INSTRUCTOR'S SOLUTIONS MANUAL
to accompany
CALCULUS

SINGLE VARIABLE

**SECOND EDITION
ALTERNATE VERSION**

Deborah Hughes-Hallett
University of Arizona

Andrew M. Gleason
Harvard University

et al.

Prepared by

Srdjan Divac
David S. Grenda
Adrian Iovita
Lee Deville
Melanie Bell
Mike Klucznik
Jian-Mei Wang
Brad Mann
Ted Pyne
Adrian Vajiac
Mihaela Vajiac
David Stephenson
Stephen A. Mallozzi
Xianboa Xu

John Wiley & Sons, Inc.
New York • Chichester • Weinheim • Brisbane • Singapore • Toronto

COVER PHOTO   © Dennis O'Clair/Tony Stone Images, Inc.

To order books or for customer service call 1-800-CALL-WILEY (225-5945).

Copyright © 2000 by John Wiley & Sons, Inc.

Excerpts from this work may be reproduced by instructors for distribution on a not-for-profit basis for testing or instructional purposes only to students enrolled in courses for which the textbook has been adopted. *Any other reproduction or translation of this work beyond that permitted by Sections 107 or 108 of the 1976 United States Copyright Act without the permission of the copyright owner is unlawful. Requests for permission or further information should be addressed to the Permissions Department, John Wiley & Sons, Inc., 605 Third Avenue, New York, NY 10158-0012.*

ISBN 0-471-36115-1

Printed in the United States of America

10 9 8 7 6 5 4 3 2 1

Printed and bound by Victor Graphics, Inc.

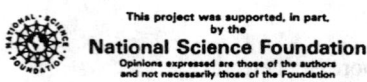

This project was supported, in part, by the National Science Foundation. Opinions expressed are those of the authors and not necessarily those of the Foundation.

CONTENTS

CHAPTER 1	1
CHAPTER 2	25
CHAPTER 3	73
CHAPTER 4	133
CHAPTER 5	203
CHAPTER 6	235
CHAPTER 7	269
CHAPTER 8	359
CHAPTER 9	419
CHAPTER 10	485
APPENDIX	569

CHAPTER ONE

Solutions for Section 1.1

1. $f(35)$ means the value of P corresponding to $t = 35$. Since t represents the number of years since 1950, we see that $f(35)$ means the population of the city in 1985. So, in 1985, the city's population was 12 million.

2. (a) $f(30) = 10$ means that the value of f at $t = 30$ was 10. In other words, the temperature at time $t = 30$ minutes was 10°C. So, 30 minutes after the object was placed outside, it had cooled to 10 °C.
   (b) The intercept a measures the value of $f(t)$ when $t = 0$. In other words, when the object was initially put outside, it had a temperature of a°C. The intercept b measures the value of t when $f(t) = 0$. In other words, at time b the object's temperature is 0 °C.

3. Generally manufacturers will produce more when prices are higher. Therefore, the first curve is a supply curve. Consumers consume less when prices are higher. Therefore, the second curve is a demand curve.

4. The price p_1 represents the maximum price any consumer would pay for the good. The quantity q_1 is the quantity of the good that could be given away if the item were free.

5. Since W is a linear function of R, we can find its slope, m, using the formula $m = \Delta W/\Delta R$. Substituting in the first two values in the table gives:
$$m = \frac{\Delta W}{\Delta R} = \frac{25 - 20}{9 - 6} = \frac{5}{3}.$$
Using the first point in the table, we have
$$W = \frac{5}{3}R + b$$
$$20 = \frac{5}{3}(6) + b$$
$$b = 10.$$
Thus, we have
$$W = \frac{5}{3}R + 10.$$
Note that we only needed two of the given points in the table to find W as a function of R.

6.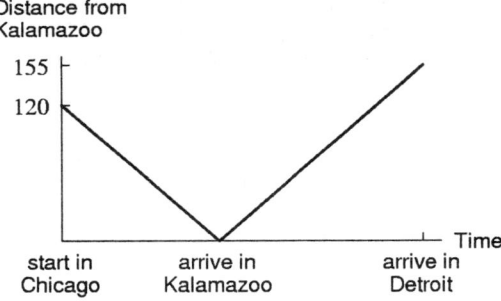

7. Since stress $\approx 2.7 \cdot 10^7$ newton/meter^2 when strain $= 0.003$, we have
$$E \approx \frac{2.7 \cdot 10^7}{0.003} = 900 \cdot 10^7 = 9 \cdot 10^9 \text{newton/meter}^2.$$

8. (a)

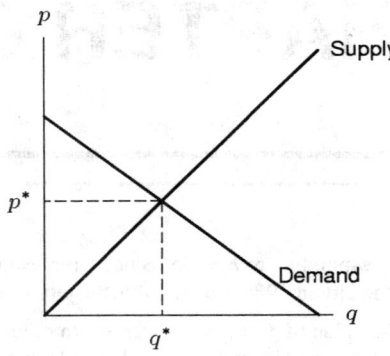

(b) If the slope of the supply curve increases, the point of intersection of the supply and demand curves moves up and to the left, as can be seen in Figure 1.1. The result is an increased equilibrium price, p_1, and a decreased equilibrium quantity, q_1. Intuitively, this makes sense since if the slope of the supply curve increases, a change in q corresponds to a larger change in p, which leads to higher prices and the corresponding lowering of quantity demanded in reaction to the higher prices.

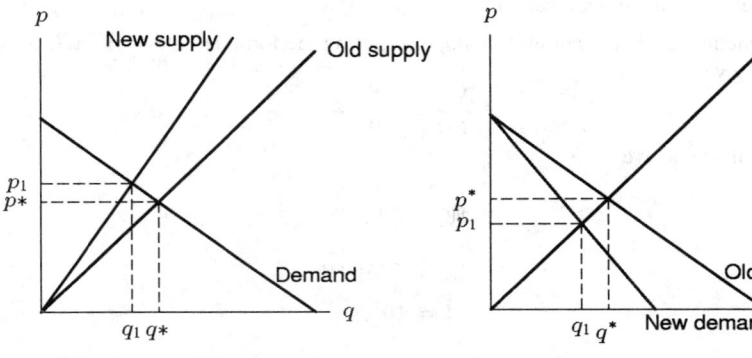

Figure 1.1   Figure 1.2

(c) If the slope of the demand curve decreases and the p-intercept is kept constant, the point of intersection of the supply and demand curves moves down and to the left, as can be seen in Figure 1.2. The result is a decreased equilibrium quantity, q_1, and a decreased equilibrium price, p_1. This follows our intuition, since if demand for a product lessens, the price and quantity purchased of the product will go down.

9.

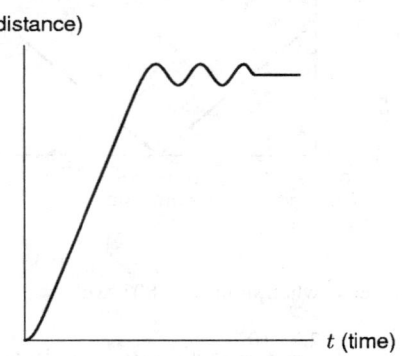

10. The plane starts far away from Laguardia, comes closer, stays at about the same constant distance from LaGuardia as it circles, then lands, when its distance is zero. See below.

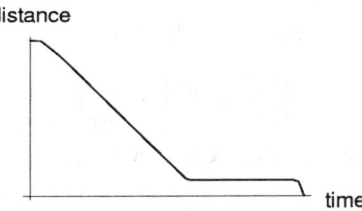

11. One possible graph is shown below.

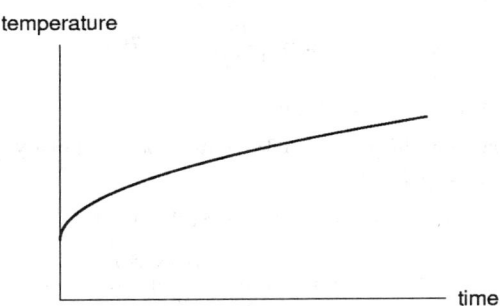

Note that any increasing and concave down graph would satisfy the stated conditions.

12.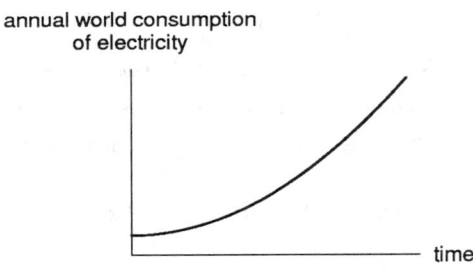

13.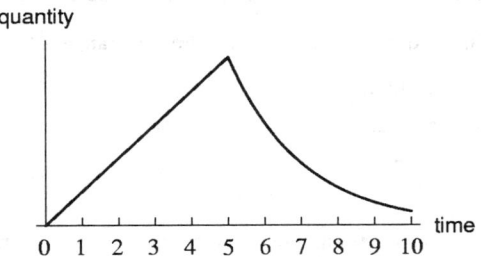

14. Since
$$\frac{g(1)}{g(0)} = \frac{g(2)}{g(1)} = \frac{g(3)}{g(2)} = \frac{g(4)}{g(3)} = 2,$$
the ratios of terms one unit apart are the same, so this appears to be an example of exponential growth. A possible formula is $g(t) = 2^t$.

15. The ratios of succeeding values of $f(t)$ are all different:

$$\frac{f(3)}{f(2)} = 2, \quad \frac{f(4)}{f(3)} = 4.5 \quad \text{and} \quad \frac{f(5)}{f(4)} = 3.$$

Thus, this is neither exponential growth nor exponential decay.

16. We see that all ratios are the same:

$$\frac{h(5)}{h(4)} = \frac{h(6)}{h(5)} = \frac{h(7)}{h(6)} = \frac{1}{2}.$$

The function is decreasing as t increases, so this appears to be an example of exponential decay. To find a formula for $h(t)$, we let

$$h(t) = a \cdot b^t, \quad \text{with } b = \frac{1}{2}.$$

Now, we solve for a:

$$2096 = a \cdot \left(\frac{1}{2}\right)^3$$

$$a = \frac{2096}{(1/2)^3} = 16768.$$

Thus, a possible formula for $h(t)$ is $h(t) = 16768(\frac{1}{2})^t$.

17. We see that $\frac{1.09}{1.06} \approx 1.03$, and therefore $h(s) = c(1.03)^s$; c must be 1. Similarly $\frac{2.42}{2.20} = 1.1$, and so $f(s) = a(1.1)^s$; $a = 2$. Lastly, $\frac{3.65}{3.47} \approx 1.05$, so $g(s) = b(1.05)^s$; $b \approx 3$.

18. If the pressure at sea level is P_0, the pressure P at altitude h is given by

$$P = P_0 \left(1 - \frac{0.4}{100}\right)^{\frac{h}{100}},$$

since we want the pressure to be multiplied by a factor of $(1 - \frac{0.4}{100}) = 0.996$ for each 100 feet we go up to make it decrease by 0.4% over that interval. At Mexico City $h = 7340$, so the pressure is

$$P = P_0(0.996)^{\frac{7340}{100}} \approx 0.745 P_0.$$

So the pressure is reduced from P_0 to approximately $0.745 P_0$, a decrease of 25.5%.

19. Since x goes from 1 to 5 and y goes from 1 to 6, the domain is $1 \leq x \leq 5$ and the range is $1 \leq y \leq 6$.

20. Since the function goes from $x = 0$ to $x = 5$ and between $y = 0$ and $y = 4$, the domain is $0 \leq x \leq 5$ and the range is $0 \leq y \leq 4$.

21. Since the function goes from $x = -2$ to $x = 2$ and from $y = -2$ to $y = 2$, the domain is $-2 \leq x \leq 2$ and the range is $-2 \leq y \leq 2$.

22. The domain is all numbers. The range is all numbers ≥ 2, since $x^2 \geq 0$ for all x.

23. The domain is all numbers except $x = 2$, since division by 0 must be avoided. The range is all numbers except 0.

24. The domain is all x-values, as the denominator is never zero. The range is $0 < y \leq \frac{1}{2}$.

25. The graph shows a concave up function.

26. The graph shows a concave down function.

27. This graph is neither concave up or down.

28. The graph is concave up.

29. (a) This is the graph of a linear function, which increases at a constant rate, and thus corresponds to $k(t)$, which increases by 0.3 over each interval of 1.
    (b) This graph is concave down, so it corresponds to a function whose increases are getting smaller, as is the case with $h(t)$, whose increases are 10, 9, 8, 7, and 6.
    (c) This graph is concave up, so it corresponds to a function whose increases are getting bigger, as is the case with $g(t)$, whose increases are 1, 2, 3, 4, and 5.

Solutions for Section 1.2

1. Figure 1.3 shows the appropriate graphs. Note that asymptotes are shown as dashed lines and x- or y-intercepts are shown as filled circles.

(a)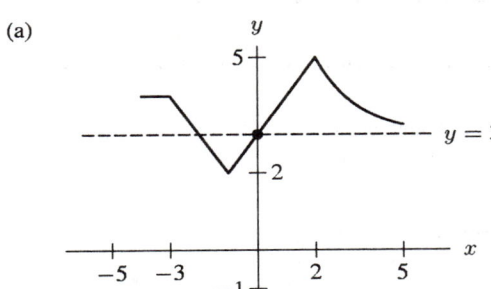

(b)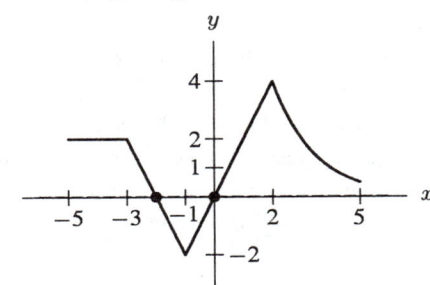

(c)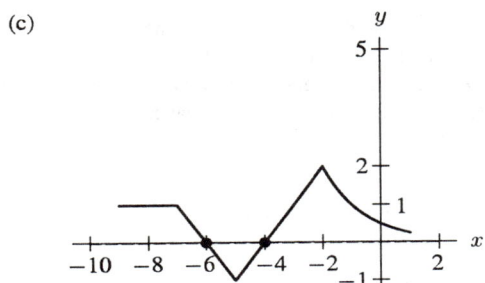

(d)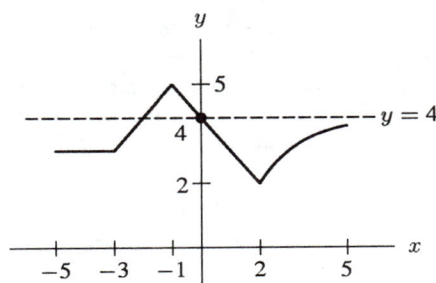

Figure 1.3

2. $f(g(x)) = f(\ln(x + 3)) = 2\ln(x + 3) + 1$.
3. $g(f(x)) = g(2x + 1) = \ln(2x + 1 + 3) = \ln(2x + 4)$.
4. $f(h(x)) = f(e^{4x+7}) = 2e^{4x+7} + 1$
5. $h(f(x)) = h(2x + 1) = e^{4(2x+1)+7} = e^{8x+11}$
6. $g(h(x)) = g(e^{4x+7}) = \ln(e^{4x+7} + 3)$
7. $h(g(x)) = h(\ln(x + 3)) = e^{4\ln(x+3)+7} = e^{\ln(x+3)^4 + 7} = e^{\ln(x+3)^4} e^7 = e^7(x + 3)^4$
8. $g(h(x) - 3) = g(e^{4x+7} - 3) = \ln(e^{4x+7} - 3 + 3) = \ln(e^{4x+7}) = 4x + 7$
9. $f(g(h(x))) = f(g(e^{4x+7})) = f(\ln(e^{4x+7} + 3)) = 2\ln(e^{4x+7} + 3) + 1$
10. $h(g(f(x))) = h(g(2x + 1)) = h(\ln(2x + 1 + 3)) = h(\ln(2x + 4))$
    $= e^{4\ln(2x+4)+7} = e^{\ln(2x+4)^4 + 7} = e^{\ln(2x+4)^4} e^7 = e^7(2x + 4)^4$.
11. The function is even since $f(-x) = f(x)$.
12. The function is even since $f(-x) = f(x)$.
13. The function is odd since $f(-x) = -f(x)$.
14. The function is neither even nor odd.
15.
$$f(-x) = (-x)^6 + (-x)^3 + 1 = x^6 - x^3 + 1.$$
Since $f(-x) \neq f(x)$ and $f(-x) \neq -f(x)$, this function is neither even nor odd.

16.
$$f(-x) = (-x)^3 + (-x)^2 + (-x) = -x^3 + x^2 - x.$$
Since $f(-x) \neq f(x)$ and $f(-x) \neq -f(x)$, this function is neither even nor odd.

6 CHAPTER ONE /SOLUTIONS

17. (a) We have $g(x) = f(x/2)$, and we can only evaluate $f(x/2)$ if $x/2$ is one of the x-values 0, 1, 2, 3, 4, 5, 6 in the table of values for f. So we can only evaluate $g(x)$ if x is one of 0, 2, 4, 6, 8, 10, 12.
    (b)
    TABLE 1.1

x	0	2	4	6	8	10	12
$g(x)$	0	2	6	12	20	30	42

    (c) The graph of g is obtained from that of f by a horizontal stretch by a factor of 2, so its graph will be twice as wide as f's.

18. $\sin x^2$ is by convention $\sin(x^2)$, which means you square the x first and then take the sine.
    $\sin^2 x = (\sin x)^2$ means find $\sin x$ and then square it.
    $\sin(\sin x)$ means find $\sin x$ and then take the sine of that.
    Expressing each as a composition: If $f(x) = \sin x$ and $g(x) = x^2$, then
    $\sin x^2 = f(g(x))$
    $\sin^2 x = g(f(x))$
    $\sin(\sin x) = f(f(x))$.

19. (a) We determine the amplitude of y by looking at the coefficient of the cosine term. Here, the coefficient is 1, so the amplitude of y is 1. Note that the constant term does not affect the amplitude.
    (b) We know that the cosine function $\cos x$ repeats itself at $x = 2\pi$, so the function $\cos(3x)$ must repeat itself when $3x = 2\pi$, or at $x = 2\pi/3$. So the period of y is $2\pi/3$. Here as well the constant term has no effect.
    (c) The graph of y is shown in the figure below.

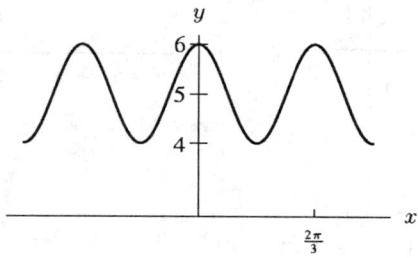

20. (a) $h(t) = 2\cos(t - \pi/2)$
    (b) $f(t) = 2\cos t$
    (c) $g(t) = 2\cos(t + \pi/2)$

21. This graph is a cosine curve with period 6π and amplitude 5, so it is given by $f(x) = 5\cos\left(\dfrac{x}{3}\right)$.

22. This graph is an inverted sine curve with amplitude 4 and period π, so it is given by $f(x) = -4\sin(2x)$.

23. This graph is a sine curve with period 8π, amplitude 2, and midline at 2, so it is given by $f(x) = 2\sin\left(\dfrac{x}{4}\right) + 2$.

24. This graph has period 6, amplitude 5 and no vertical or horizontal shift, so it is given by
$$f(x) = 5\sin\left(\frac{2\pi}{6}x\right) = 5\sin\left(\frac{\pi}{3}x\right).$$

25. This can be represented by a sine function of amplitude 3 and period 18. Thus,
$$f(x) = 3\sin\left(\frac{\pi x}{9}\right).$$

26. This graph has period 8, amplitude 3, and a vertical shift of 3 with no horizontal shift. It is given by
$$f(x) = 3 + 3\sin\left(\frac{2\pi}{8}x\right) = 3 + 3\sin\left(\frac{\pi}{4}x\right).$$

27. $f(g(1)) = f(2) \approx 0.4.$

28. $g(f(2)) \approx g(0.4) \approx 1.1.$

29. $f(f(1)) \approx f(-0.4) \approx -0.9.$

30. Computing $f(g(x))$ as in Problem 27, we get the following table. From it we graph $f(g(x))$.

x	$g(x)$	$f(g(x))$
-3	0.6	-0.5
-2.5	-1.1	-1.3
-2	-1.9	-1.2
-1.5	-1.9	-1.2
-1	-1.4	-1.3
-0.5	-0.5	-1
0	0.5	-0.6
0.5	1.4	-0.2
1	2	0.4
1.5	2.2	0.5
2	1.6	0
2.5	0.1	-0.7
3	-2.5	0.1

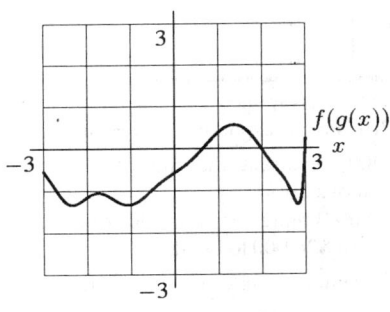

31. Using the same way to compute $g(f(x))$ as in Problem 28, we get the following table. Then we can plot the graph of $g(f(x))$.

x	$f(x)$	$g(f(x))$
-3	3	-2.6
-2.5	0.1	0.8
-2	-1	-1.4
-1.5	-1.3	-1.8
-1	-1.2	-1.7
-0.5	-1	-1.4
0	-0.8	-1
0.5	-0.6	-0.6
1	-0.4	-0.3
1.5	-0.1	0.3
2	0.3	1.1
2.5	0.9	2
3	1.6	2.2

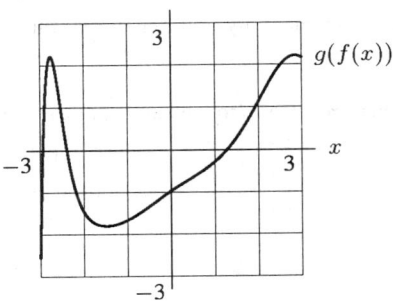

32. Using the same way to compute $f(f(x))$ as in Problem 29, we get the following table. Then we can plot the graph of $f(f(x))$.

8 CHAPTER ONE /SOLUTIONS

x	$f(x)$	$f(f(x))$
-3	3	1.6
-2.5	0.1	-0.7
-2	-1	-1.2
-1.5	-1.3	-1.3
-1	-1.2	-1.3
-0.5	-1	-1.2
0	-0.8	-1.1
0.5	-0.6	-1
1	-0.4	-0.9
1.5	-0.1	-0.8
2	0.3	-0.6
2.5	0.9	-0.4
3	1.6	0

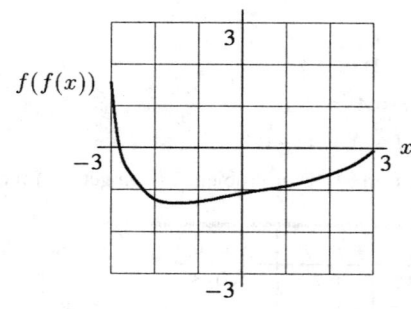

33. (a) $f(25)$ is q corresponding to $p = 25$, or, in other words, the number of items sold when the price is 25.
    (b) $f^{-1}(30)$ is p corresponding to $q = 30$, or the price at which 30 units will be sold.

34. (a) $f(10,000)$ represents the value of C corresponding to $A = 10,000$, or in other words the cost of building a 10,000 square-foot store.
    (b) $f^{-1}(20,000)$ represents the value of A corresponding to $C = 20,000$, or the area in square feet of a store which would cost $20,000 to build.

35. The function is not invertible since there are many horizontal lines which hit the function twice.

36. The function is not invertible since there are horizontal lines which hit the function more than once.

37. The function is invertible since no horizontal lines hit it more than once.

38. The function is not invertible since there are horizontal lines which hit it more than once.

39. (a) The function f tells us C in terms of q. To get its inverse, we want q in terms of C, which we find by solving for q:
$$C = 100 + 2q,$$
$$C - 100 = 2q,$$
$$q = \frac{C - 100}{2} = f^{-1}(C).$$
    (b) The inverse function tells us the number of articles that can be produced for a given cost.

40. Let n be the infant mortality of Senegal. As a function of time t, n is given by
$$n = n_0(0.90)^t.$$
To find when $n = 0.50n_0$ (so the number of cases has been reduced by 50%), we solve
$$0.50 = (0.90)^t,$$
$$\log(0.50) = t\log(0.90),$$
$$t = \frac{\log(0.50)}{\log(0.90)} \approx 6.58 \text{ years}.$$

41. (a) We have $P_0 = 1$ million, and $k = 0.02$, so $P = (1,000,000)(e^{0.02t})$.
    (b)

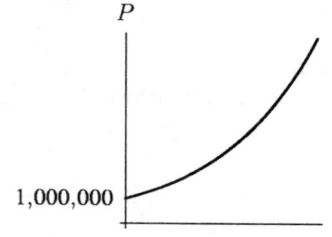

42. (a) We know the decay follows the equation
$$P = P_0 e^{-kt},$$
and that 10% of the pollution is removed after 5 hours (meaning that 90% is left). Therefore,
$$0.90 P_0 = P_0 e^{-5k}$$
$$k = -\frac{1}{5} \ln(0.90).$$

Thus, after 10 hours:
$$P = P_0 e^{-10((-0.2)\ln 0.90)}$$
$$P = P_0 (0.9)^2 = 0.81 P_0$$

so 81% of the original amount is left.

(b) We want to solve for the time when $P = 0.50 P_0$:
$$0.50 P_0 = P_0 e^{t((0.2)\ln 0.90)}$$
$$0.50 = e^{\ln(0.90^{0.2t})}$$
$$0.50 = 0.90^{0.2t}$$
$$t = \frac{5 \ln(0.50)}{\ln(0.90)} \approx 32.9 \text{ hours}.$$

(c)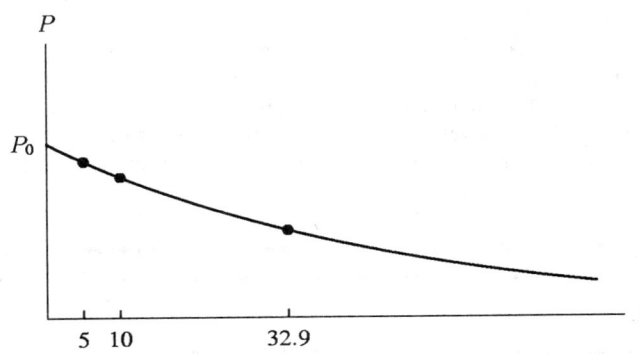

(d) When highly polluted air is filtered, there is more pollutant per liter of air to remove. If a fixed amount of air is cleaned every day, there is a higher amount of pollutant removed earlier in the process.

43. (a) The pressure P at 6198 meters is given in terms of the pressure P_0 at sea level to be
$$P = P_0 e^{-0.00012h}$$
$$= P_0 e^{(-0.00012)6198}$$
$$= P_0 e^{-0.74376}$$
$$\approx 0.4753 P_0 \quad \text{or about 47.5\% of sea level pressure.}$$

(b) At $h = 12{,}000$ meters, we have
$$P = P_0 e^{-0.00012h}$$
$$= P_0 e^{(-0.00012)12{,}000}$$
$$= P_0 e^{-1.44}$$
$$\approx 0.2369 P_0 \quad \text{or about 23.7\% of sea level pressure.}$$

44. (a) $P = 3.6(1.034)^t$

(b) Since $P = 3.6 e^{kt} = 3.6(1.034)^t$, we have
$$e^{kt} = (1.034)^t$$
$$kt = t \ln(1.034)$$
$$k = 0.0334$$

Thus, $P = 3.6 e^{0.0334t}$.

(c) The annual growth rate is 3.4%, while the continuous growth rate is 3.3%. Thus the growth rates are not equal, though for small growth rates (such as these), they are close. The annual growth rate is larger.

45. (a) A table of values for $g(x)$ is given below.

x	-1	-0.8	-0.6	-0.4	-0.2	0	0.2	0.4	0.6	0.8	1
arccos x	3.14	2.50	2.21	1.98	1.77	1.57	1.37	1.16	0.93	0.64	0

(b)

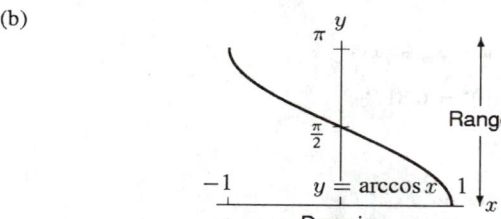

Figure 1.4

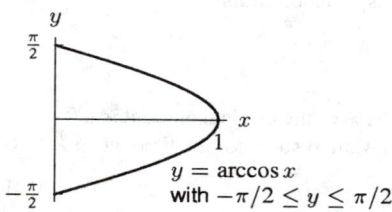

Figure 1.5

(c) The domain is $-1 \leq x \leq 1$ while the range is $0 \leq y \leq \pi$. See Figure 1.4.
(d) The domain of arccos and arcsin are the same, $-1 \leq x \leq 1$, since their inverses (sine and cosine) only take on values in this range.
(e) The domain of the original sine function was restricted to the the interval $[-\frac{\pi}{2}, \frac{\pi}{2}]$ to construct the arcsine function. Hence, the range of arcsine is also $[-\frac{\pi}{2}, \frac{\pi}{2}]$. Now, if we restrict the domain of cosine in the same way, we obtain an arccosine curve which is not a function. (See Figure 1.5.) For example, for $x = 0$, $y = \arccos x$ will have two values, $-\frac{\pi}{2}$, and $\frac{\pi}{2}$. Also, it gives no values for $x < 0$, so it is not very useful. The domain of cosine should instead be restricted to $[0, \pi]$, so that $y = \arccos x$ gives a unique y for each value of x.

Solutions for Section 1.3

1. (a) is (V), because slope is positive, vertical intercept is negative
    (b) is (IV), because slope is negative, vertical intercept is positive
    (c) is (I), because slope is 0, vertical intercept is positive
    (d) is (VI), because slope and vertical intercept are both negative
    (e) is (II), because slope and vertical intercept are both positive
    (f) is (III), because slope is positive, vertical intercept is 0

2. (a) Since $2x^5$ is a higher-degree polynomial than $200x^4$, eventually it will be larger as $x \to \infty$.
    (b) Since exponential growth functions always dominate any polynomial functions for large x, as $x \to \infty$, e^x will eventually be larger than $10x^3$.
    (c) Since x^5 dominates x^2 as x becomes large, it must be that $1/x^2$ dominates $1/x^5$ for large x. So as $x \to \infty$, x^{-2} will be larger.
    (d) Since $x^{1/2}$ is a power function and $\ln x$ is logarithmic, eventually $x^{1/2}$ will be larger than $\ln x$.

3. (a) Charge per cubic foot $= \dfrac{\Delta \$}{\Delta \text{ cu. ft.}} = \dfrac{105 - 90}{1600 - 1000} = \$0.025$/cu. ft.
    Alternatively, if we let c = cost, w = cubic feet of water, b = fixed charge, and m = cost/cubic feet, we obtain $c = b + mw$. Substituting the information given in the problem, we have
    $$90 = b + 1000m$$
    $$105 = b + 1600m.$$
    Subtracting the first equation from the second yields $15 = 600m$, so $m = 0.025$.
    (b) $c = b + 0.025w$, so $90 = b + 0.025(1000)$, which yields $b = 65$. Thus the equation is $c = 65 + 0.025w$.
    (c) We need to solve the equation $130 = 65 + 0.025w$, which yields $w = 2600$.

4. (a) $b = 0$ since the scarf is zero feet long when zero skeins of yarn have been used.
    (b) a is positive since using more yarn results in a longer scarf.
    (c) $c > a$; Because the friend's scarf has "holes", the same amount of yarn produces a longer scarf.

5. (a) The formula is $Q = Q_0 \left(\frac{1}{2}\right)^{(t/1620)}$.
   (b) The percentage left after 500 years is
   $$\frac{Q_0 \left(\frac{1}{2}\right)^{(500/1620)}}{Q_0}.$$

   The Q_0s cancel giving
   $$\left(\frac{1}{2}\right)^{(500/1620)} \approx 0.807,$$

   so 80.7% is left.

6. The population has increased by a factor of $\frac{56,000,000}{40,000,000} = 1.4$ in 10 years. Thus we have the formula
   $$P = 40,000,000(1.4)^{t/10},$$

   and $t/10$ gives the number of 10-year periods that have passed since 1980.

   In 1980, $t/10 = 0$, so we have $P = 40,000,000$.
   In 1990, $t/10 = 1$, so $P = 40,000,000(1.4) = 56,000,000$.
   In 2000, $t/10 = 2$, so $P = 40,000,000(1.4)^2 = 78,400,000$.
   To find the doubling time, solve $80,000,000 = 40,000,000(1.4)^{t/10}$, to get $t \approx 20.6$ years.

7. Let t = number of years since 1980. Then the number of vehicles, V, in millions, at time t is given by
   $$V = 170(1.04)^t$$

   and the number of people, P, in millions, at time t is given by
   $$P = 227(1.01)^t.$$

   There is an average of one vehicle per person when $\frac{V}{P} = 1$, or $V = P$. Thus, we must solve for t the equation:
   $$170(1.04)^t = 227(1.01)^t,$$

   which implies
   $$\left(\frac{1.04}{1.01}\right)^t = \frac{(1.04)^t}{(1.01)^t} = \frac{227}{170}$$

   Taking logs on both sides,
   $$t \log \frac{1.04}{1.01} = \log \frac{227}{170}.$$

   Therefore,
   $$t = \frac{\log\left(\frac{227}{170}\right)}{\log\left(\frac{1.04}{1.01}\right)} \approx 9.9 \text{ years}.$$

   So there was, according to this model, about one vehicle per person in 1990.

8. (a)

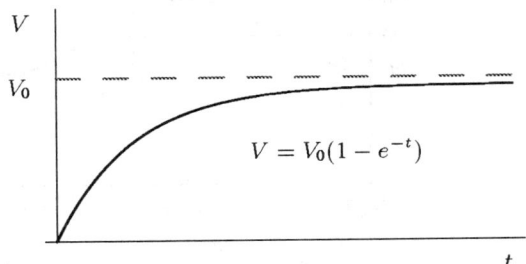

   (b) V_0 represents the terminal velocity of the raindrop or the maximum speed it can attain as it falls (although a raindrop starting at rest will never quite reach V_0 exactly).

9. Assuming the US population grows exponentially, we have

$$248.7 = 226.5e^{10k}$$
$$k = \frac{\ln(1.098)}{10} = 0.00935.$$

We want to find the time t in which

$$300 = 226.5e^{0.00935t}$$
$$t = \frac{\ln(1.324)}{0.00935} = 30 \text{ years}.$$

Thus, the population will go over 300 million around the year 2010.

10. Since the amount of strontium-90 remaining halves every 29 years, we can solve for the decay constant;

$$0.5P_0 = P_0 e^{-29k}$$
$$k = \frac{\ln(1/2)}{-29}.$$

Knowing this, we can look for the time t in which $P = 0.10P_0$, or

$$0.10P_0 = P_0 e^{\ln(0.5)t/29}$$
$$t = \frac{29\ln(0.10)}{\ln(0.5)} = 96.34 \text{ years}.$$

11. (a) D = the average depth of the water.
    (b) A = the amplitude = $15/2 = 7.5$.
    (c) Period = 12.4 hours. Thus $(B)(12.4) = 2\pi$ so $B = 2\pi/12.4 \approx 0.507$.
    (d) C is the time of a high tide.

12. (a) Beginning at time $t = 0$, the voltage will have oscillated through a complete cycle when $\cos(120\pi t) = \cos(2\pi)$, hence when $t = \frac{1}{60}$ second. The period is $\frac{1}{60}$ second.
    (b) V_0 represents the amplitude of the oscillation.
    (c)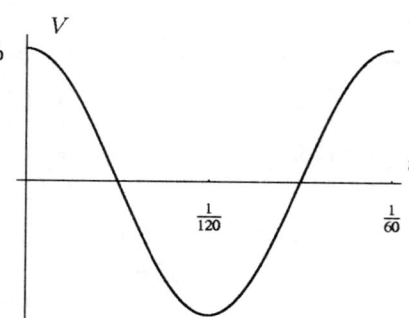

13. (a)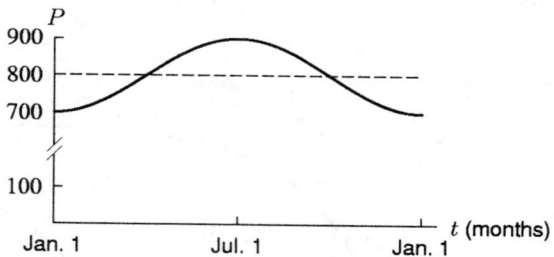

(b) Average value of population = $\frac{700+900}{2} = 800$, amplitude = $\frac{900-700}{2} = 100$, and period = 12 months, so $B = 2\pi/12 = \pi/6$. Since the population is at its minimum when $t = 0$, we use a negative cosine:

$$P = 800 - 100\cos\left(\frac{\pi t}{6}\right).$$

14. Over the one-year period, the average value is about 75° and the amplitude of the variation is about $\frac{90-60}{2} = 15°$. The function assumes its minimum value right at the beginning of the year, so we want a negative cosine function. Thus, for t in years, we have the function
$$f(t) = 75 - 15\cos\left(\frac{2\pi}{12}t\right).$$
(Many other answers are possible, depending on how you read the chart.)

15. The function R has period of π, so its graph is as shown in the figure below. The maximum value of the range is v_0^2/g and occurs when $\theta = \pi/4$.

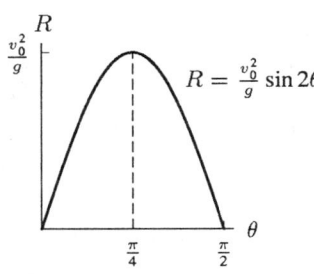

16. (a) Since the rate R varies directly with the fourth power of the radius r, we have the formula
$$R = kr^4$$
where k is a constant.

    (b) Given $R = 400$ for $r = 3$, we can determine the constant k.
$$400 = k(3)^4$$
$$400 = k(81)$$
$$k = \frac{400}{81} \approx 4.938.$$
So the formula is
$$R = 4.938r^4$$

    (c) Evaluating the formula above at $r = 5$ yields
$$R = 4.928(5)^4 = 3086.42\frac{\text{cm}^3}{\text{sec}}.$$

17. (a) The object starts at $t = 0$, when $s = v_0(0) - g(0)^2/2 = 0$. Thus it starts on the ground, with zero height.

    (b) The object hits the ground when $s = 0$. This is satisfied at $t = 0$, before it has left the ground, and at some later time t that we must solve for.
$$0 = v_0 t - gt^2/2 = t\left(v_0 - gt/2\right)$$
Thus $s = 0$ when $t = 0$ and when $v_0 - gt/2 = 0$, i.e., when $t = 2v_0/g$. The starting time is $t = 0$, so it must hit the ground at time $t = 2v_0/g$.

    (c) The object reaches its maximum height halfway between when it is released and when it hits the ground, or at
$$t = (2v_0/g)/2 = v_0/g.$$

    (d) Since we know the time at which the object reaches its maximum height, to find the height it actually reaches we just use the given formula, which tells us s at any given t. Plugging in $t = \frac{v_0}{g}$,
$$s = v_0\left(\frac{v_0}{g}\right) - \frac{1}{2}g\left(\frac{v_0^2}{g^2}\right) = \frac{v_0^2}{g} - \frac{v_0^2}{2g}$$
$$= \frac{2v_0^2 - v_0^2}{2g} = \frac{v_0^2}{2g}.$$

18. (a) If $(1, 1)$ is on the graph, we know that
$$1 = a(1)^2 + b(1) + c = a + b + c.$$

(b) If $(1, 1)$ is the vertex, then the axis of symmetry is $x = 1$, so
$$-\frac{b}{2a} = 1,$$
and thus
$$a = -\frac{b}{2}, \text{ so } b = -2a.$$
But to be the vertex, $(1, 1)$ must also be on the graph, so we know that $a + b + c = 1$. Substituting $b = -2a$, we get $-a + c = 1$, which we can rewrite as $a = c - 1$, or $c = 1 + a$.

(c) For $(0, 6)$ to be on the graph, we must have $f(0) = 6$. But $f(0) = a(0^2) + b(0) + c = c$, so $c = 6$.

(d) To satisfy all the conditions, we must first, from (c), have $c = 6$. From (b), $a = c - 1$ so $a = 5$. Also from (b), $b = -2a$, so $b = -10$. Thus the completed equation is
$$y = f(x) = 5x^2 - 10x + 6,$$
which satisfies all the given conditions.

19. (a) $R(P) = kP(L - P)$, where k is a positive constant.
(b) A possible graph is shown below.

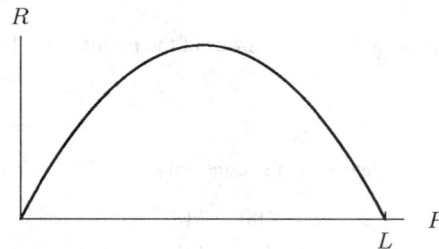

20. (a) $a(v) = \frac{1}{m}(\text{ENGINE} - \text{WIND}) = \frac{1}{m}(F_E - kv^2)$, where k is a positive constant.
(b) A possible graph is shown below.

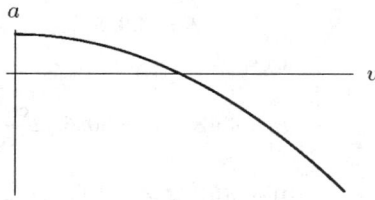

21. The graphs are shown in Figure 1.6.

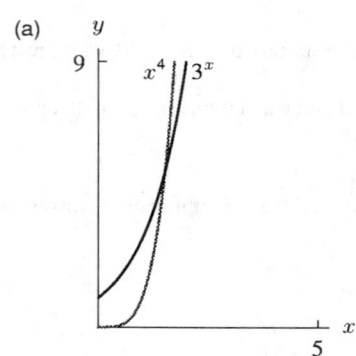 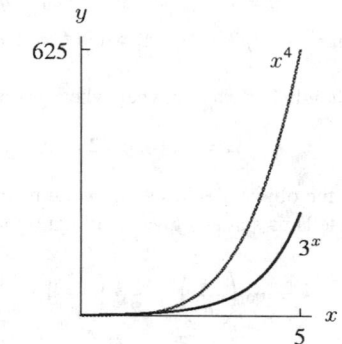 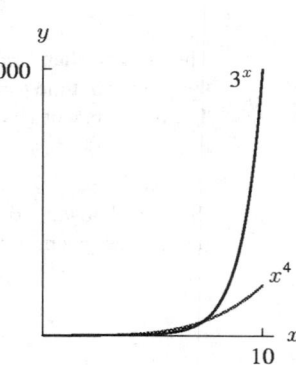

Figure 1.6

Solutions for Section 1.4

1. No, because $x - 2 = 0$ at $x = 2$.
2. Yes, because $x - 2$ is not zero on this interval.
3. Yes, because the denominator is never zero.
4. Yes, because $2x - 5$ is positive for $3 \leq x \leq 4$.
5. No, because $\sin 0 = 0$.
6. No, because $\cos(\pi/2) = 0$.
7. Yes, because $\cos \theta$ is not zero on this interval.
8. No, because $e^x - 1 = 0$ at $x = 0$.
9. The graph of g suggests that g is not continuous on any interval containing $\theta = 0$, since $g(0) = 1/2$.
10.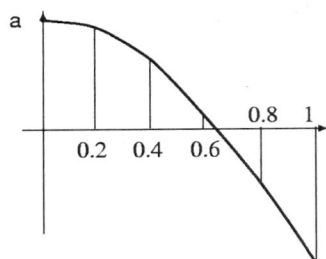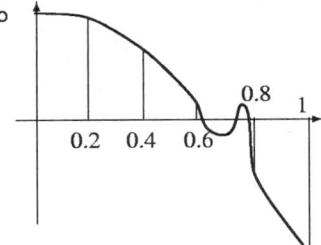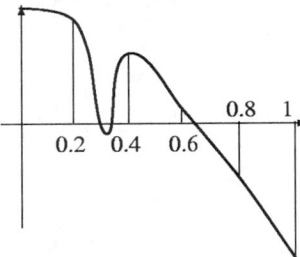

11. The functions $y(x) = \sin x$ and $z_k(x) = ke^{-x}$ for $k = 1, 2, 4, 6, 8, 10$ are shown in Figure 1.7. The values of $f(k)$ for $k = 1, 2, 4, 6, 8, 10$ are given in Table 1.2. These values can be obtained using either tracing or a numerical root finder on a calculator or computer.

    From Figure 1.7 it is clear that the smallest solution of $\sin x = ke^{-x}$ for $k = 1, 2, 4, 6$ occurs on the first period of the sine curve. For small changes in k, there are correspondingly small changes in the intersection point. For $k = 8$ and $k = 10$, the solution jumps to the second period because $\sin x < 0$ between π and 2π, but ke^{-x} is uniformly positive. Somewhere in the interval $6 \leq k \leq 8$, $f(k)$ has a discontinuity.

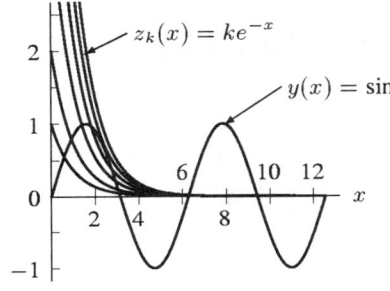

Figure 1.7

TABLE 1.2

k	$f(k)$
1	0.588
2	0.921
4	1.401
6	1.824
8	6.298
10	6.302

Solutions for Chapter 1 Review

1. A possible graph is shown below:

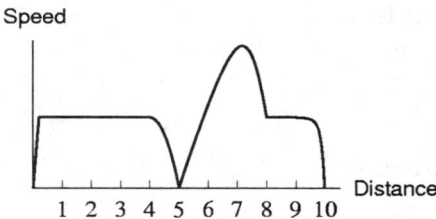

2. (a) Given the two points $(0, 32)$ and $(100, 212)$, and assuming the graph in Figure 1.8 is a line,

$$\text{Slope} = \frac{212 - 32}{100} = \frac{180}{100} = 1.8.$$

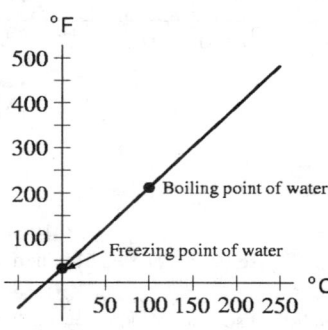

Figure 1.8

(b) The °F-intercept is $(0, 32)$, so
$$°F = 1.8(°C) + 32.$$

(c) If the temperature is 20°Celsius, then
$$°F = 1.8(20) + 32 = 68°F.$$

(d) If °F = °C, then
$$°C = 1.8°C + 32$$
$$-32 = 0.8°C$$
$$°C = -40° = °F.$$

3. Given $l - l_0 = al_0(t - t_0)$ with l_0, t_0 and a all constant,

(a) We have $l = al_0(t - t_0) + l_0 = al_0 t - al_0 t_0 + l_0$, which is a linear function of t with slope al_0 and y-intercept at $(0, -al_0 t_0 + l_0)$.

(b) If $l_0 = 100$, $t_0 = 60°F$ and $a = 10^{-5}$, then
$$l = 10^{-5}(100)t - 10^{-5}(100)(60) + 100 = 10^{-3}t + 99.94$$
$$= 0.001t + 99.94$$

(c) If the slope is positive, (as in (b)), then as the temperature rises, the length of the metal increases: it expands. If the slope were negative, then the metal would contract as the temperature rises.

4. From the figure it appears that the "seat" of the graph $y = x^3$ has been moved to the left and up by 1, to the point $(-1, 1)$. Since translation to the right by h is achieved by replacing x with $(x - h)$ and translation up by k is achieved by replacing y with $(y - k)$, the equation of our translated graph appears to be

$$y - 1 = (x - (-1))^3$$

or

$$y = (x + 1)^3 + 1.$$

Since the picture *suggests* that the graph of $y = x^3$ has been moved over and up by 1 (but does not show this explicitly), we will confirm that our equation is correct by checking the x and y intercepts (which are shown explicitly in the picture). The desired y-intercept is 2, and substituting $x = 0$ into the equation gives

$$y = (0 + 1)^3 + 1 = 2.$$

In addition, the desired x-intercept is -2, and substituting $y = 0$ into the equation gives

$$0 = (x + 1)^3 + 1$$

so

$$(x + 1)^3 = -1$$
$$x + 1 = \sqrt[3]{-1} = -1$$
$$x = -2.$$

Thus our equation does have the graph shown.

We could also have solved this problem, with less guessing and more algebra, by finding a translation of x^3 which has the given x- and y-intercepts. Let h be the horizontal, and k the vertical translation; then the formula for the translated function is

$$f(x) = (x - h)^3 + k.$$

The y-intercept of f is $f(0) = 2$, so

$$f(0) = (0 - h)^3 + k = -h^3 + k = 2.$$

Therefore $k = 2 + h^3$ and we have $f(x) = (x - h)^3 + (2 + h^3)$. The x-intercept, -2, is the x-value for which $f(x) = 0$, so

$$0 = f(-2) = (-2 - h)^3 + (2 + h^3)$$
$$= -6h^2 - 12h - 6$$
$$= -6(h + 1)^2$$

Therefore $h = -1$ and $k = 2 + h^3 = 1$.

5. Because the population is growing exponentially, the time it takes to double is the same, regardless of the population levels we are considering. For example, the population is 20,000 at time 3.7, and 40,000 at time 6.0. This represents a doubling of the population in a span of $6.0 - 3.7 = 2.3$ years.

How long does it take the population to double a second time, from 40,000 to 80,000? Looking at the graph once again, we see that the population reaches 80,000 at time $t = 8.3$. This second doubling has taken $8.3 - 6.0 = 2.3$ years, the same amount of time as the first doubling.

Further comparison of any two populations on this graph that differ by a factor of two will show that the time that separates them is 2.3 years. Similarly, during any 2.3 year period, the population will double. Thus, the doubling time is 2.3 years.

Suppose $P = P_0 a^t$ doubles from time t to time $t + d$. We now have $P_0 a^{t+d} = 2P_0 a^t$, so $P_0 a^t a^d = 2P_0 a^t$. Thus, canceling P_0 and a^t, d must be the number such that $a^d = 2$, no matter what t is.

6. (a) A linear function must change by exactly the same amount whenever x changes by some fixed quantity. While $h(x)$ decreases by 3 whenever x increases by 1, $f(x)$ and $g(x)$ fail this test, since both change by different amounts between $x = -2$ and $x = -1$ and between $x = -1$ and $x = 0$. So the only linear function is $h(x)$, so it will be given by a formula of the type: $h(x) = mx + b$. As noted, $m = -3$. Since the y-intercept of h is 31, the formula for $h(x)$ is $h(x) = 31 - 3x$.

(b) An exponential function must grow by exactly the same factor whenever x changes by some fixed quantity. Here, $g(x)$ increases by a factor of 1.5 whenever x increases by 1. Since the y-intercept of $g(x)$ is 36, $g(x)$ has the formula $g(x) = 36(1.5^x)$. The other two functions are not exponential; $h(x)$ is not because it is a linear function, and $f(x)$ is not because it both increases and decreases.

18 CHAPTER ONE /SOLUTIONS

7. $f^{-1}(75)$ is the length of the column of mercury in the thermometer when the temperature is 75°F.

8. To find a half-life, we want to find at what t value $Q = \frac{1}{2}Q_0$. Plugging this into the equation of the decay of plutonium-240, we have

$$\frac{1}{2} = e^{-0.00011t}$$

$$t = \frac{\ln(1/2)}{-0.00011} \approx 6{,}301 \text{ years.}$$

The only difference in the case of plutonium-242 is that the constant -0.00011 in the exponent is now -0.0000018. Thus, following the same procedure, the solution for t is

$$t = \frac{\ln(1/2)}{-0.0000018} \approx 385{,}081 \text{ years.}$$

9. We can solve for the growth rate k of the bacteria using the formula $P = P_0 e^{kt}$:

$$1500 = 500 e^{k(2)}$$

$$k = \frac{\ln(1500/500)}{2}.$$

Knowing the growth rate, we can find the population P at time $t = 6$:

$$P = 500 e^{(\frac{\ln 3}{2})6}$$

$$\approx 13{,}500 \text{ bacteria.}$$

10. Given the doubling time of 2 hours, $200 = 100 e^{k(2)}$, we can solve for the growth rate k using the equation:

$$2P_0 = P_0 e^{2k}$$

$$\ln 2 = 2k$$

$$k = \frac{\ln 2}{2}.$$

Using the growth rate, we wish to solve for the time t in the formula

$$P = 100 e^{\frac{\ln 2}{2} t}$$

where $P = 3{,}200$, so

$$3{,}200 = 100 e^{\frac{\ln 2}{2} t}$$

$$t = 10 \text{ hours.}$$

11. In ten years, the substance has decayed to 40% of its original mass. In another ten years, it will decay by an additional factor of 40%, so the amount remaining after 20 years will be $100 \cdot 40\% \cdot 40\% = 16$ kg.

12. Using the exponential decay equation $P = P_0 e^{-kt}$, we can solve for the substance's decay constant k:

$$(P_0 - 0.3 P_0) = P_0 e^{-20k}$$

$$k = \frac{\ln(0.7)}{-20}.$$

Knowing this decay constant, we can solve for the half-life t using the formula

$$0.5 P_0 = P_0 e^{\ln(0.7) t / 20}$$

$$t = \frac{20 \ln(0.5)}{\ln(0.7)} \approx 38.87 \text{ hours.}$$

13. (a) The line given by $(0, 2)$ and $(1, 1)$ has slope $m = \frac{2-1}{-1} = -1$ and y-intercept 2, so its equation is
$$y = -x + 2.$$
The points of intersection of this line with the parabola $y = x^2$ are given by
$$x^2 = -x + 2$$
$$x^2 + x - 2 = 0$$
$$(x + 2)(x - 1) = 0.$$
The solution $x = 1$ corresponds to the point we are already given, so the other solution, $x = -2$, gives the x-coordinate of C. When we substitute back into either equation to get y, we get the coordinates for C, $(-2, 4)$.

(b) The line given by $(0, b)$ and $(1, 1)$ has slope $m = \frac{b-1}{-1} = 1 - b$, and y-intercept at $(0, b)$, so we can write the equation for the line as we did in part (a):
$$y = (1 - b)x + b.$$
We then solve for the points of intersection with $y = x^2$ the same way:
$$x^2 = (1 - b)x + b$$
$$x^2 - (1 - b)x - b = 0$$
$$x^2 + (b - 1)x - b = 0$$
$$(x + b)(x - 1) = 0$$
Again, we have the solution at the given point $(1, 1)$, and a new solution at $x = -b$, corresponding to the other point of intersection C. Substituting back into either equation, we can find the y-coordinate for C is b^2, and thus C is given by $(-b, b^2)$. This result agrees with the particular case of part (a) where $b = 2$.

14. The period T_E of the earth is (by definition!) one year or about 365.24 days. Since the semimajor axis of the earth is 150 million km, we can use Kepler's Law to derive the constant of proportionality, k.
$$T_E = k(S_E)^{\frac{3}{2}}$$
where S_E is the earth's semimajor axis, or 150 million km.
$$365.24 = k(150)^{\frac{3}{2}}$$
$$k = \frac{365.24}{(150)^{\frac{3}{2}}} \approx 0.198.$$
Now that we know the constant of proportionality, we can use it to derive the periods of Mercury and Pluto. For Mercury,
$$T_M = (0.198)(58)^{\frac{3}{2}} \approx 87.818 \text{ days.}$$
For Pluto,
$$T_P = (0.198)(6000)^{\frac{3}{2}} \approx 92,400 \text{ days,}$$
or (converting Pluto's period to years),
$$\frac{(0.198)(6000)^{\frac{3}{2}}}{365.24} \approx 253 \text{ years.}$$

15. (a) The graph shows that $f(15)$ is approximately 48. So, the place to find find 15 million-year-old rock is about 48 meters below the Atlantic sea floor.
(b) Since f is increasing (not decreasing, since the depth axis is reversed!), f is invertible. One can see this also by noting that the graph of f is cut by a horizontal line at most once.
(c) Look at where the horizontal line through 120 intersects the graph of f and read downwards: $f^{-1}(120)$ is about 35. In practical terms, this means that at a depth of 120 meters down, the rock is 35 million years old.
(d) First, we normalize the graph of f so that time and depth are increasing from left to right and bottom to top. Points (t, d) on the graph of f correspond to points (d, t) on the graph of f^{-1}. So one could graph f^{-1} by taking points from the original graph of f reversing their coordinates, and connecting them up. However, an easier way is to change the graph of f so that the t and d axes are exchanged. (This amounts to the same thing.) When this is done,

the graph of f will be flipped over a line bisecting the 90° angle at the origin. The resulting graph (Figure 1.9) is the graph of f^{-1}.

(One cannot find the graph of f^{-1} by flipping along the line $t = d$ in this instance because t and d are in different scales.)

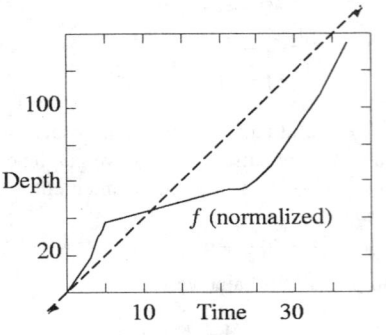
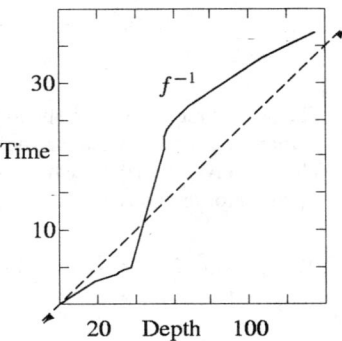

Figure 1.9: Graph of f, reflected to give that of f^{-1}

16. Looking at the given data, it seems that Galileo's hypothesis was incorrect. The first table suggests that velocity is not a linear function of distance, since the increases in velocity for each foot of distance are themselves getting smaller. Moreover, the second table suggests that velocity is instead proportional to *time*, since for each second of time, the velocity increases by 32 ft/sec.

Solutions to the Projects and Computer Algebra Investigations

1. Notice that whenever x increases by 0.5, $f(x)$ increases by 1, indicating that $f(x)$ is linear. By inspection, we see that $f(x) = 2x$.

   Similarly, $g(x)$ decreases by 1 each time x increases by 0.5. We know, therefore, that $g(x)$ is a linear function with slope $\frac{-1}{0.5} = -2$. The y-intercept is 10, so $g(x) = 10 - 2x$.

   $h(x)$ is an even function which is always positive. Comparing the values of x and $h(x)$, it appears that $h(x) = x^2$.

   $F(x)$ is an odd function that seems to vary between -1 and 1. We guess that $F(x) = \sin x$ and check with a calculator.

   $G(x)$ is also an odd function that varies between -1 and 1. Notice that $G(x) = F(2x)$, and thus $G(x) = \sin 2x$.

   Notice also that $H(x)$ is exactly 2 more than $F(x)$ for all x, so $H(x) = 2 + \sin x$.

2. (a) Compounding daily (continuously),
$$P = P_0 e^{rt}$$
$$= \$450{,}000 e^{(0.06)(213)}$$
$$\approx \$1.5977 \times 10^{11}$$

   This amounts to approximately $160 billion.

   (b) Compounding yearly,
$$A = \$450{,}000(1 + 0.06)^{213}$$
$$= \$450{,}000(1.06)^{213} \approx \$450{,}000(245555.29)$$
$$\approx \$1.10499882 \times 10^{11}$$

   This is only about $110.5 billion.

(c) We first wish to find the interest that will accrue during 1990. For 1990, the principal is 1.105×10^{11}. At 6% annual interest, during 1990 the money will earn
$$0.06 \times \$1.105 \times 10^{11} = \$6.63 \times 10^9.$$
The number of seconds in a year is
$$\left(365\frac{\text{days}}{\text{year}}\right)\left(24\frac{\text{hours}}{\text{day}}\right)\left(60\frac{\text{mins}}{\text{hour}}\right)\left(60\frac{\text{secs}}{\text{min}}\right) = 31{,}536{,}000 \text{ sec}.$$
Thus, over 1990, interest is accumulating at the rate of
$$\frac{\$6.63 \times 10^9}{31{,}536{,}000 \text{ sec}} \approx \$210.24 \text{ /sec}.$$

3. (a) A CAS gives $f(x) = (x-a)(x+a)(x+b)(x-c)$.
   (b) The graph of $f(x)$ crosses the x-axis at $x = a, = -a, x = -b, x = c$; it crosses the y-axis at a^2bc. Since the coefficient of x^4 (namely 1) is positive, the graph of f looks like that shown in Figure 1.10.

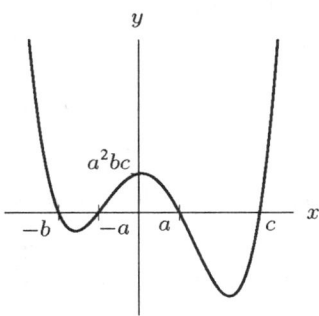

Figure 1.10: Graph of $f(x) = (x-a)(x+a)(x+b)(x-c)$

4. (a) A CAS gives $f(x) = -(x-1)^2(x-3)^3$.
   (b) The answer to part (a) shows that f has a double root at $x = 1$, so near $x = 1$, the graph of f looks like an upside-down parabola touching the x-axis at $x = 1$. Similarly, f has a triple root at $x = 3$. Near $x = 3$, the graph of f looks like the graph of $y = x^3$, flipped over the x-axis and shifted to the right by 3, so that the "seat" is at $x = 3$. See Figure 1.11.

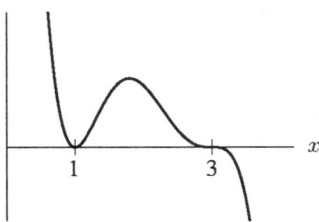

Figure 1.11: Graph of $f(x) = -(x-1)^2(x-3)^3$

5. (a) As $x \to \infty$, the term e^{6x} dominates and tends to ∞. Thus, $f(x) \to \infty$ as $x \to \infty$.
   As $x \to -\infty$, the terms of the form e^{kx}, where $k = 6, 5, 4, 3, 2, 1$, all tend to zero. Thus, $f(x) \to 16$ as $x \to -\infty$.
   (b) A CAS gives
   $$f(x) = (e^x + 1)(e^{2x} - 2)(e^x - 2)(e^{2x} + 2e^x + 4).$$
   Since e^x is always positive, the factors $(e^x + 1)$ and $(e^{2x} + 2e^x + 4)$ are never zero. The other factors each lead to a zero, so there are two zeros.
   (c) The zeros are given by
   $$e^{2x} = 2 \quad \text{so} \quad x = \frac{\ln 2}{2}$$
   $$e^x = 2 \quad \text{so} \quad x = \ln 2.$$
   Thus, one zero is twice the size of the other.

6. (a) Since $f(x) = x^2 - x$,

$$f(f(x)) = (f(x))^2 - f(x) = (x^2 - x)^2 - (x^2 - x) = x - 2x^3 + x^4.$$

Using the CAS to define the function $f(x)$, and then asking it to expand $f(f(f(x)))$, we get

$$f(f(f(x))) = -x + x^2 + 2x^3 - 5x^4 + 2x^5 + 4x^6 - 4x^7 + x^8.$$

(b) The degree of $f(f(x))$ (that is, f composed with itself 2 times) is $4 = 2^2$. The degree of $f(f(f(x)))$ (that is, f composed with itself 3 times), is $8 = 2^3$. Each time you substitute f into itself, the degree is multiplied by 2, because you are substituting in a degree 2 polynomial. So we expect the degree of $f(f(f(f(f(f(x))))))$ (that is, f composed with itself 6 times) to be $64 = 2^6$.

7. (a) A CAS or division gives

$$f(x) = \frac{x^3 - 30}{x - 3} = x^2 + 3x + 9 - \frac{3}{x - 3},$$

so $p(x) = x^2 + 3x + 9$, and $r(x) = -3$, and $q(x) = x - 3$.

(b) The vertical asymptote is $x = 3$. Near $x = 3$, the values of $p(x)$ are much smaller than the values of $r(x)/q(x)$. Thus

$$f(x) \approx \frac{-3}{x - 3} \quad \text{for } x \text{ near } 3.$$

(c) For large x, the values of $p(x)$ are much larger than the value of $r(x)/q(x)$. Thus

$$f(x) \approx x^2 + 3x + 9 \quad \text{as } x \to \infty, x \to -\infty.$$

(d) Figure 1.12 shows $f(x)$ and $y = -3/(x - 3)$ for x near 3. Figure 1.13 shows $f(x)$ and $y = x^2 + 3x + 9$ for $-20 \leq x \leq 20$. Note that in each case the graphs of f and the approximating function are close.

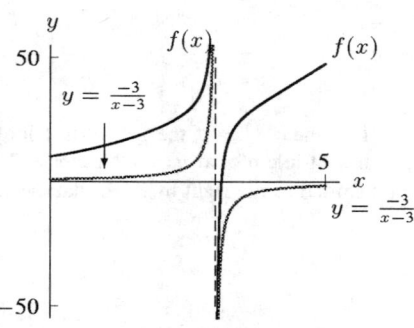

Figure 1.12: Close-up view of $f(x)$ and $y = -3/(x - 3)$

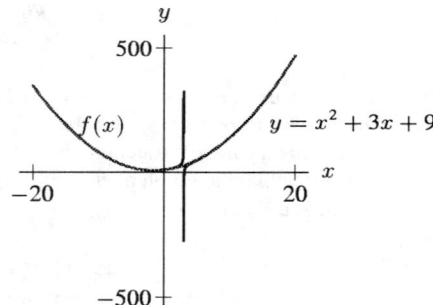

Figure 1.13: Far-away view of $f(x)$ and $y = x^2 + 3x + 9$

8. Using the trigonometric expansion capabilities of your CAS, you get something like

$$\sin(5x) = 5\cos^4(x)\sin(x) - 10\cos^2(x)\sin^3(x) + \sin^5(x).$$

Answers may vary. To get rid of the powers of cosine, use the identity $\cos^2(x) = 1 - \sin^2(x)$. This gives

$$\sin(5x) = 5\sin(x)\left(1 - \sin^2(x)\right)^2 - 10\sin^3(x)\left(1 - \sin^2(x)\right) + \sin^5(x).$$

Finally, using the CAS to simplify,

$$\sin(5x) = 5\sin(x) - 20\sin^3(x) + 16\sin^5(x).$$

9. Using the trigonometric expansion capabilities of your computer algebra system, you get something like
$$\cos(4x) = \cos^4(x) - 6\cos^2(x)\sin^2(x) + \sin^4(x).$$

Answers may vary.

(a) To get rid of the powers of cosine, use the identity $\cos^2(x) = 1 - \sin^2(x)$. This gives
$$\cos(4x) = \cos^4(x) - 6\cos^2(x)\left(1 - \cos^2(x)\right) + \left(1 - \cos^2(x)\right)^2.$$
Finally, using the CAS to simplify,
$$\cos(4x) = 1 - 8\cos^2(x) + 8\cos^4(x).$$

(b) This time we use $\sin^2(x) = 1 - \cos^2(x)$ to get rid of powers of sine. We get
$$\cos(4x) = \sin^4(x) - 6\sin^2(x)\left(1 - \sin^2(x)\right) + \left(1 - \sin^2(x)\right)^2 = 1 - 8\sin^2(x) + 8\sin^4(x).$$

Solutions to Problems on the Binomial Theorem

1. From the formula, C_k^n looks as though it could be a fraction; however, the entries we see in the first 6 rows of Pascal's triangle are all integers. Since all further entries are obtained by adding the entries above, they must all be positive integers.

    Alternatively, C_k^n represents the number of ways the term $x^{n-k}y^k$ arises when $(x+y)^n$ is multiplied out. This gives another way of seeing that C_k^n must be a positive integer.

2. (a) There seems to be a "mirror symmetry" in Pascal's triangle. Draw a vertical line down the middle of the triangle. It appears that as we move away from the center line the corresponding entries on either side of the line are equal.

    (b) Let's write this down in terms of the C_k^n for the first several rows. We already know that $C_0^n = C_1^n = 1$ for all n. The second row only has three entries. We really start to see the pattern starting in the third row. There we have $C_2^3 = C_1^3$. Then for the next row, $C_3^4 = C_1^4$. Next, $C_4^5 = C_1^5$ and $C_3^5 = C_2^5$. Finally for the sixth row we have $C_5^6 = C_1^6$ and $C_2^6 = C_4^6$. Look at how the k's are related in C_k^n on either side of the equation in these identities. In each case they add up to n.

    It looks as though we have the identity
    $$C_k^n = C_{n-k}^n.$$
    This identity holds for the cases we checked above. It also works for the middle entry in an even numbered row, for that entry corresponds to $k = n/2$ and in that case the identity just becomes
    $$C_{\frac{n}{2}}^n = C_{n-\frac{n}{2}}^n = C_{\frac{n}{2}}^n.$$
    We conjecture that the identity holds for all n and for $0 \leq k \leq n$.

    (c) The facts we know about C_k^n are the relation $C_k^n = C_{k-1}^{n-1} + C_k^{n-1}$ and the formula
    $$C_k^n = \frac{n!}{k!(n-k)!}.$$
    The first relation is not as easy to work with for our problem (try it!), but having the latter formula makes it easy to prove our conjecture. For if we write down C_{n-k}^n by substituting $n - k$ for k in the formula we, obtain
    $$C_{n-k}^n = \frac{n!}{(n-k)!(n-(n-k))!} = \frac{n!}{(n-k)!k!} = C_k^n,$$
    exactly what we wanted to show.

3. (a) For the first row we get $1 + 1 = 2$, then it's $1 + 2 + 1 = 4$, $1 + 3 + 3 + 1 = 8$, $1 + 4 + 6 + 4 + 1 = 16$, $1 + 5 + 10 + 10 + 5 + 1 = 32$, and finally $1 + 6 + 15 + 20 + 15 + 6 + 1 = 64$. The sums are 2, 4, 8, 16, 32 and 64, which we recognize as the first six powers of 2. So we might conjecture that the sum of the entries across the nth row in Pascal's triangle is 2^n. In terms of the binomial coefficients, C_k^n, our conjecture is

$$C_0^n + C_1^n + C_2^n + \cdots + C_{n-1}^n + C_n^n = 2^n.$$

(b) The most straightforward way to prove this result is to let $x = y = 1$ in the expansion of $(x + y)^n$. Then

$$2^n = (1 + 1)^n = C_0^n \cdot 1^n + C_1^n \cdot 1^{n-1} \cdot 1 + C_2^n \cdot 1^{n-2} \cdot 1^2 + \cdots + C_n^n \cdot 1^n$$
$$= C_0^n + C_1^n + C_2^n + \cdots + C_n^n.$$

Alternatively, we can prove this by induction on n. There are two steps. We first verify the statement for $n = 1$. Then we assume that the statement is true for a specific n, and using this we try to prove that it also holds for $n + 1$.

For $n = 1$ we must check the sum of the entries in the first row of Pascal's triangle. As we have already observed, in that case we have $C_0^1 + C_1^1 = 1 + 1 = 2 = 2^1$. Our formula is true for $n = 1$. We now assume that our formula holds for n (this is the induction step) and show that

$$C_0^{n+1} + C_1^{n+1} + C_2^{n+1} + \cdots + C_n^{n+1} + C_{n+1}^{n+1} = 2^{n+1}.$$

How can we use the result on adding entries in the nth row (the induction assumption) to find the sum of entries in the $(n + 1)^{\text{th}}$ row? We are well set up for this because the entries in the $(n + 1)^{\text{th}}$ row are obtained from entries in the nth row according to the formula

$$C_k^{n+1} = C_{k-1}^n + C_k^n.$$

We also know that $C_0^{n+1} = C_{n+1}^{n+1} = 1$ and $C_0^n = C_n^n = 1$. Substitute this information into the sum on the entries in the $(n + 1)^{\text{th}}$ row and group terms. We see that each entry in the nth row comes in twice except the first and last:

$$C_0^{n+1} + C_1^{n+1} + C_2^{n+1} + C_3^{n+1} + \cdots + C_n^{n+1} + C_{n+1}^{n+1} =$$
$$1 + (C_0^n + C_1^n) + (C_1^n + C_2^n) + (C_2^n + C_3^n) + \cdots + (C_{n-1}^n + C_n^n) + 1$$
$$= 1 + 1 + 2C_1^n + 2C_2^n + \cdots + 2C_{n-1}^n + 1 + 1$$
$$= 2(1 + C_1^n + C_2^n + \cdots + C_{n-1}^n + 1)$$
$$= 2 \cdot 2^n = 2^{n+1}.$$

We used the induction step to say that $1 + C_1^n + C_2^n + \cdots + C_{n-1}^n + 1 = 2^n$. We have now verified that our formula holds for $n + 1$. Therefore, by induction, it holds for all numbers and the proof is complete.

Solutions to Problems on the Completeness of Real Numbers

1. (a) (i) A lower bound is a number which is less than or equal to all the numbers in the set.
   (ii) The greatest lower bound of a set is the one which is greater than or equal to all the others.
   (b) Any nonempty set of real numbers which has a lower bound has a greatest lower bound.

2. If there were two numbers in all of the intervals, say r and r', with $r < r'$, then each interval $[a_n, b_n]$ would have to contain the interval $[r, r']$. But this means the width $|b_n - a_n|$ would always be greater than $|r - r'|$, which is greater than 0. This contradicts $|b_n - a_n| \to 0$. So our original assumption $r' \neq r$ must be false, i.e., r is unique.

3. (a) Each truncation is greater than or equal to the previous one, so x_n is greater than or equal to all the x_ks with $k \leq n$. Thus $x_n + (1/10^n) > x_k$ for all $k \leq n$.

   Now suppose that $k > n$. The furthest x_k could be from x_n is if all the digits from the $(n + 1)$-th to the k-th are 9s. In that case we would have

$$x_k = x_n + \left(9\frac{1}{10}\right)^{n+1} + \cdots + \left(9\frac{1}{10}\right)^k$$
$$= x_n + \left(\frac{1}{10}\right)^n (.99\cdots 9)$$
$$< x_n + \left(\frac{1}{10}\right)^n.$$

So $x_n + (1/10)^n$ is an upper bound for all the truncations.

(b) By the completeness axiom, the set of truncations has a least upper bound c. Since c is an upper bound for all the x_ns, we have $x_n \leq c$ for all n. Since $x_n + (1/10)^n$ is an upper bound, and c is the least upper bound, $c \leq x_n + (1/10)^n$ for all n.

CHAPTER TWO

Solutions for Section 2.1

1.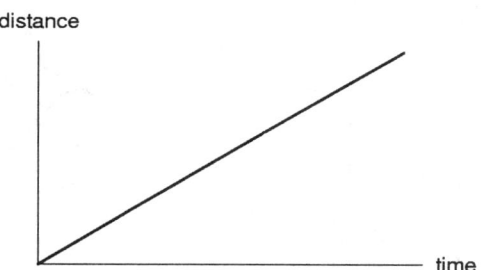

2.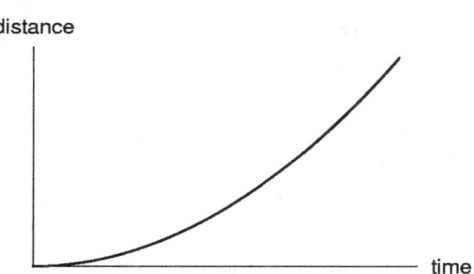

3.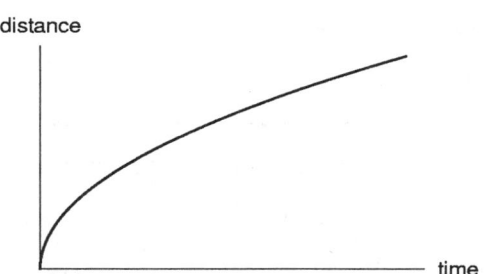

4. The slope is positive at A and D; negative at C and F. The slope is most positive at A; most negative at F.

5.

Slope	−3	−1	0	1/2	1	2
Point	F	C	E	A	B	D

6. $0 <$ slope at $C <$ slope at $B <$ slope of $AB < 1 <$ slope at A. (Note that the line $y = x$, has slope 1.)

7. Since $f(t)$ is concave down between $t = 1$ and $t = 3$, the average velocity between the two times should be less than the instantaneous velocity at $t = 1$ but greater than the instantaneous velocity at time $t = 3$, so $D < A < C$. For analogous reasons, $F < B < E$. Finally, note that f is decreasing at $t = 5$ so $E < 0$, but increasing at $t = 0$, so $D > 0$. Therefore, the ordering from smallest to greatest of the given quantities is

$$F < B < E < 0 < D < A < C.$$

8. One possibility is:

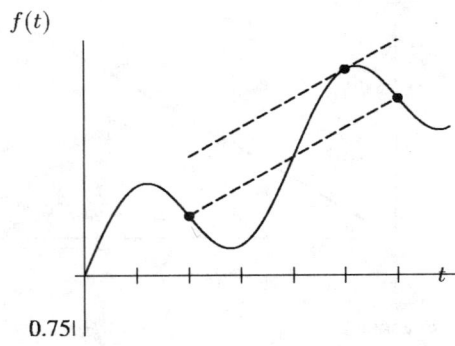

9. Using $h = 0.1, 0.01, 0.001$, we see

$$\frac{(3 + 0.1)^3 - 27}{0.1} = 27.91$$

$$\frac{(3 + 0.01)^3 - 27}{0.01} = 27.09$$

$$\frac{(3 + 0.001)^3 - 27}{0.001} = 27.009.$$

These calculations suggest that $\lim_{h \to 0} \frac{(3 + h)^3 - 27}{h} = 27$.

10. Using radians,

h	$(\cos h - 1)/h$
0.01	-0.005
0.001	-0.0005
0.0001	-0.00005

These values suggest that $\lim_{h \to 0} \frac{\cos h - 1}{h} = 0$.

11. Using $h = 0.1, 0.01, 0.001$, we see

$$\frac{7^{0.1} - 1}{0.1} = 2.148$$

$$\frac{7^{0.01} - 1}{0.01} = 1.965$$

$$\frac{7^{0.001} - 1}{0.001} = 1.948$$

$$\frac{7^{0.0001} - 1}{0.0001} = 1.946.$$

This suggests that $\lim_{h \to 0} \frac{7^h - 1}{h} \approx 1.9\ldots$

12. Using $h = 0.1, 0.01, 0.001$, we see

h	$(e^{1+h} - e)/h$
0.01	2.7319
0.001	2.7196
0.0001	2.7184

These values suggest that $\lim_{h \to 0} \dfrac{e^{1+h} - e}{h} = 2.7\ldots$. In fact, this limit is e.

13.

TABLE 2.1

x	$f(x)$	x	$f(x)$
0.1	1.3	−0.0001	0.9997
0.01	1.03	−0.001	0.997
0.001	1.003	−0.01	0.97
0.0001	1.0003	−0.1	0.7

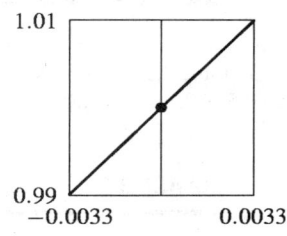

Figure 2.1

From Table 2.1, it appears the limit is 1. This is confirmed by Figure 2.1. An appropriate window is $-0.0033 < x < 0.0033, 0.99 < y < 1.01$.

14.

TABLE 2.2

x	$f(x)$	x	$f(x)$
0.1	−0.99	−0.0001	−0.99999999
0.01	−0.9999	−0.001	−0.999999
0.001	−0.999999	−0.01	−0.9999
0.0001	−0.99999999	−0.1	−0.99

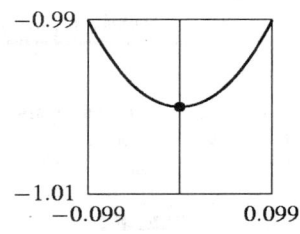

Figure 2.2

From Table 2.2, it appears the limit is -1. This is confirmed by Figure 2.2. An appropriate window is $-0.099 < x < 0.099, -1.01 < y < -0.99$.

15.

TABLE 2.3

x	$f(x)$	x	$f(x)$
0.1	0.1987	−0.0001	−0.0002
0.01	0.0200	−0.001	−0.0020
0.001	0.0020	−0.01	−0.0200
0.0001	0.0002	−0.1	−0.1987

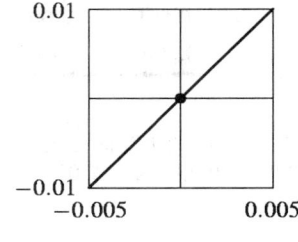

Figure 2.3

From Table 2.3, it appears the limit is 0. This is confirmed by Figure 2.3. An appropriate window is $-0.005 < x < 0.005, -0.01 < y < 0.01$.

16.

TABLE 2.4

x	$f(x)$	x	$f(x)$
0.1	0.2955	−0.0001	−0.0003
0.01	0.0300	−0.001	−0.0030
0.001	0.0030	−0.01	−0.0300
0.0001	0.0003	−0.1	−0.2955

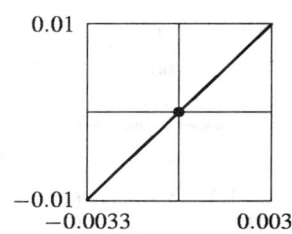

Figure 2.4

From Table 2.4, it appears the limit is 0. This is confirmed by Figure 2.4. An appropriate window is $-0.0033 < x < 0.0033, -0.01 < y < 0.01$.

17.

TABLE 2.5

x	$f(x)$	x	$f(x)$
0.1	1.9867	-0.0001	2.0000
0.01	1.9999	-0.001	2.0000
0.001	2.0000	-0.01	1.9999
0.0001	2.0000	-0.1	1.9867

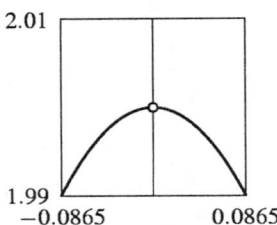

Figure 2.5

From Table 2.5, it appears the limit is 2. This is confirmed by Figure 2.5. An appropriate window is $-0.0865 < x < 0.0865$, $1.99 < y < 2.01$.

18.

TABLE 2.6

x	$f(x)$	x	$f(x)$
0.1	2.9552	-0.0001	3.0000
0.01	2.9996	-0.001	3.0000
0.001	3.0000	-0.01	2.9996
0.0001	3.0000	-0.1	2.9552

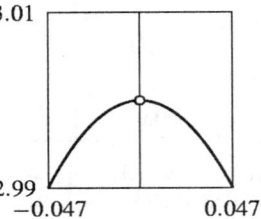

Figure 2.6

From Table 2.6, it appears the limit is 3. This is confirmed by Figure 2.6. An appropriate window is $-0.047 < x < 0.047$, $2.99 < y < 3.01$.

19.

TABLE 2.7

x	$f(x)$	x	$f(x)$
0.1	1.0517	-0.0001	1.0000
0.01	1.0050	-0.001	0.9995
0.001	1.0005	-0.01	0.9950
0.0001	1.0001	-0.1	0.9516

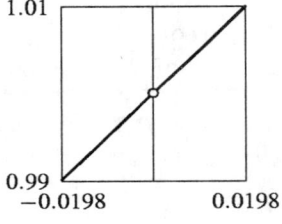

Figure 2.7

From Table 2.7, it appears the limit is 1. This is confirmed by Figure 2.7. An appropriate window is $-0.0198 < x < 0.0198$, $0.99 < y < 1.01$.

20.

TABLE 2.8

x	$f(x)$	x	$f(x)$
0.1	2.2140	-0.0001	1.9998
0.01	2.0201	-0.001	1.9980
0.001	2.0020	-0.01	1.9801
0.0001	2.0002	-0.1	1.8127

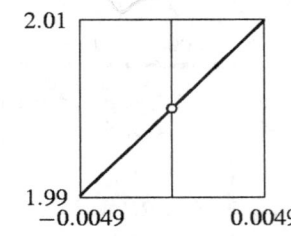

Figure 2.8

From Table 2.8, it appears the limit is 2. This is confirmed by Figure 2.8. An appropriate window is $-0.0049 < x < 0.0049$, $1.99 < y < 2.01$.

21.

TABLE 2.9

x	$f(x)$
0.1	0.0666
0.01	0.0067
0.001	0.0007
0.0001	0

x	$f(x)$
−0.0001	−0.0001
−0.001	−0.0007
−0.01	−0.0067
−0.1	−0.0666

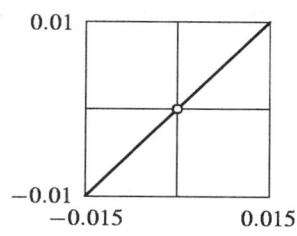

Figure 2.9

From Table 2.9, it appears the limit is 0. This is confirmed by Figure 2.9. An appropriate window is $-0.015 < x < 0.015$, $-0.01 < y < 0.01$.

22.

TABLE 2.10

x	$f(x)$
0.1	0.3365
0.01	0.0337
0.001	0.0034
0.0001	0.0004

x	$f(x)$
−0.0001	−0.0004
−0.001	−0.0034
−0.01	−0.0337
−0.1	−0.3365

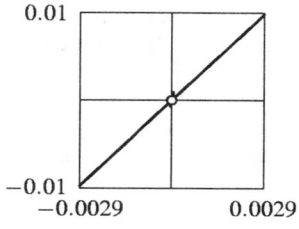

Figure 2.10

From Table 2.10, it appears the limit is 0. This is confirmed by Figure 2.10. An appropriate window is $-0.0029 < x < 0.0029$, $-0.01 < y < 0.01$.

Solutions for Section 2.2

1.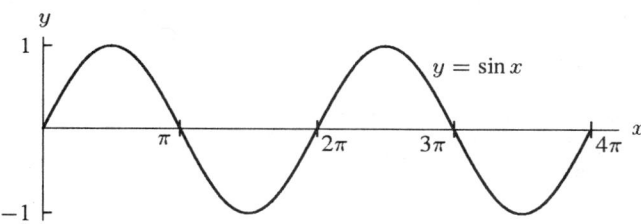

   Since $\sin x$ is decreasing for values near $x = 3\pi$, its derivative at $x = 3\pi$ is negative.

2. (a) Using a calculator we obtain the values found in the table below:

x	1	1.5	2	2.5	3
e^x	2.72	4.48	7.39	12.18	20.09

   (b) The average rate of change of $f(x) = e^x$ between $x = 1$ and $x = 3$ is

   $$\text{Average rate of change} = \frac{f(3) - f(1)}{3 - 1} = \frac{e^3 - e}{3 - 1} \approx \frac{20.09 - 2.72}{2} = 8.69.$$

   (c) First we find the average rates of change of $f(x) = e^x$ between $x = 1.5$ and $x = 2$, and between $x = 2$ and $x = 2.5$:

   $$\text{Average rate of change} = \frac{f(2) - f(1.5)}{2 - 1.5} = \frac{e^2 - e^{1.5}}{2 - 1.5} \approx \frac{7.39 - 4.48}{0.5} = 5.82$$

   $$\text{Average rate of change} = \frac{f(2.5) - f(2)}{2.5 - 2} = \frac{e^{2.5} - e^2}{2.5 - 2} \approx \frac{12.18 - 7.39}{0.5} = 9.58.$$

   Now we approximate the instantaneous rate of change at $x = 2$ by averaging these two rates:

   $$\text{Instantaneous rate of change} \approx \frac{5.82 + 9.58}{2} = 7.7.$$

30 CHAPTER TWO /SOLUTIONS

3. One possible choice of points is shown below.

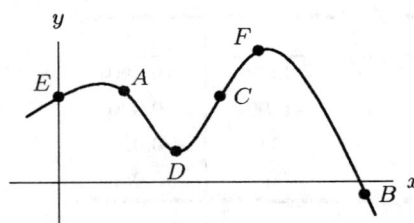

4. (a) The average rate of change from $x = a$ to $x = b$ is the slope of the line between the points on the curve with $x = a$ and $x = b$. Since the curve is concave down, the line from $x = 1$ to $x = 3$ has a greater slope than the line from $x = 3$ to $x = 5$, and so the average rate of change between $x = 1$ and $x = 3$ is greater than that between $x = 3$ and $x = 5$.
   (b) Since f is increasing, $f(5)$ is the greater.
   (c) As in part (a), f is concave down and f' is decreasing throughout so $f'(1)$ is the greater.

5.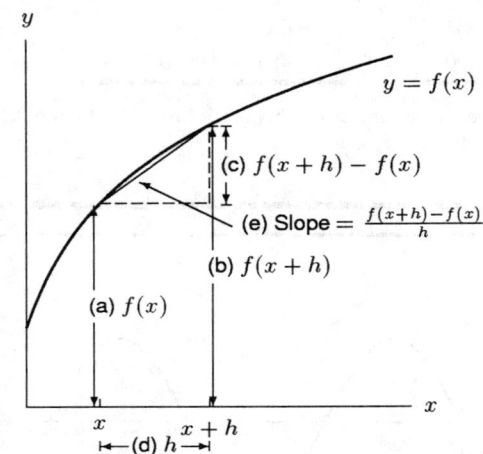

6.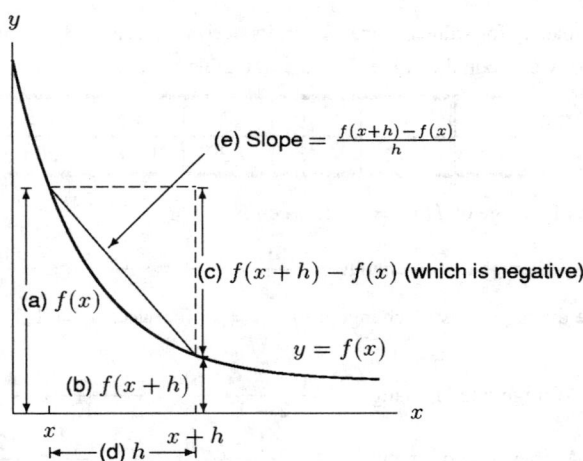

7.

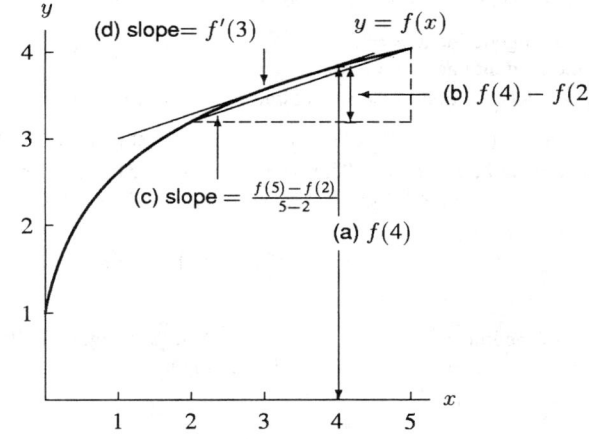

8.

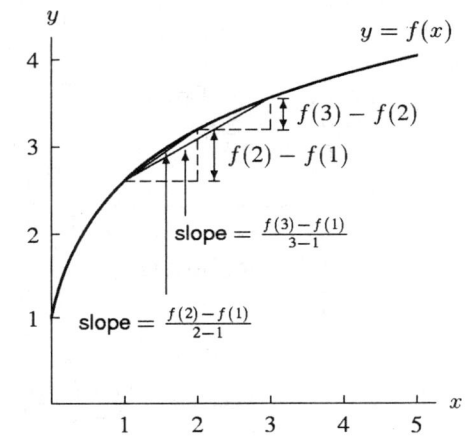

(a) $f(4) > f(3)$ since f is increasing.
(b) From the figure, it appears that $f(2) - f(1) > f(3) - f(2)$.
(c) $\dfrac{f(2) - f(1)}{2 - 1}$ represents the slope of the secant line connecting the graph at $x = 1$ and $x = 2$. This is greater than the slope of the secant line connecting the graph at $x = 1$ and $x = 3$ which is $\dfrac{f(3) - f(1)}{3 - 1}$.
(d) The function is steeper at $x = 1$ than at $x = 4$ so $f'(1) > f'(4)$.

9. Figure 2.11 shows the quantities in which we are interested.

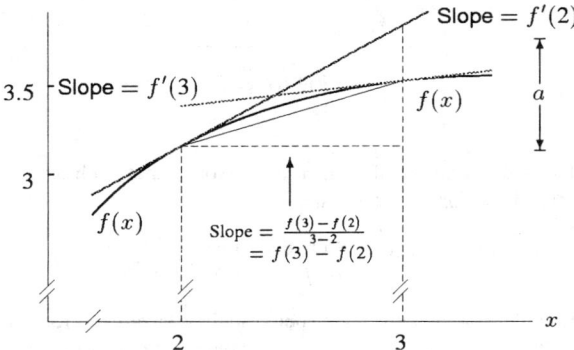

Figure 2.11

The quantities $f'(2), f'(3)$ and $f(3) - f(2)$ have the following interpretations:

- $f'(2)$ = slope of the tangent line at $x = 2$
- $f'(3)$ = slope of the tangent line at $x = 3$
- $f(3) - f(2) = \frac{f(3)-f(2)}{3-2}$ = slope of the secant line from $f(2)$ to $f(3)$.

From Figure 2.11, it is clear that $0 < f(3) - f(2) < f'(2)$. By extending the secant line past the point $(3, f(3))$, we can see that it lies above the tangent line at $x = 3$. Thus $0 < f'(3) < f(3) - f(2) < f'(2)$. From the figure, the height a appears less than 1, so $f'(2) = \frac{a}{3-2} = \frac{a}{1} < 1$.

Thus
$$0 < f'(3) < f(3) - f(2) < f'(2) < 1.$$

10. (a) $f(4)/4$ is the slope of the line connecting $(0,0)$ to $(4, f(4))$. (See Figure 2.12.)
    (b) It is clear from the picture for part (a) that $f(3)/3 > f(4)/4$.

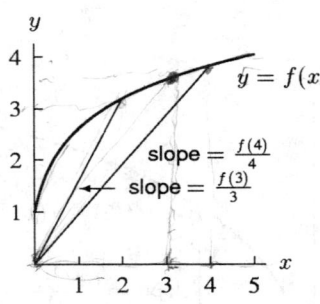

Figure 2.12

11. (a) For the line from A to B,
$$\text{Slope} = \frac{f(b) - f(a)}{b - a}.$$

(b) The tangent line at point C appears to be parallel to the line from A to B. Assuming this to be the case, the lines have the same slope.

(c) There is only one other point, labeled D in Figure 2.13, at which the tangent line is parallel to the line joining A and B.

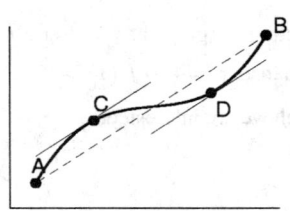

Figure 2.13

12. The derivative, $f'(2)$, is the rate of change of x^3 at $x = 2$. Notice that each time x changes by 0.001 in the table, the value of x^3 changes by 0.012. Therefore, we estimate
$$f'(2) = \frac{\text{Rate of change}}{\text{of } f \text{ at } x = 2} \approx \frac{0.012}{0.001} = 12.$$

The function values in the table look exactly linear because they have been rounded. For example, the exact value of x^3 when $x = 2.001$ is 8.012006001, not 8.012. Thus, the table can tell us only that the derivative is approximately 12. Example 5 on page 113 shows how to compute the derivative of $f(x)$ exactly.

13. Using the definition of the derivative

$$g'(-1) = \lim_{h \to 0} \frac{g(-1+h) - g(-1)}{h}$$
$$= \lim_{h \to 0} \frac{(3(-1+h)^2 + 5(-1+h)) - (3(-1)^2 + 5(-1))}{h}$$
$$= \lim_{h \to 0} \frac{(3(1 - 2h + h^2) - 5 + 5h) - (-2)}{h}$$
$$= \lim_{h \to 0} \frac{3 - 6h + 3h^2 - 3 + 5h}{h}$$
$$= \lim_{h \to 0} \frac{(-h + 3h^2)}{h} = \lim_{h \to 0} (-1 + 3h) = -1.$$

14. Using the definition of the derivative, we have

$$f'(10) = \lim_{h \to 0} \frac{f(10+h) - f(10)}{h}$$
$$= \lim_{h \to 0} \frac{5(10+h)^2 - 5(10)^2}{h}$$
$$= \lim_{h \to 0} \frac{500 + 100h + 5h^2 - 500}{h}$$
$$= \lim_{h \to 0} \frac{100h + 5h^2}{h}$$
$$= \lim_{h \to 0} \frac{h(100 + 5h)}{h}$$
$$= \lim_{h \to 0} 100 + 5h$$
$$= 100.$$

15. Using the definition of the derivative, we have

$$f'(-2) = \lim_{h \to 0} \frac{f(-2+h) - f(-2)}{h}$$
$$= \lim_{h \to 0} \frac{(-2+h)^3 - (-2)^3}{h}$$
$$= \lim_{h \to 0} \frac{(-8 + 12h - 6h^2 + h^3) - (-8)}{h}$$
$$= \lim_{h \to 0} \frac{12h - 6h^2 + h^3}{h}$$
$$= \lim_{h \to 0} \frac{h(12 - 6h + h^2)}{h}$$
$$= \lim_{h \to 0} (12 - 6h + h^2),$$

which goes to 12 as $h \to 0$. So $f'(-2) = 12$.

16.
$$f'(1) = \lim_{h \to 0} \frac{f(1+h) - f(1)}{h} = \lim_{h \to 0} \frac{((1+h)^3 + 5) - (1^3 + 5)}{h}$$
$$= \lim_{h \to 0} \frac{1 + 3h + 3h^2 + h^3 + 5 - 1 - 5}{h} = \lim_{h \to 0} \frac{3h + 3h^2 + h^3}{h}$$
$$= \lim_{h \to 0} (3 + 3h + h^2) = 3.$$

17.
$$g'(2) = \lim_{h \to 0} \frac{g(2+h) - g(2)}{h} = \lim_{h \to 0} \frac{\frac{1}{2+h} - \frac{1}{2}}{h}$$
$$= \lim_{h \to 0} \frac{2 - (2+h)}{h(2+h)2} = \lim_{h \to 0} \frac{-h}{h(2+h)2}$$
$$= \lim_{h \to 0} \frac{-1}{(2+h)2} = -\frac{1}{4}$$

18.
$$g'(2) = \lim_{h \to 0} \frac{g(2+h) - g(2)}{h} = \lim_{h \to 0} \frac{\frac{1}{(2+h)^2} - \frac{1}{2^2}}{h}$$
$$= \lim_{h \to 0} \frac{2^2 - (2+h)^2}{2^2(2+h)^2 h} = \lim_{h \to 0} \frac{4 - 4 - 4h - h^2}{4h(2+h)^2}$$
$$= \lim_{h \to 0} \frac{-4h - h^2}{4h(2+h)^2} = \lim_{h \to 0} \frac{-4 - h}{4(2+h)^2}$$
$$= \frac{-4}{4(2)^2} = -\frac{1}{4}.$$

19. As we saw in the answer to Problem 15, the slope of the tangent line to $f(x) = x^3$ at $x = -2$ is 12. When $x = -2$, $f(x) = -8$ so we know the point $(-2, -8)$ is on the tangent line. Thus the equation of the tangent line is $y = 12(x + 2) - 8 = 12x + 16$.

20. As we saw in the answer to Problem 14, the slope of the tangent line to $f(x) = 5x^2$ at $x = 10$ is 100. When $x = 10$, $f(x) = 500$ so $(10, 500)$ is a point on the tangent line. Thus $y = 100(x - 10) + 500 = 100x - 500$.

21. We know that the slope of the tangent line to $f(x) = x$ when $x = 20$ is 1. When $x = 20$, $f(x) = 20$ so $(20, 20)$ is on the tangent line. Thus the equation of the tangent line is $y = 1(x - 20) + 20 = x$.

22. First find the derivative of $f(x) = 1/x^2$ at $x = 1$.
$$f'(1) = \lim_{h \to 0} \frac{f(1+h) - f(1)}{h} = \lim_{h \to 0} \frac{\frac{1}{(1+h)^2} - \frac{1}{1^2}}{h}$$
$$= \lim_{h \to 0} \frac{1^2 - (1+h)^2}{h(1+h)^2} = \lim_{h \to 0} \frac{1 - (1 + 2h + h^2)}{h(1+h)^2}$$
$$= \lim_{h \to 0} \frac{-2h - h^2}{h(1+h)^2} = \lim_{h \to 0} \frac{-2 - h}{(1+h)^2} = -2$$

Thus the tangent line has a slope of -2 and goes through the point $(1, 1)$, and so its equation is
$$y - 1 = -2(x - 1) \quad \text{or} \quad y = -2x + 3.$$

23. We need to look at the difference quotient and take the limit as h approaches zero. The difference quotient is
$$\frac{f(3+h) - f(3)}{h} = \frac{[(3+h)^2 + 1] - 10}{h} = \frac{9 + 6h + h^2 + 1 - 10}{h} = \frac{6h + h^2}{h} = \frac{h(6+h)}{h}.$$

Since $h \neq 0$, we can divide by h in the last expression to get $6 + h$. Now the limit as h goes to 0 of $6 + h$ is clearly 6, so
$$f'(3) = \lim_{h \to 0} \frac{h(6+h)}{h} = \lim_{h \to 0} (6 + h) = 6.$$

So the slope of the tangent line at $x = 3$ is 6. Since $f(3) = 10$, the tangent line passes through $(3, 10)$, and so its equation is
$$y - 10 = 6(x - 3), \quad \text{or} \quad y = 6x - 8.$$

24.
$$f'(3) = \lim_{h \to 0} \frac{f(3+h) - f(3)}{h} = \lim_{h \to 0} \frac{(3+h)^2 + 3 + h - (3^2 + 3)}{h}$$
$$= \lim_{h \to 0} \frac{9 + 6h + h^2 + 3 + h - 9 - 3}{h} = \lim_{h \to 0} \frac{7h + h^2}{h} = \lim_{h \to 0} (7 + h) = 7.$$

Thus the slope of the tangent line is 7. Since $f(3) = 3^2 + 3 = 12$, the line goes through the point $(3, 12)$, and therefore its equation is
$$y - 12 = 7(x - 3) \quad \text{or} \quad y = 7x - 9.$$

The graph is shown below.

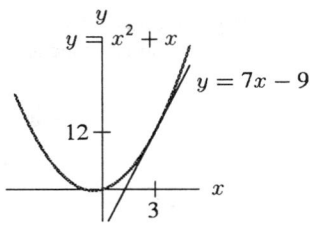

25. (a)
$$f'(0) = \lim_{h \to 0} \frac{\overbrace{\sin h}^{h \text{ in degrees}} - \overbrace{\sin 0}^{0}}{h} = \frac{\sin h}{h}.$$

To four decimal places,
$$\frac{\sin 0.2}{0.2} \approx \frac{\sin 0.1}{0.1} \approx \frac{\sin 0.01}{0.01} \approx \frac{\sin 0.001}{0.001} \approx 0.01745$$

so $f'(0) \approx 0.01745$.

(b) Consider the ratio $\frac{\sin h}{h}$. As we approach 0, the numerator, $\sin h$, will be much smaller in magnitude if h is in degrees than it would be if h were in radians. For example, if $h = 1°$ radian, $\sin h = 0.8415$, but if $h = 1$ degree, $\sin h = 0.01745$. Thus, since the numerator is smaller for h measured in degrees while the denominator is the same, we expect the ratio $\frac{\sin h}{h}$ to be smaller.

26. We want $f'(2)$. The exact answer is
$$f'(2) = \lim_{h \to 0} \frac{f(2+h) - f(2)}{h} = \lim_{h \to 0} \frac{(2+h)^{2+h} - 4}{h},$$

but we can approximate this. If $h = 0.001$, then
$$\frac{(2.001)^{2.001} - 4}{0.001} \approx 6.779$$

and if $h = 0.0001$ then
$$\frac{(2.0001)^{2.0001} - 4}{0.0001} \approx 6.773,$$

so $f'(2) \approx 6.77$.

27. Notice that we can't get all the information we want just from the graph of f for $0 \leq x \leq 2$, shown on the left in Figure 2.14. Looking at this graph, it looks as if the slope at $x = 0$ is 0. But if we zoom in on the graph near $x = 0$, we get the graph of f for $0 \leq x \leq 0.05$, shown on the right in Figure 2.14. We see that f does dip down quite a bit between $x = 0$ and $x \approx 0.11$. In fact, it now looks like $f'(0)$ is around -1. Note that since $f(x)$ is undefined for $x < 0$, this derivative only makes sense as we approach zero from the right.

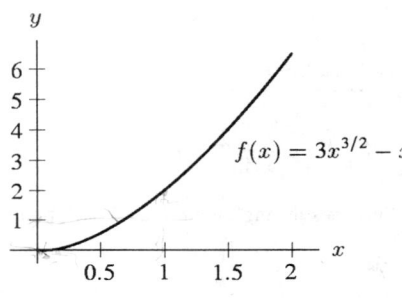
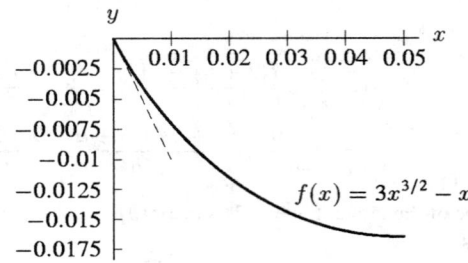

Figure 2.14

We zoom in on the graph of f near $x = 1$ to get a more accurate picture from which to estimate $f'(1)$. A graph of f for $0.7 \le x \le 1.3$ is shown in Figure 2.15. [Keep in mind that the axes shown in this graph don't cross at the origin!] Here we see that $f'(1) \approx 3.5$.

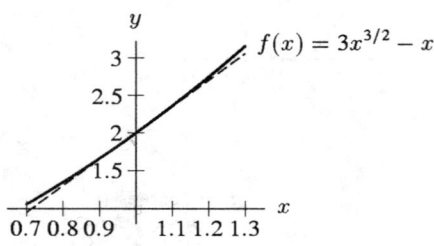

Figure 2.15

28.
$$f'(1) = \lim_{h \to 0} \frac{f(1+h) - f(1)}{h} = \lim_{h \to 0} \frac{\ln(\cos(1+h)) - \ln(\cos 1)}{h}$$

For $h = 0.001$, the difference quotient $= -1.55912$; for $h = 0.0001$, the difference quotient $= -1.55758$. The instantaneous rate of change of f therefore appears to be about -1.558 at $x = 1$.

At $x = \frac{\pi}{4}$, if we try $h = 0.0001$, then

$$\text{difference quotient} = \frac{\ln[\cos(\frac{\pi}{4} + 0.0001)] - \ln(\cos \frac{\pi}{4})}{0.0001} \approx -1.0001.$$

The instantaneous rate of change of f appears to be about -1 at $x = \frac{\pi}{4}$.

29. (a) We construct the difference quotient using erf(0) and each of the other given values:

$$\text{erf}'(0) \approx \frac{\text{erf}(1) - \text{erf}(0)}{1 - 0} = 0.84270079$$

$$\text{erf}'(0) \approx \frac{\text{erf}(0.1) - \text{erf}(0)}{0.1 - 0} = 1.1246292$$

$$\text{erf}'(0) \approx \frac{\text{erf}(0.01) - \text{erf}(0)}{0.01 - 0} = 1.128342.$$

Based on these estimates, the best estimate is $\text{erf}'(0) \approx 1.12$; the subsequent digits have not yet stabilized.

(b) Using erf(0.001), we have

$$\text{erf}'(0) \approx \frac{\text{erf}(0.001) - \text{erf}(0)}{0.001 - 0} = 1.12838$$

and so the best estimate is now 1.1283.

30. (a) **TABLE 2.11**

x	$\frac{\sinh(x+0.001)-\sinh(x)}{0.001}$	$\frac{\sinh(x+0.0001)-\sinh(x)}{0.0001}$	so $f'(0) \approx$	$\cosh(x)$
0	1.00000	1.00000	1.00000	1.00000
0.3	1.04549	1.04535	1.04535	1.04534
0.7	1.25555	1.25521	1.25521	1.25517
1	1.54367	1.54314	1.54314	1.54308

(b) It seems that they are approximately the same, i.e. the derivative of $\sinh(x) = \cosh(x)$ for $x = 0, 0.3, 0.7,$ and 1.

31. We want to approximate $P'(0)$ and $P'(2)$. Since for small h

$$P'(0) \approx \frac{P(h) - P(0)}{h},$$

if we take $h = 0.01$, we get

$$P'(0) \approx \frac{1.15(1.014)^{0.01} - 1.15}{0.01} = 0.01599 \text{ billion/year}$$
$$= 16.0 \text{ million people/year}$$

$$P'(2) \approx \frac{1.15(1.014)^{2.01} - 1.15(1.014)^2}{0.01} = 0.0164 \text{ billion/year}$$
$$= 16.4 \text{ million people/year}$$

32.

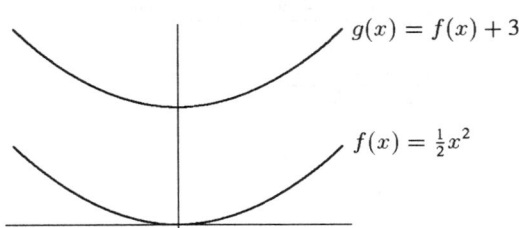

(a) From the figure above, it appears that the slopes of the tangent lines to the two graphs are the same at each x. For $x = 0$, the slopes of the tangents to the graphs of $f(x)$ and $g(x)$ at 0 are

$$f'(0) = \lim_{h \to 0} \frac{f(0+h) - f(0)}{h} \qquad g'(0) = \lim_{h \to 0} \frac{g(0+h) - g(0)}{h}$$
$$= \lim_{h \to 0} \frac{f(h) - 0}{h} \qquad\qquad = \lim_{h \to 0} \frac{g(h) - g(0)}{h}$$
$$= \lim_{h \to 0} \frac{\frac{1}{2}h^2}{h} \qquad\qquad = \lim_{h \to 0} \frac{\frac{1}{2}h^2 + 3 - 3}{h}$$
$$= \lim_{h \to 0} \frac{1}{2}h \qquad\qquad = \lim_{h \to 0} \frac{\frac{1}{2}h^2}{h}$$
$$= 0, \qquad\qquad\qquad = \lim_{h \to 0} \frac{1}{2}h$$
$$\qquad\qquad\qquad\qquad = 0.$$

For $x = 2$, the slopes of the tangents to the graphs of $f(x)$ and $g(x)$ are

$$f'(2) = \lim_{h \to 0} \frac{f(2+h) - f(2)}{h}$$
$$= \lim_{h \to 0} \frac{\frac{1}{2}(2+h)^2 - \frac{1}{2}(2)^2}{h}$$
$$= \lim_{h \to 0} \frac{\frac{1}{2}(4 + 4h + h^2) - 2}{h}$$
$$= \lim_{h \to 0} \frac{2 + 2h + \frac{1}{2}h^2 - 2}{h}$$
$$= \lim_{h \to 0} \frac{2h + \frac{1}{2}h^2}{h}$$
$$= \lim_{h \to 0} \left(2 + \frac{1}{2}h\right)$$
$$= 2,$$

$$g'(2) = \lim_{h \to 0} \frac{g(2+h) - g(2)}{h}$$
$$= \lim_{h \to 0} \frac{\frac{1}{2}(2+h)^2 + 3 - (\frac{1}{2}(2)^2 + 3)}{h}$$
$$= \lim_{h \to 0} \frac{\frac{1}{2}(2+h)^2 - \frac{1}{2}(2)^2}{h}$$
$$= \lim_{h \to 0} \frac{\frac{1}{2}(4 + 4h + h^2) - 2}{h}$$
$$= \lim_{h \to 0} \frac{2 + 2h + \frac{1}{2}(h^2) - 2}{h}$$
$$= \lim_{h \to 0} \frac{2h + \frac{1}{2}(h^2)}{h}$$
$$= \lim_{h \to 0} \left(2 + \frac{1}{2}h\right)$$
$$= 2.$$

For $x = x_0$, the slopes of the tangents to the graphs of $f(x)$ and $g(x)$ are

$$f'(x_0) = \lim_{h \to 0} \frac{f(x_0 + h) - f(x_0)}{h}$$
$$= \lim_{h \to 0} \frac{\frac{1}{2}(x_0 + h)^2 - \frac{1}{2}x_0^2}{h}$$
$$= \lim_{h \to 0} \frac{\frac{1}{2}(x_0^2 + 2x_0 h + h^2) - \frac{1}{2}x_0^2}{h}$$
$$= \lim_{h \to 0} \frac{x_0 h + \frac{1}{2}h^2}{h}$$
$$= \lim_{h \to 0} \left(x_0 + \frac{1}{2}h\right)$$
$$= x_0,$$

$$g'(x_0) = \lim_{h \to 0} \frac{g(x_0 + h) - g(x_0)}{h}$$
$$= \lim_{h \to 0} \frac{\frac{1}{2}(x_0 + h)^2 + 3 - (\frac{1}{2}(x_0)^2 + 3)}{h}$$
$$= \lim_{h \to 0} \frac{\frac{1}{2}(x_0 + h)^2 - \frac{1}{2}(x_0)^2}{h}$$
$$= \lim_{h \to 0} \frac{\frac{1}{2}(x_0^2 + 2x_0 h + h^2) - \frac{1}{2}x_0^2}{h}$$
$$= \lim_{h \to 0} \frac{x_0 h + \frac{1}{2}h^2}{h}$$
$$= \lim_{h \to 0} \left(x_0 + \frac{1}{2}h\right)$$
$$= x_0.$$

(b)
$$g'(x) = \lim_{h \to 0} \frac{g(x+h) - g(x)}{h}$$
$$= \lim_{h \to 0} \frac{f(x+h) + C - (f(x) + C)}{h}$$
$$= \lim_{h \to 0} \frac{f(x+h) - f(x)}{h}$$
$$= f'(x).$$

33. As h gets smaller, round-off error becomes important. When $h = 10^{-12}$, the quantity $2^h - 1$ is so close to 0 that the calculator rounds off the difference to 0, making the difference quotient 0. The same thing will happen when $h = 10^{-20}$.

Solutions for Section 2.3

1. The graph is that of the line $y = -2x + 2$. The slope, and hence the derivative, is -2.

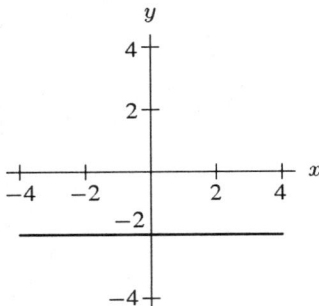

2.

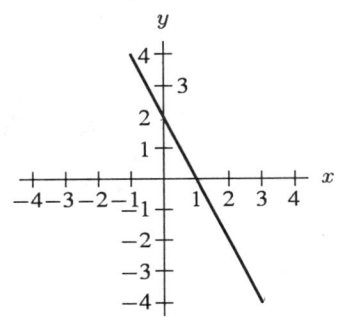

3.

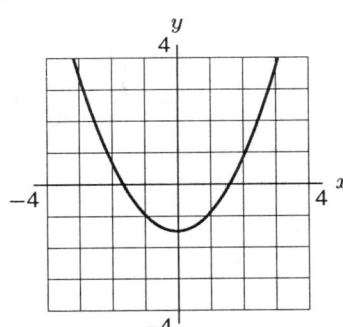

4. The slope of this curve is approximately -1 at $x = -4$ and at $x = 4$, approximately 0 at $x = -2.5$ and $x = 1.5$, and approximately 1 at $x = 0$.

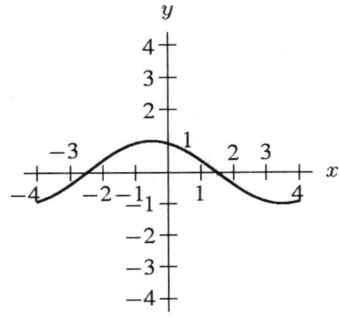

40 CHAPTER TWO /SOLUTIONS

5.
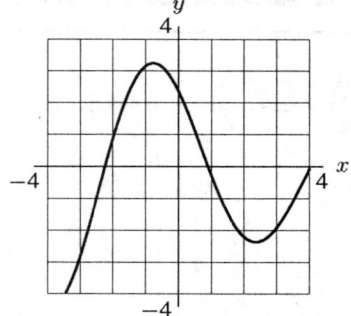

6.
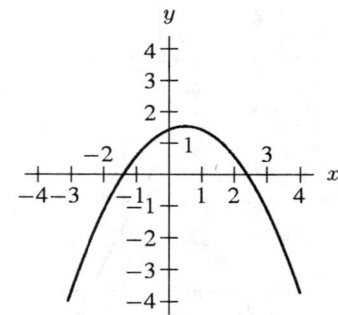

7.
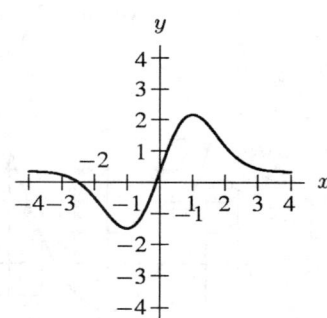

8.
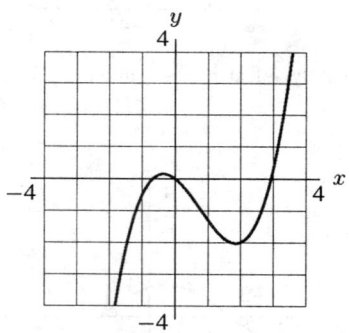

9.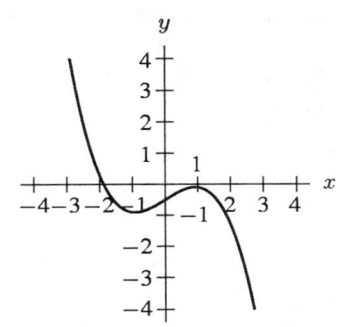

10. (a)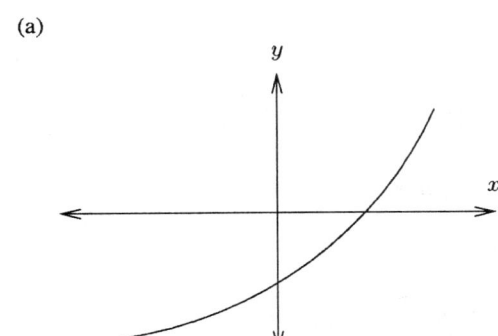
    (b)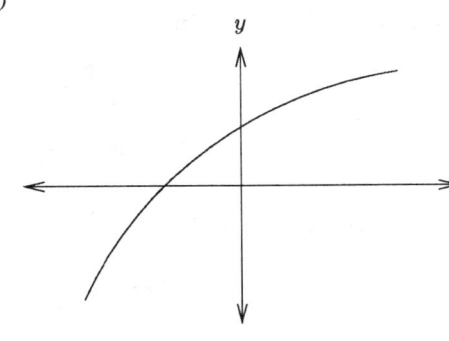
    (c)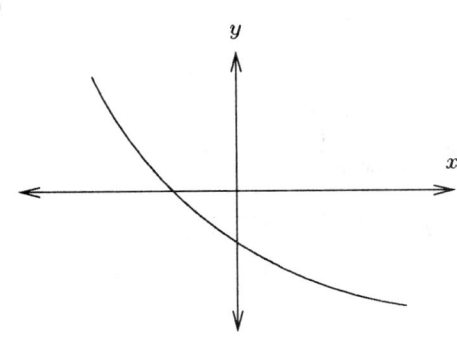
    (d)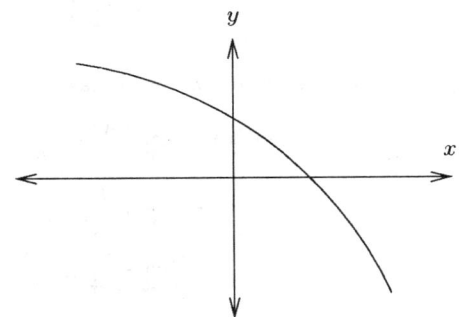

11. We know that $f'(x) \approx \dfrac{f(x+h) - f(x)}{h}$. For this problem, we'll take the average of the values obtained for $h = 1$ and $h = -1$; that's the average of $f(x+1) - f(x)$ and $f(x) - f(x-1)$ which equals $\dfrac{f(x+1) - f(x-1)}{2}$. Thus,
$f'(0) \approx f(1) - f(0) = 13 - 18 = -5.$
$f'(1) \approx [f(2) - f(0)]/2 = [10 - 18]/2 = -4.$
$f'(2) \approx [f(3) - f(1)]/2 = [9 - 13]/2 = -2.$
$f'(3) \approx [f(4) - f(2)]/2 = [9 - 10]/2 = -0.5.$
$f'(4) \approx [f(5) - f(3)]/2 = [11 - 9]/2 = 1.$
$f'(5) \approx [f(6) - f(4)]/2 = [15 - 9]/2 = 3.$
$f'(6) \approx [f(7) - f(5)]/2 = [21 - 11]/2 = 5.$
$f'(7) \approx [f(8) - f(6)]/2 = [30 - 15]/2 = 7.5.$
$f'(8) \approx f(8) - f(7) = 30 - 21 = 9.$
The rate of change of $f(x)$ is positive for $4 \leq x \leq 8$, negative for $0 \leq x \leq 3$. The rate of change is greatest at about $x = 8$.

42 CHAPTER TWO /SOLUTIONS

12. The value of $g(x)$ is increasing at a decreasing rate for $2.7 < x < 4.2$ and increasing at an increasing rate for $x > 4.2$.

$$\frac{\Delta y}{\Delta x} = \frac{7.4 - 6.0}{5.2 - 4.7} = 2.8 \quad \text{between } x = 4.7 \text{ and } x = 5.2$$

$$\frac{\Delta y}{\Delta x} = \frac{9.0 - 7.4}{5.7 - 5.2} = 3.2 \quad \text{between } x = 5.2 \text{ and } x = 5.7$$

Thus $g'(x)$ should be close to 3 near $x = 5.2$.

13. Using the definition of the derivative,

$$g'(x) = \lim_{h \to 0} \frac{g(x+h) - g(x)}{h} = \lim_{h \to 0} \frac{2(x+h)^2 - 3 - (2x^2 - 3)}{h}$$
$$= \lim_{h \to 0} \frac{2(x^2 + 2xh + h^2) - 3 - 2x^2 + 3}{h} = \lim_{h \to 0} \frac{4xh + 2h^2}{h}$$
$$= \lim_{h \to 0} (4x + 2h) = 4x.$$

14. Using the definition of the derivative,

$$k'(x) = \lim_{h \to 0} \frac{k(x+h) - k(x)}{h} = \lim_{h \to 0} \frac{\frac{1}{x+h} - \frac{1}{x}}{h} = \lim_{h \to 0} \frac{x - (x+h)}{h(x+h)x}$$
$$= \lim_{h \to 0} \frac{-h}{h(x+h)x} = \lim_{h \to 0} \frac{-1}{(x+h)x} = -\frac{1}{x^2}.$$

15. Using the definition of the derivative,

$$l'(x) = \lim_{h \to 0} \frac{\frac{1}{(x+h)^2} - \frac{1}{x^2}}{h} = \lim_{h \to 0} \frac{x^2 - (x+h)^2}{h(x+h)^2 x^2}$$
$$= \lim_{h \to 0} \frac{x^2 - (x^2 + 2xh + h^2)}{h(x+h)^2 x^2} = \lim_{h \to 0} \frac{-2xh - h^2}{h(x+h)^2 x^2}$$
$$= \lim_{h \to 0} \frac{-2x - h}{(x+h)^2 x^2} = \frac{-2x}{x^2 x^2} = -\frac{2}{x^3}.$$

16. Using the definition of the derivative, we have

$$m'(x) = \lim_{h \to 0} \frac{m(x+h) - m(x)}{h} = \lim_{h \to 0} \frac{1}{h}\left(\frac{1}{x+h+1} - \frac{1}{x+1}\right)$$
$$= \lim_{h \to 0} \frac{1}{h}\left(\frac{x+1-x-h-1}{(x+1)(x+h+1)}\right) = \lim_{h \to 0} \frac{-h}{h(x+1)(x+h+1)}$$
$$= \lim_{h \to 0} \frac{-1}{(x+1)(x+h+1)}$$
$$= \frac{-1}{(x+1)^2}.$$

17. From the given information we know that f is increasing for values of x less than -2, is decreasing between $x = -2$ and $x = 2$, and is constant for $x > 2$. Figure 2.16 shows a possible graph—yours may be different.

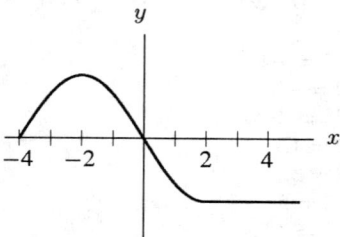

Figure 2.16

18. Figure 2.17 shows a possible graph – yours may be different.

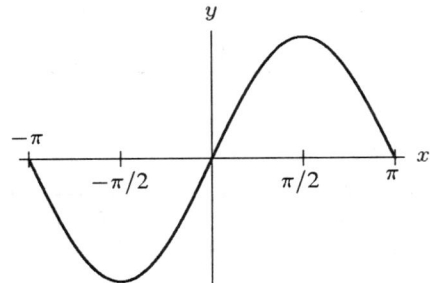

Figure 2.17

19.

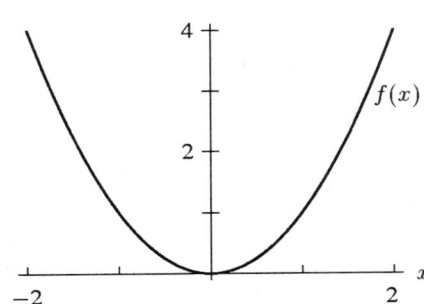 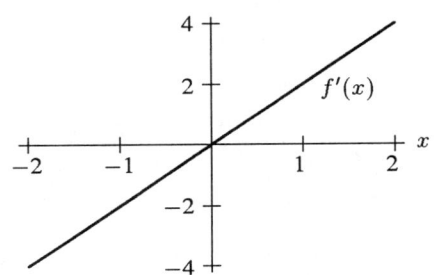

20.

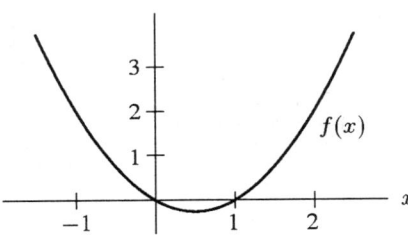 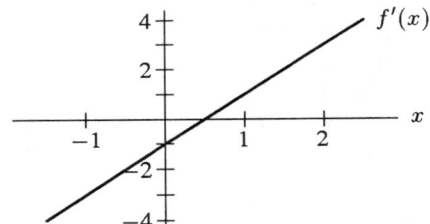

21.

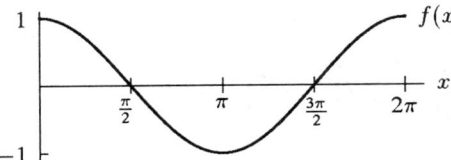 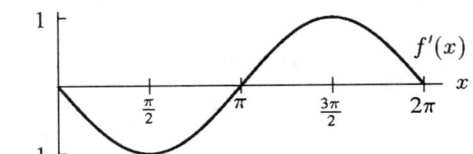

22.

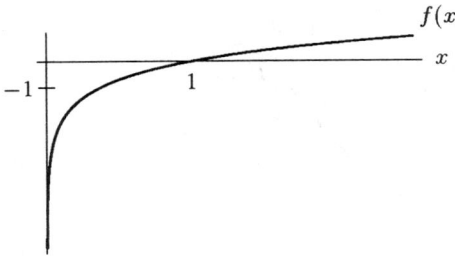 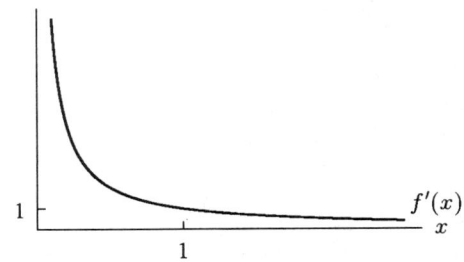

23.
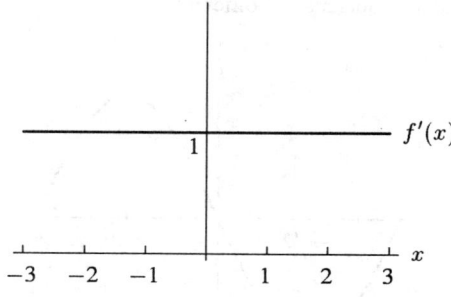

24.
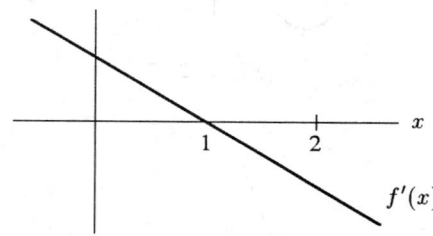

25.
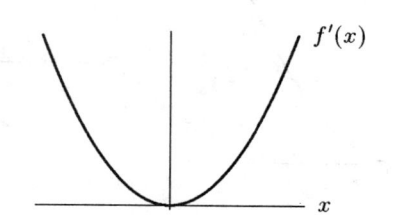

26.
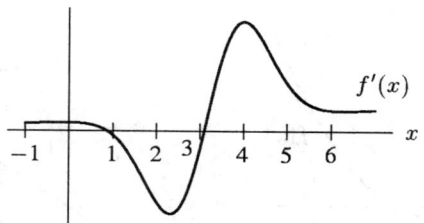

27.
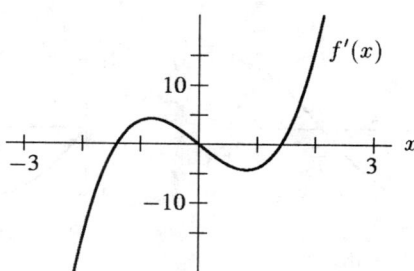

28.
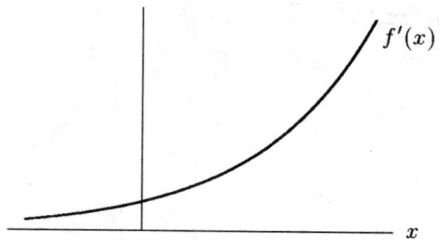

29. (a) x_3 (b) x_4 (c) x_5 (d) x_3
30. (a) Graph II
    (b) Graph I
    (c) Graph III
31. (a) $t = 3$
    (b) $t = 9$
    (c) $t = 14$
    (d)

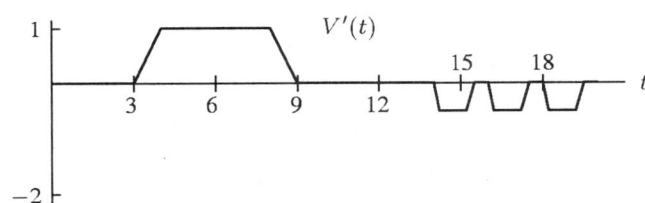

32. (a) The population varies periodically with a period of 1 year. See below.

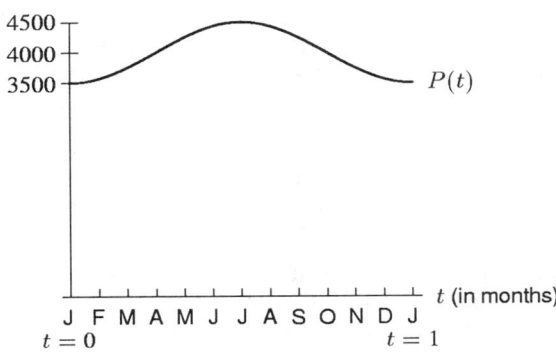

(b) The population is at a maximum on July 1st. At this time $\sin(2\pi t - \frac{\pi}{2}) = 1$, so the actual maximum population is $4000 + 500(1) = 4500$. Similarly, the population is at a minimum on January 1st. At this time, $\sin(2\pi t - \frac{\pi}{2}) = -1$, so the minimum population is $4000 + 500(-1) = 3500$.
(c) The rate of change is most positive about April 1st and most negative around October 1st.
(d) Since the population is at its maximum around July 1st, its rate of change is about 0 then.

33. If $f(x)$ is even, its graph is symmetric about the y-axis. So the tangent line to f at $x = x_0$ is the same as that at $x = -x_0$ reflected about the y-axis.

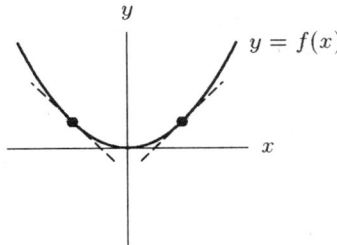 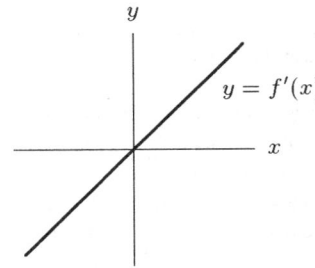

So the slopes of these two tangent lines are opposite in sign, so $f'(x_0) = -f'(-x_0)$, and f' is odd.

46 CHAPTER TWO /SOLUTIONS

34. If $g(x)$ is odd, its graph remains the same if you rotate it 180° about the origin. So the tangent line to g at $x = x_0$ is the tangent line to g at $x = -x_0$, rotated 180°.

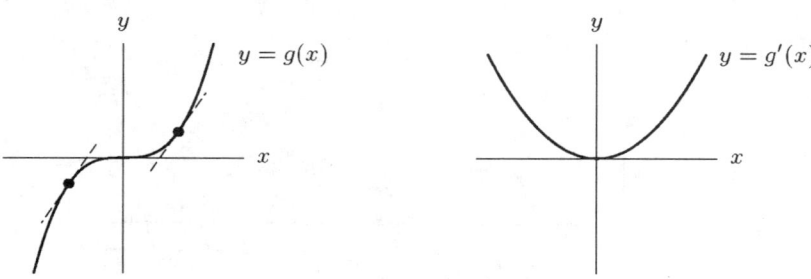

But the slope of a line stays constant if you rotate it 180°. So $g'(x_0) = g'(-x_0)$; g' is even.

Solutions for Section 2.4

1. The units of $f'(x)$ are feet/mile. The derivative, $f'(x)$, represents the rate of change of elevation with distance from the source, so if the river is flowing downhill everywhere, the elevation is always decreasing and $f'(x)$ is always negative. (In fact, there may be some stretches where the elevation is more or less constant, so $f'(x) = 0$.)

2. (Note that we are considering the average temperature of the yam, since its temperature is different at different points inside it.)
   (a) It is positive, because the temperature of the yam increases the longer it sits in the oven.
   (b) The units of $f'(20)$ are °F/min. $f'(20) = 2$ means that at time $t = 20$ minutes, the temperature T increases by approximately 2°F for each additional minute in the oven.

3. Units of $P'(t)$ are dollars/year. The practical meaning of $P'(t)$ is the rate at which the monthly payments change as the duration of the mortgage increases. Approximately, $P'(t)$ represents the change in the monthly payment if the duration is increased by one year. $P'(t)$ is negative because increasing the duration of a mortgage decreases the monthly payments.

4. Units of $C'(r)$ are dollars/percent. Approximately, $C'(r)$ means the additional amount needed to pay off the loan when the interest rate is increased by 1%. The sign of $C'(r)$ is positive, because increasing the interest rate will increase the amount it costs to pay off a loan.

5. Since B is measured in dollars and t is measured in years, dB/dt is measured in dollars per year. We can interpret dB as the extra money added to your balance in dt years. Therefore dB/dt represents how fast your balance is growing, in units of dollars/year.

6. (a) This means that investing the $1000 at 5% would yield $1649 after 10 years.
   (b) Writing $g'(r)$ as dB/dt, we see that the units of dB/dt are dollars per percent (interest). We can interpret dB as the extra money earned if interest rate is increased by dr percent. Therefore $g'(5) = \frac{dB}{dr}|_{r=5} \approx 165$ means that the balance, at 5% interest, would increase by about $165 if the interest rate were increased by 1%. In other words, $g(6) \approx g(5) + 165 = 1649 + 165 = 1814$.

7. (a) The function f takes quarts of ice cream to cost in dollars, so 200 must be the amount of ice cream, measured in quarts, and $70 must be the corresponding cost, measured in dollars. It costs $70 to produce 200 quarts of ice cream.
   (b) For similar reasons to part (a), 200 is in units of quarts. Since f' here tells how f changes with g, the 3 must be in units of dollars/quarts. After producing 200 quarts of ice cream, the cost to produce one more quart is about $3.

8. (a) If the price is $150, then 2000 items will be sold.
   (b) If the price goes up from $150 by $1 per item, about 25 fewer items will be sold. Equivalently, if the price is decreased from $150 by $1 per item, about 25 more items will be sold.

9. (a) The pressure in dynes/cm^2 at a depth of 100 meters.
   (b) The depth of water in meters giving a pressure of $1.2 \cdot 10^6$ dynes/cm^2.
   (c) The pressure at a depth of h meters plus a pressure of 20 dynes/cm^2.
   (d) The pressure at a depth of 20 meters below the diver.
   (e) The rate of increase of pressure with respect to depth, at 100 meters, in units of dynes/cm^2 per meter. Approximately, $p'(100)$ represents the increase in pressure in going from 100 meters to 101 meters.
   (f) The depth, in meters, at which the rate of change of pressure with respect to depth is 20 dynes/cm^2 per meter.

10. (a) When $t = 10$, that is, at 10 am, 3.1 cm of rain has fallen.
    (b) We are told that when 10 cm of rain has fallen, 16 hours have passed ($t = 16$); that is, 10 cm of rain has fallen by 4 pm.
    (c) The rate at which rain is falling is 0.4 cm/hr at $t = 8$, that is, at 8 am.
    (d) The units of $(f^{-1})'(5)$ are hours/cm. Thus, we are being told that when 5 cm of rain has fallen, rain is falling at a rate such that it will take 2 additional hours for another centimeter to fall.

11. (a) The derivative, dW/dt, measures the rate of change of water in the bathtub in gallons per minute.
    (b) (i) Before the plug is pulled, the rate of change of W in gallons per minute is 0 since the amount of water in the tub is not changing.
        (ii) Since dW/dt represents the rate at which the amount of water in the tub is changing, when the plug is pulled and water is leaving the tub, the sign of dW/dt is negative.
        (iii) Once all the water has drained from the tub, the amount of water in the tub is not changing, so $dW/dt = 0$.

12. (a) Negative.
    (b) $dw/dt = 0$ for t bigger than some t_0 (the time when the fire stops burning).
    (c) $|dw/dt|$ increases, so dw/dt decreases since it is negative.

13. Units of $g'(55)$ are mpg/mph. The statement $g'(55) = -0.54$ means that at 55 miles per hour the fuel efficiency (in miles per gallon, or mpg) of the car decreases at a rate of approximately one half mpg as the velocity increases by one mph.

14. (a)

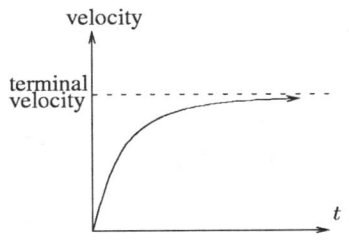

   (b) The graph should be concave down because wind resistance decreases your acceleration as you speed up, and so the slope of the graph of velocity is decreasing.
   (c) The slope represents the acceleration due to gravity.

15. (a) The company hopes that increased advertising always brings in more customers instead of turning them away. Therefore, it hopes $f'(a)$ is always positive.
    (b) If $f'(100) = 2$, it means that if the advertising budget is $100,000, each extra dollar spent on advertising will bring in $2 worth of sales. If $f'(100) = 0.5$, each dollar above $100 thousand spent on advertising will bring in $0.50 worth of sales.
    (c) If $f'(100) = 2$, then as we saw in part (b), spending slightly more than $100,000 will increase revenue by an amount greater than the additional expense, and thus more should be spent on advertising. If $f'(100) = 0.5$, then the increase in revenue is less than the additional expense, hence too much is being spent on advertising. The optimum amount to spend is an amount that makes $f'(a) = 1$. At this point, the increases in advertising expenditures just pay for themselves. If $f'(a) < 1$, too much is being spent; if $f'(a) > 1$, more should be spent.

16. Since $\frac{P(67)-P(66)}{67-66}$ is an estimate of $P'(66)$, we may think of $P'(66)$ as an estimate of $P(67) - P(66)$, and the latter is the number of people between 66 and 67 inches tall. Alternatively, since $\frac{P(66.5)-P(65.5)}{66.5-65.5}$ is a better estimate of $P'(66)$, we may regard $P'(66)$ as an estimate of the number of people of height between 65.5 and 66.5 inches. The units for $P'(x)$ are people per inch. Since there were 250 million people at the 1990 census, we might guess that there are about 200 million full-grown persons in the US whose heights are distributed between $60''(5')$ and $75''(6'3'')$. There are probably quite a few people of height $66''$–perhaps $1\frac{1}{2}$ what you'd expect from an even, or uniform, distribution–because it's nearly average. An even distribution would yield $P'(66) = \frac{200 \text{ million}}{15''} \approx 13$ million per inch–so we can expect $P'(66)$ to be perhaps $13(1.5) \approx 20$.

$P'(x)$ is never negative because $P(x)$ is never decreasing. To see this, let's look at an example involving a particular value of x, say $x = 70$. The value $P(70)$ represents the number of people whose height is less than or equal to 70 inches, and $P(71)$ represents the number of people whose height is less than or equal to 71 inches. Since everyone shorter than 70 inches is also shorter than 71 inches, $P(70) \leq P(71)$. In general, $P(x)$ is 0 for small x, and increases as x increases, and is eventually constant (for large enough x).

Solutions for Section 2.5

1. $f'(x) > 0$
   $f''(x) > 0$
2. $f'(x) = 0$
   $f''(x) = 0$
3. $f'(x) < 0$
   $f''(x) = 0$
4. $f'(x) < 0$
   $f''(x) > 0$
5. $f'(x) > 0$
   $f''(x) < 0$
6. $f'(x) < 0$
   $f''(x) < 0$
7. To measure the average acceleration over an interval, we calculate the average rate of change of velocity over the interval. The units of acceleration are ft/sec per second, or (ft/sec)/sec, written ft/sec^2.

$$\text{Average acceleration for } 0 \leq t \leq 1 = \frac{\text{Change in velocity}}{\text{Time}} = \frac{v(1) - v(0)}{1} = \frac{30 - 0}{1} = 30 \text{ ft/sec}^2$$

$$\text{Average acceleration for } 1 \leq t \leq 2 = \frac{52 - 30}{2 - 1} = 22 \text{ ft/sec}^2$$

8. (a) $dP/dt > 0$ and $d^2P/dt^2 > 0$.

   (b) $dP/dt < 0$ and $d^2P/dt^2 > 0$ (but dP/dt is close to zero).

9. (a)

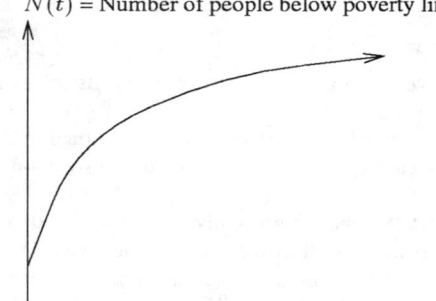

$N(t)$ = Number of people below poverty line

   (b) dN/dt is positive, since people are still slipping below the poverty line. d^2N/dt^2 is negative, since the rate at which people are slipping below the poverty line, dN/dt, is decreasing.

10. (a)

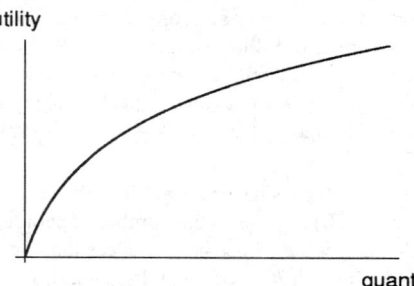

   (b) As a function of quantity, utility is increasing but at a decreasing rate; the graph is increasing but concave down. So the derivative of utility is positive, but the second derivative of utility is negative.

11. Suppose $p(t)$ is the average price level at time t. Then, if $t_0 =$ April 1991,
    "Prices are still rising" means $p'(t_0) > 0$.
    "Prices rising less fast than they were" means $p''(t_0) < 0$.
    "Prices rising not as much less fast as everybody had hoped" means $H < p''(t_0)$, where H is the rate of change in rate of change of prices that people had hoped for.

12. Since all advertising campaigns are assumed to produce an increase in sales, a graph of sales against time would be expected to have a positive slope.

    A positive second derivative means the rate at which sales are increasing is increasing. If a positive second derivative is observed during a new campaign, it is reasonable to conclude that this increase in the rate sales are increasing is caused by the new campaign–which is therefore judged a success. A negative second derivative means a decrease in the rate at which sales are increasing, and therefore suggests the new campaign is a failure.

13. (a) The EPA will say that the rate of discharge is still rising. The industry will say that the rate of discharge is increasing less quickly, and may soon level off or even start to fall.
    (b) The EPA will say that the rate at which pollutants are being discharged is levelling off, but not to zero — so pollutants will continue to be dumped in the lake. The industry will say that the rate of discharge has decreased significantly.

14. (a) $f'(0.6) \approx \dfrac{f(0.8) - f(0.6)}{0.8 - 0.6} = \dfrac{4.0 - 3.9}{0.2} = 0.5.$ $f'(0.5) \approx \dfrac{f(0.6) - f(0.4)}{0.6 - 0.4} = \dfrac{0.4}{0.2} = 2.$
    (b) Using the values of f' from part (a), we get $f''(0.6) \approx \dfrac{f'(0.6) - f'(0.5)}{0.6 - 0.5} = \dfrac{0.5 - 2}{0.2} = \dfrac{-1.5}{0.2} = -7.5.$
    (c) The maximum value of f is probably near $x = 0.8$. The minimum value of f is probably near $x = 0.3$.

15. By noting whether $f(x)$ is positive or negative, increasing or decreasing, and concave up or down at each of the given points, we get the completed Table 2.12:

TABLE 2.12

Point	f	f'	f''
A	−	0	+
B	+	0	−
C	+	−	−
D	−	+	+

16. Since f' is everywhere positive, f is everywhere increasing. Hence the greatest value of f is at x_6 and the least value of f is at x_1. Directly from the graph, we see that f' is greatest at x_3 and least at x_2. Since f'' gives the slope of the graph of f', f'' is greatest where f' is rising most rapidly, namely at x_6, and f'' is least where f' is falling most rapidly, namely at x_1.

17. (a) B (where $f', f'' > 0$) and E (where $f', f'' < 0$)
    (b) A (where $f = f' = 0$) and D (where $f' = f'' = 0$)

18. The velocity is the derivative of the distance, that is, $v(t) = s'(t)$. Therefore, we have

$$v(t) = \lim_{h \to 0} \frac{s(t+h) - s(t)}{h}$$
$$= \lim_{h \to 0} \frac{(5(t+h)^2 + 3) - (5t^2 + 3)}{h}$$
$$= \lim_{h \to 0} \frac{10th + 5h^2}{h}$$
$$= \lim_{h \to 0} \frac{h(10t + 5h)}{h} = \lim_{h \to 0} (10t + 5h) = 10t \text{ km/minute.}$$

The acceleration is the derivative of velocity, so $a(t) = v'(t)$:

$$a(t) = \lim_{h \to 0} \frac{10(t+h) - 10t}{h}$$
$$= \lim_{h \to 0} \frac{10h}{h} = 10 \text{ km/(minute)}^2.$$

Solutions for Chapter 2 Review

1.

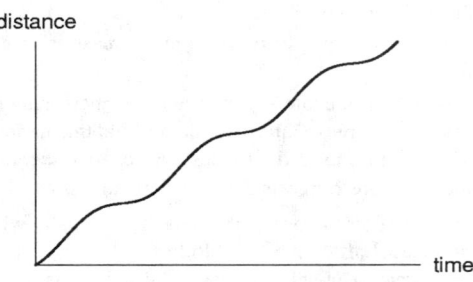

2.

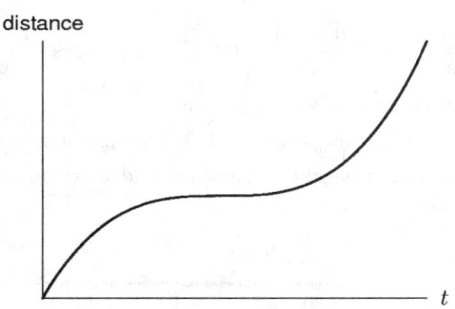

3. (a)

TABLE 2.13

x	1	1.5	2	2.5	3
$\log x$	0	0.18	0.30	0.40	0.48

(b) The average rate of change of $f(x) = \log x$ between $x = 1$ and $x = 3$ is

$$\frac{f(3) - f(1)}{3 - 1} = \frac{\log 3 - \log 1}{3 - 1} \approx \frac{0.48 - 0}{2} = 0.24$$

(c) First we find the average rates of change of $f(x) = \log x$ between $x = 1.5$ and $x = 2$, and between $x = 2$ and $x = 2.5$.

$$\frac{\log 2 - \log 1.5}{2 - 1.5} = \frac{0.30 - 0.18}{0.5} \approx 0.24$$

$$\frac{\log 2.5 - \log 2}{2.5 - 2} = \frac{0.40 - 0.30}{0.5} \approx 0.20$$

Now we approximate the instantaneous rate of change at $x = 2$ by finding the average of the above rates, i.e.

$$\left(\begin{array}{c}\text{the instantaneous rate of change} \\ \text{of } f(x) = \log x \text{ at } x = 2\end{array}\right) \approx \frac{0.24 + 0.20}{2} = 0.22.$$

4. $f'(1) = \lim\limits_{h \to 0} \dfrac{\log(1+h) - \log 1}{h} = \lim\limits_{h \to 0} \dfrac{\log(1+h)}{h}$

Evaluating $\dfrac{\log(1+h)}{h}$ for $h = 0.01, 0.001$, and 0.0001, we get $0.43214, 0.43408, 0.43427$, so $f'(1) \approx 0.43427$. The corresponding secant lines are getting steeper, because the graph of $\log x$ is concave down. We thus expect the limit to be more than 0.43427. If we consider negative values of h, the estimates are too large. We can also see this from the graph below:

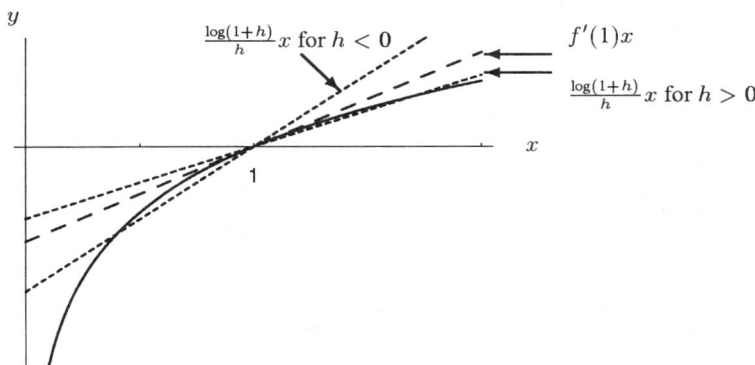

5.

x	$\ln x$	x	$\ln x$	x	$\ln x$	x	$\ln x$
0.998	-0.0020	1.998	0.6921	4.998	1.6090	9.998	2.3024
0.999	-0.0010	1.999	0.6926	4.999	1.6092	9.999	2.3025
1.000	0.0000	2.000	0.6931	5.000	1.6094	10.000	2.3026
1.001	0.0010	2.001	0.6936	5.001	1.6096	10.001	2.3027
1.002	0.0020	2.002	0.6941	5.002	1.6098	10.002	2.3028

At $x = 1$, the values of $\ln x$ are increasing by 0.001 for each increase in x of 0.001, so the derivative appears to be 1. At $x = 2$, the increase is 0.0005 for each increase of 0.001, so the derivative appears to be 0.5. At $x = 5$, $\ln x$ increases by 0.0002 for each increase of 0.001 in x, so the derivative appears to be 0.2. And at $x = 10$, the increase is 0.0001 over intervals of 0.001, so the derivative appears to be 0.1. These values suggest an inverse relationship between x and $f'(x)$, namely $f'(x) = \dfrac{1}{x}$.

6. Using the points to the right of $x = 0.5$, we obtain

$$J_0'(0.5) \approx \dfrac{J_0(0.6) - J_0(0.5)}{0.6 - 0.5} = -0.265.$$

Secondly, using the points to the left of $x = 0.5$, we obtain

$$J_0'(0.5) \approx \dfrac{J_0(0.4) - J_0(0.5)}{0.4 - 0.5} = -0.219.$$

To obtain a better estimate, we average these, giving

$$J_0'(0.5) \approx \dfrac{(-0.265 - 0.219)}{2} = -0.242.$$

7.

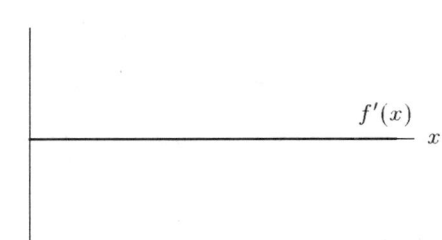

8.

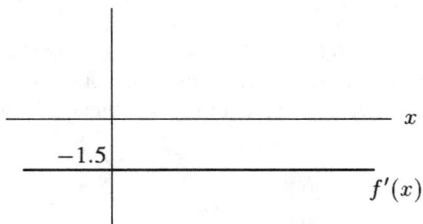

9.

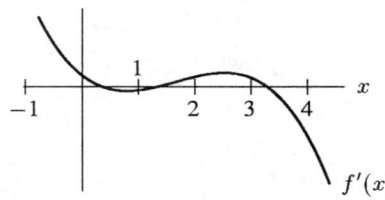

10.

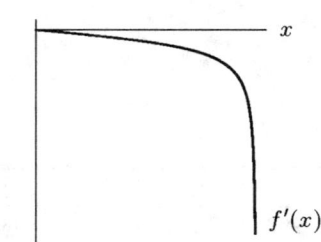

11.

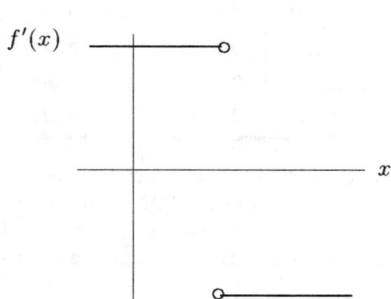

12.

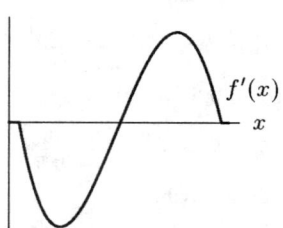

13. Using the definition of the derivative

$$f'(x) = \lim_{h \to 0} \frac{f(x+h) - f(x)}{h}$$
$$= \lim_{h \to 0} \frac{5(x+h)^2 + x + h - (5x^2 + x)}{h}$$
$$= \lim_{h \to 0} \frac{5(x^2 + 2xh + h^2) + x + h - 5x^2 - x}{h}$$
$$= \lim_{h \to 0} \frac{10xh + 5h^2 + h}{h}$$
$$= \lim_{h \to 0} (10x + 5h + 1) = 10x + 1$$

14. Using the definition of the derivative, we have

$$n'(x) = \lim_{h \to 0} \frac{n(x+h) - n(x)}{h}$$
$$= \lim_{h \to 0} \frac{1}{h}\left[\left(\frac{1}{x+h} + 1\right) - \left(\frac{1}{x} + 1\right)\right]$$
$$= \lim_{h \to 0} \frac{1}{h}\left(\frac{1}{x+h} - \frac{1}{x}\right)$$
$$= \lim_{h \to 0} \frac{x - (x+h)}{hx(x+h)}$$
$$= \lim_{h \to 0} \frac{-h}{hx(x+h)}$$
$$= \lim_{h \to 0} \frac{-1}{x(x+h)} = \frac{-1}{x^2}.$$

15. Note that f' and g' are periodic, with the same period as f and g respectively. Since g oscillates between 1 and -1 more quickly, its values change faster and therefore we would expect its derivative to reach larger values (both positive and negative); g' has a larger amplitude than f'.

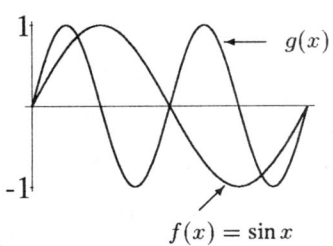
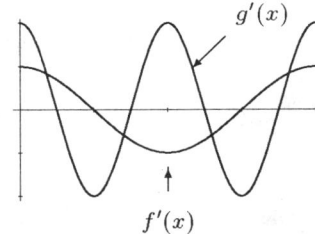

16. The slopes of the lines drawn through successive pairs of points are negative but increasing, suggesting that $f''(x) > 0$ for $1 \leq x \leq 3.3$ and that the graph of $f(x)$ is concave up.

17. (a) The yam is cooling off so T is decreasing and $f'(t)$ is negative.
(b) Since $f(t)$ is measured in degrees Fahrenheit and t is measured in minutes, df/dt must be measured in units of °F/min.

18. $f(10) = 240{,}000$ means that if the commodity costs \$10, then 240,000 units of it will be sold. $f'(10) = -29{,}000$ means that if the commodity costs \$10 now, each \$1 increase in price will cause a decline in sales of 29,000 units.

19. (a)

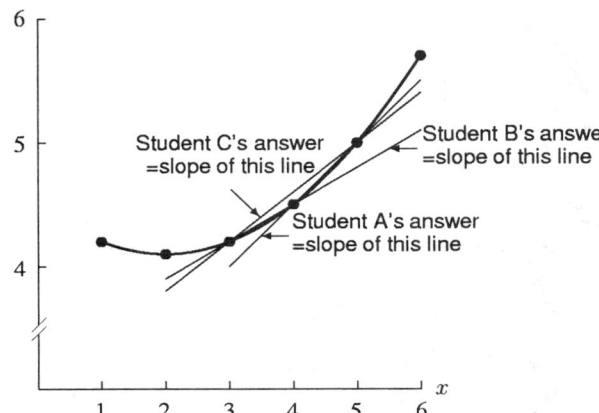

(b) The slope of f appears to be somewhere between student A's answer and student B's, so student C's answer, halfway in between, is probably the most accurate.

(c) Student A's estimate is $f'(x) \approx \frac{f(x+h)-f(x)}{h}$, while student B's estimate is $f'(x) \approx \frac{f(x)-f(x-h)}{h}$. Student C's estimate is the average of these two, or

$$f'(x) \approx \frac{1}{2}\left[\frac{f(x+h)-f(x)}{h} + \frac{f(x)-f(x-h)}{h}\right] = \frac{f(x+h)-f(x-h)}{2h}.$$

This estimate is the slope of the chord connecting $(x-h, f(x-h))$ to $(x+h, f(x+h))$. Thus, we estimate that the tangent to a curve is nearly parallel to a chord connecting points h units to the right and left, as shown below.

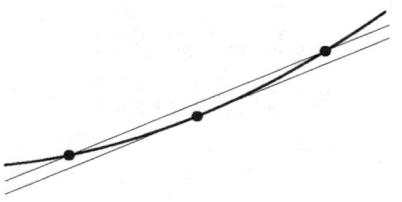

20. (a) IV, (b) III, (c) II, (d) I, (e) IV, (f) II

21. (a) See (b).
(b)

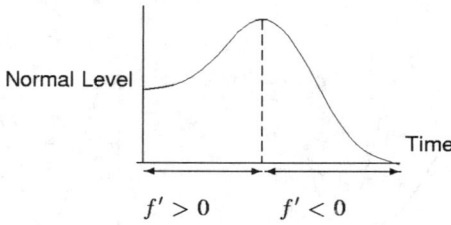

(c) f' is the rate at which the concentration is increasing or decreasing. f' is positive at the start of the disease and negative toward the end. In practice, of course, one cannot measure f' directly. Checking the value of C in blood samples taken on consecutive days would tell us

$$f(t+1) - f(t) = \frac{f(t+1) - f(t)}{(t+1) - t},$$

which is our estimate of $f'(t)$.

22. (a) The population varies periodically with a period of 12 months (i.e. one year).

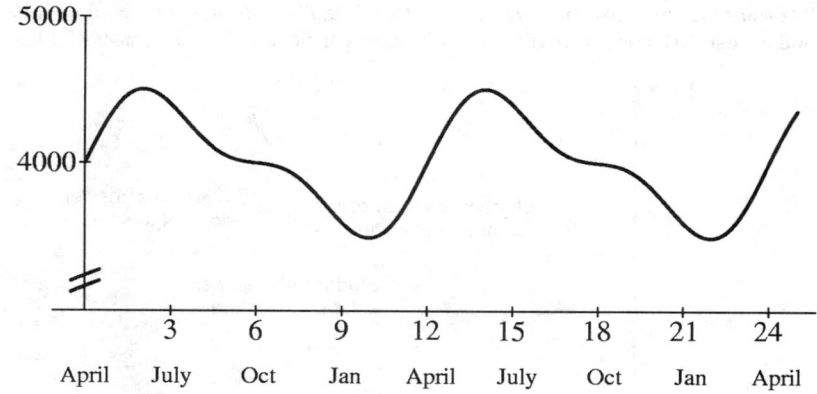

(b) The herd is largest about June 1st when there are about 4500 deer.
(c) The herd is smallest about February 1st when there are about 3500 deer.
(d) The herd grows the fastest about April 1st. The herd shrinks the fastest about July 15 and again about December 15.
(e) It grows the fastest about April 1st when the rate of growth is about 400 deer/month, i.e about 13 new fawns per day.

SOLUTIONS TO REVIEW PROBLEMS FOR CHAPTER TWO   **55**

23. (a) Slope of tangent line $= \lim_{h\to 0} \frac{\sqrt{4+h}-\sqrt{4}}{h}$. Using $h = 0.001$, $\frac{\sqrt{4.001}-\sqrt{4}}{0.001} = 0.249984$. Hence the slope of the tangent line is about 0.25.
    (b)
    $$y - y_1 = m(x - x_1)$$
    $$y - 2 = 0.25(x - 4)$$
    $$y - 2 = 0.25x - 1$$
    $$y = 0.25x + 1$$

    (c) $f(x) = kx^2$
    If $(4, 2)$ is on the graph of f, then $f(4) = 2$, so $k \cdot 4^2 = 2$. Thus $k = \frac{1}{8}$, and $f(x) = \frac{1}{8}x^2$.
    (d) To find where the graph of f crosses then line $y = 0.25x + 1$, we solve:
    $$\frac{1}{8}x^2 = 0.25x + 1$$
    $$x^2 = 2x + 8$$
    $$x^2 - 2x - 8 = 0$$
    $$(x-4)(x+2) = 0$$
    $$x = 4 \text{ or } x = -2$$
    $$f(-2) = \frac{1}{8}(4) = 0.5$$

    Therefore, $(-2, 0.5)$ is the other point of intersection. (Of course, $(4, 2)$ is a point of intersection; we know that from the start.)

24. (a) The slope of the tangent line at $(0, \sqrt{19})$ is zero: it is horizontal.
    The slope of the tangent line at $(\sqrt{19}, 0)$ is undefined: it is vertical.
    (b) The slope appears to be about $\frac{1}{2}$. (Note that when x is 2, y is about -4, but when x is 4, y is approximately -3.)

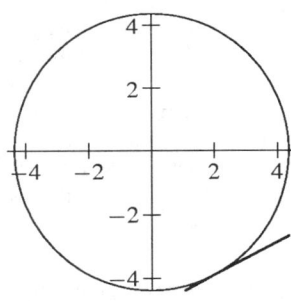

    (c) Using symmetry we can determine: Slope at $(-2, \sqrt{15})$: about $\frac{1}{2}$. Slope at $(-2, -\sqrt{15})$: about $-\frac{1}{2}$. Slope at $(2, \sqrt{15})$: about $-\frac{1}{2}$.

25. (a)

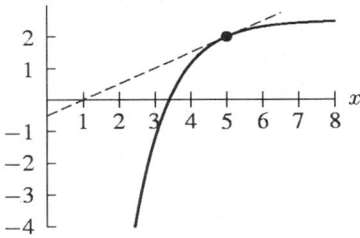

    (b) Exactly one. There can't be more than one zero because f is increasing everywhere. There does have to be one zero because f stays below its tangent line (dotted line in above graph), and therefore f must cross the x-axis.
    (c) The equation of the (dotted) tangent line is $y = \frac{1}{2}x - \frac{1}{2}$, and so it crosses the x-axis at $x = 1$. Therefore the zero of f must be between $x = 1$ and $x = 5$.

(d) $\lim_{x \to -\infty} f(x) = -\infty$, because f is increasing and concave down. Thus, as $x \to -\infty$, $f(x)$ decreases, at a faster and faster rate.

(e) Yes.

(f) No. The slope is decreasing since f is concave down, so $f'(1) > f'(5)$, i.e. $f'(1) > \frac{1}{2}$.

26. (a) The graph looks straight because the graph shows only a small part of the curve magnified greatly.

(b) The month is March: We see that about the 21st of the month there are twelve hours of daylight and hence twelve hours of night. This phenomenon (the length of the day equaling the length of the night) occurs at the equinox, midway between winter and summer. Since the length of the days is increasing, and Madrid is in the northern hemisphere, we are looking at March, not September.

(c) The slope of the curve is found from the graph to be about 0.04 (the rise is about 0.8 hours in 20 days or 0.04 hours/day). This means that the amount of daylight is increasing by about 0.04 hours (about $2\frac{1}{2}$ minutes) per calendar day, or that each day is $2\frac{1}{2}$ minutes longer than its predecessor.

27. (a) A possible graph is shown below. At first, the yam heats up very quickly, since the difference in temperature between it and its surroundings is so large. As time goes by, the yam gets hotter and hotter, its rate of temperature increase slows down, and its temperature approaches the temperature of the oven as an asymptote. The graph is thus concave down. (We are considering the average temperature of the yam, since the temperature in its center and on its surface will vary in different ways.)

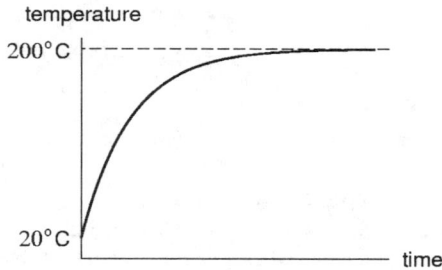

(b) If the rate of temperature increase were to remain 2°/min, in ten minutes the yam's temperature would increase 20°, from 120° to 140°. Since we know the graph is not linear, but concave down, the actual temperature is between 120° and 140°.

(c) In 30 minutes, we know the yam increases in temperature by 45° at an average rate of $45/30 = 1.5°$/min. Since the graph is concave down, the temperature at $t = 40$ is therefore between $120 + 1.5(10) = 135°$ and 140°.

(d) If the temperature increases at 2°/minute, it reaches 150° after 15 minutes, at $t = 45$. If the temperature increases at 1.5°/minute, it reaches 150° after 20 minutes, at $t = 50$. So t is between 45 and 50 mins.

Solutions to the Projects and Computer Algebra Investigations

1. (a) $S(0) = 12$ since the days are always 12 hours long at the equator.

(b) Since $S(0) = 12$ from part (a) and the formula gives $S(0) = a$, we have $a = 12$. Since $S(x)$ must be continuous at $x = x_0$, and the formula gives $S(x_0) = a + b \arcsin(1) = 12 + b\left(\frac{\pi}{2}\right)$ and also $S(x_0) = 24$, we must have $12 + b\left(\frac{\pi}{2}\right) = 24$ so $b\left(\frac{\pi}{2}\right) = 12$ and $b = \frac{24}{\pi} \approx 7.64$.

(c) $S(32°13') \approx 14.12$ and $S(46°4') \approx 15.58$.

(d)

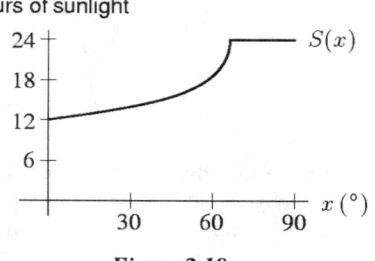

Figure 2.18

(e) The graph in Figure 2.18 appears to have a corner at $x_0 = 66°30'$. We compare the slope to the right of x_0 and to the left of x_0. To the right of S_0, the function is constant, so $S'(x) = 0$ for $x > 66°30'$.

We estimate the slope immediately to the left of x_0. We want to calculate the following:

$$\lim_{h \to 0^-} \frac{S(x_0 + h) - S(x_0)}{h}.$$

We approximate it by taking $x_0 = 66.5$ and $h = -0.1, -0.01, -0.001$:

$$\frac{S(66.49) - S(66.5)}{-0.1} \approx \frac{22.3633 - 24}{-0.1} = 16.38,$$

$$\frac{S(66.499) - S(66.5)}{-0.01} \approx \frac{23.4826 - 24}{-0.01} = 51.83,$$

$$\frac{S(66.4999) - S(66.5)}{-0.001} \approx \frac{23.8370 - 24}{-0.001} = 163.9.$$

These approximations suggest that, for $x_0 = 66.5$,

$$\lim_{h \to 0^-} \frac{S(x_0 + h) - S(x_0)}{h} \quad \text{does not exist.}$$

This evidence suggests that $S(x)$ is not differentiable at x_0. A proof requires the techniques found in Chapter 4.

2. (a) (i) Estimating derivatives using difference quotients (but other answers are possible):

$$P'(1900) \approx \frac{P(1910) - P(1900)}{10} = \frac{92.0 - 76.0}{10} = 1.6 \text{ million people per year}$$

$$P'(1945) \approx \frac{P(1950) - P(1940)}{10} = \frac{150.7 - 131.7}{10} = 1.9 \text{ million people per year}$$

$$P'(1990) \approx \frac{P(1990) - P(1980)}{10} = \frac{248.7 - 226.5}{10} = 2.22 \text{ million people per year}$$

(ii) The population growth was maximal somewhere between 1950 and 1960.

(iii) $P'(1950) \approx \frac{P(1960) - P(1950)}{10} = \frac{179.0 - 150.7}{10} = 2.83$ million people per year, so $P(1956) \approx P(1950) + P'(1950)(1956 - 1950) = 150.7 + 2.83(6) \approx 167.7$ million people.

(iv) If the growth rate between 1990 and 2000 was the same as the growth rate from 1980 to 1990, then the total population should be about 271 million people in 2000.

(b) (i) $f^{-1}(100)$ is the point in time when the population of the US was 100 million people (somewhere between 1910 and 1920).

(ii) The derivative of $f^{-1}(P)$ at $P = 100$ represents the ratio of change in time to change in population, and its units are years per million people. In other words, this derivative represents about how long it took for the population to increase by 1 million, when the population was 100 million.

(iii) Since the population increased by $105.7 - 92.0 = 13.7$ million people in 10 years, the average rate of increase is 1.37 million people per year. If the rate is fairly constant in that period, the amount of time it would take for an increase of 8 million people (100 million − 92.0 million) would be

$$\frac{8 \text{ million people}}{1.37 \text{ million people/year}} \approx 5.8 \text{ years} \approx 6 \text{ years}$$

Adding this to our starting point of 1910, we estimate that the population of the US reached 100 million around 1916, i.e. $f^{-1}(100) \approx 1916$.

(iv) Since it took 10 years between 1910 and 1920 for the population to increase by $105.7 - 92.0 = 13.7$ million people, the derivative of $f^{-1}(P)$ at $P = 100$ is approximately

$$\frac{10 \text{ years}}{13.7 \text{ million people}} = 0.73 \text{ years/million people}$$

58 CHAPTER TWO /SOLUTIONS

- (c) (i) Clearly the population of the US at any instant is an integer that varies up and down every few seconds as a child is born, a person dies, or a new immigrant arrives. So $f(t)$ has "jumps;" it is not a smooth function. But these jumps are small relative to the values of f, so f appears smooth unless we zoom in very closely on its graph (to within a few seconds).

    Major land acquisitions such as the Louisiana Purchase caused larger jumps in the population, but since the census is taken only every ten years and the territories acquired were rather sparsely populated, we cannot see these jumps in the census data.

  (ii) We can regard rate of change of the population for a particular time t as representing an estimate of how much the population will increase during the year after time t.

  (iii) Many economic indicators are treated as smooth, such as the Gross National Product, the Dow Jones Industrial Average, volumes of trading, and the price of commodities like gold. But these figures only change in increments, not continuously.

3. The CAS says the derivative is zero. This can be explained by the fact that $f(x) = \sin^2 x + \cos^2 x = 1$, so $f'(x)$ is the derivative of the constant function 1. The derivative of a constant function is zero.

4. (a) The CAS gives $f'(x) = 2\cos^2 x - 2\sin^2 x$.
   (b) Using the double angle formulas for sine and cosine, we have
   $$f(x) = 2\sin x \cos x = \sin(2x)$$
   $$f'(x) = 2\cos^2 x - 2\sin^2 x = 2(\cos^2 x - \sin^2 x) = 2\cos(2x).$$

   Thus we get
   $$\frac{d}{dx}\sin(2x) = 2\cos(2x).$$

5. (a) The first derivative is $g'(x) = -2axe^{-x^2}$, so the second derivative is
   $$g''(x) = \frac{d^2}{dx^2}e^{-ax^2} = \frac{-2a}{e^{ax^2}} + \frac{4a^2x^2}{e^{ax^2}}.$$

   (b) Both graphs get narrow as a gets larger; the graph of g'' is below the x-axis along the interval where f is concave down, and is above the x-axis where f is concave up. See Figure 2.19.

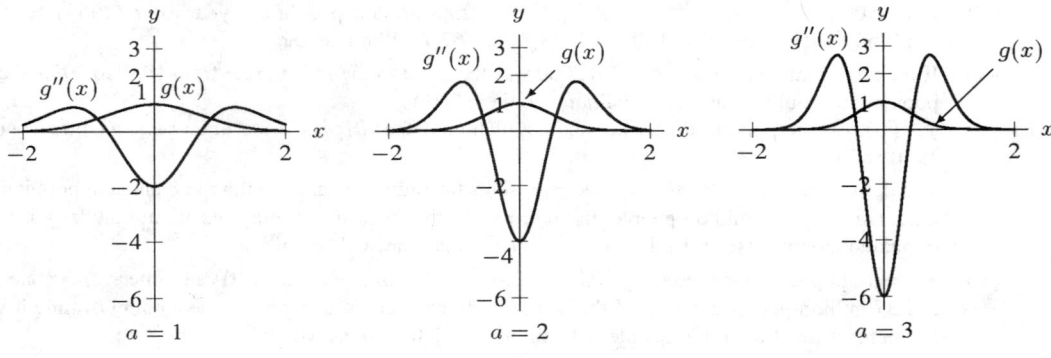

Figure 2.19

   (c) The second derivative of a function is positive when the graph of the function is concave up and negative when it is concave down.

6. (a) The computer algebra system gives
   $$\frac{d}{dx}(x^2+1)^2 = 4x(x^2+1)$$
   $$\frac{d}{dx}(x^2+1)^3 = 6x(x^2+1)^2$$
   $$\frac{d}{dx}(x^2+1)^4 = 8x(x^2+1)^3$$

   (b) The pattern suggests that
   $$\frac{d}{dx}(x^2+1)^n = 2nx(x^2+1)^{n-1}.$$
   Taking the derivative of $(x^2+1)^n$ with a CAS confirms this.

7. (a) The CAS gives the same derivative, $1/x$, in all three cases.
   (b) From the properties of logarithms, $g(x) = \ln(2x) = \ln 2 + \ln x = f(x) + \ln 2$. So the graph of g is the same shape as the graph of f, only shifted up by $\ln 2$. So the graphs have the same slope everywhere, and therefore the two functions have the same derivatve. By the same reasoning, $h(x) = f(x) + \ln 3$, so h and f have the same derivative as well.

8. (a) Using a CAS, we find
$$\frac{d}{dx}\sin x = \cos x$$
$$\frac{d}{dx}\cos x = -\sin x$$
$$\frac{d}{dx}(\sin x \cos x) = \cos^2 x - \sin^2 x.$$

   (b) The product of the derivatives of $\sin x$ and $\cos x$ is $\cos x(-\sin x) = -\cos x \sin x$. On the other hand, the derivative of the product is $\cos^2 x - \sin^2 x$, which is not the same. So no, the derivative of a product is not always equal to the product of the derivatives.

Solutions to Problems on Limits and Continuity

1. The graph in Figure 2.20 suggests that
$$\text{if } -0.05 < \theta < 0.05, \quad \text{then} \quad 0.999 < (\sin \theta)/\theta < 1.001.$$
   Thus, if θ is within 0.05 of 0, we see that $(\sin\theta)/\theta$ is within 0.001 of 1.

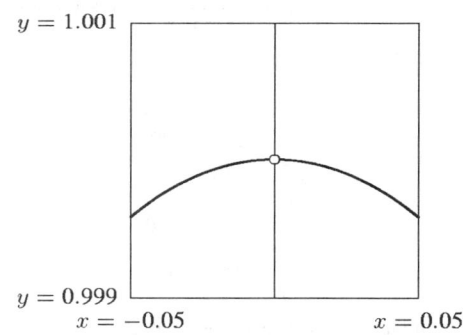

Figure 2.20: Graph of $(\sin\theta)/\theta$ with $-0.05 < \theta < 0.05$

2. The graph of $\sin\theta/\theta$ has a hole at $\theta = 0$, since division by zero is undefined. Thus, to make $g(\theta)$ continuous at $\theta = 0$ we fill in the hole. Figures 2.44 and 2.45 suggest that
$$\lim_{\theta \to 0} g(\theta) = \lim_{\theta \to 0} \frac{\sin\theta}{\theta} = 1.$$
Assuming this to be true, we define $g(0) = 1$. Then g is continuous at $\theta = 0$ since
$$\lim_{\theta \to 0} g(\theta) = 1 = g(0).$$

3. For $-0.5 \leq \theta \leq 0.5, 0 \leq y \leq 3$, the graph of $y = \dfrac{\sin(2\theta)}{\theta}$ is shown in Figure 2.21. Therefore, $\lim_{\theta \to 0} \dfrac{\sin(2\theta)}{\theta} = 2$.

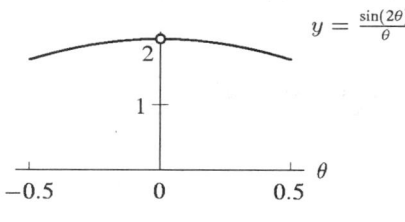

Figure 2.21

60   CHAPTER TWO /SOLUTIONS

4. For $-1 \le \theta \le 1, -1 \le y \le 1$, the graph of $y = \dfrac{\cos\theta - 1}{\theta}$ is shown in Figure 2.22. Therefore, $\lim\limits_{\theta \to 0} \dfrac{\cos\theta - 1}{\theta} = 0$.

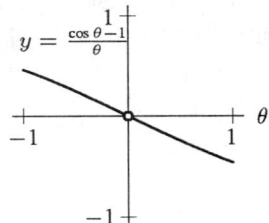

Figure 2.22

5. For $-90° \le \theta \le 90°, 0 \le y \le 0.02$, the graph of $y = \dfrac{\sin\theta}{\theta}$ is shown in Figure 2.23. Therefore, by tracing along the curve, we see that in degrees, $\lim\limits_{\theta \to 0} \dfrac{\sin\theta}{\theta} = 0.01745\ldots$.

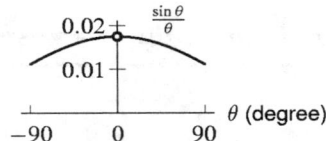

Figure 2.23

6. For $-0.5 \le \theta \le 0.5, 0 \le y \le 0.5$, the graph of $y = \dfrac{\theta}{\tan(3\theta)}$ is shown in Figure 2.24. Therefore, by tracing along the curve, we see that $\lim\limits_{\theta \to 0} \dfrac{\theta}{\tan(3\theta)} = 0.3333\ldots$.

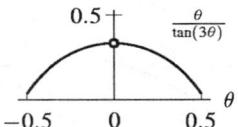

Figure 2.24

7. The limit appears to be 1; a graph and table of values is shown below.

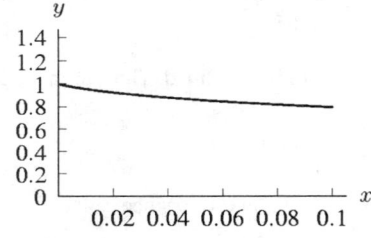

x	x^x
0.1	0.7943
0.01	0.9550
0.001	0.9931
0.0001	0.9990
0.00001	0.9999

8. (a) A possible example is $f(x) = 1/|x - 2|$ as $\lim\limits_{x \to 2} 1/|x - 2| = \infty$.
   (b) A possible example is $f(x) = -1/(x - 2)^2$ as $\lim\limits_{x \to 2} -1/(x - 2)^2 = -\infty$.

9. (a) Since $\sin(n\pi) = 0$ for $n = 1, 2, 3, \ldots$ the sequence of x-values
$$\frac{1}{\pi}, \frac{1}{2\pi}, \frac{1}{3\pi}, \ldots$$
works. These x-values $\to 0$ and are zeroes of $f(x)$.
   (b) Since $\sin(n\pi/2) = 1$ for $n = 1, 5, 9\ldots$ the sequence of x-values
$$\frac{2}{\pi}, \frac{2}{5\pi}, \frac{2}{9\pi}, \ldots$$
works.

(c) Since $\sin(n\pi)/2 = -1$ for $n = 3, 7, 11, \ldots$ the sequence of x-values
$$\frac{2}{3\pi}, \frac{2}{7\pi}, \frac{2}{11\pi} \cdots$$
works.

(d) Any two of these sequences of x-values show that if the limit were to exist, then it would have to have two (different) values: 0 and 1, or 0 and -1, or 1 and -1. Hence, the limit can not exist.

10. The statement
$$\lim_{h \to a} g(h) = K$$
means that we can make the value of $g(h)$ as close to K as we want by choosing h sufficiently close to, but not equal to, a.

In symbols, for any $\epsilon > 0$, there is a $\delta > 0$ such that
$$|g(h) - K| < \epsilon \quad \text{for all } 0 < |h - a| < \delta.$$

11. (a) **TABLE 2.14**

x	$f(x)$
2.1	4.1
2.01	4.01
2.001	4.001
2.0001	4.0001
1.9999	3.9999
1.999	3.999
1.99	3.99
1.9	3.9

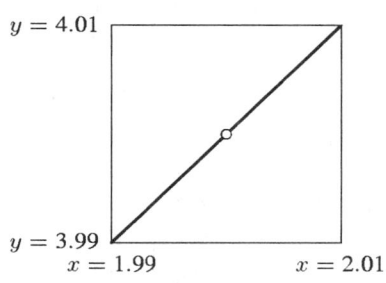

Figure 2.25

From Table 2.14, it appears the limit is 4. Figure 2.25 confirms this. An appropriate window is $1.99 < x < 2.01$, $3.99 < y < 4.01$.

(b) **TABLE 2.15**

x	$f(x)$
3.1	6.1
3.01	6.01
3.001	6.001
3.0001	6.0001
2.9999	5.9999
2.999	5.999
2.99	5.99
2.9	5.9

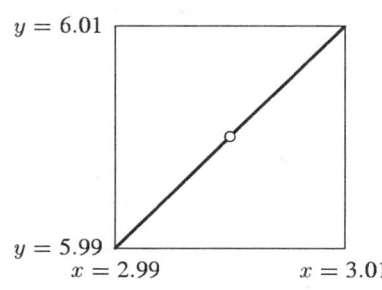

Figure 2.26

From Table 2.15, it appears the limit is 6. Figure 2.26 confirms this. An appropriate window is $2.99 < x < 3.01$, $5.99 < y < 6.01$.

(c) **TABLE 2.16**

x	$f(x)$
1.6708	−0.0500
1.5808	−0.0050
1.5718	−0.0005
1.5709	−0.0001
1.5707	0.0000
1.5698	0.0005
1.5608	0.0050
1.4708	0.0500

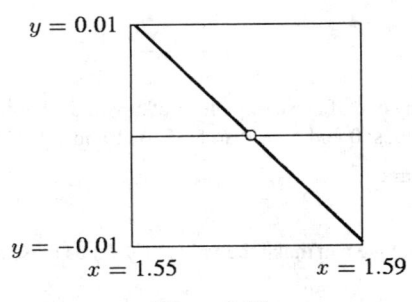

Figure 2.27

From Table 2.16, it appears the limit is 0. Figure 2.27 confirms this. An appropriate window is $1.55 < x < 1.59$, $-0.01 < y < 0.01$.

(d) **TABLE 2.17**

x	$f(x)$
1.6708	−1.2242
1.5808	−0.1250
1.5718	−0.0125
1.5709	−0.0013
1.5707	0.0012
1.5698	0.0125
1.5608	0.1249
1.4708	1.2241

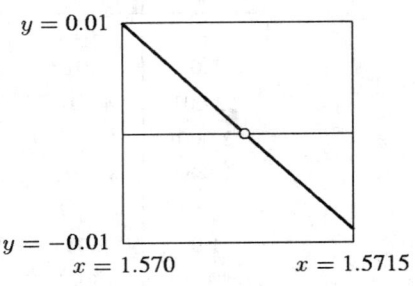

Figure 2.28

From Table 2.17, it appears the limit is 0. Figure 2.28 confirms this. An appropriate window is $1.570 < x < 1.5715$, $-0.01 < y < 0.01$.

(e) **TABLE 2.18**

x	$f(x)$
1.1	2.2140
1.01	2.0201
1.001	2.0020
1.0001	2.0002
0.9999	1.9998
0.999	1.9980
0.99	1.9801
0.9	1.8127

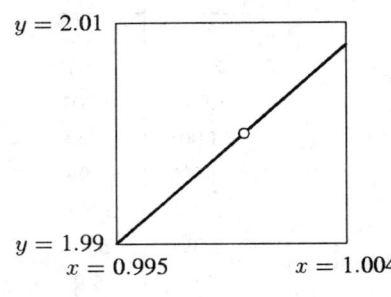

Figure 2.29

From Table 2.18, it appears the limit is 2. Figure 2.29 confirms this. An appropriate window is $0.995 < x < 1.004$, $1.99 < y < 2.01$.

(f)

TABLE 2.19

x	$f(x)$
2.1	0.5127
2.01	0.5013
2.001	0.5001
2.0001	0.5000
1.9999	0.5000
1.999	0.4999
1.99	0.4988
1.9	0.4877

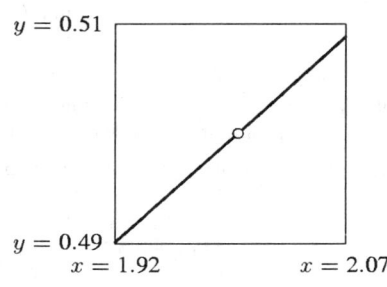

Figure 2.30

From Table 2.19, it appears the limit is $1/2$. Figure 2.30 confirms this. An appropriate window is $1.92 < x < 2.07, 0.49 < y < 0.51$.

12. (a) $f(x) = (x+3)/(2-x) = (1+3/x)/(2/x-1)$,
so $\lim_{x \to \infty} f(x) = \lim_{x \to \infty} \dfrac{1+3/x}{2/x-1} = \dfrac{\lim_{x \to \infty}(1+3/x)}{\lim_{x \to \infty}(2/x-1)} = \dfrac{1}{-1} = -1$.

(b) $f(x) = (x^2+2x-1)/(3+3x^2) = (1+2/x-1/x^2)/(3/x^2+3)$,
so $\lim_{x \to \infty} f(x) = \lim_{x \to \infty} \dfrac{1+2/x-1/x^2}{3/x^2+3} = \dfrac{\lim_{x \to \infty}(1+2/x-1/x^2)}{\lim_{x \to \infty}(3/x^2+3)} = \dfrac{1}{3}$.

(c) $f(x) = (x^2+4)/(x+3) = (x+4/x)/(1+3/x)$, so $\lim_{x \to \infty} f(x) = +\infty$.

(d) $f(x) = (2x^3 - 16x^2)/(4x^2 + 3x^3) = (2 - 16/x)/(4/x + 3)$,
so $\lim_{x \to \infty} f(x) = \lim_{x \to \infty} \dfrac{2-16/x}{4/x+3} = \dfrac{\lim_{x \to \infty}(2-16/x)}{\lim_{x \to \infty}(4/x+3)} = \dfrac{2}{3}$.

(e) $f(x) = (x^4+3x)/(x^4+2x^5) = (1/x + 3/x^4)/(1/x + 2)$,
so $\lim_{x \to \infty} f(x) = \dfrac{\lim_{x \to \infty}(1/x+3/x^4)}{\lim_{x \to \infty}(1/x+2)} = \dfrac{0}{2} = 0$.

(f) $f(x) = \dfrac{3e^x + 2}{2e^x + 3} = \dfrac{3 + 2e^{-x}}{2 + 3e^{-x}}$, so $\lim_{x \to \infty} f(x) = \dfrac{\lim_{x \to \infty}(3+2e^{-x})}{\lim_{x \to \infty}(2+3e^{-x})} = \dfrac{3}{2}$.

(g) $f(x) = \dfrac{2e^{-x} + 3}{3e^{-x} + 2}$, so $\lim_{x \to \infty} f(x) = \dfrac{\lim_{x \to \infty}(2e^{-x}+3)}{\lim_{x \to \infty}(3e^{-x}+2)} = \dfrac{3}{2}$.

13. (a) If $b = 0$, then the property says $\lim_{x \to c} 0 = 0$, which is easy to see is true.

(b) If $|f(x) - L| < \dfrac{\epsilon}{|b|}$, then multiplying by $|b|$ gives

$$|b||f(x) - L| < \epsilon.$$

Since

$$|b||f(x) - L| = |b(f(x) - L)| = |bf(x) - bL|,$$

we have

$$|bf(x) - bL| < \epsilon.$$

(c) Suppose that $\lim_{x \to c} f(x) = L$. We want to show that $\lim_{x \to c} bf(x) = bL$. If we are to have

$$|bf(x) - bL| < \epsilon,$$

then we will need

$$|f(x) - L| < \dfrac{\epsilon}{|b|}.$$

We choose δ small enough that

$$|x - c| < \delta \quad \text{implies} \quad |f(x) - L| < \dfrac{\epsilon}{|b|}.$$

By part (b), this ensures that

$$|bf(x) - bL| < \epsilon,$$

as we wanted.

64 CHAPTER TWO /SOLUTIONS

14. Suppose $\lim_{x \to c} f(x) = L_1$ and $\lim_{x \to c} g(x) = L_2$. Then we need to show that

$$\lim_{x \to c} \big(f(x) + g(x)\big) = L_1 + L_2.$$

Let $\epsilon > 0$ be given. We need to show that we can choose $\delta > 0$ so that whenever $|x - c| < \delta$, we will have $\big|\big(f(x) + g(x)\big) - (L_1 + L_2)\big| < \epsilon$. First choose $\delta_1 > 0$ so that $|x - c| < \delta_1$ implies $|f(x) - L_1| < \frac{\epsilon}{2}$; we can do this since $\lim_{x \to c} f(x) = L_1$. Similarly, choose $\delta_2 > 0$ so that $|x - c| < \delta_2$ implies $|g(x) - L_2| < \frac{\epsilon}{2}$. Now, set δ equal to the smaller of δ_1 and δ_2. Thus $|x - c| < \delta$ will make both $|x - c| < \delta_1$ and $|x - c| < \delta_2$. Then, for $|x - c| < \delta$, we have

$$|f(x) + g(x) - (L_1 + L_2)| = \big|\big(f(x) - L_1\big) + \big(g(x) - L_2\big)\big|$$
$$\leq \big|\big(f(x) - L_1\big)\big| + \big|\big(g(x) - L_2\big)\big|$$
$$\leq \frac{\epsilon}{2} + \frac{\epsilon}{2} = \epsilon.$$

This proves $\lim_{x \to c}(f(x) + g(x)) = \lim_{x \to c} f(x) + \lim_{x \to c} g(x)$, which is the result we wanted to prove.

15. (a) We need to show that for any given $\epsilon > 0$, there is a $\delta > 0$ so that $|x - c| < \delta$ implies $|f(x)g(x)| < \epsilon$. If $\epsilon > 0$ is given, choose δ_1 so that when $|x - c| < \delta_1$, we have $|f(x)| < \sqrt{\epsilon}$. This can be done since $\lim_{x \to 0} f(x) = 0$. Similarly, choose δ_2 so that when $|x - c| < \delta_2$, we have $|g(x)| < \sqrt{\epsilon}$. Then, if we take δ to be the smaller of δ_1 and δ_2, we'll have that $|x - c| < \delta$ implies both $|f(x)| < \sqrt{\epsilon}$ and $|g(x)| < \sqrt{\epsilon}$. So when $|x - c| < \delta$, we have $|f(x)g(x)| = |f(x)||g(x)| < \sqrt{\epsilon} \cdot \sqrt{\epsilon} = \epsilon$. Thus $\lim_{x \to c} f(x)g(x) = 0$.

(b) $\big(f(x) - L_1\big)\big(g(x) - L_2\big) + L_1 g(x) + L_2 f(x) - L_1 L_2$
$= f(x)g(x) - L_1 g(x) - L_2 f(x) + L_1 L_2 + L_1 g(x) + L_2 f(x) - L_1 L_2 = f(x)g(x)$.

(c) $\lim_{x \to c}\big(f(x) - L_1\big) = \lim_{x \to c} f(x) - \lim_{x \to c} L_1 = L_1 - L_1 = 0$, using the second limit property. Similarly, $\lim_{x \to c}\big(g(x) - L_2\big) = 0$.

(d) Since $\lim_{x \to c}\big(f(x) - L_1\big) = \lim_{x \to c}\big(g(x) - L_2\big) = 0$, we have that $\lim_{x \to c}\big(f(x) - L_1\big)\big(g(x) - L_2\big) = 0$ by part (a).

(e) From part (b), we have

$$\lim_{x \to c} f(x)g(x) = \lim_{x \to c} \big(\big(f(x) - L_1\big)\big(g(x) - L_2\big) + L_1 g(x) + L_2 f(x) - L_1 L_2\big)$$
$$= \lim_{x \to c}\big(f(x) - L_1\big)\big(g(x) - L_2\big) + \lim_{x \to c} L_1 g(x) + \lim_{x \to c} L_2 f(x) + \lim_{x \to c}(-L_1 L_2)$$
(using limit property 2)
$$= 0 + L_1 \lim_{x \to c} g(x) + L_2 \lim_{x \to c} f(x) - L_1 L_2$$
(using limit property 1 and part (d))
$$= L_1 L_2 + L_2 L_1 - L_1 L_2 = L_1 L_2.$$

16. (a) For any c, we have $f(c) = k$. This means $|f(x) - f(c)| = |k - k| = 0$ for any x and c. So given $\epsilon > 0$, choose any δ, for example $\delta = 1$. Then $|x - c| < \delta$ implies $|f(x) - f(c)| = 0 < \epsilon$. So $\lim_{x \to c} f(x) = f(c)$, so f is continuous everywhere.

(b) Given any $\epsilon > 0$, we can take $\delta = \epsilon$, so that when $|x - c| < \delta$, $|g(x) - g(c)| = |x - c| < \epsilon$. So, $\lim_{x \to c} g(x) = g(c)$ and g is continuous everywhere.

17. We can use the δ guaranteed by the continuity of f to be the "ϵ" for the continuity of g. That is, we can choose $\delta_1 > 0$ so that when $|x - c| < \delta_1$ we have $|g(x) - g(c)| < \delta$. But if we let $g(x) = y$ and remember $g(c) = d$, then this says that $|x - c| < \delta_1$ implies $|y - d| < \delta$, which in turn implies $|f(y) - f(d)| < \epsilon$. So, given any $\epsilon > 0$, we can find a $\delta_1 > 0$ such that $|x - c| < \delta_1$ implies $|f(y) - f(d)| = \big|f\big(g(x)\big) - f\big(g(c)\big)\big| < \epsilon$. This proves that $\lim_{x \to c} f\big(g(x)\big) = f(g(c))$, which is what it means for $f\big(g(x)\big)$ to be continuous at $x = c$.

18. By tracing on a calculator or solving equations, we find the following values of δ:
For $\epsilon = 0.2, \delta \leq 0.1$.
For $\epsilon = 0.1, \delta \leq 0.05$.
For $\epsilon = 0.02, \delta \leq 0.01$.
For $\epsilon = 0.01, \delta \leq 0.005$.
For $\epsilon = 0.002, \delta \leq 0.001$.
For $\epsilon = 0.001, \delta \leq 0.0005$.

19. By tracing on a calculator or solving equations, we find the following values of δ:
    For $\epsilon = 0.1, \delta \leq 0.46$.
    For $\epsilon = 0.01, \delta \leq 0.21$.
    For $\epsilon = 0.001, \delta < 0.1$. Thus, we can take $\delta \leq 0.09$.

20. The results of Problem 18 suggest that we can choose $\delta = \epsilon/2$. For any $\epsilon > 0$, we want to find the δ such that
$$|f(x) - 3| = |-2x + 3 - 3| = |2x| < \epsilon.$$
Then if $|x| < \delta = \epsilon/2$, it follows that $|f(x) - 3| = |2x| < \epsilon$. So $\lim_{x \to 0}(-2x + 3) = 3$.

21. For any $\epsilon > 0$, we want to find the δ such that
$$|g(x) - 2| = \left|-x^3 + 2 - 2\right| = \left|x^3\right| < \epsilon.$$
Choose $\delta = \epsilon^{1/3}$. Then if $|x| < \delta = \epsilon^{1/3}$, it follows that $|g(x) - 2| = \left|x^3\right| < \epsilon$.

22. The only change is that, instead of considering all x near c, we only consider x near to and greater than c. Thus the phrase "$|x - c| < \delta$" must be replaced by "$c < x < c + \delta$." Thus, we define
$$\lim_{x \to c^+} f(x) = L$$
to mean that for any $\epsilon > 0$ (as small as we want), there is a $\delta > 0$ (sufficiently small) such that if $c < x < c + \delta$, then $|f(x) - L| < \epsilon$.

23. The only change is that, instead of considering all x near c, we only consider x near to and less than c. Thus the phrase "$|x - c| < \delta$" must be replaced by "$c - \delta < x < c$." Thus, we define
$$\lim_{x \to c^-} f(x) = L$$
to mean that for any $\epsilon > 0$ (as small as we want), there is a $\delta > 0$ (sufficiently small) such that if $c - \delta < x < c$, then $|f(x) - L| < \epsilon$.

24. Instead of being "sufficiently close to c," we want x to be "sufficiently large." Using N to measure how large x must be, we define
$$\lim_{x \to \infty} f(x) = L$$
to mean that for any $\epsilon > 0$ (as small as we want), there is a $N > 0$ (sufficiently large) such that if $x > N$, then $|f(x) - L| < \epsilon$.

25. Since $\sin x$ is continuous everywhere, and $1/x$ is continuous except at $x = 0$, we know that $\sin(1/x)$ is continuous at each $x \neq 0$, because the composition of continuous functions is continuous. Furthermore, the product of two continuous functions is continuous, so $f(x)$ is continuous except perhaps at $x = 0$. Thus the only place we need to check continuity is $x = 0$. Since the values of the sine function go between 1 and -1, we have $|\sin(1/x)| \leq 1$ for $x \neq 0$, so
$$\left|x \sin\left(\frac{1}{x}\right)\right| \leq |x|.$$
It follows that we can make $x \sin(1/x)$ as small as we like by making x small enough. Hence $\lim_{x \to 0} (x \sin(1/x)) = 0$, so
$$\lim_{x \to 0} f(x) = \lim_{x \to 0} \left(x \sin\left(\frac{1}{x}\right)\right) = 0 = f(0).$$
So f is continuous at $x = 0$.

On the other hand, $\sin(1/x)$ has infinitely many oscillations between ϵ and 0, since it crosses the x-axis infinitely many times, at the points $x = 1/(n\pi)$, where n is an integer. So $x \sin(1/x)$ also has infinitely many oscillations, although they get smaller and smaller as $x \to 0$. Thus, f is not always increasing or always decreasing on any interval of the form $[0, \epsilon]$.

26. (a) Since $L > 0$, we take any $\epsilon > 0$ and $\epsilon < L$. Then, by the continuity of f, there is a $\delta > 0$ such that
$$|f(x) - L| < \epsilon \quad \text{for all} \quad |x - r| < \delta,$$
that is
$$L - \epsilon < f(x) < L + \epsilon \quad \text{for all} \quad r - \delta < x < r + \delta.$$
Since $\epsilon < L$, we know that $L - \epsilon > 0$, so
$$0 < f(x) < L + \epsilon \quad \text{for all} \quad r - \delta < x < r + \delta.$$
Since the interval $[a_n, b_n]$ has length 10^{-n}, we pick n so that $10^{-n} < \delta$. Then a_n is in the interval $[r - \delta, r + \delta]$; thus, $f(a_n) > 0$. In addition, $f(a_n) < 0$ because of the way a_n was originally chosen. We have a contradiction, so $f(r)$ cannot be positive.

(b) Since $L < 0$, we take any $\epsilon > 0$ and $\epsilon < |L|$. By the continuity of f, there is a $\delta > 0$ such that
$$|f(x) - L| < \epsilon \quad \text{for all} \quad |x - r| < \delta,$$
so
$$L - \epsilon < f(x) < L + \epsilon \quad \text{for all} \quad r - \delta < x < r + \delta.$$
Since $\epsilon < |L|$, we know that $L + \epsilon < 0$, so we have
$$L - \epsilon < f(x) < 0 \quad \text{for all} \quad r - \delta < x < r + \delta.$$
But, as in part (a), there is some b_n in the interval $[r - \delta, r + \delta]$. Thus, $f(b_n) < 0$. In addition, $f(b_n) > 0$ because of the way b_n was originally chosen. We again have a contradiction, so $f(r)$ cannot be negative.

(c) Since we have seen that $f(r) > 0$ leads to a contradiction, and $f(r) < 0$ leads to a contradiction, the only possibility is that $f(r) = 0$.

27. If $k = f(a)$ or $k = f(b)$, we can take $c = a$ or $c = b$. If $f(a) = f(b)$, the only possibility is $k = f(a) = f(b)$, and we take $c = a$.

Now we suppose $f(a) < k < f(b)$. The case $f(b) < k < f(a)$ is handled similarly. Consider the function
$$g(x) = f(x) - k.$$
Then g is continuous in $[a, b]$, and $g(a) = f(a) - k < 0$ and $g(b) = f(b) - k > 0$. The number c that we are looking for is a zero of g, since if $g(c) = 0$, then $f(c) = k$.

Since $g(a) < 0$ and $g(b) > 0$, we divide the interval $[a, b]$ into two (or more) equal subintervals. We pick an interval $[a_n, b_n]$ with $g(a_n) < 0$ and $g(b_n) > 0$. (If g is zero on one of the endpoints, we have found c and can stop.) Continuing this way, we construct an infinite sequence of closed intervals $[a_n, b_n]$, each one contained within the previous ones and with length tending to zero.

By the nested interval theorem, there is a number c in all of these intervals. By the continuity of g, we can prove that $g(c) = 0$, using the argument in Problem 26. Hence $f(c) = k$.

Solutions to Problems on Differentiability and Linear Approximation

1. (a) (i) Function f is not continuous at $x = 1$.
   (ii) Function f appears not differentiable at $x = 1, 2, 3$.
   (b) (i) Function g appears continuous at all x-values shown.
   (ii) Function g appears not differentiable at $x = 2, 4$.

2. We want to look at
$$\lim_{h \to 0} \frac{(h^2 + 0.0001)^{1/2} - (0.0001)^{1/2}}{h}.$$
As $h \to 0$ from positive or negative numbers, the difference quotient approaches 0. (Try evaluating it for $h = 0.001$, 0.0001, etc.) So it appears there is a derivative at $x = 0$ and that this derivative is zero. How can this be if f has a corner at $x = 0$?

The answer lies in the fact that what appears to be a corner is in fact smooth—when you zoom in, the graph of f looks like a straight line with slope 0! See Figure 2.31.

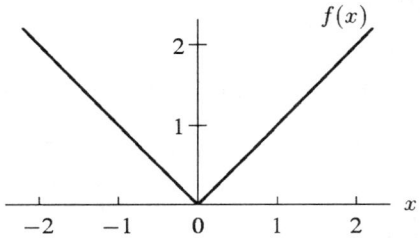 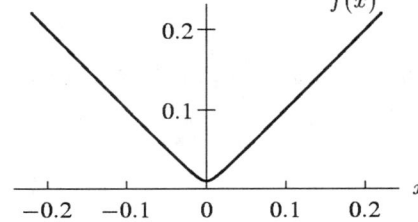

Figure 2.31: Close-ups of $f(x) = (x^2 + 0.0001)^{1/2}$ showing differentiability at $x = 0$

3. Yes, f is differentiable at $x = 0$, since its graph does not have a "corner" at $x = 0$. See below.

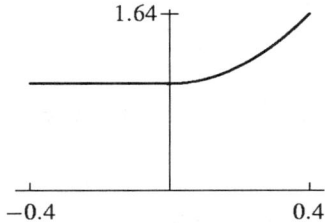

Another way to see this is by computing:
$$\lim_{h \to 0} \frac{f(h) - f(0)}{h} = \lim_{h \to 0} \frac{(h + |h|)^2}{h} = \lim_{h \to 0} \frac{h^2 + 2h|h| + |h|^2}{h}.$$
Since $|h|^2 = h^2$, we have:
$$\lim_{h \to 0} \frac{f(h) - f(0)}{h} = \lim_{h \to 0} \frac{2h^2 + 2h|h|}{h} = \lim_{h \to 0} 2(h + |h|) = 0.$$
So f is differentiable at 0 and $f'(0) = 0$.

4. As we can see in Figure 2.32, f oscillates infinitely often between the x-axis and the line $y = 2x$ near the origin. This means a line from $(0, 0)$ to a point $(h, f(h))$ on the graph of f alternates between slope 0 (when $f(h) = 0$) and slope 2 (when $f(h) = 2h$) infinitely often as h tends to zero. Therefore, there is no limit of the slope of this line as h tends to zero, and thus there is no derivative at the origin. Another way to see this is by noting that
$$\lim_{h \to 0} \frac{f(h) - f(0)}{h} = \lim_{h \to 0} \frac{h \sin(\frac{1}{h}) + h}{h} = \lim_{h \to 0} \left(\sin\left(\frac{1}{h}\right) + 1 \right)$$
does not exist, since $\sin(\frac{1}{h})$ does not have a limit as h tends to zero. Thus, f is not differentiable at $x = 0$.

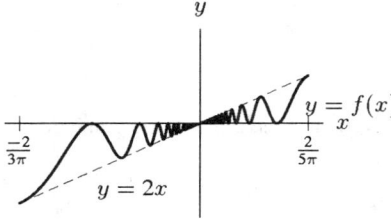

Figure 2.32

5. We can see from Figure 2.33 that the graph of f oscillates infinitely often between the curves $y = x^2$ and $y = -x^2$ near the origin. Thus the slope of the line from $(0,0)$ to $(h, f(h))$ oscillates between h (when $f(h) = h^2$ and $\frac{f(h)-0}{h-0} = h$) and $-h$ (when $f(h) = -h^2$ and $\frac{f(h)-0}{h-0} = -h$) as h tends to zero. So, the limit of the slope as h tends to zero is 0, which is the derivative of f at the origin. Another way to see this is to observe that

$$\lim_{h \to 0} \frac{f(h) - f(0)}{h} = \lim_{h \to 0} \left(\frac{h^2 \sin(\frac{1}{h})}{h} \right)$$
$$= \lim_{h \to 0} h \sin(\frac{1}{h})$$
$$= 0,$$

since $\lim_{h \to 0} h = 0$ and $-1 \leq \sin(\frac{1}{h}) \leq 1$ for any h. Thus f is differentiable at $x = 0$, and $f'(0) = 0$.

Figure 2.33

6. (a) The graph of Q against t does not have a break at $t = 0$, so Q appears to be continuous at $t = 0$. See below.

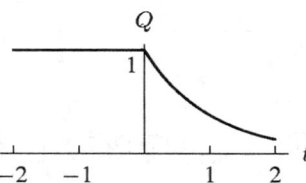

(b) The slope dQ/dt is zero for $t < 0$, and negative for all $t > 0$. At $t = 0$, there appears to be a corner, which does not disappear as you zoom in, suggesting that I is defined for all times t except $t = 0$.

7. (a) Notice that B is a linear function of r for $r \leq r_0$ and a reciprocal for $r > r_0$. The constant B_0 is the value of B at $r = r_0$ and the maximum value of B.

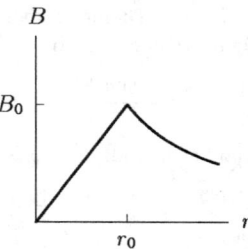

(b) B is continuous at $r = r_0$ because there is no break in the graph there. Using the formula for B, we have

$$\lim_{r \to r_0^-} B = \frac{r_0}{r_0} B_0 = B_0 \quad \text{and} \quad \lim_{r \to r_0^+} B = \frac{r_0}{r_0} B_0 = B_0.$$

(c) The function B is not differentiable at $r = r_0$ because the graph has a corner there. The slope is positive for $r < r_0$ and the slope is negative for $r > r_0$.

8. (a) Since
$$\lim_{r \to r_0^-} E = kr_0$$
and
$$\lim_{r \to r_0^+} E = \frac{kr_0^2}{r_0} = kr_0$$
and
$$E(r_0) = kr_0,$$
we see that E is continuous at r_0.

(b) The function E is not differentiable at $r = r_0$ because the graph has a corner there. The slope is positive for $r < r_0$ and the slope is negative for $r > r_0$.

(c)
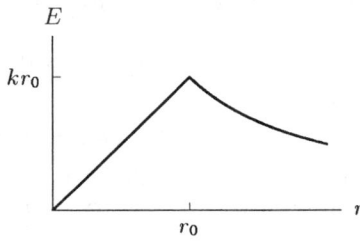

9.
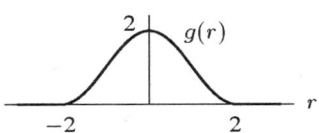

(a) The graph of $g(r)$ does not have a break or jump at $r = 2$, and so $g(r)$ is continuous there. This is confirmed by the fact that
$$g(2) = 1 + \cos(\pi 2/2) = 1 + (-1) = 0$$
so the value of $g(r)$ as you approach $r = 2$ from the left is the same as the value when you approach $r = 2$ from the right.

(b) The graph of $g(r)$ does not have a corner at $r = 2$, even after zooming in, so $g(r)$ appears to be differentiable at $r = 0$. This is confirmed by the fact that $\cos(\pi r/2)$ is at the bottom of a trough at $r = 2$, and so its slope is 0 there. Thus the slope to the left of $r = 2$ is the same as the slope to the right of $r = 2$.

10. (a) The graph of ϕ does not have a break at $y = 0$, and so ϕ appears to be continuous there. See figure below.

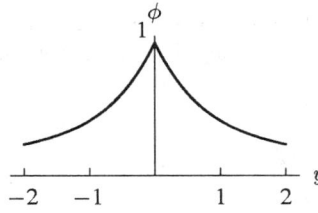

(b) The graph of ϕ has a corner at $y = 0$ which does not disappear as you zoom in. Therefore ϕ appears not be differentiable at $y = 0$.

11. (a) Since $f(4) = \sqrt{4} = 2$ and $f'(4) = 1/4$, the tangent line approximation is
$$f(x) \approx f(4) + f'(4)(x-4)$$
$$\approx 2 + \frac{1}{4}(x-4).$$

See Figure 2.34.

(b) For $x = 4.1$, the true value is
$$f(4.1) = \sqrt{4.1} = 2.02485...,$$
whereas the approximation is
$$f(4.1) \approx 2 + \frac{1}{4}(4.1-4) = 2.025.$$
Thus, the approximation differs from the true value by about 0.00015.

(c) For $x = 16$ the true value is:
$$f(16) = \sqrt{16} = 4$$
whereas the approximation is
$$f(16) \approx 2 + \frac{1}{4}(16-4) = 5.$$
Thus, the approximation differs from the true value by 1.

(d) The tangent line is a good approximation to the graph near $x = 4$, but not necessarily far away. Of course, there's no reason to expect that the curve will look like the tangent line if we go too far away, and usually it doesn't. (See Figure 2.34.) The problem is that we have traveled too far from the place where the curve looks like a line with slope $1/4$.

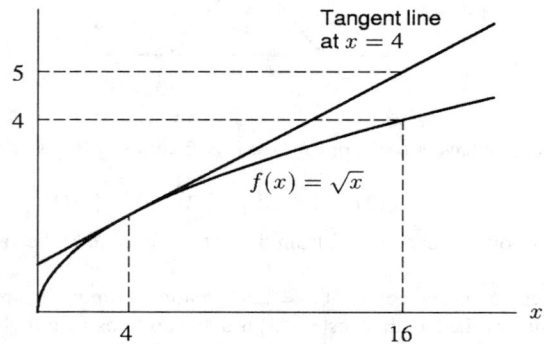

Figure 2.34: Local linearization: Approximating $f(x) = \sqrt{x}$ by its tangent line at $x = 4$

12. The graph of x^2 is concave up and lies above its tangent line; therefore, the linearization will always be too small. The graph of $\sqrt{x}$ is concave down and lies below its tangent line, and therefore the linearization will be too large.

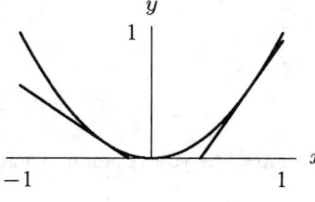 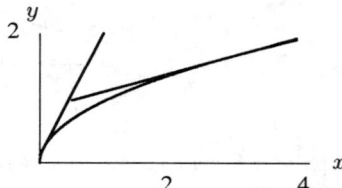

13. Since $f(1) = 1$ and we showed that $f'(1) = 2$, the local linearization is
$$f(x) \approx 1 + 2(x-1) = 2x - 1.$$

14. We know that the local linearization is $y = 2x - 1$. The fact that $f(x)$ is differentiable tells us that

$$\lim_{x \to 0} \frac{E(x)}{x-1} = \lim_{x \to 0} \frac{x^2 - (2x-1)}{x-1} = 0.$$

Suppose we take $\epsilon = 1/10$, then there is a δ such that

$$\left| \frac{x^2 - (2x-1)}{x-1} \right| < \frac{1}{10} \quad \text{for all} \quad |x-1| < \delta.$$

Thus

$$-\frac{1}{10}(x-1) < x^2 - (2x-1) < \frac{1}{10}(x-1) \quad \text{for all} \quad 1 - \delta < x < 1 + \delta.$$

To find this δ, we observe that

$$\left| \frac{x^2 - (2x-1)}{x-1} \right| = \left| \frac{x^2 - 2x + 1}{x-1} \right| = \left| \frac{(x-1)^2}{x-1} \right| = |x-1|.$$

Therefore we can take $\delta = 1/10$. Then

$$\left| \frac{x^2 - (2x-1)}{x-1} \right| < \frac{1}{10} \quad \text{for all} \quad |x-1| < \frac{1}{10}$$

so

$$-\frac{1}{10}(x-1) < |x^2 - (2x-1)| < \frac{1}{10}(x-1).$$

15. We want to show that

$$\lim_{x \to a} \frac{f(x) - f(a)}{x - a} = L.$$

Substituting for $f(x)$ we have

$$\lim_{x \to a} \frac{f(x) - f(a)}{x - a} = \lim_{x \to a} \frac{f(a) + L(x-a) + E_L(x) - f(a)}{x - a}$$

$$= \lim_{x \to a} \left(L + \frac{E_L(x)}{x - a} \right) = L + \lim_{x \to 0} \frac{E_L(x)}{x - a} = L.$$

Thus, we have shown that f is differentiable at $x = a$ and that its derivative is L, that is, $f'(a) = L$.

CHAPTER THREE

Solutions for Section 3.1

1. The derivative, $f'(x)$, is defined as
$$f'(x) = \lim_{h \to 0} \frac{f(x+h) - f(x)}{h}.$$
   If $f(x) = 7$, then
$$f'(x) = \lim_{h \to 0} \frac{7-7}{h} = \lim_{h \to 0} \frac{0}{h} = 0.$$

2. (a) $f(x) = -3x + 2, g(x) = 2x + 1$.
$$\begin{aligned}k(x) &= f(x) + g(x) \\ &= (-3x + 2) + (2x + 1) \\ &= -x + 3 \\ k'(x) &= -1.\end{aligned}$$
   Also, $f'(x) = -3, g'(x) = 2$, so $f'(x) + g'(x) = -3 + 2 = -1$, confirming that $k'(x) = f'(x) + g'(x)$.

   (b)
$$\begin{aligned}j(x) &= f(x) - g(x) \\ &= (-3x + 2) - (2x + 1) \\ &= -5x + 1 \\ j'(x) &= -5.\end{aligned}$$
   Also, $f'(x) - g'(x) = -3 - 2 = -5$, confirming that $j'(x) = f'(x) - g'(x)$.

3. (a) $f(x) = 5x - 3, g(x) = -2x + 1$.
   $f[g(x)] = f(-2x + 1) = 5(-2x + 1) - 3 = -10x + 5 - 3 = -10x + 2$.
   So, $\frac{d}{dx}[f[g(x)]] = -10$.

   (b) $f'(x) = 5, g'(x) = -2$, and the derivative of $f[g(x)] = -10 = f'(x)g'(x)$. So one might speculate that if f and g are linear functions, then the derivative of $f[g(x)]$ is $f'(x)g'(x)$. This is true, and it can be proved as follows. Consider general linear functions f and g:
$$\begin{aligned}f(x) &= m_1 x + b_1 \\ f'(x) &= m_1 \\ g(x) &= m_2 x + b_2 \\ g'(x) &= m_2.\end{aligned}$$
   Then
$$f[g(x)] = m_1(m_2 x + b_2) + b_1 = m_1 m_2 x + m_1 b_2 + b_1,$$
   and
$$\frac{d}{dx}\left(f[g(x)]\right) = m_1 m_2 = f'(x)g'(x).$$

4. $y' = 11x^{10}$.
5. $y' = 12x^{11}$.
6. $y' = 11x^{-12}$.
7. $y' = 3.2x^{2.2}$.

74   CHAPTER THREE /SOLUTIONS

8. $y' = -12x^{-13}$.
9. $y' = \frac{4}{3}x^{1/3}$.
10. $y' = \frac{3}{4}x^{-1/4}$.
11. $y' = -\frac{3}{4}x^{-7/4}$.
12. $f'(x) = -4x^{-5}$.
13. $f'(x) = \frac{1}{4}x^{-3/4}$.
14. $f'(x) = ex^{e-1}$.
15. $y' = 6x^{1/2} - \frac{5}{2}x^{-1/2}$.
16. $y' = 18x^2 + 8x - 2$.
17. $y' = -12x^3 - 12x^2 - 6$.
18. $y' = 15t^4 - \frac{5}{2}t^{-1/2} - \frac{7}{t^2}$.
19. $y' = 6t - \frac{6}{t^{3/2}} + \frac{2}{t^3}$.
20. $y' = 2z - \frac{1}{2z^2}$.
21. $y = x + \frac{1}{x}$, so $y' = 1 - \frac{1}{x^2}$.
22. $g(z) = z^5 + 5z^4 - z$
    $g'(z) = 5z^4 + 20z^3 - 1$.
23. $f(t) = \frac{1}{t^2} + \frac{1}{t} - \frac{1}{t^4} = t^{-2} + t^{-1} - t^{-4}$
    $f'(t) = -2t^{-3} - t^{-2} + 4t^{-5}$.
24. $y = \frac{\theta}{\sqrt{\theta}} - \frac{1}{\sqrt{\theta}} = \sqrt{\theta} - \frac{1}{\sqrt{\theta}}$
    $y' = \frac{1}{2\sqrt{\theta}} + \frac{1}{2\theta^{3/2}}$.
25. The functions whose derivatives don't exist at $x = 0$ are in problems 8, 10, 11, 12, 13, and 15.
26. $y' = \frac{1}{2}x^{-1/2}$.   (power rule)
27. So far, we can only take the derivative of powers of x and the sums of constant multiples of powers of x. Since we cannot write $\sqrt{x+3}$ in this form, we cannot yet take its derivative.
28. $y' = 6x$.   (power rule and sum rule)
29. $y' = -\frac{2}{3z^3}$.   (power rule and sum rule)
30. We cannot write $\frac{1}{3x^2+4}$ as the sum of powers of x multiplied by constants.
31. The x is in the exponent and we haven't learned how to handle that yet.
32. $y' = -\frac{1}{6x^{3/2}}$.   (power rule and sum rule)
33. $g'(x) = \frac{12}{\sqrt[6]{x^5}} - \frac{18}{\sqrt[3]{x^5}}$, using the power and sum rules.
34. $g'(x) = \pi x^{(\pi-1)} + \pi x^{-(\pi+1)}$, by the power and sum rules.
35. $f'(t) = 6t^2 - 8t + 3$ and $f''(t) = 12t - 8$.
36.
$$f'(x) = 12x^2 + 12x - 23 \geq 1$$
$$12x^2 + 12x - 24 \geq 0$$
$$12(x^2 + x - 2) \geq 0$$
$$12(x+2)(x-1) \geq 0.$$

Hence $x \geq 1$ or $x \leq -2$.

37. Decreasing means $f'(x) < 0$:
$$f'(x) = 4x^3 - 12x^2 = 4x^2(x-3),$$
so $f'(x) < 0$ when $x < 3$ and $x \neq 0$. Concave up means $f''(x) > 0$:
$$f''(x) = 12x^2 - 24x = 12x(x-2)$$
so $f''(x) > 0$ when
$$12x(x-2) > 0$$
$$x < 0 \quad \text{or} \quad x > 2.$$
So, both conditions hold for $x < 0$ or $2 < x < 3$.

38. The graph increases when $dy/dx > 0$:
$$\frac{dy}{dx} = 5x^4 - 5 > 0$$
$$5(x^4 - 1) > 0 \quad \text{so} \quad x^4 > 1 \quad \text{so} \quad x > 1 \text{ or } x < -1.$$
The graph is concave up when $d^2y/dx^2 > 0$:
$$\frac{d^2y}{dx^2} = 20x^3 > 0 \quad \text{so} \quad x > 0.$$
We need values of x where $\{x > 1 \text{ or } x < -1\}$ AND $\{x > 0\}$, which implies $x > 1$. Thus, both conditions hold for all values of x larger than 1.

39.
$$f'(x) = -8 + 2\sqrt{2}x$$
$$f'(r) = -8 + 2\sqrt{2}r = 4$$
$$r = \frac{12}{2\sqrt{2}} = 3\sqrt{2}.$$

40. (a) Since the power of x will go down by one every time you take a derivative (until the exponent is zero after which the derivative will be zero), we can see immediately that $f^{(8)}(x) = 0$.
    (b) $f^{(7)}(x) = 7 \cdot 6 \cdot 5 \cdot 4 \cdot 3 \cdot 2 \cdot 1 \cdot x^0 = 5040$.

41.
$$f'(x) = 6x^2 - 4x \quad \text{so} \quad f'(1) = 6 - 4 = 2.$$
Thus the equation of the tangent line is $(y - 1) = 2(x - 1)$ or $y = 2x - 1$.

42. If $f(x) = x^n$, then $f'(x) = nx^{n-1}$. This means $f'(1) = n \cdot 1^{n-1} = n \cdot 1 = n$, because any power of 1 equals 1.

43. Since $f(x) = ax^n$, $f'(x) = anx^{n-1}$. We know that $f'(2) = (an)2^{n-1} = 3$, and $f'(4) = (an)4^{n-1} = 24$. Therefore,
$$\frac{f'(4)}{f'(2)} = \frac{24}{3}$$
$$\frac{(an)4^{n-1}}{(an)2^{n-1}} = \left(\frac{4}{2}\right)^{n-1} = 8$$
$$2^{n-1} = 8, \text{ and thus } n = 4.$$
Substituting $n = 4$ into the expression for $f'(2)$, we get $3 = a(4)(8)$, or $a = 3/32$.

44. Yes. To see why, we substitute $y = x^n$ into the equation $13x\frac{dy}{dx} = y$. We first calculate $\frac{dy}{dx} = \frac{d}{dx}(x^n) = nx^{n-1}$. The differential equation becomes
$$13x(nx^{n-1}) = x^n$$
But $13x(nx^{n-1}) = 13n(x \cdot x^{n-1}) = 13nx^n$, so we have
$$13n(x^n) = x^n$$
This equality must hold for all x, so we get $13n = 1$, so $n = 1/13$. Thus, $y = x^{1/13}$ is a solution.

45. The slopes of the tangent lines to $y = x^2 - 2x + 4$ are given by $y' = 2x - 2$. A line through the origin has equation $y = mx$. So, at the tangent point, $x^2 - 2x + 4 = mx$ where $m = y' = 2x - 2$.

$$x^2 - 2x + 4 = (2x - 2)x$$
$$x^2 - 2x + 4 = 2x^2 - 2x$$
$$-x^2 + 4 = 0$$
$$-(x+2)(x-2) = 0$$
$$x = 2, -2.$$

Thus, the points of tangency are $(2, 4)$ and $(-2, 12)$. The lines through these points and the origin are $y = 2x$ and $y = -6x$, respectively. Graphically, this can be seen in Figure 3.1:

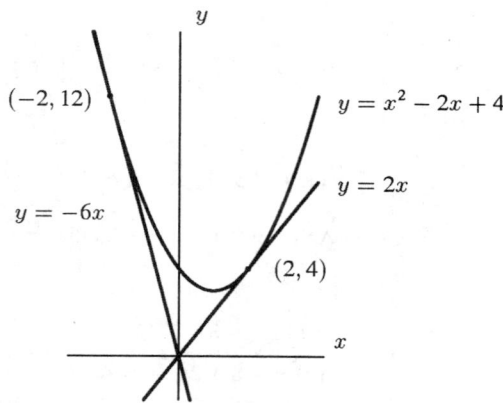

Figure 3.1

46. (a) Velocity $v(t) = \frac{dy}{dt} = \frac{d}{dt}(1250 - 16t^2) = -32t$.
    Since $t \geq 0$, the ball's velocity is negative. This is reasonable, since its height y is decreasing.
    (b) Acceleration $a(t) = \frac{dv}{dt} = \frac{d}{dt}(-32t) = -32$.
    So its acceleration is the negative constant -32.
    (c) The ball hits the ground when its height $y = 0$. This gives

    $$1250 - 16t^2 = 0$$
    $$t = \pm 8.84 \text{ seconds}$$

    We discard $t = -8.84$ because time t is nonnegative. So the ball hits the ground 8.84 seconds after its release, at which time its velocity is

    $$v(8.84) = -32(8.84) = -282.88 \text{ feet/sec} = -192.84 \text{ mph}.$$

47. $\frac{dF}{dr} = -\frac{2GMm}{r^3}$.

48. (a) $T = 2\pi\sqrt{\frac{l}{g}} = \frac{2\pi}{\sqrt{g}}\left(l^{\frac{1}{2}}\right)$, so $\frac{dT}{dl} = \frac{2\pi}{\sqrt{g}}\left(\frac{1}{2}l^{-\frac{1}{2}}\right) = \frac{\pi}{\sqrt{gl}}$.
    (b) Since $\frac{dT}{dl}$ is positive, the period T increases as the length l increases.

49. $f(x) = \frac{1}{x} = x^{-1}$ $f'(x) = -x^{-2} = -\frac{1}{x^2}$.
    The tangent line at $x = 1$ will have slope $f'(1) = -1$. See Figure 3.2. Using the point $(1, 1)$ which lies on the line, we obtain the equation $y = -x + 2$. We approximate $f(2)$ by using the y-value corresponding to $x = 2$, so $f(2) \approx 0$.
    Similarly, the tangent line at $x = 100$ will have slope $f'(100) = \frac{-1}{(100)^2} = -0.0001$. The equation of the line is then $y = -0.0001x + 0.02$. The approximate value of $f(2)$ predicted by this tangent line is $f(2) \approx 0.0198$.

The actual value of $f(2)$ is $\frac{1}{2}$, so the approximation from $x = 100$ is better than that from $x = 1$. This is because the slope changes less between $x = 2$ and $x = 100$ than it does between $x = 1$ and $x = 2$.

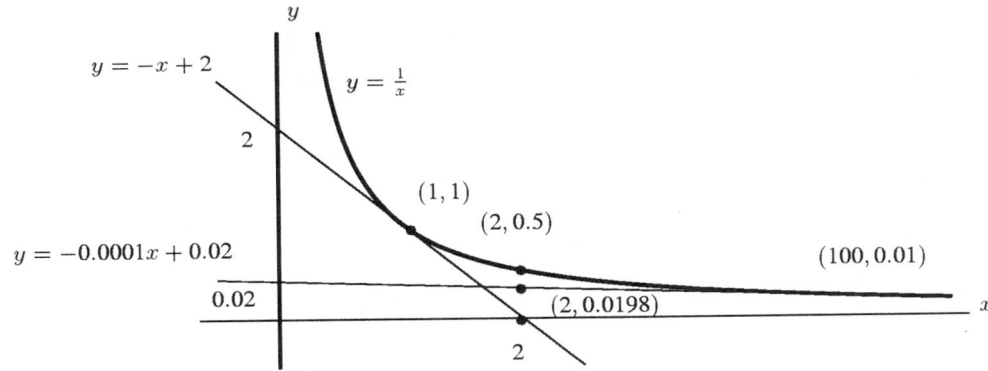

Figure 3.2

50. (a) $A = \pi r^2$
    $\frac{dA}{dr} = 2\pi r$.
    (b) This is the formula for the circumference of a circle.
    (c) $A'(r) \approx \frac{A(r+h)-A(r)}{h}$ for small h. When $h > 0$, the numerator of the difference quotient denotes the area of the region contained between the inner circle (radius r) and the outer circle (radius $r + h$). See figure below. As h approaches 0, this area can be approximated by the product of the circumference of the inner circle and the "width" of the region, i.e., h. Dividing this by the denominator, h, we get $A' =$ the circumference of the circle with radius r.

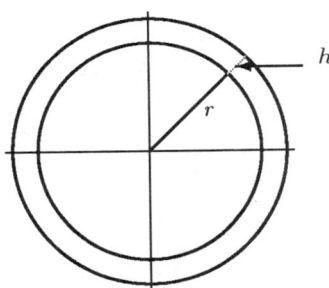

We can also think about the derivative of A as the rate of change of area for a small change in radius. If the radius increases by a tiny amount, the area will increase by a thin ring whose area is simply the circumference at that radius times the small amount. To get the rate of change, we divide by the small amount and obtain the circumference.

51. $V = \frac{4}{3}\pi r^3$
    $\frac{dV}{dr} = 4\pi r^2 =$ surface area of a sphere.

    Our reasoning is similar to that of Problem 50. The difference quotient $\frac{V(r+h)-V(r)}{h}$ is the volume between two spheres divided by the change in radius. Furthermore, when h is very small (and consequently $V(r + h) \approx V(r)$) this volume is like a coating of paint of depth h applied to the surface of the sphere. The volume of the paint is about $h \cdot$ (Surface Area) for small h: dividing by h gives back the surface area. Also, thinking about the derivative as the rate of change of the function for a small change in the variable, the answer seems clear. If you increase the radius of a sphere the tiniest amount, the volume will increase by a very thin layer whose volume will be the surface area at that radius multiplied by that tiniest amount.

52. (a)
$$\frac{d(x^{-1})}{dx} = \lim_{h \to 0} \frac{(x+h)^{-1} - x^{-1}}{h} = \lim_{h \to 0} \frac{1}{h}\left[\frac{1}{x+h} - \frac{1}{x}\right]$$
$$= \lim_{h \to 0} \frac{1}{h}\left[\frac{x - (x+h)}{x(x+h)}\right] = \lim_{h \to 0} \frac{1}{h}\left[\frac{-h}{x(x+h)}\right]$$
$$= \lim_{h \to 0} \frac{-1}{x(x+h)} = \frac{-1}{x^2} = -1x^{-2}.$$

$$\frac{d(x^{-3})}{dx} = \lim_{h \to 0} \frac{(x+h)^{-3} - x^{-3}}{h}$$
$$= \lim_{h \to 0} \frac{1}{h}\left[\frac{1}{(x+h)^3} - \frac{1}{x^3}\right]$$
$$= \lim_{h \to 0} \frac{1}{h}\left[\frac{x^3 - (x+h)^3}{x^3(x+h)^3}\right]$$
$$= \lim_{h \to 0} \frac{1}{h}\left[\frac{x^3 - (x^3 + 3hx^2 + 3h^2x + h^3)}{x^3(x+h)^3}\right]$$
$$= \lim_{h \to 0} \frac{1}{h}\left[\frac{-3hx^2 - 3xh^2 - h^3}{x^3(x+h)^3}\right]$$
$$= \lim_{h \to 0} \frac{-3x^2 - 3xh - h^2}{x^3(x+h)^3}$$
$$= \frac{-3x^2}{x^6} = -3x^{-4}.$$

(b) For clarity, let $n = -k$, where k is a positive integer. So $x^n = x^{-k}$.
$$\frac{d(x^{-k})}{dx} = \lim_{h \to 0} \frac{(x+h)^{-k} - x^{-k}}{h}$$
$$= \lim_{h \to 0} \frac{1}{h}\left[\frac{1}{(x+h)^k} - \frac{1}{x^k}\right]$$
$$= \lim_{h \to 0} \frac{1}{h}\left[\frac{x^k - (x+h)^k}{x^k(x+h)^k}\right]$$
$$= \lim_{h \to 0} \frac{1}{h}\left[\frac{x^k - x^k - khx^{k-1} - \overbrace{\ldots - h^k}^{\text{terms involving } h^2 \text{ and higher powers of } h}}{x^k(x+h)^k}\right]$$
$$= \frac{-kx^{k-1}}{x^k(x)^k} = \frac{-k}{x^{k+1}} = -kx^{-(k+1)} = -kx^{-k-1}.$$

Solutions for Section 3.2

1. $f'(x) = 2e^x + 2x$.
2. $y' = 10t + 4e^t$.
3. $y' = (\ln 5)5^x$.
4. $f'(x) = (\ln 2)2^x + 2(\ln 3)3^x$.
5. $y' = 10x + (\ln 2)2^x$.
6. $f'(x) = 12e^x + (\ln 11)11^x$.
7. $\frac{dy}{dx} = 4(\ln 10)10^x - 3x^2$.
8. $\frac{dy}{dx} = 3 - 2(\ln 4)4^x$.

9. $\frac{dy}{dx} = \frac{1}{3}(\ln 3)3^x - \frac{33}{2}(x^{-\frac{3}{2}}).$

10. $f'(x) = ex^{e-1}.$

11. $f(x) = e^{1+x} = e^1 \cdot e^x.$ Then, since e^1 is just a constant,
    $f'(x) = e \cdot e^x = e^{1+x}.$

12. $f(t) = e^t \cdot e^2.$ Then, since e^2 is just a constant, $f'(t) = \frac{d}{dt}(e^t e^2) = e^2 \frac{d}{dt} e^t = e^2 e^t = e^{t+2}.$

13. $y = e^\theta e^{-1}$ $y' = \frac{d}{d\theta}(e^\theta e^{-1}) = e^{-1}\frac{d}{d\theta}e^\theta = e^\theta e^{-1} = e^{\theta - 1}.$

14. $z' = (\ln 4)e^x.$

15. $z' = (\ln 4)^2 4^x.$

16. $f'(z) = (2\ln 3)z + (\ln 4)e^z.$

17. $f'(t) = (\ln(\ln 3))(\ln 3)^t.$

18. $f'(x) = 3x^2 + 3^x \ln 3$

19. $\frac{dy}{dx} = 5 \cdot 5^t \ln 5 + 6 \cdot 6^t \ln 6$

20. $\frac{dy}{dx} = \pi^x \ln \pi$

21. $f'(x) = \pi^2 x^{(\pi^2 - 1)} + (\pi^2)^x \ln(\pi^2)$

22. $y' = 2x + (\ln 2)2^x.$

23. $y' = \frac{1}{2}x^{-\frac{1}{2}} - \ln\frac{1}{2}(\frac{1}{2})^x = \frac{1}{2\sqrt{x}} + \ln 2 (\frac{1}{2})^x.$

24. We can take the derivative of the sum $x^2 + 2^x$, but not the product.

25. Once again, this is a product of two functions, 2^x and $\frac{1}{x}$, each of which we can take the derivative of; but we don't know how to take the derivative of the product.

26. Since $y = e^5 e^x$, $y' = e^5 e^x = e^{x+5}.$

27. $y = e^{5x} = (e^5)^x$, so $y' = \ln(e^5) \cdot (e^5)^x = 5e^{5x}.$

28. The exponent is x^2, and we haven't learned what to do about that yet.

29. $f'(z) = (\ln \sqrt{4})(\sqrt{4})^z = (\ln 2)2^z.$

30. We can't use our rules if the exponent is $\sqrt{\theta}$.

31.
$$\frac{dP}{dt} = 35{,}000 \cdot (\ln 0.98)(0.98^t).$$

At $t = 23$, this is $35{,}000(\ln 0.98)(0.98^{23}) \approx -444.3\frac{\text{people}}{\text{year}}$. (Note: the negative sign indicates that the population is decreasing.)

32. Since $P = 1 \cdot (1.05)^t$, $\frac{dP}{dt} = \ln(1.05)1.05^t.$ When $t = 10$,

$$\frac{dP}{dt} = (\ln 1.05)(1.05)^{10} \approx \$0.07947/\text{year} \approx 7.95\text{¢}/\text{year}.$$

33. $\frac{dV}{dt} = 75(1.35)^t \ln 1.35 \approx 22.5(1.35)^t.$

34. (a) $V(4) = 25(0.85)^4 = 25(0.522) = 13{,}050.$ Thus the value of the car after 4 years is $13,050.
    (b) We have a function of the form $f(t) = Ca^t.$ We know that such functions have a derivative of the form $(C \ln a) \cdot a^t.$ Thus, $V'(t) = 25(0.85)^t \cdot \ln 0.85 = -4.063(0.85)^t.$ The units would be the change in value (in thousands of dollars) with respect to time (in years), or thousands of dollars/year.
    (c) $V'(4) = -4.063(0.85)^4 = -4.063(0.522) = -2.121.$ This means that at the end of the fourth year, the value of the car is decreasing by $2121 per year.
    (d) $V(t)$ is a positive decreasing function, so that the value of the automobile is positive and decreasing. $V'(t)$ is a negative function whose magnitude is decreasing, meaning the value of the automobile is always dropping, but the yearly loss of value is less as time goes on. The graphs of $V(t)$ and $V'(t)$ confirm that the value of the car decreases with time. What they do not take into account are the *costs* associated with owning the vehicle. At some time, t, it is likely that the costs of owning the vehicle will outweigh its value. At that time, it may no longer be worthwhile to keep the car.

35. (a) $f(x) = 1 - e^x$ crosses the x-axis where $0 = 1 - e^x$, which happens when $e^x = 1$, so $x = 0$. Since $f'(x) = -e^x$, $f'(0) = -e^0 = -1$.
    (b) $y = -x$
    (c) The negative of the reciprocal of -1 is 1, so the equation of the normal line is $y = x$.

36. Since $y = 2^x$, $y' = (\ln 2)2^x$. At $(0, 1)$, the tangent line has slope $\ln 2$ so its equation is $y = (\ln 2)x + 1$. At c, $y = 0$, so $0 = (\ln 2)c + 1$, thus $c = -\frac{1}{\ln 2}$.

37.
$$g(x) = ax^2 + bx + c \qquad f(x) = e^x$$
$$g'(x) = 2ax + b \qquad f'(x) = e^x$$
$$g''(x) = 2a \qquad f''(x) = e^x$$

So, using $g''(0) = f''(0)$, etc., we have $2a = 1$, $b = 1$, and $c = 1$, and thus $g(x) = \frac{1}{2}x^2 + x + 1$, as shown in the figure below.

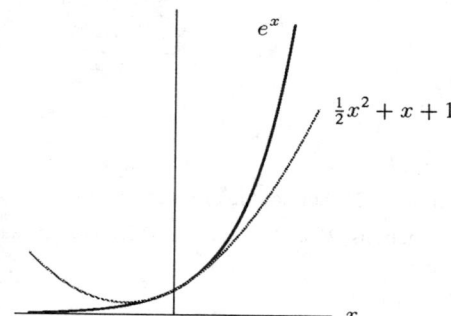

The two functions do look very much alike near $x = 0$. They both increase for large values of x, but e^x increases much more quickly. For very negative values of x, the quadratic goes to ∞ whereas the exponential goes to 0. By choosing a function whose first few derivatives agreed with the exponential when $x = 0$, we got a function which looks like the exponential for x-values near 0.

38. The derivative of e^x is $\frac{d}{dx}(e^x) = e^x$. Thus the tangent line at $x = 0$, has slope $e^0 = 1$, and the tangent line is $y = x + 1$. A function which is always concave up will always stay above any of its tangent lines. Thus $e^x \geq x + 1$ for all x, as shown in the figure below.

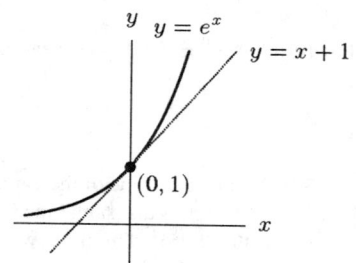

39. The equation $2^x = 2x$ has solutions $x = 1$ and $x = 2$. (Check this by substituting these values into the equation). The graph below suggests that these are the only solutions, but how can we be sure?

Let's look at the slope of the curve $f(x) = 2^x$, which is $f'(x) = (\ln 2)2^x \approx (0.693)2^x$, and the slope of the line $g(x) = 2x$ which is 2. At $x = 1$, the slope of $f(x)$ is less than 2; at $x = 2$, the slope of $f(x)$ is more than 2. Since the slope of $f(x)$ is always increasing, there can be no other point of intersection. (If there were another point of intersection, the graph f would have to "turn around".)

Here's another way of seeing this. Suppose $g(x)$ represents the position of a car going a steady 2 mph, while $f(x)$ represents a car which starts ahead of g (because the graph of f is above g) and is initially going slower than g. The car f is first overtaken by g. All the while, however, f is speeding up until eventually it overtakes g again. Notice that the two cars will only meet twice (corresponding to the two intersections of the curve): once when g overtakes f and once when f overtakes g.

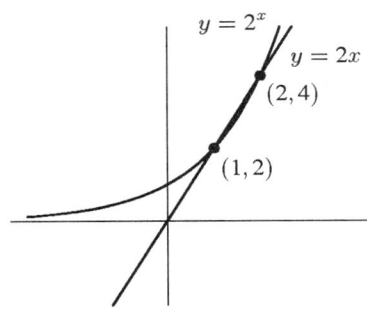

Solutions for Section 3.3

1. By the product rule, $f'(x) = 2x(x^3 + 5) + x^2(3x^2) = 2x^4 + 3x^4 + 10x = 5x^4 + 10x$. Alternatively, $f'(x) = (x^5 + 5x^2)' = 5x^4 + 10x$. The two answers should, and do, match.

2.
$$f'(x) = (\ln 2)2^x 3^x + (\ln 3)2^x 3^x = (\ln 2 + \ln 3)(2^x \cdot 3^x) = \ln(2 \cdot 3)(2 \cdot 3)^x = (\ln 6)6^x$$
or, since $2^x \cdot 3^x = (2 \cdot 3)^x = 6^x$,
$$f'(x) = (6^x)' = (\ln 6)(6^x).$$

The two answers should, and do, match.

3. $f'(x) = x \cdot e^x + e^x \cdot 1 = e^x(x + 1)$.

4. $y' = 2^x + x(\ln 2)2^x = 2^x(1 + x \ln 2)$.

5. $y' = \frac{1}{2\sqrt{x}}2^x + \sqrt{x}(\ln 2)2^x$.

6.
$$f'(x) = (x^2 - x^{\frac{1}{2}}) \cdot 3^x (\ln 3) + 3^x \left(2x - \frac{1}{2}x^{-\frac{1}{2}}\right)$$
$$= 3^x \left[(\ln 3)(x^2 - x^{\frac{1}{2}}) + \left(2x - \frac{1}{2\sqrt{x}}\right)\right].$$

7. It is easier to do this by multiplying it out first, rather than using the product rule first:
$$z = s^4 - s, \quad z' = 4s^3 - 1.$$

8. $\frac{dy}{dt} = 2te^t + (t^2 + 3)e^t = e^t(t^2 + 2t + 3)$.

9. $w' = (3t^2 + 5)(t^2 - 7t + 2) + (t^3 + 5t)(2t - 7)$.

10. $y' = (3t^2 - 14t)e^t + (t^3 - 7t^2 + 1)e^t = (t^3 - 4t^2 - 14t + 1)e^t$.

11. $f'(x) = \dfrac{e^x \cdot 1 - x \cdot e^x}{(e^x)^2} = \dfrac{e^x(1-x)}{(e^x)^2} = \dfrac{1-x}{e^x}.$

12. $g'(x) = \dfrac{50xe^x - 25x^2 e^x}{e^{2x}} = \dfrac{50x - 25x^2}{e^x}.$

13. $g'(w) = \dfrac{3.2w^{2.2}(5^w) - (\ln 5)(w^{3.2})5^w}{5^{2w}} = \dfrac{3.2w^{2.2} - w^{3.2}(\ln 5)}{5^w}.$

14. $h'(t) = \dfrac{(1)(t-4) - (1)(t+4)}{(t-4)^2} = \dfrac{t-4-t-4}{(t-4)^2} = \dfrac{-8}{(t-4)^2}.$

15. $\dfrac{dz}{dt} = \dfrac{3(5t+2) - (3t+1)5}{(5t+2)^2} = \dfrac{15t+6-15t-5}{(5t+2)^2} = \dfrac{1}{(5t+2)^2}.$

16. $z' = \dfrac{(2t+5)(t+3) - (t^2+5t+2)}{(t+3)^2} = \dfrac{t^2+6t+13}{(t+3)^2}.$

17. Divide and then differentiate
$$f(x) = x + \dfrac{3}{x}$$
$$f'(x) = 1 - \dfrac{3}{x^2}.$$

18. $w = y^2 - 6y + 7.\quad w' = 2y - 6, y \neq 0.$

19. $y' = \dfrac{\frac{1}{2\sqrt{t}}(t^2+1) - \sqrt{t}(2t)}{(t^2+1)^2}.$

20. Notice that you can cancel a z out of the numerator and denominator to get
$$f(z) = \dfrac{3z}{5z+7}, \quad z \neq 0$$

Then
$$f'(z) = \dfrac{(5z+7)3 - 3z(5)}{(5z+7)^2}$$
$$= \dfrac{15z+21-15z}{(5z+7)^2}$$
$$= \dfrac{21}{(5z+7)^2}, z \neq 0.$$

[If you used the quotient rule correctly without canceling the z out first, your answer should simplify to this one, but it is usually a good idea to simplify as much as possible before differentiating.]

21. $w'(x) = \dfrac{17e^x(2^x) - (\ln 2)(17e^x)2^x}{2^{2x}} = \dfrac{17e^x(2^x)(1 - \ln 2)}{2^{2x}} = \dfrac{17e^x(1 - \ln 2)}{2^x}.$

22. $h'(p) = \dfrac{2p(3+2p^2) - 4p(1+p^2)}{(3+2p^2)^2} = \dfrac{6p+4p^3-4p-4p^3}{(3+2p^2)^2} = \dfrac{2p}{(3+2p^2)^2}.$

23.
$$f'(x) = \dfrac{(2+3x+4x^2)(1) - (1+x)(3+8x)}{(2+3x+4x^2)^2}$$
$$= \dfrac{2+3x+4x^2 - 3 - 11x - 8x^2}{(2+3x+4x^2)^2}$$
$$= \dfrac{-4x^2 - 8x - 1}{(2+3x+4x^2)^2}.$$

24.
$$f'(x) = 3(2x-5) + 2(3x+8) = 12x + 1$$
$$f''(x) = 12.$$

25. Using the product rule, we have
$$f'(x) = e^{-x} - xe^{-x}$$
$$f''(x) = -e^{-x} - e^{-x} + xe^{-x} = e^{-x}(x-2).$$
Since $e^{-x} > 0$, for all x, we have $f''(x) < 0$ if $x - 2 < 0$, that is, $x < 2$.

26. Using the quotient rule, we have
$$g'(x) = \frac{0 - 1(2x)}{(x^2+1)^2} = \frac{-2x}{(x^2+1)^2}$$
$$g''(x) = \frac{-2(x^2+1)^2 + 2x(4x^3 + 4x)}{(x^2+1)^4}$$
$$= \frac{-2(x^2+1)^2 + 8x^2(x^2+1)}{(x^2+1)^4}$$
$$= \frac{-2(x^2+1) + 8x^2}{(x^2+1)^3}$$
$$= \frac{2(3x^2 - 1)}{(x^2+1)^3}.$$

Since $(x^2+1)^3 > 0$ for all x, we have $g''(x) < 0$ if $(3x^2 - 1) < 0$, or when
$$3x^2 < 1$$
$$-\frac{1}{\sqrt{3}} < x < \frac{1}{\sqrt{3}}.$$

27.
$$f(t) = \frac{1}{e^t}$$
$$f'(t) = \frac{e^t \cdot 0 - e^t \cdot 1}{(e^t)^2}$$
$$= \frac{-1}{e^t} = -e^{-t}.$$

28. $f(x) = e^x \cdot e^x$
$f'(x) = e^x \cdot e^x + e^x \cdot e^x = 2e^{2x}.$

29.
$$f(x) = e^x e^{2x}$$
$$f'(x) = e^x(e^{2x})' + (e^x)'e^{2x}$$
$$= 2e^x e^{2x} + e^x e^{2x} \text{ (from Problem 28)}$$
$$= 3e^{3x}.$$

30. Since $\frac{d}{dx}e^{2x} = 2e^{2x}$ and $\frac{d}{dx}e^{3x} = 3e^{3x}$, we might guess that $\frac{d}{dx}e^{4x} = 4e^{4x}$.

31. (a) Although the answer you would get by using the quotient rule is equivalent, the answer looks simpler in this case if you just use the product rule:
$$\frac{d}{dx}\left(\frac{e^x}{x}\right) = \frac{d}{dx}\left(e^x \cdot \frac{1}{x}\right) = \frac{e^x}{x} - \frac{e^x}{x^2}$$
$$\frac{d}{dx}\left(\frac{e^x}{x^2}\right) = \frac{d}{dx}\left(e^x \cdot \frac{1}{x^2}\right) = \frac{e^x}{x^2} - \frac{2e^x}{x^3}$$
$$\frac{d}{dx}\left(\frac{e^x}{x^3}\right) = \frac{d}{dx}\left(e^x \cdot \frac{1}{x^3}\right) = \frac{e^x}{x^3} - \frac{3e^x}{x^4}.$$

(b) $\frac{d}{dx}\frac{e^x}{x^n} = \frac{e^x}{x^n} - \frac{ne^x}{x^{n+1}}.$

32.

$$\frac{d(x^2)}{dx} = \frac{d}{dx}(x \cdot x)$$
$$= x\frac{d(x)}{dx} + x\frac{d(x)}{dx}$$
$$= 2x.$$

$$\frac{d(x^3)}{dx} = \frac{d}{dx}(x^2 \cdot x)$$
$$= x^2\frac{d(x)}{dx} + x\frac{d(x^2)}{dx}$$
$$= x^2\frac{d(x)}{dx} + x\left[x\frac{d(x)}{dx} + x\frac{d(x)}{dx}\right]$$
$$= x^2\frac{d(x)}{dx} + x^2\frac{d(x)}{dx} + x^2\frac{d(x)}{dx}$$
$$= 3x^2.$$

33. Since
$$x^{1/2} \cdot x^{1/2} = x,$$
we differentiate to obtain
$$\frac{d}{dx}(x^{1/2}) \cdot x^{1/2} + x^{1/2} \cdot \frac{d}{dx}(x^{1/2}) = 1.$$

Now solve for $d(x^{1/2})/dx$:

$$2x^{1/2}\frac{d}{dx}(x^{1/2}) = 1$$
$$\frac{d}{dx}(x^{1/2}) = \frac{1}{2x^{1/2}}.$$

34. (a) We have $h'(2) = f'(2) + g'(2) = 5 - 2 = 3$.
   (b) We have $h'(2) = f'(2)g(2) + f(2)g'(2) = 5(4) + 3(-2) = 14$.
   (c) We have $h'(2) = \frac{f'(2)g(2) - f(2)g'(2)}{(g(2))^2} = \frac{5(4) - 3(-2)}{4^2} = \frac{26}{16} = \frac{13}{8}$.

35. (a) $G'(z) = F'(z)H(z) + H'(z)F(z)$, so
   $G'(3) = F'(3)H(3) + H'(3)F(3) = 4 \cdot 1 + 3 \cdot 5 = 19$.
   (b) $G'(w) = \frac{F'(w)H(w) - H'(w)F(w)}{[H(w)]^2}$, so $G'(3) = \frac{4(1) - 3(5)}{1^2} = -11$.

36. $f'(x) = 10x^9 e^x + x^{10} e^x$ is of the form $g'h + h'g$, where

$$g(x) = x^{10}, \ g'(x) = 10x^9$$

and
$$h(x) = e^x, \ h'(x) = e^x.$$

Therefore, using the product rule, let $f = g \cdot h$, with $g(x) = x^{10}$ and $h(x) = e^x$. Thus

$$f(x) = x^{10}e^x.$$

37. (a) $f(140) = 15{,}000$ says that 15,000 skateboards are sold when the cost is $140 per board.
   $f'(140) = -100$ means that if the price is increased from $140, roughly speaking, every dollar of increase will decrease the total sales by 100 boards.
   (b) $\frac{dR}{dp} = \frac{d}{dp}(p \cdot q) = \frac{d}{dp}(p \cdot f(p)) = f(p) + pf'(p)$.
   So,

$$\left.\frac{dR}{dp}\right|_{p=140} = f(140) + 140f'(140)$$
$$= 15{,}000 + 140(-100) = 1000.$$

   (c) From (b) we see that $\left.\frac{dR}{dp}\right|_{p=140} = 1000 > 0$. This means that the revenue will increase by about $1000 if the price is raised by $1.

38. We want dR/dr_1. Solving for R:
$$\frac{1}{R} = \frac{1}{r_1} + \frac{1}{r_2} = \frac{r_2 + r_1}{r_1 r_2}, \text{ which gives } R = \frac{r_1 r_2}{r_2 + r_1}.$$

So, thinking of r_2 as a constant and using the quotient rule,
$$\frac{dR}{dr_1} = \frac{r_2(r_2 + r_1) - r_1 r_2(1)}{(r_2 + r_1)^2} = \frac{r_2^2}{(r_1 + r_2)^2}.$$

39. (a) If the museum sells the painting and invests the proceeds $P(t)$ at time t, then t years have elapsed since 2000, and the time span up to 2020 is $20 - t$. This is how long the proceeds $P(t)$ are earning interest in the bank. Each year the money is in the bank it earns 5% interest, which means the amount in the bank is multiplied by a factor of 1.05. So, at the end of $(20 - t)$ years, the balance is given by
$$B(t) = P(t)(1 + 0.05)^{20-t} = P(t)(1.05)^{20-t}.$$

(b)
$$B(t) = P(t)(1.05)^{20}(1.05)^{-t} = (1.05)^{20}\frac{P(t)}{(1.05)^t}.$$

(c) By the quotient rule,
$$B'(t) = (1.05)^{20}\left[\frac{P'(t)(1.05)^t - P(t)(1.05)^t \ln 1.05}{(1.05)^{2t}}\right].$$

So,
$$B'(10) = (1.05)^{20}\left[\frac{5000(1.05)^{10} - 150{,}000(1.05)^{10}\ln 1.05}{(1.05)^{20}}\right]$$
$$= (1.05)^{10}(5000 - 150{,}000 \ln 1.05)$$
$$\approx -3776.63.$$

40. Note first that $f(v)$ is in $\frac{\text{liters}}{\text{km}}$, and v is in $\frac{\text{km}}{\text{hour}}$.

(a) $g(v) = \frac{1}{f(v)}$. (This is in $\frac{\text{km}}{\text{liter}}$.) Differentiating gives
$$g'(v) = \frac{-f'(v)}{(f(v))^2}.$$

So,
$$g(80) = \frac{1}{0.05} = 20 \tfrac{\text{km}}{\text{liter}}.$$
$$g'(80) = \frac{-0.0005}{(0.05)^2} = -\frac{1}{5} \tfrac{\text{km}}{\text{liter}} \text{ for each } 1 \tfrac{\text{km}}{\text{hr}} \text{ increase in speed.}$$

(b) $h(v) = v \cdot f(v)$. (This is in $\frac{\text{km}}{\text{hour}} \cdot \frac{\text{liters}}{\text{km}} = \frac{\text{liters}}{\text{hour}}$.) Differentiating gives
$$h'(v) = f(v) + v \cdot f'(v),$$

so
$$h(80) = 80(0.05) = 4 \tfrac{\text{liters}}{\text{hr}}.$$
$$h'(80) = 0.05 + 80(0.0005) = 0.09 \tfrac{\text{liters}}{\text{hr}} \text{ for each } 1 \tfrac{\text{km}}{\text{hr}} \text{ increase in speed.}$$

(c) Part (a) tells us that at 80 km/hr, the car can go 20 km on 1 liter. Since the first derivative evaluated at this velocity is negative, this implies that as velocity increases, fuel efficiency decreases, i.e., at higher velocities the car will not go as far on 1 liter of gas. Part (b) tells us that at 80 km/hr, the car uses 4 liters in an hour. Since the first derivative evaluated at this velocity is positive, this means that at higher velocities, the car will use more gas per hour.

41. Assume for $g(x) \neq f(x)$, $g'(x) = g(x)$ and $g(0) = 1$. Then for

$$h(x) = \frac{g(x)}{e^x}$$

$$h'(x) = \frac{g'(x)e^x - g(x)e^x}{(e^x)^2} = \frac{e^x(g'(x) - g(x))}{(e^x)^2} = \frac{g'(x) - g(x)}{e^x}.$$

But, since $g(x) = g'(x)$, $h'(x) = 0$, so $h(x)$ is constant. Thus, the ratio of $g(x)$ to e^x is constant. Since $\frac{g(0)}{e^0} = \frac{1}{1} = 1$, $\frac{g(x)}{e^x}$ must equal 1 for all x. Thus $g(x) = e^x = f(x)$ for all x, so f and g are the same function.

42. (a) $f'(x) = (x-2) + (x-1)$.
   (b) Think of f as the product of two factors, with the first as $(x-1)(x-2)$. (The reason for this is that we have already differentiated $(x-1)(x-2)$).

$$f(x) = [(x-1)(x-2)](x-3).$$

Now $f'(x) = [(x-1)(x-2)]'(x-3) + [(x-1)(x-2)](x-3)'$
Using the result of a):

$$f'(x) = [(x-2) + (x-1)](x-3) + [(x-1)(x-2)] \cdot 1$$
$$= (x-2)(x-3) + (x-1)(x-3) + (x-1)(x-2).$$

   (c) Because we have already differentiated $(x-1)(x-2)(x-3)$, rewrite f as the product of two factors, the first being $(x-1)(x-2)(x-3)$:

$$f(x) = [(x-1)(x-2)(x-3)](x-4)$$

Now $f'(x) = [(x-1)(x-2)(x-3)]'(x-4) + [(x-1)(x-2)(x-3)](x-4)'$.

$$f'(x) = [(x-2)(x-3) + (x-1)(x-3) + (x-1)(x-2)](x-4)$$
$$+ [(x-1)(x-2)(x-3)] \cdot 1$$
$$= (x-2)(x-3)(x-4) + (x-1)(x-3)(x-4)$$
$$+ (x-1)(x-2)(x-4) + (x-1)(x-2)(x-3).$$

From the solutions above, one can observe that when f is a product, its derivative is obtained by differentiating each factor in turn (leaving the other factors alone), and adding the results.

43. From the answer to Problem 42, we find that

$$f'(x) = (x-r_1)(x-r_2)\cdots(x-r_{n-1}) \cdot 1$$
$$+ (x-r_1)(x-r_2)\cdots(x-r_{n-2}) \cdot 1 \cdot (x-r_n)$$
$$+ (x-r_1)(x-r_2)\cdots(x-r_{n-3}) \cdot 1 \cdot (x-r_{n-1})(x-r_n)$$
$$+ \cdots + 1 \cdot (x-r_2)(x-r_3)\cdots(x-r_n)$$
$$= f(x)\left(\frac{1}{x-r_1} + \frac{1}{x-r_2} + \cdots + \frac{1}{x-r_n}\right).$$

44. (a) We can approximate $\frac{d}{dx}[F(x)G(x)H(x)]$ using the large rectangular solids by which our original cube is increased:

Volume of whole − volume of original solid = change in volume.

$F(x+h)G(x+h)H(x+h) - F(x)G(x)H(x) = $ change in volume.

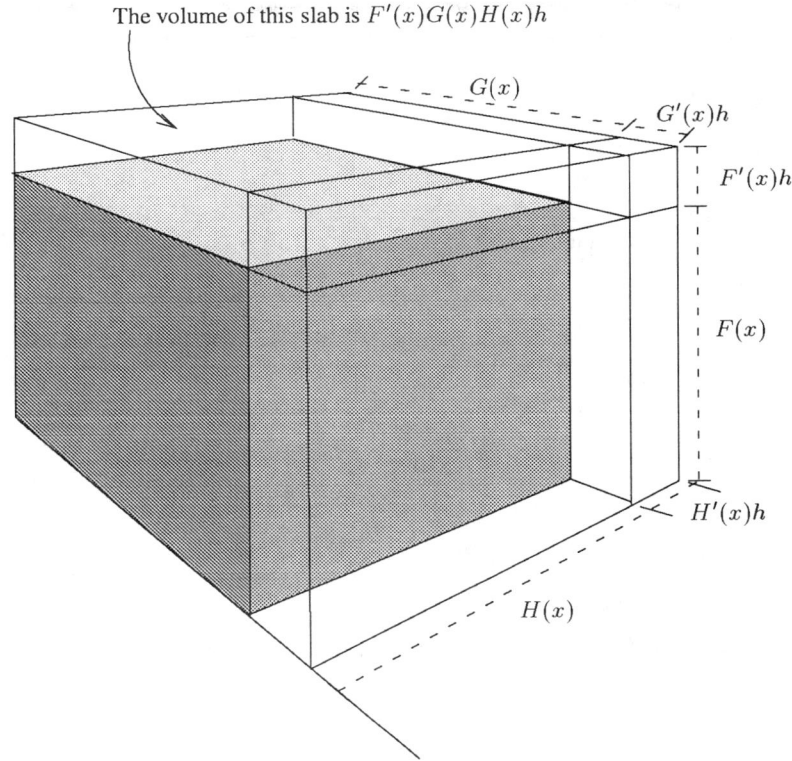

The volume of this slab is $F'(x)G(x)H(x)h$

As in the book, we will ignore the <u>smaller</u> regions which are added (the long, thin rectangular boxes and the small cube in the corner.) This can be justified by recognizing that as $h \to 0$, these volumes will shrink much faster than the volumes of the big slabs and will therefore be insignificant. (Note that these smaller regions have an h^2 or h^3 in the formulas of their volumes.) Then we can approximate the change in volume above by:

$$F(x+h)G(x+h)H(x+h) - F(x)G(x)H(x) \approx F'(x)G(x)H(x)h \quad \text{(top slab)}$$
$$+ F(x)G'(x)H(x)h \quad \text{(front slab)}$$
$$+ F(x)G(x)H'(x)h \quad \text{(other slab)}.$$

Dividing by h gives

$$\frac{F(x+h)G(x+h)H(x+h) - F(x)G(x)H(x)}{h}$$
$$\approx F'(x)G(x)H(x) + F(x)G'(x)H(x) + F(x)G(x)H'(x).$$

Letting $h \to 0$
$$(FGH)' = F'GH + FG'H + FGH'.$$

(b) Verifying,

$$\frac{d}{dx}[(F(x) \cdot G(x)) \cdot H(x)] = (F \cdot G)'(H) + (F \cdot G)(H)'$$
$$= [F'G + FG']H + FGH'$$
$$= F'GH + FG'H + FGH'$$

as before.

(c) From the answer to (b), we observe that the derivative of a product is obtained by differentiating each factor in turn (leaving the other factors alone), and adding the results. So, in general,

$$(f_1 \cdot f_2 \cdot f_3 \cdot \ldots \cdot f_n)' = f_1'f_2f_3 \cdots f_n + f_1f_2'f_3 \cdots f_n + \cdots + f_1 \cdots f_{n-1}f_n'.$$

Solutions for Section 3.4

1. $f'(x) = 99(x+1)^{98} \cdot 1 = 99(x+1)^{98}$.
2. $f'(x) = \dfrac{1}{2}(1-x^2)^{-\frac{1}{2}}(-2x) = \dfrac{-x}{\sqrt{1-x^2}}$.
3. $w' = 100(t^2+1)^{99}(2t) = 200t(t^2+1)^{99}$.
4. $w' = 100(t^3+1)^{99}(3t^2) = 300t^2(t^3+1)^{99}$.
5. $w' = 100(\sqrt{t}+1)^{99}\left(\dfrac{1}{2\sqrt{t}}\right) = \dfrac{50}{\sqrt{t}}(\sqrt{t}+1)^{99}$.
6. $f'(t) = (e^{3t})(3) = 3e^{3t}$.
7. $f'(x) = (\ln 2)2^{(x+2)}$.
8. $g'(x) = 2(\ln 3)3^{(2x+7)}$.
9. $k'(x) = 4(x^3+e^x)^3(3x^2+e^x)$.
10. $z'(x) = \dfrac{(\ln 2)2^x}{3\sqrt[3]{(2^x+5)^2}}$.
11. $y' = \dfrac{\frac{2^z}{2\sqrt{z}} - (\sqrt{z})(\ln 2)(2^z)}{2^{2z}} = \dfrac{1 - 2z\ln 2}{2^{z+1}\sqrt{z}}$.
12. $w' = \dfrac{3}{2}\sqrt{x^2 \cdot 5^x}[2x(5^x) + (\ln 5)(x^2)(5^x)] = \dfrac{3}{2}x^2\sqrt{5^{3x}}(2 + x\ln 5)$.
13. $y' = \dfrac{3}{2}e^{\frac{3}{2}w}$.
14. $y' = -4e^{-4t}$.
15. $y' = \dfrac{3s^2}{2\sqrt{s^3+1}}$.
16. $w' = \dfrac{1}{2\sqrt{s}}e^{\sqrt{s}}$.
17. $y' = 1 \cdot e^{-t^2} + te^{-t^2}(-2t)$.
18. $f'(z) = \dfrac{1}{2\sqrt{z}}e^{-z} - \sqrt{z}e^{-z}$.
19. We can write this as $f(z) = \sqrt{z}e^{-z}$, in which case it is the same as problem 18. So $f'(z) = \dfrac{1}{2\sqrt{z}}e^{-z} - \sqrt{z}e^{-z}$.
20. $z' = 5 \cdot \ln 2 \cdot 2^{5t-3}$.
21. $f'(t) = 1 \cdot e^{5-2t} + te^{5-2t}(-2) = e^{5-2t}(1-2t)$.
22. $f'(z) = -2(e^z+1)^{-3} \cdot e^z = \dfrac{-2e^z}{(e^z+1)^3}$.
23. $f'(\theta) = -1(1+e^{-\theta})^{-2}(e^{-\theta})(-1) = \dfrac{e^{-\theta}}{(1+e^{-\theta})^2}$.
24. $f'(x) = 6(e^{5x})(5) + (e^{-x^2})(-2x) = 30e^{5x} - 2xe^{-x^2}$.
25.
$$f'(w) = (e^{w^2})(10w) + (5w^2+3)(e^{w^2})(2w)$$
$$= 2we^{w^2}(5 + 5w^2 + 3)$$
$$= 2we^{w^2}(5w^2 + 8).$$
26. $w' = (2t+3)(1-e^{-2t}) + (t^2+3t)(2e^{-2t})$.
27. $f(y) = \left[10^{(5-y)}\right]^{\frac{1}{2}} = 10^{\frac{5}{2}-\frac{1}{2}y}$
$f'(y) = (\ln 10)\left(10^{\frac{5}{2}-\frac{1}{2}y}\right)\left(-\dfrac{1}{2}\right) = -\dfrac{1}{2}(\ln 10)(10^{\frac{5}{2}-\frac{1}{2}y})$.

28. $f'(x) = e^{-(x-1)^2} \cdot (-2)(x-1)$.

29. $f'(y) = e^{e^{(y^2)}} \left[(e^{y^2})(2y)\right] = 2ye^{[e^{(y^2)}+y^2]}$.

30. $f'(t) = 2(e^{-2e^{2t}})(-2e^{2t})2 = -8(e^{-2e^{2t}+2t})$.

31.

$$f(x) = 6e^{5x} + e^{-x^2} \qquad\qquad f'(x) = 30e^{5x} - 2xe^{-x^2}$$
$$f(1) = 6e^5 + e^{-1} \qquad\qquad f'(1) = 30e^5 - 2(1)e^{-1}$$

$$y - y_1 = m(x - x_1)$$
$$y - (6e^5 + e^{-1}) = (30e^5 - 2e^{-1})(x-1)$$
$$y - (6e^5 + e^{-1}) = (30e^5 - 2e^{-1})x - (30e^5 - 2e^{-1})$$
$$y = (30e^5 - 2e^{-1})x - 30e^5 + 2e^{-1} + 6e^5 + e^{-1}$$
$$\approx 4451.66x - 3560.81.$$

32. The graph is concave down when $f''(x) < 0$.

$$f'(x) = e^{-x^2}(-2x)$$
$$f''(x) = \left[e^{-x^2}(-2x)\right](-2x) + e^{-x^2}(-2)$$
$$= \frac{4x^2}{e^{x^2}} - \frac{2}{e^{x^2}}$$
$$= \frac{4x^2 - 2}{e^{x^2}} < 0$$

The graph is concave down when $4x^2 < 2$. This occurs when $x^2 < \frac{1}{2}$, or $-\frac{1}{\sqrt{2}} < x < \frac{1}{\sqrt{2}}$.

33.
$$f'(x) = [10(2x+1)^9(2)][(3x-1)^7] + [(2x+1)^{10}][7(3x-1)^6(3)]$$
$$= (2x+1)^9(3x-1)^6[20(3x-1) + 21(2x+1)]$$
$$= [(2x+1)^9(3x-1)^6](102x+1)$$
$$f''(x) = [9(2x+1)^8(2)(3x-1)^6 + (2x+1)^9(6)(3x-1)^5(3)](102x+1)$$
$$\qquad + (2x+1)^9(3x-1)^6(102).$$

34. (a) $H(x) = F(G(x))$
$H(4) = F(G(4)) = F(2) = 1$
(b) $H(x) = F(G(x))$
$H'(x) = F'(G(x)) \cdot G'(x)$
$H'(4) = F'(G(4)) \cdot G'(4) = F'(2) \cdot 6 = 5 \cdot 6 = 30$
(c) $H(x) = G(F(x))$
$H(4) = G(F(4)) = G(3) = 4$
(d) $H(x) = G(F(x))$
$H'(x) = G'(F(x)) \cdot F'(x)$
$H'(4) = G'(F(4)) \cdot F'(4) = G'(3) \cdot 7 = 8 \cdot 7 = 56$
(e) $H(x) = \frac{F(x)}{G(x)}$
$H'(x) = \frac{G(x) \cdot F'(x) - F(x) \cdot G'(x)}{[G(x)]^2}$
$H'(4) = \frac{G(4) \cdot F'(4) - F(4) \cdot G'(4)}{[G(4)]^2} = \frac{2 \cdot 7 - 3 \cdot 6}{2^2} = \frac{14-18}{4} = \frac{-4}{4} = -1$

35. (a) Since $h'(x) = f'(g(x)) \cdot g'(x)$, we have
$$h'(2) = f'(g(2)) \cdot g'(2) = f'(5) \cdot g'(2) = \pi\sqrt{2}.$$

(b) Since $h'(x) = g'(f(x)) \cdot f'(x)$, we have
$$h'(2) = g'(f(2)) \cdot f'(2) = g'(5) \cdot f'(2) = 7e.$$

(c) We have $h'(x) = f'(f(x)) \cdot f'(x)$, so
$$h'(2) = f'(f(2)) \cdot f'(2) = f'(5) \cdot f'(2) = \pi e.$$

36. (a) If
$$p(x) = k(2x),$$
then
$$p'(x) = k'(2x) \cdot 2.$$
When $x = \frac{1}{2}$,
$$p'\left(\frac{1}{2}\right) = k'\left(2 \cdot \frac{1}{2}\right)(2) = 2 \cdot 2 = 4.$$

(b) If
$$q(x) = k(x+1),$$
then
$$q'(x) = k'(x+1) \cdot 1.$$
When $x = 0$,
$$q'(0) = k'(0+1)(1) = 2 \cdot 1 = 2.$$

(c) If
$$r(x) = k\left(\frac{1}{4}x\right),$$
then
$$r'(x) = k'\left(\frac{1}{4}x\right) \cdot \frac{1}{4}.$$
When $x = 4$,
$$r'(4) = k'\left(\frac{1}{4}4\right)\frac{1}{4} = 2 \cdot \frac{1}{4} = \frac{1}{2}.$$

37. (a) Differentiating $g(x) = \sqrt{f(x)} = (f(x))^{1/2}$, we have
$$g'(x) = \frac{1}{2}(f(x))^{-1/2} \cdot f'(x) = \frac{f'(x)}{2\sqrt{f(x)}}$$
$$g'(1) = \frac{f'(1)}{2\sqrt{f(1)}} = \frac{3}{2\sqrt{4}} = \frac{3}{4}.$$

(b) Differentiating $h(x) = f(\sqrt{x})$, we have
$$h'(x) = f'(\sqrt{x}) \cdot \frac{1}{2\sqrt{x}}$$
$$h'(1) = f'(\sqrt{1}) \cdot \frac{1}{2\sqrt{1}} = \frac{f'(1)}{2} = \frac{3}{2}.$$

38. Yes. To see why, simply plug $x = \sqrt[3]{2t+5}$ into the expression $3x^2\frac{dx}{dt}$ and evaluate it. To do this, first we calculate $\frac{dx}{dt}$. By the chain rule,
$$\frac{dx}{dt} = \frac{d}{dt}(2t+5)^{\frac{1}{3}} = \frac{2}{3}(2t+5)^{-\frac{2}{3}} = \frac{2}{3}[(2t+5)^{\frac{1}{3}}]^{-2}.$$
But since $x = (2t+5)^{\frac{1}{3}}$, we have (by substitution)
$$\frac{dx}{dt} = \frac{2}{3}x^{-2}.$$
It follows that $3x^2\frac{dx}{dt} = 3x^2\left(\frac{2}{3}x^{-2}\right) = 2.$

39. We see that $m'(x)$ is nearly of the form $f'(g(x)) \cdot g'(x)$ where

$$f(g) = e^g \quad \text{and} \quad g(x) = x^6,$$

but $g'(x)$ is off by a multiple of 6. Therefore, using the chain rule, let

$$m(x) = \frac{f(g(x))}{6} = \frac{e^{(x^6)}}{6}.$$

40. (a) $P(12) = 10e^{0.6(12)} = 10e^{7.2} \approx 13{,}394$ zebra mussels. There are 13,394 zebra mussels in the area after 12 months.
    (b) We differentiate to find $P'(t)$, and then substitute in to find $P'(12)$:

$$P'(t) = 10(e^{0.6t})(0.6) = 6e^{0.6t}$$
$$P'(12) = 6e^{0.6(12)} \approx 8{,}037 \text{ mussels/month}.$$

The population is growing at a rate of approximately 8037 zebra mussels per month.

41. (a)
$$\frac{dQ}{dt} = \frac{d}{dt}e^{-0.000121t}$$
$$= -0.000121e^{-0.000121t}$$

(b)
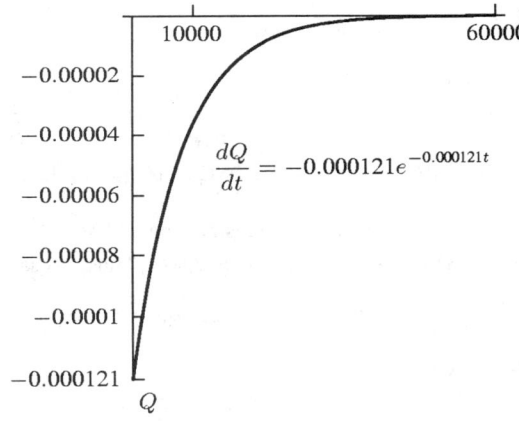

42. (a)
$$\frac{dH}{dt} = \frac{d}{dt}(40 + 30e^{-2t}) = 30(-2)e^{-2t} = -60e^{-2t}.$$

(b) Since e^{-2t} is always positive, $\frac{dH}{dt} < 0$; this makes sense because the temperature of the soda is decreasing.

(c) The magnitude of $\frac{dH}{dt}$ is

$$\left|\frac{dH}{dt}\right| = \left|-60e^{-2t}\right| = 60e^{-2t} \leq 60 = \left|\frac{dH}{dt}\right|_{t=0},$$

since $e^{-2t} \leq 1$ for all $t \geq 0$ and $e^0 = 1$. This is just saying that at the moment that the can of soda is put in the refrigerator (at $t = 0$), the temperature difference between the soda and the inside of the refrigerator is the greatest, so the temperature of the soda is dropping the quickest.

43. (a) $\frac{dB}{dt} = P\left(1 + \frac{r}{100}\right)^t \ln\left(1 + \frac{r}{100}\right)$. The expression $\frac{dB}{dt}$ tells us how fast the amount of money in the bank is changing with respect to time for fixed initial investment P and interest rate r.

(b) $\frac{dB}{dr} = Pt\left(1 + \frac{r}{100}\right)^{t-1} \frac{1}{100}$. The expression $\frac{dB}{dr}$ indicates how fast the amount of money changes with respect to the interest rate r, assuming fixed initial investment P and time t.

92 CHAPTER THREE /SOLUTIONS

44. The ripple's area and radius are related by $A(t) = \pi[r(t)]^2$. Taking derivatives and using the chain rule gives
$$\frac{dA}{dt} = \pi \cdot 2r \frac{dr}{dt}.$$
We know that $dr/dt = 10$ cm/sec, so when $r = 20$ cm we have
$$\frac{dA}{dt} = \pi \cdot 2 \cdot 20 \cdot 10 \text{cm}^2/\text{sec} = 400\pi \text{cm}^2/\text{sec}.$$

45. (a)
$$\frac{dm}{dv} = \frac{d}{dv}\left[m_0\left(1 - \frac{v^2}{c^2}\right)^{-1/2}\right]$$
$$= m_0\left(-\frac{1}{2}\right)\left(1 - \frac{v^2}{c^2}\right)^{-3/2}\left(-\frac{2v}{c^2}\right)$$
$$= \frac{m_0 v}{c^2}\frac{1}{\sqrt{\left(1 - \frac{v^2}{c^2}\right)^3}}.$$

(b) $\dfrac{dm}{dv}$ represents the rate of change of mass with respect to the speed v.

46. (a) For $t < 0$, $I = \dfrac{dQ}{dt} = 0$.
   For $t > 0$, $I = \dfrac{dQ}{dt} = -\dfrac{Q_0}{RC}e^{-t/RC}$.
   (b) For $t > 0$, $t \to 0$ (that is, as $t \to 0^+$),
$$I = -\frac{Q_0}{RC}e^{-t/RC} \to -\frac{Q_0}{RC}.$$
   Since $I = 0$ just to the left of $t = 0$ and $I = -Q_0/RC$ just to the right of $t = 0$, it is not possible to define I at $t = 0$.
   (c) Q is not differentiable at $t = 0$ because there is no tangent line at $t = 0$.

47. The time constant for Q is the time, T_Q, such that $Q = Q_0/e$. Thus, T_Q satisfies
$$\frac{Q_0}{e} = Q_0 e^{-T_Q/RC}.$$
Canceling Q_0 and taking natural logs gives
$$e^{-T_Q/RC} = \frac{1}{e} = e^{-1}$$
$$\frac{-T_Q}{RC} = -1$$
$$T_Q = RC.$$
To find $I = dQ/dt$, differentiate Q:
$$I = \frac{dQ}{dt} = \frac{-Q_0}{RC}e^{-t/RC}.$$
Since the exponent of e is unchanged, so is the time constant. We know that the initial current is
$$I_0 = \frac{-Q_0}{RC}.$$
If T_I is the time constant for I, we know
$$\frac{1}{e}\left(\frac{-Q_0}{RC}\right) = \frac{-Q_0}{RC}e^{-T_I/RC}.$$
Canceling $-Q_0/RC$ gives
$$\frac{1}{e} = e^{-T_I/RC}.$$
This is the same equation as the one we solved for T_Q, so
$$T_I = RC.$$

48. Let f have a zero of multiplicity m at $x = a$ so that
$$f(x) = (x-a)^m h(x), \quad h(a) \neq 0.$$
Differentiating this expression gives
$$f'(x) = (x-a)^m h'(x) + m(x-a)^{(m-1)} h(x)$$
and both terms in the sum are zero when $x = a$ so $f'(a) = 0$. Taking another derivative gives
$$f''(x) = (x-a)^m h''(x) + 2m(x-a)^{(m-1)} h'(x) + m(m-1)(x-a)^{(m-2)} h(x).$$
Again, each term in the sum contains a factor of $(x-a)$ to some positive power, so at $x = a$ this will evaluate to 0. Differentiating repeatedly, all derivatives will have positive integer powers of $(x-a)$ until the mth and will therefore vanish. However,
$$f^{(m)}(a) = m!h(a) \neq 0.$$

Solutions for Section 3.5

1.
TABLE 3.1

x	$\cos x$	Difference Quotient	$-\sin x$
0	1.0	-0.0005	0.0
0.1	0.995	-0.10033	-0.099833
0.2	0.98007	-0.19916	-0.19867
0.3	0.95534	-0.296	-0.29552
0.4	0.92106	-0.38988	-0.38942
0.5	0.87758	-0.47986	-0.47943
0.6	0.82534	-0.56506	-0.56464

2. $r'(\theta) = \cos\theta - \sin\theta$.
3. $s'(\theta) = -\sin\theta\sin\theta + \cos\theta\cos\theta = \cos^2\theta - \sin^2\theta = \cos 2\theta$.
4. $t'(\theta) = \dfrac{-\sin\theta\sin\theta - \cos\theta\cos\theta}{\sin^2\theta} = -\dfrac{(\sin^2\theta + \cos^2\theta)}{\sin^2\theta} = -\dfrac{1}{\sin^2\theta}$.
5. $z' = -4\sin(4\theta)$.
6. $f'(x) = \cos(3x) \cdot 3 = 3\cos(3x)$.
7. $w' = e^t \cos(e^t)$.
8. $f'(x) = (2x)(\cos x) + x^2(-\sin x) = 2x\cos x - x^2\sin x$.
9. $f'(x) = (e^{\cos x})(-\sin x) = -\sin x e^{\cos x}$.
10. $f'(y) = (\cos y)e^{\sin y}$.
11.
$$f(x) = (1 - \cos x)^{\frac{1}{2}}$$
$$f'(x) = \frac{1}{2}(1 - \cos x)^{-\frac{1}{2}}(-(-\sin x))$$
$$= \frac{\sin x}{2\sqrt{1 - \cos x}}.$$
12. $f'(x) = [-\sin(\sin x)](\cos x)$.
13. $f'(x) = \dfrac{\cos x}{\cos^2(\sin x)}$.
14. $k'(x) = \frac{3}{2}\sqrt{\sin(2x)}(2\cos(2x)) = 3\cos(2x)\sqrt{\sin(2x)}$.

15. $h'(x) = (\ln 2)2^{\sin x} \cos x$.
16. $w' = (\ln 2)(2^{2\sin x + e^x})(2\cos x + e^x)$.
17. $z' = e^{\cos\theta} - \theta(\sin\theta)e^{\cos\theta}$.
18. $f'(x) = 2 \cdot [\sin(3x)] + 2x[\cos(3x)] \cdot 3 = 2\sin(3x) + 6x\cos(3x)$
19. $f'(x) = 2\cos(2x)\sin(3x) + 3\sin(2x)\cos(3x)$.
20. $y' = e^{\theta}\sin(2\theta) + 2e^{\theta}\cos(2\theta)$.
21. $f'(x) = (e^{-2x})(-2)(\sin x) + (e^{-2x})(\cos x) = -2\sin x(e^{-2x}) + (e^{-2x})(\cos x) = e^{-2x}[\cos x - 2\sin x]$.
22. $z' = \dfrac{\cos t}{2\sqrt{\sin t}}$.
23. $y' = 5\sin^4\theta \cos\theta$.
24. $g'(z) = \dfrac{e^z}{\cos^2(e^z)}$.
25. $z' = \dfrac{-3e^{-3\theta}}{\cos^2(e^{-3\theta})}$.
26. $w' = (-\cos\theta)e^{-\sin\theta}$.
27. $h'(t) = 1 \cdot (\cos t) + t(-\sin t) + \dfrac{1}{\cos^2 t} = \cos t - t\sin t + \dfrac{1}{\cos^2 t}$.
28. $f'(\alpha) = -\sin\alpha + 3\cos\alpha$
29. $f'(\theta) = 2\theta\sin\theta + \theta^2\cos\theta + 2\cos\theta - 2\theta\sin\theta - 2\cos\theta = \theta^2\cos\theta$.
30. The pattern in the table below allows us to generalize and say that the $(4n)^{\text{th}}$ derivative of $\cos x$ is $\cos x$, i.e.,

$$\frac{d^4 y}{dx^4} = \frac{d^8 y}{dx^8} = \cdots = \frac{d^{4n} y}{dx^{4n}} = \cos x.$$

Thus we can say that $d^{48}y/dx^{48} = \cos x$. From there we differentiate twice more to obtain $d^{50}y/dx^{50} = -\cos x$.

n	1	2	3	4	$\cdots$	48	49	50
n^{th} derivative	$-\sin x$	$-\cos x$	$\sin x$	$\cos x$		$\cos x$	$-\sin x$	$-\cos x$

31. We see that $q'(x)$ is of the form

$$\frac{g(x) \cdot f'(x) - f(x) \cdot g'(x)}{(g(x))^2},$$

with $f(x) = e^x$ and $g(x) = \sin x$. Therefore, using the quotient rule, let

$$q(x) = \frac{f(x)}{g(x)} = \frac{e^x}{\sin x}.$$

32. (a) $v(t) = \dfrac{dy}{dt} = \dfrac{d}{dt}(15 + \sin(2\pi t)) = 2\pi\cos(2\pi t)$.

(b)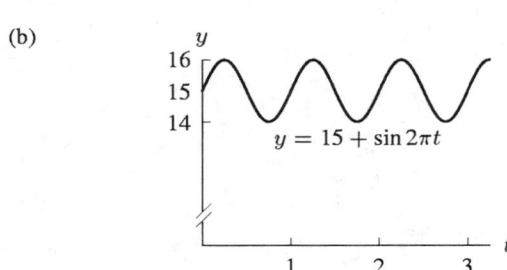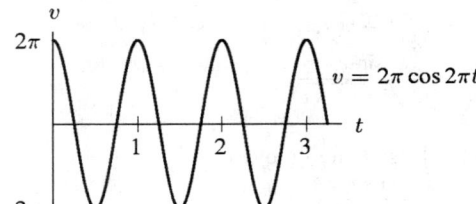

33. (a) $\frac{dy}{dt} = -\frac{4.9\pi}{6}\sin\left(\frac{\pi}{6}t\right)$. It represents the rate of change of the depth of the water.

(b) $\frac{dy}{dt}$ is zero where the tangent line to the curve y is horizontal. $\frac{dy}{dt} = 0$ occurs when $\sin(\frac{\pi}{6}t) = 0$, or at $t = 6, 12, 18$ and 24 (6 am, noon, 6 pm, and midnight). When $\frac{dy}{dt} = 0$, the depth of the water is no longer changing. Therefore, it has either just finished rising or just finished falling, and we know that the harbor's level is at a maximum or a minimum.

34. (a) Differentiating, we find

$$\text{Rate of change of voltage with time} = \frac{dV}{dt} = -120\pi \cdot 156\sin(120\pi t)$$
$$= -18720\pi \sin(120\pi t) \text{ volts per second.}$$

(b) The rate of change of voltage with time is zero when $\sin(120\pi t) = 0$. This occurs when $120\pi t$ equals any multiple of π. For example, $\sin(120\pi t) = 0$ when $120\pi t = \pi$, or at $t = 1/120$ seconds. Since there are an infinite number of multiples of π, there are many times when the rate of change dV/dt is zero.

(c) The maximum value of the rate of change is $18720\pi = 58810.6$ volts/sec.

35. (a) When $\sqrt{\frac{k}{m}}t = \frac{\pi}{2}$ the spring is farthest from the equilibrium position. This occurs at time $t = \frac{\pi}{2}\sqrt{\frac{m}{k}}$
$v = A\sqrt{\frac{k}{m}}\cos\left(\sqrt{\frac{k}{m}}t\right)$, so the maximum velocity occurs when $t = 0$
$a = -A\frac{k}{m}\sin\left(\sqrt{\frac{k}{m}}t\right)$, so the maximum acceleration occurs when $\sqrt{\frac{k}{m}}t = \frac{3\pi}{2}$, which is at time $t = \frac{3\pi}{2}\sqrt{\frac{m}{k}}$

(b) $T = \frac{2\pi}{\sqrt{k/m}} = 2\pi\sqrt{\frac{m}{k}}$

(c) $\frac{dT}{dm} = \frac{2\pi}{\sqrt{k}} \cdot \frac{1}{2}m^{-\frac{1}{2}} = \frac{\pi}{\sqrt{km}}$

Since $\frac{dT}{dm} > 0$, an increase in the mass causes the period to increase.

36. The tangent lines to $f(x) = \sin x$ have slope $\frac{d}{dx}(\sin x) = \cos x$. The tangent line at $x = 0$ has slope $f'(0) = \cos 0 = 1$ and goes through the point $(0, 0)$. Consequently, its equation is $y = g(x) = x$. The approximate value of $\sin\frac{\pi}{6}$ given by this equation is then $g(\frac{\pi}{6}) = \frac{\pi}{6} \approx 0.524$.

Similarly, the tangent line at $x = \frac{\pi}{3}$ has slope $f'(\frac{\pi}{3}) = \cos\frac{\pi}{3} = \frac{1}{2}$ and goes through the point $(\frac{\pi}{3}, \frac{\sqrt{3}}{2})$. Consequently, its equation is $y = h(x) = \frac{1}{2}x + \frac{3\sqrt{3}-\pi}{6}$. The approximate value of $\sin\frac{\pi}{6}$ given by this equation is then $h(\frac{\pi}{6}) = \frac{6\sqrt{3}-\pi}{12} \approx 0.604$.

The actual value of $\sin\frac{\pi}{6}$ is $\frac{1}{2}$, so the approximation from 0 is better than that from $\frac{\pi}{3}$. This is because the slope of the function changes less between $x = 0$ and $x = \frac{\pi}{6}$ than it does between $x = \frac{\pi}{6}$ and $x = \frac{\pi}{3}$. This is illustrated below.

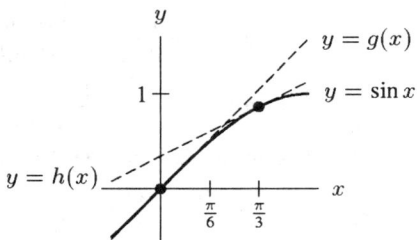

37. Using the triangle OSL in Figure 3.3, we label the distance x.

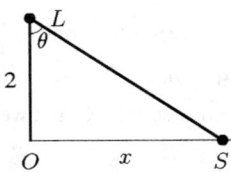

Figure 3.3

We want to calculate $dx/d\theta$. First we must find x as a function of θ. From the triangle, we see
$$\frac{x}{2} = \tan\theta \quad \text{so} \quad x = 2\tan\theta.$$
Thus,
$$\frac{dx}{d\theta} = \frac{2}{\cos^2\theta}.$$

38. (a) Using triangle OPD in the figure below, we see
$$\frac{OD}{a} = \cos\theta \quad \text{so} \quad OD = a\cos\theta$$
$$\frac{PD}{a} = \sin\theta \quad \text{so} \quad PD = a\sin\theta.$$

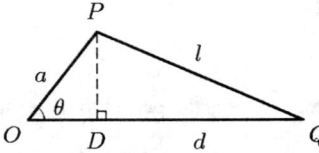

Using triangle PQD, we have
$$(PD)^2 + d^2 = l^2$$
so
$$a^2\sin^2\theta + d^2 = l^2, \quad d = \sqrt{l^2 - a^2\sin^2\theta}.$$
Thus,
$$x = OD + DQ$$
$$= a\cos\theta + \sqrt{l^2 - a^2\sin^2\theta}.$$

(b) Differentiating, regarding a and l as constants,
$$\frac{dx}{d\theta} = -a\sin\theta + \frac{1}{2}\frac{(-2a^2\sin\theta\cos\theta)}{\sqrt{l^2 - a^2\sin^2\theta}}$$
$$= -a\sin\theta - \frac{a^2\sin\theta\cos\theta}{\sqrt{l^2 - a^2\sin^2\theta}}.$$

We want to find dx/dt. Using the chain rule and the fact that $d\theta/dt = 2$, we have
$$\frac{dx}{dt} = \frac{dx}{d\theta} \cdot \frac{d\theta}{dt} = 2\frac{dx}{d\theta}.$$

(i) Substituting $\theta = \pi/2$, we have
$$\frac{dx}{dt} = 2\frac{dx}{d\theta}\bigg|_{\theta=\pi/2} = -2a\sin\left(\frac{\pi}{2}\right) - 2\frac{a^2\sin(\frac{\pi}{2})\cos(\frac{\pi}{2})}{\sqrt{l^2 - a^2\sin^2(\frac{\pi}{2})}}$$
$$= -2a \text{ cm/sec}.$$

(ii) Substituting $\theta = \pi/4$, we have
$$\frac{dx}{dt} = 2\frac{dx}{d\theta}\bigg|_{\theta=\pi/4} = -2a\sin\left(\frac{\pi}{4}\right) - 2\frac{a^2\sin(\frac{\pi}{4})\cos(\frac{\pi}{4})}{\sqrt{l^2 - a^2\sin^2(\frac{\pi}{4})}}$$
$$= -a\sqrt{2} - \frac{a^2}{\sqrt{l^2 - a^2/2}} \text{ cm/sec}.$$

39. (a) If $f(x) = \sin x$, then

$$f'(x) = \lim_{h \to 0} \frac{\sin(x+h) - \sin x}{h}$$
$$= \lim_{h \to 0} \frac{(\sin x \cos h + \sin h \cos x) - \sin x}{h}$$
$$= \lim_{h \to 0} \frac{\sin x(\cos h - 1) + \sin h \cos x}{h}$$
$$= \sin x \lim_{h \to 0} \frac{\cos h - 1}{h} + \cos x \lim_{h \to 0} \frac{\sin h}{h}.$$

(b) $\frac{\cos h - 1}{h} \to 0$ and $\frac{\sin h}{h} \to 1$, as $h \to 0$. Thus, $f'(x) = \sin x \cdot 0 + \cos x \cdot 1 = \cos x$.

(c) Similarly,

$$g'(x) = \lim_{h \to 0} \frac{\cos(x+h) - \cos x}{h}$$
$$= \lim_{h \to 0} \frac{(\cos x \cos h - \sin x \sin h) - \cos x}{h}$$
$$= \lim_{h \to 0} \frac{\cos x(\cos h - 1) - \sin x \sin h}{h}$$
$$= \cos x \lim_{h \to 0} \frac{\cos h - 1}{h} - \sin x \lim_{h \to 0} \frac{\sin h}{h}$$
$$= -\sin x.$$

40. (a) Sector OAQ is a sector of a circle with radius $\frac{1}{\cos \theta}$ and angle $\Delta \theta$. Thus its area is the left side of the inequality. Similarly, the area of Sector OBR is the right side of the equality. The area of the triangle OQR is $\frac{1}{2}\Delta \tan \theta$ since it is a triangle with base $\Delta \tan \theta$ (the segment QR) and height 1 (if you turn it sideways, it is easier to see this). Thus, using the given fact about areas (which is also clear from looking at the picture), we have

$$\frac{\Delta \theta}{2\pi} \cdot \pi \left(\frac{1}{\cos \theta}\right)^2 \leq \frac{1}{2} \cdot \Delta(\tan \theta) \leq \frac{\Delta \theta}{2\pi} \cdot \pi \left(\frac{1}{\cos(\theta + \Delta \theta)}\right)^2.$$

(b) Dividing the inequality through by $\frac{\Delta \theta}{2}$ and canceling the π's gives:

$$\left(\frac{1}{\cos \theta}\right)^2 \leq \frac{\Delta \tan \theta}{\Delta \theta} \leq \left(\frac{1}{\cos(\theta + \Delta \theta)}\right)^2$$

Then as $\Delta \theta \to 0$, the right and left sides both tend towards $\left(\frac{1}{\cos \theta}\right)^2$ while the middle (which is the difference quotient for tangent) tends to $(\tan \theta)'$. Thus, the derivative of tangent is "squeezed" between two values heading towards the same thing and must, itself, also tend to that value. Therefore, $(\tan \theta)' = \left(\frac{1}{\cos \theta}\right)^2$.

(c) Take the identity $\sin^2 \theta + \cos^2 \theta = 1$ and divide through by $\cos^2 \theta$ to get $(\tan \theta)^2 + 1 = \left(\frac{1}{\cos \theta}\right)^2$. Differentiating with respect to θ yields:

$$2(\tan \theta) \cdot (\tan \theta)' = 2\left(\frac{1}{\cos \theta}\right) \cdot \left(\frac{1}{\cos \theta}\right)'$$
$$2\left(\frac{\sin \theta}{\cos \theta}\right) \cdot \left(\frac{1}{\cos \theta}\right)^2 = 2\left(\frac{1}{\cos \theta}\right) \cdot (-1)\left(\frac{1}{\cos \theta}\right)^2 (\cos \theta)'$$
$$2\frac{\sin \theta}{\cos^3 \theta} = (-1)2\frac{1}{\cos^3 \theta}(\cos \theta)'$$
$$-\sin \theta = (\cos \theta)'.$$

(d)

$$\frac{d}{d\theta}\left(\sin^2 \theta + \cos^2 \theta\right) = \frac{d}{d\theta}(1)$$
$$2\sin \theta \cdot (\sin \theta)' + 2\cos \theta \cdot (\cos \theta)' = 0$$
$$2\sin \theta \cdot (\sin \theta)' + 2\cos \theta \cdot (-\sin \theta) = 0$$
$$(\sin \theta)' - \cos \theta = 0$$
$$(\sin \theta)' = \cos \theta.$$

Solutions for Section 3.6

1. $f'(t) = \frac{2t}{t^2+1}$.
2. $f'(x) = \frac{-1}{1-x} = \frac{1}{x-1}$.
3. Since $\ln(e^{2x}) = 2x$, the derivative $f'(x) = 2$.
4. Since $e^{\ln(e^{2x^2+3})} = e^{2x^2+3}$, the derivative $f'(x) = 4xe^{2x^2+3}$.
5. $f'(z) = -1(\ln z)^{-2} \cdot \frac{1}{z} = \frac{-1}{z(\ln z)^2}$.
6. $f'(\theta) = \frac{-\sin\theta}{\cos\theta} = -\tan\theta$.
7. $f'(x) = \frac{1}{1-e^{-x}} \cdot -e^{-x}(-1) = \frac{e^{-x}}{1-e^{-x}}$.
8. $f'(\alpha) = \frac{1}{\sin\alpha} \cdot \cos\alpha = \frac{\cos\alpha}{\sin\alpha}$.
9. $f'(x) = \frac{1}{e^x+1} \cdot e^x$.
10. $f'(t) = \frac{1}{\ln t} \cdot \frac{1}{t} = \frac{1}{t \ln t}$.
11. $f'(x) = \frac{1}{e^{7x}} \cdot (e^{7x})7 = 7$.
    (Note also that $\ln(e^{7x}) = 7x$ implies $f'(x) = 7$.)
12. Note that $f(x) = e^{\ln x} \cdot e^1 = x \cdot e = ex$. So $f'(x) = e$. (Remember, e is just a constant.) You might also use the chain rule to get:
    $$f'(x) = e^{(\ln x)+1} \cdot \frac{1}{x}.$$
    [Are the two answers the same? Of course they are, since
    $$e^{(\ln x)+1}\left(\frac{1}{x}\right) = e^{\ln x} \cdot e\left(\frac{1}{x}\right) = xe\left(\frac{1}{x}\right) = e.]$$

13. $f'(w) = \frac{1}{\cos(w-1)}[-\sin(w-1)] = -\tan(w-1)$.
    [This could be done easily using the answer from Problem 6 and the chain rule.]
14. $f(t) = \ln t$ (because $\ln e^x = x$ or because $e^{\ln t} = t$), so $f'(t) = \frac{1}{t}$.
15. $f'(y) = \dfrac{2y}{\sqrt{1-y^4}}$.
16. $g'(t) = \dfrac{3}{(3t-4)^2+1}$.
17. $g(\alpha) = \alpha$, so $g'(\alpha) = 1$.
18. $g'(t) = e^{\arctan(3t^2)}\left(\dfrac{1}{1+(3t^2)^2}\right)(6t) = e^{\arctan(3t^2)}\left(\dfrac{6t}{1+9t^4}\right)$.
19. $g'(t) = \dfrac{-\sin(\ln t)}{t}$.
20. $h'(z) = (\ln 2)z^{(\ln 2 - 1)}$.
21. $h'(w) = \arcsin w + \dfrac{w}{\sqrt{1-w^2}}$.
22. $f'(x) = -\sin(\arcsin(x+1))\left(\dfrac{1}{\sqrt{1-(x+1)^2}}\right) = \dfrac{-(x+1)}{\sqrt{1-(x+1)^2}}$.
23. Differentiating
    $$f'(x) = \frac{1}{x^2+1} \cdot 2x = 2x(x^2+1)^{-1}$$
    $$f''(x) = 2(x^2+1)^{-1} - 2x(x^2+1)^{-2} \cdot 2x$$
    $$= \frac{2}{(x^2+1)} - \frac{4x^2}{(x^2+1)^2} = \frac{2x^2+2}{(x^2+1)^2} - \frac{4x^2}{(x^2+1)^2}$$
    $$= \frac{2(1-x^2)}{(x^2+1)^2}.$$
    Since $(x^2+1)^2 > 0$ for all x, we see that $f''(0) > 0$ for $1-x^2 > 0$ or $x^2 < 1$. That is, $\ln(x^2+1)$ is concave up on the interval $-1 < x < 1$.

24. Let
$$g(x) = \arcsin x$$
so
$$\sin[g(x)] = x.$$
Differentiating,
$$\cos[g(x)] \cdot g'(x) = 1$$
$$g'(x) = \frac{1}{\cos[g(x)]}$$
Using the fact that $\sin^2 \theta + \cos^2 \theta = 1$, and $\cos[g(x)] \geq 0$, since $-\frac{\pi}{2} \leq g(x) \leq \frac{\pi}{2}$, we get
$$\cos[g(x)] = \sqrt{1 - (\sin[g(x)])^2}.$$
Therefore,
$$g'(x) = \frac{1}{\sqrt{1 - (\sin[g(x)])^2}}$$
Since $\sin[g(x)] = x$, we have
$$g'(x) = \frac{1}{\sqrt{1 - x^2}}, -1 < x < 1.$$

25. Let
$$g(x) = \log x.$$
Then
$$10^{g(x)} = x.$$
Differentiating,
$$(\ln 10)[10^{g(x)}]g'(x) = 1$$
$$g'(x) = \frac{1}{(\ln 10)[10^{g(x)}]}$$
$$g'(x) = \frac{1}{(\ln 10)x}.$$

26. pH $= 2 = -\log x$ means $\log x = -2$ so $x = 10^{-2}$. Rate of change of pH with hydrogen ion concentration is
$$\frac{d}{dx}\text{pH} = -\frac{d}{dx}(\log x) = \frac{-1}{x(\ln 10)} = -\frac{1}{(10^{-2})\ln 10} = -43.4$$

27. (a) $P = 10.8(0.998)^{10} \approx 10.59$ million.
    (b)
    $$\frac{dP}{dt} = 10.8(\ln 0.998)(0.998)^t$$
    so
    $$\left.\frac{dP}{dt}\right|_{t=10} = 10.8(\ln 0.998)(0.998)^{10} \approx -0.02 \text{ million/year}.$$
    Thus in 2000, Hungary's population will be decreasing by 20,000 people per year.

28. The closer you look at the function, the more it begins to look like a line with slope equal to the derivative of the function at $x = 0$. Hence, functions whose derivatives at $x = 0$ are equal will look the same there.

    The following functions look like the line $y = x$ since, in all cases, $y' = 1$ at $x = 0$.

    $y = x$ $y' = 1$
    $y = \sin x$ $y' = \cos x$
    $y = \tan x$ $y' = \frac{1}{\cos^2 x}$
    $y = \ln(x+1)$ $y' = \frac{1}{x+1}$

The following functions look like the line $y = 0$ since, in all cases, $y' = 0$ at $x = 0$.

$y = x^2$ $y' = 2x$
$y = x \sin x$ $y' = x \cos x + \sin x$
$y = x^3$ $y' = 3x^2$
$y = \frac{1}{2} \ln(x^2 + 1)$ $y' = 2x \cdot \frac{1}{2} \cdot \frac{1}{x^2+1} = \frac{x}{x^2+1}$
$y = 1 - \cos x$ $y' = \sin x$

The following functions look like the line $x = 0$ since, in all cases, as $x \to 0^+$, the slope $y' \to \infty$.

$y = \sqrt{x}$ $y' = \frac{1}{2\sqrt{x}}$
$y = \sqrt{\frac{x}{x+1}}$ $y' = \frac{(x+1)-x}{(x+1)^2} \cdot \frac{1}{2} \cdot \frac{1}{\sqrt{\frac{x}{x+1}}} = \frac{1}{2(x+1)^2} \cdot \sqrt{\frac{x+1}{x}}$
$y = \sqrt{2x - x^2}$ $y' = (2 - 2x)\frac{1}{2} \cdot \frac{1}{\sqrt{2x-x^2}} = \frac{1-x}{\sqrt{2x-x^2}}$

29. (a)
$$f'(x) = \frac{1}{1+x^2} + \frac{1}{1+\frac{1}{x^2}} \cdot \left(-\frac{1}{x^2}\right)$$
$$= \frac{1}{1+x^2} + \left(-\frac{1}{x^2+1}\right)$$
$$= \frac{1}{1+x^2} - \frac{1}{1+x^2}$$
$$= 0$$

(b) f is a constant function. Checking at a few values of x,

TABLE 3.2

x	arctan x	arctan x^{-1}	$f(x) = \arctan x + \arctan x^{-1}$
1	0.785392	0.7853982	1.5707963
2	1.1071487	0.4636476	1.5707963
3	1.2490458	0.3217506	1.5707963

30. (a) $y = \ln x$, $y' = \frac{1}{x}$; $f'(1) = \frac{1}{1} = 1$.
$y - y_1 = m(x - x_1)$, $y - 0 = 1(x - 1)$; $y = g(x) = x - 1$.
(b) $g(1.1) = 1.1 - 1 = 0.1$; $g(2) = 2 - 1 = 1$.
(c) $f(1.1)$ and $f(2)$ are below $g(x) = x - 1$. $f(0.9)$ and $f(0.5)$ are also below $g(x)$. This would be true for any approximation of this function by a tangent line since f is concave down ($f''(x) = -\frac{1}{x^2} < 0$ for all $x \neq 0$). See figure below. Thus, for a given x-value, the y-value given by the function is always below the value given by the tangent line.

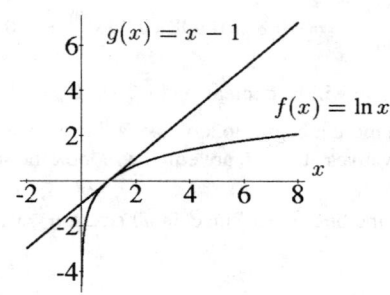

31. (a) Let $g(x) = ax^2 + bx + c$ be our quadratic and $f(x) = \ln x$. For the best approximation, we want to find a quadratic with the same value as $\ln x$ at $x = 1$ and the same first and second derivatives as $\ln x$ at $x = 1$.
$g'(x) = 2ax + b, g''(x) = 2a, f'(x) = \frac{1}{x}, f''(x) = -\frac{1}{x^2}$.

$$g(1) = a(1)^2 + b(1) + c \quad f(1) = 0$$
$$g'(1) = 2a(1) + b \quad f'(1) = 1$$
$$g''(1) = 2a \quad f''(1) = -1$$

Thus, we obtain the equations

$$a + b + c = 0$$
$$2a + b = 1$$
$$2a = -1$$

We find $a = -\frac{1}{2}, b = 2$ and $c = -\frac{3}{2}$. Thus our approximation is:

$$g(x) = -\frac{1}{2}x^2 + 2x - \frac{3}{2}$$

(b) From the graph below, we notice that around $x = 1$, the value of $f(x) = \ln x$ and the value of $g(x) = -\frac{1}{2}x^2 + 2x - \frac{3}{2}$ are very close.

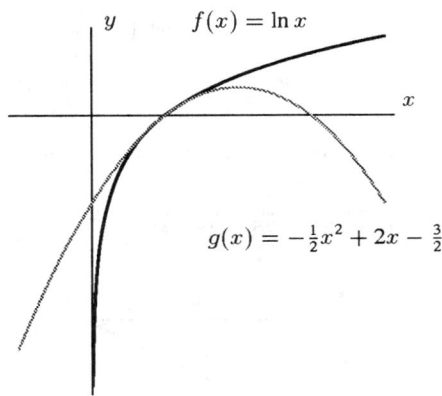

(c) $g(1.1) = 0.095 \quad g(2) = 0.5$
Compare with $f(1.1) = 0.0953, f(2) = 0.693$.

32. We differentiate $F = k/r^2$ with respect to t using the chain rule to give

$$\frac{dF}{dt} = -\frac{2k}{r^3} \cdot \frac{dr}{dt}.$$

We know that $k = 10^{13}$ newton $\cdot$ km^2 and that the rocket is moving at 0.2 km/sec when $r = 10^4$ km. In other words, $dr/dt = 0.2$ km/sec when $r = 10^4$. Substituting gives

$$\frac{dF}{dt} = -\frac{2 \cdot 10^{13}}{(10^4)^3} \cdot 0.2 = -4 \text{ newtons/sec}.$$

33. (a) Using Pythagoras' theorem, we see

$$z^2 = 0.5^2 + x^2$$

so

$$z = \sqrt{0.25 + x^2}.$$

(b) We want to calculate dz/dt. Using the chain rule, we have
$$\frac{dz}{dt} = \frac{dz}{dx} \cdot \frac{dx}{dt} = \frac{2x}{2\sqrt{0.25 + x^2}} \frac{dx}{dt}.$$

Because the train is moving at 0.8 km/hr, we know that
$$\frac{dx}{dt} = 0.8 \text{ km/hr}.$$

At the moment we are interested in $z = 1$ km so
$$1^2 = 0.25 + x^2$$

giving
$$x = \sqrt{0.75} = 0.866 \text{ km}.$$

Therefore
$$\frac{dz}{dt} = \frac{2(0.866)}{2\sqrt{0.25 + 0.75}} \cdot 0.8 = 0.866 \cdot 0.8 = 0.693 \text{ km/min}.$$

(c) We want to know $d\theta/dt$, where θ is as shown in Figure 3.4. Since
$$\frac{x}{0.5} = \tan\theta$$

we know
$$\theta = \arctan\left(\frac{x}{0.5}\right),$$

so
$$\frac{d\theta}{dt} = \frac{1}{1 + (x/0.5)^2} \cdot \frac{1}{0.5} \frac{dx}{dt}.$$

We know that $dx/dt = 0.8$ km/min and, at the moment we are interested in, $x = \sqrt{0.75}$. Substituting gives
$$\frac{d\theta}{dt} = \frac{1}{1 + 0.75/0.25} \cdot \frac{1}{0.5} \cdot 0.8 = 0.4 \text{ radians/min}.$$

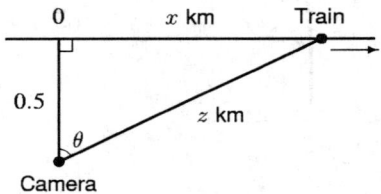

Figure 3.4

34. If V is the volume of the balloon and r is its radius, then
$$V = \frac{4}{3}\pi r^3.$$

We want to know the rate at which air is being blown into the balloon, which is the rate at which the volume is increasing, dV/dt. We are told that
$$\frac{dr}{dt} = 2 \text{ cm/sec} \quad \text{when} \quad r = 10 \text{ cm}.$$

Using the chain rule, we have
$$\frac{dV}{dt} = \frac{dV}{dr} \cdot \frac{dr}{dt} = 4\pi r^2 \frac{dr}{dt}.$$

Substituting gives
$$\frac{dV}{dt} = 4\pi(10)^2 2 = 800\pi = 2513.3 \text{ cm}^3/\text{sec}.$$

35. (a) Assuming that $T(1) = 98.6 - 2 = 96.6$, we get
$$96.6 = 68 + 30.6e^{-k \cdot 1}$$
$$28.6 = 30.6e^{-k}$$
$$0.935 = e^{-k}.$$

So
$$k = -\ln(0.935) \approx 0.067.$$

(b) We're looking for a value of t which gives $T'(t) = -1$. First we find $T'(t)$:
$$T(t) = 68 + 30.6e^{-0.067t}$$
$$T'(t) = (30.6)(-0.067)e^{-0.067t} \approx -2e^{-0.067t}.$$

Setting this equal to $-1°$F per hour gives
$$-1 = -2e^{-0.067t}$$
$$\ln(0.5) = -0.067t$$
$$t = -\frac{\ln(0.5)}{0.067} \approx 10.3.$$

Thus, when $t \approx 10.3$ hours, we have $T'(t) \approx -1°$F per hour.

(c) The coroner's rule of thumb predicts that in 24 hours the body temperature will decrease 25°F, to about 73.6°F. The formula predicts a temperature of
$$T(24) = 68 + 30.6e^{-0.067 \cdot 24} \approx 74.1°\text{F}.$$

36. (a) Since the elevator is descending at 30 ft/sec, its height from the ground is given by $h(t) = 300 - 30t$, for $0 \leq t \leq 10$.

(b) From the triangle in the figure,
$$\tan \theta = \frac{h(t) - 100}{150} = \frac{300 - 30t - 100}{150} = \frac{200 - 30t}{150}.$$

Therefore
$$\theta = \arctan\left(\frac{200 - 30t}{150}\right)$$

and
$$\frac{d\theta}{dt} = \frac{1}{1 + \left(\frac{200-30t}{150}\right)^2} \cdot \left(\frac{-30}{150}\right) = -\frac{1}{5}\left(\frac{150^2}{150^2 + (200 - 30t)^2}\right).$$

Notice that $\frac{d\theta}{dt}$ is always negative, which is reasonable since θ decreases as the elevator descends.

(c) If we want to know when θ changes (decreases) the fastest, we want to find out when $d\theta/dt$ has the largest magnitude. This will occur when the denominator, $150^2 + (200 - 30t)^2$, in the expression for $d\theta/dt$ is the smallest, or when $200 - 30t = 0$. This occurs when $t = \frac{200}{30}$ seconds, and so $h(\frac{200}{30}) = 100$ feet, i.e., when the elevator is at the level of the observer.

Solutions for Section 3.7

1. We differentiate implicitly both sides of the equation with respect to x.
$$2x + 2y\frac{dy}{dx} = 0,$$
$$\frac{dy}{dx} = -\frac{2x}{2y} = -\frac{x}{y}.$$

2. We differentiate implicitly both sides of the equation with respect to x.
$$2x + \left(y + x\frac{dy}{dx}\right) - 3y^2\frac{dy}{dx} = y^2 + x(2y)\frac{dy}{dx},$$
$$x\frac{dy}{dx} - 3y^2\frac{dy}{dx} - 2xy\frac{dy}{dx} = y^2 - y - 2x,$$
$$\frac{dy}{dx} = \frac{y^2 - y - 2x}{x - 3y^2 - 2xy}.$$

104 CHAPTER THREE /SOLUTIONS

3. We differentiate implicitly both sides of the equation with respect to x.
$$x^{1/2} = 5y^{1/2}$$
$$\frac{1}{2}x^{-1/2} = \frac{5}{2}y^{-1/2}\frac{dy}{dx}$$
$$\frac{dy}{dx} = \frac{\frac{1}{2}x^{-1/2}}{\frac{5}{2}y^{-1/2}} = \frac{1}{5}\sqrt{\frac{y}{x}} = \frac{1}{25}.$$

We can also obtain this answer by realizing that the original equation represents part of the line $x = 25y$ which has slope $1/25$.

4. We differentiate implicitly both sides of the equation with respect to x.
$$x^{\frac{1}{2}} + y^{\frac{1}{2}} = 25,$$
$$\frac{1}{2}x^{-\frac{1}{2}} + \frac{1}{2}y^{-\frac{1}{2}}\frac{dy}{dx} = 0,$$
$$\frac{dy}{dx} = -\frac{\frac{1}{2}x^{-\frac{1}{2}}}{\frac{1}{2}y^{-\frac{1}{2}}} = -\frac{x^{-\frac{1}{2}}}{y^{-\frac{1}{2}}} = -\frac{\sqrt{y}}{\sqrt{x}} = -\sqrt{\frac{y}{x}}.$$

5. We differentiate implicitly both sides of the equation with respect to x.
$$\ln x + \ln(y^2) = 3$$
$$\frac{1}{x} + \frac{1}{y^2}(2y)\frac{dy}{dx} = 0$$
$$\frac{dy}{dx} = \frac{-1/x}{2y/y^2} = -\frac{y}{2x}.$$

6. We differentiate implicitly both sides of the equation with respect to x.
$$e^{x^2} + \ln y = 0$$
$$2xe^{x^2} + \frac{1}{y}\frac{dy}{dx} = 0$$
$$\frac{dy}{dx} = -2xye^{x^2}.$$

7. We differentiate implicitly both sides of the equation with respect to x.
$$\arctan(x^2 y) = xy^2$$
$$\frac{1}{1+x^4y^2}(2xy + x^2\frac{dy}{dx}) = y^2 + 2xy\frac{dy}{dx}$$
$$2xy + x^2\frac{dy}{dx} = [1+x^4y^2][y^2 + 2xy\frac{dy}{dx}]$$
$$\frac{dy}{dx}[x^2 - (1+x^4y^2)(2xy)] = (1+x^4y^2)y^2 - 2xy$$
$$\frac{dy}{dx} = \frac{y^2 + x^4y^4 - 2xy}{x^2 - 2xy - 2x^5y^3}.$$

8. We differentiate implicitly both sides of the equation with respect to x.
$$\ln y + x\frac{1}{y}\frac{dy}{dx} + 3y^2\frac{dy}{dx} = \frac{1}{x}$$
$$\frac{x}{y}\frac{dy}{dx} + 3y^2\frac{dy}{dx} = \frac{1}{x} - \ln y$$
$$\frac{dy}{dx}\left(\frac{x}{y} + 3y^2\right) = \frac{1 - x\ln y}{x}$$
$$\frac{dy}{dx}\left(\frac{x + 3y^3}{y}\right) = \frac{1 - x\ln y}{x}$$
$$\frac{dy}{dx} = \frac{(1 - x\ln y)}{x} \cdot \frac{y}{(x + 3y^3)}$$

9. We differentiate implicitly both sides of the equation with respect to x.

$$\cos(xy)\left(y + x\frac{dy}{dx}\right) = 2$$

$$y\cos(xy) + x\cos(xy)\frac{dy}{dx} = 2$$

$$\frac{dy}{dx} = \frac{2 - y\cos(xy)}{x\cos(xy)}.$$

10. $\frac{2}{3}x^{-1/3} + \frac{2}{3}y^{-1/3} \cdot \frac{dy}{dx} = 0, \frac{dy}{dx} = -\frac{x^{-1/3}}{y^{-1/3}} = -\frac{y^{1/3}}{x^{1/3}}.$

11. We differentiate implicitly both sides of the equation with respect to x.

$$e^{\cos y}(-\sin y)\frac{dy}{dx} = 3x^2 \arctan y + x^3 \frac{1}{1+y^2}\frac{dy}{dx}$$

$$\frac{dy}{dx}\left(-e^{\cos y}\sin y - \frac{x^3}{1+y^2}\right) = 3x^2 \arctan y$$

$$\frac{dy}{dx} = \frac{3x^2 \arctan y}{-e^{\cos y}\sin y - x^3(1+y^2)^{-1}}.$$

12. Using the relation $\cos^2 y + \sin^2 y = 1$, the equation becomes:
$1 = y + 2$ or $y = -1$. Hence, $\frac{dy}{dx} = 0$.

13. First, we must find the slope of the tangent, i.e. $\left.\frac{dy}{dx}\right|_{(1,-1)}$. Differentiating implicitly, we have:

$$y^2 + x(2y)\frac{dy}{dx} = 0,$$

$$\frac{dy}{dx} = -\frac{y^2}{2xy} = -\frac{y}{2x}.$$

Substitution yields $\left.\frac{dy}{dx}\right|_{(1,-1)} = -\frac{-1}{2} = \frac{1}{2}$. Using the point-slope formula for a line, we have that the equation for the tangent line is $y + 1 = \frac{1}{2}(x - 1)$ or $y = \frac{1}{2}x - \frac{3}{2}$.

14. First we must find the slope of the tangent, $\frac{dy}{dx}$, at $(1, e^2)$. Differentiating implicitly, we have:

$$\frac{1}{xy}\left(x\frac{dy}{dx} + y\right) = 2$$

$$\frac{dy}{dx} = \frac{2xy - y}{x}.$$

Evaluating dy/dx at $(1, e^2)$ yields $(2(1)e^2 - e^2)/1 = e^2$. Using the point-slope formula for the equation of the line, we have:

$$y - e^2 = e^2(x - 1),$$

or

$$y = e^2 x.$$

15. First, we must find the slope of the tangent, $\left.\frac{dy}{dx}\right|_{(4,2)}$. Implicit differentiation yields:

$$2y\frac{dy}{dx} = \frac{2x(xy - 4) - x^2\left(x\frac{dy}{dx} + y\right)}{(xy - 4)^2}.$$

Given the complexity of the above equation, we first want to substitute 4 for x and 2 for y (the coordinates of the point where we are constructing our tangent line), then solve for $\dfrac{dy}{dx}$. Substitution yields:

$$2\cdot 2\frac{dy}{dx} = \frac{(2\cdot 4)(4\cdot 2 - 4) - 4^2\left(4\frac{dy}{dx} + 2\right)}{(4\cdot 2 - 4)^2} = \frac{8(4) - 16(4\frac{dy}{dx} + 2)}{16} = -4\frac{dy}{dx}.$$

$$4\frac{dy}{dx} = -4\frac{dy}{dx},$$

Solving for $\dfrac{dy}{dx}$, we have:

$$\frac{dy}{dx} = 0.$$

The tangent is a horizontal line through $(4, 2)$, hence its equation is $y = 2$.

16. First, we must find the slope of the tangent, $\left.\dfrac{dy}{dx}\right|_{(a,0)}$. We differentiate implicitly, obtaining:

$$\frac{2}{3}x^{-\frac{1}{3}} + \frac{2}{3}y^{-\frac{1}{3}}\frac{dy}{dx} = 0,$$

$$\frac{dy}{dx} = -\frac{\frac{2}{3}x^{-\frac{1}{3}}}{\frac{2}{3}y^{-\frac{1}{3}}} = -\frac{\sqrt[3]{y}}{\sqrt[3]{x}}.$$

Substitution yields, $\left.\dfrac{dy}{dx}\right|_{(a,0)} = \dfrac{\sqrt[3]{0}}{\sqrt[3]{a}} = 0$. The tangent is a horizontal line through $(a, 0)$, hence its equation is $y = 0$.

17. $y = x^{\frac{m}{n}}$. Taking n^{th} powers of both sides of this expression yields $(y)^n = (x^{\frac{m}{n}})^n$, or $y^n = x^m$.

$$\frac{d}{dx}(y^n) = \frac{d}{dx}(x^m)$$

$$ny^{n-1}\frac{dy}{dx} = mx^{m-1}$$

$$\frac{dy}{dx} = \frac{m}{n}\frac{x^{m-1}}{y^{n-1}}$$

$$= \frac{m}{n}\frac{x^{m-1}}{(x^{m/n})^{n-1}}$$

$$= \frac{m}{n}\frac{x^{m-1}}{x^{m-\frac{m}{n}}}$$

$$= \frac{m}{n}x^{(m-1)-(m-\frac{m}{n})} = \frac{m}{n}x^{\frac{m}{n}-1}.$$

18. (a) If $x = 4$ then $16 + y^2 = 25$, so $y = \pm 3$. We find $\dfrac{dy}{dx}$ implicitly:

$$2x + 2y\frac{dy}{dx} = 0$$

$$\frac{dy}{dx} = -\frac{x}{y}$$

So the slope at $(4, 3)$ is $-\frac{4}{3}$ and at $(4, -3)$ is $\frac{4}{3}$. The tangent lines are:

$$(y - 3) = -\frac{4}{3}(x - 4) \quad \text{and} \quad (y + 3) = \frac{4}{3}(x - 4)$$

(b) The normal lines have slopes that are the negative of the reciprocal of the slopes of the tangent lines. Thus,

$$(y - 3) = \frac{3}{4}(x - 4) \quad \text{so} \quad y = \frac{3}{4}x$$

and

$$(y + 3) = -\frac{3}{4}(x - 4) \quad \text{so} \quad y = -\frac{3}{4}x$$

are the normal lines.

(c) These lines meet at the origin, which is the center of the circle.

19. (a) Taking derivatives implicitly, we get
$$\frac{2}{25}x + \frac{2}{9}y\frac{dy}{dx} = 0$$
$$\frac{dy}{dx} = \frac{-9x}{25y}$$

(b) The slope is not defined anywhere along the line $y = 0$. This ellipse intersects that line in two places, $(-5, 0)$ and $(5, 0)$. (These are, of course, the "ends" of the ellipse where the tangent is vertical.)

20. (a) Solving for $\frac{dy}{dx}$ by implicit differentiation yields
$$3x^2 + 3y^2\frac{dy}{dx} - y^2 - 2xy\frac{dy}{dx} = 0$$
$$\frac{dy}{dx} = \frac{y^2 - 3x^2}{3y^2 - 2xy}.$$

(b) We can approximate the curve near $x = 1, y = 2$ by its tangent line. The tangent line will have slope $\frac{(2)^2 - 3(1)^2}{3(2)^2 - 2(1)(2)} = \frac{1}{8} = 0.125$. Thus its equation is
$$y = 0.125x + 1.875$$

Using the y-values of the tangent line to approximate the y-values of the curve, we get:

x	0.96	0.98	1	1.02	1.04
approximate y	1.995	1.9975	2.000	2.0025	2.005

(c) When $x = 0.96$, we get the equation $0.96^3 + y^3 - 0.96y^2 = 5$, whose solution by numerical methods is 1.9945, which is close to the one above.

(d) The tangent line is horizontal when $\frac{dy}{dx}$ is zero and vertical when $\frac{dy}{dx}$ is undefined. These will occur when the numerator is zero and when the denominator is zero, respectively.

Thus, we know that the tangent is horizontal when $y^2 - 3x^2 = 0 \Rightarrow y = \pm\sqrt{3}x$. To find the points that satisfy this condition, we substitute back into the original equation for the curve:
$$x^3 + y^3 - xy^2 = 5$$
$$x^3 \pm 3\sqrt{3}x^3 - 3x^3 = 5$$
$$x^3 = \frac{5}{\pm 3\sqrt{3} - 2}$$
So $x \approx 1.1609$ or $x \approx -0.8857$.

Substituting,
$$y = \pm\sqrt{3}x \text{ so } y \approx 2.0107 \quad \text{or} \quad y \approx 1.5341.$$

Thus, the tangent line is horizontal at $(1.1609, 2.0107)$ and $(-0.8857, 1.5341)$.

Also, we know that the tangent is vertical whenever $3y^2 - 2xy = 0$, that is, when $y = \frac{2}{3}x$ or $y = 0$. Substituting into the original equation for the curve gives us $x^3 + (\frac{2}{3}x)^3 - (\frac{2}{3})^2 x^3 = 5$. This means $x^3 \approx 5.8696$, so $x \approx 1.8039, y \approx 1.2026$. The other vertical tangent is at $y = 0, x = \sqrt[3]{5}$.

21. The slope of the tangent to the curve $y = x^2$ at $x = 1$ is 2 so the equation of such a tangent will be of the form $y = 2x + c$. As the tangent must pass through $(1, 1)$, $c = -1$ and so the required tangent is $y = 2x - 1$.

Any circle centered at $(8, 0)$ will be of the form
$$(x - 8)^2 + y^2 = R^2.$$

The slope of this curve at (x, y) is given by implicit differentiation:
$$2(x - 8) + 2yy' = 0$$

or
$$y' = \frac{8 - x}{y}$$

For the tangent to the parabola to be tangential to the circle we need

$$\frac{8-x}{y} = 2$$

so that at the point of contact of the circle and the line the coordinates are given by (x, y) when $y = 4 - x/2$. Substituting into the equation of the tangent line gives $x = 2$ and $y = 3$. From this we conclude that $R^2 = 45$ so that the equation of the circle is

$$(x-8)^2 + y^2 = 45.$$

22. Let the point of intersection of the tangent line with the smaller circle be (x_1, y_1) and the point of intersection with the larger be (x_2, y_2). Let the tangent line be $y = mx + c$. Then at (x_1, y_1) and (x_2, y_2) the slopes of $x^2 + y^2 = 1$ and $y^2 + (x-3)^2 = 4$ are also m. The slope of $x^2 + y^2 = 1$ is found by implicit differentiation: $2x + 2yy' = 0$ so $y' = -x/y$. Similarly, the slope of $y^2 + (x-3)^2 = 4$ is $y' = -(x-3)/y$. Thus,

$$m = \frac{y_2 - y_1}{x_2 - x_1} = -\frac{x_1}{y_1} = -\frac{(x_2 - 3)}{y_2},$$

where $y_1 = \sqrt{1 - x_1^2}$ and $y_2 = \sqrt{4 - (x_2 - 3)^2}$. The positive values for y_1 and y_2 follow from Figure 3.5 and from our choice of $m > 0$. We obtain

$$\frac{x_1}{\sqrt{1 - x_1^2}} = \frac{x_2 - 3}{\sqrt{4 - (x_2 - 3)^2}}$$

$$\frac{x_1^2}{1 - x_1^2} = \frac{(x_2 - 3)^2}{4 - (x_2 - 3)^2}$$

$$x_1^2[4 - (x_2 - 3)^2] = (1 - x_1^2)(x_2 - 3)^2$$

$$4x_1^2 - (x_1^2)(x_2 - 3)^2 = (x_2 - 3)^2 - x_1^2(x_2 - 3)^2$$

$$4x_1^2 = (x_2 - 3)^2$$

$$2|x_1| = |x_2 - 3|.$$

From the picture $x_1 < 0$ and $x_2 < 3$. This gives $x_2 = 2x_1 + 3$ and $y_2 = 2y_1$. From

$$\frac{y_2 - y_1}{x_2 - x_1} = -\frac{x_1}{y_1},$$

substituting $y_1 = \sqrt{1 - x_1^2}$, $y_2 = 2y_1$ and $x_2 = 2x_1 + 3$ gives

$$x_1 = -\frac{1}{3}.$$

From $x_2 = 2x_1 + 3$ we get $x_2 = 7/3$. In addition, $y_1 = \sqrt{1 - x_1^2}$ gives $y_1 = 2\sqrt{2}/3$, and finally $y_2 = 2y_1$ gives $y_2 = 4\sqrt{2}/3$.

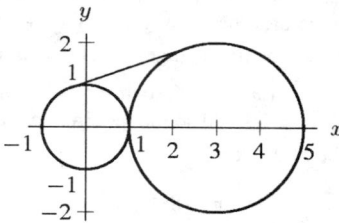

Figure 3.5

Solutions for Section 3.8

1. With $f(x) = 1/x$, we see that the tangent line approximation to f near $x = 1$ is
$$f(x) \approx f(1) + f'(1)(x-1),$$
which becomes
$$\frac{1}{x} \approx 1 + f'(1)(x-1).$$
Since $f'(x) = -1/x^2$, $f'(1) = -1$. Thus our formula reduces to
$$\frac{1}{x} \approx 1 - (x-1) = 2 - x.$$
This is the local linearization of $1/x$ near $x = 1$.

2. With $f(x) = 1/(\sqrt{1+x})$, we see that the tangent line approximation to f near $x = 0$ is
$$f(x) \approx f(0) + f'(0)(x-0),$$
which becomes
$$\frac{1}{\sqrt{1+x}} \approx 1 + f'(0)x.$$
Since $f'(x) = (-1/2)(1+x)^{-3/2}$, $f'(0) = -1/2$. Thus our formula reduces to
$$\frac{1}{\sqrt{1+x}} \approx 1 - x/2.$$
This is the local linearization of $\frac{1}{\sqrt{1+x}}$ near $x = 0$.

3. Let $f(x) = e^{-x}$. Then $f'(x) = -e^{-x}$. So $f(0) = 1$, $f'(0) = -e^0 = -1$. Therefore, $e^{-x} \approx f(0) + f'(0)x = 1 - x$.

4. If \$P were deposited, then Pe^{rt} would be the balance after t years if interest were compounded continuously at the nominal rate r (see Section 3.3). If interest were compounded n times a year, then the balance would be $P(1 + \frac{r}{n})^{nt}$. The local linearization $e^{rt} \approx 1 + rt$ tells us that for small values of t, say $t = \frac{1}{n}$,
$$Pe^{r(\frac{1}{n})} \approx P\left(1 + r\left(\frac{1}{n}\right)\right) = P(1 + \frac{r}{n}).$$
In other words, the balance after one compounding period is approximately the same whether interest is compounded n times a year or continuously, and is given by local linearization of $Pe^{rt} \approx P(1 + rt)$ at $t =$ compounding period $(1/n)$.

5. (a) Let $f(x) = (1+x)^k$. Then $f'(x) = k(1+x)^{k-1}$. Since
$$f(x) \approx f(0) + f'(0)(x-0)$$
is the tangent line approximation, and $f(0) = 1$, $f'(0) = k$, for small x we get
$$f(x) \approx 1 + kx.$$
(b) Since $\sqrt{1.1} = (1+0.1)^{1/2} \approx 1 + (1/2)0.1 = 1.05$ by the above method, this estimate is about right.
(c) The real answer is less than 1.05. Since $(1.05)^2 = (1+0.05)^2 = 1 + 2(1)(0.05) + (0.05)^2 = 1.1 + (0.05)^2 > 1.1$, we have $(1.05)^2 > 1.1$ Therefore
$$\sqrt{1.1} < 1.05.$$
Graphically, we can see this because the graph of $\sqrt{1+x}$ is concave down, so it bends below its tangent line. (See Figure 3.8 on page 140.) Therefore the true value ($\sqrt{1.1}$) which is on the curve is below the approximate value (1.05) which is on the tangent line.

6. Since $f'(a) > 0$ and $g'(a) < 0$, l'Hopital's rule tells us that
$$\lim_{x \to a} \frac{f(x)}{g(x)} = \frac{f'(a)}{g'(a)} < 0.$$

7. Since $f'(a) < 0$ and $g'(a) < 0$, l'Hopital's rule tells us that

$$\lim_{x \to a} \frac{f(x)}{g(x)} = \frac{f'(a)}{g'(a)} > 0.$$

8. Here $f(a) = g(a) = f'(a) = g'(a) = 0$, and $f''(a) > 0$ and $g''(a) < 0$.

$$\lim_{x \to a} \frac{f(x)}{g(x)} = \lim_{x \to a} \frac{f'(x)}{g'(x)} = \frac{f''(a)}{g''(a)} < 0$$

9. Note that $f(0) = g(0) = 0$ and $f'(0) = g'(0)$. Since $x = 0$ looks like a point of inflection for each curve, $f''(0) = g''(0) = 0$. Therefore, applying l'Hopital's rule successively gives us

$$\lim_{x \to 0} \frac{f(x)}{g(x)} = \lim_{x \to 0} \frac{f'(x)}{g'(x)} = \lim_{x \to 0} \frac{f''(x)}{g''(x)} = \lim_{x \to 0} \frac{f'''(x)}{g'''(x)}.$$

Now notice how the concavity of f changes: for $x < 0$, it is concave up, so $f''(x) > 0$, and for $x > 0$ it is concave down, so $f''(x) < 0$. Thus $f''(x)$ is a decreasing function at 0 and so $f'''(0)$ is negative. Similarly, for $x < 0$, we see g is concave down and for $x > 0$ it is concave up, so $g''(x)$ is increasing at 0 and so $g'''(0)$ is positive. Consequently,

$$\lim_{x \to 0} \frac{f(x)}{g(x)} = \lim_{x \to 0} \frac{f'''(x)}{g'''(0)} < 0.$$

10. Let $f(x) = \ln x$ and $g(x) = 1/x$ so $f'(x) = 1/x$ and $g'(x) = -1/x^2$ and

$$\lim_{x \to 0^+} \frac{\ln x}{1/x} = \lim_{x \to 0^+} \frac{1/x}{-1/x^2} = \lim_{x \to 0^+} \frac{x}{-1} = 0.$$

11. (a) You know that the numerator approaches zero as x goes to zero and the denominator goes to zero even faster, so you should expect that the limit will not exist. You can verify this by substituting several values of x close to zero. Alternatively, using l'Hopital's rule, we have

$$\lim_{x \to 0} \frac{\sin x}{x^2} = \lim_{x \to 0} \frac{\cos x}{2x}$$

which is not defined.

(b) Here the numerator goes to zero faster than the denominator does, so you should expect that the limit will be zero. Using l'Hopital's rule, we have

$$\lim_{x \to 0} \frac{\sin^2 x}{x} = \lim_{x \to 0} \frac{2 \sin x \cos x}{1} = 0.$$

(c) Here the denominator goes to zero more slowly than x does, so you should expect that the numerator will dominate and the limit should be zero. With l'Hopital's rule,

$$\lim_{x \to 0} \frac{\sin x}{x^{1/3}} = \lim_{x \to 0} \frac{\cos x}{\frac{1}{3} x^{-2/3}} = 0.$$

(d) Here the numerator goes to zero more slowly than $\sin x$ does, so the x in the denominator should dominate. Therefore, we should expect that the limit will not exist. Using l'Hopital's rule,

$$\lim_{x \to 0} \frac{(\sin x)^{1/3}}{x} = \lim_{x \to 0} \frac{\frac{1}{3}(\sin x)^{-2/3} \cos x}{1}$$

which is not defined.

12. (a) Since $f'(x) = 3\cos(3x)$, we have $f'(0) = 3$.
(b) Since $g'(x) = 5$, we have $g'(0) = 5$.
(c) Since $f(x) = \sin 3x$ and $g(x) = 5x$ are both 0 at $x = 0$, we apply l'Hopital's rule to obtain

$$\lim_{x \to 0} \frac{\sin(3x)}{5x} = \frac{f'(0)}{g'(0)} = \frac{3}{5}.$$

13. The larger power dominates. Using l'Hopital's rule

$$\lim_{x\to\infty} \frac{x^5}{0.1x^7} = \lim_{x\to\infty} \frac{5x^4}{0.7x^6} = \lim_{x\to\infty} \frac{20x^3}{4.2x^5}$$
$$= \lim_{x\to\infty} \frac{60x^2}{21x^4} = \lim_{x\to\infty} \frac{120x}{84x^3} = \lim_{x\to\infty} \frac{120}{252x^2} = 0$$

so $0.1x^7$ dominates.

14. We apply l'Hopital's rule twice to the ratio $50x^2/0.01x^3$:

$$\lim_{x\to\infty} \frac{50x^2}{0.01x^3} = \lim_{x\to\infty} \frac{100x}{0.03x^2} = \lim_{x\to\infty} \frac{100}{0.06x} = 0.$$

Since the limit is 0, we see that $0.01x^3$ is much larger than $50x^2$ as $x \to \infty$.

15. The power function dominates. Using l'Hopital's rule

$$\lim_{x\to\infty} \frac{\ln(x+3)}{x^{0.2}} = \lim_{x\to\infty} \frac{\frac{1}{(x+3)}}{0.2x^{-0.8}} = \lim_{x\to\infty} \frac{x^{0.8}}{0.2(x+3)}.$$

Using l'Hopital's rule again gives

$$\lim_{x\to\infty} \frac{x^{0.8}}{0.2(x+3)} = \lim_{x\to\infty} \frac{0.8x^{-0.2}}{0.2} = 0,$$

so $x^{0.2}$ dominates.

16. The exponential dominates. After 10 applications of l'Hopital's rule

$$\lim_{x\to\infty} \frac{x^{10}}{e^{0.1x}} = \lim_{x\to\infty} \frac{10x^9}{0.1e^{0.1x}} = \cdots = \lim_{x\to\infty} \frac{10!}{(0.1)^{10}e^{0.1x}} = 0.$$

so $e^{0.1x}$ dominates.

17. Observe that both $f(4)$ and $g(4)$ are zero. Also, $f'(4) = 1.4$ and $g'(4) = -0.7$, so by l'Hopital's rule,

$$\lim_{x\to 4} \frac{f(x)}{g(x)} = \frac{f'(4)}{g'(4)} = \frac{1.4}{-0.7} = -2.$$

18. (a) Let $f(x) = \sin(2x)$ and $g(x) = x$. Observe that $f(1) = \sin 2 \neq 0$ and $g(1) = 1 \neq 0$. Therefore l'Hopital's rule does not apply. However,

$$\lim_{x\to 1} \frac{\sin 2x}{x} = \frac{\sin 2}{1} = 0.909297.$$

(b) Let $f(x) = \cos x$ and $g(x) = x$. Observe that since $f(0) = 1$, l'Hopital's rule does not apply. But since $g(0) = 0$,

$$\lim_{x\to 0} \frac{\cos x}{x} \quad \text{does not exist.}$$

(c) Let $f(x) = e^{-x}$ and $g(x) = \sin x$. Observe that as x increases, $f(x)$ approaches 0 but $g(x)$ oscillates between -1 and 1. Since $g(x)$ does not approach 0 in the limit, l'Hopital's rule does not apply. Because $g(x)$ is in the denominator and oscillates through 0 forever, the limit does not exist.

19. We want to find $\lim_{x\to\infty} f(x)$, which we do by three applications of l'Hopital's rule:

$$\lim_{x\to\infty} \frac{2x^3 + 5x^2}{3x^3 - 1} = \lim_{x\to\infty} \frac{6x^2 + 10x}{9x^2} = \lim_{x\to\infty} \frac{12x + 10}{18x} = \lim_{x\to\infty} \frac{12}{18} = \frac{2}{3}.$$

So the line $y = 2/3$ is the horizontal asymptote.

20. The local linearization of e^x near $x = 0$ is $1 + 1x$ so

$$e^x \approx 1 + x.$$

Squaring this yields, for small x,

$$e^{2x} = (e^x)^2 \approx (1+x)^2 = 1 + 2x + x^2.$$

Local linearization of e^{2x} directly yields

$$e^{2x} \approx 1 + 2x$$

for small x. The two approximations are consistent because they agree: the tangent line approximation to $1 + 2x + x^2$ is just $1 + 2x$.

The first approximation is more accurate. One can see this numerically or by noting that the approximation for e^{2x} given by $1 + 2x$ is really the same as approximating e^y at $y = 2x$. Since the other approximation approximates e^y at $y = x$, which is twice as close to 0 and therefore a better general estimate, it's more likely to be correct.

21. (a) Let $f(x) = 1/(1 + x)$. Then $f'(x) = -1/(1 + x)^2$ by the chain rule. So $f(0) = 1$, and $f'(0) = -1$. Therefore, for x near 0, $1/(1 + x) \approx f(0) + f'(0)x = 1 - x$.
   (b) We know that for small y, $1/(1 + y) \approx 1 - y$. Let $y = x^2$; when x is small, so is $y = x^2$. Hence, for small x, $1/(1 + x^2) \approx 1 - x^2$.
   (c) Since the linearization of $1/(1 + x^2)$ is the line $y = 1$, and this line has a slope of 0, the derivative of $1/(1 + x^2)$ is zero at $x = 0$.

22. The local linearizations of $f(x) = e^x$ and $g(x) = \sin x$ near $x = 0$ are

$$f(x) = e^x \approx 1 + x$$

and

$$g(x) = \sin x \approx x.$$

Thus, the local linearization of $e^x \sin x$ is the local linearization of the product:

$$e^x \sin x \approx (1 + x)x = x + x^2 \approx x.$$

We therefore know that the derivative of $e^x \sin x$ at $x = 0$ must be 1. Similarly, using the local linearization of $1/(1 + x)$ near $x = 0$, $1/(1 + x) \approx 1 - x$, we have

$$\frac{e^x \sin x}{1 + x} = (e^x)(\sin x)\left(\frac{1}{1 + x}\right) \approx (1 + x)(x)(1 - x) = x - x^3$$

so the local linearization of the triple product $\dfrac{e^x \sin x}{1 + x}$ at $x = 0$ is simply x. And therefore the derivative of $\dfrac{e^x \sin x}{1 + x}$ at $x = 0$ is 1.

23. Note that

$$[f(x)g(x)]' = \lim_{h \to 0} \frac{f(x + h)g(x + h) - f(x)g(x)}{h}.$$

We use the hint: For small h, $f(x + h) \approx f(x) + f'(x)h$, and $g(x + h) \approx g(x) + g'(x)h$. Therefore

$$\begin{aligned}
f(x + h)g(x + h) - f(x)g(x) &\approx [f(x) + hf'(x)][g(x) + hg'(x)] - f(x)g(x) \\
&= f(x)g(x) + hf'(x)g(x) + hf(x)g'(x) \\
&\quad + h^2 f'(x)g'(x) - f(x)g(x) \\
&= hf'(x)g(x) + hf(x)g'(x) + h^2 f'(x)g'(x).
\end{aligned}$$

Therefore

$$\begin{aligned}
\lim_{h \to 0} \frac{f(x + h)g(x + h) - f(x)g(x)}{h} &= \lim_{h \to 0} \frac{hf'(x)g(x) + hf(x)g'(x) + h^2 f'(x)g'(x)}{h} \\
&= \lim_{h \to 0} \frac{h\left(f'(x)g(x) + f(x)g'(x) + hf'(x)g'(x)\right)}{h} \\
&= \lim_{h \to 0} \left(f'(x)g(x) + f(x)g'(x) + hf'(x)g'(x)\right) \\
&= f'(x)g(x) + f(x)g'(x).
\end{aligned}$$

A more complete derivation can be given using the error term discussed in the section on Differentiability and Linear Approximation in Chapter 2. Adapting the notation of that section to this problem, we write

$$f(x+h) = f(x) + f'(x)h + E_f(h) \quad \text{and} \quad g(x+h) = g(x) + g'(x)h + E_g(h),$$

where $\lim_{h \to 0} \frac{E_f(h)}{h} = \lim_{h \to 0} \frac{E_g(h)}{h} = 0$. (This implies that $\lim_{h \to 0} E_f(h) = \lim_{h \to 0} E_g(h) = 0$.)

We have

$$\frac{f(x+h)g(x+h) - f(x)g(x)}{h} = \frac{f(x)g(x)}{h} + f(x)g'(x) + f'(x)g(x) + f(x)\frac{E_g(h)}{h} + g(x)\frac{E_f(h)}{h}$$
$$+ f'(x)g'(x)h + f'(x)E_g(h) + g'(x)E_f(h) + \frac{E_f(h)E_g(h)}{h} - \frac{f(x)g(x)}{h}$$

The terms $f(x)g(x)/h$ and $-f(x)g(x)/h$ cancel out. All the remaining terms on the right, with the exception of the second and third terms, go to zero as $h \to 0$. Thus, we have

$$[f(x)g(x)]' = \lim_{h \to 0} \frac{f(x+h)g(x+h) - f(x)g(x)}{h} = f(x)g'(x) + f'(x)g(x).$$

24. Note that

$$[f(g(x))]' = \lim_{h \to 0} \frac{f(g(x+h)) - f(g(x))}{h}.$$

Using the local linearizations of f and g, we get that

$$f(g(x+h)) - f(g(x)) \approx f\big(g(x) + g'(x)h\big) - f(g(x))$$
$$\approx f\big(g(x)\big) + f'(g(x))g'(x)h - f(g(x))$$
$$= f'(g(x))g'(x)h.$$

Therefore,

$$[f(g(x))]' = \lim_{h \to 0} \frac{f(g(x+h)) - f(g(x))}{h}$$
$$= \lim_{h \to 0} \frac{f'(g(x))g'(x)h}{h}$$
$$= \lim_{h \to 0} f'(g(x))g'(x) = f'(g(x))g'(x).$$

A more complete derivation can be given using the error term discussed in the section on Differentiability and Linear Approximation in Chapter 2. Adapting the notation of that section to this problem, we write

$$f(z+k) = f(z) + f'(z)k + E_f(k) \quad \text{and} \quad g(x+h) = g(x) + g'(x)h + E_g(h),$$

where $\lim_{h \to 0} \frac{E_g(h)}{h} = \lim_{k \to 0} \frac{E_f(k)}{k} = 0$.

Now we let $z = g(x)$ and $k = g(x+h) - g(x)$. Then we have $k = g'(x)h + E_g(h)$. Thus,

$$\frac{f(g(x+h)) - f(g(x))}{h} = \frac{f(z+k) - f(z)}{h}$$
$$= \frac{f(z) + f'(z)k + E_f(k) - f(z)}{h} = \frac{f'(z)k + E_f(k)}{h}$$
$$= \frac{f'(z)g'(x)h + f'(z)E_g(h)}{h} + \frac{E_f(k)}{k} \cdot \left(\frac{k}{h}\right)$$
$$= f'(z)g'(x) + \frac{f'(z)E_g(h)}{h} + \frac{E_f(k)}{k}\left[\frac{g'(x)h + E_g(h)}{h}\right]$$
$$= f'(z)g'(x) + \frac{f'(z)E_g(h)}{h} + \frac{g'(x)E_f(k)}{k} + \frac{E_g(h) \cdot E_f(k)}{h \cdot k}$$

Now, if $h \to 0$ then $k \to 0$ as well, and all the terms on the right except the first go to zero, leaving us with the term $f'(z)g'(x)$. Substituting $g(x)$ for z, we obtain

$$[f(g(x))]' = \lim_{h \to 0} \frac{f(g(x+h)) - f(g(x))}{h} = f'(g(x))g'(x).$$

Solutions for Chapter 3 Review

1. We wish to find the slope $m = dy/dx$. To do this, we can implicitly differentiate the given formula in terms of x:

$$x^2 + 3y^2 = 7$$
$$2x + 6y\frac{dy}{dx} = \frac{d}{dx}(7) = 0$$
$$\frac{dy}{dx} = \frac{-2x}{6y} = \frac{-x}{3y}.$$

Thus, at $(2, -1)$, $m = -(2)/3(-1) = 2/3$.

2. Taking derivatives implicitly, we find

$$\frac{dy}{dx} + \cos y \frac{dy}{dx} + 2x = 0$$
$$\frac{dy}{dx} = \frac{-2x}{1 + \cos y}$$

So, at the point $x = 3, y = 0$,

$$\frac{dy}{dx} = \frac{(-2)(3)}{1 + \cos 0} = \frac{-6}{2} = -3.$$

3. Taking the values of f, f', g, and g' from the table we get:
   (a) $h(4) = f(g(4)) = f(3) = 1.$
   (b) $h'(4) = f'(g(4))g'(4) = f'(3) \cdot 1 = 2.$
   (c) $h(4) = g(f(4)) = g(4) = 3.$
   (d) $h'(4) = g'(f(4))f'(4) = g'(4) \cdot 3 = 3.$
   (e) $h'(4) = \big(f(4)g'(4) - g(4)f'(4)\big)/f^2(4) = -5/16.$
   (f) $h'(4) = f(4)g'(4) + g(4)f'(4) = 13.$

4. (a) $H'(2) = r'(2)s(2) + r(2)s'(2) = -1 \cdot 1 + 4 \cdot 3 = 11.$
   (b) $H'(2) = \dfrac{r'(2)}{2\sqrt{r(2)}} = \dfrac{-1}{2\sqrt{4}} = -\dfrac{1}{4}.$
   (c) $H'(2) = r'(s(2))s'(2) = r'(1) \cdot 3$, but we don't know $r'(1)$.
   (d) $H'(2) = s'(r(2))r'(2) = s'(4)r'(2) = -3.$

5. When we zoom in on the origin, we find that two functions are not defined there. The other functions all look like straight lines through the origin. The only way we can tell them apart is their slope.

    The following functions all have slope 0 and are therefore indistinguishable:
$\sin x - \tan x$, $\frac{x^2}{x^2+1}$, $x - \sin x$, and $\frac{1-\cos x}{\cos x}$.

    These functions all have slope 1 at the origin, and are thus indistinguishable:
$\arcsin x$, $\frac{\sin x}{1+\sin x}$, $\arctan x$, $e^x - 1$, $\frac{x}{x+1}$, and $\frac{x}{x^2+1}$.

    Now, $\frac{\sin x}{x} - 1$ and $-x \ln x$ both are undefined at the origin, so they are distinguishable from the other functions. In addition, while $\frac{\sin x}{x} - 1$ has a slope that approaches zero near the origin, $-x \ln x$ becomes vertical near the origin, so they are distinguishable from each other.

    Finally, $x^{10} + \sqrt[10]{x}$ is the only function defined at the origin and with a vertical tangent there, so it is distinguishable from the others.

6. It makes sense to define the angle between two curves to be the angle between their tangent lines. (The tangent lines are the best linear approximations to the curves). See Figure 3.6. The functions $\sin x$ and $\cos x$ are equal at $x = \frac{\pi}{4}$.

$$\text{For } f_1(x) = \sin x, \quad f_1'\left(\frac{\pi}{4}\right) = \cos\left(\frac{\pi}{4}\right) = \frac{\sqrt{2}}{2}$$
$$\text{For } f_2(x) = \cos x, \quad f_2'\left(\frac{\pi}{4}\right) = -\sin\left(\frac{\pi}{4}\right) = -\frac{\sqrt{2}}{2}.$$

Using the point $(\frac{\pi}{4}, \frac{\sqrt{2}}{2})$ for each tangent line we get $y = \frac{\sqrt{2}}{2}x + \frac{\sqrt{2}}{2}(1 - \frac{\pi}{4})$ and $y = -\frac{\sqrt{2}}{2}x + \frac{\sqrt{2}}{2}(1 + \frac{\pi}{4})$, respectively.

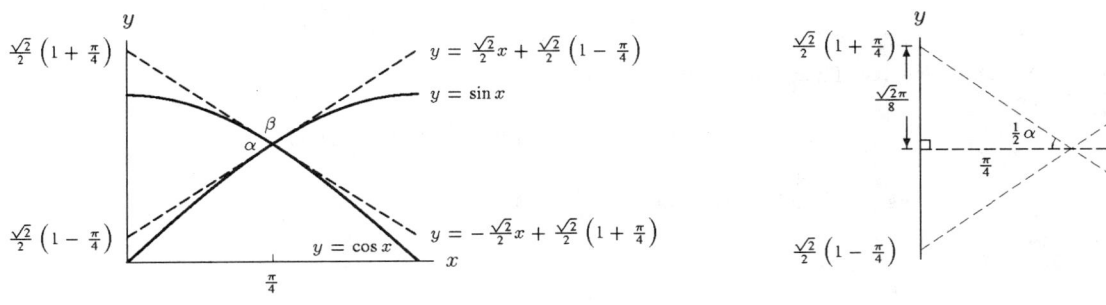

Figure 3.6 **Figure 3.7**

There are two possibilities of how to define the angle between the tangent lines, indicated by α and β above. The choice is arbitrary, so we will solve for both. To find the angle, α, we consider the triangle formed by these two lines and the y-axis. See Figure 3.7.

$$\tan\left(\frac{1}{2}\alpha\right) = \frac{\sqrt{2}\pi/8}{\pi/4} = \frac{\sqrt{2}}{2}$$

$$\frac{1}{2}\alpha = 0.61548 \text{ radians}$$

$$\alpha = 1.231 \text{ radians, or } 70.5°.$$

Now let us solve for β, the other possible measure of the angle between the two tangent lines. Since α and β are supplementary, $\beta = \pi - 1.231 = 1.909$ radians, or $109.4°$.

7. (a) $\dfrac{dg}{dr} = GM\dfrac{d}{dr}\left(\dfrac{1}{r^2}\right) = GM\dfrac{d}{dr}(r^{-2}) = GM(-2)r^{-3} = -\dfrac{2GM}{r^3}$.

   (b) $\dfrac{dg}{dr}$ is the rate of change of acceleration due to the pull of gravity. The further away from the center of the earth, the weaker the pull of gravity is. So g is decreasing and therefore its derivative, $\dfrac{dg}{dr}$, is negative.

   (c) By part (a),

   $$\left.\frac{dg}{dr}\right|_{r=6400} = -\left.\frac{2GM}{r^3}\right|_{r=6400}$$
   $$= -\frac{2(6.67 \times 10^{-20})(6 \times 10^{24})}{(6400)^3}$$
   $$\approx -3.05 \times 10^{-6}.$$

   (d) It is reasonable to assume that g is a constant near the surface of the earth.

8. The population of Mexico is given by the formula
$$M = 84(1 + 0.026)^t = 84(1.026)^t \text{ million}$$

and that of the US by
$$U = 250(1 + 0.007)^t = 250(1.007)^t \text{ million},$$

where t is measured in years ($t = 0$ corresponds to the year 1990). So,

$$\left.\frac{dM}{dt}\right|_{t=0} = 84\frac{d}{dt}(1.026)^t\bigg|_{t=0} = 84(1.026)^t \ln(1.026)\bigg|_{t=0} \approx 2.156$$

and

$$\left.\frac{dU}{dt}\right|_{t=0} = 250\frac{d}{dt}(1.007)^t\bigg|_{t=0} = 250(1.007)^t \ln(1.007)\bigg|_{t=0} \approx 1.744$$

Since $\left.\dfrac{dM}{dt}\right|_{t=0} > \left.\dfrac{dU}{dt}\right|_{t=0}$, the population of Mexico was growing faster in 1990.

9. (a) If the distance $s(t) = 20e^{\frac{t}{2}}$, then the velocity, $v(t)$, is given by

$$v(t) = s'(t) = \left(20e^{\frac{t}{2}}\right)' = \left(\frac{1}{2}\right)\left(20e^{\frac{t}{2}}\right) = 10e^{\frac{t}{2}}.$$

(b) Observing the differentiation in (a), we note that

$$s'(t) = v(t) = \frac{1}{2}\left(20e^{\frac{t}{2}}\right) = \frac{1}{2}s(t).$$

Substituting $s(t)$ for $20e^{\frac{t}{2}}$, we obtain $s'(t) = \frac{1}{2}s(t)$.

10. (a)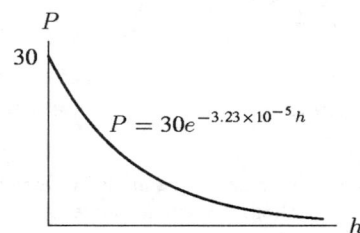

(b)
$$\frac{dP}{dh} = 30e^{-3.23 \times 10^{-5} h}(-3.23 \times 10^{-5})$$

so

$$\left.\frac{dP}{dh}\right|_{h=0} = -30(3.23 \times 10^{-5}) = -9.69 \times 10^{-4}$$

Hence, at $h = 0$, the slope of the tangent line is -9.69×10^{-4}, so the equation of the tangent line is

$$y - 30 = (-9.69 \times 10^{-4})(h - 0)$$
$$y = (-9.69 \times 10^{-4})h + 30.$$

(c) The rule of thumb says

$$\left(\begin{array}{c}\text{Drop in pressure from}\\ \text{sea level to height } h\end{array}\right) = \frac{h}{1000}$$

But since the pressure at sea level is 30 inches of mercury, this drop in pressure is also $(30 - P)$, so

$$30 - P = \frac{h}{1000}$$

giving

$$P = 30 - 0.001h.$$

(d) The equations in (b) and (c) are almost the same: both have P intercepts of 30, and the slopes are almost the same $(9.69 \times 10^{-4} \approx 0.001)$. The rule of thumb calculates values of P which are very close to the tangent lines, and therefore yields values very close to the curve.

(e) The tangent line is slightly below the curve, and the rule of thumb line, having a slightly more negative slope, is slightly below the tangent line (for $h > 0$). Thus, the rule of thumb values are slightly smaller.

11.
$$\frac{dy}{dt} = -7.5(0.507)\sin(0.507t) = -3.80\sin(0.507t)$$

(a) When $t = 6$, $\frac{dy}{dt} = -3.80\sin(0.507 \cdot 6) = -0.38$ meters/hour. So it's falling at 0.38 meters/hour.
(b) When $t = 9$, $\frac{dy}{dt} = -3.80\sin(0.507 \cdot 9) = 3.76$ meters/hour. So it's rising at 3.76 meters/hour.
(c) When $t = 12$, $\frac{dy}{dt} = -3.80\sin(0.507 \cdot 12) = 0.75$ meters/hour. So it's rising at 0.75 meters/hour.
(d) When $t = 18$, $\frac{dy}{dt} = -3.80\sin(0.507 \cdot 18) = -1.12$ meters/hour. So it's falling at 1.12 meters/hour.

12. Since we're given that the instantaneous rate of change of T at $t = 30$ is 2, we want to choose a and b so that the derivative of T agrees with this value. Differentiating, $T'(t) = ab \cdot e^{-bt}$. Then we have

$$2 = T'(30) = abe^{-30b} \text{ or } e^{-30b} = \frac{2}{ab}$$

We also know that at $t = 30$, $T = 120$, so

$$120 = T(30) = 200 - ae^{-30b} \text{ or } e^{-30b} = \frac{80}{a}$$

Thus $\frac{80}{a} = e^{-30b} = \frac{2}{ab}$, so $b = \frac{1}{40} = 0.025$ and $a = 169.36$.

13. (a) Differentiating, we see

$$v = \frac{dy}{dt} = -2\pi\omega y_0 \sin(2\pi\omega t)$$

$$a = \frac{dv}{dt} = -4\pi^2\omega^2 y_0 \cos(2\pi\omega t).$$

(b) We have

$$y = y_0 \cos(2\pi\omega t)$$
$$v = -2\pi\omega y_0 \sin(2\pi\omega t)$$
$$a = -4\pi^2\omega^2 y_0 \cos(2\pi\omega t).$$

So

Amplitude of y is $|y_0|$,

Amplitude of v is $|2\pi\omega y_0| = 2\pi\omega|y_0|$,

Amplitude of a is $|4\pi^2\omega^2 y_0| = 4\pi^2\omega^2|y_0|$.

The amplitudes are different (provided $2\pi\omega \neq 1$). The periods of the three functions are all the same, namely $1/\omega$.

(c) Looking at the answer to part (a), we see

$$\frac{d^2 y}{dt^2} = a = -4\pi^2\omega^2 \left(y_0 \cos(2\pi\omega t)\right)$$
$$= -4\pi^2\omega^2 y.$$

So we see that

$$\frac{d^2 y}{dt^2} + 4\pi^2\omega^2 y = 0.$$

14. (a) Since $\lim_{t \to \infty} e^{-0.1t} = 0$, we see that $\lim_{t \to \infty} \frac{1000000}{1 + 5000e^{-0.1t}} = 1000000$. Thus, in the long run, close to 1,000,000 people will have had the disease. This can be seen in the figure below.

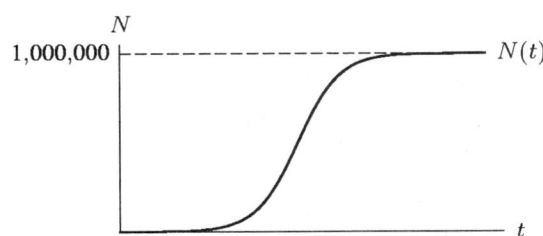

(b) The rate at which people fall sick is given by the first derivative $N'(t)$.
$N'(t) \approx \frac{\Delta N}{\Delta t}$, where $\Delta t = 1$ day.

$$N'(t) = \frac{500{,}000{,}000}{e^{0.1t}(1 + 5000e^{-0.1t})^2} = \frac{500{,}000{,}000}{e^{0.1t} + 25{,}000{,}000e^{-0.1t} + 10^4}$$

Graphing this we see that the maximum value of $N'(t)$ is approximately 25,000. Therefore the maximum number of people to fall sick on any given day is 25,000.

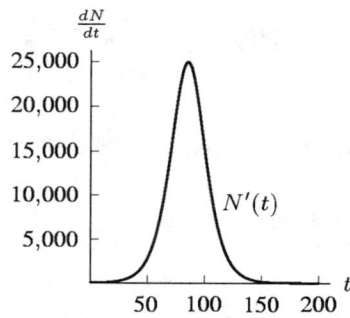

15. We are given that the volume is increasing at a constant rate $\frac{dV}{dt} = 400$. The radius r is related to the volume by the formula $V = \frac{4}{3}\pi r^3$. By implicit differentiation, we have

$$\frac{dV}{dt} = \frac{4}{3}\pi 3r^2 \frac{dr}{dt} = 4\pi r^2 \frac{dr}{dt}$$

Plugging in $\frac{dV}{dt} = 400$ and $r = 10$, we have

$$400 = 400\pi \frac{dr}{dt}$$

so $\frac{dr}{dt} = \frac{1}{\pi} \approx 0.32 \mu$m/day.

16. The radius r is related to the volume by the formula $V = \frac{4}{3}\pi r^3$. By implicit differentiation, we have

$$\frac{dV}{dt} = \frac{4}{3}\pi 3r^2 \frac{dr}{dt} = 4\pi r^2 \frac{dr}{dt}.$$

The surface area of a sphere is $4\pi r^2$, so we have

$$\frac{dV}{dt} = s \cdot \frac{dr}{dt},$$

but since $\frac{dV}{dt} = \frac{1}{3}s$ was given, we have

$$\frac{dr}{dt} = \frac{1}{3}.$$

17. Using Pythagoras' theorem, we see that the distance x between the aircraft's current position and the point 2 miles directly above the ground station are related to s by the formula $x = (s^2 - 2^2)^{1/2}$. See Figure 3.8. The speed along the aircraft's constant altitude flight path is

$$\frac{dx}{dt} = \left(\frac{1}{2}\right)(s^2 - 4)^{-1/2}(2s)\left(\frac{ds}{dt}\right) = \frac{s}{x}\frac{ds}{dt}.$$

When $s = 4.6$ and $ds/dt = 210$,

$$\frac{dx}{dt} = \frac{4.6}{\sqrt{(4.6)^2 - 4}} 210$$

$$= \frac{966}{\sqrt{21.16 - 4}}$$

$$= \frac{966}{4.14} \approx 233.2 \text{ miles/hour}.$$

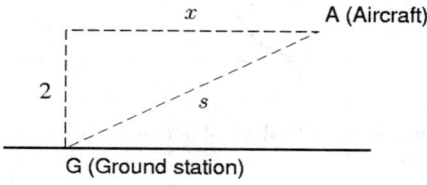

Figure 3.8

18. We want to find dP/dV. Solving $PV = k$ for P gives

$$P = k/V$$

so,

$$\frac{dP}{dV} = -\frac{k}{V^2}.$$

19. (a) Since $V = k/P$, the volume decreases.
  (b) Since $PV = k$ and $P = 2$ when $V = 10$, we have $k = 20$, so

$$V = \frac{20}{P}.$$

We think of both P and V as functions of time, so by the chain rule

$$\frac{dV}{dt} = \frac{dV}{dP}\frac{dP}{dt},$$

$$\frac{dV}{dt} = -\frac{20}{P^2}\frac{dP}{dt}.$$

We know that $dP/dt = 0.05$ atm/min when $P = 2$ atm, so

$$\frac{dV}{dt} = -\frac{20}{2^2} \cdot (0.05) = -0.25 \text{ cm}^3/\text{min}.$$

20. (a) If $y = \ln x$, then

$$y' = \frac{1}{x}$$

$$y'' = -\frac{1}{x^2}$$

$$y''' = \frac{2}{x^3}$$

$$y'''' = -\frac{3 \cdot 2}{x^4}$$

and so

$$y^{(n)} = (-1)^{n+1}(n-1)!x^{-n}.$$

  (b) If $y = xe^x$, then

$$y' = xe^x + e^x$$

$$y'' = xe^x + 2e^x$$

$$y''' = xe^x + 3e^x$$

so that

$$y^{(n)} = xe^x + ne^x.$$

  (c) If $y = e^x \cos x$, then

$$y' = e^x(\cos x - \sin x)$$

$$y'' = -2e^x \sin x$$

$$y''' = e^x(-2\cos x - 2\sin x)$$

$$y^{(4)} = -4e^x \cos x$$

$$y^{(5)} = e^x(-4\cos x + 4\sin x)$$

$$y^{(6)} = 8e^x \sin x.$$

Combining these results we get

$$y^{(n)} = (-4)^{(n-1)/4}e^x(\cos x - \sin x), \quad n = 4m+1, \quad m = 0, 1, 2, 3, \ldots$$

$$y^{(n)} = -2(-4)^{(n-2)/4}e^x \sin x, \quad n = 4m+2, \quad m = 0, 1, 2, 3, \ldots$$

$$y^{(n)} = -2(-4)^{(n-3)/4}e^x(\cos x + \sin x), \quad n = 4m+3, \quad m = 0, 1, 2, 3, \ldots$$

$$y^{(n)} = (-4)^{(n/4)}e^x \cos x, \quad n = 4m, \quad m = 1, 2, 3, \ldots.$$

Solutions to the Projects and Computer Algebra Investigations

1. Let $r = i/100$. (For example if $i = 5\%$, $r = 0.05$.) Then the balance, $\$B$, after t years is given by
$$B = P(1+r)^t,$$
where $\$P$ is the original deposit. If we are doubling our money, then $B = 2P$, so we wish to solve for t in the equation $2P = P(1+r)^t$. This is equivalent to
$$2 = (1+r)^t.$$
Taking natural logarithms of both sides and solving for t yields
$$\ln 2 = t \ln(1+r),$$
$$t = \frac{\ln 2}{\ln(1+r)}.$$
We now approximate $\ln(1+r)$ near $r = 0$. Let $f(r) = \ln(1+r)$. Then $f'(r) = 1/(1+r)$. Thus, $f(0) = 0$ and $f'(0) = 1$, so
$$f(r) \approx f(0) + f'(0)r$$
becomes
$$\ln(1+r) \approx r.$$
Therefore,
$$t = \frac{\ln 2}{\ln(1+r)} \approx \frac{\ln 2}{r} = \frac{100 \ln 2}{i} \approx \frac{70}{i},$$
as claimed. We expect this approximation to hold for small values of i; it turns out that values of i up to 10 give good enough answers for most everyday purposes.

2. (a) (i) Set $f(x) = \sin x$, so $f'(x) = \cos x$. Guess $x_0 = 3$. Then
$$x_1 = 3 - \frac{\sin 3}{\cos 3} \approx 3.1425$$
$$x_2 = x_1 - \frac{\sin x_1}{\cos x_1} \approx 3.1415926533,$$
which is correct to one billionth!

   (ii) Newton's method uses the tangent line at $x = 3$, i.e. $y - \sin 3 = \cos(3)(x-3)$. Around $x = 3$, however, $\sin x$ is almost linear, since the second derivative $\sin''(\pi) = 0$. Thus using the tangent line to get an approximate value for the root gives us a very good approximation.

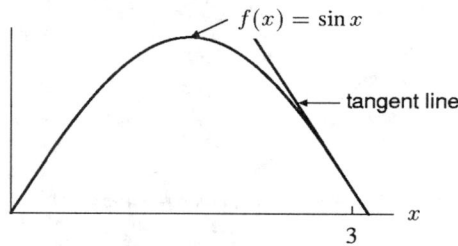

   (iii) For $f(x) = \sin x$, we have
$$f(3) = 0.14112$$
$$f(4) = -0.7568,$$
so there is a root in $[3, 4]$. We now continue bisecting:
$$[3, 3.5] : f(3.5) = -0.35078 \text{ (bisection 1)}$$
$$[3, 3.25] : f(3.25) = -0.10819 \text{ (bisection 2)}$$
$$[3.125, 3.25] : f(3.125) = 0.01659 \text{ (bisection 3)}$$
$$[3.125, 3.1875] : f(3.1875) = -0.04584 \text{ (bisection 4)}$$

   We continue this process; after 11 bisections, we know the root lies between 3.1411 and 3.1416, which still is not as good an approximation as what we get from Newton's method in just two steps.

(b) (i) We have $f(x) = \sin x - \frac{2}{3}x$ and $f'(x) = \cos x - \frac{2}{3}$.

Using $x_0 = 0.904$,

$$x_1 = 0.904 - \frac{\sin(0.904) - \frac{2}{3}(0.904)}{\cos(0.904) - \frac{2}{3}} \approx 4.704,$$

$$x_2 = 4.704 - \frac{\sin(4.704) - \frac{2}{3}(4.704)}{\cos(4.704) - \frac{2}{3}} \approx -1.423,$$

$$x_3 = -1.433 - \frac{\sin(-1.423) - \frac{2}{3}(-1.423)}{\cos(-1.423) - \frac{2}{3}} \approx -1.501,$$

$$x_4 = -1.499 - \frac{\sin(-1.501) - \frac{2}{3}(-1.501)}{\cos(-1.501) - \frac{2}{3}} \approx -1.496,$$

$$x_5 = -1.496 - \frac{\sin(-1.496) - \frac{2}{3}(-1.496)}{\cos(-1.496) - \frac{2}{3}} \approx -1.496.$$

Using $x_0 = 0.905$,

$$x_1 = 0.905 - \frac{\sin(0.905) - \frac{2}{3}(0.905)}{\cos(0.905) - \frac{2}{3}} \approx 4.643,$$

$$x_2 = 4.643 - \frac{\sin(4.643) - \frac{2}{3}(4.643)}{\cos(4.643) - \frac{2}{3}} \approx -0.918,$$

$$x_3 = -0.918 - \frac{\sin(-0.918) - \frac{2}{3}(-0.918)}{\cos(-0.918) - \frac{2}{3}} \approx -3.996,$$

$$x_4 = -3.996 - \frac{\sin(-3.996) - \frac{2}{3}(-3.996)}{\cos(-3.996) - \frac{2}{3}} \approx -1.413,$$

$$x_5 = -1.413 - \frac{\sin(-1.413) - \frac{2}{3}(-1.413)}{\cos(-1.413) - \frac{2}{3}} \approx -1.502,$$

$$x_6 = -1.502 - \frac{\sin(-1.502) - \frac{2}{3}(-1.502)}{\cos(-1.502) - \frac{2}{3}} \approx -1.496.$$

Now using $x_0 = 0.906$,

$$x_1 = 0.906 - \frac{\sin(0.906) - \frac{2}{3}(0.906)}{\cos(0.906) - \frac{2}{3}} \approx 4.584,$$

$$x_2 = 4.584 - \frac{\sin(4.584) - \frac{2}{3}(4.584)}{\cos(4.584) - \frac{2}{3}} \approx -0.509,$$

$$x_3 = -0.510 - \frac{\sin(-0.509) - \frac{2}{3}(-0.509)}{\cos(-0.509) - \frac{2}{3}} \approx .207,$$

$$x_4 = -1.300 - \frac{\sin(.207) - \frac{2}{3}(.207)}{\cos(.207) - \frac{2}{3}} \approx -0.009,$$

$$x_5 = -1.543 - \frac{\sin(-0.009) - \frac{2}{3}(-0.009)}{\cos(-0.009) - \frac{2}{3}} \approx 0,$$

(ii) Starting with 0.904 and 0.905 yields the same value, but the two paths to get to the root are very different. Starting with 0.906 leads to a different root. Our starting points were near the maximum value of f. Consequently, a small change in x_0 makes a large change in x_1.

3. (a) A CAS gives $f'(x) = 1$.
   (b) By the chain rule,
   $$f'(x) = \cos(\arcsin x) \cdot \frac{1}{\sqrt{1-x^2}}.$$
   Now $\cos t = \pm\sqrt{1 - \sin^2 t}$. Furthermore, if $-\pi/2 \leq t \leq \pi/2$ then $\cos t > 0$, so we take the positive square root and get $\cos t = \sqrt{1 - \sin^2 t}$. Since $-\pi/2 \leq \arcsin x \leq \pi/2$ for all x in the domain of arcsin, we have
   $$\cos(\arcsin x) = \sqrt{1 - (\sin(\arcsin x))^2} = \sqrt{1 - x^2},$$
   so
   $$\frac{d}{dx}\sin(\arcsin(x)) = \sqrt{1-x^2} \cdot \frac{1}{\sqrt{1-x^2}} = 1.$$
   (c) Since $\sin(\arcsin(x)) = x$, its derivative is 1.

4. (a) A CAS gives $g'(r) = 0$.
   (b) Using the product rule,
   $$g'(r) = \frac{d}{dr}(2^{-2r})4^r + 2^{-2r}\frac{d}{dr}(4^r) = -2\ln 2 \cdot 2^{-2r}4^r + 2^{-2r}\ln 4 \cdot 4^r$$
   $$= -\ln 4 \cdot 2^{-2r}4^r + \ln 4 \cdot 2^{-2r}4^r = (-\ln 4 + \ln 4)2^{-2r}4^r = 0 \cdot 2^{-2r}4^r = 0.$$
   (c) By the laws of exponents, $4^r = (2^2)^r = 2^{2r}$, so $2^{-2r}4^r = 2^{-2r}2^{2r} = 2^0 = 1$. Therefore, its derivative is zero.

5. (a) A CAS gives $h'(t) = 0$
   (b) By the chain rule
   $$h'(t) = \frac{\frac{d}{dt}\left(1 - \frac{1}{t}\right)}{1 - \frac{1}{t}} + \frac{\frac{d}{dt}\left(\frac{t}{t-1}\right)}{\frac{t}{t-1}} = \frac{\frac{1}{t^2}}{\frac{t-1}{t}} + \frac{\frac{1}{t-1} - \frac{t}{(t-1)^2}}{\frac{t}{t-1}}$$
   $$= \frac{1}{t^2 - t} + \frac{(t-1) - t}{t^2 - t} = \frac{1}{t^2 - t} + \frac{-1}{t^2 - t} = 0.$$
   (c) The expression inside the first logarithm is $1 - (1/t) = (t-1)/t$. Using the property $\log A + \log B = \log(AB)$, we get
   $$\ln\left(1 - \frac{1}{t}\right) + \ln\left(\frac{t}{t-1}\right) = \ln\left(\frac{t-1}{t}\right) + \ln\left(\frac{t}{t-1}\right)$$
   $$= \ln\left(\frac{t-1}{t} \cdot \frac{t-1}{t}\right) = \ln 1 = 0.$$
   Thus $h(t) = 0$, so $h'(t) = 0$ also.

6. (a) Answers from different computer algebra systems may be in different forms. One form is:
   $$\frac{d}{dx}(x+1)^x = x(x+1)^{x-1} + (x+1)^x \ln(x+1)$$
   $$\frac{d}{dx}(\sin x)^x = x\cos x(\sin x)^{x-1} + (\sin x)^x \ln(\sin x)$$
   (b) Both the answers in part (a) follow the general rule:
   $$\frac{d}{dx}f(x)^x = xf'(x)\left(f(x)\right)^{x-1} + \left(f(x)\right)^x \ln(f(x)).$$
   (c) Applying this rule to $g(x)$, we get
   $$\frac{d}{dx}(\ln x)^x = x(1/x)(\ln x)^{x-1} + (\ln x)^x \ln(\ln x) = (\ln x)^{x-1} + (\ln x)^x \ln(\ln x).$$
   This agrees with the answer given by the computer algebra system.
   (d) We can write $f(x) = e^{\ln(f(x))}$. So
   $$\left(f(x)\right)^x = (e^{\ln(f(x))})^x = e^{x\ln(f(x))}.$$

Therefore, using the chain rule and the product rule,

$$\frac{d}{dx}\left(f(x)\right)^x = \frac{d}{dx}\left(x\ln(f(x))\right)\cdot e^{x\ln(f(x))} = \left(\ln(f(x)) + x\frac{d}{dx}\ln(f(x))\right)e^{x\ln(f(x))}$$

$$= \left(\ln(f(x)) + x\frac{f'(x)}{f(x)}\right)\left(f(x)\right)^x = \ln(f(x))\left(f(x)\right)^x + xf'(x)\left(f(x)\right)^{x-1}$$

$$= xf'(x)\left(f(x)\right)^{x-1} + \left(f(x)\right)^x \ln(f(x)).$$

Solutions to Practice Problems on Differentiation

1. $f'(t) = 6t - 4$.
2. $y' = 17 + 12x^{-1/2}$.
3. $g'(x) = -\frac{1}{2}(5x^4 + 2)$.
4. Dividing gives $g(t) = t^2 + k/t$ so $g'(t) = 2t - \frac{k}{t^2}$.
5. The power rule gives $f'(x) = 20x^3 - \frac{2}{x^3}$.
6. Using the quotient rule gives $\dfrac{dz}{dt} = \dfrac{(2t+3)(t+1) - (t^2+3t+1)}{(t+1)^2}$.

   or

   $\dfrac{dz}{dt} = \dfrac{t^2 + 2t + 2}{(t+1)^2}$.

7. $\dfrac{dy}{dx} = \dfrac{2e^{2x}(x^2+1) - e^{2x}(2x)}{(x^2+1)^2} = \dfrac{2e^{2x}(x^2+1-x)}{(x^2+1)^2}$

8. Either notice that $f(x) = \dfrac{x^2 + 3x + 2}{x+1}$ can be written as $f(x) = \dfrac{(x+2)(x+1)}{x+1}$ which reduces to $f(x) = x + 2$, giving $f'(x) = 1$, or use the quotient rule which gives

$$f'(x) = \frac{(x+1)(2x+3) - (x^2+3x+2)}{(x+1)^2}$$

$$= \frac{2x^2 + 5x + 3 - x^2 - 3x - 2}{(x+1)^2}$$

$$= \frac{x^2 + 2x + 1}{(x+1)^2}$$

$$= \frac{(x+1)^2}{(x+1)^2}$$

$$= 1.$$

9. $y' = 2\left(\dfrac{x^2+2}{3}\right)\left(\dfrac{2x}{3}\right) = \dfrac{4}{9}x\left(x^2+2\right)$

10. $g'(\theta) = 2\sin(2\theta)\cos(2\theta)\cdot 2 - \pi = 4\sin(2\theta)\cos(2\theta) - \pi$

11. $\dfrac{d}{dx}\sin(2-3x) = \cos(2-3x)\dfrac{d}{dx}(2-3x) = -3\cos(2-3x)$.

12. Using the chain rule gives $R'(x) = 3\pi\sin(\pi x)$.

13. $f(z) = \dfrac{z}{3} + \dfrac{1}{3}z^{-1} = \dfrac{1}{3}\left(z + z^{-1}\right)$, so $f'(z) = \dfrac{1}{3}\left(1 - z^{-2}\right) = \dfrac{1}{3}\left(\dfrac{z^2-1}{z^2}\right)$.

14. $q'(r) = \dfrac{3(5r+2) - 3r(5)}{(5r+2)^2} = \dfrac{15r + 6 - 15r}{(5r+2)^2} = \dfrac{6}{(5r+2)^2}$

15.
$$h'(z) = \frac{1}{2}\left(\frac{\sin(2z)}{\cos(2z)}\right)^{-1/2}\left[\frac{2\cos(2z)\cos(2z) - \sin(2z)(-2\sin(2z))}{\cos^2(2z)}\right]$$
$$= \left(\frac{\cos(2z)}{\sin(2z)}\right)^{1/2}\left[\frac{\cos^2(2z) + \sin^2(2z)}{\cos^2(2z)}\right]$$
$$= \frac{(\cos(2z))^{1/2}}{(\sin(2z))^{1/2}\cos^2(2z)} = \frac{1}{\sqrt{\sin(2z)}\sqrt{\cos^3(2z)}}$$

16. $\dfrac{dy}{dx} = \ln x + x\left(\dfrac{1}{x}\right) - 1 = \ln x$

17. $j'(x) = \dfrac{ae^{ax}}{(e^{ax} + b)}$

18. $\dfrac{dy}{dx} = 2(\ln x + \ln 2) + 2x\left(\dfrac{1}{x}\right) - 2 = 2(\ln x + \ln 2) = 2\ln(2x)$

19. $g'(\theta) = \dfrac{\cos(\tan \theta)}{\cos^2 \theta}$

20. $w'(x) = \dfrac{2x}{\cos^2(x^2)}$

21. $f'(x) = \cos(\cos x + \sin x)(\cos x - \sin x)$

22. $j'(x) = -\sin\left(\sin^{-1} x\right) \cdot \left[\dfrac{1}{\sqrt{1-x^2}}\right] = -\dfrac{x}{\sqrt{1-x^2}}$

23. $k'(\alpha) = (5\sin^4 \alpha \cos \alpha)\cos^3 \alpha + \sin^5 \alpha(3\cos^2 \alpha(-\sin \alpha)) = 5\sin^4 \alpha \cos^4 \alpha - 3\sin^6 \alpha \cos^2 \alpha$

24. $f'(w) = -2\cos w \sin w - \sin(w^2)(2w) = -2(\cos w \sin w + w\sin(w^2))$

25. $g'(t) = -4(3 + \sqrt{t})^{-2}\left(\dfrac{1}{2}t^{-1/2}\right) = \dfrac{-2}{\sqrt{t}(3 + \sqrt{t})^2}$

26. $g'(t) = \dfrac{(t+4) - (t-4)}{(t+4)^2} = \dfrac{8}{(t+4)^2}$.

27. $y' = \dfrac{-(3e^{3x} + 2x)}{(e^{3x} + x^2)^2}$.

28. $h'(w) = 5(w^4 - 2w)^4(4w^3 - 2)$

29. $q'(\theta) = \dfrac{1}{2}(4\theta^2 - \sin^2(2\theta))^{-1/2}(8\theta - 2\sin(2\theta)(2\cos(2\theta))) = \dfrac{4\theta - 2\sin(2\theta)\cos(2\theta)}{\sqrt{4\theta^2 - \sin^2(2\theta)}}$

30. $g'(t) = 4(t\cos t + \tan^3(t^5))^3 \cdot \left(\cos t - t\sin t + 3\tan^2(t^5)\left[\dfrac{5t^4}{\cos^2(t^5)}\right]\right)$

31. Using the product and chain rules gives $h'(w) = 3w^2 \ln(10w) + w^3 \dfrac{10}{10w} = 3w^2 \ln(10w) + w^2$.

32. Using the chain rule gives $f'(x) = \dfrac{\cos x - \sin x}{\sin x + \cos x}$.

33. Note that $g(x) = \arcsin(\sin \pi x) = \pi x$.
    Thus, $g'(x) = \pi$.

34. Using the chain rule gives $r'(t) = \dfrac{2}{\sqrt{1 - 4t^2}}$.

35. We can write $w(r) = (r^4 + 1)^{1/2}$, so
$$w'(r) = \dfrac{1}{2}(r^4 + 1)^{-1/2}(4r^3) = \dfrac{2r^3}{\sqrt{r^4 + 1}}.$$

SOLUTIONS TO PRACTICE PROBLEMS ON DIFFERENTIATION   **125**

36. $h'(w) = 6w^{-4} + \dfrac{3}{2}w^{-1/2}$

37. We can write $h(x) = \left(\dfrac{x^2+9}{x+3}\right)^{1/2}$, so

$$h'(x) = \dfrac{1}{2}\left(\dfrac{x^2+9}{x+3}\right)^{-1/2}\left[\dfrac{2x(x+3) - (x^2+9)}{(x+3)^2}\right] = \dfrac{1}{2}\sqrt{\dfrac{x+3}{x^2+9}}\left[\dfrac{x^2+6x-9}{(x+3)^2}\right].$$

38. Using the power and quotient rules gives

$$f'(x) = \dfrac{1}{2}\left(\dfrac{1-\sin x}{1-\cos x}\right)^{-1/2}\left[\dfrac{-\cos x(1-\cos x) - (1-\sin x)\sin x}{(1-\cos x)^2}\right]$$

$$= \dfrac{1}{2}\sqrt{\dfrac{1-\cos x}{1-\sin x}}\left[\dfrac{-\cos x(1-\cos x) - (1-\sin x)\sin x}{(1-\cos x)^2}\right]$$

$$= \dfrac{1}{2}\sqrt{\dfrac{1-\cos x}{1-\sin x}}\left[\dfrac{1-\cos x - \sin x}{(1-\cos x)^2}\right].$$

39. Using the chain rule gives

$$T'(u) = \left[\dfrac{1}{1+\left(\dfrac{u}{1+u}\right)^2}\right]\left[\dfrac{(1+u)-u}{(1+u)^2}\right]$$

$$= \dfrac{(1+u)^2}{(1+u)^2+u^2}\left[\dfrac{1}{(1+u)^2}\right]$$

$$= \dfrac{1}{1+2u+2u^2}.$$

40. Using the product rule and factoring gives $\dfrac{dw}{dz} = 2^{-4z}\left[-4\ln(2)\sin(\pi z) + \pi\cos(\pi z)\right].$

41. Using the product rule gives $v'(t) = 2te^{-ct} - ce^{-ct}t^2 = (2t - ct^2)e^{-ct}.$

42. This is the sum of an exponential function and a power function, so $f'(x) = \ln(\pi)\pi^x + \pi x^{\pi-1}.$

43. Using the quotient rule gives

$$f'(x) = \dfrac{1+\ln x - x\left(\dfrac{1}{x}\right)}{(1+\ln x)^2}$$

$$= \dfrac{\ln x}{(1+\ln x)^2}.$$

44. The quotient rule gives $G'(x) = \dfrac{2\sin x\cos x(\cos^2 x + 1) + 2\sin x\cos x(\sin^2 x + 1)}{(\cos^2 x + 1)^2}$

or, using $\sin^2 x + \cos^2 x = 1$,
$G'(x) = \dfrac{6\sin x\cos x}{(\cos^2 x + 1)^2}.$

45. Since $\ln\left[\left(\dfrac{1-\cos t}{1+\cos t}\right)^4\right] = 4\ln\left[\left(\dfrac{1-\cos t}{1+\cos t}\right)\right]$ we have

$$a'(t) = 4\left(\dfrac{1+\cos t}{1-\cos t}\right)\left[\dfrac{\sin t(1+\cos t) + \sin t(1-\cos t)}{(1+\cos t)^2}\right]$$

$$= \left[\dfrac{1+\cos t}{1-\cos t}\right]\left[\dfrac{8\sin t}{(1+\cos t)^2}\right]$$

$$= \dfrac{8\sin t}{1-\cos^2 t}$$

$$= \dfrac{8}{\sin t}.$$

126 CHAPTER THREE /SOLUTIONS

46. Note that $f(x) = kx$ so, $f'(x) = k$.
47. Using the chain rule gives $R'(\theta) = 3\cos(3\theta)e^{\sin(3\theta)}$.
48. $f'(x) = (\ln \pi)\pi^x$.
49. $y' = (\ln \pi)\pi^{(x+2)}$.
50. $g(x) = \pi e^{\pi x}$.
51. Using the chain rule, $g'(\theta) = (\cos\theta)e^{\sin\theta}$.
52. $f(\theta) = (2^{-1})^\theta = (\frac{1}{2})^\theta$ so $f'(\theta) = (\ln\frac{1}{2})2^{-\theta}$.
53. $f'(x) = 2e^{2x}[x^2 + 5^x] + e^{2x}[2x + (\ln 5)5^x] = e^{2x}[2x^2 + 2x + (\ln 5 + 2)5^x]$.
54. $h'(x) = (\ln 2)(3e^{3x})2^{e^{3x}} = 3e^{3x}2^{e^{3x}}\ln 2$.
55. $h'(t) = \dfrac{(-1)(4+t) - (4-t)}{(4+t)^2} = -\dfrac{8}{(4+t)^2}$.
56. $\dfrac{d}{dy}\left(\dfrac{y}{\cos y + a}\right) = \dfrac{\cos y + a - y(-\sin y)}{(\cos y + a)^2} = \dfrac{\cos y + a + y\sin y}{(\cos y + a)^2}$.
57. $h'(z) = \dfrac{-8b^4 z}{(a+z^2)^5}$.
58. $p'(t) = 4e^{4t+2}$.
59. $h'(z) = (\ln(\ln 2))(\ln 2)^z$.
60. $j'(x) = \dfrac{3x^2}{a} + \dfrac{2ax}{b} - c$
61. $f'(x) = -\sin(\arctan 3x)\left(\dfrac{1}{1+(3x)^2}\right)(3) = \dfrac{-3\sin(\arctan 3x)}{1+9x^2}$.
62. $f'(x) = 6x(e^x - 4) + (3x^2 + \pi)e^x = 6xe^x - 24x + 3x^2 e^x + \pi e^x$.
63. $\dfrac{d}{dt}e^{(1+3t)^2} = e^{(1+3t)^2}\dfrac{d}{dt}(1+3t)^2 = e^{(1+3t)^2}\cdot 2(1+3t)\cdot 3 = 6(1+3t)e^{(1+3t)^2}$.
64. $\dfrac{d}{dz}\left(\dfrac{z^2+1}{\sqrt{z}}\right) = \dfrac{d}{dz}(z^{\frac{3}{2}} + z^{-\frac{1}{2}}) = \dfrac{3}{2}z^{\frac{1}{2}} - \dfrac{1}{2}z^{-\frac{3}{2}} = \dfrac{\sqrt{z}}{2}(3 - z^{-2})$.
65. $h'(r) = \dfrac{d}{dr}\left(\dfrac{r^2}{2r+1}\right) = \dfrac{(2r)(2r+1) - 2r^2}{(2r+1)^2} = \dfrac{2r(r+1)}{(2r+1)^2}$.
66. $g'(x) = \dfrac{d}{dx}(2x - x^{-1/3} + 3^x - e) = 2 + \dfrac{1}{3x^{\frac{4}{3}}} + 3^x \ln 3$.
67. $f'(t) = \dfrac{d}{dt}\left(2te^t - \dfrac{1}{\sqrt{t}}\right) = 2e^t + 2te^t + \dfrac{1}{2t^{3/2}}$.
68.
$$\dfrac{dw}{dz} = \dfrac{(-3)(5+3z) - (5-3z)(3)}{(5+3z)^2}$$
$$= \dfrac{-15 - 9z - 15 + 9z}{(5+3z)^2} = \dfrac{-30}{(5+3z)^2}$$

69. $g'(w) = \dfrac{d}{dw}\left(\dfrac{1}{2^w + e^w}\right) = -\dfrac{2^w \ln 2 + e^w}{(2^w + e^w)^2}$.
70. $\dfrac{d}{dy}\ln\ln(2y^3) = \dfrac{1}{\ln(2y^3)}\dfrac{1}{2y^3}6y^2 = \dfrac{3}{y\ln(2y^3)}$.
71. $f'(x) = \dfrac{3x^2}{9}(3\ln x - 1) + \dfrac{x^3}{9}\left(\dfrac{3}{x}\right) = x^2 \ln x - \dfrac{x^2}{3} + \dfrac{x^2}{3} = x^2 \ln x$.

SOLUTIONS TO PRACTICE PROBLEMS ON DIFFERENTIATION **127**

72. $g'(x) = \dfrac{d}{dx}\left(x^k + k^x\right) = kx^{k-1} + k^x \ln k.$

73. $r'(\theta) = \dfrac{d}{d\theta}\sin[(3\theta - \pi)^2] = \cos[(3\theta - \pi)^2] \cdot 2(3\theta - \pi) \cdot 3 = 6(3\theta - \pi)\cos[(3\theta - \pi)^2].$

74. $s'(\theta) = \dfrac{d}{d\theta}\sin^2(3\theta - \pi) = 6\cos(3\theta - \pi)\sin(3\theta - \pi).$

75. $h'(t) = \dfrac{1}{e^{-t} - t}\left(-e^{-t} - 1\right).$

76.
$$\dfrac{d}{d\theta}\left(\dfrac{\sin(5-\theta)}{\theta^2}\right) = \dfrac{\cos(5-\theta)(-1)\theta^2 - \sin(5-\theta)(2\theta)}{\theta^4}$$
$$= -\dfrac{\theta\cos(5-\theta) + 2\sin(5-\theta)}{\theta^3}.$$

77. $w'(\theta) = \dfrac{1}{\sin^2 \theta} - \dfrac{2\theta \cos \theta}{\sin^3 \theta}$

78. $g'(x) = \dfrac{d}{dx}\left(x^{\frac{1}{2}} + x^{-1} + x^{-\frac{3}{2}}\right) = \dfrac{1}{2}x^{-\frac{1}{2}} - x^{-2} - \dfrac{3}{2}x^{-\frac{5}{2}}.$

79. $s'(x) = \dfrac{d}{dx}\left(\arctan(2-x)\right) = \dfrac{-1}{1+(2-x)^2}.$

80. $r'(\theta) = \dfrac{d}{d\theta}\left(e^{\left(e^\theta + e^{-\theta}\right)}\right) = e^{\left(e^\theta + e^{-\theta}\right)}\left(e^\theta - e^{-\theta}\right).$

81. Using the chain rule, we get:
$$m'(n) = \cos(e^n) \cdot (e^n)$$

82. Using the chain rule we get:
$$k'(\alpha) = e^{\tan(\sin \alpha)}(\tan(\sin \alpha))' = e^{\tan(\sin \alpha)} \cdot \dfrac{1}{\cos^2(\sin \alpha)} \cdot \cos \alpha.$$

83. Here we use the product rule, and then the chain rule, and then the product rule.
$$g'(t) = \cos(\sqrt{t}e^t) + t(\cos \sqrt{t}e^t)' = \cos(\sqrt{t}e^t) + t(-\sin(\sqrt{t}e^t) \cdot (\sqrt{t}e^t)')$$
$$= \cos(\sqrt{t}e^t) - t\sin(\sqrt{t}e^t) \cdot \left(\sqrt{t}e^t + \dfrac{1}{2\sqrt{t}}e^t\right)$$

84. $f'(r) = e(\tan 2 + \tan r)^{e-1}(\tan 2 + \tan r)' = e(\tan 2 + \tan r)^{e-1}\left(\dfrac{1}{\cos^2 r}\right)$

85. $y' = 0$

86.
$$\dfrac{dy}{dz} = 3(x^2+5)^2(2x)(3x^3-2)^2 + (x^2+5)^3[2(3x^3-2)(9x^2)]$$
$$= 3(2x)(x^2+5)^2(3x^3-2)[(3x^3-2) + (x^2+5)(3x)]$$
$$= 6x(x^2+5)^2(3x^3-2)[6x^3 + 15x - 2]$$

87. $\dfrac{d}{dx}xe^{\tan x} = e^{\tan x} + xe^{\tan x}\dfrac{1}{\cos^2 x}.$

88. $\dfrac{dy}{dx} = 2e^{2x}\sin^2(3x) + e^{2x}(2\sin(3x)\cos(3x)3) = 2e^{2x}\sin(3x)(\sin(3x) + 3\cos(3x))$

89. $g'(x) = \dfrac{6x}{1+\left(3x^2+1\right)^2} = \dfrac{6x}{9x^4+6x^2+2}$

90. $\dfrac{dy}{dx} = (\ln 2)2^{\sin x}\cos x \cdot \cos x + 2^{\sin x}(-\sin x) = 2^{\sin x}\left((\ln 2)\cos^2 x - \sin x\right)$

91. $h(x) = ax \cdot \ln e = ax$, so $h'(x) = a$.
92. $k'(x) = a$
93. $f'(\theta) = ke^{k\theta}$
94. $N'(\theta) = k$
95. Using the product rule and factoring gives $f'(t) = e^{-4kt}(\cos t - 4k \sin t)$.
96. Using the chain rule gives $f'(x) = 5\ln(a)a^{5x}$.
97. Using the quotient rule gives
$$f'(x) = \frac{(-2x)(a^2 + x^2) - (2x)(a^2 - x^2)}{(a^2 + x^2)^2}$$
$$= \frac{-4a^2 x}{(a^2 + x^2)^2}.$$

98. Using the quotient rule gives
$$w'(r) = \frac{2ar(b + r^3) - 3r^2(ar^2)}{(b + r^3)^2}$$
$$= \frac{2abr - ar^4}{(b + r^3)^2}.$$

99. Using the quotient rule gives
$$f'(s) = \frac{-2s\sqrt{a^2 + s^2} - \frac{s}{\sqrt{a^2+s^2}}(a^2 - s^2)}{(a^2 + s^2)}$$
$$= \frac{-2s(a^2 + s^2) - s(a^2 - s^2)}{(a^2 + s^2)^{3/2}}$$
$$= \frac{-2a^2 s - 2s^3 - a^2 s + s^3}{(a^2 + s^2)^{3/2}}$$
$$= \frac{-3a^2 s - s^3}{(a^2 + s^2)^{3/2}}.$$

100. Using the product rule gives $h'(t) = ke^{kt}(\sin at + \cos bt) + e^{kt}(a \cos at - b \sin bt)$.
101. Using the product rule gives
$$H'(t) = 2ate^{-ct} - c(at^2 + b)e^{-ct}$$
$$= (-cat^2 + 2at - bc)e^{-ct}.$$

102. $\dfrac{d}{d\theta}\sqrt{a^2 - \sin^2\theta} = \dfrac{1}{2\sqrt{a^2 - \sin^2\theta}}(-2\sin\theta\cos\theta) = -\dfrac{\sin\theta\cos\theta}{\sqrt{a^2 - \sin^2\theta}}$.

103. $\dfrac{dy}{dx} = \dfrac{1}{1 + \left(\frac{2}{x}\right)^2}\left(\dfrac{-2}{x^2}\right) = \dfrac{-2}{x^2 + 4}$.

104. Using the chain rule gives $r'(t) = \dfrac{\cos(\frac{t}{k})}{\sin(\frac{t}{k})}\left(\dfrac{1}{k}\right)$.

105. $g'(u) = \dfrac{ae^{au}}{a^2 + b^2}$

106. Since $g(w) = 5(a^2 - w^2)^{-2}$, $g'(w) = -10(a^2 - w^2)^{-3}(-2w) = \dfrac{20w}{(a^2 - w^2)^3}$

107.
$$\frac{dy}{dx} = \frac{(e^x + e^{-x})(e^x + e^{-x}) - (e^x - e^{-x})(e^x - e^{-x})}{(e^x + e^{-x})^2}$$
$$= \frac{(e^x + e^{-x})^2 - (e^x - e^{-x})^2}{(e^x + e^{-x})^2} = \frac{(e^{2x} + 2 + e^{-2x}) - (e^{2x} - 2 + e^{-2x})}{(e^x + e^{-x})^2}$$
$$= \frac{4}{(e^x + e^{-x})^2}$$

108.
$$\frac{dy}{dx} = \frac{(ae^{ax} + ae^{-ax})(e^{ax} + e^{-ax}) - (e^{ax} - e^{-ax})(ae^{ax} - ae^{-ax})}{(e^{ax} + e^{-ax})^2}$$
$$= \frac{a(e^{ax} + e^{-ax})^2 - a(e^{ax} - e^{-ax})^2}{(e^{ax} + e^{-ax})^2}$$
$$= \frac{a[(e^{2ax} + 2 + e^{-2ax}) - (e^{2ax} - 2 + e^{-2ax})]}{(e^{ax} + e^{-ax})^2}$$
$$= \frac{4a}{(e^{ax} + e^{-ax})^2}$$

109. $f'(x) = \dfrac{d}{dx}(2 - 4x - 3x^2)(6x^e - 3\pi) = (-4 - 6x)(6x^e - 3\pi) + (2 - 4x - 3x^2)(6ex^{e-1})$.

110. $f'(t) = 4(\sin(2t) - \cos(3t))^3[2\cos(2t) + 3\sin(3t)]$

111. Since $\cos^2 y + \sin^2 y = 1$, we have $s(y) = \sqrt[3]{1+3} = \sqrt[3]{4}$. Thus $s'(y) = 0$.

112.
$$f'(x) = (-2x + 6x^2)(6 - 4x + x^7) + (4 - x^2 + 2x^3)(-4 + 7x^6)$$
$$= (-12x + 44x^2 - 24x^3 - 2x^8 + 6x^9) + (-16 + 4x^2 - 8x^3 + 28x^6 - 7x^8 + 14x^9)$$
$$= -16 - 12x + 48x^2 - 32x^3 + 28x^6 - 9x^8 + 20x^9$$

113.
$$h'(x) = \left(-\frac{1}{x^2} + \frac{2}{x^3}\right)(2x^3 + 4) + \left(\frac{1}{x} - \frac{1}{x^2}\right)(6x^2)$$
$$= -2x + 4 - \frac{4}{x^2} + \frac{8}{x^3} + 6x - 6$$
$$= 4x - 2 - 4x^{-2} + 8x^{-3}$$

114. Note: $f(z) = (5z)^{1/2} + 5z^{1/2} + 5z^{-1/2} - \sqrt{5}z^{-1/2} + \sqrt{5}$, so $f'(z) = \dfrac{5}{2}(5z)^{-1/2} + \dfrac{5}{2}z^{-1/2} - \dfrac{5}{2}z^{-3/2} + \dfrac{\sqrt{5}}{2}z^{-3/2}$.

115. (a) $f(x) = x^2 - 4g(x)$
$f'(x) = 2x - 4g'(x)$
$f'(2) = 2(2) - 4(-4) = 4 + 16 = 20$
(b) $f(x) = \dfrac{x}{g(x)}$
$f'(x) = \dfrac{g(x) - xg'(x)}{(g(x))^2}$
$f'(2) = \dfrac{g(2) - 2g'(2)}{(g(2))^2} = \dfrac{3 - 2(-4)}{(3)^2} = \dfrac{11}{9}$
(c) $f(x) = x^2 g(x)$
$f'(x) = 2xg(x) + x^2 g'(x)$
$f'(2) = 2(2)(3) + (2)^2(-4) = 12 - 16 = -4$
(d) $f(x) = (g(x))^2$
$f'(x) = 2g(x) \cdot g'(x)$
$f'(2) = 2(3)(-4) = -24$

(e) $f(x) = x\sin(g(x))$
$f'(x) = \sin(g(x)) + x\cos(g(x)) \cdot g'(x)$
$f'(2) = \sin(g(2)) + 2\cos(g(2)) \cdot g'(2)$
$= \sin 3 + 2\cos(3) \cdot (-4)$
$= \sin 3 - 8\cos 3$

(f) $f(x) = x^2 \ln(g(x))$
$f'(x) = 2x\ln(g(x)) + x^2(\frac{g'(x)}{g(x)})$
$f'(2) = 2(2)\ln 3 + (2)^2(\frac{-4}{3})$
$= 4\ln 3 - \frac{16}{3}$

116. (a) $f(x) = x^2 - 4g(x)$
$f(2) = 4 - 4(3) = -8$
$f'(2) = 20$
Thus, we have a point $(2, -8)$ and slope $m = 20$. This gives
$$-8 = 2(20) + b$$
$$b = -48, \quad \text{so}$$
$$y = 20x - 48.$$

(b) $f(x) = \dfrac{x}{g(x)}$
$f(2) = \dfrac{2}{3}$
$f'(2) = \dfrac{11}{9}$
Thus, we have point $(2, \frac{2}{3})$ and slope $m = \frac{11}{9}$. This gives
$$\frac{2}{3} = (\frac{11}{9})(2) + b$$
$$b = \frac{2}{3} - \frac{22}{9} = \frac{-16}{9}, \quad \text{so}$$
$$y = \frac{11}{9}x - \frac{16}{9}.$$

(c) $f(x) = x^2 g(x)$
$f(2) = 4 \cdot g(2) = 4(3) = 12$
$f'(2) = -4$
Thus, we have point $(2, 12)$ and slope $m = -4$. This gives
$$12 = 2(-4) + b$$
$$b = 20, \quad \text{so}$$
$$y = -4x + 20.$$

(d) $f(x) = (g(x))^2$
$f(2) = (g(2))^2 = (3)^2 = 9$
$f'(2) = -24$
Thus, we have point $(2, 9)$ and slope $m = -24$. This gives
$$9 = 2(-24) + b$$
$$b = 57, \quad \text{so}$$
$$y = -24x + 57.$$

(e) $f(x) = x\sin(g(x))$
$f(2) = 2\sin(g(2)) = 2\sin 3$
$f'(2) = \sin 3 - 8\cos 3$
We will use a decimal approximation for $f(2)$ and $f'(2)$, so the point $(2, 2\sin 3) \approx (2, 0.28)$ and $m \approx 8.06$. Thus,
$$0.28 = 2(8.06) + b$$
$$b = -15.84, \quad \text{so}$$
$$y = 8.06x - 15.84.$$

(f) $f(x) = x^2 \ln g(x)$
$f(2) = 4 \ln g(2) = 4 \ln 3 \approx 4.39$
$f'(2) = 4 \ln 3 - \dfrac{16}{3} \approx -0.94$.
Thus, we have point $(2, 4.39)$ and slope $m = -0.94$. This gives
$$4.39 = 2(-0.94) + b$$
$$b = 6.27, \quad \text{so}$$
$$y = -0.94x + 6.27.$$

117.
$$y + \frac{x\,dy}{dx} - 1 - \frac{3\,dy}{dx} = 0$$
$$(x - 3)\frac{dy}{dx} = 1 - y$$
$$\frac{dy}{dx} = \frac{1 - y}{x - 3}$$

118.
$$12x + 8y\frac{dy}{dx} = 0$$
$$\frac{dy}{dx} = \frac{-12x}{8y} = \frac{-3x}{2y}$$

119.
$$2ax - 2by\frac{dy}{dx} = 0$$
$$\frac{dy}{dx} = \frac{-2ax}{-2by} = \frac{ax}{by}$$

120.
$$2xy + x^2\frac{dy}{dx} - 2\frac{dy}{dx} = 0$$
$$(x^2 - 2)\frac{dy}{dx} = -2xy$$
$$\frac{dy}{dx} = \frac{-2xy}{(x^2 - 2)}$$

121.
$$3x^2 + 3y^2\frac{dy}{dx} - 8xy - 4x^2\frac{dy}{dx} = 0$$
$$(3y^2 - 4x^2)\frac{dy}{dx} = 8xy - 3x^2$$
$$\frac{dy}{dx} = \frac{8xy - 3x^2}{3y^2 - 4x^2}$$

122.
$$a\cos(ay)\frac{dy}{dx} - b\sin(bx) = y + x\frac{dy}{dx}$$
$$(a\cos(ay) - x)\frac{dy}{dx} = y + b\sin(bx)$$
$$\frac{dy}{dx} = \frac{y + b\sin(bx)}{a\cos(ay) - x}$$

CHAPTER FOUR

Solutions for Section 4.1

1.

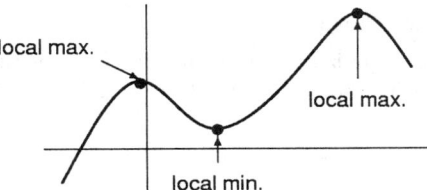

2. The critical points of f are zeros of f'. Just to the left of the first critical point $f' > 0$, so f is increasing. Immediately to the right of the first critical point $f' < 0$, so f is decreasing. Thus, the first point must be a maximum. To the left of the second critical point, $f' < 0$, and to its right, $f' > 0$; hence it is a minimum. On either side of the last critical point, $f' > 0$, so it is neither a maximum nor a minimum. See the figure below.

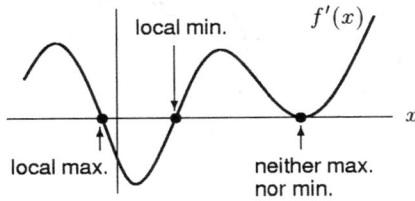

3. (a) A critical point occurs when $f'(x) = 0$. Since $f'(x)$ changes sign between $x = 2$ and $x = 3$, between $x = 6$ and $x = 7$, and between $x = 9$ and $x = 10$, we expect critical points at around $x = 2.5$, $x = 6.5$, and $x = 9.5$.
   (b) Since $f'(x)$ goes from positive to negative at $x \approx 2.5$, a local maximum should occur there. Similarly, $x \approx 6.5$ is a local minimum and $x \approx 9.5$ a local maximum.

4. (a)   (b)

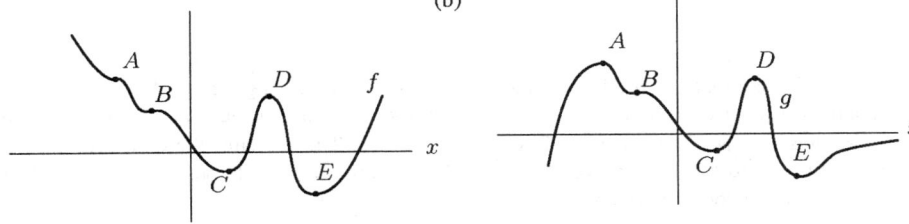

5. To find inflection points of the function f we must find points where f'' changes sign. However, because f'' is the derivative of f', any point where f'' changes sign will be a local maximum or minimum on the graph of f'.

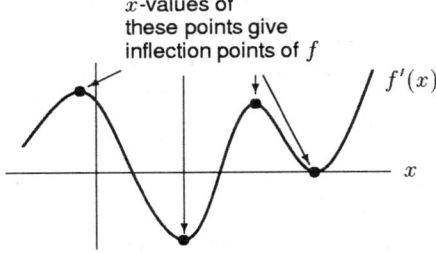

6.

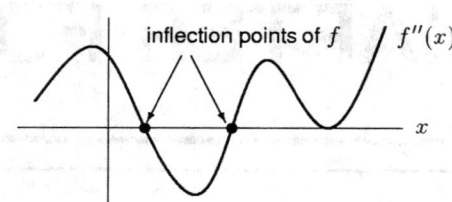

The inflection points of f are the points where f'' changes sign.

7. From the graph of $f(x)$ in the figure below, we see that the function must have two inflection points. We calculate $f'(x) = 4x^3 + 3x^2 - 6x$, and $f''(x) = 12x^2 + 6x - 6$. Solving $f''(x) = 0$ we find that:

$$x_1 = -1 \quad \text{and} \quad x_2 = \frac{1}{2}.$$

Since $f''(x) > 0$ for $x < x_1$, $f''(x) < 0$ for $x_1 < x < x_2$, and $f''(x) > 0$ for $x_2 < x$, it follows that both points are inflection points.

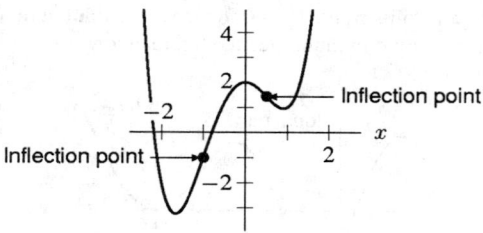

8.

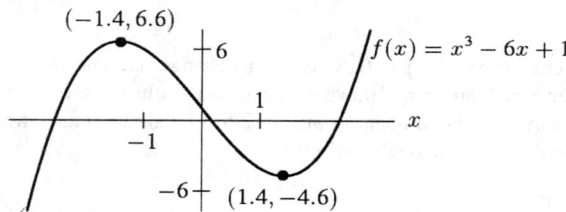

The graph of f above appears to be increasing for $x < -1.4$, decreasing for $-1.4 < x < 1.4$, and increasing for $x > 1.4$. There is a local maximum near $x = -1.4$ and local minimum near $x = 1.4$. The derivative of f is $f'(x) = 3x^2 - 6$. Thus $f'(x) = 0$ when $x^2 = 2$, that is $x = \pm\sqrt{2}$. This explains the critical points near $x = \pm 1.4$. Since $f'(x)$ changes from positive to negative at $x = -\sqrt{2}$, and from negative to positive at $x = \sqrt{2}$, there is a local maximum at $x = -\sqrt{2}$ and a local minimum at $x = \sqrt{2}$.

9.

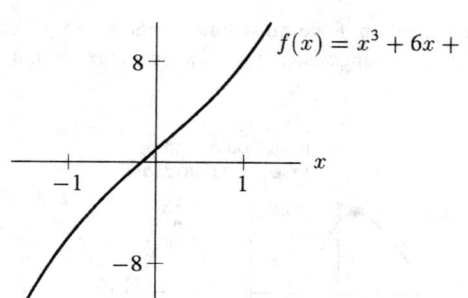

The graph of f in above appears to be increasing for all x, with no critical points. Since $f'(x) = 3x^2 + 6$ and $x^2 \geq 0$ for all x, we have $f'(x) > 0$ for all x. That explains why f is increasing for all x.

10.

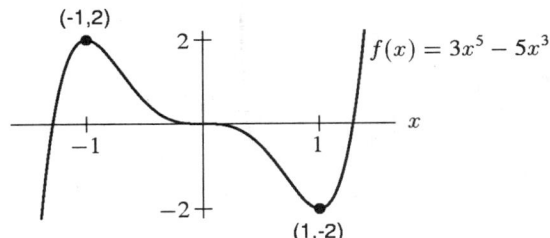

The graph of f above appears to be increasing for $x < -1$, decreasing for $-1 < x < 1$ although it is flat at $x = 0$, and increasing for $x > 1$. There are critical points at $x = -1$ and $x = 1$, and apparently also at $x = 0$. Since $f'(x) = 15x^4 - 15x^2 = 15x^2(x^2 - 1)$, we have $f'(x) = 0$ at $x = 0, -1, 1$. Notice that although $f'(0) = 0$, making $x = 0$ a critical point, there is no change in sign of $f'(x)$ at $x = 0$; the only sign changes are at $x = \pm 1$. Thus the graph of f must alternate increasing/decreasing for $x < -1, -1 < x < 1, x > 1$, just as we described.

11.

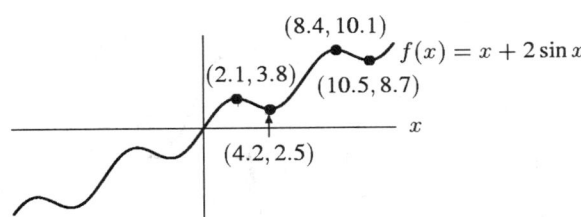

The graph of f above looks like a climbing sine curve, alternately increasing and decreasing, with more time spent increasing than decreasing. Here $f'(x) = 1 + 2\cos x$, so $f'(x) = 0$ when $\cos x = -1/2$; this occurs when

$$x = \pm\frac{2\pi}{3}, \pm\frac{4\pi}{3}, \pm\frac{8\pi}{3}, \pm\frac{10\pi}{3}, \pm\frac{14\pi}{3}, \pm\frac{16\pi}{3}\ldots$$

Since $f'(x)$ changes sign at each of these values, the graph of f must alternate increasing/decreasing. However, the distance between values of x for critical points alternates between $(2\pi)/3$ and $(4\pi)/3$, with $f'(x) > 0$ on the intervals of length $(4\pi)/3$. For example, $f'(x) > 0$ on the interval $(4\pi)/3 < x < (8\pi)/3$. As a result, f is increasing on the intervals of length $(4\pi/3)$ and decreasing on the intervals of length $(2\pi/3)$.

12.

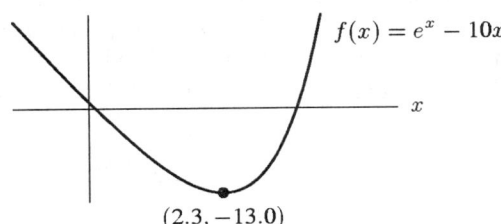

The graph of f above appears to be decreasing for $x < 2.3$ (almost like a straight line for $x < 0$), and increasing sharply for $x > 2.3$. Here $f'(x) = e^x - 10$, so $f'(x) = 0$ when $e^x = 10$, that is $x = \ln 10 = 2.302\ldots$ This is the only place where $f'(x)$ changes sign, and it is a minimum of f. Notice that e^x is small for $x < 0$ so $f'(x) \approx -10$ for $x < 0$, which means the graph looks like a straight line of slope -10 for $x < 0$. However, e^x gets large quickly for $x > 0$, so $f'(x)$ gets large quickly for $x > \ln 10$, meaning the graph increases sharply there.

13.

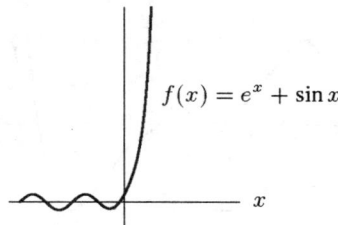

The graph of f above looks like $\sin x$ for $x < 0$ and e^x for $x > 0$. In particular, there are no waves for $x > 0$. We have $f'(x) = \cos x + e^x$, and so the critical points of f occur at those values of x for which $\cos x = -e^x$. Since $e^x > 1$ for all $x > 0$, we know immediately that there are no critical points at positive values of x. The specific locations of the critical points at $x < 0$ must be determined numerically; the first few are $x \approx -1.7, -4.7, -7.9$. For $x < 0$, the quantity e^x is small so that the graph looks like the graph of $\sin x$. For $x > 0$, we have $f'(x) > 0$ since $-1 \le \cos x$ and $e^x > 1$. Thus, the graph is increasing for all $x > 0$ and there are no such waves.

14.

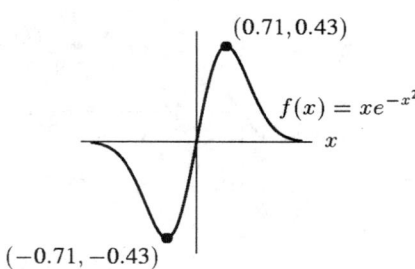

The graph of f above appears to be asymptotic to the x-axis from below for large negative x, decreasing to a global minimum at about $x = -0.71$, increasing to a global maximum at about $x = 0.71$ (passing through the origin along the way), and then decreasing asymptotically to the x-axis from above.

We have $f'(x) = e^{-x^2} + xe^{-x^2}(-2x) = e^{-x^2}(1 - 2x^2)$. Since $e^{-x^2} > 0$ for all x, the sign of $f'(x)$ is the same as the sign of $(1 - 2x^2)$. Thus $f'(x)$ changes sign at $x = \pm 1/\sqrt{2} \approx \pm 0.71$, going from negative to positive to negative, which explains the critical points and increasing/decreasing behavior described. Note that $xe^{-x^2} = x/e^{x^2}$ clearly approaches 0 as $x \to \pm\infty$, since e^{x^2} is much larger than x when $|x|$ is large. Thus the graph is asymptotic to the x-axis as $x \to \pm\infty$. Note also that the sign of $f(x) = xe^{-x^2}$ is the same as x, so $f(x) < 0$ for $x < 0$ and $f(x) > 0$ for $x > 0$. Since the graph increases from $x = 0$ to $x = 0.71$ and then decreases, $x = 1/\sqrt{2}$ is the maximum point for $x \ge 0$. Since $f(x) < 0$ for $x < 0$, $x = 1/\sqrt{2}$ is a global maximum. The global minimum at $x = -1/\sqrt{2}$ can be explained similarly.

15.

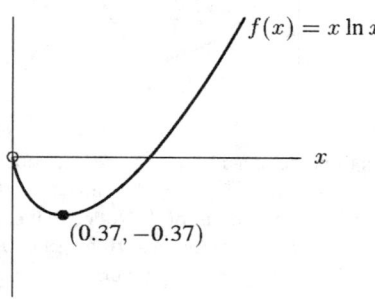

The graph of f above appears to be decreasing for $0 < x < 0.37$, and then increasing for $x > 0.37$. We have $f'(x) = \ln x + x(1/x) = \ln x + 1$, so $f'(x) = 0$ when $\ln x = -1$, that is, $x = e^{-1} \approx 0.37$. This is the only place where f' changes sign and $f'(1) = 1 > 0$, so the graph must decrease for $0 < x < e^{-1}$ and increase for $x > e^{-1}$. Thus, there is a local minimum at $x = e^{-1}$.

16. (8) The graph of $f(x) = x^3 - 6x + 1$ appears to be concave up for $x > 0$ and concave down for $x < 0$, with a point of inflection at $x = 0$. This is because $f''(x) = 6x$ is negative for $x < 0$ and positive for $x > 0$.
   (9) Same answer as number 8.
   (10) There appear to be three points of inflection at about $x = \pm 0.7$ and $x = 0$. This is because $f''(x) = 60x^3 - 30x = 30x(2x^2 - 1)$, which changes sign at $x = 0$ and $x = \pm 1/\sqrt{2}$.
   (11) There appear to be points of inflection equally spaced about 3 units apart. This is because $f''(x) = -2\sin x$, which changes sign at $x = 0, \pm\pi, \pm 2\pi, \ldots$.
   (12) The graph appears to be concave up for all x. This is because $f''(x) = e^x > 0$ for all x.
   (13) The graph appears to be concave up for all $x > 0$, and has almost periodic changes in concavity for $x < 0$. This is because for $x > 0$, $f''(x) = e^x - \sin x > 0$, and for $x < 0$, since e^x is small, $f''(x)$ changes sign at approximately the same values of x as $\sin x$.
   (14) There appears to be a point of inflection for some $x < -0.71$, for $x = 0$, and for some $x > 0.71$. This is because $f'(x) = e^{-x^2}(1 - 2x^2)$ so
   $$f''(x) = e^{-x^2}(-4x) + (1 - 2x^2)e^{-x^2}(-2x)$$
   $$= e^{-x^2}(4x^3 - 6x).$$
   Since $e^{-x^2} > 0$, this means $f''(x)$ has the same sign as $(4x^3 - 6x) = 2x(2x^2 - 3)$. Thus $f''(x)$ changes sign at $x = 0$ and $x = \pm\sqrt{3/2} \approx \pm 1.22$.
   (15) The graph appears to be concave up for all x. This is because $f'(x) = 1 + \ln x$, so $f''(x) = 1/x$, which is greater than 0 for all $x > 0$.

17. Figure 4.1 contains the graph of $f(x) = x^2 + \cos x$.

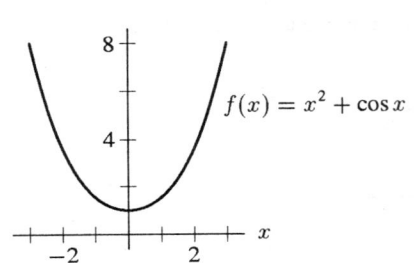

Figure 4.1

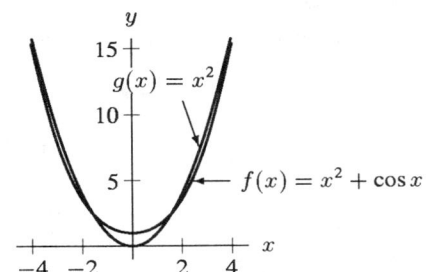

Figure 4.2

The graph looks like a parabola with no waves because $f''(x) = 2 - \cos x$, which is always positive. Thus, the graph of f is concave up everywhere; there are no waves. If you plot the graph of $f(x)$ together with the graph of $g(x) = x^2$, you see that the graph of f does wave back and forth across the graph of g, but never enough to change the concavity of f. See Figure 4.2.

18. First, we wish to have $f'(6) = 0$, since $f(6)$ should be a local minimum:
$$f'(x) = 2x + a = 0$$
$$x = -\frac{a}{2} = 6$$
$$a = -12.$$

Next, we need to have $f(6) = -5$, since the point $(6, -5)$ is on the graph of $f(x)$. We can substitute $a = -12$ into our equation for $f(x)$ and solve for b:
$$f(x) = x^2 - 12x + b$$
$$f(6) = 36 - 72 + b = -5$$
$$b = 31.$$

Thus, $f(x) = x^2 - 12x + 31$.

19. We wish to have $f'(3) = 0$. Differentiating to find $f'(x)$ and then solving $f'(3) = 0$ for a gives:

$$f'(x) = x(ae^{ax}) + 1(e^{ax}) = e^{ax}(ax + 1)$$
$$f'(3) = e^{3a}(3a + 1) = 0$$
$$3a + 1 = 0$$
$$a = -\frac{1}{3}.$$

Thus, $f(x) = xe^{-x/3}$.

20. Using the product rule on the function $f(x) = axe^{bx}$, we have $f'(x) = ae^{bx} + abxe^{bx} = ae^{bx}(1 + bx)$. We want $f(\frac{1}{3}) = 1$, and since this is to be a maximum, we require $f'(\frac{1}{3}) = 0$. These conditions give

$$f(1/3) = a(1/3)e^{b/3} = 1,$$
$$f'(1/3) = ae^{b/3}(1 + b/3) = 0.$$

Since $ae^{(1/3)b}$ is non-zero, we can divide both sides of the second equation by $ae^{(1/3)b}$ to obtain $0 = 1 + \frac{b}{3}$. This implies $b = -3$. Plugging $b = -3$ into the first equation gives us $a(\frac{1}{3})e^{-1} = 1$, or $a = 3e$. How do we know we have a maximum at $x = \frac{1}{3}$ and not a minimum? Since $f'(x) = ae^{bx}(1 + bx) = (3e)e^{-3x}(1 - 3x)$, and $(3e)e^{-3x}$ is always positive, it follows that $f'(x) > 0$ when $x < \frac{1}{3}$ and $f'(x) < 0$ when $x > \frac{1}{3}$. Since f' is positive to the left of $x = \frac{1}{3}$ and negative to the right of $x = \frac{1}{3}$, $f(\frac{1}{3})$ is a local maximum.

21. Since f is differentiable everywhere, f' must be zero (not undefined) at any critical points; thus, $f'(3) = 0$. Since f has exactly one critical point, f' may change sign only at $x = 3$. Thus f is always increasing or always decreasing for $x < 3$ and for $x > 3$. Using the information in parts (a) through (d), we determine whether $x = 3$ is a local minimum, local maximum, or neither.

(a) $x = 3$ is a local maximum because $f(x)$ is increasing when $x < 3$ and decreasing when $x > 3$.

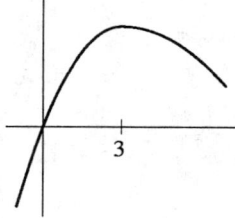

(b) $x = 3$ is a local minimum because $f(x)$ heads to infinity to either side of $x = 3$.

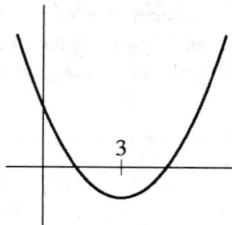

(c) $x = 3$ is neither a local minimum nor maximum, as $f(1) < f(2) < f(4) < f(5)$.

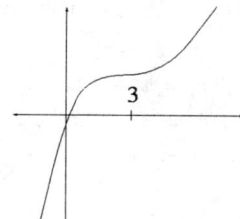

(d) $x = 3$ is a local minimum because $f(x)$ is decreasing to the left of $x = 3$ and must increase to the right of $x = 3$, as $f(3) = 1$ and eventually $f(x)$ must become close to 3.

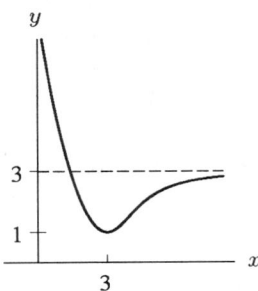

22. (a) This is one of many possible graphs.

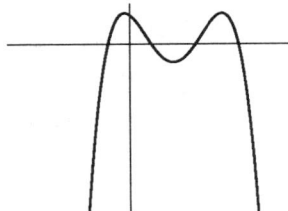

(b) Since f must have a bump between each pair of zeros, f could have at most four zeros.
(c) f could well have no zeros at all. To see this, consider the graph of the above function shifted vertically downwards.
(d) f must have at least two inflection points. Since f has 3 maxima or minima, it has 3 critical points. Consequently f' will have 3 corresponding zeros. Between each consecutive pair of these zeroes f' must have a local maximum or minimum. Thus f' will have one local maximum and one local minimum, which implies that f'' will have two zeros. These values, where the second derivative is zero, correspond to points of inflection on the graph of f.
(e) The 3 critical points are zeros of f', so degree(f') ≥ 3. Thus degree(f) ≥ 4.
(f) For example:
$$f(x) = \frac{-2}{15}(x+1)(x-1)(x-3)(x-5)$$
will look something like the graph in part (a). Many other answers are possible.

23. (a) Since the volume of water in the container is proportional to its depth, and the volume is increasing at a constant rate,
$$d(t) = \text{Depth at time } t = Kt,$$
where K is some positive constant. So the graph is linear, as shown in Figure 4.3. Since initially no water is in the container, we have $d(0) = 0$, and the graph starts from the origin.

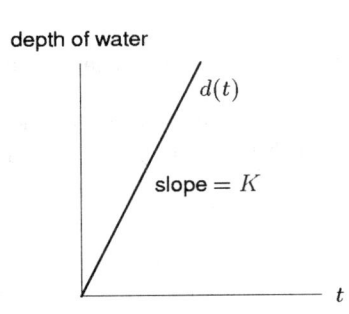

Figure 4.3

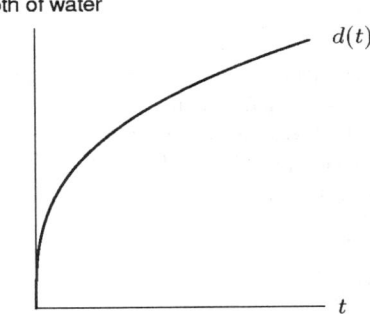

Figure 4.4

(b) As time increases, the additional volume needed to raise the water level by a fixed amount increases. Thus, although the depth, $d(t)$, of water in the cone at time t, continues to increase, it does so more and more slowly. This means $d'(t)$ is positive but decreasing, i.e., $d(t)$ is concave down. See Figure 4.4.

24.

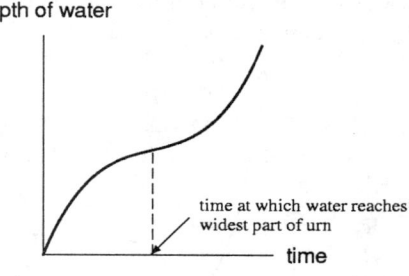

25.

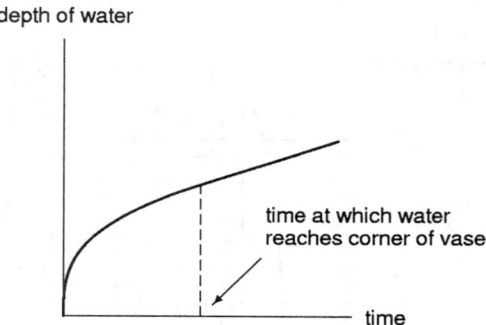

26. (a) From the graph of $P(t) = \dfrac{2000}{1 + e^{(5.3-0.4t)}}$ in Figure 4.5, we see that the population levels off at about 2000 rabbits.

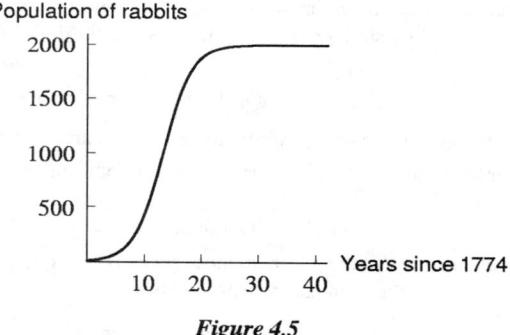

Figure 4.5

(b) The population appears to have been growing fastest when there were about 1000 rabbits, about 13 years after Captain Cook left them there.

(c) The rabbits reproduce quickly, so their population initially grew very rapidly. Limited food and space availability and perhaps predators on the island probably account for the population being unable to grow past 2000.

27.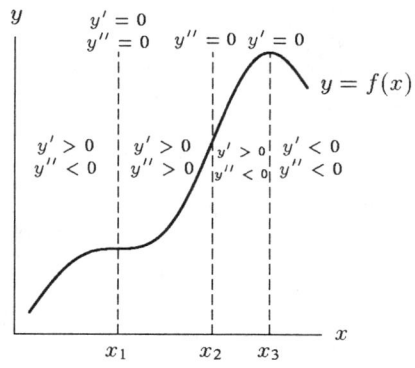

28.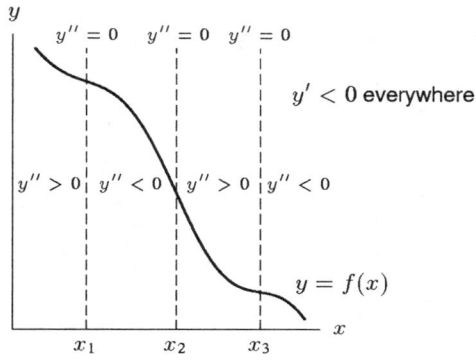

29.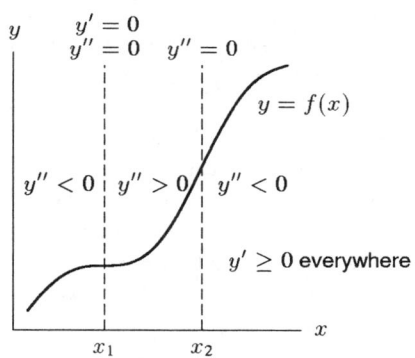

30.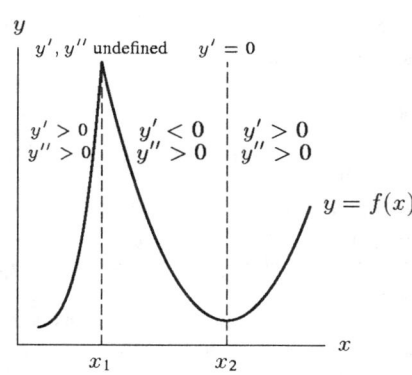

31.

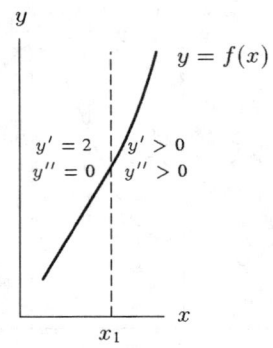

32.

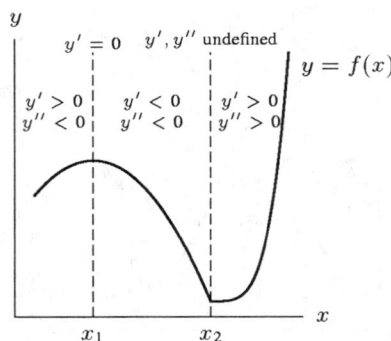

33. (a) When a number grows larger, its reciprocal grows smaller. Therefore, since f is increasing near x_0, we know that g (its reciprocal) must be decreasing. Another argument can be made using derivatives. We know that (since f is increasing) $f'(x) > 0$ near x_0. We also know (by the chain rule) that $g'(x) = (f(x)^{-1})' = -\frac{f'(x)}{f(x)^2}$. Since both $f'(x)$ and $f(x)^2$ are positive, this means $g'(x)$ is negative, which in turn means $g(x)$ is decreasing near $x = x_0$.

(b) Since f has a local maximum near x_1, $f(x)$ increases as x nears x_1, and then $f(x)$ decreases as x exceeds x_1. Thus the reciprocal of f, g, decreases as x nears x_1 and then increases as x exceeds x_1. Thus g has a local minimum at $x = x_1$. To put it another way, since f has a local maximum at $x = x_1$, we know $f'(x_1) = 0$. Since $g'(x) = -\frac{f'(x)}{f(x)^2}$, $g'(x_1) = 0$. To the left of x_1, $f'(x_1)$ is positive, so $g'(x)$ is negative. To the right of x_1, $f'(x_1)$ is negative, so $g'(x)$ is positive. Therefore, g has a local minimum at x_1.

(c) Since f is concave down at x_2, we know $f''(x_2) < 0$. We also know (from above) that

$$g''(x_2) = \frac{2f'(x_2)^2}{f(x_2)^3} - \frac{f''(x_2)}{f(x_2)^2} = \frac{1}{f(x_2)^2}\left(\frac{2f'(x_2)^2}{f(x_2)} - f''(x_2)\right).$$

Since $\frac{1}{f(x_2)^2} > 0$, $2f'(x_2)^2 > 0$, and $f(x_2) > 0$ (as f is assumed to be everywhere positive), we see that $g''(x_2)$ is positive. Thus g is concave up at x_2.

Note that for the first two parts of the problem, we didn't need to require f to be positive (only non-zero). However, it was necessary here.

34. (a) Since $f''(x) > 0$ and $g''(x) > 0$ for all x, then $f''(x) + g''(x) > 0$ for all x, so $f(x) + g(x)$ is concave up for all x.

(b) Nothing can be concluded about the concavity of $(f + g)(x)$. For example, if $f(x) = ax^2$ and $g(x) = bx^2$ with $a > 0$ and $b < 0$, then $(f + g)''(x) = a + b$. So $f + g$ is either always concave up, always concave down, or a straight line, depending on whether $a > |b|$, $a < |b|$, or $a = |b|$. More generally, it is even possible that $(f + g)(x)$ may have one or more changes in concavity.

(c) It is possible to have infinitely many changes in concavity. Consider $f(x) = x^2 + \cos x$ and $g(x) = -x^2$. Since $f''(x) = 2 - \cos x$, we see that $f(x)$ is concave up for all x. Clearly $g(x)$ is concave down for all x. However, $f(x) + g(x) = \cos x$, which changes concavity an infinite number of times.

Solutions for Section 4.2

1. (a) Let $p(x) = x^3 - ax$, and suppose $a < 0$. Then $p'(x) = 3x^2 - a > 0$ for all x, so $p(x)$ is always increasing.
   (b) Now suppose $a > 0$. We have $p'(x) = 3x^2 - a = 0$ when $x^2 = a/3$, i.e., when $x = \sqrt{a/3}$ and $x = -\sqrt{a/3}$. We also have $p''(x) = 6x$; so $x = \sqrt{a/3}$ is a local minimum since $6\sqrt{a/3} > 0$, and $x = -\sqrt{a/3}$ is a local maximum since $-6\sqrt{a/3} < 0$.
   (c) Case 1: $a < 0$
   In this case, $p(x)$ is always increasing. We have $p''(x) = 6x > 0$ if $x > 0$, meaning the graph is concave up for $x > 0$. Furthermore, $6x < 0$ if $x < 0$, meaning the graph is concave down for $x < 0$. Thus, $x = 0$ is an inflection point.
   Case 2: $a > 0$
   We have

   $$p\left(\sqrt{\tfrac{a}{3}}\right) = \left(\sqrt{\tfrac{a}{3}}\right)^3 - a\sqrt{\tfrac{a}{3}} = \frac{a\sqrt{a}}{\sqrt{27}} - \frac{a\sqrt{a}}{\sqrt{3}} = -\frac{2a\sqrt{a}}{3\sqrt{3}} < 0,$$

   and $p\left(-\sqrt{\tfrac{a}{3}}\right) = -\frac{a\sqrt{a}}{\sqrt{27}} + \frac{a\sqrt{a}}{\sqrt{3}} = -p\left(\sqrt{\tfrac{a}{3}}\right) > 0.$

   $$p'(x) = 3x^2 - a \begin{cases} = 0 & \text{if } |x| = \sqrt{\tfrac{a}{3}}; \\ > 0 & \text{if } |x| > \sqrt{\tfrac{a}{3}}; \\ < 0 & \text{if } |x| < \sqrt{\tfrac{a}{3}}. \end{cases}$$

   So p is increasing for $x < -\sqrt{a/3}$, decreasing for $-\sqrt{a/3} < x < \sqrt{a/3}$, and increasing for $x > \sqrt{a/3}$. Since $p''(x) = 6x$, the graph of $p(x)$ is concave down for values of x less than zero and concave up for values greater than zero. Graphs of $p(x)$ for $a < 0$ and $a > 0$ are found in Figures 4.6 and 4.7, respectively.

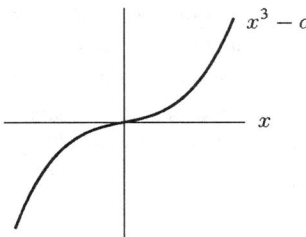
Figure 4.6: $p(x)$ for $a < 0$

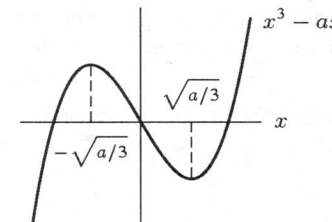
Figure 4.7: $p(x)$ for $a > 0$

2. (a) We have $p'(x) = 3x^2 - a$, and (see solution to Problem 1)

p increasing	p decreasing	p increasing
$x = -\sqrt{\tfrac{a}{3}}$		$x = \sqrt{\tfrac{a}{3}}$

   Local maximum: $p(-\sqrt{\tfrac{a}{3}}) = \frac{-a\sqrt{a}}{\sqrt{27}} + \frac{a\sqrt{a}}{\sqrt{3}} = +\frac{2a\sqrt{a}}{3\sqrt{3}}$
   Local minimum: $p(\sqrt{\tfrac{a}{3}}) = -p(-\sqrt{\tfrac{a}{3}}) = -\frac{2a\sqrt{a}}{3\sqrt{3}}$

   (b) Increasing the value of a moves the critical points of p away from the y-axis, and moves the critical values away from the x-axis. Thus, the "bumps" get further apart and higher. At the same time, increasing the value of a spreads the zeros of p further apart (while leaving the one at the origin fixed).

(c)

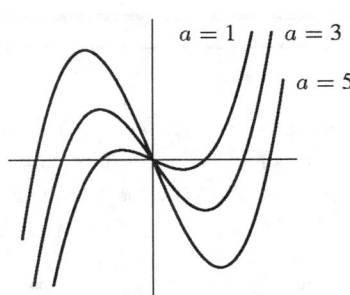

3. We have $f(x) = x^2 + 2ax = x(x + 2a) = 0$ when $x = 0$ or $x = -2a$.

$$f'(x) = 2x + 2a = 2(x + a) \begin{cases} = 0 & \text{when } x = -a \\ > 0 & \text{when } x > -a \\ < 0 & \text{when } x < -a. \end{cases}$$

See figure below. Furthermore, $f''(x) = 2$, so that $f(-a) = -a^2$ is a global minimum, and the graph is always concave up.

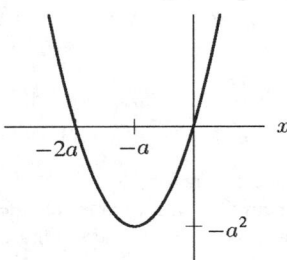

Increasing $|a|$ stretches the graph horizontally. Also, the critical value (the value of f at the critical point) drops further beneath the x-axis. Letting $a < 0$ would reflect the graph shown through the y-axis.

4. Since $\lim_{t \to \infty} N = a$, we have $a = 200{,}000$. Note that while $N(t)$ will never actually reach 200,000, it will become arbitrarily close to 200,000. Since N represents the number of people, it makes sense to round up long before $t \to \infty$. When $t = 1$, we have $N = 0.1(200{,}000) = 20{,}000$ people, so plugging into our formula gives

$$N(1) = 20{,}000 = 200{,}000 \left(1 - e^{-k(1)}\right).$$

Solving for k gives

$$0.1 = 1 - e^{-k}$$
$$e^{-k} = 0.9$$
$$k = -\ln 0.9 \approx 0.105.$$

5. $T(t) = $ the temperature at time $t = a(1 - e^{-kt}) + b$.

   (a) Since at time $t = 0$ the yam is at 20°C, we have

   $$T(0) = 20° = a\left(1 - e^0\right) + b = a(1 - 1) + b = b.$$

   Thus $b = 20°$C. Now, common sense tells us that after a period of time, the yam will heat up to about 200°, or oven temperature. Thus the temperature T should approach 200° as the time t grows large:

   $$\lim_{t \to \infty} T(t) = 200°\text{C} = a(1 - 0) + b = a + b.$$

   Since $a + b = 200°$, and $b = 20°$C, this means $a = 180°$C.

   (b) Since we're talking about how quickly the yam is heating up, we need to look at the derivative, $T'(t) = ake^{-kt}$:

   $$T'(t) = (180)ke^{-kt}.$$

   We know $T'(0) = 2°$C/min, so

   $$2 = (180)ke^{-k(0)} = (180)(k).$$

   So $k = (2°\text{C/min})/180°\text{C} = \tfrac{1}{90}\text{min}^{-1}$.

6. We begin by finding the intercepts, which occur where $f(x) = 0$, that is

$$x - k\sqrt{x} = 0$$
$$\sqrt{x}(\sqrt{x} - k) = 0$$

so $x = 0$ or $\sqrt{x} = k$, $x = k^2$.

So 0 and k^2 are the x-intercepts. Now we find the location of the critical points by setting $f'(x)$ equal to 0:

$$f'(x) = 1 - k\left(\frac{1}{2}x^{-(1/2)}\right) = 1 - \frac{k}{2\sqrt{x}} = 0.$$

This means

$$1 = \frac{k}{2\sqrt{x}}, \quad \text{so} \quad \sqrt{x} = \frac{1}{2}k, \quad \text{and} \quad x = \frac{1}{4}k^2.$$

We can use the second derivative to verify that $x = \frac{k^2}{4}$ is a local minimum. $f''(x) = 1 + \frac{k}{4x^{3/2}}$ is positive for all $x > 0$. So the critical point, $x = \frac{1}{4}k^2$, is 1/4 of the way between the x-intercepts, $x = 0$ and $x = k^2$. Since $f''(x) = \frac{1}{4}kx^{-3/2}$, $f''(\frac{1}{4}k^2) = 2/k^2 > 0$, this critical point is a minimum.

7. (a) $f'(x) = 4x^3 + 2ax = 2x(2x^2 + a)$; so $x = 0$ and $x = \pm\sqrt{-a/2}$ (if $\pm\sqrt{-a/2}$ is real, i.e. if $-a/2 \geq 0$) are critical points.
   (b) $x = 0$ is a critical point for any value of a. In order to guarantee that $x = 0$ is the only critical point, the factor $2x^2 + a$ should not have a root other than possibly $x = 0$. This means $a \geq 0$, since $2x^2 + a$ has only one root ($x = 0$) for $a = 0$, and no roots for $a > 0$. There is no restriction on the constant b.
   Now $f''(x) = 12x^2 + 2a$ and $f''(0) = 2a$.
   If $a > 0$, then by the second derivative test, $f(0)$ is a local minimum.
   If $a = 0$, then $f(x) = x^4 + b$, which has a local minimum at $x = 0$.
   So $x = 0$ is a local minimum when $a \geq 0$.
   (c) Again, b will have no effect on the location of the critical points. In order for $f'(x) = 2x(2x^2 + a)$ to have three different roots, the constant a has to be negative. Let $a = -2c^2$, for some $c > 0$. Then
   $$f'(x) = 4x(x^2 - c^2) = 4x(x - c)(x + c).$$
   The critical points of f are $x = 0$ and $x = \pm c = \pm\sqrt{-a/2}$.
   To the left of $x = -c$, $f'(x) < 0$.
   Between $x = -c$ and $x = 0$, $f'(x) > 0$.
   Between $x = 0$ and $x = c$, $f'(x) < 0$.
   To the right of $x = c$, $f'(x) > 0$.
   So, $f(-c)$ and $f(c)$ are local minima and $f(0)$ is a local maximum.
   (d) For $a \geq 0$, there is exactly one critical point, $x = 0$. For $a < 0$ there are exactly three different critical points. These exhaust all the possibilities. (Notice that the value of b is irrelevant here.)

8. Graphs of $y = e^{-ax}\sin(bx)$ for $b = 1$ and various values of a are shown in Figure 4.8. The parameter a controls the amplitude of the oscillations.

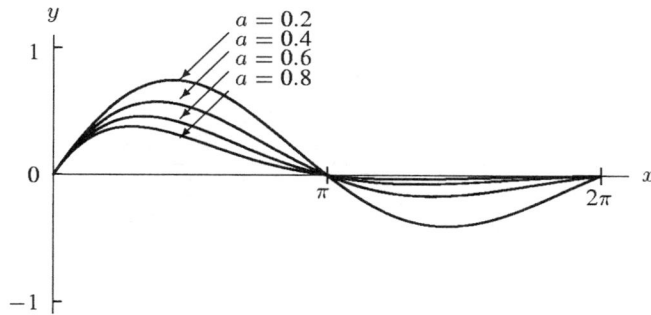

Figure 4.8

9.

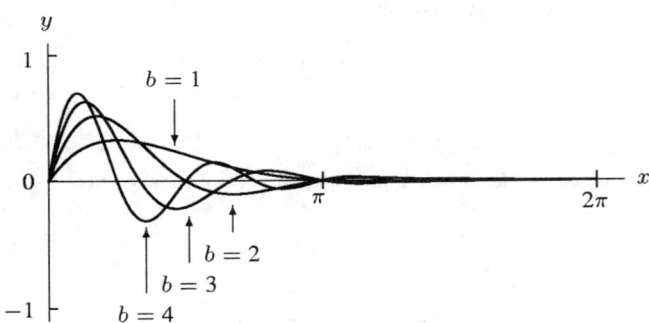

The larger the value of b, the narrower the humps and more humps per given region there are in the graph.

10. Graphs of $y = xe^{-bx}$ for $b = 1, 2, 3, 4$ are shown below. All the graphs rise at first, passing through the origin, reach a maximum and then decay toward 0. If b is small, the graph rises longer and to a higher maximum before the decay begins.

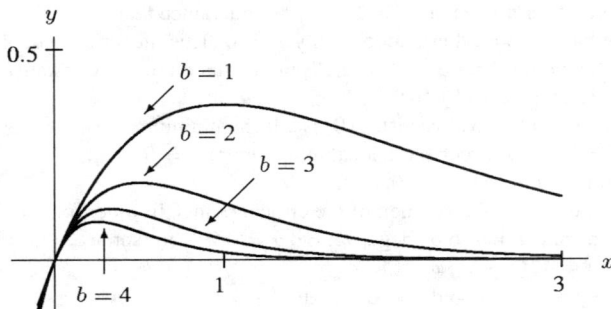

11. Since
$$\frac{dy}{dx} = (1 - bx)e^{-bx},$$
we see
$$\frac{dy}{dx} = 0 \quad \text{at} \quad x = \frac{1}{b}.$$
The critical point has coordinates $(1/b, 1/(be))$. If b is small, the x and y-coordinates of the critical point are both large, indicating a higher maximum further to the right. See figure below.

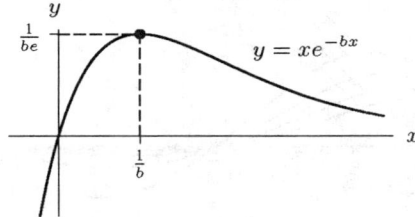

12. Since $f'(x) = abe^{-bx}$, we have $f'(x) > 0$ for all x. Therefore, f is increasing for all x. Since $f''(x) = -ab^2 e^{-bx}$, we have $f''(x) < 0$ for all x. Therefore, f is concave down for all x.

13. Let $f(x) = Ae^{-Bx^2}$. Since
$$f(x) = Ae^{-Bx^2} = Ae^{-\frac{(x-0)^2}{(1/B)}},$$
this is just the family of curves $y = e^{-\frac{(x-a)^2}{b}}$ multiplied by a constant A. This family of curves is discussed in the text; here, $a = 0, b = \frac{1}{B}$. When $x = 0, y = Ae^0 = A$, so A determines the y-intercept. A also serves to flatten or stretch the graph of e^{-Bx^2} vertically. Since $f'(x) = -2ABxe^{-Bx^2}$, $f(x)$ has a critical point at $x = 0$. For $B > 0$, the graphs are bell-shaped curves centered at $x = 0$, and $f(0) = A$ is a global maximum.

To find the inflection points of f, we solve $f''(x) = 0$. Since $f'(x) = -2ABxe^{-Bx^2}$,
$$f''(x) = -2ABe^{-Bx^2} + 4AB^2x^2e^{-Bx^2}.$$
Since e^{-Bx^2} is always positive, $f''(x) = 0$ when
$$-2AB + 4AB^2x^2 = 0$$
$$x^2 = \frac{2AB}{4AB^2}$$
$$x = \pm\sqrt{\frac{1}{2B}}.$$
These are points of inflection, since the second derivative changes sign here. Thus for large values of B, the inflection points are close to $x = 0$, and for smaller values of B the inflection points are further from $x = 0$. Therefore B affects the width of the graph.

In the graphs in Figure 4.9, A is held constant, and variations in B are shown.

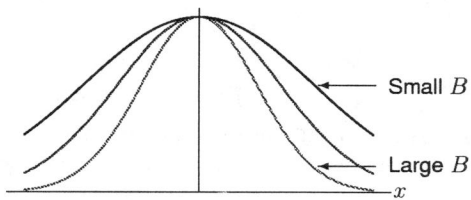

Figure 4.9: $f(x) = Ae^{-Bx^2}$ for varying B

14. (a) Let $f(x) = axe^{-bx}$. To find the local maxima and local minima of f, we solve
$$f'(x) = ae^{-bx} - abxe^{-bx} = ae^{-bx}(1 - bx) \begin{cases} = 0 & \text{if } x = 1/b \\ < 0 & \text{if } x > 1/b \\ > 0 & \text{if } x < 1/b. \end{cases}$$
Therefore, f is increasing ($f' > 0$) for $x < 1/b$ and decreasing ($f' > 0$) for $x > 1/b$. A local maximum occurs at $x = 1/b$. There are no local minima. To find the points of inflection, we write
$$f''(x) = -abe^{-bx} + ab^2xe^{-bx} - abe^{-bx}$$
$$= -2abe^{-bx} + ab^2xe^{-bx}$$
$$= ab(bx - 2)e^{-bx},$$
so $f'' = 0$ at $x = 2/b$. Therefore, f is concave up for $x < 2/b$ and concave down for $x > 2/b$, and the inflection point is $x = 2/b$.

(b) Varying a stretches or flattens the graph but does not affect the critical point $x = 1/b$ and the inflection point $x = 2/b$. Since the critical and inflection points are depend on b, varying b will change these points, as well as the maximum $f(1/b) = a/be$. For example, an increase in b will shift the critical and inflection points to the left, and also lower the maximum value of f.

(c)

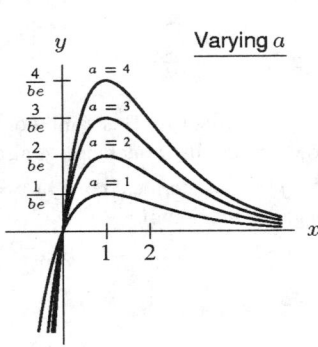
Varying a

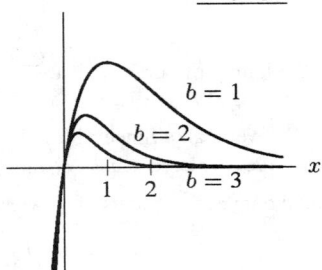
Varying b

15. (a) Figures 4.10- 4.13 show graphs of $f(x) = x^2 + \cos(kx)$ for various values of k. For $k = 0.5$ and $k = 1$, the graphs look like parabolas. For $k = 3$, there is some waving in the parabola, which becomes more noticeable if $k = 5$. The waving begins to happen at about $k = 1.5$.

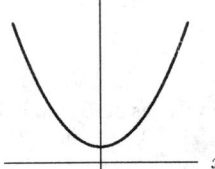
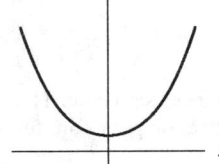
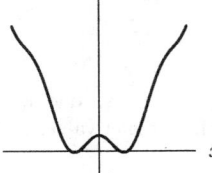
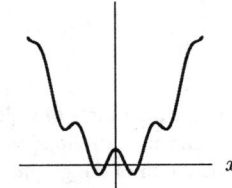

Figure 4.10: $k = 0.5$ **Figure 4.11**: $k = 1$ **Figure 4.12**: $k = 3$ **Figure 4.13**: $k = 5$

(b) Differentiating, we have
$$f'(x) = 2x - k\sin(kx)$$
$$f''(x) = 2 - k^2\cos(kx).$$

If $k^2 \leq 2$, then $f''(x) \geq 2 - 2\cos(kx) \geq 0$, since $\cos(kx) \leq 1$. Thus, the graph is always concave up if $k \leq \sqrt{2}$. If $k^2 > 2$, then $f''(x)$ changes sign whenever $\cos(kx) = 2/k^2$, which occurs for infinitely many values of x, since $0 < 2/k^2 < 1$.

(c) Since $f'(x) = 2x - k\sin(kx)$, we want to find all points where
$$2x - k\sin(kx) = 0.$$

Since
$$-1 \leq \sin(kx) \leq 1,$$
$f'(x) \neq 0$ if $x > k/2$ or $x < -k/2$. Thus, all the roots of $f'(x)$ must be in the interval $-k/2 \leq x \leq k/2$. The roots occur where the line $y = 2x$ intersects the curve $y = k\sin(kx)$, and there are only a finite number of such points for $-k/2 \leq x \leq k/2$.

16. (a)

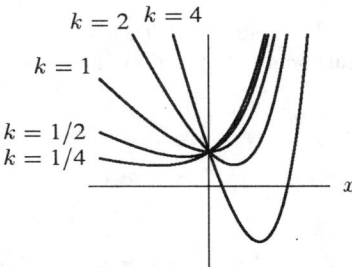

Figure 4.14

Figure 4.14 suggests that each graph decreases to a local minimum and then increases sharply. The local minimum appears to move to the right as k increases. It appears to move up until $k = 1$, and then to move back down.

(b) $f'(x) = e^x - k = 0$ when $x = \ln k$. Since $f'(x) < 0$ for $x < \ln k$ and $f'(x) > 0$ for $x > \ln k$, f is decreasing to the left of $x = \ln k$ and increasing to the right, so f reaches a local minimum at $x = \ln k$.

(c) The minimum value of f is
$$f(\ln k) = e^{\ln k} - k(\ln k) = k - k \ln k.$$

Since we want to maximize the expression $k - k \ln k$, we can imagine a function $g(k) = k - k \ln k$. To maximize this function we simply take its derivative and find the critical points. Differentiating, we obtain
$$g'(k) = 1 - \ln k - k(1/k) = -\ln k.$$

Thus $g'(k) = 0$ when $k = 1$, $g'(k) > 0$ for $k < 1$, and $g'(k) < 0$ for $k > 1$. Thus $k = 1$ is a local maximum for $g(k)$. That is, the largest global minimum for f occurs when $k = 1$.

17. (a) The larger the value of $|A|$, the steeper the graph (for the same x-value).
    (b) The graph is shifted horizontally by B. The shift is to the left for positive B, to the right for negative B. There is a vertical asymptote at $x = -B$.
    (c)

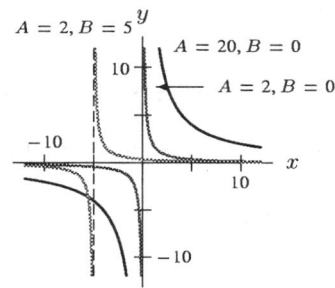

18. (a) Since
$$U = b\left(\frac{a^2 - ax}{x^2}\right) = 0 \quad \text{when} \quad x = a,$$

the x-intercept is $x = a$. There is a vertical asymptote at $x = 0$ and a horizontal asymptote at $U = 0$.

(b) Setting $dU/dx = 0$, we have
$$\frac{dU}{dx} = b\left(-\frac{2a^2}{x^3} + \frac{a}{x^2}\right) = b\left(\frac{-2a^2 + ax}{x^3}\right) = 0.$$

So the critical point is
$$x = 2a.$$

When $x = 2a$,
$$U = b\left(\frac{a^2}{4a^2} - \frac{a}{2a}\right) = -\frac{b}{4}.$$

The second derivative of U is
$$\frac{d^2 U}{dx^2} = b\left(\frac{6a^2}{x^4} - \frac{2a}{x^3}\right).$$

When we evaluate this at $x = 2a$, we get
$$\frac{d^2 U}{dx^2} = b\left(\frac{6a^2}{(2a)^4} - \frac{2a}{(2a)^3}\right) = \frac{b}{8a^2} > 0.$$

Since $d^2 U/dx^2 > 0$ at $x = 2a$, we see that the point $(2a, -b/4)$ is a local minimum.

(c)

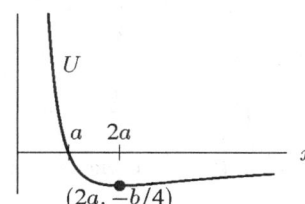

19. Both U and F have asymptotes at $x = 0$ and the x-axis. In Problem 18 we saw that U has intercept $(a, 0)$ and local minimum $(2a, -b/4)$. Differentiating U gives

$$F = b\left(\frac{2a^2}{x^3} - \frac{a}{x^2}\right).$$

Since

$$F = b\left(\frac{2a^2 - ax}{x^3}\right) = 0 \quad \text{for} \quad x = 2a,$$

F has one intercept: $(2a, 0)$. Differentiating again to find the critical points:

$$\frac{dF}{dx} = b\left(-\frac{6a^2}{x^4} + \frac{2a}{x^3}\right) = b\left(\frac{-6a^2 + 2ax}{x^4}\right) = 0,$$

so $x = 3a$. When $x = 3a$,

$$F = b\left(\frac{2a^2}{27a^3} - \frac{a}{9a^2}\right) = -\frac{b}{27a}.$$

By the first or second derivative test, $x = 3a$ is a local minimum of F. See figure below.

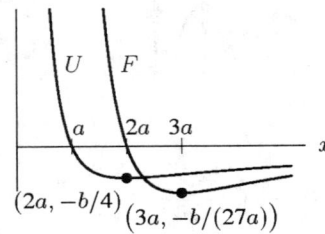

20. (a) The force is zero where

$$f(r) = -\frac{A}{r^2} + \frac{B}{r^3} = 0$$
$$Ar^3 = Br^2$$
$$r = \frac{B}{A}.$$

The vertical asymptote is $r = 0$ and the horizontal asymptote is the r-axis.

(b) To find critical points, we differentiate and set $f'(r) = 0$:

$$f'(r) = \frac{2A}{r^3} - \frac{3B}{r^4} = 0$$
$$2Ar^4 = 3Br^3$$
$$r = \frac{3B}{2A}.$$

Thus, $r = 3B/(2A)$ is the only critical point. Since $f'(r) < 0$ for $r < 3B/(2A)$ and $f'(r) > 0$ for $r > 3B/(2A)$, we see that $r = 3B/(2A)$ is a local minimum. At that point,

$$f\left(\frac{3B}{2A}\right) = -\frac{A}{9B^2/4A^2} + \frac{B}{27B^3/8A^3} = -\frac{4A^3}{27B^2}.$$

Differentiating again, we have

$$f''(r) = -\frac{6A}{r^4} + \frac{12B}{r^5} = -\frac{6}{r^5}(Ar - 2B).$$

So $f''(r) < 0$ where $r > 2B/A$ and $f''(r) > 0$ when $r < 2B/A$. Thus, $r = 2B/A$ is the only point of inflection. At that point

$$f\left(\frac{2B}{A}\right) = -\frac{A}{4B^2/A^2} + \frac{B}{8B^3/A^3} = -\frac{A^3}{8B^2}.$$

(c)

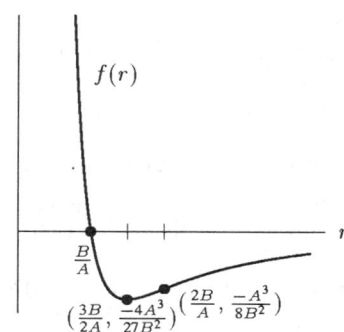

(d) (i) Increasing B means that the r-values of the zero, the minimum, and the inflection point increase, while the $f(r)$ values of the minimum and the point of inflection decrease in magnitude. See Figure 4.15.

(ii) Increasing A means that the r-values of the zero, the minimum, and the point of inflection decrease, while the $f(r)$ values of the minimum and the point of inflection increase in magnitude. See Figure 4.16.

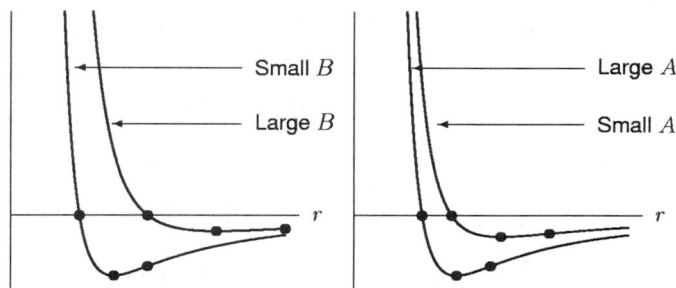

Figure 4.15: Increasing B **Figure 4.16:** Increasing A

Solutions for Section 4.3

1.

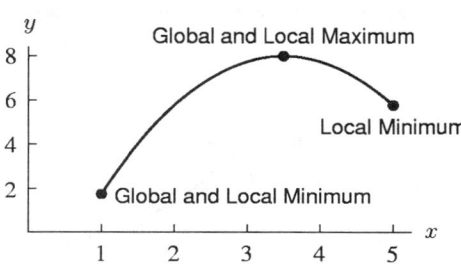

2.

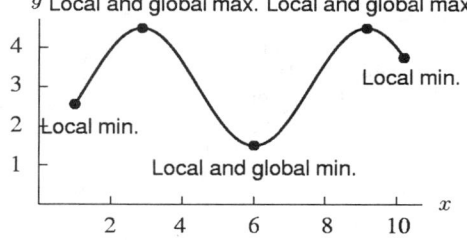

The global maximum is achieved at the two local maxima, which are at the same height.

3. (a) We have $f'(x) = 10x^9 - 10 = 10(x^9 - 1)$. This is zero when $x = 1$, so $x = 1$ is a critical point of f. For values of x less than 1, x^9 is less than 1, and thus $f'(x)$ is negative when $x < 1$. Similarly, $f'(x)$ is positive for $x > 1$. Thus $f(1) = -9$ is a local minimum.

We also consider the endpoints $f(0) = 0$ and $f(2) = 1004$. Since $f'(0) < 0$ and $f'(2) > 0$, we see $x = 0$ and $x = 2$ are local maxima.

(b) Comparing values of f shows that the global minimum is at $x = 1$, and the global maximum is at $x = 2$.

4. (a) $f'(x) = 1 - 1/x$. This is zero only when $x = 1$. Now $f'(x)$ is positive when $1 < x \leq 2$, and negative when $0.1 < x < 1$. Thus $f(1) = 1$ is a local minimum. The endpoints $f(0.1) \approx 2.4026$ and $f(2) \approx 1.3069$ are local maxima.

(b) Comparing values of f shows that $x = 0.1$ gives the global maximum and $x = 1$ gives the global minimum.

5. (a)
$$f(x) = \sin^2 x - \cos x \quad \text{for } 0 \leq x \leq \pi$$
$$f'(x) = 2\sin x \cos x + \sin x = (\sin x)(2\cos x + 1)$$

$f'(x) = 0$ when $\sin x = 0$ or when $2\cos x + 1 = 0$. Now, $\sin x = 0$ when $x = 0$ or when $x = \pi$. On the other hand, $2\cos x + 1 = 0$ when $\cos x = -1/2$, which happens when $x = 2\pi/3$. So the critical points are $x = 0$, $x = 2\pi/3$, and $x = \pi$.

Note that $\sin x > 0$ for $0 < x < \pi$. Also, $2\cos x + 1 < 0$ if $2\pi/3 < x \leq \pi$ and $2\cos x + 1 > 0$ if $0 < x < 2\pi/3$. Therefore,

$$f'(x) < 0 \quad \text{for} \quad \frac{2\pi}{3} < x < \pi$$
$$f'(x) > 0 \quad \text{for} \quad 0 < x < \frac{2\pi}{3}.$$

Thus f has a local maximum at $x = 2\pi/3$ and local minima at $x = 0$ and $x = \pi$.

(b) We have
$$f(0) = [\sin(0)]^2 - \cos(0) = -1$$
$$f\left(\frac{2\pi}{3}\right) = \left[\sin\left(\frac{2\pi}{3}\right)\right]^2 - \cos\frac{2\pi}{3} = 1.25$$
$$f(\pi) = [\sin(\pi)]^2 - \cos(\pi) = 1.$$

Thus the global maximum is at $x = 2\pi/3$, and the global minimum is at $x = 0$.

6. Since the function is positive, the graph lies above the x-axis. If there is a global maximum at $x = 3$, $t'(x)$ must be positive, then negative. Since $t'(x)$ and $t''(x)$ have the same sign for $x < 3$, they must both be positive, and thus the graph must be increasing and concave up. Since $t'(x)$ and $t''(x)$ have opposite signs for $x > 3$ and $t'(x)$ is negative, $t''(x)$ must again be positive and the graph must be decreasing and concave up. A possible sketch of $y = t(x)$ is shown in the figure below.

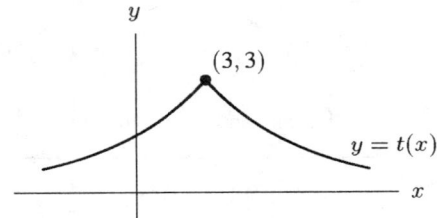

7. Here is one possible graph of g:

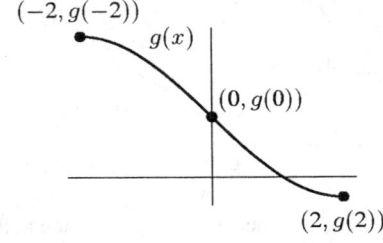

(a) From left to right, the graph of $g(x)$ starts "flat", decreases slowly at first then more rapidly, most rapidly at $x = 0$. The graph then continues to decrease but less and less rapidly until flat again at $x = 2$. The graph should exhibit symmetry about the point $(0, g(0))$.

(b) The graph has an inflection point at $(0, g(0))$ where the slope changes from negative and decreasing to negative and increasing.

(c) The function has a global maximum at $x = -2$ and a global minimum at $x = 2$.

(d) Since the function is decreasing over the interval $-2 \leq x \leq 2$

$$g(-2) = 5 > g(0) > g(2).$$

Since the function appears symmetric about $(0, g(0))$, we have

$$g(-2) - g(0) = g(0) - g(2).$$

8. (a) We know that $h''(x) < 0$ for $-2 \leq x < -1$, $h''(-1) = 0$, and $h''(x) > 0$ for $x > -1$. Thus, $h'(x)$ decreases to its minimum value at $x = -1$, which we know to be zero, and then increases; it is never negative.

(b) Since $h'(x)$ is non-negative for $-2 \leq x \leq 1$, we know that $h(x)$ is never decreasing on $[-2, 1]$. So a global maximum must occur at the right hand endpoint of the interval.

(c) The graph below shows a function that is increasing on the interval $-2 \leq x \leq 1$ with a horizontal tangent and an inflection point at $(-1, 2)$.

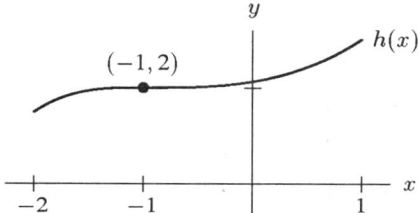

9. We want to maximize the height, y, of the grapefruit above the ground, as shown in the figure below. Using the derivative we can find exactly when the grapefruit is at the highest point. We can think of this in two ways. By common sense, at the peak of the grapefruit's flight, the velocity, dy/dt, must be zero. Alternately, we are looking for a global maximum of y, so we look for critical points where $dy/dt = 0$. We have

$$\frac{dy}{dt} = -32t + 50 = 0 \quad \text{and so} \quad t = \frac{-50}{-32} \approx 1.56 \text{ sec.}$$

Thus, we have the time at which the height is a maximum; the maximum value of y is then

$$y \approx -16(1.56)^2 + 50(1.56) + 5 = 44.1 \text{ feet.}$$

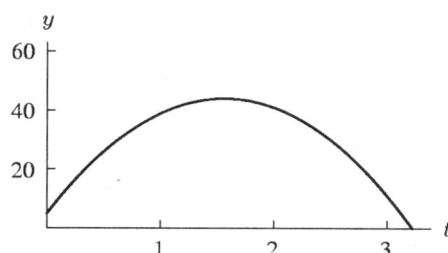

10. (a) We have

$$T(D) = \left(\frac{C}{2} - \frac{D}{3}\right)D^2 = \frac{CD^2}{2} - \frac{D^3}{3},$$

and

$$\frac{dT}{dD} = CD - D^2 = D(C - D).$$

Since, by this formula, dT/dD is zero when $D = 0$ or $D = C$, negative when $D > C$, and positive when $D < C$, we have (by the first derivative test) that the temperature change is maximized when $D = C$.

(b) The sensitivity is $dT/dD = CD - D^2$; its derivative is $d^2T/dD^2 = C - 2D$, which is zero if $D = C/2$, negative if $D > C/2$, and positive if $D < C/2$. Thus by the first derivative test the sensitivity is maximized at $D = C/2$.

154 CHAPTER FOUR /SOLUTIONS

11. We have that $v(r) = a(R-r)r^2 = aRr^2 - ar^3$, and $v'(r) = 2aRr - 3ar^2 = 2ar(R - \frac{3}{2}r)$, which is zero if $r = \frac{2}{3}R$, or if $r = 0$, and so $v(r)$ has critical points there.
    $v''(r) = 2aR - 6ar$, and thus $v''(0) = 2aR > 0$, which by the second derivative test implies that v has a minimum at $r = 0$. $v''(\frac{2}{3}R) = 2aR - 4aR = -2aR < 0$, and so by the second derivative test v has a maximum at $r = \frac{2}{3}R$. In fact, this is a global max of $v(r)$ since $v(0) = 0$ and $v(R) = 0$ at the endpoints.

12. We look for critical points of M:
$$\frac{dM}{dx} = \frac{1}{2}wL - wx.$$
Now $dM/dx = 0$ when $x = L/2$. At this point $d^2M/dx^2 = -w$ so this point is a local maximum. The graph of $M(x)$ is a parabola opening downwards, so the local maximum is also the global maximum.

13.
$$\frac{dE}{d\theta} = \frac{(\mu+\theta)(1-2\mu\theta) - (\theta - \mu\theta^2)}{(\mu+\theta)^2} = \frac{\mu(1 - 2\mu\theta - \theta^2)}{(\mu+\theta)^2}.$$
Now $dE/d\theta = 0$ when $\theta = -\mu \pm \sqrt{1+\mu^2}$. Since $\theta > 0$, the only possible critical point is when $\theta = -\mu + \sqrt{\mu^2+1}$. Differentiating again gives $E'' < 0$ at this point and so it is a local maximum. Since $E(\theta)$ is continuous for $\theta > 0$ and $E(\theta)$ has only one critical point, the local maximum is the global maximum.

14. A graph of F against θ is shown below.

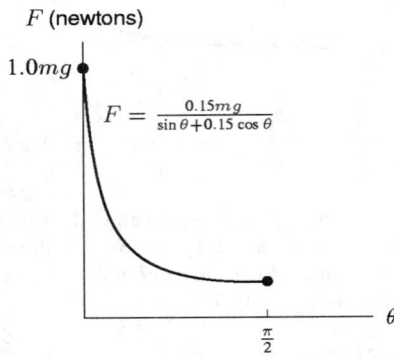

Taking the derivative:
$$\frac{dF}{d\theta} = -\frac{mg\mu(\cos\theta - \mu\sin\theta)}{(\sin\theta + \mu\cos\theta)^2}.$$
At a critical point, $dF/d\theta = 0$, so
$$\cos\theta - \mu\sin\theta = 0$$
$$\tan\theta = \frac{1}{\mu}$$
$$\theta = \arctan\left(\frac{1}{\mu}\right).$$

If $\mu = 0.15$, then $\theta = \arctan(1/0.15) = 1.422 \approx 81.5°$. To calculate the maximum and minimum values of F, we evaluate at this critical point and the endpoints:

$$\text{At } \theta = 0, \quad F = \frac{0.15mg}{\sin 0 + 0.15\cos 0} = 1.0mg \text{ newtons.}$$

$$\text{At } \theta = 1.422, \quad F = \frac{0.15mg}{\sin(1.422) + 0.15\cos(1.422)} = 0.148mg \text{ newtons.}$$

$$\text{At } \theta = \pi/2, \quad F = \frac{0.15mg}{\sin(\frac{\pi}{2}) + 0.15\cos(\frac{\pi}{2})} = 0.15mg \text{ newtons.}$$

Thus, the maximum value of F is $1.0mg$ newtons when $\theta = 0$ (her arm is vertical) and the minimum value of F is $0.148mg$ newtons is when $\theta = 1.422$ (her arm is close to horizontal). See Figure 4.17.

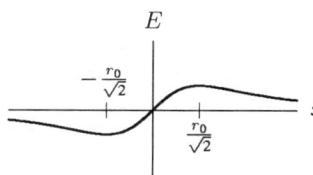

Figure 4.17

15. The domain for E is all real x. Note $E \to 0$ as $x \to \pm\infty$. The critical points occur where $dE/dx = 0$. The derivative is

$$\frac{dE}{dx} = \frac{k}{\left(x^2 + r_0^2\right)^{3/2}} - \frac{3}{2} \cdot \frac{kx(2x)}{\left(x^2 + r_0^2\right)^{5/2}}$$

$$= \frac{k\left(x^2 + r_0^2 - 3x^2\right)}{\left(x^2 + r_0^2\right)^{5/2}}$$

$$= \frac{k\left(r_0^2 - 2x^2\right)}{\left(x^2 + r_0^2\right)^{5/2}}.$$

So $dE/dx = 0$ where

$$r_0^2 - 2x^2 = 0$$

$$x = \pm \frac{r_0}{\sqrt{2}}.$$

Looking at the formula for dE/dx shows

$$\frac{dE}{dx} > 0 \text{ for } -\frac{r_0}{\sqrt{2}} < x < \frac{r_0}{\sqrt{2}}$$

$$\frac{dE}{dx} < 0 \text{ for } x < -\frac{r_0}{\sqrt{2}}$$

$$\frac{dE}{dx} < 0 \text{ for } x > \frac{r_0}{\sqrt{2}}.$$

Therefore, $x = -r_0/\sqrt{2}$ gives the minimum value of E and $x = r_0/\sqrt{2}$ gives the maximum value of E.

16. Since $I(t)$ is a periodic function with period $2\pi/w$, it is enough to consider $I(t)$ for $0 \leq wt \leq 2\pi$. Differentiating, we find

$$\frac{dI}{dt} = -w\sin(wt) + \sqrt{3}w\cos(wt).$$

At a critical point

$$-w\sin(wt) + \sqrt{3}w\cos(wt) = 0$$

$$\sin(wt) = \sqrt{3}\cos(wt)$$

$$\tan(wt) = \sqrt{3}.$$

So $wt = \pi/3$ or $4\pi/3$, or these values plus multiples of 2π. Substituting into I, we see

At $wt = \dfrac{\pi}{3}$: $\quad I = \cos\left(\dfrac{\pi}{3}\right) + \sqrt{3}\sin\left(\dfrac{\pi}{3}\right) = \dfrac{1}{2} + \sqrt{3}\cdot\left(\dfrac{\sqrt{3}}{2}\right) = 2.$

At $wt = \dfrac{4\pi}{3}$: $\quad I = \cos\left(\dfrac{4\pi}{3}\right) + \sqrt{3}\sin\left(\dfrac{4\pi}{3}\right) = -\dfrac{1}{2} - \sqrt{3}\cdot\left(\dfrac{\sqrt{3}}{2}\right) = -2.$

Thus, the maximum value is 2 amps and the minimum is -2 amps.

17. (a) We want to find where $x > 2\ln x$, which is the same as solving $x - 2\ln x > 0$. Let $f(x) = x - 2\ln x$. Then $f'(x) = 1 - \dfrac{2}{x}$, which implies that $x = 2$ is the only critical point of f. Since $f'(x) < 0$ for $x < 2$ and $f'(x) > 0$ for $x > 2$, by the first derivative test we see that f has a local and global minimum at $x = 2$. Since $f(2) = 2 - 2\ln 2 \approx 0.61$, then for all $x > 0$, $f(x) \geq f(2) > 0$. Thus $f(x)$ is always positive, which means $x > 2\ln x$ for any $x > 0$.

   (b) We've shown that $x > 2\ln x = \ln(x^2)$ for all $x > 0$. Since e^x is an increasing function, $e^x > e^{\ln x^2} = x^2$, so $e^x > x^2$ for all $x > 0$.

   (c) Let $f(x) = x - 3\ln x$. Then $f'(x) = 1 - \dfrac{3}{x} = 0$ at $x = 3$. By the first derivative test, f has a local minimum at $x = 3$. But, $f(3) \approx -0.295$, which is less than zero. Thus $3\ln x > x$ at $x = 3$. So, x is not less than $3\ln x$ for all $x > 0$.

   (One could also see this by plugging in $x = e$: since $3\ln e = 3$, $x < 3\ln x$ when $x = e$.)

18. Let $y = e^{-x^2}$. Since $y' = -2xe^{-x^2}$, y is increasing for $x < 0$ and decreasing for $x > 0$. Hence $y = e^0 = 1$ is a global maximum.

    When $x = \pm 0.3$, $y = e^{-0.09} \approx 0.9139$, which is a global minimum on the given interval. Thus $e^{-0.09} \leq y \leq 1$ for $|x| \leq 0.3$.

19. Let $y = \ln(1+x)$. Since $y' = 1/(1+x)$, y is increasing for all $x \geq 0$. The lower bound is at $x = 0$, so, $\ln(1) = 0 \leq y$. There is no upper bound.

20. Let $y = \ln(1+x^2)$. Then $y' = 2x/(1+x^2)$. Since the denominator is always positive, the sign of y' is determined by the numerator $2x$. Thus $y' > 0$ when $x > 0$, and $y' < 0$ when $x < 0$, and we have a local (and global) minimum for y at $x = 0$. Since $y(-1) = \ln 2$ and $y(2) = \ln 5$, the global maximum is at $x = 2$. Thus $0 \leq y \leq \ln 5$, or (in decimals) $0 \leq y < 1.61$. (Note that our upper bound has been rounded *up* from 1.6094.)

21. Let $y = x^3 - 4x^2 + 4x$. To locate the critical points, we solve $y' = 0$. Since $y' = 3x^2 - 8x + 4 = (3x-2)(x-2)$, the critical points are $x = 2/3$ and $x = 2$. To find the global minimum and maximum on $0 \leq x \leq 4$, we check the critical points and the endpoints: $y(0) = 0$; $y(2/3) = 32/27$; $y(2) = 0$; $y(4) = 16$. Thus, the global minimum is at $x = 0$ and $x = 2$, the global maximum is at $x = 4$, and $0 \leq y \leq 16$.

22. The graph of $y = x + \sin x$ in Figure 4.18 suggests that the function is nondecreasing over the entire interval. You can confirm this by looking at the derivative:
$$y' = 1 + \cos x$$

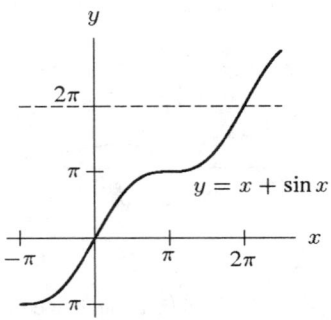

Figure 4.18: Graph of $y = x + \sin x$

Since $\cos x \geq -1$, we have $y' \geq 0$ everywhere, so y never decreases. This means that a lower bound for y is 0 (its value at the left endpoint of the interval) and an upper bound is 2π (its value at the right endpoint). That is, if $0 \leq x \leq 2\pi$:
$$0 \leq y \leq 2\pi.$$

These are the best bounds for y over the interval.

23. (a) To maximize benefit (surviving young), we pick 10, because that's the highest point of the benefit graph.
    (b) To optimize the vertical distance between the curves, we can either do it by inspection or note that the slopes of the two curves will be the same where the difference is maximized. Either way, one gets approximately 9.

24. (a) At higher speeds, more energy is used so the graph rises to the right. The initial drop is explained by the fact that the energy it takes a bird to fly at very low speeds is greater than that needed to fly at a slightly higher speed. When it flies slightly faster, the amount of energy consumed decreases. But when it flies at very high speeds, the bird consumes a lot more energy (this is analogous to our swimming in a pool).
    (b) $f(v)$ measures energy per second; $a(v)$ measures energy per meter. A bird traveling at rate v will in 1 second travel v meters, and thus will consume $v \cdot a(v)$ joules of energy in that 1 second period. Thus $v \cdot a(v)$ represents the energy consumption per second, and so $f(v) = v \cdot a(v)$.
    (c) Since $v \cdot a(v) = f(v)$, $a(v) = f(v)/v$. But this ratio has the same value as the slope of a line passing from the origin through the point $(v, f(v))$ on the curve (see figure). Thus $a(v)$ is minimal when the slope of this line is minimal. To find the value of v minimizing $a(v)$, we solve $a'(v) = 0$. By the quotient rule,
    $$a'(v) = \frac{vf'(v) - f(v)}{v^2}.$$

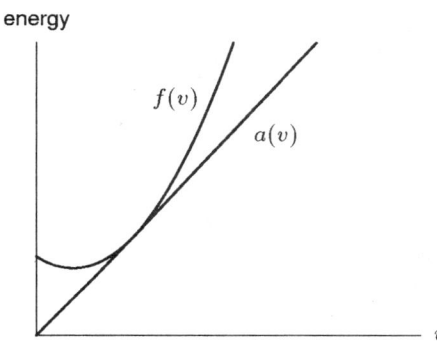

Thus $a'(v) = 0$ when $vf'(v) = f(v)$, or when $f'(v) = f(v)/v = a(v)$. Since $a(v)$ is represented by the slope of a line through the origin and a point on the curve, $a(v)$ is minimized when this line is tangent to $f(v)$, so that the slope $a(v)$ equals $f'(v)$.
    (d) The bird should minimize $a(v)$ assuming it wants to go from one particular point to another, i.e. where the distance is set. Then minimizing $a(v)$ minimizes the total energy used for the flight.

25. (a) Figure 4.19 contains the graph of total drag, plotted on the same coordinate system with induced and parasite drag. It was drawn by adding the vertical coordinates of Induced and Parasite drag.

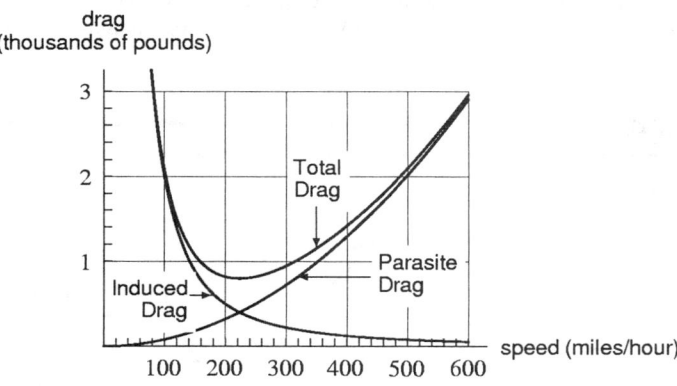

Figure 4.19

(b) Airspeeds of approximately 160 mph and 320 mph each result in a total drag of 1000 pounds. Since two distinct airspeeds are associated with a single total drag value, the total drag function does not have an inverse. The parasite and induced drag functions do have inverses, because they are strictly increasing and strictly decreasing functions, respectively.

(c) To conserve fuel, fly the at the airspeed which minimizes total drag. This is the airspeed corresponding to the lowest point on the total drag curve in part (a): that is, approximately 220 mph.

26. (a) To obtain $g(v)$, which is in gallons per mile, we need to divide $f(v)$ (in gallons per hour) by v (in miles per hour). Thus, $g(v) = f(v)/v$.
(b) By inspecting the graph, we see that $f(v)$ is minimized at approximately 220 mph.
(c) Note that a point on the graph of $f(v)$ has the coordinates $(v, f(v))$. The line passing through this point and the origin $(0, 0)$ has

$$\text{Slope} = \frac{f(v) - 0}{v - 0} = \frac{f(v)}{v} = g(v).$$

So minimizing $g(v)$ corresponds to finding the line of minimum slope from the family of lines which pass through the origin $(0, 0)$ and the point $(v, f(v))$ on the graph of $f(v)$. This line is the unique member of the family which is tangent to the graph of $f(v)$. The value of v corresponding to the point of tangency will minimize $g(v)$. This value of v will satisfy $f(v)/v = f'(v)$. From the graph in Figure 4.20, we see that $v \approx 300$ mph.

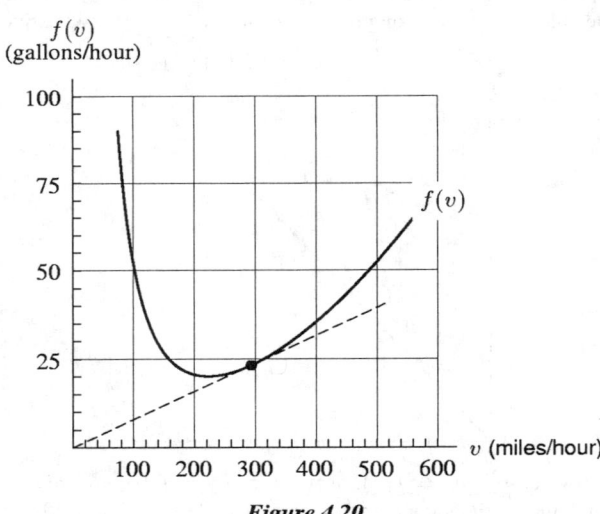

Figure 4.20

(d) The pilot's goal with regard to $f(v)$ and $g(v)$ would depend on the purpose of the flight, and might even vary within a given flight. For example, if the mission involved aerial surveillance or banner-towing over some limited area, or if the plane was flying a holding pattern, then the pilot would want to minimize $f(v)$ so as to remain aloft as long as possible. In a more normal situation where the purpose was economical travel between two fixed points, then the minimum net fuel expenditure for the trip would result from minimizing $g(v)$.

Solutions for Section 4.4

1.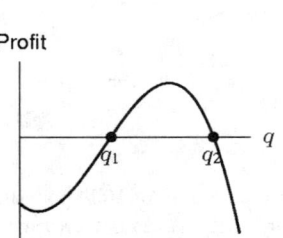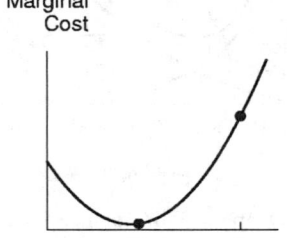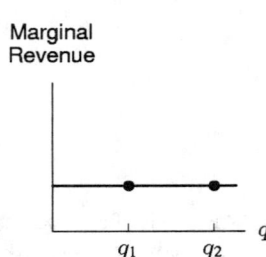

2. (a) $\pi(q)$ is maximized when $R(q) > C(q)$ and they are as far apart as possible:

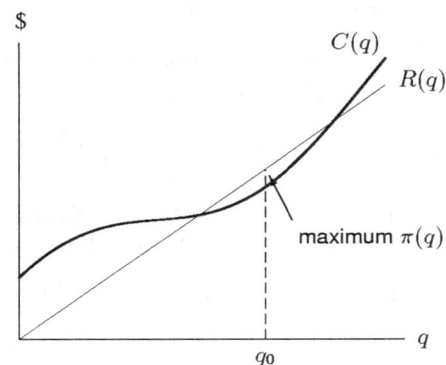

(b) $\pi'(q_0) = R'(q_0) - C'(q_0) = 0$ implies that $C'(q_0) = R'(q_0) = p$.

Graphically, the slopes of the two curves at q_0 are equal. This is plausible because if $C'(q_0)$ were greater than p or less than p, the maximum of $\pi(q)$ would be to the left or right of q_0, respectively. In economic terms, if the cost were rising more quickly than revenues, the profit would be maximized at a lower quantity (and if the cost were rising more slowly, at a higher quantity).

(c)

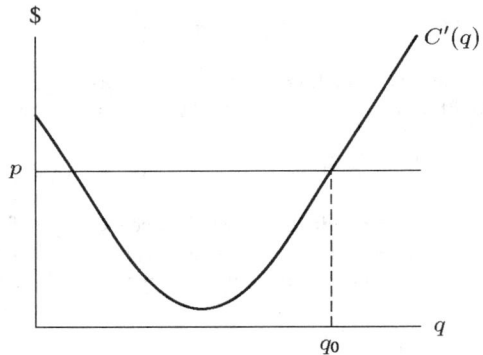

3. (a) $C(0)$ represents the fixed costs before production, that is, the cost of producing zero units, incurred for initial investments in equipment, etc.
   (b) The marginal cost decreases slowly, and then increases as quantity produced increases. See Problem 1, graph (b).
   (c) Concave down implies decreasing marginal cost, while concave up implies increasing marginal cost.
   (d) An inflection point of the cost function is (locally) the point of maximum or minimum marginal cost.
   (e) One would think that the more of an item you produce, the less it would cost to produce extra items. In economic terms, one would expect the marginal cost of production to decrease, so we would expect the cost curve to be concave down. In practice, though, it eventually becomes more expensive to produce more items, because workers and resources may become scarce as you increase production. Hence after a certain point, the marginal cost may rise again. This happens in oil production, for example.

4. (a) We know that Profit = Revenue − Cost, so differentiating with respect to q gives:

   Marginal Profit = Marginal Revenue − Marginal Cost.

   We see from the figure in the problem that just to the left of $q = a$, marginal revenue is less than marginal cost, so marginal profit is negative there. To the right of $q = a$ marginal revenue is greater than marginal cost, so marginal profit is positive there. At $q = a$ marginal profit changes from negative to positive. This means that profit is decreasing to the left of a and increasing to the right. The point $q = a$ corresponds to a local minimum of profit, and does not maximize profit. It would be a terrible idea for the company to set its production level at $q = a$.

   (b) We see from the figure in the problem that just to the left of $q = b$ marginal revenue is greater than marginal cost, so marginal profit is positive there. Just to the right of $q = b$ marginal revenue is less than marginal cost, so marginal profit is negative there. At $q = b$ marginal profit changes from positive to negative. This means that profit is increasing to the left of b and decreasing to the right. The point $q = b$ corresponds to a local maximum of profit. In fact, since the area between the MC and MR curves in the figure in the text between $q = a$ and $q = b$ is bigger than the area between $q = 0$ and $q = a$, $q = b$ is in fact a global maximum.

5. (a) The fixed cost is 0 because $C(0) = 0$.
   (b) Profit, $\pi(q)$, is equal to money from sales, $7q$, minus total cost to produce those items, $C(q)$.

   $$\pi = 7q - 0.01q^3 + 0.6q^2 - 13q$$
   $$\pi' = -0.03q^2 + 1.2q - 6$$

   $$\pi' = 0 \text{ if } q = \frac{-1.2 \pm \sqrt{(1.2)^2 - 4(0.03)(6)}}{-0.06} \approx 5.9 \text{ or } 34.1.$$

   Now $\pi'' = -0.06q + 1.2$, so $\pi''(5.9) > 0$ and $\pi''(34.1) < 0$. This means $q = 5.9$ is a local min and $q = 34.1$ a local max. We now evaluate the endpoint, $\pi(0) = 0$, and the points nearest $q = 34.1$ with integer q-values:

   $$\pi(35) = 7(35) - 0.01(35)^3 + 0.6(35)^2 - 13(35) = 245 - 148.75 = 96.25,$$

   $$\pi(34) = 7(34) - 0.01(34)^3 + 0.6(34)^2 - 13(34) = 238 - 141.44 = 96.56.$$

   So the (global) maximum profit is $\pi(34) = 96.56$. The money from sales is \$238, the cost to produce the items is \$141.44, resulting in a profit of \$96.56.
   (c) The money from sales is equal to price×quantity sold. If the price is raised from \$7 by \$$x$ to \$$(7+x)$, the result is a reduction in sales from 34 items to $(34 - 2x)$ items. So the result of raising the price by \$$x$ is to change the money from sales from $(7)(34)$ to $(7+x)(34-2x)$ dollars. If the production level is fixed at 34, then the production costs are fixed at \$141.44, as found in part (b), and the profit is given by:

   $$\pi(x) = (7+x)(34-2x) - 141.44$$

   This expression gives the profit as a function of change in price x, rather than as a function of quantity as in part (b). We set the derivative of π with respect to x equal to zero to find the change in price that maximizes the profit:

   $$\frac{d\pi}{dx} = (1)(34 - 2x) + (7+x)(-2) = 20 - 4x = 0$$

   So $x = 5$, and this must give a maximum for $\pi(x)$ since the graph of π is a parabola which opens downwards. The profit when the price is \$12 $(= 7 + x = 7 + 5)$ is thus $\pi(5) = (7+5)(34 - 2(5)) - 141.44 = \$146.56$. This is indeed higher than the profit when the price is \$7, so the smart thing to do is to raise the price by \$5.

6. For each month,

   $$\text{Profit} = \text{Revenue} - \text{Cost}$$
   $$\pi = pq - wL = pcK^\alpha L^\beta - wL$$

   The variable on the right is L, so at the maximum

   $$\frac{d\pi}{dL} = \beta pcK^\alpha L^{\beta-1} - w = 0$$

   Now $\beta - 1$ is negative, since $0 < \beta < 1$, so $1 - \beta$ is positive and we can write

   $$\frac{\beta pcK^\alpha}{L^{1-\beta}} = w$$

   giving

   $$L = \left(\frac{\beta pcK^\alpha}{w}\right)^{\frac{1}{1-\beta}}$$

   Since $\beta - 1$ is negative, when L is just above 0, the quantity $L^{\beta-1}$ is huge and positive, so $d\pi/dL > 0$. When L is large, $L^{\beta-1}$ is small, so $d\pi/dL < 0$. Thus the value of L we have found gives a global maximum, since it is the only critical point.

7. (a) $N = 100 + 20x$, graphed in Figure 4.21.

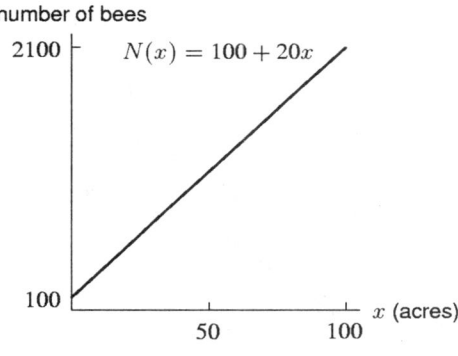

Figure 4.21

(b) $N'(x) = 20$ and its graph is just a horizontal line. This means that rate of increase of the number of bees with acres of clover is constant — each acre of clover brings 20 more bees.

On the other hand, $N(x)/x = 100/x + 20$ means that the average number of bees per acre of clover approaches 20 as more acres are put under clover. See Figure 4.22. As x increases, $100/x$ decreases to 0, so $N(x)/x$ approaches 20 (i.e. $N(x)/x \to 20$). Since the total number of bees is 20 per acre plus the original 100, the average number of bees per acre is 20 plus the 100 shared out over x acres. As x increases, the 100 are shared out over more acres, and so its contribution to the average becomes less. Thus the average number of bees per acre approaches 20 for large x.

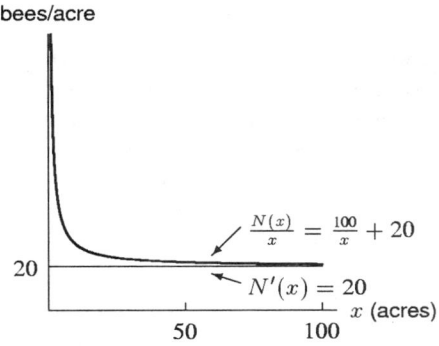

Figure 4.22

8. This question implies that the line from the origin to the point $(x, R(x))$ has some relationship to $r(x)$. The slope of this line is $R(x)/x$, which is $r(x)$. So the point x_0 at which $r(x)$ is maximal will also be the point at which the slope of this line is maximal. The question claims that the line from the origin to $(x_0, R(x_0))$ will be tangent to the graph of $R(x)$. We can understand this by trying to see what would happen if it were otherwise.

If the line from the origin to $(x_0, R(x_0))$ intersects the graph of $R(x)$, but is not tangent to the graph of $R(x)$ at x_0, then there are points of this graph on both sides of the line — and, in particular, there is some point x_1 such that the line from the origin to $(x_1, R(x_1))$ has larger slope than the line to $(x_0, R(x_0))$. (See the graph below.) But we picked x_0 so that no other line had larger slope, and therefore no such x_1 exists. So the original supposition is false, and the line from the origin to $(x_0, R(x_0))$ is tangent to the graph of $R(x)$.

(a) See (b).

(b)

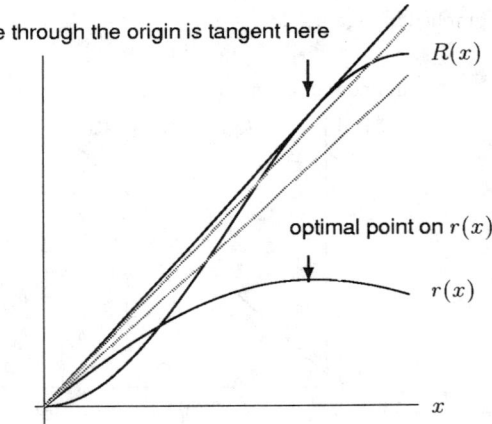

(c)
$$r(x) = \frac{R(x)}{x}$$
$$r'(x) = \frac{xR'(x) - R(x)}{x^2}$$

So when $r(x)$ is maximized $0 = xR'(x) - R(x)$, the numerator of $r'(x)$, or $R'(x) = R(x)/x = r(x)$. i.e. when $r(x)$ is maximized, $r(x) = R'(x)$.

Let us call the x-value at which the maximum of r occurs x_m. Then the line passing through $R(x_m)$ and the origin is $y = x \cdot R(x_m)/x_m$. Its slope is $R(x_m)/x_m$, which also happens to be $r(x_m)$. In the previous paragraph, we showed that at x_m, this is also equal to the slope of the tangent to $R(x)$. So, the line through the origin is the tangent line.

9. (a) The value of MC is the slope of the tangent to the curve at q_0. See Figure 4.23.
   (b) The line from the curve to the origin joins $(0,0)$ and $(q_0, C(q_0))$, so its slope is $C(q_0)/q_0 = a(q_0)$.
   (c) Figure 4.24 shows that the line whose slope is the minimum $a(q)$ is tangent to the curve $C(q)$. This line, therefore, also has slope MC, so $a(q) = MC$ at the q making $a(q)$ minimum.

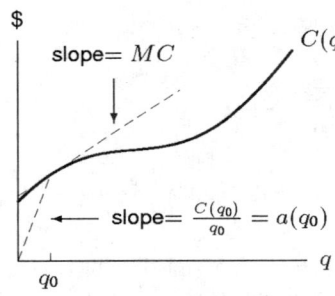

Figure 4.23

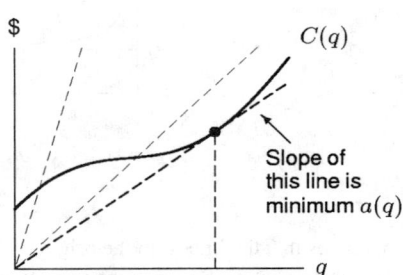

Figure 4.24

10. (a) $a(q) = C(q)/q$, so $C(q) = 0.01q^3 - 0.6q^2 + 13q$.
    (b) Taking the derivative of $C(q)$ gives an expression for the marginal cost:
    $$C'(q) = MC(q) = 0.03q^2 - 1.2q + 13.$$

    To find the smallest MC we take its derivative and find the value of q that makes it zero. So: $MC'(q) = 0.06q - 1.2 = 0$ when $q = 1.2/0.06 = 20$. This value of q must give a minimum because the graph of $MC(q)$ is a parabola opening upwards. Therefore the minimum marginal cost is $MC(20) = 1$. So the marginal cost is at a minimum when the additional cost per item is $1.

(c) $a'(q) = 0.02q - 0.6$
Setting $a'(q) = 0$ and solving for q gives $q = 30$ as the quantity at which the average is minimized, since the graph of a is a parabola which opens upwards. The minimum average cost is $a(30) = 4$ dollars per item.

(d) The marginal cost at $q = 30$ is $MC(30) = 0.03(30)^2 - 1.2(30) + 13 = 4$. This is the same as the average cost at this quantity. Note that since $a(q) = C(q)/q$, we have $a'(q) = (qC'(q) - C(q))/q^2$. At a critical point, q_0, of $a(q)$, we have

$$0 = a'(q_0) = \frac{q_0 C'(q_0) - C(q_0)}{q_0^2},$$

so $C'(q_0) = C(q_0)/q_0 = a(q_0)$. Therefore $C'(30) = a(30) = 4$ dollars per item.

Another way to see why the marginal cost at $q = 30$ must equal the minimum average cost $a(30) = 4$ is to view $C'(30)$ as the approximate cost of producing the 30th or 31st good. If $C'(30) < a(30)$, then producing the 31st good would lower the average cost, i.e. $a(31) < a(30)$. If $C'(30) > a(30)$, then producing the 30th good would raise the average cost, i.e. $a(30) > a(29)$. Since $a(30)$ is the global minimum, we must have $C'(30) = a(30)$.

11. (a) Differentiating $C(q)$ gives

$$C'(q) = \frac{K}{a} q^{(1/a)-1}, \quad C''(q) = \frac{K}{a}\left(\frac{1}{a} - 1\right) q^{(1/a)-2}.$$

If $a > 1$, then $C''(q) < 0$, so C is concave down.

(b) We have

$$a(q) = \frac{C(q)}{q} = \frac{Kq^{1/a} + F}{q}$$

$$C'(q) = \frac{K}{a} q^{(1/a)-1}$$

so $a(q) = C'(q)$ means

$$\frac{Kq^{1/a} + F}{q} = \frac{K}{a} q^{(1/a)-1}.$$

Solving,

$$Kq^{1/a} + F = \frac{K}{a} q^{1/a}$$

$$K\left(\frac{1}{a} - 1\right) q^{1/a} = F$$

$$q = \left[\frac{Fa}{K(1-a)}\right]^a.$$

Solutions for Section 4.5

1. We take the derivative, set it equal to 0, and solve for x:

$$\frac{dt}{dx} = \frac{1}{6} - \frac{1}{4} \cdot \frac{1}{2}\left((2000-x)^2 + 600^2\right)^{-1/2} \cdot 2(2000-x) = 0$$

$$(2000 - x) = \frac{2}{3}\left((2000-x)^2 + 600^2\right)^{1/2}$$

$$(2000 - x)^2 = \frac{4}{9}\left((2000-x)^2 + 600^2\right)$$

$$\frac{5}{9}(2000 - x)^2 = \frac{4}{9} \cdot 600^2$$

$$2000 - x = \sqrt{\frac{4}{5} \cdot 600^2} = \frac{1200}{\sqrt{5}}$$

$$x = 2000 - \frac{1200}{\sqrt{5}} \text{ feet.}$$

Note that $2000 - (1200/\sqrt{5}) \approx 1463$ feet, as given in the example.

2. Call the stacks A and B. (See below.) Assume that A corresponds to k_1, and B corresponds to k_2.

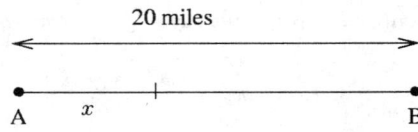

Suppose the point where the concentration of deposit is a minimum occurs at a distance of x miles from stack A. We want to find x such that
$$S = \frac{k_1}{x^2} + \frac{k_2}{(20-x)^2} = k_2\left(\frac{7}{x^2} + \frac{1}{(20-x)^2}\right)$$
is a minimum, which is the same thing as minimizing $f(x) = 7x^{-2} + (20-x)^{-2}$ since k_2 is nonnegative.

We have
$$f'(x) = -14x^{-3} - 2(20-x)^{-3}(-1) = \frac{-14}{x^3} + \frac{2}{(20-x)^3} = \frac{-14(20-x)^3 + 2x^3}{x^3(20-x)^3}.$$

Thus we want to find x such that $-14(20-x)^3 + 2x^3 = 0$, which implies $2x^3 = 14(20-x)^3$. That's equivalent to $x^3 = 7(20-x)^3$, or $\frac{20-x}{x} = (1/7)^{1/3} \approx 0.523$. Solving for x, we have $20 - x = 0.523x$, whence $x = 20/1.523 \approx 13.13$.

To verify that this minimizes f, we take the second derivative:
$$f''(x) = 42x^{-4} + 6(20-x)^{-4} = \frac{42}{x^4} + \frac{6}{(20-x)^4} > 0$$
for any $0 < x < 20$, so by the second derivative test the concentration is minimized 13.13 miles from A.

3. We only consider $\lambda > 0$. For such λ, the value of $v \to \infty$ as $\lambda \to \infty$ and as $\lambda \to 0^+$. Thus, v does not have a maximum velocity. It will have a minimum velocity. To find it, we set $dv/d\lambda = 0$:
$$\frac{dv}{d\lambda} = k\frac{1}{2}\left(\frac{\lambda}{c} + \frac{c}{\lambda}\right)^{-1/2}\left(\frac{1}{c} - \frac{c}{\lambda^2}\right) = 0.$$

Solving, and remembering that $\lambda > 0$, we obtain
$$\frac{1}{c} - \frac{c}{\lambda^2} = 0$$
$$\frac{1}{c} = \frac{c}{\lambda^2}$$
$$\lambda^2 = c^2,$$
so
$$\lambda = c.$$

Thus, we have one critical point. Since
$$\frac{dv}{d\lambda} < 0 \quad \text{for } \lambda < c$$
and
$$\frac{dv}{d\lambda} > 0 \quad \text{for } \lambda > c,$$
the first derivative test tells us that we have a local minimum of v at $x = c$. Since $\lambda = c$ is the only critical point, it gives the global minimum. Thus the minimum value of v is
$$v = k\sqrt{\frac{c}{c} + \frac{c}{c}} = \sqrt{2}k.$$

4. Let w and l be the width and length, respectively, of the rectangular area you wish to enclose. Then
$$w + w + l = 100 \text{ feet}$$
$$l = 100 - 2w$$
$$\text{Area} = w \cdot l = w(100 - 2w) = 100w - 2w^2$$

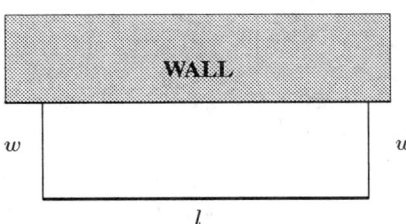

To maximize area, we solve $A' = 0$ to find critical points. This gives $A' = 100 - 4w = 0$, so $w = 25, l = 50$. So the area is $25 \cdot 50 = 1250$ square feet. This is a local maximum by the second derivative test because $A'' = -4 < 0$. Since the graph of A is a parabola, the local maximum is in fact a global maximum.

5. (a) Suppose the height of the box is h. The box has six sides, four with area xh and two, the top and bottom, with area x^2. Thus,
$$4xh + 2x^2 = A.$$
So
$$h = \frac{A - 2x^2}{4x}.$$
Then, the volume, V, is given by
$$V = x^2 h = x^2 \left(\frac{A - 2x^2}{4x}\right) = \frac{x}{4}\left(A - 2x^2\right)$$
$$= \frac{A}{4}x - \frac{1}{2}x^3.$$

(b) The graph is shown below. We are assuming A is a positive constant. Also, we have drawn the whole graph, but we should only consider $V > 0, x > 0$ as V and x are lengths.

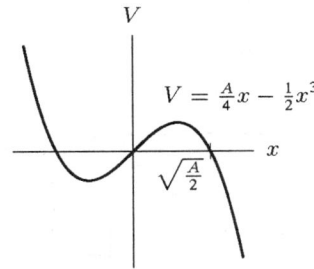

(c) To find the maximum, we differentiate, regarding A as a constant:
$$\frac{dV}{dx} = \frac{A}{4} - \frac{3}{2}x^2.$$
So $dV/dx = 0$ if
$$\frac{A}{4} - \frac{3}{2}x^2 = 0$$
$$x = \pm\sqrt{\frac{A}{6}}.$$
For a real box, we must use $x = \sqrt{A/6}$. The graph makes it clear that this value of x gives the maximum. Evaluating at $x = \sqrt{A/6}$, we get
$$V = \frac{A}{4}\sqrt{\frac{A}{6}} - \frac{1}{2}\left(\sqrt{\frac{A}{6}}\right)^3 = \frac{A}{4}\sqrt{\frac{A}{6}} - \frac{1}{2} \cdot \frac{A}{6}\sqrt{\frac{A}{6}} = \left(\frac{A}{6}\right)^{3/2}.$$

6. Consider the rectangle of sides x and y shown in the figure below.

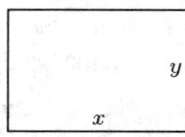

The total area is $xy = 3000$, so $y = 3000/x$. Suppose the left and right edges and the lower edge have the shrubs and the top edge has the fencing. The total cost is

$$C = 25(x + 2y) + 10(x)$$
$$= 35x + 50y.$$

Since $y = 3000/x$, this reduces to

$$C(x) = 35x + 50(3000/x) = 35x + 150{,}000/x.$$

Therefore, $C'(x) = 35 - 150{,}000/x^2$. We set this to 0 to find the critical points:

$$35 - \frac{150{,}000}{x^2} = 0$$
$$\frac{150{,}000}{x^2} = 35$$
$$x^2 = 4285.71$$
$$x \approx 65.5 \text{ ft}$$

so that

$$y = 3000/x \approx 45.8 \text{ ft}.$$

Since $C(x) \to \infty$ as $x \to 0^+$ and $x \to \infty$, $x = 65.5$ is a minimum. The minimum total cost is then

$$C(65.5) \approx \$4583.$$

7. Consider the diagram shown below where the pool has dimensions x by y and the deck extends 5 feet at either side and 10 feet at the ends of the pool.

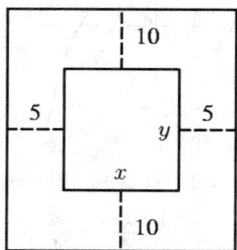

The dimensions of the plot of land containing the pool are then $(x + 5 + 5)$ by $(y + 10 + 10)$. The area of the land is then

$$A = (x + 10)(y + 20),$$

which is to be minimized. We also are told that the area of the pool is $xy = 1800$, so

$$y = 1800/x$$

and

$$A = (x + 10)\left(\frac{1800}{x} + 20\right)$$
$$= 1800 + 20x + \frac{18000}{x} + 200.$$

We find dA/dx and set it to zero to get

$$\frac{dA}{dx} = 20 - \frac{18000}{x^2} = 0$$
$$20x^2 = 18000$$
$$x^2 = 900$$
$$x = 30 \text{ feet.}$$

Since $A \to \infty$ as $x \to 0^+$ and as $x \to \infty$, this critical point must be a global minimum. Also, $y = 1800/30 = 60$ feet. The plot of land is therefore $(30 + 10) = 40$ by $(60 + 20) = 80$ feet.

8. From the triangle shown below, we see that

$$\left(\frac{w}{2}\right)^2 + \left(\frac{h}{2}\right)^2 = 30^2$$
$$w^2 + h^2 = 4(30)^2 = 3600.$$

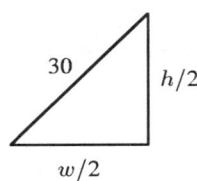

The strength, S, of the beam is given by
$$S = kwh^2,$$
for some constant k. To make S a function of only one variable, substitute for h^2, giving

$$S = kw(3600 - w^2) = k(3600w - w^3).$$

Differentiating and setting $dS/dw = 0$,
$$\frac{dS}{dw} = k(3600 - 3w^2) = 0.$$

Solving for w gives
$$w = \sqrt{1200} = 34.64 \text{ cm,}$$

so
$$h^2 = 3600 - w^2 = 3600 - 1200 = 2400$$
$$h = \sqrt{2400} = 48.99 \text{ cm.}$$

Thus, $w = 34.64$ cm and $h = 48.99$ cm give a critical point. To check that this is a local maximum, we compute

$$\frac{d^2S}{dw^2} = -6w < 0 \quad \text{for} \quad w > 0.$$

Since $d^2S/dw^2 < 0$, we see that $w = 34.64$ cm is a local maximum. It is the only critical point, so it is a global maximum.

9. If the illumination is represented by I, then we know that

$$I = \frac{k \cos \theta}{r^2}.$$

See Figure 4.25.

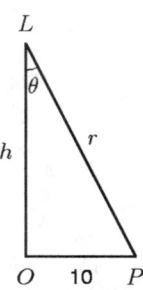

Figure 4.25

Since $r^2 = h^2 + 10^2$ and $\cos\theta = h/r = h/\sqrt{h^2 + 10^2}$, we have

$$I = \frac{kh}{(h^2 + 10^2)^{3/2}}.$$

To find the height at which I is maximized, we differentiate

$$\frac{dI}{dh} = \frac{k}{(h^2 + 10^2)^{3/2}} - \frac{3kh(2h)}{2(h^2 + 10^2)^{5/2}} = \frac{k(h^2 + 10^2) - 3kh^2}{(h^2 + 10^2)^{5/2}} = \frac{k(10^2 - 2h^2)}{(h^2 + 10^2)^{5/2}}.$$

Setting $dI/dh = 0$ gives

$$10^2 - 2h^2 = 0$$
$$h = \sqrt{50} \text{ meters.}$$

Since $dI/dh > 0$ for $0 \leq h < \sqrt{50}$ and $dI/dh < 0$ for $h > \sqrt{50}$, we know that I is a maximum when $h = \sqrt{50}$ meters.

10. We see that the width of the tunnel is $2r$. The area of the rectangle is then $(2r)h$. The area of the semicircle is $(\pi r^2)/2$. The cross-sectional area, A, is then

$$A = 2rh + \frac{1}{2}\pi r^2$$

and the perimeter, P, is

$$P = 2h + 2r + \pi r.$$

From $A = 2rh + (\pi r^2)/2$ we get

$$h = \frac{A}{2r} - \frac{\pi r}{4}.$$

Thus,

$$P = 2\left(\frac{A}{2r} - \frac{\pi r}{4}\right) + 2r + \pi r = \frac{A}{r} + 2r + \frac{\pi r}{2}.$$

We now have the perimeter in terms of r and the constant A. Differentiating, we obtain

$$\frac{dP}{dr} = -\frac{A}{r^2} + 2 + \frac{\pi}{2}.$$

To find the critical points we set $P' = 0$:

$$-\frac{A}{r^2} + \frac{\pi}{2} + 2 = 0$$
$$\frac{r^2}{A} = \frac{2}{4 + \pi}$$
$$r = \sqrt{\frac{2A}{4 + \pi}}.$$

Substituting this back into our expression for h, we have

$$h = \frac{A}{2} \cdot \frac{\sqrt{4+\pi}}{\sqrt{2A}} - \frac{\pi}{4} \cdot \frac{\sqrt{2A}}{\sqrt{4+\pi}}.$$

Since $P \to \infty$ as $r \to 0^+$ and as $r \to \infty$, this critical point must be a global minimum. Notice that the h-value simplifies to

$$h = \sqrt{\frac{2A}{4+\pi}} = r.$$

11. The distance from a given point on the parabola (x, x^2) to $(1, 0)$ is given by
$$D = \sqrt{(x-1)^2 + (x^2 - 0)^2}.$$
Minimizing this is equivalent to minimizing $d = (x-1)^2 + x^4$. (We can ignore the square root if we are only interested in minimizing because the square root is smallest when the thing it is the square root of is smallest.) To minimize d, we find its critical points by solving $d' = 0$. Since $d = (x-1)^2 + x^4 = x^2 - 2x + 1 + x^4$,
$$d' = 2x - 2 + 4x^3 = 2(2x^3 + x - 1).$$
By graphing $d' = 2(2x^3 + 2x - 1)$ on a calculator, we see that it has only 1 root, $x \approx 0.59$. This must give a minimum because $d \to \infty$ as $x \to -\infty$ and as $x \to +\infty$, and d has only one critical point. This is confirmed by the second derivative test: $d'' = 12x^2 + 2 = 2(6x^2 + 1)$, which is always positive. Thus the point $(0.59, 0.59^2) \approx (0.59, 0.35)$ is approximately the closest point of $y = x^2$ to $(1, 0)$.

12. Any point on the curve can be written (x, x^2). The distance between such a point and $(3, 0)$ is given by
$$s(x) = \sqrt{(3-x)^2 + (0-x^2)^2} = \sqrt{(3-x)^2 + x^4}.$$
Plotting this function in Figure 4.26, we see that there is a minimum near $x = 1$.

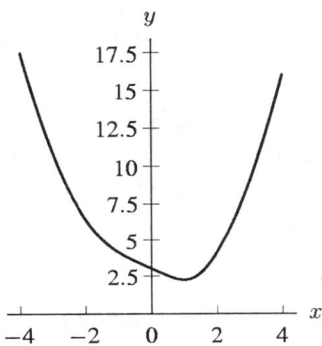

Figure 4.26

To find the value of x that minimizes the distance we can instead minimize the function $Q = s^2$ (the derivative is simpler). Then we have
$$Q(x) = (3-x)^2 + x^4.$$
Differentiating $Q(x)$ gives
$$\frac{dQ}{dx} = -6 + 2x + 4x^3.$$
Plotting the function $4x^3 + 2x - 6$ shows that there is one real solution at $x = 1$, which can be verified by substitution; the required coordinates are therefore $(1, 1)$. Because $Q''(x) = 2 + 12x^2$ is always positive, $x = 1$ is indeed the minimum. See Figure 4.27.

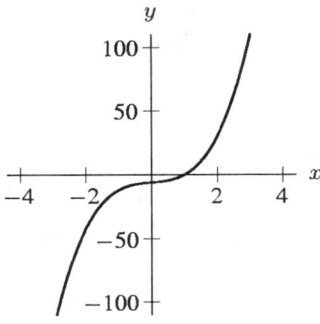

Figure 4.27

13. Let the sides of the rectangle have lengths a and b. We shall look for the minimum of the square s of the length of either diagonal, i.e. $s = a^2 + b^2$. The area is $A = ab$, so $b = A/a$. This gives

$$s(a) = a^2 + \frac{A^2}{a^2}.$$

To find the minimum squared length we need to find the critical points of s. Differentiating s with respect to a gives

$$\frac{ds}{da} = 2a + (-2)A^2 a^{-3} = 2a\left(1 - \frac{A^2}{a^4}\right).$$

The derivative $ds/da = 0$ when $a = \sqrt{A}$, that is when $a = b$ and so the rectangle is a square. Because $\frac{d^2s}{da^2} = 2\left(1 + \frac{3A^2}{a^4}\right) > 0$, this is a minimum.

14. Let x equal the number of chairs ordered in excess of 300, so $0 \leq x \leq 100$.

$$\text{Revenue} = R = (90 - 0.25x)(300 + x)$$
$$= 27{,}000 - 75x + 90x - 0.25x^2 = 27{,}000 + 15x - 0.25x^2$$

At a critical point $dR/dx = 0$. Since $dR/dx = 15 - 0.5x$, we have $x = 30$, and the maximum revenue is $\$27{,}225$ since the graph of R is a parabola which opens downwards. The minimum is $\$0$ (when no chairs are sold).

15. If v is the speed of the boat in miles per hour, then

$$\text{Cost of fuel per hour (in \$/hour)} = kv^3,$$

where k is the constant of proportionality. To find k, use the information that the boat uses $\$100$ worth of fuel per hour when cruising at 10 miles per hour: $100 = k10^3$, so $k = 100/10^3 = 0.1$. Thus,

$$\text{Cost of fuel per hour (in \$/hour)} = 0.1v^3.$$

From the given information, we also have

$$\text{Cost of other operations (labor, maintenance, etc.) per hour (in \$/hour)} = 675.$$

So

$$\text{Total Cost per hour (in \$/hour)} = \text{Cost of fuel (in \$/hour)} + \text{Cost of other (in \$/hour)}$$
$$= 0.1v^3 + 675.$$

However, we want to find the Cost per *mile*, which is the Total Cost per *hour* divided by the number of miles that the ferry travels in one hour. Since v is the speed in miles/hour at which the ferry travels, the number of miles that the ferry travels in one hour is simply v miles. Let $C = $ Cost per *mile*. Then

$$\text{Cost per \emph{mile} (in \$/mile)} = \frac{\text{Total Cost per \emph{hour} (in \$/hour)}}{\text{Distance traveled per hour (in miles/hour)}}$$

$$C = \frac{0.1v^3 + 675}{v} = 0.1v^2 + \frac{675}{v}.$$

We also know that $0 < v < \infty$. To find the speed at which Cost per *mile* is minimized, set

$$\frac{dC}{dv} = 2(0.1)v - \frac{675}{v^2} = 0$$

so

$$2(0.1)v = \frac{675}{v^2}$$
$$v^3 = \frac{675}{2(0.1)} = 3375$$
$$v = 15 \text{ miles/hour}.$$

Since

$$\frac{d^2C}{dv^2} = 0.2 + \frac{2(675)}{v^3} > 0$$

for $v > 0$, $v = 15$ gives a local minimum for C by the second-derivative test. Since this is the only critical point for $0 < v < \infty$, it must give a global minimum.

16. (a) We have
$$x^{1/x} = e^{\ln(x^{1/x})} = e^{(1/x)\ln x}.$$

Thus
$$\frac{d(x^{1/x})}{dx} = \frac{d(e^{(1/x)\ln x})}{dx} = \frac{d(\frac{1}{x}\ln x)}{dx} e^{(1/x)\ln x}$$
$$= \left(-\frac{\ln x}{x^2} + \frac{1}{x^2}\right) x^{1/x}$$
$$= \frac{x^{1/x}}{x^2}(1 - \ln x) \begin{cases} = 0 & \text{when } x = e \\ < 0 & \text{when } x > e \\ > 0 & \text{when } x < e. \end{cases}$$

Hence $e^{1/e}$ is the global maximum for $x^{1/x}$, by the first derivative test.

(b) Since $x^{1/x}$ is increasing for $0 < x < e$ and decreasing for $x > e$, and 2 and 3 are the closest integers to e, either $2^{1/2}$ or $3^{1/3}$ is the maximum for $n^{1/n}$. We have $2^{1/2} \approx 1.414$ and $3^{1/3} \approx 1.442$, so $3^{1/3}$ is the maximum.

(c) Since $e < 3 < \pi$, and $x^{1/x}$ is decreasing for $x > e$, $3^{1/3} > \pi^{1/\pi}$.

17. (a) If, following the hint, we set $f(x) = (a+x)/2 - \sqrt{ax}$, then $f(x)$ represents the difference between the arithmetic and geometric means for some fixed a and any $x > 0$. We can find where this difference is minimized by solving $f'(x) = 0$. Since $f'(x) = \frac{1}{2} - \frac{1}{2}\sqrt{a}x^{-1/2}$, if $f'(x) = 0$ then $\frac{1}{2}\sqrt{a}x^{-1/2} = \frac{1}{2}$, or $x = a$. Since $f''(x) = \frac{1}{4}\sqrt{a}x^{-3/2}$ is positive for all positive x, by the second derivative test $f(x)$ has a minimum at $x = a$, and $f(a) = 0$. Thus $f(x) = (a+x)/2 - \sqrt{ax} \geq 0$ for all $x > 0$, which means $(a+x)/2 \geq \sqrt{ax}$. This means that the arithmetic mean is greater than the geometric mean unless $a = x$, in which case the two means are equal.

Alternatively, and without using calculus, we obtain
$$\frac{a+b}{2} - \sqrt{ab} = \frac{a - 2\sqrt{ab} + b}{2}$$
$$= \frac{(\sqrt{a} - \sqrt{b})^2}{2} \geq 0,$$

and again we have $(a+b)/2 \geq \sqrt{ab}$.

(b) Following the hint, set $f(x) = \frac{a+b+x}{3} - \sqrt[3]{abx}$. Then $f(x)$ represents the difference between the arithmetic and geometric means for some fixed a, b and any $x > 0$. We can find where this difference is minimized by solving $f'(x) = 0$. Since $f'(x) = \frac{1}{3} - \frac{1}{3}\sqrt[3]{ab}x^{-2/3}$, $f'(x) = 0$ implies that $\frac{1}{3}\sqrt[3]{ab}x^{-2/3} = \frac{1}{3}$, or $x = \sqrt{ab}$. Since $f''(x) = \frac{2}{9}\sqrt[3]{ab}x^{-5/3}$ is positive for all positive x, by the second derivative test $f(x)$ has a minimum at $x = \sqrt{ab}$. But
$$f(\sqrt{ab}) = \frac{a+b+\sqrt{ab}}{3} - \sqrt[3]{ab\sqrt{ab}} = \frac{a+b+\sqrt{ab}}{3} - \sqrt{ab} = \frac{a+b-2\sqrt{ab}}{3}.$$

By the first part of this problem, we know that $\frac{a+b}{2} - \sqrt{ab} \geq 0$, which implies that $a + b - 2\sqrt{ab} \geq 0$. Thus $f(\sqrt{ab}) = \frac{a+b-2\sqrt{ab}}{3} \geq 0$. Since f has a maximum at $x = \sqrt{ab}$, $f(x)$ is always nonnegative. Thus $f(x) = \frac{a+b+x}{3} - \sqrt[3]{abx} \geq 0$, so $\frac{a+b+c}{3} \geq \sqrt[3]{abc}$. Note that equality holds only when $a = b = c$. (Part (b) may also be done without calculus, but it's harder than (a).)

18. Let x be as indicated in the figure in the text. Then the distance from S to Town 1 is $\sqrt{1+x^2}$ and the distance from S to Town 2 is $\sqrt{(4-x)^2 + 4^2} = \sqrt{x^2 - 8x + 32}$.

$$\text{Total length of pipe} = f(x) = \sqrt{1+x^2} + \sqrt{x^2 - 8x + 32}.$$

We want to look for critical points of f. The easiest way is to graph f and see that it has a local minimum at about $x = 0.8$ miles. Alternatively, we can use the formula:

$$f'(x) = \frac{2x}{2\sqrt{1+x^2}} + \frac{2x-8}{2\sqrt{x^2-8x+32}}$$
$$= \frac{x}{\sqrt{1+x^2}} + \frac{x-4}{\sqrt{x^2-8x+32}}$$
$$= \frac{x\sqrt{x^2-8x+32} + (x-4)\sqrt{1+x^2}}{\sqrt{1+x^2}\sqrt{x^2-8x+32}} = 0.$$

$f'(x)$ is equal to zero when the numerator is equal to zero.

$$x\sqrt{x^2 - 8x + 32} + (x - 4)\sqrt{1 + x^2} = 0$$
$$x\sqrt{x^2 - 8x + 32} = (4 - x)\sqrt{1 + x^2}.$$

Squaring both sides and simplifying, we get

$$x^2(x^2 - 8x + 32) = (x^2 - 8x + 16)(14x^2)$$
$$x^4 - 8x^3 + 32x^2 = x^4 - 8x^3 + 17x^2 - 8x + 16$$
$$15x^2 + 8x - 16 = 0,$$
$$(3x + 4)(5x - 4) = 0.$$

So $x = 4/5$. (Discard $x = -4/3$ since we are only interested in x between 0 and 4, between the two towns.) Using the second derivative test, we can verify that $x = 4/5$ is a local minimum.

19. (a) The distance the pigeon flies over water is

$$\overline{BP} = \frac{\overline{AB}}{\sin \theta} = \frac{500}{\sin \theta},$$

and over land is

$$\overline{PL} = \overline{AL} - \overline{AP} = 2000 - \frac{500}{\tan \theta} = 2000 - \frac{500 \cos \theta}{\sin \theta}.$$

Therefore the energy required is

$$E = 2e \left(\frac{500}{\sin \theta}\right) + e \left(2000 - \frac{500 \cos \theta}{\sin \theta}\right)$$
$$= 500e \left(\frac{2 - \cos \theta}{\sin \theta}\right) + 2000e, \quad \text{for} \quad \arctan\left(\frac{500}{2000}\right) \leq \theta \leq \frac{\pi}{2}.$$

(b) Notice that E and the function $f(\theta) = \frac{2 - \cos \theta}{\sin \theta}$ must have the same critical points since the graph of E is just a stretch and a vertical shift of the graph of f. The graph of $\frac{2 - \cos \theta}{\sin \theta}$ for $\arctan(\frac{500}{2000}) \leq \theta \leq \frac{\pi}{2}$ in Figure 4.28 shows that E has precisely one critical point, and that a minimum for E occurs at this point.

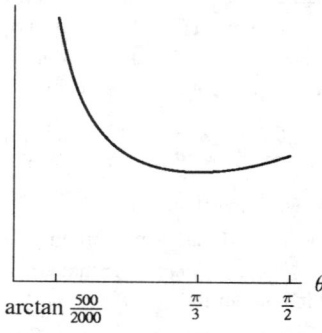

Figure 4.28: Graph of $f(\theta) = \frac{2-\cos\theta}{\sin\theta}$ for $\arctan(\frac{500}{2000}) \leq \theta \leq \frac{\pi}{2}$

To find the critical point θ, we solve $f'(\theta) = 0$ or

$$E' = 0 = 500e \left(\frac{\sin\theta \cdot \sin\theta - (2 - \cos\theta) \cdot \cos\theta}{\sin^2 \theta}\right)$$
$$= 500e \left(\frac{1 - 2\cos\theta}{\sin^2 \theta}\right).$$

Therefore $1 - 2\cos\theta = 0$ and so $\theta = \pi/3$.

(c) Letting $a = \overline{AB}$ and $b = \overline{AL}$, our formula for E becomes

$$E = 2e \left(\frac{a}{\sin\theta}\right) + e \left(b - \frac{a \cos\theta}{\sin\theta}\right)$$
$$= ea \left(\frac{2 - \cos\theta}{\sin\theta}\right) + eb, \quad \text{for} \quad \arctan\left(\frac{a}{b}\right) \leq \theta \leq \frac{\pi}{2}.$$

Again, the graph of E is just a stretch and a vertical shift of the graph of $\dfrac{2 - \cos \theta}{\sin \theta}$. Thus, the critical point $\theta = \pi/3$ is independent of e, a, and b. But the maximum of E *on the domain* $\arctan(a/b) \leq \theta \leq \frac{\pi}{2}$ is dependent on the ratio $b = \dfrac{\overline{AB}}{\overline{AL}}$. In other words, the optimal angle is $\theta = \pi/3$ provided $\arctan(a/b) \leq \frac{\pi}{3}$; otherwise, the optimal angle is $\arctan(a/b)$, which means the pigeon should fly over the lake for the entire trip—this occurs when $a/b > 1.733$.

20. We want to maximize the viewing angle, which is $\theta = \theta_1 - \theta_2$.

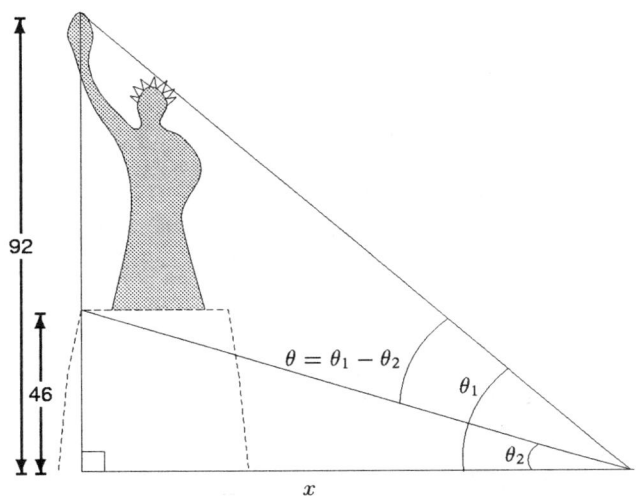

Now
$$\tan(\theta_1) = \frac{92}{x} \quad \text{so } \theta_1 = \arctan\left(\frac{92}{x}\right)$$
$$\tan(\theta_2) = \frac{46}{x} \quad \text{so } \theta_2 = \arctan\left(\frac{46}{x}\right).$$

Then
$$\theta = \arctan\left(\frac{92}{x}\right) - \arctan\left(\frac{46}{x}\right) \quad \text{for} \quad x > 0.$$

We look for critical points of the function by computing $d\theta/dx$:
$$\frac{d\theta}{dx} = \frac{1}{1 + (92/x)^2}\left(\frac{-92}{x^2}\right) - \frac{1}{1 + (46/x)^2}\left(\frac{-46}{x^2}\right)$$
$$= \frac{-92}{x^2 + 92^2} - \frac{-46}{x^2 + 46^2}$$
$$= \frac{-92(x^2 + 46^2) + 46(x^2 + 92^2)}{(x^2 + 92^2) \cdot (x^2 + 46^2)}$$
$$= \frac{46(4232 - x^2)}{(x^2 + 92^2) \cdot (x^2 + 46^2)}.$$

Setting $d\theta/dx = 0$ gives
$$x^2 = 4232$$
$$x = \pm\sqrt{4232}.$$

Since $x > 0$, the critical point is $x = \sqrt{4232} \approx 65.1$ meters. To verify that this is indeed where θ attains a maximum, we note that $d\theta/dx > 0$ for $0 < x < \sqrt{4232}$ and $d\theta/dx < 0$ for $x > \sqrt{4232}$. By the First Derivative Test, θ attains a maximum at $x = \sqrt{4232} \approx 65.1$.

21. (a) Since $RB' = x$ and $A'R = c - x$, we have
$$AR = \sqrt{a^2 + (c-x)^2} \quad \text{and} \quad RB = \sqrt{b^2 + x^2}.$$

See Figure 4.29.

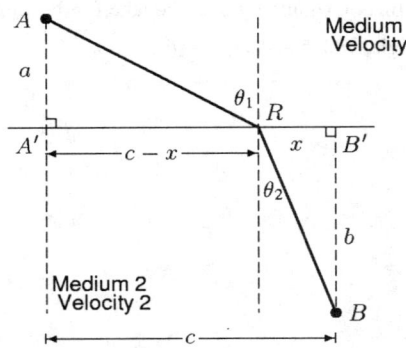

Figure 4.29

The time traveled, T, is given by
$$T = \text{Time } AR + \text{Time } RB = \frac{\text{Distance } AR}{v_1} + \frac{\text{Distance } RB}{v_2}$$
$$= \frac{\sqrt{a^2 + (c-x)^2}}{v_1} + \frac{\sqrt{b^2 + x^2}}{v_2}.$$

(b) Let us calculate dT/dx:
$$\frac{dT}{dx} = \frac{-2(c-x)}{2v_1\sqrt{a^2 + (c-x)^2}} + \frac{2x}{2v_2\sqrt{b^2 + x^2}}.$$

At the minimum $dT/dx = 0$, so
$$\frac{c-x}{v_1\sqrt{a^2 + (c-x)^2}} = \frac{x}{v_2\sqrt{b^2 + x^2}}.$$

But we have
$$\sin\theta_1 = \frac{c-x}{\sqrt{a^2 + (c-x)^2}} \quad \text{and} \quad \sin\theta_2 = \frac{x}{\sqrt{b^2 + x^2}}.$$

Therefore, setting $dT/dx = 0$ tells us that
$$\frac{\sin\theta_1}{v_1} = \frac{\sin\theta_2}{v_2}$$

which gives
$$\frac{\sin\theta_1}{\sin\theta_2} = \frac{v_1}{v_2}.$$

22. We know that the time taken is given by
$$T = \frac{\sqrt{a^2 + (c-x)^2}}{v_1} + \frac{\sqrt{b^2 + x^2}}{v_2}$$
$$\frac{dT}{dx} = \frac{-(c-x)}{v_1\sqrt{a^2 + (c-x)^2}} + \frac{x}{v_2\sqrt{b^2 + x^2}}.$$

Differentiating again gives
$$\frac{d^2T}{dx^2} = \frac{1}{v_1\sqrt{a^2+(c-x)^2}} + \frac{(c-x)(-2(c-x))}{2v_1(a^2+(c-x)^2)^{3/2}} + \frac{1}{v_2\sqrt{b^2+x^2}} - \frac{x(2x)}{2v_2(b^2+x^2)^{3/2}}$$
$$= \frac{a^2 + (c-x)^2 - (c-x)^2}{v_1(a^2+(c-x)^2)^{3/2}} + \frac{b^2+x^2-x^2}{v_2(b^2+x^2)^{3/2}}$$
$$= \frac{a^2}{v_1(a^2+(c-x)^2)^{3/2}} + \frac{b^2}{v_2(b^2+x^2)^{3/2}}.$$

This expression for d^2T/dx^2 shows that for any value of x, a, c, v_1, and v_2 with $v_1, v_2 > 0$, we have $d^2T/dx^2 > 0$. Thus, any critical point must be a local minimum. Since there is only one critical point, it must be a global minimum.

Solutions for Section 4.6

1. Substitute $x = 0$ into the formula for $\sinh x$. This yields
$$\sinh 0 = \frac{e^0 - e^{-0}}{2} = \frac{1-1}{2} = 0.$$

2. Substituting $-x$ for x in the formula for $\sinh x$ gives
$$\sinh(-x) = \frac{e^{-x} - e^{-(-x)}}{2} = \frac{e^{-x} - e^x}{2} = -\frac{e^x - e^{-x}}{2} = -\sinh x.$$

3. The graph of $\sinh x$ in the text suggests that
$$\text{As } x \to \infty, \quad \sinh x \to \frac{1}{2}e^x.$$
$$\text{As } x \to -\infty, \quad \sinh x \to -\frac{1}{2}e^{-x}.$$

   Using the facts that
$$\text{As } x \to \infty, \quad e^{-x} \to 0,$$
$$\text{As } x \to -\infty, \quad e^x \to 0,$$

   we can obtain the same results analytically:
$$\text{As } x \to \infty, \quad \sinh x = \frac{e^x - e^{-x}}{2} \to \frac{1}{2}e^x.$$
$$\text{As } x \to -\infty, \quad \sinh x = \frac{e^x - e^{-x}}{2} \to -\frac{1}{2}e^{-x}.$$

4. First we observe that
$$\sinh(2x) = \frac{e^{2x} - e^{-2x}}{2}.$$

   Now let's calculate
$$(\sinh x)(\cosh x) = \left(\frac{e^x - e^{-x}}{2}\right)\left(\frac{e^x + e^{-x}}{2}\right)$$
$$= \frac{(e^x)^2 - (e^{-x})^2}{4}$$
$$= \frac{e^{2x} - e^{-2x}}{4}$$
$$= \frac{1}{2}\sinh(2x).$$

   Thus, we see that
$$\sinh(2x) = 2\sinh x \cosh x.$$

5. First, we observe that
$$\cosh(2x) = \frac{e^{2x} + e^{-2x}}{2}.$$

Now let's use the fact that $e^x \cdot e^{-x} = 1$ to calculate
$$\cosh^2 x = \left(\frac{e^x + e^{-x}}{2}\right)^2$$
$$= \frac{(e^x)^2 + 2e^x \cdot e^{-x} + (e^{-x})^2}{4}$$
$$= \frac{e^{2x} + 2 + e^{-2x}}{4}.$$

Similarly, we have
$$\sinh^2 x = \left(\frac{e^x - e^{-x}}{2}\right)^2$$
$$= \frac{(e^x)^2 - 2e^x \cdot e^{-x} + (e^{-x})^2}{4}$$
$$= \frac{e^{2x} - 2 + e^{-2x}}{4}.$$

Thus, to obtain $\cosh(2x)$, we need to add (rather than subtract) $\cosh^2 x$ and $\sinh^2 x$, giving
$$\cosh^2 x + \sinh^2 x = \frac{e^{2x} + 2 + e^{-2x} + e^{2x} - 2 + e^{-2x}}{4}$$
$$= \frac{2e^{2x} + 2e^{-2x}}{4}$$
$$= \frac{e^{2x} + e^{-2x}}{2}$$
$$= \cosh(2x).$$

Thus, we see that the identity relating $\cosh(2x)$ to $\cosh x$ and $\sinh x$ is
$$\cosh(2x) = \cosh^2 x + \sinh^2 x.$$

6. Using the formula for $\sinh x$ and the fact that $d(e^{-x})/dx = -e^{-x}$, we see that
$$\frac{d}{dx}\left(\frac{e^x - e^{-x}}{2}\right) = \frac{e^x + e^{-x}}{2} = \cosh x.$$

7. Using the chain rule, $\dfrac{d}{dx}\left(\cosh(2x)\right) = \left(\sinh(2x)\right) \cdot 2 = 2\sinh(2x)$.

8. Using the chain rule, $\dfrac{d}{dz}\left(\sinh(3z+5)\right) = \cosh(3z+5) \cdot 3 = 3\cosh(3z+5)$.

9. Using the chain rule twice, $\dfrac{d}{dt}\left(\cosh(e^{t^2})\right) = \sinh(e^{t^2}) \cdot e^{t^2} \cdot 2t = 2te^{t^2}\sinh(e^{t^2})$.

10. Using the chain rule twice,
$$\frac{d}{dy}\left(\sinh\left(\sinh(3y)\right)\right) = \cosh\left(\sinh(3y)\right) \cdot \cosh(3y) \cdot 3$$
$$= 3\cosh(3y) \cdot \cosh\left(\sinh(3y)\right).$$

11. Using the chain rule,
$$\frac{d}{d\theta}\left(\ln\left(\cosh(1+\theta)\right)\right) = \frac{1}{\cosh(1+\theta)} \cdot \sinh(1+\theta) = \frac{\sinh(1+\theta)}{\cosh(1+\theta)} = \tanh(1+\theta).$$

12. For $-5 \leq x \leq 5$, we have the graphs of $y = a\cosh(x/a)$ shown below.

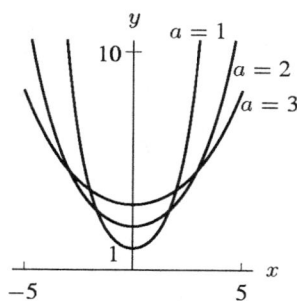

Increasing the value of a makes the graph flatten out and raises the minimum value. The minimum value of y occurs at $x = 0$ and is given by

$$y = a\cosh\left(\frac{0}{a}\right) = a\left(\frac{e^{0/a} + e^{-0/a}}{2}\right) = a.$$

13. (a) The graph in Figure 4.30 looks like the graph of $y = \cosh x$, with the minimum at about $(0.5, 6.3)$.

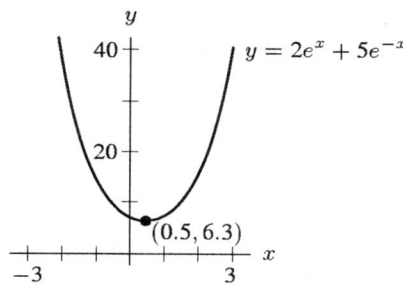

Figure 4.30

(b) We want to write

$$y = 2e^x + 5e^{-x} = A\cosh(x - c) = \frac{A}{2}e^{x-c} + \frac{A}{2}e^{-(x-c)}$$
$$= \frac{A}{2}e^x e^{-c} + \frac{A}{2}e^{-x}e^{c}$$
$$= \left(\frac{Ae^{-c}}{2}\right)e^x + \left(\frac{Ae^{c}}{2}\right)e^{-x}.$$

Thus, we need to choose A and c so that

$$\frac{Ae^{-c}}{2} = 2 \quad \text{and} \quad \frac{Ae^{c}}{2} = 5.$$

Dividing gives

$$\frac{Ae^c}{Ae^{-c}} = \frac{5}{2}$$
$$e^{2c} = 2.5$$
$$c = \frac{1}{2}\ln 2.5 \approx 0.458.$$

Solving for A gives

$$A = \frac{4}{e^{-c}} = 4e^c \approx 6.325.$$

Thus,

$$y = 6.325\cosh(x - 0.458).$$

Rewriting the function in this way shows that the graph in part (a) is the graph of $\cosh x$ shifted to the right by 0.458 and stretched vertically by a factor of 6.325.

14. We want to show that for any A, B with $A > 0$, $B > 0$, we can find K and c such that

$$y = Ae^x + Be^{-x} = \frac{Ke^{(x-c)} + Ke^{-(x-c)}}{2}$$
$$= \frac{K}{2}e^x e^{-c} + \frac{K}{2}e^{-x}e^c$$
$$= \left(\frac{Ke^{-c}}{2}\right)e^x + \left(\frac{Ke^c}{2}\right)e^{-x}.$$

Thus, we want to find K and c such that

$$\frac{Ke^{-c}}{2} = A \quad \text{and} \quad \frac{Ke^c}{2} = B.$$

Dividing, we have

$$\frac{Ke^c}{Ke^{-c}} = \frac{B}{A}$$
$$e^{2c} = \frac{B}{A}$$
$$c = \frac{1}{2}\ln\left(\frac{B}{A}\right).$$

If $A > 0$, $B > 0$, then there is a solution for c. Substituting to find K, we have

$$\frac{Ke^{-c}}{2} = A$$
$$K = 2Ae^c = 2Ae^{(\ln(B/A))/2}$$
$$= 2Ae^{\ln\sqrt{B/A}} = 2A\sqrt{\frac{B}{A}} = 2\sqrt{AB}.$$

Thus, if $A > 0$, $B > 0$, there is a solution for K also.

The fact that $y = Ae^x + Be^{-x}$ can be rewritten in this way shows that the graph of $y = Ae^x + Be^{-x}$ is the graph of $\cosh x$, shifted over by c and stretched (or shrunk) vertically by a factor of K.

15. (a) The graphs are shown in Figures 4.31–4.36.

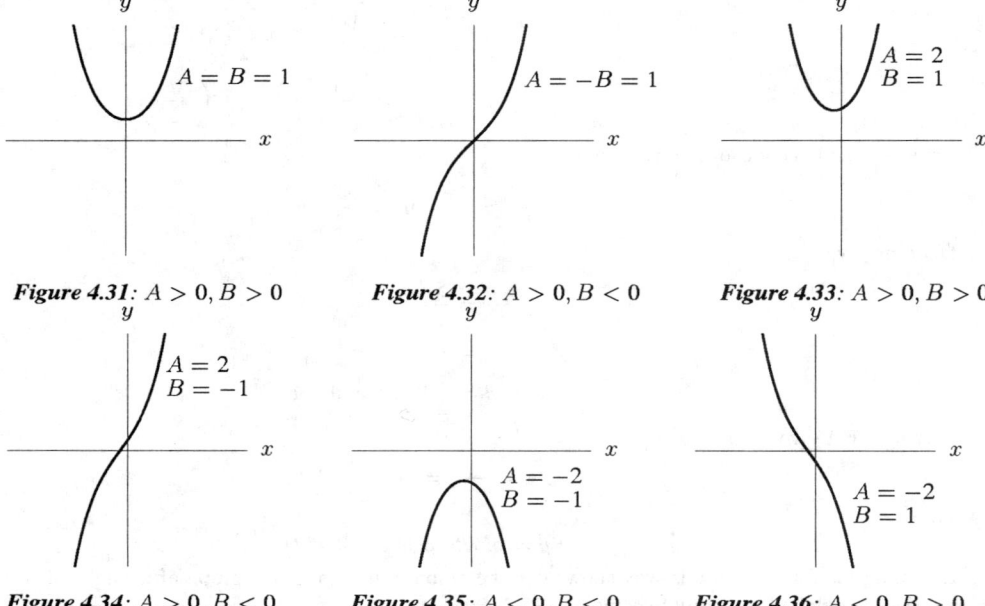

Figure 4.31: $A > 0, B > 0$

Figure 4.32: $A > 0, B < 0$

Figure 4.33: $A > 0, B > 0$

Figure 4.34: $A > 0, B < 0$

Figure 4.35: $A < 0, B < 0$

Figure 4.36: $A < 0, B > 0$

(b) If A and B have the same sign, the graph is U-shaped. If A and B are both positive, the graph opens upward. If A and B are both negative, the graph opens downward.

(c) If A and B have different signs, the graph appears to be everywhere increasing (if $A > 0, B < 0$) or decreasing (if $A < 0, B > 0$).

(d) The function appears to have a local maximum if $A < 0$ and $B < 0$, and a local minimum if $A > 0$ and $B > 0$.
To justify this, calculate the derivative

$$\frac{dy}{dx} = Ae^x - Be^{-x}.$$

Setting $dy/dx = 0$ gives

$$Ae^x - Be^{-x} = 0$$
$$Ae^x = Be^{-x}$$
$$e^{2x} = \frac{B}{A}.$$

This equation has a solution only if B/A is positive, that is, if A and B have the same sign. In that case,

$$2x = \ln\left(\frac{B}{A}\right)$$
$$x = \frac{1}{2}\ln\left(\frac{B}{A}\right).$$

This value of x gives the only critical point.
To determine whether the critical point is a local maximum or minimum, we use the first derivative test. Since

$$\frac{dy}{dx} = Ae^x - Be^{-x},$$

we see that:
If $A > 0, B > 0$, we have $dy/dx > 0$ for large positive x and $dy/dx < 0$ for large negative x, so there is a local minimum.
If $A < 0, B < 0$, we have $dy/dx < 0$ for large positive x and $dy/dx > 0$ for large negative x, so there is a local maximum.

16. (a) Since the cosh function is even, the height, y, is the same at $x = -T/w$ and $x = T/w$. The height at these endpoints is

$$y = \frac{T}{w}\cosh\left(\frac{w}{T}\cdot\frac{T}{w}\right) = \frac{T}{w}\cosh 1 = \frac{T}{w}\left(\frac{e^1 + e^{-1}}{2}\right).$$

At the lowest point, $x = 0$, and the height is

$$y = \frac{T}{w}\cosh 0 = \frac{T}{w}.$$

Thus the "sag" in the cable is given by

$$\text{Sag} = \frac{T}{w}\left(\frac{e + e^{-1}}{2}\right) - \frac{T}{w} = \frac{T}{w}\left(\frac{e + e^{-1}}{2} - 1\right) \approx 0.54\frac{T}{w}.$$

(b) To show that the differential equation is satisfied, take derivatives

$$\frac{dy}{dx} = \frac{T}{w}\cdot\frac{w}{T}\sinh\left(\frac{wx}{T}\right) = \sinh\left(\frac{wx}{T}\right)$$
$$\frac{d^2y}{dx^2} = \frac{w}{T}\cosh\left(\frac{wx}{T}\right).$$

Therefore, using the fact that $1 + \sinh^2 a = \cosh^2 a$ and that cosh is always positive, we have:

$$\frac{w}{T}\sqrt{1+\left(\frac{dy}{dx}\right)^2} = \frac{w}{T}\sqrt{1+\sinh^2\left(\frac{wx}{T}\right)} = \frac{w}{T}\sqrt{\cosh^2\left(\frac{wx}{T}\right)}$$
$$= \frac{w}{T}\cosh\left(\frac{wx}{T}\right).$$

So

$$\frac{w}{T}\sqrt{1+\left(\frac{dy}{dx}\right)^2} = \frac{d^2y}{dx^2}.$$

180 CHAPTER FOUR /SOLUTIONS

17.

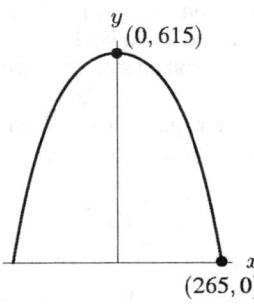

We know $x = 0$ and $y = 615$ at the top of the arch, so
$$615 = b - a\cosh(0/a) = b - a.$$

This means $b = a + 615$. We also know that $x = 265$ and $y = 0$ where the arch hits the ground, so
$$0 = b - a\cosh(265/a) = a + 615 - a\cosh(265/a).$$

We can solve this equation numerically on a calculator and get $a \approx 100$, which means $b \approx 715$. This results in the equation
$$y \approx 715 - 100\cosh\left(\frac{x}{100}\right).$$

Solutions for Chapter 4 Review

1.

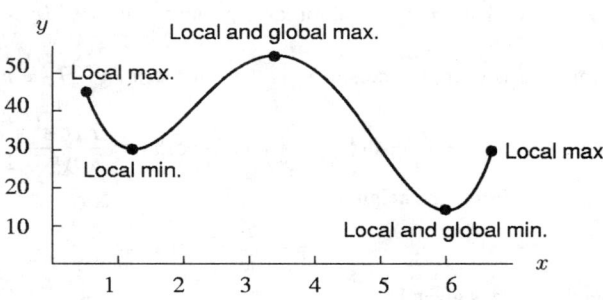

2.

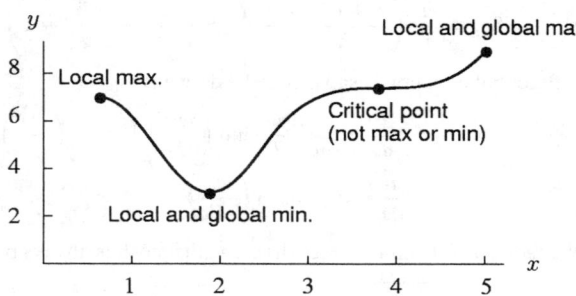

3. (a) Increasing for $x > 0$, decreasing for $x < 0$.
    (b) $f(0)$ is a local and global minimum, and f has no global maximum.
4. (a) Increasing for all x.
    (b) No maxima or minima.
5. (a) Decreasing for $x < 0$, increasing for $0 < x < 4$, and decreasing for $x > 4$.
    (b) $f(0)$ is a local minimum, and $f(4)$ is a local maximum.

6. (a) Decreasing for $x < -1$, increasing for $-1 < x < 0$, decreasing for $0 < x < 1$, and increasing for $x > 1$.
(b) $f(-1)$ and $f(1)$ are local minima, $f(0)$ is a local maximum.

7. (a) We wish to investigate the behavior of $f(x) = x^3 - 3x^2$ on the interval $-1 \leq x \leq 3$. We find:
$$f'(x) = 3x^2 - 6x = 3x(x-2)$$
$$f''(x) = 6x - 6 = 6(x-1)$$

(b) The critical points of f are $x = 2$ and $x = 0$ since $f'(x) = 0$ at those points. Using the second derivative test, we find that $x = 0$ is a local maximum since $f'(0) = 0$ and $f''(0) = -6 < 0$, and that $x = 2$ is a local minimum since $f'(2) = 0$ and $f''(2) = 6 > 0$.

(c) There is an inflection point at $x = 1$ since f'' changes sign at $x = 1$.

(d) At the critical points, $f(0) = 0$ and $f(2) = -4$.
At the endpoints: $f(-1) = -4, f(3) = 0$.
So the global maxima are $f(0) = 0$ and $f(3) = 0$, while the global minima are $f(-1) = -4$ and $f(2) = -4$.

(e)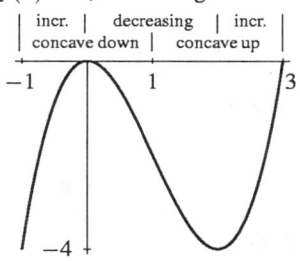

8. (a) First we find f' and f''; $f'(x) = 1 + \cos x$ and $f''(x) = -\sin x$.
(b) The critical point of f is $x = \pi$, since $f'(\pi) = 0$.
(c) Since f'' changes sign at $x = \pi$, it means that $x = \pi$ is an inflection point.
(d) Evaluating f at the critical point and endpoints, we find $f(0) = 0$, $f(\pi) = \pi$, $f(2\pi) = 2\pi$,. Therefore, the global maximum is $f(2\pi) = 2\pi$, and the global minimum is $f(0) = 0$. Note that $x = \pi$ isn't a local maximum or minimum of f, and that the second derivative test is inconclusive here.

(e)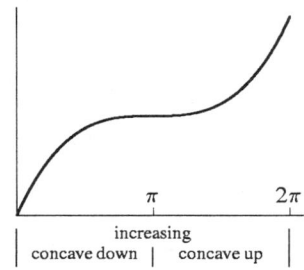

9. (a) First we find f' and f'':
$$f'(x) = -e^{-x}\sin x + e^{-x}\cos x$$
$$f''(x) = e^{-x}\sin x - e^{-x}\cos x$$
$$-e^{-x}\cos x - e^{-x}\sin x$$
$$= -2e^{-x}\cos x$$

(b) The critical points are $x = \pi/4, 5\pi/4$, since $f'(x) = 0$ here.
(c) The inflection points are $x = \pi/2, 3\pi/2$, since f'' changes sign at these points.
(d) At the endpoints, $f(0) = 0$, $f(2\pi) = 0$. So we have $f(\pi/4) = (e^{-\pi/4})(\sqrt{2}/2)$ as the global maximum; $f(5\pi/4) = -e^{-5\pi/4}(\sqrt{2}/2)$ as the global minimum.

(e)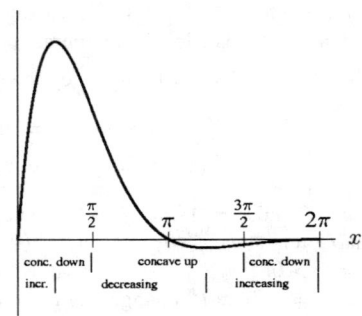

10. (a) We first find f' and f'':
$$f'(x) = -\frac{2}{3}x^{-\frac{5}{3}} + \frac{1}{3}x^{-\frac{2}{3}} = \frac{1}{3}x^{-\frac{5}{3}}(x-2)$$
$$f''(x) = \frac{10}{9}x^{-\frac{8}{3}} - \frac{2}{9}x^{-\frac{5}{3}} = -\frac{2}{9}x^{-\frac{8}{3}}(x-5)$$

(b) Critical point: $x = 2$.
(c) There are no inflection points, since f'' does not change sign on the interval $1.2 \leq x \leq 3.5$.
(d) At the endpoints, $f(1.2) \approx 1.94821$ and $f(3.5) \approx 1.95209$. So, the global minimum is $f(2) \approx 1.88988$ and the global maximum is $f(3.5) \approx 1.95209$.

(e)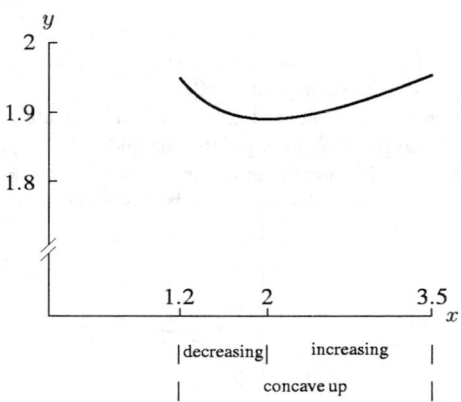

11. The polynomial $f(x)$ behaves like $2x^3$ as x goes to ∞. Therefore, $\lim_{x \to \infty} f(x) = \infty$ and $\lim_{x \to -\infty} f(x) = -\infty$.

We have $f'(x) = 6x^2 - 18x + 12 = 6(x-2)(x-1)$, which is zero when $x = 1$ or $x = 2$.

Also, $f''(x) = 12x - 18 = 6(2x-3)$, which is zero when $x = 3/2$. For $x < 3/2$, $f''(x) < 0$; for $x > 3/2$, $f''(x) > 0$. Thus $x = 3/2$ is an inflection point.

The critical points are $x = 1$ and $x = 2$, and $f(1) = 6$, $f(2) = 5$. By the second derivative test, $f''(1) = -6 < 0$, so $x = 1$ is a local maximum; $f''(2) = 6 > 0$, so $x = 2$ is a local minimum.

Now we can draw the diagrams below.

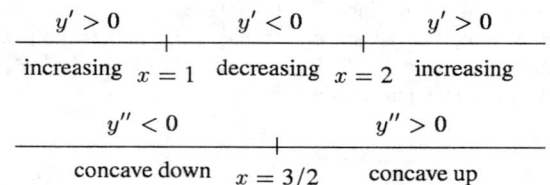

The graph of $f(x) = 2x^3 - 9x^2 + 12x + 1$ is shown below. It has no global maximum or minimum.

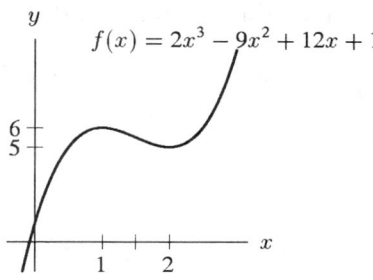

12. If we divide the denominator and numerator of $f(x)$ by x^2 we have

$$\lim_{x \to \pm\infty} \frac{4x^2}{x^2 + 1} = \lim_{x \to \pm\infty} \frac{4}{1 + \frac{1}{x^2}} = 4$$

since

$$\lim_{x \to \pm\infty} \frac{1}{x^2} = 0.$$

Using the quotient rule we get

$$f'(x) = \frac{(x^2 + 1)8x - 4x^2(2x)}{(x^2 + 1)^2} = \frac{8x}{(x^2 + 1)^2},$$

which is zero when $x = 0$, positive when $x > 0$, and negative when $x < 0$. Thus $f(x)$ has a local minimum when $x = 0$, with $f(0) = 0$.

Because $f'(x) = 8x/(x^2 + 1)^2$, the quotient rule implies that

$$f''(x) = \frac{(x^2 + 1)^2 8 - 8x[2(x^2 + 1)2x]}{(x^2 + 1)^4}$$

$$= \frac{8x^2 + 8 - 32x^2}{(x^2 + 1)^3} = \frac{8(1 - 3x^2)}{(x^2 + 1)^3}.$$

The denominator is always positive, so $f''(x) = 0$ when $x = \pm\sqrt{1/3}$, positive when $-\sqrt{1/3} < x < \sqrt{1/3}$, and negative when $x > \sqrt{1/3}$ or $x < -\sqrt{1/3}$. This gives the diagram

```
        y' < 0           |           y' > 0
      decreasing       x = 0       increasing
```

```
    y'' < 0      |      y'' > 0      |      y'' < 0
  concave down   |    concave up     |   concave down
            x = -√(1/3)         x = √(1/3)
```

and the graph of f looks like:

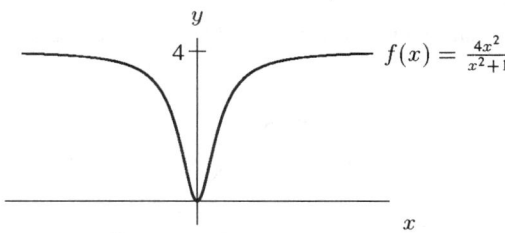

with inflection points $x = \pm\sqrt{1/3}$, a global minimum at $x = 0$, and no local or global maxima (since $f(x)$ never equals 4).

13. As $x \to -\infty$, $e^{-x} \to \infty$, so $xe^{-x} \to -\infty$. Thus $\lim_{x \to -\infty} xe^{-x} = -\infty$.

As $x \to \infty$, $\frac{x}{e^x} \to 0$, since e^x grows much more quickly than x. Thus $\lim_{x \to \infty} xe^{-x} = 0$.

Using the product rule,
$$f'(x) = e^{-x} - xe^{-x} = (1-x)e^{-x},$$
which is zero when $x = 1$, negative when $x > 1$, and positive when $x < 1$. Thus $f(1) = 1/e^1 = 1/e$ is a local maximum.

Again, using the product rule,
$$\begin{aligned}f''(x) &= -e^{-x} - e^{-x} + xe^{-x} \\ &= xe^{-x} - 2e^{-x} \\ &= (x-2)e^{-x},\end{aligned}$$
which is zero when $x = 2$, positive when $x > 2$, and negative when $x < 2$, giving an inflection point at $(2, \frac{2}{e^2})$. With the above, we have the following diagram:

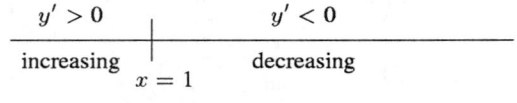

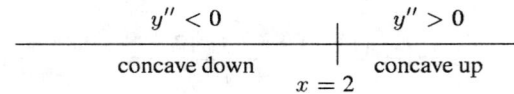

The graph of f is shown below.

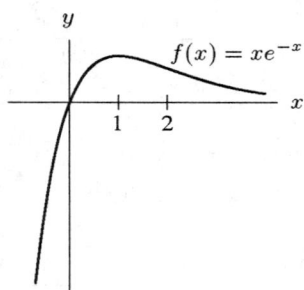

and $f(x)$ has one global maximum at $1/e$ and no local or global minima.

14. We first compute f' and f'':
$$\begin{aligned}f'(x) &= -xe^{-x^2/2} \\ f''(x) &= -e^{-x^2/2} + x^2 e^{-x^2/2} \\ &= e^{-x^2/2}(x^2 - 1)\end{aligned}$$

$x = 0$ is a critical point, since $f'(0) = 0$. Since $f''(0) = -1$, by the second derivative test $f(0) = 1$ is a local maximum. $x = 1$ and $x = -1$ are inflection points, because $f''(x)$ changes sign there.

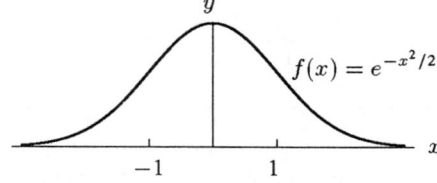

15. $\lim_{x \to \infty} f(x) = +\infty$, and $\lim_{x \to -\infty} f(x) = -\infty$.
There are no asymptotes.
$f'(x) = 3x^2 + 6x - 9 = 3(x+3)(x-1)$. Critical points are $x = -3, x = 1$.
$f''(x) = 6(x+1)$.

x		-3		-1		1	
f'	$+$	0	$-$	$-$	$-$	0	$+$
f''	$-$	$-$	$-$	0	$+$	$+$	$+$
f	⤴		⤵		⤷		⤶

Thus, $x = -1$ is an inflection point. $f(-3) = 12$ is a local maximum; $f(1) = -20$ is a local minimum. There are no global maxima or minima.

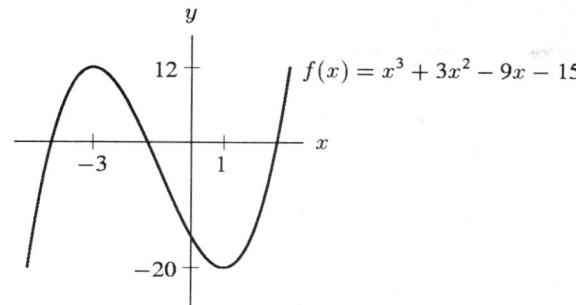

16. $\lim_{x \to +\infty} f(x) = +\infty$, and $\lim_{x \to -\infty} f(x) = -\infty$.
There are no asymptotes.
$f'(x) = 5x^4 - 45x^2 = 5x^2(x^2 - 9) = 5x^2(x+3)(x-3)$.
The critical points are $x = 0, x = \pm 3$. f' changes sign at 3 and -3 but not at 0.
$f''(x) = 20x^3 - 90x = 10x(2x^2 - 9)$. f'' changes sign at $0, \pm 3/\sqrt{2}$.
So, inflection points are at $x = 0, x = \pm 3/\sqrt{2}$.

x		-3		$-3/\sqrt{2}$		0		$3/\sqrt{2}$		3	
f'	$+$	0	$-$		$-$	0	$-$		$-$	0	$+$
f''	$-$	$-$	$-$	0	$+$	0	$-$	0	$+$	$+$	$+$
f	⤴		⤵		⤷		⤵		⤷		⤶

Thus, $f(-3)$ is a local maximum; $f(3)$ is a local minimum. There are no global maxima or minima.

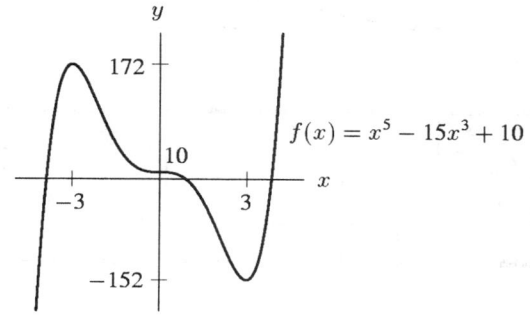

17. $\lim_{x \to +\infty} f(x) = +\infty$, and $\lim_{x \to 0^+} f(x) = +\infty$.
Hence, $x = 0$ is a vertical asymptote.
$f'(x) = 1 - \dfrac{2}{x} = \dfrac{x-2}{x}$, so $x = 2$ is the only critical point.
$f''(x) = \dfrac{2}{x^2}$, which can never be zero. So there are no inflection points.

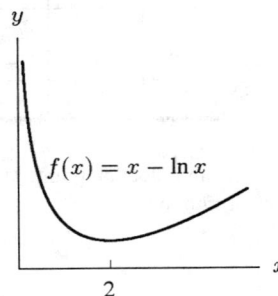

x		2	
f'	−	0	+
f''	+	+	+
f	↘⌣		↗⌣

Thus, $f(2)$ is a local and global minimum.

18. $\lim_{x \to +\infty} f(x) = +\infty$, $\lim_{x \to -\infty} f(x) = 0$.
$y = 0$ is the horizontal asymptote.
$f'(x) = 2xe^{5x} + 5x^2 e^{5x} = xe^{5x}(5x + 2)$.
Thus, $x = -\dfrac{2}{5}$ and $x = 0$ are the critical points.

$$f''(x) = 2e^{5x} + 2xe^{5x} \cdot 5 + 10xe^{5x} + 25x^2 e^{5x}$$
$$= e^{5x}(25x^2 + 20x + 2).$$

So, $x = \dfrac{-2 \pm \sqrt{2}}{5}$ are inflection points.

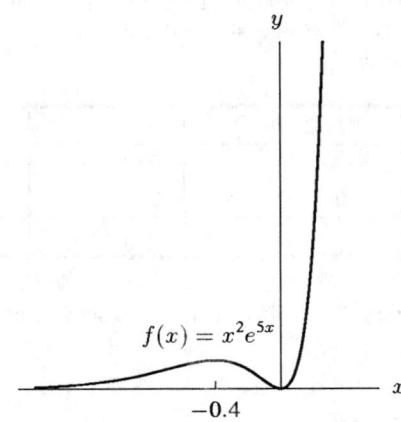

x		$\frac{-2-\sqrt{2}}{5}$		$-\frac{2}{5}$		$\frac{-2+\sqrt{2}}{5}$		0	
f'	+	+	+	0	−	−	−	0	+
f''	+	0	−	−	−	0	+	+	+
f	↗⌣		↗⌢		↘⌢		↘⌣		↗⌣

So, $f(-\frac{2}{5})$ is a local maximum; $f(0)$ is a local and global minimum.

19. Since $\lim_{x \to -\infty} f(x) = \lim_{x \to +\infty} f(x) = 0$, $y = 0$ is a horizontal asymptote.

$f'(x) = -2xe^{-x^2}$. So, $x = 0$ is the only critical point.
$f''(x) = -2(e^{-x^2} + x(-2x)e^{-x^2}) = 2e^{-x^2}(2x^2 - 1) = 2e^{-x^2}(\sqrt{2}x - 1)(\sqrt{2}x + 1)$.
Thus, $x = \pm 1/\sqrt{2}$ are inflection points.

TABLE 4.1

x		$-1/\sqrt{2}$		0		$1/\sqrt{2}$	
f'	+	+	+	0	−	−	−
f''	+	0	−	−	−	0	+
f	↗⌣		↗⌢		↘⌢		↘⌣

Thus, $f(0) = 1$ is a local and global maximum.

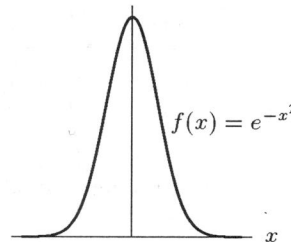

$f(x) = e^{-x^2}$

20. $\lim_{x \to +\infty} f(x) = \lim_{x \to -\infty} f(x) = 1$.

Thus, $y = 1$ is a horizontal asymptote. Since $x^2 + 1$ is never 0, there are no vertical asymptotes.

$$f'(x) = \frac{2x(x^2 + 1) - x^2(2x)}{(x^2 + 1)^2} = \frac{2x}{(x^2 + 1)^2}.$$

So, $x = 0$ is the only critical point.

$$f''(x) = \frac{2(x^2 + 1)^2 - 2x \cdot 2(x^2 + 1) \cdot 2x}{(x^2 + 1)^4}$$
$$= \frac{2(x^2 + 1 - 4x^2)}{(x^2 + 1)^3}$$
$$= \frac{2(1 - 3x^2)}{(x^2 + 1)^3}.$$

So, $x = \pm \frac{1}{\sqrt{3}}$ are inflection points.

TABLE 4.2

x		$\frac{-1}{\sqrt{3}}$		0		$\frac{1}{\sqrt{3}}$	
f'	−	−	−	0	+	+	+
f''	−	0	+	+	+	0	−
f	↘⌢		↘⌣		↗⌣		↗⌢

Thus, $f(0) = 0$ is a local and global minimum. A graph of $f(x)$ can be found in Figure 4.37.

188 CHAPTER FOUR /SOLUTIONS

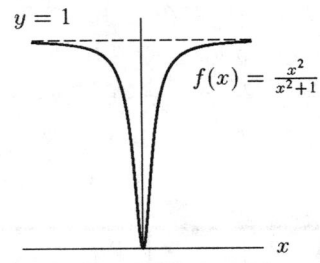

Figure 4.37

21. The critical points of f occur where f' is zero. These two points are indicated in the figure below.

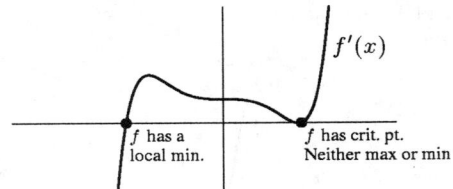

Note that the point labeled as a local minimum of f is not a critical point of f'.

22. (a)

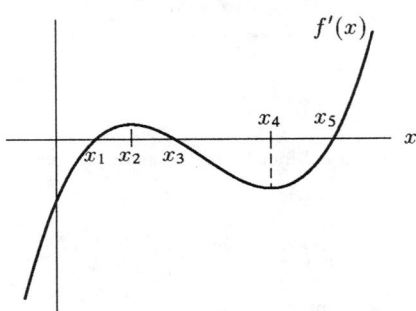

(b) $f'(x)$ changes sign at x_1, x_3, and x_5.
(c) $f'(x)$ has local extrema at x_2 and x_4.

23. The local maxima and minima of f correspond to places where f' is zero and changes sign or, possibly, to the endpoints of intervals in the domain of f. The points at which f changes concavity correspond to local maxima and minima of f'. The change of sign of f', from positive to negative corresponds to a maximum of f and change of sign of f' from negative to positive corresponds to a minimum of f.

24. The volume is given by $V = x^2 y$. The surface area is given by

$$S = 2x^2 + 4xy$$
$$= 2x^2 + 4xV/x^2 = 2x^2 + 4V/x.$$

To find the dimensions which minimize the area, find x such that $dS/dx = 0$:

$$\frac{dS}{dx} = 4x - \frac{4V}{x^2} = 0$$
$$x^3 = V.$$

Solving for x gives $x = \sqrt[3]{V} = y$. To see that this gives a minimum, note that for small x, $S \approx 4V/x$ is decreasing. For large x, $S \approx 2x^2$ is increasing. Since there is only one critical point, it must give a global minimum. Therefore, when the width equals the height, the surface area is minimized.

25. Volume: $V = x^2 y$,
Surface: $S = x^2 + 4xy = x^2 + 4xV/x^2 = x^2 + 4V/x$.
To find the dimensions which minimize the area, find x such that $dS/dx = 0$.

$$\frac{dS}{dx} = 2x - \frac{4V}{x^2} = 0,$$

so

$$x^3 = 2V,$$

and solving for x gives $x = \sqrt[3]{2V}$. To see that this gives a minimum, note that for small x, $S \approx 4V/x$ is decreasing. For large x, $S \approx x^2$ is increasing. Since there is only one critical point, this must give a global minimum. Using x to find y gives $y = V/x^2 = V/(2V)^{2/3} = \sqrt[3]{V/4}$.

26.

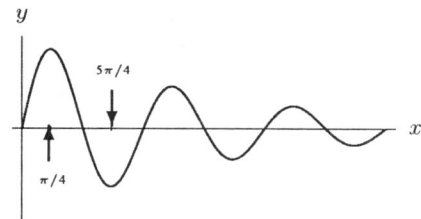

Letting $f(x) = e^{-x} \sin x$, we have

$$f'(x) = -e^{-x} \sin x + e^{-x} \cos x.$$

Solving $f'(x) = 0$, we get $\sin x = \cos x$. This means $x = \arctan(1) = \pi/4$, and $\pi/4$ plus multiples of π, are the critical points of $f(x)$. By evaluating $f(x)$ at the points $k\pi + \pi/4$, where k is an integer, we can find:

$$e^{-5\pi/4} \sin(5\pi/4) \le e^{-x} \sin x \le e^{-\pi/4} \sin(\pi/4),$$

since $f(0) = 0$ at the endpoint. So

$$-0.014 \le e^{-x} \sin x \le 0.322.$$

27. Let $f(x) = x \sin x$. Then $f'(x) = x \cos x + \sin x$.
$f'(x) = 0$ when $x = 0$, $x \approx 2$, and $x \approx 5$. The latter two estimates we can get from the graph of $f'(x)$.

Zooming in (or using some other approximation method), we can find the zeros of $f'(x)$ with more precision. They are (approximately) 0, 2.029, and 4.913. We check the endpoints and critical points for the global maximum and minimum.

$$f(0) = 0, \qquad f(2\pi) = 0,$$
$$f(2.029) \approx 1.8197, \qquad f(4.914) \approx -4.814.$$

Thus for $0 \le x \le 2\pi$, $-4.81 \le f(x) \le 1.82$.

28.

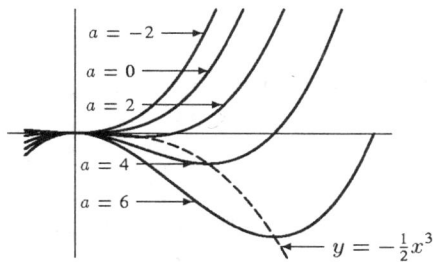

To solve for the critical points, we set $\frac{dy}{dx} = 0$. Since $\frac{d}{dx}(x^3 - ax^2) = 3x^2 - 2ax$, we want $3x^2 - 2ax = 0$, so $x = 0$ or $x = \frac{2}{3}a$. At $x = 0$, we have $y = 0$. This first critical point is independent of a and lies on the curve $y = -\frac{1}{2}x^3$. At $x = \frac{2}{3}a$, we calculate $y = -\frac{4}{27}a^3 = -\frac{1}{2}\left(\frac{2}{3}a\right)^3$. Thus the second critical point also lies on the curve $y = -\frac{1}{2}x^3$.

29. (a) We have $g'(t) = \frac{t(1/t) - \ln t}{t^2} = \frac{1 - \ln t}{t^2}$, which is zero if $t = e$, negative if $t > e$, and positive if $t < e$, since $\ln t$ is increasing. Thus $g(e) = \frac{1}{e}$ is a global maximum for g. Since $t = e$ was the only point at which $g'(t) = 0$, there is no minimum.

(b) Now $\ln t/t$ is increasing for $0 < t < e$, $\ln 1/1 = 0$, and $\ln 5/5 \approx 0.322 < \ln(e)/e$. Thus, for $1 < t < e$, $\ln t/t$ increases from 0 to above $\ln 5/5$, so there must be a t between 1 and e such that $\ln t/t = \ln 5/5$. For $t > e$, there is only one solution to $\ln t/t = \ln 5/5$, namely $t = 5$, since $\ln t/t$ is decreasing for $t > e$. For $0 < t < 1$, $\ln t/t$ is negative and so cannot equal $\ln 5/5$. Thus $\ln x/x = \ln t/t$ has exactly two solutions.

(c) The graph of $\ln t/t$ intersects the horizontal line $y = \ln 5/5$, at $x = 5$ and $x \approx 1.75$.

30. (a)

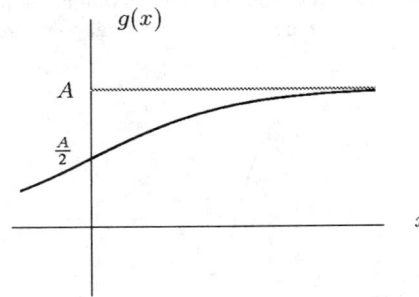

A is the limiting value of $g(x)$ for large values of x.

(b) Calculating $g(x) + g(-x)$ directly, we have

$$g(x) + g(-x) = \frac{A}{1 + e^{-Cx}} + \frac{A}{1 + e^{Cx}}$$
$$= \frac{A(1 + e^{Cx}) + A(1 + e^{-Cx})}{(1 + e^{-Cx})(1 + e^{Cx})}$$
$$= \frac{A(2 + e^{Cx} + e^{-Cx})}{1 + 1 + e^{-Cx} + e^{Cx}} = A.$$

This is the sum of two mirror-image functions. Their sum is A for any value of x.

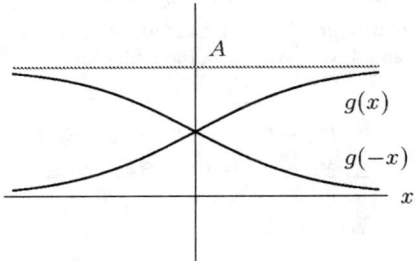

(c) If C is increased, then the slope near $x = 0$ increases.

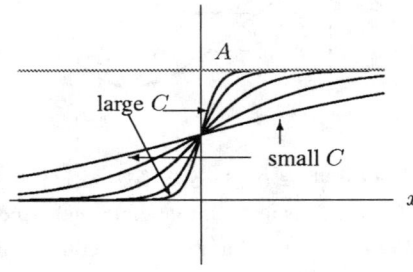

(d) Consider a horizontal shift of $g(x)$. Replacing x by $x - b$ in the function $g(x)$, we get a shift to the right by the distance b.

$$g(x - b) = \frac{A}{1 + e^{-C(x-b)}}$$
$$= \frac{A}{1 + e^{-Cx+bC}}$$
$$= \frac{A}{1 + e^{bC}e^{-Cx}}$$

Since e^{bC} is a constant, we can rename it B. Thus $g(x - b) = \dfrac{A}{1 + Be^{-Cx}}$, that is, a horizontal shift can be written in terms of B instead of b.

31. (a) x-intercept: $(a, 0)$, y-intercept: $(0, \frac{1}{a^2+1})$

(b) Area $= \frac{1}{2}(a)(\frac{1}{a^2+1}) = \frac{a}{2(a^2+1)}$

(c)
$$A = \frac{a}{2(a^2 + 1)}$$
$$A' = \frac{2(a^2 + 1) - a(4a)}{4(a^2 + 1)^2}$$
$$= \frac{2(1 - a^2)}{4(a^2 + 1)^2}$$
$$= \frac{(1 - a^2)}{2(a^2 + 1)^2}.$$

If $A' = 0$, then $a = \pm 1$. We only consider positive values of a, and we note that A' changes sign from positive to negative at $a = 1$. Hence $a = 1$ is a local maximum of A which is a global maximum because $A' < 0$ for all $a > 1$ and $A' > 0$ for $0 < a < 1$.

(d) $A = \frac{1}{2}(1)(\frac{1}{2}) = \frac{1}{4}$

(e) Set $\frac{a}{2(a^2+a)} = \frac{1}{3}$ and solve for a:

$$5a = 2a^2 + 2$$
$$2a^2 - 5a + 2 = 0$$
$$(2a - 1)(a - 2) = 0.$$

32.

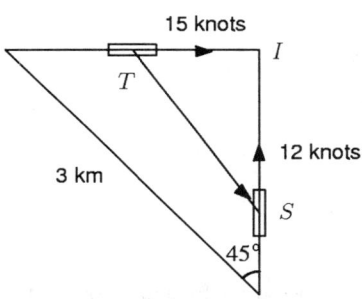

Figure 4.38: Position of the tanker and ship

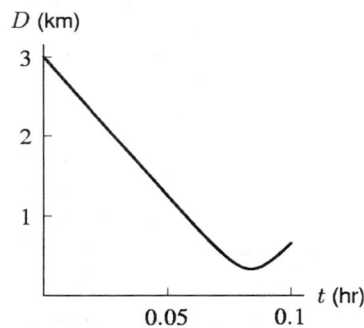

Figure 4.39: Distance between the ship at S and the tanker at T

Suppose t is the time, in hours, since the ships were 3 km apart. Then $\overline{TI} = \frac{3\sqrt{2}}{2} - (15)(1.85)t$ and $\overline{SI} = \frac{3\sqrt{2}}{2} - (12)(1.85)t$. So the distance, $D(t)$, in km, between the ships at time t is

$$D(t) = \sqrt{\left(\frac{3\sqrt{2}}{2} - 27.75t\right)^2 + \left(\frac{3\sqrt{2}}{2} - 22.2t\right)^2}.$$

Differentiating gives

$$\frac{dD}{dt} = \frac{-55.5\left(\frac{3}{\sqrt{2}} - 27.75t\right) - 44.4\left(\frac{3}{\sqrt{2}} - 22.2t\right)}{2\sqrt{\left(\frac{3}{\sqrt{2}} - 27.75t\right)^2 + \left(\frac{3}{\sqrt{2}} - 22.2t\right)^2}}.$$

Solving $dD/dt = 0$ gives a critical point at $t = 0.0839$ hours when the ships will be approximately 331 meters apart. So the ships do not need to change course. Alternatively, tracing along the curve in Figure 4.39 gives the same result. Note that this is after the eastbound ship crosses the path of the northbound ship.

33. (a) The concavity changes at t_1 and t_3, as shown below.

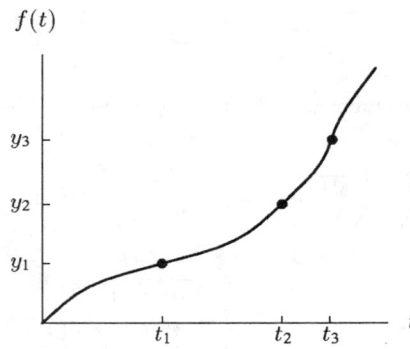

(b) $f(t)$ grows most quickly where the vase is skinniest (at y_3) and most slowly where the vase is widest (at y_1). The diameter of the widest part of the vase looks to be about 4 times as large as the diameter at the skinniest part. Since the area of a cross section is given by πr^2, where r is the radius, the ratio between areas of cross sections at these two places is about 4^2, so the growth rates are in a ratio of about 1 to 16 (the wide part being 16 times slower).

34.
$$r(\lambda) = a(\lambda)^{-5}(e^{b/\lambda} - 1)^{-1}$$
$$r'(\lambda) = a(-5\lambda^{-6})(e^{b/\lambda} - 1)^{-1} + a(\lambda^{-5})\left(\frac{b}{\lambda^2}e^{b/\lambda}\right)(e^{b/\lambda} - 1)^{-2}$$

$(0.96, 3.13)$ is a maximum, so $r'(0.96) = 0$ implies that the following holds, with $\lambda = 0.96$:

$$5\lambda^{-6}(e^{b/\lambda} - 1)^{-1} = \lambda^{-5}\left(\frac{b}{\lambda^2}e^{b/\lambda}\right)(e^{b/\lambda} - 1)^{-2}$$
$$5\lambda(e^{b/\lambda} - 1) = be^{b/\lambda}$$
$$5\lambda e^{b/\lambda} - 5\lambda = be^{b/\lambda}$$
$$5\lambda e^{b/\lambda} - be^{b/\lambda} = 5\lambda$$
$$\left(\frac{5\lambda - b}{5\lambda}\right)e^{b/\lambda} = 1$$
$$\frac{4.8 - b}{4.8}e^{b/0.96} - 1 = 0.$$

Using Newton's method, or some other approximation method, we search for a root. The root should be near 4.8. Using our initial guess, we get $b \approx 4.7665$. At $\lambda = 0.96, r = 3.13$, so

$$3.13 = \frac{a}{0.96^5(e^{b/0.96} - 1)} \quad \text{or}$$
$$a = 3.13(0.96)^5(e^{b/0.96} - 1)$$
$$\approx 363.23.$$

As a check, we try $r(4) \approx 0.155$, which looks about right on the given graph.

Solutions to the Projects and Computer Algebra Investigations

1. (a) To find the critical points we must first take a derivative of y.

$$\frac{dy}{dx} = -1((1-x^2)^2 + 2ax^2)^{-2}((1-x^2)^2 + 2ax^2)'$$

$$= -((1-x^2)^2 + 2ax^2)^{-2}(2(1-x^2)(-2x) + 4ax) = -\frac{4x^3 - 4x + 4ax}{\left((1-x^2)^2 + 2ax^2\right)^2}.$$

Set $\dfrac{dy}{dx} = 0$. Thus,

$$4x^3 - 4x + 4ax = 0$$
$$4x\left(x^2 - 1 + a\right) = 0.$$

So, we have critical points at $x = 0$ and $x = +\sqrt{1-a}$. Note that we never take $x = -\sqrt{1-a}$ since $x \geq 0$ was assumed. Also, $x = +\sqrt{1-a}$ only exists for $a < 1$.

What kind of critical point is $x = 0$? First, note that the denominator of $\dfrac{dy}{dx}$ is always positive. Since the numerator is made up of odd powers of x, it will change sign at $x = 0$ regardless of the value of a. The sign of the numerator, $-4x(x^2 - 1 + a)$, does depend on the value of a. If $a > 1$, the factor $(x^2 - 1 + a)$ is positive when $x = 0$, so the sign of $\dfrac{dy}{dx}$ switches from positive to negative across $x = 0$, hence we get a local maximum. If $0 < a < 1$, the factor $(x^2 - 1 + a)$ is negative when $x = 0$, so the sign of $\dfrac{dy}{dx}$ switches from negative to positive across $x = 0$ and we get a local minimum. When $a = 1$, the function is $y = 1/(1 + x^4)$ which has a local maximum at the origin.

What kind of critical point is $x = \sqrt{1-a}$? Again, look at the numerator of $\dfrac{dy}{dx}$. For x slightly greater than $\sqrt{1-a}$, the quantity $-4x(x^2 - 1 + a)$ will be negative. For x slightly less than $\sqrt{1-a}$, the quantity $-4x(x^2 - 1 + a)$ will be positive. Hence, there is a local maximum at $x = \sqrt{1-a}$.

For $0 < a < 1$, both critical points exist, making the family more interesting. If $a > 1$, then $\sqrt{1-a}$ does not exist and we get only a critical point at $x = 0$. For $a = 1$, $\sqrt{1-a} = 0$ and again there is only one critical point at $x = 0$.

(b) At the critical point where $x = \sqrt{1-a}$, we calculate the y-coordinate.

$$y = \frac{1}{(1-(1-a))^2 + 2a(1-a)} = \frac{1}{a^2 + 2a - 2a^2} = \frac{1}{2a - a^2}.$$

When a is small, we have $\sqrt{1-a} \approx 1$ and $2a - a^2 \approx 2a$ (since a^2 is small relative to a). Thus, we get a critical point very close to $\left(1, \dfrac{1}{2a}\right)$.

(c) For $x > 0$, $(1-x^2)^2 + 2ax^2 > (1-x^2)^2$. So, $y = \dfrac{1}{(1-x^2)^2 + 2ax^2} < \dfrac{1}{(1-x^2)^2}$.

(d) The curves are sketched below.

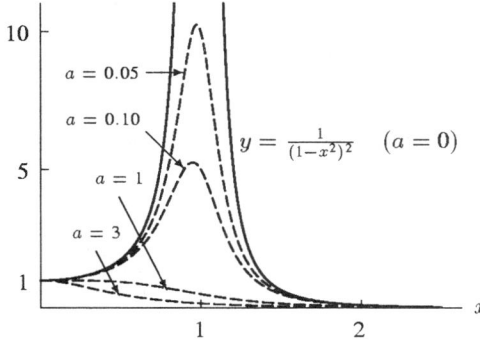

(e) For $0 < a < 1$, the smaller the value of a, the higher the peak of the graph will be near $x = 1$. When $a > 1$, there is no peak; the function is strictly decreasing.

2.

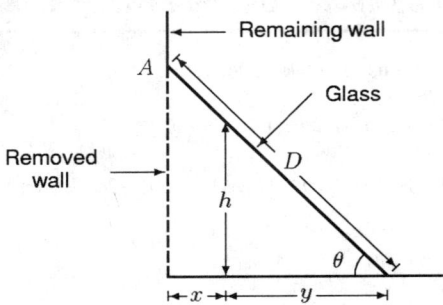

Figure 4.40: A Cross-section of the Projected Greenhouse

Suppose that the glass is at an angle θ (as shown in Figure 4.40), that the length of the wall is l, and that the glass has dimensions D ft by l ft. Since your parents will spend a fixed amount, the area of the glass, say k ft^2, is fixed:

$$Dl = k.$$

The width of the extension is $D\cos\theta$. If h is the height of your tallest parent, he or she can walk in a distance of x, and

$$\frac{h}{y} = \tan\theta, \quad \text{so} \quad y = \frac{h}{\tan\theta}.$$

Thus,

$$x = D\cos\theta - y = D\cos\theta - \frac{h}{\tan\theta} \quad \text{for } 0 < \theta < \frac{\pi}{2}.$$

We maximize x since doing so maximizes the usable area:

$$\frac{dx}{d\theta} = -D\sin\theta + \frac{h}{(\tan\theta)^2} \cdot \frac{1}{(\cos\theta)^2} = 0$$

$$\sin^3\theta = \frac{h}{D}$$

$$\theta = \arcsin\left(\left(\frac{h}{D}\right)^{1/3}\right).$$

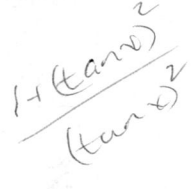

This is the only critical point, and $x \to 0$ when $\theta \to 0$ and when $\theta \to \pi/2$. Thus, the critical point is a global maximum. Since

$$\cos\theta = \sqrt{1 - \sin^2\theta} = \sqrt{1 - \left(\frac{h}{D}\right)^{2/3}},$$

the maximum value of x is

$$x = D\cos\theta - \frac{h}{\tan\theta} = D\cos\theta - \frac{h\cos\theta}{\sin\theta}$$

$$= \left(D - \frac{h}{\sin\theta}\right)\cos\theta = \left(D - \frac{h}{(h/D)^{1/3}}\right) \cdot \left(1 - \left(\frac{h}{D}\right)^{2/3}\right)^{1/2}$$

$$= (D - h^{2/3}D^{1/3}) \cdot \left(1 - \frac{h^{2/3}}{D^{2/3}}\right)^{1/2}$$

$$= D\left(1 - \frac{h^{2/3}}{D^{2/3}}\right) \cdot \left(1 - \frac{h^{2/3}}{D^{2/3}}\right)^{1/2} = D\left(1 - \frac{h^{2/3}}{D^{2/3}}\right)^{3/2}.$$

This means

$$\text{Maximum Usable Area} = lx$$

$$= lD\left(1 - \frac{h^{2/3}}{D^{2/3}}\right)^{3/2}$$

$$= k\left(1 - \left(\frac{hl}{k}\right)^{2/3}\right)^{3/2}$$

3. Since $a(q) = C(q)/q$, we have $C(q) = a(q) \cdot q$. Thus $C'(q) = a'(q)q + a(q)$, and so
$$C'(q_0) = a'(q_0)q_0 + a(q_0).$$

Since t_1 is the line tangent to $a(q)$ at $q = q_0$, the slope of t_1 is $a'(q_0)$, and the equation of t_1 is
$$y = a(q_0) + a'(q_0) \cdot (q - q_0) = a'(q_0) \cdot q + \big(a(q_0) - a'(q_0) \cdot q_0\big).$$

Thus the y-intercept of t_1 is given by $a(q_0) - a'(q_0)q_0$, and the equation of the line t_2 is
$$y = 2 \cdot a'(q_0) \cdot q + \big(a(q_0) - a'(q_0) \cdot q_0\big)$$

since t_2 has twice the slope of t_1. Let's compute the y-value on t_2 when $q = q_0$:
$$y = 2 \cdot a'(q_0) \cdot q_0 + \big(a(q_0) - a'(q_0) \cdot q_0\big) = a'(q_0)q_0 + a(q_0) = C'(q_0).$$

Hence $C'(q_0)$ is given by the point on t_2 where $q = q_0$.

4. (a) If the water is colder, the air warms it. If the water is warmer, the air cools it.
   (b) See Figure 4.41.

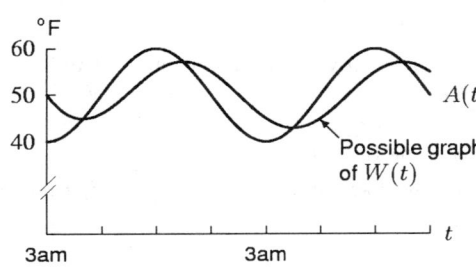

Figure 4.41   Figure 4.42

   (c) When the graphs intersect, the water temperature is not changing (since it is exactly the same as the air temperature). In other words, the derivative of the water temperature will be zero. Thus, the extrema of the water temperature occur at these intersections.
   (d) The greater $A(t) - W(t)$ is, the greater the rate of change of the water temperature.
   (e) When $A(t) - W(t)$ reaches a maximum or minimum, the rate at which $W(t)$ changes reaches a maximum or minimum. In other words, $W'(t)$ is at a maximum or minimum and therefore $W''(t) = 0$ and $W''(t)$ changes sign. Thus $W(t)$ has inflection points whenever $A(t) - W(t)$ reaches a maximum or minimum.
   (f) See Figure 4.42 above.

5. (a) Since $k > 0$, we have $\lim_{t \to \infty} e^{-kt} = 0$. Thus
$$\lim_{t \to \infty} P = \lim_{t \to \infty} \frac{L}{1 + Ce^{-kt}} = \frac{L}{1 + C \cdot 0} = L.$$

The constant L is called the camjing capacity of the environment because it represents the long-run population in the environment.

   (b) Using a CAS, we find
$$\frac{d^2 P}{dt^2} = -\frac{LCk^2 e^{-kt}(1 - Ce^{-kt})}{(1 + Ce^{-kt})^3}.$$

Thus, $d^2 P/dt^2 = 0$ when
$$1 - Ce^{-kt} = 0$$
$$t = -\frac{\ln(1/c)}{k}.$$

Since e^{-kt} and $(1 + Ce^{-kt})$ are both always positive, the sign of d^2P/dt^2 is negative when $(1 - Ce^{-kt}) > 0$, that is, for $t > -\ln(1/C)/k$. Similarly, the sign of d^2p/dt^2 is positive when $(1 - Ce^{-kt}) < 0$, that is, for $t < -\ln(1/C)/k$. Thus, there is an inflection point at $t = -\ln(1/C)/k$.

For $t = -\ln(1/C)/k$,
$$P = \frac{L}{1 + Ce^{\ln(1/C)}} = \frac{L}{1 + C(1/C)} = \frac{L}{2}.$$

Thus, the inflection point occurs where $P = L/2$.

6. (a) The graph is has a jump discontinuity whose position depends on a. The function is increasing, and the slope at a given x-value seems to be the same for all values of a. See Figure 4.43.

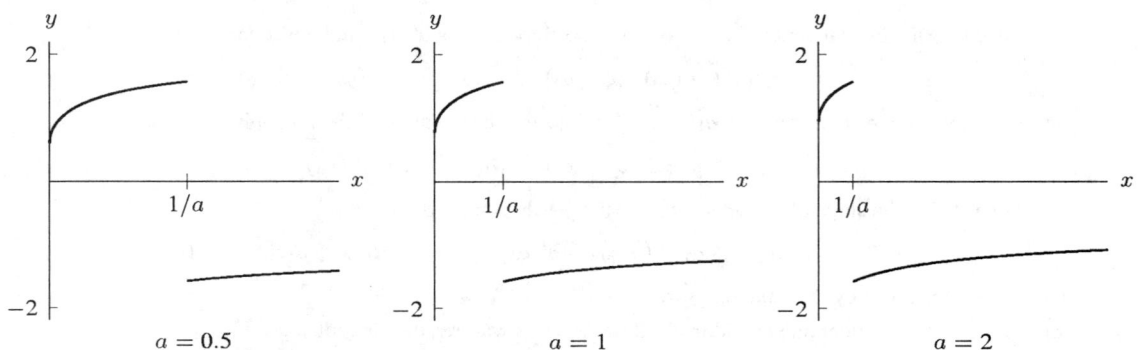

Figure 4.43

(b) Most computer algebra systems will give a fairly complicated answer for the derivative. Here is one example; others may be different.
$$\frac{dy}{dx} = \frac{\sqrt{x} + \sqrt{a}\sqrt{ax}}{2x\left(1 + a + 2\sqrt{a}\sqrt{x} + x + ax - 2\sqrt{ax}\right)}.$$

When we graph the derivative, it appears that we get the same graph for all values of a. See Figure 4.44.

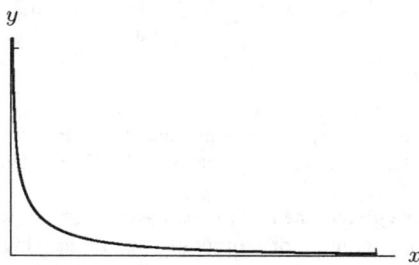

Figure 4.44

(c) Since a and x are positive, we have $\sqrt{ax} = \sqrt{a}\sqrt{x}$. We can use this to simplify the expression we found for the derivative:
$$\frac{dy}{dx} = \frac{\sqrt{x} + \sqrt{a}\sqrt{ax}}{2x\left(1 + a + 2\sqrt{a}\sqrt{x} + x + ax - 2\sqrt{ax}\right)}$$
$$= \frac{\sqrt{x} + \sqrt{a}\sqrt{a}\sqrt{x}}{2x\left(1 + a + 2\sqrt{a}\sqrt{x} + x + ax - 2\sqrt{a}\sqrt{x}\right)}$$
$$= \frac{\sqrt{x} + a\sqrt{x}}{2x\left(1 + a + x + ax\right)} = \frac{(1+a)\sqrt{x}}{2x(1+a)(1+x)} = \frac{\sqrt{x}}{2x(1+x)}$$

Since a has canceled out, the derivative is independent of a. This explains why all the graphs look the same in part (b). (In fact they are not exactly the same, because $f'(x)$ is undefined where $f(x)$ has its jump discontinuity. The point at which this happens changes with a.)

7. (a) A CAS gives
$$\frac{d}{dx}\text{arcsinh}\,x = \frac{1}{\sqrt{1+x^2}}$$

(b) Differentiating both sides of $\sinh(\text{arcsinh}\,x) = x$, we get
$$\cosh(\text{arcsinh}\,x)\frac{d}{dx}(\text{arcsinh}\,x) = 1$$
$$\frac{d}{dx}(\text{arcsinh}\,x) = \frac{1}{\cosh(\text{arcsinh}\,x)}.$$

Since $\cosh^2 x - \sinh^2 x = 1$, $\cosh x = \pm\sqrt{1 + \sinh^2 x}$. Furthermore, since $\cosh x > 0$ for all x, we take the positive square root, so $\cosh x = \sqrt{1 + \sinh^2 x}$. Therfore, $\cosh(\text{arcsinh } x) = \sqrt{1 + (\sinh(\text{arcsinh } x))^2} = \sqrt{1 + x^2}$. Thus

$$\frac{d}{dx}\text{arcsinh } x = \frac{1}{\sqrt{1 + x^2}}.$$

8. (a) A CAS gives
$$\frac{d}{dx}\text{arcosh } x = \frac{1}{\sqrt{x^2 - 1}}, \quad x \geq 0.$$

   (b) Differentiating both sides of $\cosh(\text{arccosh } x) = x$, we get

$$\sinh(\text{arccosh } x)\frac{d}{dx}(\text{arccosh } x) = 1$$

$$\frac{d}{dx}(\text{arccosh } x) = \frac{1}{\sinh(\text{arccosh } x)}.$$

Since $\cosh^2 x - \sinh^2 x = 1$, $\sinh x = \pm\sqrt{\cosh^2 x - 1}$. If $x \geq 0$, then $\sinh x \geq 0$, so we take the positive square root. So $\sinh x = \sqrt{\cosh^2 x - 1}$, $x \geq 0$. Therefore, $\sinh(\text{arccosh } x) = \sqrt{(\cosh(\text{arccosh } x))^2 - 1} = \sqrt{x^2 - 1}$, $x \geq 0$. Thus

$$\frac{d}{dx}\text{arccosh } x = \frac{1}{\sqrt{x^2 - 1}}.$$

9. (a) Using a computer algebra system or differentiating by hand, we get

$$f'(x) = \frac{1}{2\sqrt{a + x}(\sqrt{a} + \sqrt{x})} - \frac{\sqrt{a + x}}{2\sqrt{x}(\sqrt{a} + \sqrt{x})^2}.$$

Simplifying gives

$$f'(x) = \frac{-a + \sqrt{a}\sqrt{x}}{2\left(\sqrt{a} + \sqrt{x}\right)^2 \sqrt{x}\sqrt{a + x}}.$$

The denominator of the derivative is always positive if $x > 0$, and the numerator is zero when $x = a$. Writing the numerator as $\sqrt{a}(\sqrt{x} - \sqrt{a})$, we see that the derivative changes from negative to positive at $x = a$. Thus, by the first derivative test, the function has a local minimum at $x = a$.

   (b) As a increases, the local minimum moves to the right. See Figure 4.45. This is consistent with what we found in part (a), since the local minimum is at $x = a$.

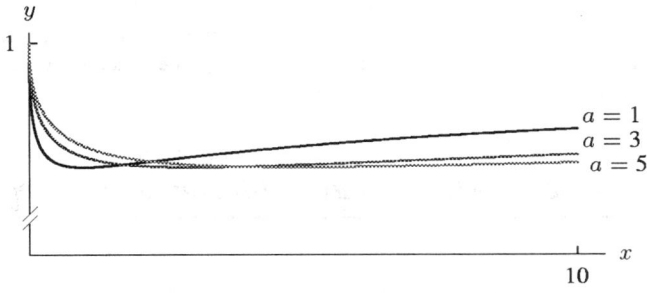

Figure 4.45

   (c) Using a computer algebra system to find the second derivative when $a = 2$, we get

$$f''(x) = \frac{4\sqrt{2} + 12\sqrt{x} + 6x^{3/2} - 3\sqrt{2}x^2}{4\left(\sqrt{2} + \sqrt{x}\right)^3 x^{3/2}(2 + x)^{3/2}}.$$

Using the computer algebra system again to solve $f''(x) = 0$, we find that it has one zero at $x = 4.6477$. Graphing the second derivative, we see that it goes from positive to negative at $x = 4.6477$, so this is an inflection point.

198 CHAPTER FOUR /SOLUTIONS

10. (a) Different CASs give different answers. (In fact, their answers could be more complicated than what you get by hand.) One possible answer is

$$\frac{dy}{dx} = \frac{\tan\left(\frac{x}{2}\right)}{2\sqrt{\frac{1-\cos x}{1+\cos x}}}.$$

(b) The graph in Figure 4.46 is a step function:

$$f(x) = \begin{cases} 1/2 & 2n\pi < x < (2n+1)\pi \\ -1/2 & (2n+1)\pi < x < (2n+2)\pi. \end{cases}$$

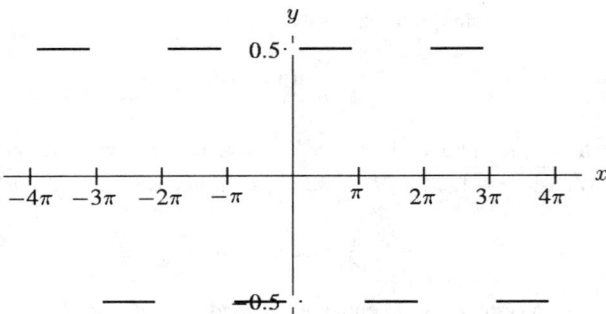

Figure 4.46

Figure 4.46, which shows the graph in disconnected line segments, is correct. However, unless you select certain graphing options in your CAS, it may join up the segments. Use the double angle formula $\cos(x) = \cos^2(x/2) - \sin^2(x/2)$ to simplify the answer in part (a). We find

$$\frac{dy}{dx} = \frac{\tan(x/2)}{2\sqrt{\frac{1-\cos x}{1+\cos x}}} = \frac{\tan(x/2)}{2\sqrt{\frac{1-\cos(2\cdot(x/2))}{1+\cos(2\cdot(x/2))}}} = \frac{\tan(x/2)}{2\sqrt{\frac{1-\cos^2(x/2)+\sin^2(x/2)}{1+\cos^2(x/2)-\sin^2(x/2)}}}$$

$$= \frac{\tan(x/2)}{2\sqrt{\frac{2\sin^2(x/2)}{2\cos^2(x/2)}}} = \frac{\tan(x/2)}{2\sqrt{\tan^2(x/2)}} = \frac{\tan(x/2)}{2\,|\tan(x/2)|}$$

Thus, $dy/dx = 1/2$ when $\tan(x/2) > 0$, i.e. when $0 < x < \pi$ (more generally, when $2n\pi < x < (2n+1)\pi$), and $dy/dx = -1/2$ when $\tan(x/2) < 0$, i.e., when $\pi < x < 2\pi$ (more generally, when $(2n+1)\pi < x < (2n+2)\pi$, where n is any integer).

Solutions to Problems on the Theorems about Continuous and Differentiable Functions

1. Let $f(x) = \sin x$ and $g(x) = x$. Then $f(0) = 0$ and $g(0) = 0$. Also $f'(x) = \cos x$ and $g'(x) = 1$, so for all $x \geq 0$ we have $f'(x) \leq g'(x)$. So the graphs of f and g both go through the origin and the graph of f climbs slower than the graph of g. Thus the graph of f is below the graph of g for $x \geq 0$ by the Racetrack Principle. In other words, $\sin x \leq x$ for $x \geq 0$.

2. Let $g(x) = \ln x$ and $h(x) = x - 1$. For $x \geq 1$, we have $g'(x) = 1/x \leq 1 = h'(x)$. Since $g(1) = h(1)$, the Racetrack Principle with $a = 1$ says that $g(x) \leq h(x)$ for $x \geq 1$, that is, $\ln x \leq x - 1$ for $x \geq 1$. For $0 < x \leq 1$, we have $h'(x) = 1 \leq 1/x = g'(x)$. Since $g(1) = h(1)$, the Racetrack Principle with $b = 1$ says that $g(x) \leq h(x)$ for $0 < x \leq 1$, that is, $\ln x \leq x - 1$ for $0 < x \leq 1$.

3.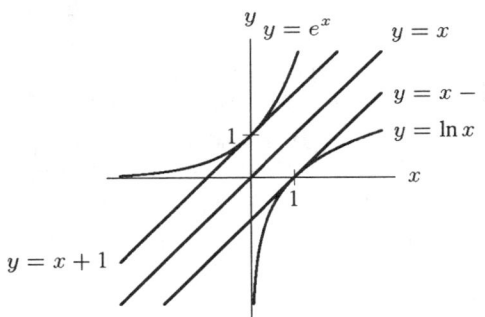

Graphical solution: If f and g are inverse functions then the graph of g is just the graph of f reflected through the line $y = x$. But e^x and $\ln x$ are inverse functions, and so are the functions $x + 1$ and $x - 1$. Thus the equivalence is clear from the figure.

Algebraic solution: If $x > 0$ and
$$x + 1 \leq e^x,$$
then, replacing x by $x - 1$, we have
$$x \leq e^{x-1}.$$
Taking logarithms, and using the fact that ln is an increasing function, gives
$$\ln x \leq x - 1.$$
We can also go in the opposite direction, which establishes the equivalence.

4. Since $f''(t) \leq 7$ for $0 \leq t \leq 2$, if we apply the Racetrack Principle with $a = 0$ to the functions $f'(t) - f'(0)$ and $7t$, both of which go through the origin, we get
$$f'(t) - f'(0) \leq 7t \quad \text{for } 0 \leq t \leq 2.$$

The left side of this inequality is the derivative of $f(t) - f'(0)t$, so if we apply the Racetrack Principle with $a = 0$ again, this time to the functions $f(t) - f'(0)t$ and $(7/2)t^2 + 3$, both of which have the value 3 at $t = 0$, we get
$$f(t) - f'(0)t \leq \frac{7}{2}t^2 + 3 \quad \text{for } 0 \leq t \leq 2.$$

That is,
$$f(t) \leq 3 + 4t + \frac{7}{2}t^2 \quad \text{for } 0 \leq t \leq 2.$$

In the same way, we can show that the lower bound on the acceleration, $5 \leq f''(t)$ leads to:
$$f(t) \geq 3 + 4t + \frac{5}{2}t^2 \quad \text{for } 0 \leq t \leq 2.$$

If we substitute $t = 2$ into these two inequalities, we get bounds on the position at time 2:
$$21 \leq f(2) \leq 25.$$

5. If f is continuous then $-f$ is continuous also. So $-f$ has a global maximum at some point $x = c$. Thus $-f(x) \leq -f(c)$ for all x in $[a, b]$. Hence $f(x) \geq f(c)$ for all x in $[a, b]$. So f has a global minimum at $x = c$.

6. The Decreasing Function Theorem is: Suppose that f is continuous on $[a, b]$ and differentiable on (a, b). If $f'(x) < 0$ on (a, b), then f is decreasing on $[a, b]$. If $f'(x) \leq 0$ on (a, b), then f is nonincreasing on $[a, b]$.

To prove the theorem, we note that if f is decreasing then $-f$ is increasing and vice-versa. Similarly, if f is nonincreasing, then $-f$ is nondecreasing. Thus if $f'(x) < 0$, then $-f'(x) > 0$, so $-f$ is increasing, which means f is decreasing. And if $f'(x) \leq 0$, then $-f'(x) \geq 0$, so $-f$ is nondecreasing, which means f is nonincreasing.

7. Consider the function $f(x) = h(x) - g(x)$. Since $f'(x) = h'(x) - g'(x) \geq 0$, we know that f is nondecreasing by the Increasing Function Theorem. This means $f(x) \leq f(b)$ for $a \leq x \leq b$. However, $f(b) = h(b) - g(b) = 0$, so $f(x) \leq 0$, which means $h(x) \leq g(x)$.

8. If $f'(x) = 0$, then both $f'(x) \geq 0$ and $f'(x) \leq 0$. By the Increasing and Decreasing Function Theorems, f is both nondecreasing and nonincreasing, so f is constant.

9. Let $h(x) = f(x) - g(x)$. Then $h'(x) = f'(x) - g'(x) = 0$ for all x in (a, b). Hence, by the Constant Function Theorem, there is a constant C such that $h(x) = C$ on (a, b). Thus $f(x) = g(x) + C$.

10. We will show $f(x) = Ce^x$ by deducing that $f(x)/e^x$ is a constant. By the Constant Function Theorem, we need only show the derivative of $g(x) = f(x)/e^x$ is zero. By the quotient rule (since $e^x \neq 0$), we have
$$g'(x) = \frac{f'(x)e^x - e^x f(x)}{(e^x)^2}.$$
Since $f'(x) = f(x)$, we simplify and obtain
$$g'(x) = \frac{f(x)e^x - e^x f(x)}{(e^x)^2} = \frac{0}{e^{2x}} = 0,$$
which is what we needed to show.

11. (a)
$$\left(\frac{e-c}{d-c}\right) \text{slope}(c, e) + \left(\frac{d-e}{d-c}\right) \text{slope}(e, d)$$
$$= \left(\frac{e-c}{d-c}\right)\left(\frac{f(e)-f(c)}{e-c}\right) + \left(\frac{d-e}{d-c}\right)\left(\frac{f(d)-f(e)}{d-e}\right)$$
$$= \frac{f(e)-f(c)}{d-c} + \frac{f(d)-f(e)}{d-c} = \frac{f(d)-f(c)}{d-c} = \text{slope}(c, d)$$

If $\text{slope}(c, e) \leq \text{slope}(e, d)$, then
$$\left(\frac{e-c}{d-c}\right) \text{slope}(c, e) + \left(\frac{d-e}{d-c}\right) \text{slope}(e, d) \leq \left(\frac{e-c}{d-c}\right) \text{slope}(e, d) + \left(\frac{d-e}{d-c}\right) \text{slope}(e, d)$$
$$= \left(\frac{e-c}{d-c} + \frac{d-e}{d-c}\right) \text{slope}(e, d) = \text{slope}(e, d)$$

and
$$\left(\frac{e-c}{d-c}\right) \text{slope}(c, e) + \left(\frac{d-e}{d-c}\right) \text{slope}(e, d) \geq \left(\frac{e-c}{d-c}\right) \text{slope}(c, e) + \left(\frac{d-e}{d-c}\right) \text{slope}(c, e)$$
$$= \left(\frac{e-c}{d-c} + \frac{d-e}{d-c}\right) \text{slope}(c, e) = \text{slope}(c, e)$$

Thus $\text{slope}(c, d)$ lies between $\text{slope}(c, e)$ and $\text{slope}(e, d)$. The argument if $\text{slope}(c, e) \geq \text{slope}(e, d)$ is similar.

(b) Bisect the interval $[a_1, b_1]$. At least one of the two halves will have slope less than or equal to $\text{slope}(a_1, b_1)$ by (a). Let $[a_2, b_2]$ be one such half. Bisect $[a_2, b_2]$. At least one half will have slope less than or equal to $\text{slope}(a_2, b_2) \leq \text{slope}(a_1, b_1)$. Let $[a_3, b_3]$ be one such half. Continue.

(c) If $c = a_n$ for some n, then $\text{slope}(c, b_n) = \text{slope}(a_n, b_n) \leq \text{slope}(a_1, b_1)$. If $c = b_n$, then $\text{slope}(a_n, c) = \text{slope}(a_n, b_n) \leq \text{slope}(a_1, b_1)$. If $a_n < c < b_n$, then it follows from (a) that one of $\text{slope}(a_n, c)$ and $\text{slope}(c, b_n)$ is less than or equal to $\text{slope}(a_n, b_n)$, which is less than or equal to $\text{slope}(a_1, b_1)$.

(d) Let $\epsilon = -\text{slope}(a_1, b_1)$. For x sufficiently close to c, we have $\text{slope}(x, c) > f'(c) - \epsilon \geq 0 + \text{slope}(a_1, b_1)$ and $\text{slope}(c, x) > f'(c) - \epsilon \geq 0 + \text{slope}(a_1, b_1)$. Thus, for n sufficiently large, both a_n and b_n will be sufficiently close to c to ensure that $\text{slope}(a_n, c) > \text{slope}(a_1, b_1)$ and $\text{slope}(c, b_n) > \text{slope}(a_1, b_1)$.

(e) Since we already know that f is nondecreasing on (a, b), we only need to show that $f(a) \leq f(x)$ and $f(x) \leq f(b)$ for all x in (a, b). Suppose that $f(a) > f(d)$ for some d in (a, b). Let $\epsilon = f(a) - f(d)$. Then for all x sufficiently close to a, we have $f(x) > f(a) - \epsilon = f(d)$. This contradicts the fact that f is nondecreasing on (a, b). The argument for the other inequality is similar.

(f) Suppose $f'(x) > 0$. We must show that if $a_1 < b_1$, then $f(a_1) < f(b_1)$. Since $f'(x) > 0$ implies $f'(x) \geq 0$, we already know that f is nondecreasing. Thus, if $a_1 < x < b_1$, then $f(a_1) \leq f(x) \leq f(b_1)$. If $f(a_1) = f(b_1)$, then $f(x) = f(a_1) = f(b_1)$, so f is constant on $[a_1, b_1]$. We know the derivative of a constant function is zero, so $f'(x) = 0$ on $[a_1, b_1]$, contradicting $f'(x) > 0$. So we must have $f(a_1) < f(b_1)$, which means f is increasing on $[a, b]$.

12. Apply the Racetrack Principle to the functions $f(x) - f(a)$ and $M(x-a)$; we can do this since $f(a) - f(a) = M(a-a)$ and $f'(x) \leq M$. We conclude that $f(x) - f(a) \leq M(x-a)$. Similarly, apply the Racetrack Principle to the functions $m(x-a)$ and $f(x) - f(a)$ to obtain $m(x-a) \leq f(x) - f(a)$. If we substitute $x = b$ into these inequalities we get
$$m(b-a) \leq f(b) - f(a) \leq M(b-a).$$
Now, divide by $b - a$.

13. (a) Since $f''(x) \geq 0$, $f'(x)$ is nondecreasing on (a, b). Thus $f'(c) \leq f'(x)$ for $c \leq x < b$ and $f'(x) \leq f'(c)$ for $a < x \leq c$.
    (b) Let $g(x) = f(c) + f'(c)(x - c)$ and $h(x) = f(x)$. Then $g(c) = f(c) = h(c)$, and $g'(x) = f'(c)$ and $h'(x) = f'(x)$. If $c \leq x < b$, then $g'(x) \leq h'(x)$, and if $a < x \leq c$, then $g'(x) \geq h'(x)$, by (a). By the Racetrack Principle, $g(x) \leq h(x)$ for $c \leq x < b$ and for $a < x \leq c$, as we wanted.

14. (a) We have
$$F(b) - F(a) = ((F(b) - F(x_{n-1})) + (F(x_{n-1}) - F(x_{n-2})) + \cdots + (F(x_1) - F(a)).$$
By the Mean Value Theorem applied to the interval $[x_i, x_{i+1}]$, there is a number c_i in (x_i, x_{i+1}) such that
$$F'(c_i) = \frac{F(x_{i+1}) - F(x_i)}{x_{i+1} - x_i}.$$
Since $F' = f$, this implies
$$F(x_{i+1}) - F(x_i) = f(c_i)(x_{i+1} - x_i) = f(c_i)\Delta x_i.$$
Thus
$$F(b) - F(a) = ((F(b) - F(x_{n-1})) + (F(x_{n-1}) - F(x_{n-2})) + \cdots + (F(x_1) - F(a))$$
$$= \sum_{i=0}^{n-1} f(c_i)\Delta x_i.$$

   (b) Since f is continuous on $[a, b]$, it is integrable. Therefore we can choose a subdivision so that the lower and upper sums for that subdivision are as close as we like. Since both $\int_a^b f(x)\,dx$ and the Riemann sum constructed in part (a) are between the lower sum and upper sum, this means that we can make the Riemann sum in part (a) as close as we like to $\int_a^b f(x)\,dx$ by choosing the appropriate subdivision. But the value of the Riemann sum is $F(b) - F(a)$. Therefore we must have $F(b) - F(a) = \int_a^b f(x)\,dx$.

15. Since f is continuous, $\lim_{x \to c} f(x) = f(c)$. Let $\epsilon = M - f(c)$, and choose δ such that
$$|f(x) - f(c)| < \epsilon$$
if $|x - c| < \delta$. Then $f(x) < f(c) + \epsilon = M$ for all x in $[a, b]$ such that $c - \delta < x < c + \delta$.

16. (a) Suppose there were no such upper bound. Let m be the midpoint of $[a, b]$. If both of the subintervals $[a, m]$ and $[m, b]$ have upper bounds, then the larger of the two upper bounds would be an upper bound for the original interval. Therefore, at least one of the half-subintervals has no upper bound; choose one and call it $[a_1, b_1]$.
    (b) Continue bisecting so that at each stage there is no upper bound for f on $[a_n, b_n]$. The bisected intervals close in on a number c.
    (c) Let U be any number larger than $f(c)$. Since f is continuous, $f(x) < U$ for all x sufficiently close to U, which was shown in Problem 15. Thus $f(x) < U$ on $[a_n, b_n]$ for all sufficiently large n, which contradicts our choice of the intervals $[a_n, b_n]$. Therefore, we were wrong when we assumed no upper bound existed for f on the interval $[a, b]$.

17. (a) We have $y < y + 1$. Since $y \geq 0$, we have $y + 1 > 0$, so dividing both sides by $y + 1$ we get $y/(y+1) < 1$.
    (b) It follows from the theorem on properties of continuous functions on page 89 that g is continuous. It follows from part (a) that g is bounded above by 1.
    (c) Let $g(x) = x/(x+1)$. Then,
$$g'(x) = \frac{1}{(x+1)^2} > 0 \quad \text{for all } x \geq 0.$$
As a result, $g(y_1) \leq g(y_2)$ implies $y_1 \leq y_2$.
    (d) Let c be the point where g has a global maximum. Then $g(x) \leq g(c)$ for all x in $[a, b]$, hence $f(x)/(1 + f(x)) \leq f(c)/(1 + f(c))$, and hence, by part (c), $f(x) \leq f(c)$ for all x in $[a, b]$. Hence f has a global maximum on $[a, b]$ at $x = c$.
    (e) If f is continuous, then so is $|f|$. By what we have shown, $|f|$ has a global maximum U on $[a, b]$. But if $|f(x)| \leq U$, then $f(x) \leq |f(x)| \leq U$ also, so f is also bounded above.

18. (a) If both the global minimum and the global maximum are at the endpoints, then $f(x) = 0$ everywhere in $[a, b]$, since $f(a) = f(b) = 0$. In that case $f'(x) = 0$ everywhere as well, so any point in (a, b) will do for c.
   (b) Suppose that either the global maximum or the global minimum occurs at an interior point of the interval. Let c be that point. Then c must be a local extremum of f, so, by the theorem concerning local extrema on page 202, we have $f'(c) = 0$, as required.

19. (a) The equation of the secant line is
$$y = f(a) + \frac{f(b) - f(a)}{b - a}(x - a),$$
thus
$$g(x) = f(x) - f(a) - \frac{f(b) - f(a)}{b - a}(x - a).$$
   (b) The figure shows that $g(a) = g(b) = 0$. (You can also easily check this from the formula.) By Rolle's Theorem, there must be a point c in (a, b) where $g'(c) = 0$.
   (c) We have
$$g'(x) = f'(x) - \frac{f(b) - f(a)}{b - a}.$$
So from $g'(c) = 0$, we get
$$f'(c) = \frac{f(b) - f(a)}{b - a},$$
as required.

CHAPTER FIVE

Solutions for Section 5.1

1. (a) Lower estimate $= (45)(2) + (16)(2) + (0)(2) = 122$ feet.
   Upper estimate $= (88)(2) + (45)(2) + (16)(2) = 298$ feet.
   (b)
   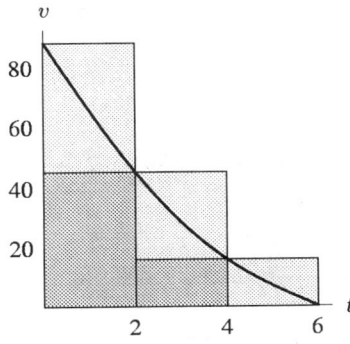

2. To find the distance the car moved before stopping, we estimate the distance traveled for each two-second interval. Since speed decreases throughout, we know that the left-handed sum will be an overestimate to the distance traveled and the right-hand sum an underestimate. Applying the formulas for these sums with $\Delta t = 2$ gives:

   $$\text{LEFT} = 2(100 + 80 + 50 + 25 + 10) = 530 \text{ ft.}$$
   $$\text{RIGHT} = 2(80 + 50 + 25 + 10 + 0) = 330 \text{ ft.}$$

   (a) The best estimate of the distance traveled will be the average of these two estimates, or

   $$\text{Best estimate} = \frac{530 + 330}{2} = 430 \text{ ft.}$$

   (b) All we can be sure of is that the distance traveled lies between the upper and lower estimates calculated above. In other words, all the black-box data tells us for sure is that the car traveled between 330 and 530 feet before stopping. So we can't be completely sure about whether it hit the skunk or not.

3. (a) Note that 15 minutes equals 0.25 hours. Lower estimate $= 11(0.25) + 10(0.25) = 5.25$ miles. Upper estimate $= 12(0.25) + 11(0.25) = 5.75$ miles.
   (b) Lower estimate $= 11(0.25) + 10(0.25) + 10(0.25) + 8(0.25) + 7(0.25) + 0(0.25) = 11.5$ miles. Upper estimate $= 12(0.25) + 11(0.25) + 10(0.25) + 10(0.25) + 8(0.25) + 7(0.25) = 14.5$ miles.
   (c) The difference between Roger's pace at the beginning and the end of his run is 12 mph. If the time between the measurements is h, then the difference between the upper and lower estimates is $12h$. We want $12h < 0.1$, so

   $$h < \frac{0.1}{12} \approx 0.0083 \text{ hours} = 30 \text{ seconds}$$

   Thus Jeff would have to measure Roger's pace every 30 seconds.

4. (a) An overestimate is 7 tons. An underestimate is 5 tons.
   (b) An overestimate is $7 + 8 + 10 + 13 + 16 + 20 = 74$ tons. An underestimate is $5 + 7 + 8 + 10 + 13 + 16 = 59$ tons.
   (c) If measurements are made every Δt months, then the error is $|f(6) - f(0)| \cdot \Delta t$. So for this to be less than 1 ton, we need $(20 - 5) \cdot \Delta t < 1$, or $\Delta t < 1/15$. So measurements every 2 days or so will guarantee an error in over- and underestimates of less than 1 ton.

5. Using $\Delta t = 2$,

$$\text{Left-hand sum} = v(0) \cdot 2 + v(2) \cdot 2 + v(4) \cdot 2$$
$$= 1(2) + 5(2) + 17(2)$$
$$= 46$$
$$\text{Right-hand sum} = v(2) \cdot 2 + v(4) \cdot 2 + v(6) \cdot 2$$
$$= 5(2) + 17(2) + 37(2)$$
$$= 118$$
$$\text{Average} = \frac{46 + 118}{2} = 82$$
$$\text{Distance traveled} \approx 82 \text{ meters}.$$

6. Using $\Delta t = 0.2$, our upper estimate is

$$\frac{1}{1+0}(0.2) + \frac{1}{1+0.2}(0.2) + \frac{1}{1+0.4}(0.2) + \frac{1}{1+0.6}(0.2) + \frac{1}{1+0.8}(0.2) \approx 0.75.$$

The lower estimate is

$$\frac{1}{1+0.2}(0.2) + \frac{1}{1+0.4}(0.2) + \frac{1}{1+0.6}(0.2) + \frac{1}{1+0.8}(0.2) \frac{1}{1+1}(0.2) \approx 0.65.$$

Since v is a decreasing function, the bug has crawled more than 0.65 meters, but less than 0.75 meters. We average the two to get a better estimate:

$$\frac{0.65 + 0.75}{2} = 0.70 \text{ meters}.$$

7. Using whole grid squares, we can overestimate the area as $3+3+3+3+2+1 = 15$, and we can underestimate the area as $1+2+2+1+0+0 = 6$. Using triangles as in Figure 5.1, we can overestimate the area as $2+2\frac{7}{8}+3+2\frac{1}{2}+1\frac{1}{2}+\frac{3}{4} = 12\frac{5}{8}$ and we can underestimate the area as $1\frac{1}{2}+2\frac{1}{4}+2\frac{1}{2}+2+1+\frac{1}{4} = 9\frac{1}{2}$. It also appears from the graph that our upper estimate is closer than the lower estimate to the actual area, so we can further estimate the area to be a little greater than $\frac{1}{2}(9\frac{1}{2} + 12\frac{5}{8}) = 11\frac{1}{16}$.

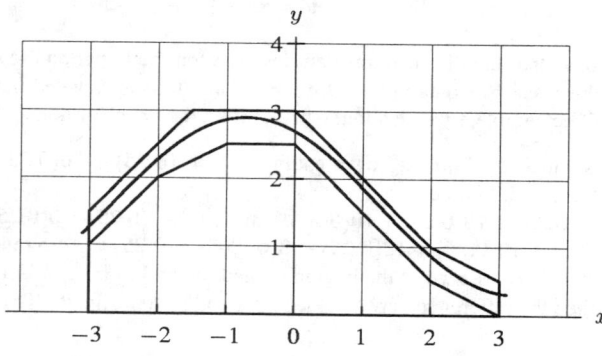

Figure 5.1: Estimating the Area

8. (a) We want the error to be less than 0.1, so take Δx such that $|f(1) - f(0)|\Delta x < 0.1$, giving

$$\Delta x < \frac{0.1}{|e^{-\frac{1}{2}} - 1|} \approx 0.25$$

so take $\Delta x = 0.25$ or $n = 4$. Then the left sum $= 0.9016$, and the right sum $= 0.8033$, so a reasonable estimate for the area is $(0.9016 + 0.8033)/2 = 0.8525$. Certainly 0.85 is within 0.1 of the actual answer.

(b) Take Δx smaller. To have an error of at most E, you need Δx such that

$$|f(1) - f(0)|\Delta x < E.$$

This means

$$\Delta x < \frac{E}{|e^{-\frac{1}{2}} - 1|} \approx \frac{E}{0.39}.$$

Using n equal subdivisions, we have

$$\Delta x = \frac{b-a}{n} = \frac{1-0}{n} = \frac{1}{n}.$$

Thus, to approximate the shaded area with an error $< E$ requires $n > \frac{0.39}{E}$ subdivisions.

9. Just counting the squares (each of which has area 10), and allowing for the broken squares, we can see that the area under the curve from 0 to 6 is between 140 and 150. Hence the distance traveled is between 140 and 150 meters.

10. (a) An upper estimate is $9.81 + 8.03 + 6.53 + 5.38 + 4.41 = 34.16$ m/sec. A lower estimate is $8.03 + 6.53 + 5.38 + 4.41 + 3.61 = 27.96$ m/sec.

(b) The average is $\frac{1}{2}(34.16 + 27.96) = 31.06$ m/sec. Because the graph of acceleration is concave up, this estimate is too high, as can be seen in the figure to the right. The area of the shaded region is the average of the areas of the rectangles $ABFE$ and $CDFE$.

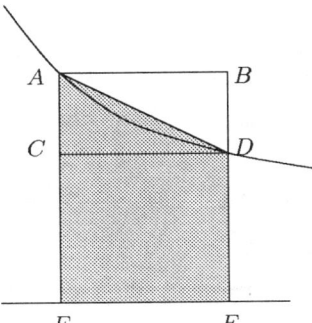

11. (a) When the aircraft is climbing at v ft/min, it takes $1/v$ minutes to climb 1 foot. Therefore

$$\text{Lower estimate} = \left(\frac{1\,\text{min}}{925\,\text{ft}}\right)(1000\,\text{ft}) + \left(\frac{1\,\text{min}}{875\,\text{ft}}\right)(1000\,\text{ft}) + \cdots + \left(\frac{1\,\text{min}}{490\,\text{ft}}\right)(1000\,\text{ft})$$
$$\approx 14.73\,\text{minutes}.$$
$$\text{Upper estimate} = \left(\frac{1\,\text{min}}{875\,\text{ft}}\right)(1000\,\text{ft}) + \left(\frac{1\,\text{min}}{830\,\text{ft}}\right)(1000\,\text{ft}) + \cdots + \left(\frac{1\,\text{min}}{440\,\text{ft}}\right)(1000\,\text{ft})$$
$$\approx 15.93\,\text{minutes}.$$

Note: The Pilot Operating Manual for this aircraft gives 16 minutes as the estimated time required to climb to 10,000 ft.

(b) The difference between upper and lower sums with $\Delta x = 500$ ft would be

$$\text{Difference} = \left(\frac{1\,\text{min}}{440\,\text{ft}} - \frac{1\,\text{min}}{925\,\text{ft}}\right)(500\,\text{ft}) = 0.60\,\text{minutes}.$$

Solutions for Section 5.2

1.

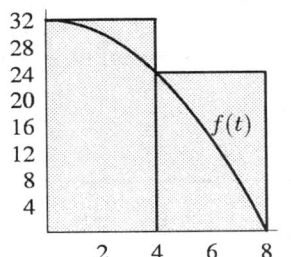

Figure 5.2: Left Sum, $\Delta t = 4$

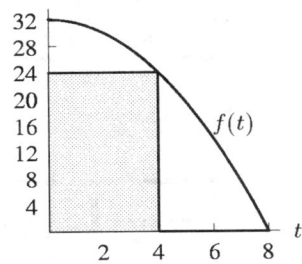

Figure 5.3: Right Sum, $\Delta t = 4$

(a) Left-hand sum $= 32(4) + 24(4) = 224$.
(b) Right-hand sum $= 24(4) + 0(4) = 96$.

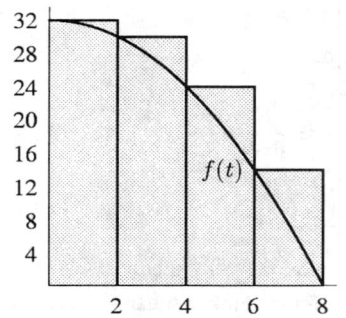

Figure 5.4: Left Sum, $\Delta t = 2$

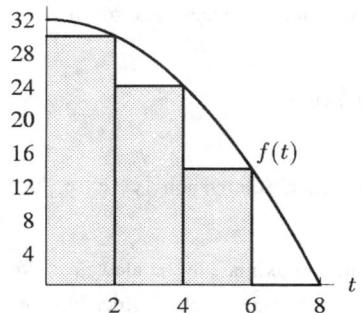

Figure 5.5: Right Sum, $\Delta t = 2$

(c) Left-hand sum $= 32(2) + 30(2) + 24(2) + 14(2) = 200$.
(d) Right-hand sum $= 30(2) + 24(2) + 14(2) + 0(2) = 136$.

2. Since we have 5 subdivisions,
$$\Delta x = \frac{b-a}{n} = \frac{7-3}{5} = 0.8.$$

The interval begins at $x = 3$ and ends at $x = 7$. Table 5.1 gives the value of $f(x)$ at the pertinent points.

TABLE 5.1

x	3.0	3.8	4.6	5.4	6.2	7.0
$f(x)$	$\frac{1}{1+3.0}$	$\frac{1}{1+3.8}$	$\frac{1}{1+4.6}$	$\frac{1}{1+5.4}$	$\frac{1}{1+6.2}$	$\frac{1}{1+7.0}$

So a right-hand sum is
$$\frac{1}{1+3.8}(0.8) + \frac{1}{1+4.6}(0.8) + \cdots + \frac{1}{1+7.0}(0.8).$$

3.

n	2	10	50	250
Left-hand Sum	0.0625	0.2025	0.2401	0.248004
Right-hand Sum	0.5625	0.3025	0.2601	0.252004

The sums seem to be converging to $\frac{1}{4}$. Since x^3 is monotone on $[0, 1]$, the true value is between 0.248004 and 0.252004.

4.

n	2	10	50	250
Left-hand Sum	1.34076	1.07648	1.01563	1.00314
Right-hand Sum	0.55536	0.91940	0.98421	0.99686

The sums seem to be converging to 1. Since $\cos x$ is monotone on $[0, \pi/2]$, the true value is between 1.00314 and .99686.

5.

n	2	10	50	250
Left-hand Sum	-0.394991	-0.0920539	-0.0429983	-0.0335556
Right-hand Sum	0.189470	0.0248382	-0.0196199	-0.0288799

There is no obvious guess as to what the limiting sum is. Moreover, since $\sin(t^2)$ is *not* monotonic on $[2, 3]$, we cannot be sure that the true value is between -0.0335556 and -0.0288799.

6.

n	2	10	50	250
Left-hand Sum	1.14201	1.38126	1.44565	1.45922
Right-hand Sum	2.00115	1.55309	1.48002	1.46610

There is no obvious guess as to what the limiting sum is. We can only observe that since e^{t^2} is monotonic on $[0, 1]$, the true value is between 1.45922 and 1.46610.

7.

n	2	10	50	250
Left-hand Sum	-0.52336	1.31159	1.49798	1.52526
Right-hand Sum	1.27721	1.6717	1.57000	1.53966

There is no obvious guess as to what the limiting sum is. Moreover, since $\sin(1/x)$ is *not* monotonic on $[0.2, 3]$, we cannot be sure that the true value is between 1.52526 and 1.53966.

8.

n	2	10	50	250
Left-hand Sum	1.41856	1.90525	2.02064	2.04445
Right-hand Sum	2.91856	2.20525	2.08064	2.05645

There is no obvious guess as to what the limiting sum is. We can only observe that since x^x is monotonic on $[1, 2]$, the true value is between 2.04445 and 2.05645.

9. Left-hand sum gives: $1^2(1/4) + (1.25)^2(1/4) + (1.5)^2(1/4) + (1.75)^2(1/4) = 1.96875$.
Right-hand sum gives: $(1.25)^2(1/4) + (1.5)^2(1/4) + (1.75)^2(1/4) + (2)^2(1/4) = 2.71875$.

We estimate the value of the integral by taking the average of these two sums, which is 2.34375. Since x^2 is monotonic on $1 \leq x \leq 2$, the true value of the integral lies between 1.96875 and 2.71875. Thus the most our estimate could be off is 0.375. We expect it to be much closer. (And it is—the true value of the integral is $7/3 \approx 2.333$.)

10. (a) $\int_0^6 (x^2 + 1)\, dx = 78$

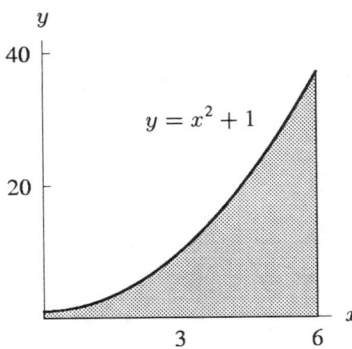

(b) Using $n = 3$, we have

$$\text{Left-hand sum} = f(0) \cdot 2 + f(2) \cdot 2 + f(4) \cdot 2$$
$$= 1(2) + 5(2) + 17(2) = 46.$$

This sum is an underestimate. See Figure 5.6.

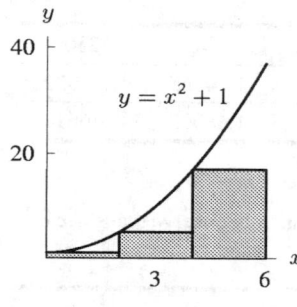

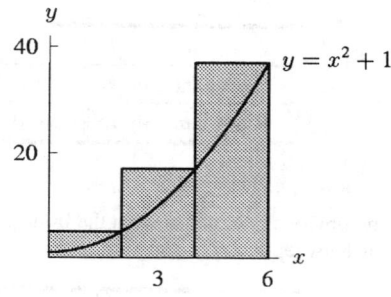

Figure 5.6　　　　　　　　　　　Figure 5.7

(c)
$$\text{Right-hand sum} = f(2) \cdot 2 + f(4) \cdot 2 + f(6) \cdot 2$$
$$= 5(2) + 17(2) + 37(2)$$
$$= 118.$$

This sum is an overestimate. See Figure 5.7.

11. We take $\Delta t = 20$. Then:
$$\text{Left-hand sum} = 1.2(20) + 2.8(20) + 4.0(20) + 4.7(20) + 5.1(20)$$
$$= 356.$$
$$\text{Right-hand sum} = 2.8(20) + 4.0(20) + 4.7(20) + 5.1(20) + 5.2(20)$$
$$= 436.$$
$$\int_0^{100} f(t)\,dt \approx \text{Average} = \frac{356 + 436}{2} = 396.$$

12. Since $\cos t \geq 0$ for $0 \leq t \leq \pi/2$, the area is given by
$$\text{Area} = \int_0^{\pi/2} \cos t\,dt = 1.$$

13. The graph of $y = 7 - x^2$ has intercepts $x = \pm\sqrt{7}$. See Figure 5.8. Therefore we have
$$\text{Area} = \int_{-\sqrt{7}}^{\sqrt{7}} (7 - x^2)\,dx = 24.7.$$

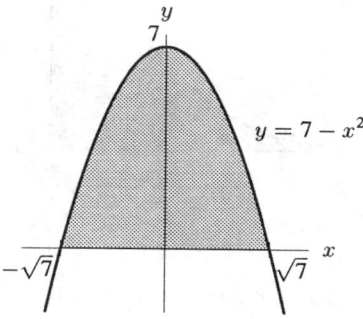

Figure 5.8

14. Since $\cos\sqrt{x} > 0$ for $0 \leq x \leq 2$, the area is given by

$$\text{Area} = \int_0^2 \cos\sqrt{x}\, dx = 1.1.$$

15. The graph of $y = e^x$ is above the line $y = 1$ for $0 \leq x \leq 2$. See Figure 5.9. Therefore

$$\text{Area} = \int_0^2 (e^x - 1)\, dx = 4.39.$$

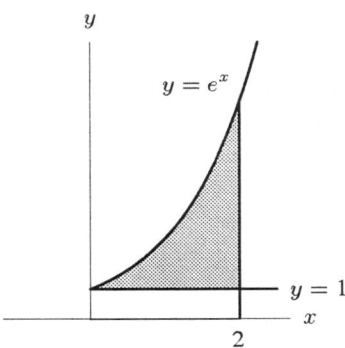

Figure 5.9

16. Since $x^3 \leq x^2$ for $0 \leq x \leq 1$, we have

$$\text{Area} = \int_0^1 (x^2 - x^3)\, dx = 0.0833.$$

17. Since $x^{1/2} \leq x^{1/3}$ for $0 \leq x \leq 1$, we have

$$\text{Area} = \int_0^1 (x^{1/3} - x^{1/2})\, dx = 0.0833.$$

18. The areas we computed are shaded in Figure 5.10. Since $y = x^2$ and $y = x^{1/2}$ are inverse functions, their graphs are reflections about the line $y = x$. Similarly, $y = x^3$ and $y = x^{1/3}$ are inverse functions and their graphs are reflections about the line $y = x$. Therefore, the two shaded areas in Figure 5.10 are equal.

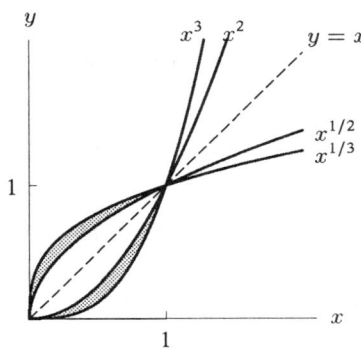

Figure 5.10

19. (a)

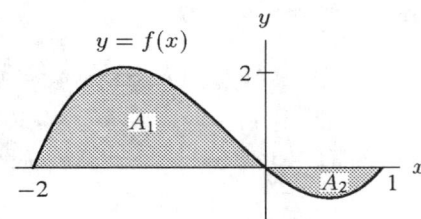

(b) $A_1 = \int_{-2}^{0} f(x)\,dx = 2.667.$

$A_2 = -\int_{0}^{1} f(x)\,dx = 0.417.$

So total area $= A_1 + A_2 \approx 3.084$. Note that while A_1 and A_2 are accurate to 3 decimal places, the quoted value for $A_1 + A_2$ is accurate only to 2 decimal places.

(c) $\int_{-2}^{1} f(x)\,dx = A_1 - A_2 = 2.250.$

20. $\int_{0}^{4} \cos\sqrt{x}\,dx = 0.80 = \text{Area } A_1 - \text{Area } A_2$

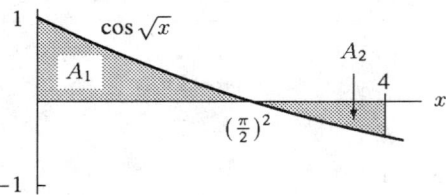

21.

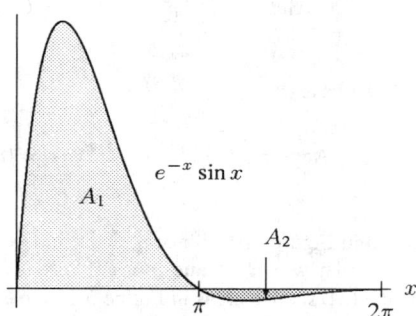

Figure 5.11

Looking at the graph of $e^{-x}\sin x$ for $0 \le x \le 2\pi$ in Figure 5.11, we see that the area, A_1, below the curve for $0 \le x \le \pi$ is much greater than the area, A_2, above the curve for $\pi \le x \le 2\pi$. Thus, the integral is

$$\int_{0}^{2\pi} e^{-x}\sin x\,dx = A_1 - A_2 > 0.$$

22. We have $\Delta x = 2/500 = 1/250$. The formulas for the left- and right-hand Riemann sums give us that

$$\text{Left} = \Delta x[f(-1) + f(-1+\Delta x) + \ldots + f(1-2\Delta x) + f(1-\Delta x)]$$
$$\text{Right} = \Delta x[f(-1+\Delta x) + f(-1+2\Delta x) + \ldots + f(1-\Delta x) + f(1)].$$

Subtracting these yields

$$\text{Right} - \text{Left} = \Delta x[f(1) - f(-1)] = \frac{1}{250}[6-2] = \frac{4}{250} = \frac{2}{125}.$$

23. The graph given shows that f is positive for $0 \leq t \leq 1$. Since the graph is contained within a rectangle of height 100 and length 1, the answers -98.35 and 100.12 are both either too small or too large to represent $\int_0^1 f(t)\,dt$. Since the graph of f is above the horizontal line $y = 80$ for $0 \leq t \leq 0.95$, the best estimate is 93.47 and not 71.84.

24. (a) $\int_{-3}^{0} f(x)\,dx = -2.$

    (b) $\int_{-3}^{4} f(x)\,dx = \int_{-3}^{0} f(x)\,dx + \int_{0}^{3} f(x)\,dx + \int_{3}^{4} f(x)\,dx = -2 + 2 - \dfrac{A}{2} = -\dfrac{A}{2}.$

25. (a) We have $\Delta x = (1-0)/n = 1/n$ and $x_i = 0 + i \cdot \Delta x = i/n$. So we get

    $$\text{Right-hand sum} = \sum_{i=1}^{n} (x_i)^4 \Delta x = \sum_{i=1}^{n} \left(\dfrac{i}{n}\right)^4 \left(\dfrac{1}{n}\right) = \sum_{i=1}^{n} \dfrac{i^4}{n^5}.$$

    (b) The CAS gives

    $$\text{Right-hand sum} = \sum_{i=1}^{n} \dfrac{i^4}{n^5} = \dfrac{6n^4 + 15n^3 + 10n^2 - 1}{30n^4}.$$

    (The results may look slightly different depending on the CAS you use.)

    (c) Using a CAS or by hand, we get

    $$\lim_{n \to \infty} \dfrac{6n^4 + 15n^3 + 10n^2 - 1}{30n^4} = \lim_{n \to \infty} \dfrac{6n^4}{30n^4} = \dfrac{1}{5}.$$

    The numerator is dominated by the highest power term, which is $6n^4$, so when n is large, the ratio behaves like $6n^4/30n^4 = 1/5$ as $n \to \infty$. Thus we see that

    $$\int_0^1 x^4 \, dx = \dfrac{1}{5}.$$

26. (a) A Riemann sum with n subdivisions of $[0, 1]$ has $\Delta x = 1/n$ and $x_i = i/n$. Thus,

    $$\text{Right-hand sum} = \sum_{i=1}^{n} \left(\dfrac{i}{n}\right)^5 \left(\dfrac{1}{n}\right) = \sum_{i=1}^{n} \dfrac{i^5}{n^6}.$$

    (b) A CAS gives

    $$\text{Right-hand sum} = \sum_{i=1}^{n} \dfrac{i^5}{n^6} = \dfrac{2n^4 + 6n^3 + 5n^2 - 1}{12n^4}.$$

    (c) Taking the limit by hand or using a CAS gives

    $$\lim_{n \to \infty} \dfrac{2n^4 + 6n^3 + 5n^2 - 1}{12n^4} = \lim_{n \to \infty} \dfrac{2n^4}{12n^4} = \dfrac{1}{6}.$$

    The numerator is dominated by the highest power term, which is $2n^4$, so the ratio behaves like $2n^4/12n^4 = 1/6$, as $n \to \infty$. Thus we see that

    $$\int_0^1 x^5 \, dx = \dfrac{1}{6}.$$

27. We have
    $$\Delta x = \dfrac{4}{3} = \dfrac{b-a}{n} \quad \text{and} \quad n = 3, \quad \text{so} \quad b - a = 4 \quad \text{or} \quad b = a + 4.$$

    The function, $f(x)$, is squaring something. Since it is a left-hand sum, $f(x)$ could equal x^2 with $a = 2$ and $b = 6$ (note that $2 + 3(\tfrac{4}{3})$ gives the right-hand endpoint of the last interval). Or, $f(x)$ could possibly equal $(x + 2)^2$ with $a = 0$ and $b = 4$. Other answers are possible.

28. (a) If the interval $1 \leq t \leq 2$ is divided into n equal subintervals of length $\Delta t = 1/n$, the subintervals are given by

$$1 \leq t \leq 1 + \frac{1}{n}, \ 1 + \frac{1}{n} \leq t \leq 1 + \frac{2}{n}, \ldots, 1 + \frac{n-1}{n} \leq t \leq 2.$$

The left-hand sum is given by

$$\text{Left sum} = \sum_{r=0}^{n-1} f\left(1 + \frac{r}{n}\right) \frac{1}{n} = \sum_{r=0}^{n-1} \frac{1}{1 + r/n} \cdot \frac{1}{n} = \sum_{r=0}^{n-1} \frac{1}{n+r}$$

and the right-hand sum is given by

$$\text{Right sum} = \sum_{r=1}^{n} f\left(1 + \frac{r}{n}\right) \frac{1}{n} = \sum_{r=1}^{n} \frac{1}{n+r}.$$

Since $f(t) = 1/t$ is decreasing in the interval $1 \leq t \leq 2$, we know that the right-hand sum is less than $\int_1^2 1/t \, dt$ and the left-hand sum is larger than this integral. Thus we have

$$\sum_{r=1}^{n} \frac{1}{n+r} < \int_1^t \frac{1}{t} \, dt < \sum_{r=0}^{n-1} \frac{1}{n+r}.$$

(b) Subtracting the sums gives

$$\sum_{r=0}^{n-1} \frac{1}{n+r} - \sum_{r=1}^{n} \frac{1}{n+r} = \frac{1}{n} - \frac{1}{2n} = \frac{1}{2n}.$$

(c) Here we need to find n such that

$$\frac{1}{2n} \leq 5 \times 10^{-6}, \quad \text{so} \quad n \geq \frac{1}{10} \times 10^6 = 10^5.$$

Solutions for Section 5.3

1. (a) One small box on the graph corresponds to moving at 750 ft/min for 15 seconds, which corresponds to a distance of 187.5 ft. Estimating the area beneath the velocity curves, we find:
   Distance traveled by car 1 $\approx$ 5.5 boxes = 1031.25 ft.
   Distance traveled by car 2 $\approx$ 3 boxes = 562.5 ft.

(b) The two cars will have gone the same distance when the areas beneath their velocity curves are equal. Since the two areas overlap, they are equal when the two shaded regions have equal areas, at $t \approx 1.6$ minutes. See Figure 5.12.

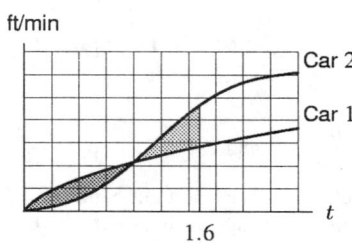

Figure 5.12

2. (a) At 3 pm, the car is traveling with a velocity of about 67 mph, while the truck has a velocity of 50 mph. Because the car is ahead of the truck at 3 pm and is traveling at a greater velocity, the distance between the car and the truck is increasing at this time. If d_{car} and d_{truck} represent the distance traveled by the car and the truck respectively, then

$$\text{distance apart} = d_{\text{car}} - d_{\text{truck}}.$$

The rate of change of the distance apart is given by its derivative:

$$(\text{distance apart})' = (d_{\text{car}})' - (d_{\text{truck}})'$$
$$= v_{\text{car}} - v_{\text{truck}}$$

At 3 pm, we get (distance apart)$' = 67$ mph $- 50$ mph $= 17$ mph. Thus, at 3 pm the car is traveling with a velocity 17 mph greater than the truck's velocity, and the distance between them is increasing at 17 miles per hour.

(b) At 2 pm, the car's velocity is greatest. Because the truck's velocity is constant, $v_{\text{car}} - v_{\text{truck}}$ will be largest when the car's velocity is largest. Thus, at 2 pm the distance between the car and the truck is increasing fastest—i.e., the car is pulling away at the greatest rate.

(Note: This only takes into account the time when the truck is moving. When the truck <u>isn't</u> moving from 12:00 to 1:00 the car pulls away from the truck at an even greater rate.)

3. (a)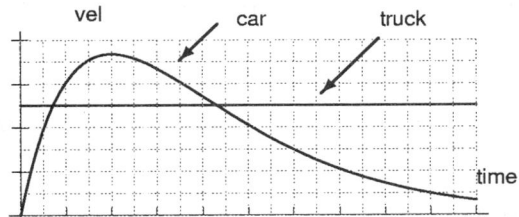

(b) The graphs intersect twice, at about 0.7 hours and 4.3 hours. At each intersection point, the velocity of the car is equal to the velocity of the truck, so $v_{\text{car}} = v_{\text{truck}}$. From the time they start until 0.7 hours later, the truck is traveling at a greater velocity than the car, so the truck is ahead of the car and is pulling farther away. At 0.7 hours they are traveling at the same velocity, and after 0.7 hours the car is traveling faster than the truck, so that the car begins to gain on the truck. Thus, at 0.7 hours the truck is farther from the car than it is immediately before or after 0.7 hours.

Similarly, because the car's velocity is greater than the truck's after 0.7 hours, it will catch up with the truck and eventually pass and pull away from the truck until 4.3 hours, at which point the two are again traveling at the same velocity. After 4.3 hours the truck travels faster than the car, so that it now gains on the car. Thus, 4.3 hours represents the point where the car is farthest ahead of the truck.

4. The units of measurement are meters per second (which are units of velocity).

5. The units of measurement are dollars.

6. The units of measurement are foot-pounds (which are units of work).

7. For any t, consider the interval $[t, t + \Delta t]$. During this interval, oil is leaking out at an approximately constant rate of $f(t)$ gallons/minute. Thus, the amount of oil which has leaked out during this interval can be expressed as

$$\text{Amount of oil leaked} = \text{Rate} \times \text{Time} = f(t)\,\Delta t$$

and the units of $f(t)\,\Delta t$ are gallons/minute $\times$ minutes $=$ gallons. The total amount of oil leaked is obtained by adding all these amounts between $t = 0$ and $t = 60$. (An hour is 60 minutes.) The sum of all these infinitesimal amounts is the integral

$$\text{Total amount of oil leaked, in gallons} = \int_0^{60} f(t)\,dt.$$

8. Average value $= \dfrac{1}{2-0}\displaystyle\int_0^2 (1+t)\,dt = \dfrac{1}{2}(4) = 2.$

9. Sketch the graph of f on $1 \leq x \leq 3$. The integral is the area under the curve, which is a trapezoidal area. So the average value is

$$\frac{1}{3-1}\int_1^3 (4x+7)\,dx = \frac{1}{2}\cdot\frac{11+19}{2}\cdot 2 = \frac{30}{2} = 15.$$

10. Average value $= \dfrac{1}{10-0}\displaystyle\int_0^{10} e^t\,dt = \dfrac{1}{10}(22025) = 2202.5$

11. The integral represents the area below the graph of $f(x)$ but above the x-axis.

    (a) Since each square has area 1, by counting squares and half-squares we find
    $$\int_1^6 f(x)\,dx = 8.5.$$

    (b) The average value is $\dfrac{1}{6-1}\int_1^6 f(x)\,dx = \dfrac{8.5}{5} = \dfrac{17}{10} = 1.7.$

12. The total number of "worker-hours" is equal to the area under the curve. The total area is about 14.5 boxes. Since each box represents (10 workers)(8 hours) = 80 worker-hours, the total area is 1160 worker-hours. At $10 per hour, the total cost is $11,600.

13. The time period 9am to 5pm is represented by the time $t = 0$ to $t = 8$ and $t = 24$ to $t = 32$. The area under the curve, or total number of worker-hours for these times, is about 9 boxes or $9(80) = 720$ worker-hours. The total cost for 9am to 5pm is $(720)(10) = \$7200$. The area under the rest of the curve is about 5.5 boxes, or $5.5(80) = 440$ worker-hours. The total cost for this time period is $(440)(15) = \$6600$. The total cost is about $7200 + 6600 = \$13,800$.

14. The area under the curve represents the number of cubic feet of storage times the number of days the storage was used. This area is given by
    $$\text{Area under graph} = \text{Area of rectangle} + \text{Area of triangle}$$
    $$= 30 \cdot 10{,}000 + \frac{1}{2} \cdot 30(30{,}000 - 10{,}000)$$
    $$= 600{,}000.$$

    Since the warehouse charges $5 for every 10 cubic feet of storage used for a day, the company will have to pay $(5)(60{,}000) = \$300{,}000$.

15.
$$\text{Average value} = \frac{1}{b-a}\int_a^b f(x)\,dx$$
$$= \frac{1}{b-a}\int_a^b 2\,dx = \frac{1}{b-a} \cdot \left(\begin{array}{c}\text{Area of rectangle} \\ \text{of height 2 and base } b - a\end{array}\right)$$
$$= \frac{1}{b-a}[2(b-a)] = 2.$$

16. (a) Since $f(x) = \sin x$ over $[0, \pi]$ is between 0 and 1, the average of $f(x)$ must itself be between 0 and 1. Furthermore, since the graph of $f(x)$ is concave down on this interval, the average value must be greater than the average height of the triangle shown in the figure, namely, 0.5.

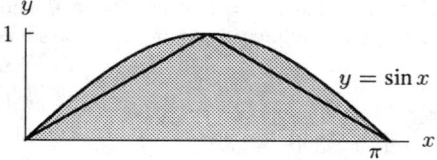

    (b) Average $= \dfrac{1}{\pi - 0}\int_0^\pi \sin x\,dx = 0.64.$

17. (a) Average value $= \int_0^1 \sqrt{1-x^2}\,dx = 0.79.$

(b) The area between the graph of $y = 1 - x$ and the x-axis is 0.5. Because the graph of $y = \sqrt{1 - x^2}$ is concave down, it lies above the line $y = 1 - x$, so its average value is above 0.5. See figure below.

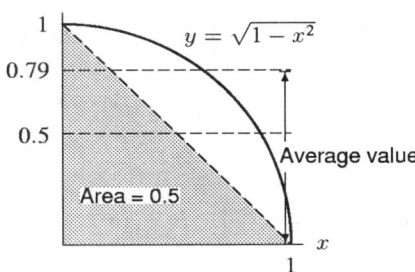

18. Since the average value is given by
$$\text{Average value} = \frac{1}{b-a} \int_a^b f(x)\,dx,$$
the units for dx inside the integral are canceled by the units for $1/(b-a)$ outside the integral, leaving only the units for $f(x)$. This is as it should be, since the average value of f should be measured in the same units as $f(x)$.

19. (a) At the end of one hour $t = 60$, and $H = 22°C$.
    (b)
$$\text{Average temperature} = \frac{1}{60} \int_0^{60} (20 + 980e^{-0.1t})\,dt$$
$$= \frac{1}{60}(10976) = 183°C.$$

    (c) Average temperature at beginning and end of hour $= (1000 + 22)/2 = 511°C$. The average found in part (b) is smaller than the average of these two temperatures because the bar cools quickly at first and so spends less time at high temperatures. Alternatively, the graph of H against t is concave up.

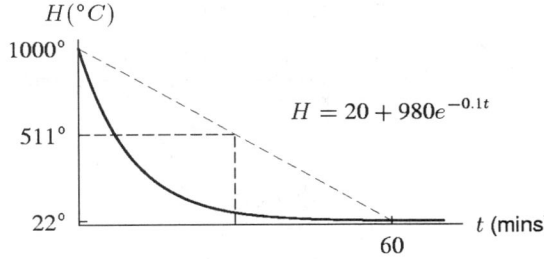

20. Since $t = 0$ in 1965 and $t = 35$ in 2000, we want:
$$\text{Average Value} = \frac{1}{35 - 0} \int_0^{35} 225(1.15)^t\,dt$$
$$= \frac{1}{35}(212{,}787) = \$6080.$$

21. (a) Since $t = 0$ to $t = 31$ covers January:
$$\left(\begin{array}{c}\text{Average number of}\\\text{daylight hours in January}\end{array}\right) = \frac{1}{31} \int_0^{31} \left[12 + 2.4\sin(0.0172(t - 80))\right]\,dt.$$

Using left and right sums with $n = 100$ gives
$$\text{Average} \approx \frac{306}{31} \approx 9.9 \text{ hours}.$$

(b) Assuming it is not a leap year, the last day of May is $t = 151 (= 31 + 28 + 31 + 30 + 31)$ and the last day of June is $t = 181 (= 151 + 30)$. Again finding the integral numerically:

$$\begin{pmatrix} \text{Average number of} \\ \text{daylight hours in June} \end{pmatrix} = \frac{1}{30} \int_{151}^{181} \left[12 + 2.4 \sin(0.0172(t-80)) \right] dt$$

$$\approx \frac{431}{30} \approx 14.4 \text{ hours.}$$

(c)

$$(\text{Average for whole year}) = \frac{1}{365} \int_0^{365} \left[12 + 2.4 \sin(0.0172(t-80)) \right] dt$$

$$\approx \frac{4381}{365} \approx 12.0 \text{ hours.}$$

(d) The average over the whole year should be 12 hours, as computed in (c). Since Madrid is in the northern hemisphere, the average for a winter month, such as January, should be less than 12 hours (it is 9.9 hours) and the average for a summer month, such as June, should be more than 12 hours (it is 14.4 hours).

22. Notice that the area of a square on the graph represents $\frac{10}{6}$ miles. At $t = 1/3$ hours, $v = 0$. The area between the curve v and the t-axis over the interval $0 \leq t \leq 1/3$ is $-\int_0^{1/3} v \, dt \approx \frac{5}{3}$. Since v is negative here, she is moving toward the lake. At $t = \frac{1}{3}$, she is about $5 - \frac{5}{3} = \frac{10}{3}$ miles from the lake. Then, as she moves away from the lake, v is positive for $\frac{1}{3} \leq t \leq 1$. At $t = 1$,

$$\int_0^1 v \, dt = \int_0^{1/3} v \, dt + \int_{1/3}^1 v \, dt \approx -\frac{5}{3} + 8 \cdot \frac{10}{6} = \frac{35}{3},$$

and the cyclist is about $5 + \frac{35}{3} = \frac{50}{3} = 16\frac{2}{3}$ miles from the lake. Since, starting from the moment $t = \frac{1}{3}$, she moves away from the lake, the cyclist will be farthest from the lake at $t = 1$. The maximal distance equals $16\frac{2}{3}$ miles.

23. We know that the the integral of F, and therefore the work, can be obtained by computing the areas in Figure 5.13.

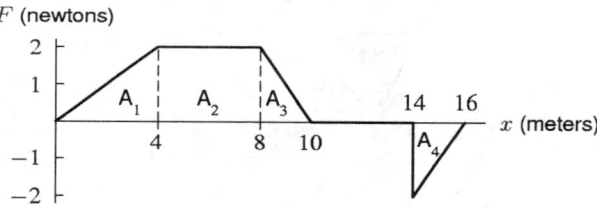

Figure 5.13

$$W = \int_0^{16} F(x) \, dx = \text{Area above } x\text{-axis} - \text{Area below } x\text{-axis}$$

$$= A_1 + A_2 + A_3 - A_4$$

$$= \frac{1}{2} \cdot 4 \cdot 2 + 4 \cdot 2 + \frac{1}{2} \cdot 2 \cdot 2 - \frac{1}{2} \cdot 2 \cdot 2$$

$$= 12 \text{ newton} \cdot \text{meters.}$$

24. (a) Average value of $f = \frac{1}{5} \int_0^5 f(x) \, dx$.

(b) Average value of $|f| = \frac{1}{5} \int_0^5 |f(x)| \, dx = \frac{1}{5} (\int_0^2 f(x) \, dx - \int_2^5 f(x) \, dx)$.

25. We'll show that in terms of the average value of f,

$$\text{I} > \text{II} = \text{IV} > \text{III}$$

Using Problem 24 (a) on page 227,

$$\begin{aligned}\frac{\text{Average value}}{\text{of } f \text{ on II}} &= \frac{\int_0^2 f(x)\,dx}{2} = \frac{\frac{1}{2}\int_{-2}^2 f(x)\,dx}{2} \\ &= \frac{\int_{-2}^2 f(x)\,dx}{4} \\ &= \text{Average value of } f \text{ on IV.}\end{aligned}$$

Since f is decreasing on [0,5], the average value of f on the interval $[0,c]$, where $0 \le c \le 5$, is decreasing as a function of c. The larger the interval the more low values of f are included. Hence

$$\begin{array}{c}\text{Average value of } f \\ \text{on } [0,1]\end{array} > \begin{array}{c}\text{Average value of } f \\ \text{on } [0,2]\end{array} > \begin{array}{c}\text{Average value of } f \\ \text{on } [0,5]\end{array}$$

Solutions for Section 5.4

1. We find the changes in $f(x)$ between any two values of x by counting the area between the curve of $f'(x)$ and the x-axis. Since $f'(x)$ is linear throughout, this is quite easy to do. From $x = 0$ to $x = 1$, we see that $f'(x)$ outlines a triangle of area $1/2$ below the x-axis (the base is 1 and the height is 1). By the Fundamental Theorem,

$$\int_0^1 f'(x)\,dx = f(1) - f(0),$$

so

$$f(0) + \int_0^1 f'(x)\,dx = f(1)$$

$$f(1) = 2 - \frac{1}{2} = \frac{3}{2}$$

Similarly, between $x = 1$ and $x = 3$ we can see that $f'(x)$ outlines a rectangle below the x-axis with area -1, so $f(2) = 3/2 - 1 = 1/2$. Continuing with this procedure (note that at $x = 4$, $f'(x)$ becomes positive), we get the table below.

x	0	1	2	3	4	5	6
$f(x)$	2	3/2	1/2	$-1/2$	-1	$-1/2$	1/2

2. (a) The amount leaked between $t = 0$ and $t = 2$ is $\int_0^2 R(t)\,dt$.

   (b)

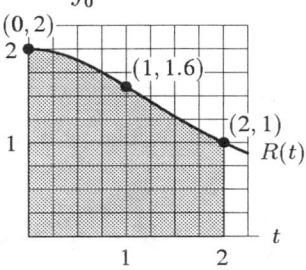

   (c) The rectangular boxes on the diagram each have area $\frac{1}{16}$. Of these 45 are wholly beneath the curve, hence the area under the curve is certainly more than $\frac{45}{16} > 2.81$. There are 9 more partially beneath the curve, and so the desired area is completely covered by 54 boxes. Therefore the area is less than $\frac{54}{16} < 3.38$.

   These are very safe estimates but far apart. We can do much better by estimating what fractions of the broken boxes are beneath the curve. Using this method, we can estimate the area to be about 3.2, which corresponds to 3.2 gallons leaking over two hours.

3. Since $F(0) = 0$, $F(b) = \int_0^b f(t)\,dt$. For each b we determine $F(b)$ graphically as follows:
$F(0) = 0$
$F(1) = F(0) + \text{Area of } 1 \times 1 \text{ rectangle} = 0 + 1 = 1$
$F(2) = F(1) + \text{Area of triangle }(\frac{1}{2} \cdot 1 \cdot 1) = 1 + 0.5 = 1.5$
$F(3) = F(2) + \text{Negative of area of triangle} = 1.5 - 0.5 = 1$
$F(4) = F(3) + \text{Negative of area of rectangle} = 1 - 1 = 0$
$F(5) = F(4) + \text{Negative of area of rectangle} = 0 - 1 = -1$
$F(6) = F(5) + \text{Negative of area of triangle} = -1 - 0.5 = -1.5$
The graph of $F(t)$, for $0 \leq t \leq 6$, is shown in Figure 5.14.

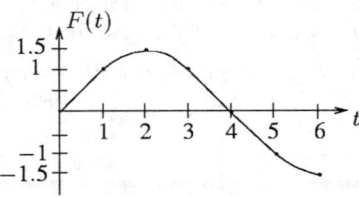

Figure 5.14

4. By the Fundamental Theorem,
$$f(1) - f(0) = \int_0^1 f'(x)\,dx,$$
Since $f'(x)$ is negative for $0 \leq x \leq 1$, this integral must be negative and so $f(1) < f(0)$.

5. First rewrite each of the quantities in terms of f', since we have the graph of f'. If A_1 and A_2 are the positive areas shown in Figure 5.15:

$$f(3) - f(2) = \int_2^3 f'(t)\,dt = -A_1$$

$$f(4) - f(3) = \int_3^4 f'(t)\,dt = -A_2$$

$$\frac{f(4) - f(2)}{2} = \frac{1}{2}\int_2^4 f'(t)\,dt = -\frac{A_1 + A_2}{2}$$

Since Area $A_1 >$ Area A_2,
$$A_2 < \frac{A_1 + A_2}{2} < A_1$$

so
$$-A_1 < -\frac{A_1 + A_2}{2} < -A_2$$

and therefore
$$f(3) - f(2) < \frac{f(4) - f(2)}{2} < f(4) - f(3).$$

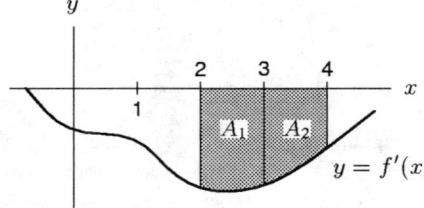

Figure 5.15

6. Change in income $= \int_0^{12} r(t)\, dt = \int_0^{12} 40(1.002)^t\, dt = \$485.80$

7. If $H(t)$ is the temperature of the coffee at time t, by the Fundamental Theorem of Calculus

$$\text{Change in temperature} = H(10) - H(0) = \int_0^{10} H'(t)\, dt = \int_0^{10} -7e^{-0.1t}\, dt.$$

Therefore,

$$H(10) = H(0) + \int_0^{10} -7(0.9^t)\, dt \approx 90 - 44.2 = 45.8°C.$$

8. (a) Quantity used $= \int_0^5 f(t)\, dt$.
   (b) Using a left sum, our approximation is

   $$32e^{0.05(0)} + 32e^{0.05(1)} + 32e^{0.05(2)} + 32e^{0.05(3)} + 32e^{0.05(4)} = 177.27.$$

   Since f is an increasing function, this represents an underestimate.
   (c) Each term is a lower estimate of one year's consumption of oil.

9. (a) Over the interval $[-1, 3]$, we estimate that the total change of the population is about 1.5, by counting boxes between the curve and the x-axis; we count about 1.5 boxes below the x-axis from $x = -1$ to $x = 1$ and about 3 above from $x = 1$ to $x = 3$. So the average rate of change is just the total change divided by the length of the interval, that is $1.5/4 = 0.375$ thousand/hour.
   (b) We can estimate the total change of the algae population by counting boxes between the curve and the x-axis. Here, there is about 1 box above the x-axis from $x = -3$ to $x = -2$, about 0.75 of a box below the x-axis from $x = -2$ to $x = -1$, and a total change of about 1.5 boxes thereafter (as discussed in part (a)). So the total change is about $1 - 0.75 + 1.5 = 1.75$ thousands of algae.

10.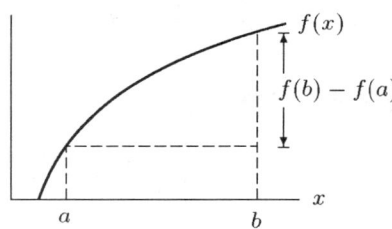

11.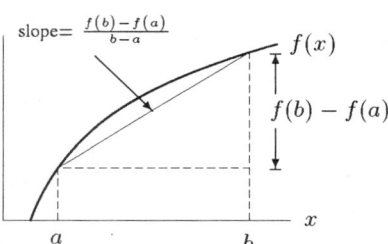

12.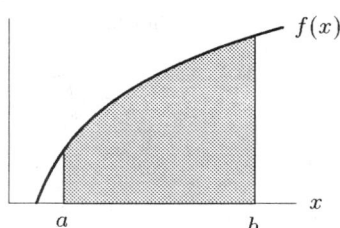

13.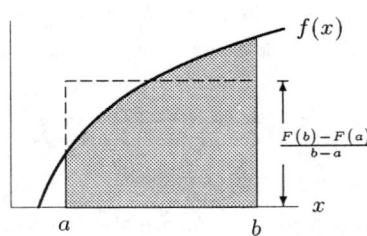

Note that we are using the interpretation of the definite integral as the length of the interval times the average value of the function on that interval, which we developed in Section 5.3.

14. Note that $\int_a^b g(x)\,dx = \int_a^b g(t)\,dt$. Thus, we have

$$\int_a^b \bigl(f(x)+g(x)\bigr)\,dx = \int_a^b f(x)\,dx + \int_a^b g(x)\,dx = 8+2 = 10.$$

15. Note that $\int_a^b (g(x))^2\,dx = \int_a^b (g(t))^2\,dt$. Thus, we have

$$\int_a^b \bigl((f(x))^2 - (g(x))^2\bigr)\,dx = \int_a^b (f(x))^2\,dx - \int_a^b (g(x))^2\,dx = 12-3 = 9.$$

16. We have

$$\int_a^b (f(x))^2\,dx - \left(\int_a^b f(x)\,dx\right)^2 = 12 - 8^2 = -52.$$

17. Note that $\int_a^b f(z)\,dz = \int_a^b f(x)\,dx$. Thus, we have

$$\int_a^b cf(z)\,dz = c\int_a^b f(z)\,dz = 8c.$$

18. We write

$$\int_a^b \bigl(c_1 g(x) + (c_2 f(x))^2\bigr)\,dx = \int_a^b \bigl(c_1 g(x) + c_2^2 (f(x))^2\bigr)\,dx$$
$$= \int_a^b c_1 g(x)\,dx + \int_a^b c_2^2 (f(x))^2\,dx$$
$$= c_1 \int_a^b g(x)\,dx + c_2^2 \int_a^b (f(x))^2\,dx$$
$$= c_1(2) + c_2^2(12) = 2c_1 + 12c_2^2.$$

19. The graph of $y = f(x-5)$ is the graph of $y = f(x)$ shifted to the right by 5. Since the limits of integration have also shifted by 5 (to $a+5$ and $b+5$), the areas corresponding to $\int_{a+5}^{b+5} f(x-5)\,dx$ and $\int_a^b f(x)\,dx$ are the same. Thus,

$$\int_{a+5}^{b+5} f(x-5)\,dx = \int_a^b f(x)\,dx = 8.$$

20. (a) $\frac{1}{\sqrt{2\pi}} \int_1^3 e^{-\frac{x^2}{2}} dx$

$= \frac{1}{\sqrt{2\pi}} \int_0^3 e^{-\frac{x^2}{2}} dx - \frac{1}{\sqrt{2\pi}} \int_0^1 e^{-\frac{x^2}{2}} dx$

$\approx 0.4987 - 0.3413 = 0.1574.$

(b) $\left(\text{by symmetry of } e^{x^2/2}\right)$ $\frac{1}{\sqrt{2\pi}} \int_{-2}^3 e^{-\frac{x^2}{2}} dx = \frac{1}{\sqrt{2\pi}} \int_{-2}^0 e^{-\frac{x^2}{2}} dx + \frac{1}{\sqrt{2\pi}} \int_0^3 e^{-\frac{x^2}{2}} dx$

$= \frac{1}{\sqrt{2\pi}} \int_0^2 e^{-\frac{x^2}{2}} dx + \frac{1}{\sqrt{2\pi}} \int_0^3 e^{-\frac{x^2}{2}} dx$

$\approx 0.4772 + 0.4987 = 0.9759.$

21. (a) $\int_{-1}^1 e^{x^2} dx > 0$, since $e^{x^2} > 0$, and $\int_{-1}^1 e^{x^2} dx$ represents the area below the curve $y = e^{x^2}$.

(b) Looking at the figure below, we see that $\int_0^1 e^{x^2} dx$ represents the area under the curve. This area is clearly greater than zero, but it is less than e since it fits inside a rectangle of width 1 and height e (with room to spare). Thus

$$0 < \int_0^1 e^{x^2} dx < e < 3.$$

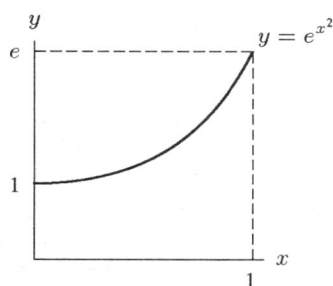

22. (a) The integrand is positive, so the integral can't be negative.
    (b) The integrand ≥ 0. If the integral $= 0$, then the integrand must be identically 0, which isn't true.

23. (a) 0, since the integrand is an odd function and the limits are symmetric around 0.
    (b) 0, since the integrand is an odd function and the limits are symmetric around 0.

24. By the given property, $\int_a^a f(x) dx = -\int_a^a f(x) dx$, so $2\int_a^a f(x) dx = 0$. Thus $\int_a^a f(x) dx = 0$.

25. We know that the average value of $v(x) = 4$, so

$$\frac{1}{6-1} \int_1^6 v(x) dx = 4, \text{ and thus } \int_1^6 v(x) dx = 20.$$

Similarly, we are told that

$$\frac{1}{8-6} \int_6^8 v(x) dx = 5, \text{ so } \int_6^8 v(x) dx = 10.$$

The average value for $1 \leq x \leq 8$ is given by

$$\text{Average value} = \frac{1}{8-1} \int_1^8 v(x) dx = \frac{1}{7}\left(\int_1^6 v(x) dx + \int_6^8 v(x) dx\right) = \frac{20+10}{7} = \frac{30}{7}.$$

Solutions for Chapter 5 Review

1. (a) We calculate the right- and left-hand sums as follows:

$$\text{Left} = 2[80 + 52 + 28 + 10] = 340 \text{ ft.}$$
$$\text{Right} = 2[52 + 28 + 10 + 0] = 180 \text{ ft.}$$

Our best estimate will be the average of these two sums,

$$\text{Best} = \frac{\text{Left} + \text{Right}}{2} = \frac{340 + 180}{2} = 260 \text{ ft.}$$

(b) Since v is decreasing throughout,

$$\text{Left} - \text{Right} = \Delta t \cdot [f(0) - f(8)]$$
$$= 80 \Delta t.$$

Since our best estimate is the average of Left and Right, the maximum error is $(80)\Delta t/2$. For $(80)\Delta t/2 \leq 20$, we must have $\Delta t \leq 1/2$. In other words, we must measure the velocity every 0.5 second.

2. Using the Riemann formulas for the left- and right-hand sums, we get Table 5.2:

TABLE 5.2

n	Left	Right
5	8.546891	7.498199
25	8.244542	8.034803
500	8.149759	8.139272

To best approximate the definite integral, we will average the left- and right-hand sums involving the highest n. Thus,

$$\int_{-1}^{2} 5e^{-x^2} \, dx \approx \frac{\text{Left}(500) + \text{Right}(500)}{2} = \frac{8.149759 + 8.139272}{2} = 8.144516.$$

3. $\int_{0}^{20} f(x) \, dx$ is equal to the area shaded in the figure below. We estimate the area by counting boxes. There are about 15 boxes and each box represents 4 square units, so the area shaded is about 60. We have

$$\int_{0}^{20} f(x) \, dx \approx 60.$$

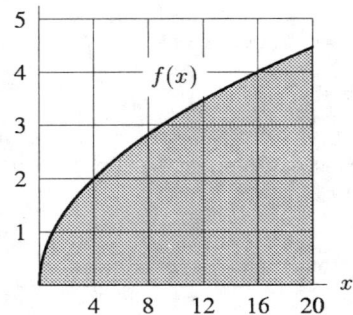

4. By counting squares and fractions of squares, we find that the area under the graph appears to be around 310 (miles/hour) sec, within about 10. So the distance traveled was about $310 \left(\frac{5280}{3600}\right) \approx 455$ feet, within about $10 \left(\frac{5280}{3600}\right) \approx 15$ feet. (Note that 455 feet is about 0.086 miles)

5. Distance traveled $= \int_0^{1.1} \sin(t^2)\, dt \approx 0.40.$

6. The x intercepts of $y = x^2 - 9$ are $x = -3$ and $x = 3$, and since the graph is below the x axis on the interval $[-3, 3]$.

$$\text{Area} = -\int_{-3}^{3} (x^2 - 9)\, dx = 36.00.$$

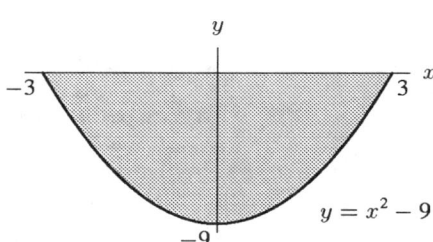

7. Since x intercepts are $x = 0, \pi, 2\pi, \ldots,$

$$\text{Area} = \int_0^{\pi} \sin x \, dx = 2.00.$$

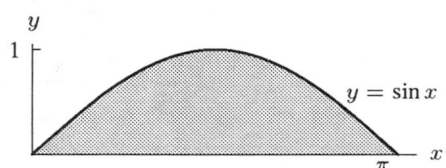

8. The x intercepts of $y = 4 - x^2$ are $x = -2$ and $x = 2$, and the graph is above the x-axis on the interval $[-2, 2]$.

$$\text{Area} = \int_{-2}^{2} (4 - x^2)\, dx = 10.67.$$

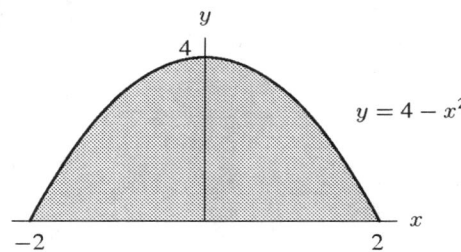

9. Since the θ intercepts of $y = \sin \theta$ are $\theta = 0, \pi, 2\pi, \ldots,$

$$\text{Area} = \int_0^{\pi} 1\, d\theta - \int_0^{\pi} \sin \theta \, d\theta = \pi - 2 \approx 1.14.$$

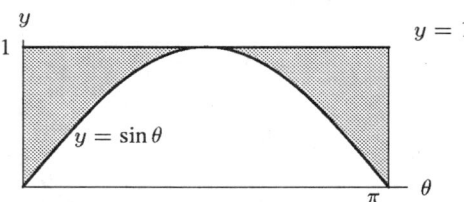

10. Using properties of the definite integral, we have:

$$\int_2^5 (2f(x) + 3)\, dx = 17$$

$$2\int_2^5 f(x)\, dx + 3\int_2^5 1\, dx = 17$$

$$2\int_2^5 f(x)\, dx + 3 \cdot 3 = 17$$

$$2\int_2^5 f(x)\, dx = 8$$

$$\int_2^5 f(x)\, dx = 4.$$

11. All the integrals have positive values, since $f \geq 0$. The integral in (ii) is about one-half the integral in (i), due to the apparent symmetry of f. The integral in (iv) will be much larger than the integral in (i), since the two peaks of f^2 rise to 10,000. The integral in (iii) will be smaller than half of the integral in (i), since the peaks in $f^{1/2}$ will only rise to 10. So

$$\int_0^2 (f(x))^{1/2}\, dx < \int_0^1 f(x)\, dx < \int_0^2 f(x)\, dx < \int_0^2 (f(x))^2\, dx.$$

12.

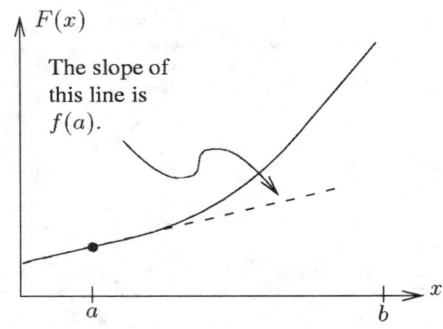

13. By the FTC, we know that $\int_a^b f(x)\, dx = F(b) - F(a)$.

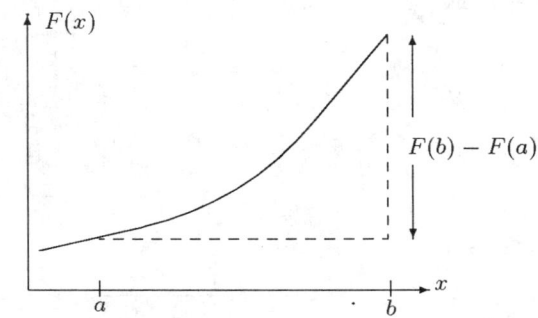

14.

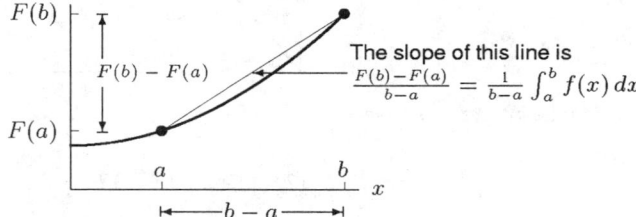

15. (a) Clearly, the points where $x = \sqrt{\pi}, \sqrt{2\pi}, \sqrt{3\pi}, \sqrt{4\pi}$ are where the graph intersects the x-axis because $f(x) = \sin(x^2) = 0$ where x is the square root of some multiple of π.

(b) Let $f(x) = \sin(x^2)$, and let $A, B, C,$ and D be the areas of the regions indicated in the figure below. Then we see that $A > B > C > D$.

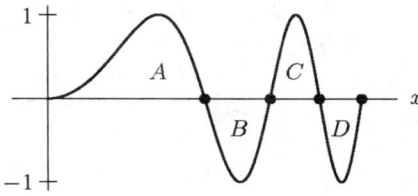

Note that
$$\int_0^{\sqrt{\pi}} f(x)\,dx = A, \quad \int_0^{\sqrt{2\pi}} f(x)\,dx = A - B,$$
$$\int_0^{\sqrt{3\pi}} f(x)\,dx = A - B + C, \quad \text{and} \quad \int_0^{\sqrt{4\pi}} f(x)\,dx = A - B + C - D.$$

It follows that
$$\int_0^{\sqrt{\pi}} f(x)\,dx = A > \int_0^{\sqrt{3\pi}} f(x)\,dx = A - (B - C) = A - B + C >$$
$$\int_0^{\sqrt{4\pi}} f(x)\,dx = A - B + C - D > \int_0^{\sqrt{2\pi}} f(x)\,dx = (A - B) > 0.$$

And thus the ordering is $n = 1$, $n = 3$, $n = 4$, and $n = 2$ from largest to smallest. All the numbers are positive.

16. (a) If the level first becomes acceptable at time t_1, then $R_0 = 4R(t_1)$, or
$$\frac{1}{4}R_0 = R_0 e^{-0.004 t_1}$$
$$\frac{1}{4} = e^{-0.004 t_1}.$$

Taking natural logs on both sides yields
$$\ln \frac{1}{4} = -0.004 t_1$$
$$t_1 = \frac{\ln \frac{1}{4}}{-0.004} \approx 346.574 \text{ hours}.$$

(b) The acceptable limit is 0.6 millirems/hour, so from the above, $R_0 = 4(0.6) = 2.4$. Since the rate at which radiation is emitted is $R(t) = R_0 e^{-0.004 t}$,
$$\text{Total radiation emitted} = \int_0^{346.574} 2.4 e^{-0.004 t}\,dt.$$

Approximating the integral using left and right sums with 200 subdivisions, we find that about 450 millirems were emitted over that interval.

17. (a)
$$\text{Average population} = \frac{1}{10}\int_0^{10} 67.38(1.026)^t\,dt$$

Evaluating the integral numerically gives

$$\text{Average population} \approx 76.8 \text{ million}$$

(b) In 1980, $t = 0$, and $P = 67.38(1.026)^0 = 67.38$.
In 1990, $t = 10$, and $P = 67.38(1.026)^{10} = 87.10$.
Average = $\frac{1}{2}(67.38 + 87.10) = 77.24$ million.

(c) If P had been linear, the average value found in (a) would have been the one we found in (b). Since the population graph is concave up, it is below the secant line. Thus, the actual values of P are less than the corresponding values on the secant line, and so the average found in (a) is smaller than that in (b).

18. (a) V, since the slope is constant.
(b) IV, since the net area under this curve is the most negative.
(c) III, since the area under the curve is largest.
(d) II, since the steepest ascent at $t = 0$ occurs on this curve.
(e) III, since average velocity is (total distance)/5, and III moves the largest total distance.
(f) I, since average acceleration is $\frac{1}{5}\int_0^5 v'(t)\,dt = \frac{1}{5}(v(5) - v(0))$, and in I, the velocity increases the most from start ($t = 0$) to finish ($t = 5$).

226 CHAPTER FIVE /SOLUTIONS

19. (a) For the first twelve months, the total number of appliances sold is
$$7 + 9 + 11 + \cdots + 29 = 216,$$
so the average number $= \frac{216}{12} = 18$ appliances per month.
(b) Average $= \frac{1}{12} \int_0^{12} (2t + 5)\, dt$, which by the Fundamental Theorem is given by $\frac{1}{12}(t^2 + 5t)\Big|_0^{12} = \frac{204}{12} = 17$ appliances per month.
(c) They are close, but not equal. Using integration gives an underestimate of the true value. This is because the function $2t + 5$ equals the rate at which she sells only at the end of each month; see the figure below.
(d) The integral is easier to calculate than the sum, particularly for a large number of months.
(e) The rectangles in the figure below represent the true answer. The triangles (portions of the rectangles above the sales line) represent the error.

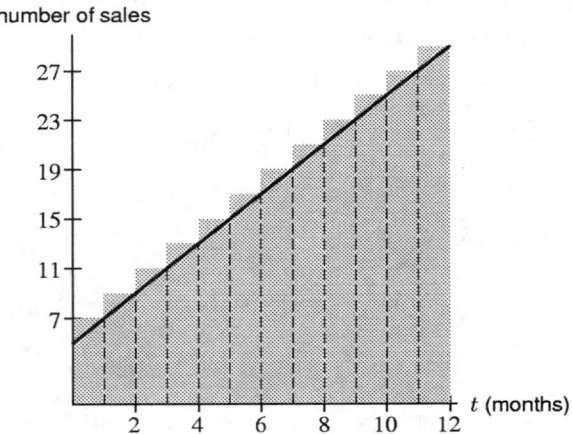

20. Let C be the rate of the flow through the hose. At $t = t_0$, the volume of water in the tank is equal to the area under the lower curve (flow rate through the hose) minus the area under the upper curve (flow rate through the hole) in the region to the left of the vertical line $t = t_0$. Since the overlap of these regions cancels, the volume is also equal to $5C$ (that's the area under the lower curve from $t = 0$ to $t = 5$) minus the region bounded by the upper curve, the horizontal line of height C, the vertical line $t = t_0$, and the vertical line $t = 5$. If $t_0 > 15$, movement of the vertical line $t = t_0$ doesn't change the area of the latter region, so the difference becomes constant. Thus the volume of water in the tank becomes constant, and the physical system is in a steady state.

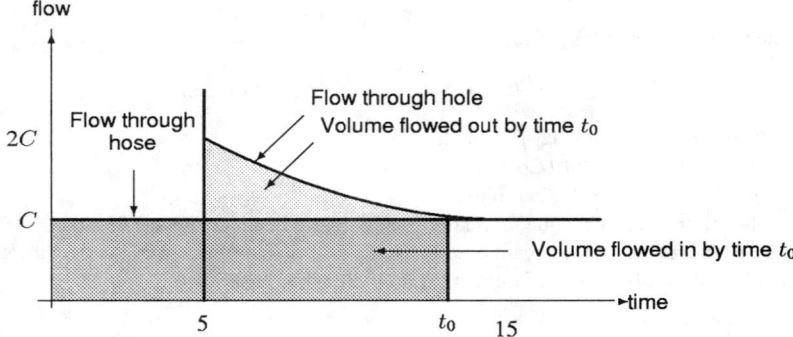

21. (a) About 300 meter^3/sec.
(b) About 250 meter^3/sec.
(c) Looking at the graph, we can see that the 1996 flood reached its maximum just between March and April, for a high of about 1250 meter^3/sec. Similarly, the 1957 flood reached its maximum in mid-June, for a maximum flow rate of 3500 meter^3/sec.
(d) The 1996 flood lasted about 1/3 of a month, or about 10 days. The 1957 flood lasted about 4 months.

(e) The area under the controlled flood graph is about 2/3 box. Each box represents 500 meter^3/sec for one month. Since

$$1 \text{ month} = 30\frac{\text{days}}{\text{month}} \cdot 24\frac{\text{hours}}{\text{day}} \cdot 60\frac{\text{minutes}}{\text{hour}} \cdot 60\frac{\text{seconds}}{\text{minute}}$$
$$= 2.592 \cdot 10^6 \approx 3 \cdot 10^6 \text{seconds},$$

each box represents

$$\text{Flow} \approx (500 \text{ meter}^3/\text{sec}) \cdot (2.6 \cdot 10^6 \text{ sec}) = 13 \cdot 10^8 \text{ meter}^3 \text{of water.}$$

So, for the artificial flood,

$$\text{Additional flow} \approx \frac{2}{3} \cdot 13 \cdot 10^8 = 9 \cdot 10^8 \text{ meter}^3 \approx 10^9 \text{ meter}^3.$$

(f) The 1957 flood released a volume of water represented by about 12 boxes above the 250 meter/sec baseline. Thus, for the natural flood,

$$\text{Additional flow} \approx 12 \cdot 15 \cdot 10^8 = 1.8 \cdot 10^{10} \approx 2 \cdot 10^{10} \text{ meter}^3.$$

So, the natural flood was nearly 20 times larger than the controlled flood and lasted much longer.

22. (a) The acceleration is positive for $0 \leq t < 40$ and for a tiny period before $t = 60$, since the slope is positive over these intervals. Just to the left of $t = 40$, it looks like the acceleration is approaching 0. Between $t = 40$ and a moment just before $t = 60$, the acceleration is negative.
   (b) The maximum altitude was about 500 feet, when t was a little greater than 40 (here we are estimating the area under the graph for $0 \leq t \leq 42$).
   (c) The acceleration is greatest when the slope of the velocity is most positive. This happens just before $t = 60$, where the magnitude of the velocity is plunging and the direction of the acceleration is positive, or up.
   (d) The deceleration is greatest when the slope of the velocity is most negative. This happens just after $t = 40$.
   (e) After the Montgolfier Brothers hit their top climbing speed (at $t = 40$), they suddenly stopped climbing and started to fall. This suggests some kind of catastrophe—the flame going out, the balloon ripping, etc. (In actual fact, in their first flight in 1783, the material covering their balloon, held together by buttons, ripped and the balloon landed in flames.)
   (f) The total change in altitude for the Montgolfiers and their balloon is the definite integral of their velocity, or the total area under the given graph (counting the part after $t = 42$ as negative, of course). As mentioned before, the total area of the graph for $0 \leq t \leq 42$ is about 500. The area for $t > 42$ is about 220. So subtracting, we see that the balloon finished 280 feet or so higher than where it began.

23. (a) The mouse changes direction (when its velocity is zero) at about times 17, 23, and 27.
    (b) The mouse is moving most rapidly to the right at time 10 and most rapidly to the left at time 40.
    (c) The mouse is farthest to the right when the integral of the velocity, $\int_0^t v(t)\,dt$, is most positive. Since the integral is the sum of the areas above the axis minus the areas below the axis, the integral is largest when the velocity is zero at about 17 seconds. The mouse is farthest to the left of center when the integral is most negative at 40 seconds.
    (d) The mouse's speed decreases during seconds 10 to 17, from 20 to 23 seconds, and from 24 seconds to 27 seconds.
    (e) The mouse is at the center of the tunnel at any time t for which the integral from 0 to t of the velocity is zero. This is true at time 0 and again somewhere around 35 seconds.

24. (a) For $-2 \leq x \leq 2$, f is symmetrical about the y-axis, so $\int_{-2}^{0} f(x)\,dx = \int_{0}^{2} f(x)\,dx$ and $\int_{-2}^{2} f(x)\,dx = 2\int_{0}^{2} f(x)\,dx$.
    (b) For any function f, $\int_{0}^{2} f(x)\,dx = \int_{0}^{5} f(x)\,dx - \int_{2}^{5} f(x)\,dx$.
    (c) Note that $\int_{-2}^{0} f(x)\,dx = \frac{1}{2}\int_{-2}^{2} f(x)\,dx$, so $\int_{0}^{5} f(x)\,dx = \int_{-2}^{5} f(x)\,dx - \int_{-2}^{0} f(x)\,dx = \int_{-2}^{5} f(x)\,dx - \frac{1}{2}\int_{-2}^{2} f(x)\,dx$.

25. (a) We know that $\int_{2}^{5} f(x)\,dx = \int_{0}^{5} f(x)\,dx - \int_{0}^{2} f(x)\,dx$. By symmetry, $\int_{0}^{2} f(x)\,dx = \frac{1}{2}\int_{-2}^{2} f(x)\,dx$, so $\int_{2}^{5} f(x)\,dx = \int_{0}^{5} f(x)\,dx - \frac{1}{2}\int_{-2}^{2} f(x)\,dx$.
    (b) $\int_{2}^{5} f(x)\,dx = \int_{-2}^{5} f(x)\,dx - \int_{-2}^{2} f(x)\,dx = \int_{-2}^{5} f(x)\,dx - 2\int_{-2}^{0} f(x)\,dx$.
    (c) Using symmetry again, $\int_{0}^{2} f(x)\,dx = \frac{1}{2}\left(\int_{-2}^{5} f(x)\,dx - \int_{2}^{5} f(x)\,dx\right)$.

228 CHAPTER FIVE /SOLUTIONS

26. In (a), $f'(1)$ is the slope of a tangent line at $x = 1$, which is negative. As for (c), the rate of change in $f(x)$ is given by $f'(x)$, and the average value of this over $0 \leq x \leq a$ is

$$\frac{1}{a-0} \int_0^a f'(x)\,dx = \frac{f(a) - f(0)}{a - 0}.$$

This is the slope of the line through the points $(0, 1)$ and $(a, 0)$, which is less negative that the tangent line at $x = 1$. Therefore, (a) < (c) < 0. The quantity (b) is $\left(\int_0^a f(x)\,dx\right)/a$ and (d) is $\int_0^a f(x)\,dx$, which is the net area under the graph of f (counting the area as negative for f below the x-axis). Since $a > 1$ and $\int_0^a f(x)\,dx > 0$, we have $0 <$(b)$<$(d). Therefore

$$(a) < (c) < (b) < (d).$$

27. For $[0, t_1]$, velocity is constant so $a = F = 0$ and work is 0.
For $[t_1, t_2]$, velocity is positive so distance moved is positive. Since velocity is increasing, $a > 0$ so $F > 0$. Thus, Work > 0.
For $[t_2, t_3]$, velocity is constant so $a = 0$ and Work $= 0$.
For $[t_3, t_4]$, velocity is positive so distance moved is positive.
Since velocity is decreasing, $a < 0$ so $F < 0$. Thus, Work < 0.

Solutions to the Projects and Computer Algebra Investigations

1. (a) CO_2 is being taken out of the water during the day and returned at night. The pond must therefore contain some plants. (The data is in fact from pond water containing both plants and animals.)

(b) Suppose t is the number of hours past dawn. The graph in Figure 5.60 of the text shows that the CO_2 content changes at a greater rate for the first 6 hours of daylight, $0 < t < 6$, than it does for the final 6 hours of daylight, $6 < t < 12$. It turns out that plants photosynthesize more vigorously in the morning than in the afternoon. Similarly, CO_2 content changes more rapidly in the first half of the night, $12 < t < 18$, than in the 6 hours just before dawn, $18 < t < 24$. The reason seems to be that at night plants quickly use up most of the sugar that they synthesized during the day, and then their respiration rate is inhibited. So the constant rate hypothesis is false, if we assume plants are the main cause of CO_2 changes in the pond.

(c) The question asks about the total quantity of CO_2 in the pond, rather than the rate at which it is changing. We will let $f(t)$ denote the CO_2 content of the pond water (in mmol/l) at t hours past dawn. Then Figure 5.60 of the text is a graph of the derivative $f'(t)$. There are 2.600 mmol/l of CO_2 in the water at dawn, so $f(0) = 2.600$.

The CO_2 content $f(t)$ decreases during the 12 hours of daylight, $0 < t < 12$, when $f'(t) < 0$, and then $f(t)$ increases for the next 12 hours. Thus, $f(t)$ is at a minimum when $t = 12$, at dusk. By the Fundamental Theorem,

$$f(12) = f(0) + \int_0^{12} f'(t)\,dt = 2.600 + \int_0^{12} f'(t)\,dt.$$

We must approximate the definite integral by a Riemann sum. From the graph in Figure 5.60 of the text, we estimate the values of the function $f'(t)$ in Table 5.3.

TABLE 5.3 Rate, $f'(t)$, at which CO_2 is entering or leaving water

t	$f'(t)$	t	$f'(t)$	t	$f'(t)$	t	$f'(t)$	t	$f'(t)$	t	$f'(t)$
0	0.000	4	−0.039	8	−0.026	12	0.000	16	0.035	20	0.020
1	−0.042	5	−0.038	9	−0.023	13	0.054	17	0.030	21	0.015
2	−0.044	6	−0.035	10	−0.020	14	0.045	18	0.027	22	0.012
3	−0.041	7	−0.030	11	−0.008	15	0.040	19	0.023	23	0.005

The left Riemann sum with $n = 12$ terms, corresponding to $\Delta t = 1$, gives

$$\int_0^{12} f'(t)\,dt \approx (0.000)(1) + (-0.042)(1) + (-0.044)(1) + \cdots + (-0.008)(1) = -0.346.$$

At 12 hours past dawn, the CO_2 content of the pond water reaches its lowest level, which is approximately

$$2.600 - 0.346 = 2.254 \text{ mmol/l}.$$

(d) The increase in CO_2 during the 12 hours of darkness equals

$$f(24) - f(12) = \int_{12}^{24} f'(t)\,dt.$$

Using Riemann sums to estimate this integral, we find that about 0.306 mmol/l of CO_2 was released into the pond during the night. In part (c) we calculated that about 0.346 mmol/l of CO_2 was absorbed from the pond during the day. If the pond is in equilibrium, we would expect the daytime absorption to equal the nighttime release. These quantities are sufficiently close (0.346 and 0.306) that the difference could be due to measurement error, or to errors from the Riemann sum approximation.

If the pond is in equilibrium, the area between the rate curve in Figure 5.60 of the text and the t-axis for $0 \le t \le 12$ will equal the area between the rate curve and the t-axis for $12 \le t \le 24$. In this experiment the areas do look approximately equal.

(e) We must evaluate

$$f(b) = f(0) + \int_0^b f'(t)\,dt = 2.600 + \int_0^b f'(t)\,dt$$

for the values $b = 0, 3, 6, 9, 12, 15, 18, 21, 24$. Left Riemann sums with $\Delta t = 1$ give the values for the CO_2 content in Table 5.4. The graph is shown in Figure 5.16.

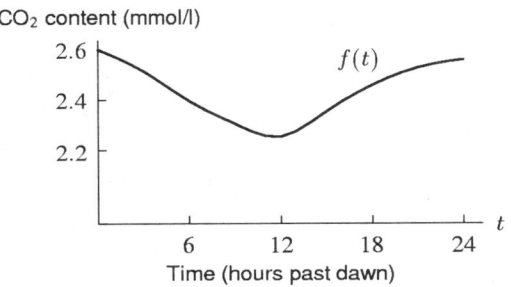

Figure 5.16: CO_2 content in pond water throughout the day

TABLE 5.4 CO_2 content throughout the day

b (hours after dawn)	0	3	6	9	12	15	18	21	24
$f(b)$ (CO_2 content)	2.600	2.514	2.396	2.305	2.254	2.353	2.458	2.528	2.560

2. Let $v = f(t)$. If we can show that $\int_a^b f(t)\,dt$ is the change in position of the particle over the time interval $a < t < b$, then $\dfrac{1}{b-a}\int_a^b f(t)\,dt$ should give us the average velocity. But

$$\int_a^b f(t)\,dt = \int_0^b f(t)\,dt - \int_0^a f(t)\,dt = s(b) - s(a),$$

where $s(b) = \int_0^b f(t)\,dt$ and $s(a) = \int_0^a f(t)\,dt$ are positions of the particle at time $t = b$ and $t = a$, respectively. So $\int_a^b f(t)\,dt$ is indeed the change of position over $a < t < b$.

3. (a) Since the length of the interval of integration is $2 - 1 = 1$, the width of each subdivision is $\Delta t = 1/n$. Thus the endpoints of the subdivision are

$$t_0 = 1, \quad t_1 = 1 + \Delta t = 1 + \frac{1}{n}, \quad t_2 = 1 + 2\Delta t = 1 + \frac{2}{n}, \ldots,$$

$$t_i = 1 + i\Delta t = 1 + \frac{i}{n}, \ldots, \quad t_{n-1} = 1 + (n-1)\Delta t = 1 + \frac{n-1}{n}.$$

Thus, since the integrand is $f(t) = t$,

$$\text{Left-hand sum} = \sum_{i=0}^{n-1} f(t_i)\Delta t = \sum_{i=0}^{n-1} t_i \Delta t = \sum_{i=0}^{n-1} \left(1 + \frac{i}{n}\right) \frac{1}{n} = \sum_{i=0}^{n-1} \frac{n+i}{n^2}.$$

(b) The CAS finds the formula for the Riemann sum

$$\sum_{i=0}^{n-1} \frac{n+i}{n^2} = \frac{\frac{(-1+n)n}{2} + n^2}{n^2} = \frac{3}{2} - \frac{1}{2n}.$$

(c) Taking the limit as $n \to \infty$

$$\lim_{n \to \infty} \left(\frac{3}{2} - \frac{1}{2n}\right) = \lim_{n \to \infty} \frac{3}{2} - \lim_{n \to \infty} \frac{1}{2n} = \frac{3}{2} + 0 = \frac{3}{2}.$$

(d) The shape under the graph of $y = t$ between $t = 1$ and $t = 2$ is a trapezoid of width 1, height 1 on the left and 2 on the right. So its area is $1 \cdot (1+2)/2 = 3/2$. This is the same answer we got by computing the definite integral.

4. (a) Since the length of the interval of integration is $2 - 1 = 1$, the width of each subdivision is $\Delta t = 1/n$. Thus the endpoints of the subdivision are

$$t_0 = 1, \quad t_1 = 1 + \Delta t = 1 + \frac{1}{n}, \quad t_2 = 1 + 2\Delta t = 1 + \frac{2}{n}, \ldots,$$

$$t_i = 1 + i\Delta t = 1 + \frac{i}{n}, \ldots, \quad t_{n-1} = 1 + (n-1)\Delta t = 1 + \frac{n-1}{n}.$$

Thus, since the integrand is $f(t) = t^2$,

$$\text{Left-hand sum} = \sum_{i=0}^{n-1} f(t_i)\Delta t = \sum_{i=0}^{n-1} t_i^2 \Delta t = \sum_{i=0}^{n-1} \left(1 + \frac{i}{n}\right)^2 \frac{1}{n} = \sum_{i=0}^{n-1} \frac{(n+i)^2}{n^3}.$$

(b) Using a CAS to find the sum, we get

$$\sum_{i=0}^{n-1} \frac{(n+i)^2}{n^3} = \frac{(-1+2n)(-1+7n)}{6n^2} = \frac{7}{3} + \frac{1}{6n^2} - \frac{3}{2n}.$$

(c) Taking the limit as $n \to \infty$

$$\lim_{n \to \infty} \left(\frac{7}{3} + \frac{1}{6n^2} - \frac{3}{2n}\right) = \lim_{n \to \infty} \frac{7}{3} + \lim_{n \to \infty} \frac{1}{6n^2} - \lim_{n \to \infty} \frac{3}{2n} = \frac{7}{3} + 0 + 0 = \frac{7}{3}.$$

(d) We have calculated $\int_1^2 t^2\, dt$ using Reimann sums. Since t^2 is above the t-axis between $t = 1$ and $t = 2$, this integral is the area; so the area is 7/3.

5. (a) Since the length of the interval of integration is π, the width of each subdivision is $\Delta x = \pi/n$. Thus the endpoints of the subdivision are

$$x_0 = 1, \quad x_1 = 0 + \Delta x = \frac{\pi}{n}, \quad x_2 = 0 + 2\Delta x = \frac{2\pi}{n}, \ldots,$$

$$x_i = 0 + i\Delta x = \frac{i\pi}{n}, \quad \ldots, \quad x_n = 0 + n\Delta x = \frac{n\pi}{n} = \pi.$$

Thus, since the integrand is $f(x) = \sin x$,

$$\text{Right-hand sum} = \sum_{i=1}^{n} f(x_i)\Delta x = \sum_{i=1}^{n} \sin(x_i)\Delta x = \sum_{i=1}^{n} \sin\left(\frac{i\pi}{n}\right) \frac{\pi}{n}.$$

(b) If the CAS can evaluate this sum, we get

$$\sum_{i=1}^{n} \sin\left(\frac{i\pi}{n}\right) \frac{\pi}{n} = \frac{\pi \cot(\pi/2n)}{n} = \frac{\pi \cos(\pi/2n)}{n \sin(\pi/2n)}.$$

(c) Using the computer algebra system, we find that
$$\lim_{n\to\infty} \frac{\pi \cos(\pi/2n)}{n \sin(\pi/2n)} = 2.$$

(d) The computer algebra system gives
$$\int_0^\pi \sin x \, dx = 2.$$

6. (a) A CAS gives
$$\int_a^b \sin(cx) \, dx = \frac{\cos(ac)}{c} - \frac{\cos(bc)}{c}.$$

(b) If $F(x)$ is an antiderivative of $\sin(cx)$, then the Fundamental Theorem of Calculus says that
$$\int_a^b \sin(cx) \, dx = F(b) - F(a).$$

Comparing this with the answer to part (a), we see that
$$F(b) - F(a) = \frac{\cos(ac)}{c} - \frac{\cos(bc)}{c} = \left(-\frac{\cos(cb)}{c}\right) - \left(-\frac{\cos(ca)}{c}\right).$$

This suggests that
$$F(x) = -\frac{\cos(cx)}{c}.$$

Taking the derivative confirms this:
$$\frac{d}{dx}\left(-\frac{\cos(cx)}{c}\right) = \sin(cx).$$

7. (a) Different systems may give different answers. A typical answer is
$$\int_a^c \frac{x}{1+bx^2} \, dx = \frac{\ln\left(\frac{|c^2 b + 1|}{|a^2 b + 1|}\right)}{2b}.$$

Some CASs may not have the absolute values in the answer; since $b > 0$, the answer is correct without the absolute values.

(b) Using the properties of logarithms, we can rewrite the answer to part (a) as
$$\int_a^c \frac{x}{1+bx^2} \, dx = \frac{\ln|c^2 b + 1| - \ln|a^2 b + 1|}{2b} = \frac{\ln|c^2 b + 1|}{2b} - \frac{\ln|a^2 b + 1|}{2b}.$$

If $F(x)$ is an antiderivative of $x/(1+bx^2)$, then the Fundamental Theorem of Calculus says that
$$\int_a^c \frac{x}{1+bx^2} \, dx = F(c) - F(a).$$

Thus
$$F(c) - F(a) = \frac{\ln|c^2 b + 1|}{2b} - \frac{\ln|a^2 b + 1|}{2b} = \frac{\ln|c^2 b + 1|}{2b} - \frac{\ln|a^2 b + 1|}{2b}.$$

This suggests that
$$F(x) = \frac{\ln|1+bx^2|}{2b} = \frac{\ln(1+bx^2)}{2b}.$$

(Since $b > 0$, we know $|1+bx^2| = 1+bx^2$.) Taking the derivative confirms this:
$$\frac{d}{dx}\left(\frac{\ln(1+bx^2)}{2b}\right) = \frac{x}{1+bx^2}.$$

Solutions to Problems on the Definite Integral

1. The statement is rarely true. The graph of almost any non-linear monotonic function, such as x^{10} for $0 < x < 1$, should provide convincing geometric evidence. Furthermore, if the statement were true, then (LHS+RHS)/2 would always give the exact value of the definite integral. This is not true.

2. As illustrated in Figure 5.17, the left- and right-hand sums are both equal to $(4\pi) \cdot 3 = 12\pi$, while the integral is smaller. Thus we have:
$$\int_0^{4\pi} (2 + \cos x)\, dx < \text{Left-hand sum} = \text{Right-hand sum}.$$

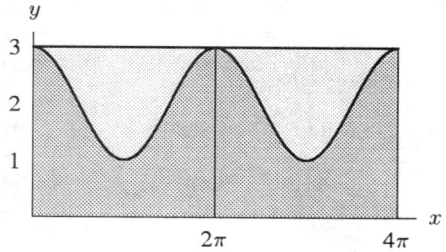

Figure 5.17: Integral vs. Left- and Right-Hand Sums

3.

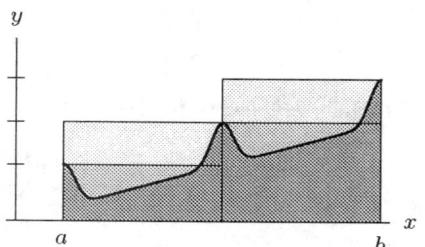

Figure 5.18: Integral vs. Left- and Right-Hand Sums

4. Since x^2 is increasing on $[0, 5]$, the left-hand sum is the lower sum and the right-hand sum is the upper sum. For $n = 1300$, LHS ≈ 41.618 (rounding down) and RHS ≈ 41.715 (rounding up). Since the left and right sums differ by 0.097, their average must be within 0.0485 of the true value, so $\int_0^5 x^2\, dx = 41.667$ to the required accuracy.

5. Since the integrand is increasing on $[1, 2]$, the left-hand sum is the lower sum and the right-hand sum is the upper sum. For $n = 30$, LHS ≈ 2.852 (rounding down) and RHS ≈ 2.919 (rounding up). Since the left and right sums differ by 0.067, their average must be within 0.0335 of the true value, so $\int_1^2 2^x\, dx = 2.886$ to the required accuracy.

6. Since the integrand is decreasing on $[1, 4]$, the left-hand sum is the upper sum and the right-hand sum is the lower sum. For $n = 20$, LHS ≈ 1.249 (rounding up) and RHS ≈ 1.179 (rounding down). Since the left and right sums differ by 0.07, their average value must be within 0.035 of the true value, so $\int_1^4 \frac{1}{\sqrt{1+x^2}}\, dx = 1.214$ to the required accuracy.

7. Since the integrand is increasing on $[1, 1.5]$, the left-hand sum is the lower sum and the right-hand sum is the upper sum. For $n = 10$, LHS ≈ 0.465 (rounding down) and RHS ≈ 0.474 (rounding up). Since the left and right sums differ by 0.009, their average must be within 0.0045 of the true value, so $\int_1^{1.5} \sin x\, dx = 0.470$ to the required accuracy.

8. Since the integrand is increasing on $[0, \pi/4]$, the left-hand sum is the lower sum and the right-hand sum is the upper sum. For $n = 10$, LHS ≈ 0.865 (rounding down) and RHS ≈ 0.899 (rounding up). Since the left and right sums differ by 0.034, their average must be within 0.017 of the true value, so $\int_0^{\frac{\pi}{4}} \frac{d\theta}{\cos \theta} = 0.882$ to the required accuracy.

9. Since the integrand is increasing on $[-2, -1]$, the left-hand sum is the lower sum and the right-hand sum is the upper sum. For $n = 10$, LHS ≈ 0.0045 (rounding down) and RHS ≈ 0.0276 (rounding up). Since the left and right sums differ by 0.0231, their average must be within 0.01155 of the true value, so $\int_{-2}^{-1} \cos^3 y\, dy = 0.016$ to the required accuracy.

10. Since the integrand is increasing on $[1, 5]$, the left-hand sum is the lower sum and the right-hand sum is the upper sum. For $n = 110$, LHS ≈ 4.810 (rounding down) and RHS ≈ 4.905 (rounding up). Since the left and right sums differ by 0.095, their average must be within 0.0475 of the true value, so $\int_1^5 (\ln x)^2 \, dx = 4.858$ to the required accuracy.

11. Since the integrand is increasing on $[1.1, 1.7]$, the left-hand sum is the lower sum and the right-hand sum is the upper sum. For $n = 20$, LHS ≈ 0.825 (rounding down) and RHS ≈ 0.905 (rounding up). Since the left and right sums differ by 0.08, their average must be within 0.04 of the true value, so $\int_{1.1}^{1.7} e^t \ln t \, dt = 0.865$ to the required accuracy.

12. The integrand is increasing on $[-3, 0]$ and decreasing on $[0, 3]$. Thus, to obtain a lower sum, we must take left-hand endpoints on $[-3, 0]$ and right-hand endpoints on $[0, 3]$. For $n = 140$, Upper sum ≈ 1.816 (rounding up) and Lower Sum ≈ 1.728 (rounding down). Since the lower and upper sums differ by 0.088, their average must be within 0.044 of the true value, so $\int_{-3}^{3} e^{-t^2} dt = 2 \int_0^3 e^{-t^2} dt = 2(0.886) = 1.772$, to the required accuracy.

13. Let $a = x_0 < x_1 < x_2 < \cdots < x_{n-1} < x_n$ be the endpoints of the subdivision. The lower sum is

$$\sum_{i=1}^{n} m_i \Delta x_i$$

and the upper sum is

$$\sum_{i=1}^{n} M_i \Delta x_i,$$

where m_i is the greatest lower bound for f on $[x_{i-1}, x_i]$ and M_i is the least upper bound for f on the same interval. Since a lower bound for a set must be less than or equal to an upper bound for the same set, $m_i \leq M_i$. Thus the lower sum is less than or equal to the upper sum.

14. (a) Let $m_{i,1}$ be the greatest lower bound for f on $[x_{i-1}, y]$ and $m_{i,2}$ be the greatest lower bound for f on $[y, x_i]$. Since m_i is a lower bound for f on the interval $[x_{i-1}, x_i]$, it is certainly a lower bound for f on the smaller interval $[x_{i-1}, y]$; so it less than or equal to $m_{i,1}$, since that is the greatest lower bound for f on $[x_{i-1}, y]$. By a similarly argument, $m_i \leq m_{i,2}$.

    (b) The lower sum for $a = x_0 < x_1 < \cdots < x_{n-1} < x_n = b$ is

$$\sum_{k=1}^{n} m_i(x_k - x_{k-1}).$$

Adding y subdivides $[x_{i-1}, x_i]$ into two pieces: $[x_{i-1}, y]$ and $[y, x_i]$. The lower sum constructed using the new subdivision has the same terms as the old one, except that the term $m_i(x_i - x_{i-1})$ in the old one is replaced by $m_{i,1}(y - x_{i,1}) + m_{i,2}(x_i - y)$. By part (a), $m_i \leq m_{i,1}$ and $m_i \leq m_{i,2}$, so

$$m_i(x_i - x_{i-1}) = m_i(x_i - y) + m_i(y - x_{i-1}) \leq m_{i,2}(x_i - y) + m_{i,1}(y - x_{i-1}).$$

So the term in the first lower sum is less than or equal to the corresponding terms in the second. Since all the other terms are the same, the first lower sum is less than or equal to the second.

    (c) In part (b) we showed that a lower sum for a particular subdivision was less than or equal to the lower sum for the subdivision that you get by adding just one point. But this shows that it is less than the lower sum for any refinement, because you can get from one subdivision to a refinement by successively adding one point at a time.

15. (a) Let $M_{i,1}$ be the least upper bound for f on $[x_{i-1}, y]$ and $M_{i,2}$ be the least upper bound for f on $[y, x_i]$. Since M_i is an upper bound for f on the interval $[x_{i-1}, x_i]$, it is certainly an upper bound for f on the smaller interval $[x_{i-1}, y]$; so it greater than or equal to $M_{i,1}$, since that is the least upper bound for f on $[x_{i-1}, y]$. By a similarly argument, $M_i \geq M_{i,2}$.

    (b) The upper sum for $a = x_0 < x_1 < \cdots < x_{n-1} < x_n = b$ is

$$\sum_{k=1}^{n} M_i(x_k - x_{k-1}).$$

Adding y subdivides $[x_{i-1}, x_i]$ into two pieces: $[x_{i-1}, y]$ and $[y, x_i]$. The upper sum constructed using the new subdivision has the same terms as the old one, except that the term $M_i(x_i - x_{i-1})$ in the old one is replaced by $M_{i,1}(y - x_{i,1}) + M_{i,2}(x_i - y)$. By part (a), $M_i \geq M_{i,1}$ and $M_i \geq M_{i,2}$, so

$$M_i(x_i - x_{i-1}) = M_i(x_i - y) + M_i(y - x_{i-1}) \geq M_{i,2}(x_i - y) + M_{i,1}(y - x_{i-1}).$$

So the term in the first upper sum is greater than or equal to the corresponding terms in the second. Since all the other terms are the same, the first upper sum is greater than or equal to the second.

(c) In part (b) we showed that a upper sum for a particular subdivision was greater than or equal to the upper sum for the subdivision that you get by adding just one point. But this shows that it is greater than the upper sum for any refinement, because you can get from one subdivision to a refinement by successively adding one point at a time.

16. Form a new subdivision whose set of endpoints is the union of the sets of endpoints of the two given subdivisions. Since it's set of endpoints contains the set of endpoints from each of the two, it is a refinement of each.

17. Let A be a lower sum and let B be an upper sum. Using Problem 16, choose a subdivision which is a refinement of both the subdivisions used in these sums, and let A' and B' be the lower and upper sums corresponding to this subdivision. By Problem 14, $A \leq A'$, and by Problem 15, $B' \leq B$. Finally, by Problem 13, $A' \leq B'$. So

$$A \leq A' \leq B' \leq B,$$

hence $A \leq B$.

18. (a) Let $\epsilon = L - U$. If $L > U$, then ϵ is positive. Since L is the least upper bound of all the lower sums, we can find lower sums arbitrarily close to L; in particular, there must be a lower sum within $\epsilon/3$ of L. By similar reasoning, there must be an upper sum within $\epsilon/3$ of U. Hence

$$\text{Lower Sum} \geq L - \frac{\epsilon}{3}$$

and

$$\text{Upper Sum} \leq U + \frac{\epsilon}{3}.$$

Since $L = U + \epsilon$, we have

$$L - \frac{\epsilon}{3} > U + \frac{\epsilon}{3},$$

thus

$$\text{Lower Sum} > \text{Upper Sum}.$$

(b) By Problem 17, every lower sum is less than or equal to every upper sum. Hence the conclusion of part (a) is impossible. So the hypothesis that $L > U$ is incorrect; we must have $L \leq U$.

19. Consider the subdivision of $[a, b]$ that consists of one subinterval, namely, $[a, b]$ itself. The greatest lower bound for f on $[a, b]$ is greater than or equal to m, and the length of the only subinterval in the subdivision is $b - a$, so the lower sum for this subdivision is greater than or equal to $m(b - a)$. Since every lower sum is an lower estimate for $\int_a^b f(x)\,dx$, we have

$$m(b - a) \leq \text{Lower Sum} \leq \int_a^b f(x)\,dx.$$

20. (a) By putting together the subdivision of $[a, c]$ corresponding to ℓ_1 and the subdivision of $[c, b]$ corresponding to ℓ_2, we obtain a subdivision of $[a, b]$. The lower sum corresponding to this subdivision is $\ell_1 + \ell_2$.

(b) Given a subdivision of $[a, b]$, with lower sum equal to ℓ, we can refine it by inserting c as one of the endpoints. The resulting subdivision is a union of a subdivision of $[a, c]$, with corresponding lower sum ℓ_1, and a subdivision of $[c, b]$, with corresponding sum ℓ_2. Thus its lower sum is $\ell_1 + \ell_2$. Since it is a refinement of the subdivision corresponding to ℓ, we have $\ell \leq \ell_1 + \ell_2$, by Problem 14.

(c) By part (b), every lower sum on $[a, b]$ is less than or equal to a lower sum on $[a, c]$ plus a lower sum on $[c, b]$. Thus $L_1 + L_2$ is an upper bound for the lower sum on $[a, b]$, and so, since L is the least upper bound, $L \leq L_1 + L_2$.

On the other hand, by part (a), adding any lower sum l_1 on $[a, c]$ to any lower sum l_2 on $[c, b]$ yields a lower sum on $[a, b]$. So L is an upper bound for $\ell_1 + \ell_2$. Thus we have $l_1 + l_2 \leq L \leq L_1 + L_2$. But an l_1 can be chosen as close as we like to L_1, since L_1 is a least upper bound, and similarly for l_2 and L_2. So $l_1 + l_2$ can be made arbitrarily close to $L_1 + L_2$. This, combined with $l_1 + l_2 \leq L \leq L_1 + L_2$, means $L = L_1 + L_2$.

CHAPTER SIX

Solutions for Section 6.1

1.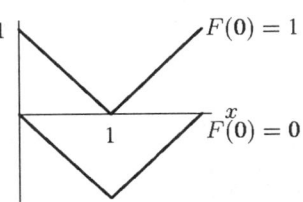

2.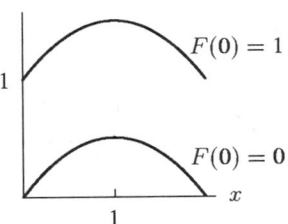

3.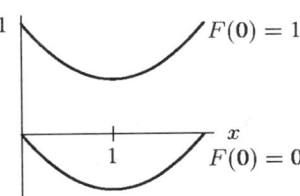

4.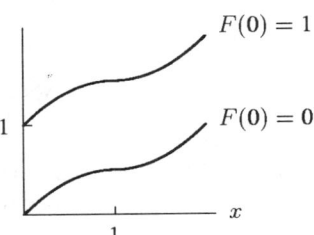

5. (a) Critical points of $F(x)$ are the zeros of f: $x = 1$ and $x = 3$.
   (b) $F(x)$ has a local minimum at $x = 1$ and a local maximum at $x = 3$.
   (c)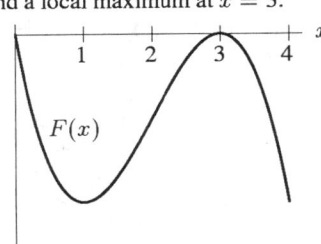

   Notice that the graph could also be above or below the x-axis at $x = 3$.

6. (a) Critical points of $F(x)$ are $x = -1$, $x = 1$ and $x = 3$.
   (b) $F(x)$ has a local minimum at $x = -1$, a local maximum at $x = 1$, and a local minimum at $x = 3$.
   (c)

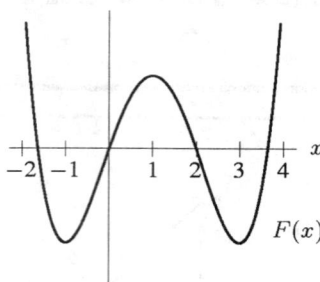

7. (a) Critical points of $F(x)$ are $x = 1$ and $x = 4$.
   (b) $F(x)$ has a local maximum at $x = 1$ and neither at $x = 4$.
   (c) A possible graph of $F(x)$ is in Figure 6.1. We've taken $F(4) = 0$ since the areas in the graph of f between $x = 0$ and $x = 1$, and between $x = 1$ and $x = 4$, appear equal.

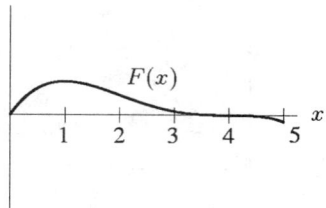

Figure 6.1

8.

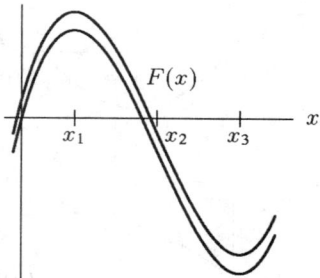

Note that since $f(x_1) = 0$ and $f'(x_1) < 0$, $F(x_1)$ is a local maximum; since $f(x_3) = 0$ and $f'(x_3) > 0$, $F(x_3)$ is a local minimum. Also, since $f'(x_2) = 0$ and f changes from decreasing to increasing about $x = x_2$, F has an inflection point at $x = x_2$.

9.

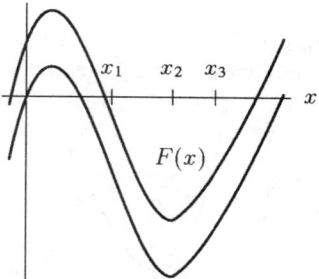

Note that since $f(x_2) = 0$, $f'(x_2) > 0$, so $F(x_2)$ is a local minimum. Since $f'(x_1) = 0$ and f changes from decreasing to increasing at $x = x_1$, F has an inflection point at $x = x_1$.

10.

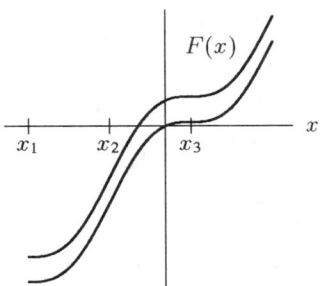

Note that since $f(x_1) = 0$, $F(x_1)$ is either a local minimum or a point of inflection; it is impossible to tell which from the graph. Since $f'(x_3) = 0$, and f' changes sign around $x = x_3$, $F(x_3)$ is an inflection point. Also, since $f'(x_2) = 0$ and f changes from increasing to decreasing about $x = x_2$, F has another inflection point at $x = x_2$.

11. (a) We know that $\int_0^3 f'(x)dx = f(3) - f(0)$ from the Fundamental Theorem of Calculus. From the graph of f' we can see that $\int_0^3 f'(x)dx = 2 - 1 = 1$ by subtracting areas between f' and the x-axis. Since $f(0) = 0$, we find that $f(3) = 1$. Similar reasoning gives $f(7) = \int_0^7 f'(x)dx = 2 - 1 + 2 - 4 + 1 = 0$.
   (b) We have $f(0) = 0$, $f(2) = 2$, $f(3) = 1$, $f(4) = 3$, $f(6) = -1$, and $f(7) = 0$. So the graph, beginning at $x = 0$, starts at zero, increases to 2 at $x = 2$, decreases to 1 at $x = 3$, increases to 3 at $x = 4$, then passes through a zero as it decreases to -1 at $x = 6$, and finally increases to 0 at 7. Thus, there are three zeroes: $x = 0$, $x = 5.5$, and $x = 7$.
   (c)

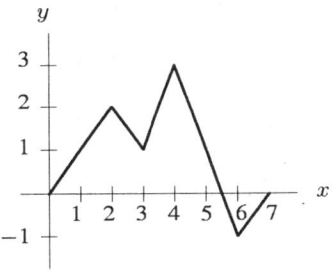

12. (a) Starting at $x = 3$, we are given that $f(3) = 0$. Moving to the left on the interval $2 < x < 3$, we have $f'(x) = -1$, so $f(2) = f(3) - (1)(-1) = 1$. On the interval $0 < x < 2$, we have $f'(x) = 1$, so

$$f(0) = f(2) + 1(-2) = -1.$$

Moving to the right from $x = 3$, we know that $f'(x) = 2$ on $3 < x < 4$. So $f(4) = f(3) + 2 = 2$. On the interval $4 < x < 6$, $f'(x) = -2$ so

$$f(6) = f(4) + 2(-2) = -2.$$

On the interval $6 < x < 7$, we have $f'(x) = 1$, so

$$f(7) = f(6) + 1 = -2 + 1 = -1.$$

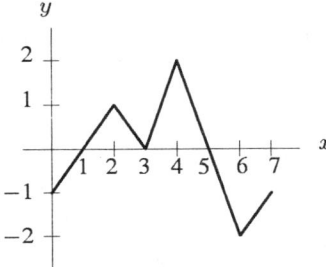

(b) In part (a) we found that $f(0) = -1$ and $f(7) = -1$.

(c) The integral $\int_0^7 f'(x)\,dx$ is given by the sum

$$\int_0^7 f'(x)\,dx = (1)(2) + (-1)(1) + (2)(1) + (-2)(2) + (1)(1) = 0.$$

Alternatively, knowing $f(7)$ and $f(0)$ and using the Fundamental Theorem of Calculus, we have

$$\int_0^7 f'(x)\,dx = f(7) - f(0) = -1 - (-1) = 0.$$

13. Looking at the graph of g' below, we see that the critical points of g occur when $x = 15$ and $x = 40$, since $g'(x) = 0$ at these values. Inflection points of g occur when $x = 10$ and $x = 20$, because $g'(x)$ has a local maximum or minimum at these values. Knowing these four key points, we sketch the graph of $g(x)$ as follows.

   We start at $x = 0$, where $g(0) = 50$. Since g' is negative on the interval $[0, 10]$, the value of $g(x)$ is decreasing there. At $x = 10$ we have

$$g(10) = g(0) + \int_0^{10} g'(x)\,dx$$
$$= 50 - (\text{area of shaded trapezoid } T_1)$$
$$= 50 - \left(\frac{10+20}{2} \cdot 10\right) = -100.$$

Similarly,

$$g(15) = g(10) + \int_{10}^{15} g'(x)\,dx$$
$$= -100 - (\text{area of triangle } T_2)$$
$$= -100 - \frac{1}{2}(5)(20) = -150.$$

Continuing,

$$g(20) = g(15) + \int_{15}^{20} g'(x)\,dx = -150 + \frac{1}{2}(5)(10) = -125,$$

and

$$g(40) = g(20) + \int_{20}^{40} g'(x)\,dx = -125 + \frac{1}{2}(20)(10) = -25.$$

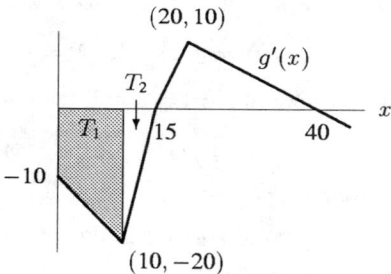

We now find concavity of $g(x)$ in the intervals $[0, 10]$, $[10, 15]$, $[15, 20]$, $[20, 40]$ by checking whether $g'(x)$ increases or decreases in these same intervals. If $g'(x)$ increases, then $g(x)$ is concave up; if $g'(x)$ decreases, then $g(x)$ is concave down. Thus we finally have our graph of $g(x)$:

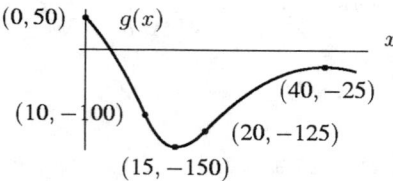

14. The critical points are at $(0,5)$, $(2,21)$, $(4,13)$, and $(5,15)$. A graph is given below.

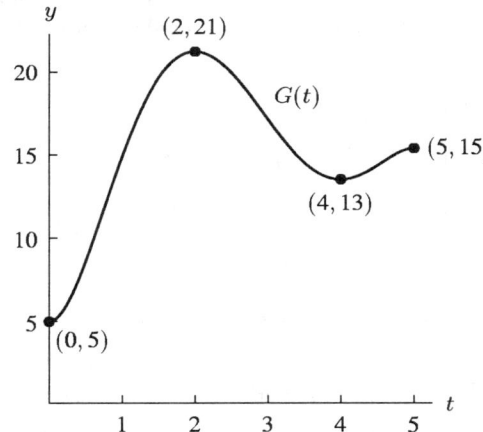

15. Between time $t = 0$ and time $t = B$, the velocity of the cork is always positive, which means the cork is moving upwards. At time $t = B$, the velocity is zero, and so the cork has stopped moving altogether. Since shortly thereafter the velocity of the cork becomes negative, the cork will next begin to move downwards. Thus when $t = B$ the cork has risen as far as it ever will, and is riding on top of the crest of the wave.

From time $t = B$ to time $t = D$, the velocity of the cork is negative, which means it is falling. When $t = D$, the velocity is again zero, and the cork has ceased to fall. Thus when $t = D$ the cork is riding on the bottom of the trough of the wave.

Since the cork is on the crest at time B and in the trough at time D, it is probably midway between crest and trough when the time is midway between B and D. Thus at time $t = C$ the cork is moving through the equilibrium position on its way down. (The equilibrium position is where the cork would be if the water were absolutely calm.) By symmetry, $t = A$ is the time when the cork is moving through the equilibrium position on the way up.

Since acceleration is the derivative of velocity, points where the acceleration is zero would be critical points of the velocity function. Since point A (a maximum) and point C (a minimum) are critical points, the acceleration is zero there.

A possible graph of the height of the cork is shown below. The horizontal axis represents a height equal to the average depth of the ocean at that point (the equilibrium position of the cork).

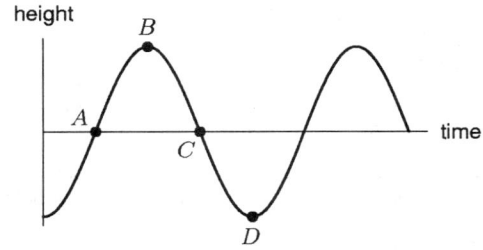

16.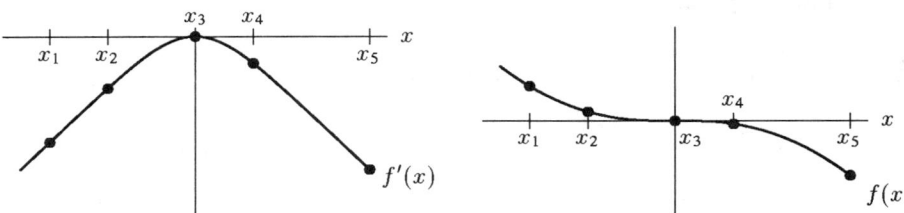

(a) $f(x)$ is greatest at x_1.
(b) $f(x)$ is least at x_5.
(c) $f'(x)$ is greatest at x_3..
(d) $f'(x)$ is least at x_5.
(e) $f''(x)$ is greatest at x_1.
(f) $f''(x)$ is least at x_5.

17. Both $F(x)$ and $G(x)$ have roots at $x = 0$ and $x = 4$. Both have a critical point (which is a local maximum) at $x = 2$. However, since the area under $g(x)$ between $x = 0$ and $x = 2$ is larger than the area under $f(x)$ between $x = 0$ and $x = 2$, the y-coordinate of $G(x)$ at 2 will be larger than the y-coordinate of $F(x)$ at 2. See below.

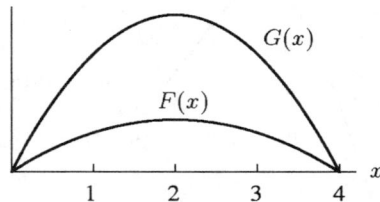

18. (a) Suppose $Q(t)$ is the amount of water in the reservoir at time t. Then

$$Q'(t) = \begin{array}{c}\text{Rate at which water} \\ \text{in reservoir is changing}\end{array} = \begin{array}{c}\text{Inflow} \\ \text{rate}\end{array} - \begin{array}{c}\text{Outflow} \\ \text{rate}\end{array}$$

Thus the amount of water in the reservoir is increasing when the inflow curve is above the outflow, and decreasing when it is below. This means that $Q(t)$ is a maximum where the curves cross in July 1993 (as shown in Figure 6.2), and $Q(t)$ is decreasing fastest when the outflow is farthest above the inflow curve, which occurs about October 1993 (see Figure 6.2).

To estimate values of $Q(t)$, we use the Fundamental Theorem which says that the change in the total quantity of water in the reservoir is given by

$$Q(t) - Q(\text{Jan'93}) = \int_{\text{Jan93}}^{t} (\text{inflow rate} - \text{outflow rate})\, dt$$

or $$Q(t) = Q(\text{Jan'93}) + \int_{\text{Jan93}}^{t} (\text{inflow rate} - \text{outflow rate})\, dt.$$

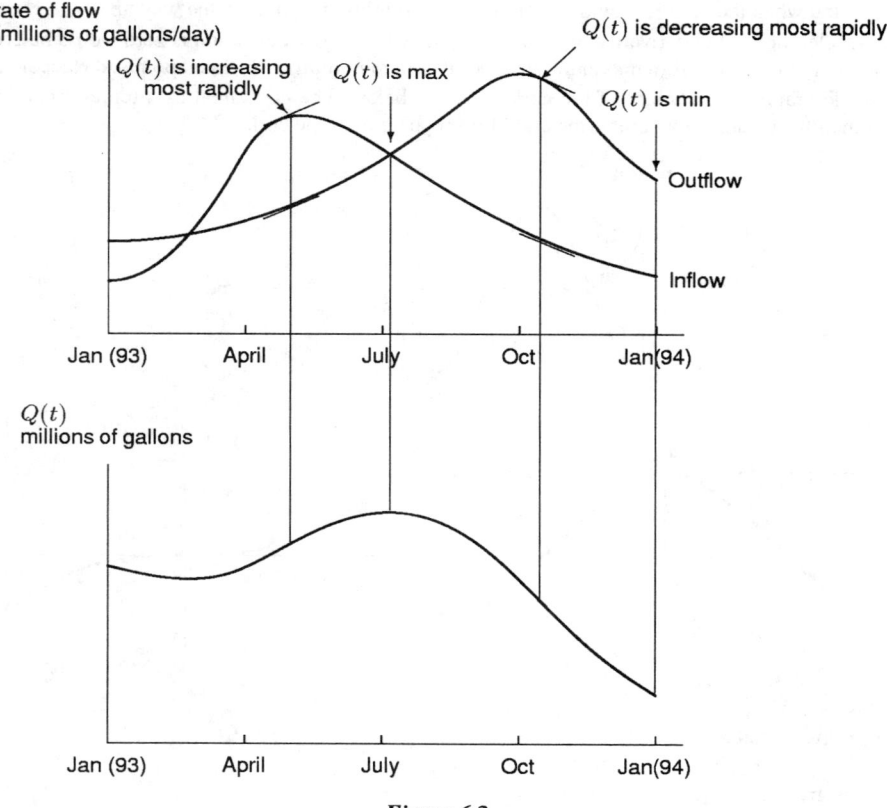

Figure 6.2

(b) See Figure 6.2. Maximum in July 1993. Minimum in Jan 1994.
(c) See Figure 6.2. Increasing fastest in May 1993. Decreasing fastest in Oct 1993.
(d) In order for the water to be the same as Jan '93 the total amount of water which has flowed into the reservoir must be 0. Referring to Figure 6.3, we have

$$\int_{Jan93}^{July94} (\text{inflow} - \text{outflow})dt = -A_1 + A_2 - A_3 + A_4 = 0$$

giving $A_1 + A_3 = A_2 + A_4$

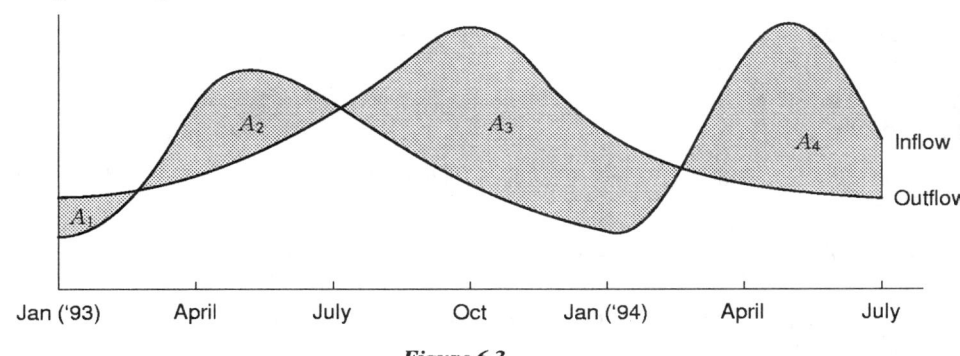

Figure 6.3

Solutions for Section 6.2

1. $5x$
2. $\frac{5}{2}x^2$
3. $\frac{1}{3}x^3$
4. $\frac{1}{3}t^3 + \frac{1}{2}t^2$
5. $\sin t$
6. $\frac{2}{3}z^{\frac{3}{2}}$
7. $\ln|z|$
8. $-\frac{1}{t}$
9. $-\frac{1}{2z^2}$
10. e^z
11. $-\cos t$
12. $\frac{2}{3}t^3 + \frac{3}{4}t^4 + \frac{4}{5}t^5$
13. $\frac{t^4}{4} - \frac{t^3}{6} - \frac{t^2}{2}$
14. $\frac{y^5}{5} + \ln|y|$
15. $\frac{5}{2}x^2 - \frac{2}{3}x^{\frac{3}{2}}$

16. $\dfrac{t^2+1}{t} = t + \dfrac{1}{t}$, which has antiderivative $\dfrac{t^2}{2} + \ln|t|$

17. $-\cos 2\theta$

18. $e^t + 5\dfrac{1}{5}e^{5t} = e^t + e^{5t}$

19. $\frac{1}{3}(t+1)^3$

20. $\dfrac{5^x}{\ln 5}$

21. $\sin t + \tan t$

22. $f(x) = 3$, so $F(x) = 3x + C$. $F(0) = 0$ implies that $3 \cdot 0 + C = 0$, so $C = 0$. Thus $F(x) = 3x$ is the only possibility.

23. $f(x) = 2x$, so $F(x) = x^2 + C$. $F(0) = 0$ implies that $0^2 + C = 0$, so $C = 0$. Thus $F(x) = x^2$ is the only possibility.

24. $f(x) = -7x$, so $F(x) = \frac{-7x^2}{2} + C$. $F(0) = 0$ implies that $-\frac{7}{2} \cdot 0^2 + C = 0$, so $C = 0$. Thus $F(x) = -7x^2/2$ is the only possibility.

25. $f(x) = \frac{1}{4}x$, so $F(x) = \frac{x^2}{8} + C$. $F(0) = 0$ implies that $\frac{1}{8} \cdot 0^2 + C = 0$, so $C = 0$. Thus $F(x) = x^2/8$ is the only possibility.

26. $f(x) = x^2$, so $F(x) = \frac{x^3}{3} + C$. $F(0) = 0$ implies that $\frac{0^3}{3} + C = 0$, so $C = 0$. Thus $F(x) = \frac{x^3}{3}$ is the only possibility.

27. $f(x) = x^{1/2}$, so $F(x) = \frac{2}{3}x^{3/2} + C$. $F(0) = 0$ implies that $\frac{2}{3} \cdot 0^{3/2} + C = 0$, so $C = 0$. Thus $F(x) = \frac{2}{3}x^{3/2}$ is the only possibility.

28. $f(x) = 2 + 4x + 5x^2$, so $F(x) = 2x + 2x^2 + \frac{5}{3}x^3 + C$. $F(0) = 0$ implies that $C = 0$. Thus $F(x) = 2x + 2x^2 + \frac{5}{3}x^3$ is the only possibility.

29. $f(x) = \sin x$, so $F(x) = -\cos x + C$. $F(0) = 0$ implies that $-\cos 0 + C = 0$, so $C = 1$. Thus $F(x) = -\cos x + 1$ is the only possibility.

30. $\dfrac{3x^2}{2} + C$

31. $2t^2 + 7t + C$

32. $\sin\theta + C$

33. $5e^z + C$

34. $\dfrac{x^2}{2} + 2x^{1/2} + C$

35. $-\cos t + C$

36. $\pi x + \dfrac{x^{12}}{12} + C$

37. $\displaystyle\int \left(t^{3/2} + t^{-3/2}\right) dt = \dfrac{2t^{5/2}}{5} - 2t^{-1/2} + C$

38. $\sin(x+1) + C$

39. $\frac{1}{2}e^{2r} + C$

40. $\displaystyle\int \dfrac{1}{e^z} dz = \int e^{-z} dz = -e^{-z} + C$

41. $\displaystyle\int \left(y - \dfrac{1}{y}\right)^2 dy = \int \left(y^2 - 2 + \dfrac{1}{y^2}\right) dy = \dfrac{y^3}{3} - 2y - \dfrac{1}{y} + C$

42. $\ln|x+1| + C$

43. $\frac{1}{2}\ln|2x-1| + C$, since $\dfrac{d}{dx}\ln|2x-1| = 2\left(\dfrac{1}{2x-1}\right)$.

44. $e^{5+x} + \frac{1}{5}e^{5x} + C$, since $\dfrac{d}{dx}(e^{5x}) = 5e^{5x}$.

45. $\frac{1}{2}\sin 2x + 2\cos x + C$, since $\dfrac{d}{dx}(\sin 2x) = 2\cos 2x$.

46. $\int_2^5 (x^3 - \pi x^2) \, dx = \left(\frac{x^4}{4} - \frac{\pi x^3}{3} \right) \Big|_2^5 = \frac{609}{4} - 39\pi \approx 29.728.$

47. $\int_0^1 \sin \theta \, d\theta = -\cos \theta \Big|_0^1 = 1 - \cos 1 \approx 0.460.$

48. Since $\frac{1 + y^2}{y} = \frac{1}{y} + y,$

$$\int_1^2 \frac{1 + y^2}{y} \, dy = \left(\ln|y| + \frac{y^2}{2} \right) \Big|_1^2 = \ln 2 + \frac{3}{2} \approx 2.193.$$

49. $\int_0^2 \left(\frac{x^3}{3} + 2x \right) dx = \left(\frac{x^4}{12} + x^2 \right) \Big|_0^2 = \frac{4}{3} + 4 = 16/3 \approx 5.333.$

50. $\int_0^{\pi/4} (\sin t + \cos t) \, dt = (-\cos t + \sin t) \Big|_0^{\pi/4} = \left(-\frac{\sqrt{2}}{2} + \frac{\sqrt{2}}{2} \right) - (-1 + 0) = 1.$

51. $\int_{-3}^{-1} \frac{2}{r^3} \, dr = -r^{-2} \Big|_{-3}^{-1} = -1 + \frac{1}{9} = -8/9 \approx -0.889.$

52. $\int_0^1 2e^x \, dx = 2e^x \Big|_0^1 = 2e - 2 \approx 3.437.$

53. Since $(\tan x)' = \frac{1}{\cos^2 x},$ $\int_0^{\pi/4} \frac{1}{\cos^2 x} \, dx = \tan x \Big|_0^{\pi/4} = \tan \frac{\pi}{4} - \tan 0 = 1.$

54. $\int 2^x \, dx = \frac{1}{\ln 2} 2^x + C,$ since $\frac{d}{dx} 2^x = \ln 2 \cdot 2^x,$ so

$$\int_{-1}^1 2^x \, dx = \frac{1}{\ln 2} \left[2^x \Big|_{-1}^1 \right] = \frac{3}{2 \ln 2} \approx 2.164.$$

55. Since $\frac{d}{dx} \sin 2x = 2 \cos 2x,$

$$\int_0^{\frac{\pi}{6}} (\sin x + \cos 2x) \, dx = \left(-\cos x + \frac{1}{2} \sin 2x \right) \Big|_0^{\frac{\pi}{6}}$$
$$= \left(-\cos \frac{\pi}{6} + \frac{1}{2} \sin \frac{\pi}{3} \right) - \left(-\cos 0 + \frac{1}{2} \sin 0 \right)$$
$$= 1 + \frac{1}{2} \sin \frac{\pi}{3} - \cos \frac{\pi}{6}$$
$$= 1 + \frac{\sqrt{3}}{4} - \frac{\sqrt{3}}{2}$$
$$= 1 - \frac{\sqrt{3}}{4}$$
$$\approx 0.567.$$

56. The graph crosses the x-axis where

$$7 - 8x + x^2 = 0$$
$$(x - 7)(x - 1) = 0;$$

so $x = 1$ and $x = 7$. See Figure 6.4. The parabola opens upward and the region is below the x-axis, so

$$\text{Area} = -\int_1^7 (7 - 8x + x^2) \, dx$$
$$= -\left(7x - 4x^2 + \frac{x^3}{3} \right) \Big|_1^7 = 36.$$

244 CHAPTER SIX /SOLUTIONS

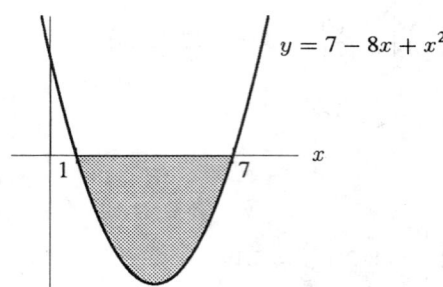

Figure 6.4

57. The graph is shown in the figure below. Since $\cos\theta \geq \sin\theta$ for $0 \leq \theta \leq \pi/4$, we have

$$\text{Area} = \int_0^{\pi/4} (\cos\theta - \sin\theta)\,d\theta$$
$$= (\sin\theta + \cos\theta)\Big|_0^{\pi/4}$$
$$= \frac{1}{\sqrt{2}} + \frac{1}{\sqrt{2}} - 1 = \sqrt{2} - 1.$$

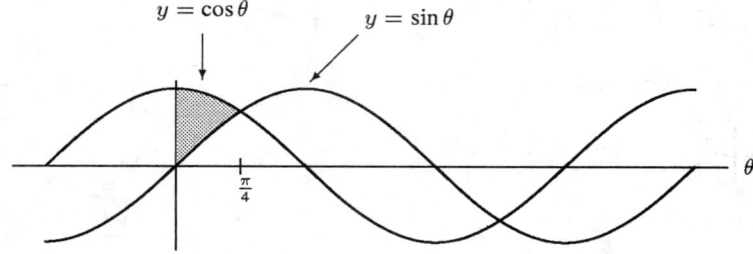

58. Since the graph of $y = e^x$ is above the graph of $y = \cos x$ (see the figure below), we have

$$\text{Area} = \int_0^1 (e^x - \cos x)\,dx$$
$$= \int_0^1 e^x\,dx - \int_0^1 \cos x\,dx$$
$$= e^x\Big|_0^1 - \sin x\Big|_0^1$$
$$= e^1 - e^0 - \sin 1 + \sin 0$$
$$= e - 1 - \sin 1.$$

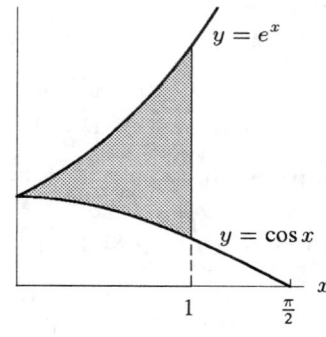

59. The graph of $y = x^2 - c^2$ has x-intercepts of $x = \pm c$. See the figure below. The shaded area is given by

$$\text{Area} = -\int_{-c}^{c} (x^2 - c^2)\, dx$$
$$= -2\int_{0}^{c} (x^2 - c^2)\, dx$$
$$= -2\left(\frac{x^3}{3} - c^2 x\right)\bigg|_0^c = -2\left(\frac{c^3}{3} - c^3\right) = \frac{4}{3}c^3.$$

We want c to satisfy $(4c^3)/3 = 36$, so $c = 3$.

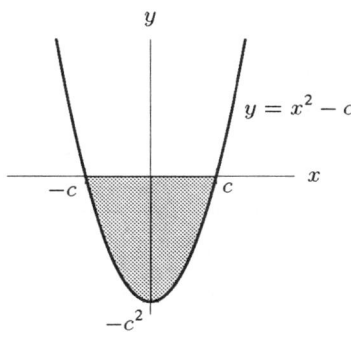

60. The average value of $v(x)$ on the interval $1 \le x \le c$ is

$$\frac{1}{c-1}\int_1^c \frac{6}{x^2}\,dx = \frac{1}{c-1}\left(-\frac{6}{x}\right)\bigg|_1^c = \frac{1}{c-1}\left(\frac{-6}{c} + 6\right) = \frac{6}{c}.$$

Since $\dfrac{1}{c-1}\int_1^c \dfrac{6}{x^2}\,dx = 1$, we have $\dfrac{6}{c} = 1$, so $c = 6$.

61. (a) The average value of $f(t) = \sin t$ over $0 \le t \le 2\pi$ is given by the formula

$$\text{Average} = \frac{1}{2\pi - 0}\int_0^{2\pi} \sin t\, dt$$
$$= \frac{1}{2\pi}(-\cos t)\bigg|_0^{2\pi}$$
$$= \frac{1}{2\pi}(-\cos 2\pi - (-\cos 0)) = 0.$$

We can check this answer by looking at the graph of $\sin t$ below. The area below the curve and above the t-axis over the interval $0 \le t \le \pi$, A_1, is the same as the area above the curve but below the t-axis over the interval $\pi \le t \le 2\pi$, A_2. When we take the integral of $\sin t$ over the entire interval $0 \le t \le 2\pi$, we get $A_1 - A_2 = 0$.

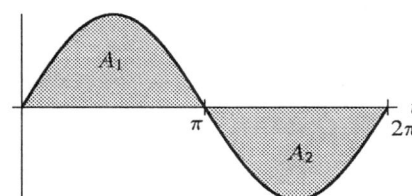

(b) Since

$$\int_0^{\pi} \sin t\, dt = -\cos t\bigg|_0^{\pi} = -\cos\pi - (-\cos 0) = -(-1) - (-1) = 2,$$

the average value of $\sin t$ on $0 \le t \le \pi$ is given by

$$\text{Average value} = \frac{1}{\pi}\int_0^{\pi} \sin t\, dt = \frac{2}{\pi}.$$

62. Since $C'(x) = 4000 + 10x$ we want to evaluate the indefinite integral

$$\int (4000 + 10x)\,dx = 4000x + 5x^2 + K$$

where K is a constant. Thus $C(x) = 5x^2 + 4000x + K$, and the fixed cost of 1,000,000 riyal means that $C(0) = 1,000,000 = K$. Therefore, the total cost is

$$C(x) = 5x^2 + 4000x + 1,000,000.$$

Since $C(x)$ depends on x^2, the square of the depth drilled, costs will increase dramatically when x grows large.

63. (a)

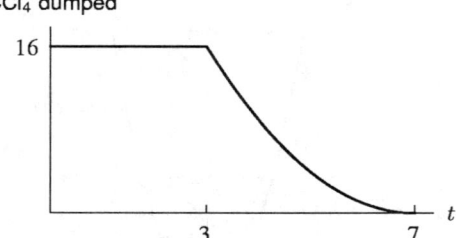

(b) 7 years, because $t^2 - 14t + 49 = (t-7)^2$ indicates that the rate of flow was zero after 7 years.

(c)

$$\text{Area under the curve} = 3(16) + \int_3^7 (t^2 - 14t + 49)\,dt$$

$$= 48 + \left(\frac{1}{3}t^3 - 7t^2 + 49t\right)\bigg|_3^7$$

$$= 48 + \frac{343}{3} - 343 + 343 - 9 + 63 - 147$$

$$= \frac{208}{3} = 69\frac{1}{3} \text{ cubic yards.}$$

Solutions for Section 6.3

1. $y = \int (x^3 + 5)\,dx = \dfrac{x^4}{4} + 5x + C$

2. $y = \int \left(8x + \dfrac{1}{x}\right) dx = 4x^2 + \ln|x| + C$

3. $W = \int 4\sqrt{t}\,dt = \dfrac{8}{3}t^{3/2} + C$

4. $r = \int 3\sin p\,dp = -3\cos p + C$

5. $y = \int (6x^2 + 4x)\,dx = 2x^3 + 2x^2 + C$. If $y(2) = 10$, then $2(2)^3 + 2(2)^2 + C = 10$ and $C = 10 - 16 - 8 = -14$.
Thus, $y = 2x^3 + 2x^2 - 14$.

6. $P = \int 10e^t\,dt = 10e^t + C$. If $P(0) = 25$, then $10e^0 + C = 25$ so $C = 15$. Thus, $P = 10e^t + 15$.

7. $s = \int (-32t + 100)\,dt = -16t^2 + 100t + C$. If $s = 50$ when $t = 0$, then $-16(0)^2 + 100(0) + C = 50$, so $C = 50$.
Thus $s = -16t^2 + 100t + 50$.

8. Since $y = x + \sin x - \pi$, we differentiate to see that $dy/dx = 1 + \cos x$, so y satisfies the differential equation. To show that it also satisfies the initial condition, we check that $y(\pi) = 0$:
$$y = x + \sin x - \pi$$
$$y(\pi) = \pi + \sin \pi - \pi = 0.$$

9. We differentiate $y = xe^{-x} + 2$ using the product rule to obtain
$$\frac{dy}{dx} = x\left(e^{-x}(-1)\right) + (1)e^{-x} + 0$$
$$= -xe^{-x} + e^{-x}$$
$$= (1-x)e^{-x},$$
and so $y = xe^{-x} + 2$ satisfies the differential equation. We now check that $y(0) = 2$:
$$y = xe^{-x} + 2$$
$$y(0) = 0e^{0} + 2 = 2.$$

10. (a) $y = \int (2x + 1)\, dx$, so the solution is $y = x^2 + x + C$.
    (b)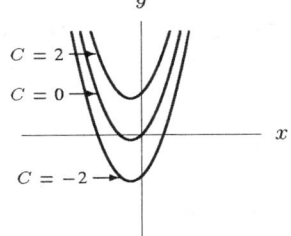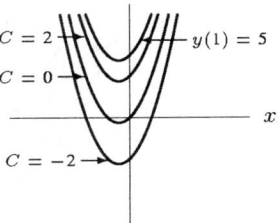
    (c) At $y(1) = 5$, we have $1^2 + 1 + C = 5$ and so $C = 3$. Thus we have the solution $y = x^2 + x + 3$.

11.
$$\frac{dy}{dt} = k\sqrt{t} = kt^{1/2}$$
$$y = \frac{2}{3}kt^{3/2} + C.$$
Since $y = 0$ when $t = 0$, we have $C = 0$, so
$$y = \frac{2}{3}kt^{3/2}.$$

12. Since the car's acceleration is constant, a graph of its velocity against time t is linear, as shown below.

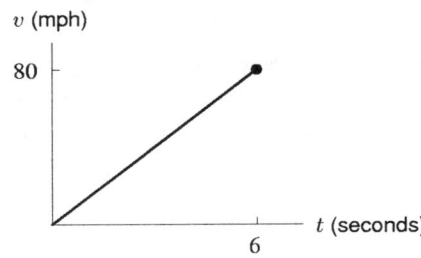

The acceleration is just the slope of this line:
$$\frac{dv}{dt} = \frac{80 - 0 \text{ mph}}{6 \text{ sec}} = \frac{40}{3} = 13.33 \frac{\text{mph}}{\text{sec}}.$$
To convert our units into ft/sec^2,
$$\frac{40}{3} \cdot \frac{\text{mph}}{\text{sec}} \cdot \frac{5280 \text{ ft}}{1 \text{ mile}} \cdot \frac{1 \text{ hour}}{3600 \text{ sec}} = 19.55 \frac{\text{ft}}{\text{sec}^2}$$

13. (a)

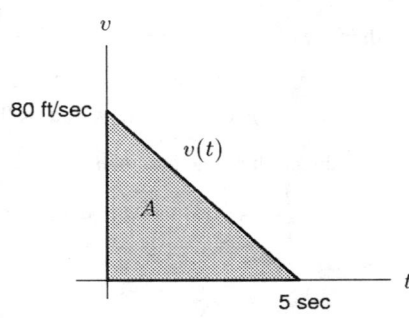

(b) The total distance is represented by the shaded region A, the area under the graph of $v(t)$.

(c) The area A, a triangle, is given by

$$A = \frac{1}{2}(\text{base})(\text{height}) = \frac{1}{2}(5\,\text{sec})(80\,\text{ft/sec}) = 200\,\text{ft}.$$

(d) Using integration and the Fundamental Theorem of Calculus, we have $A = \int_0^5 v(t)\,dt$ or $A = s(5) - s(0)$, where $s(t)$ is an antiderivative of $v(t)$.

We have that $a(t)$, the acceleration, is constant: $a(t) = k$ for some constant k. Therefore $v(t) = kt + C$ for some constant C. We have $80 = v(0) = k(0) + C = C$, so that $v(t) = kt + 80$. Putting in $t = 5$, $0 = v(5) = (k)(5) + 80$, or $k = -80/5 = -16$.

Thus $v(t) = -16t + 80$, and an antiderivative for $v(t)$ is $s(t) = -8t^2 + 80t + C$. Since the total distance traveled at $t = 0$ is 0, we have $s(0) = 0$ which means $C = 0$. Finally, $A = \int_0^5 v(t)\,dt = s(5) - s(0) = (-8(5)^2 + (80)(5)) - (-8(0)^2 + (80)(0)) = 200\,\text{ft}$, which agrees with the previous part.

14. Since the acceleration is constant, a graph of the velocity versus time looks like this:

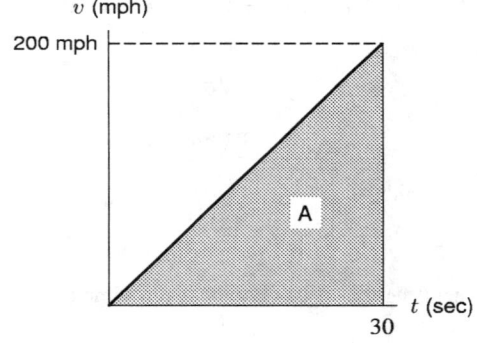

The distance traveled in 30 seconds, which is how long the runway must be, is equal to the area represented by A. We have $A = \frac{1}{2}(\text{base})(\text{height})$. First we convert the required velocity into miles per second.

$$200\,\text{mph} = \frac{200\,\text{miles}}{\text{hour}}\left(\frac{1\,\text{hour}}{60\,\text{minutes}}\right)\left(\frac{1\,\text{minute}}{60\,\text{seconds}}\right)$$

$$= \frac{200}{3600}\,\frac{\text{miles}}{\text{second}}$$

$$= \frac{1}{18}\,\text{miles/second}.$$

Therefore $A = \frac{1}{2}(30\,\text{sec})(200\,\text{mph}) = \frac{1}{2}(30\,\text{sec})\left(\frac{1}{18}\,\text{miles/sec}\right) = \frac{5}{6}$ miles.

15. (a) Since the velocity is constantly decreasing, and $v(6) = 0$, the car stops after 6 seconds.

t (sec)	0	0.5	1	1.5	2	2.5	3	3.5	4	4.5	5	5.5	6
$v(t)$ (ft/sec)	30	27.5	25	22.5	20	17.5	15	12.5	10	7.5	5	2.5	0

(b) Over the interval $a \leq t \leq a + \frac{1}{2}$, the left-hand velocity is $v(a)$, and the right-hand velocity is $v(a + \frac{1}{2})$. Since we are considering half-second intervals, $\Delta t = \frac{1}{2}$, and $n = 12$. The left sum is 97.5 ft., and the right sum is 82.5 ft.

(c) Area A in the figure below represents distance traveled.

$$A = \frac{1}{2}(\text{base})(\text{height}) = \frac{1}{2} \cdot 6 \cdot 30 = 90 \text{ ft.}$$

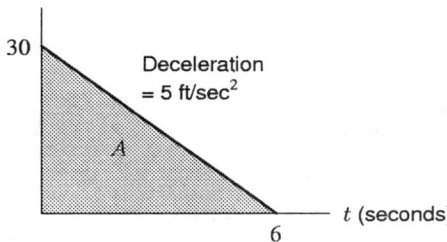

(d) The velocity is constantly decreasing at a rate of 5 ft/sec per second, i.e. after each second the velocity has dropped by 5 units. Therefore $v(t) = 30 - 5t$.

An antiderivative for $v(t)$ is $s(t)$, where $s(t) = 30t - \frac{5}{2}t^2$. Thus by the Fundamental Theorem of Calculus, the distance traveled $= s(6) - s(0) = (30(6) - \frac{5}{2}(6)^2) - (30(0) - \frac{5}{2}(0)^2) = 90$ ft. Since $v(t)$ is decreasing, the left-hand sum in part (b) overestimates the distance traveled, while the right-hand sum underestimates it.

The area A is equal to the average of the left-hand and right-hand sums: 90 ft $= \frac{1}{2}(97.5 \text{ ft} + 82.5 \text{ ft})$. The left-hand sum is an overestimate of A; the right-hand sum is an underestimate.

16. (a) To find the height of the balloon, we integrate its velocity with respect to time:

$$h(t) = \int v(t)\,dt$$
$$= \int (-32t + 40)\,dt$$
$$= -32\frac{t^2}{2} + 40t + C.$$

Since at $t = 0$, we have $h = 30$, we can solve for C to get $C = 30$, giving us a height of

$$h(t) = -16t^2 + 40t + 30.$$

(b) To find the average velocity between $t = 1.5$ and $t = 3$, we find the total displacement and divide by time.

$$\text{Average velocity} = \frac{h(3) - h(1.5)}{3 - 1.5} = \frac{6 - 54}{1.5} = -32 \text{ ft/sec.}$$

The balloon's average velocity is 32 ft/sec downward.

(c) First, we must find the time when $h(t) = 6$. Solving the equation $-16t^2 + 40t + 30 = 6$, we get

$$6 = -16t^2 + 40t + 30$$
$$0 = -16t^2 + 40t + 24$$
$$0 = 2t^2 - 5t - 3$$
$$0 = (2t + 1)(t - 3).$$

Thus, $t = -1/2$ or $t = 3$. Since $t = -1/2$ makes no physical sense, we use $t = 3$ to calculate the balloon's velocity. At $t = 3$, we have a velocity of $v(3) = -32(3) + 40 = -56$ ft/sec. So the balloon's velocity is 56 ft/sec downward at the time of impact.

17. (a)

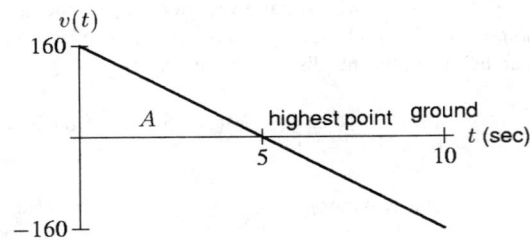

(b) The highest point is at $t = 5$ seconds. The object hits the ground at $t = 10$ seconds, since by symmetry if the object takes 5 seconds to go up, it takes 5 seconds to come back down.

(c) The maximum height is the distance traveled when going up, which is represented by the area A of the triangle above the time axis.

$$\text{Area} = \frac{1}{2}(160 \text{ ft/sec})(5 \text{ sec}) = 400 \text{ feet}.$$

(d) The slope of the line is -32, so $v(t) = -32t + 160$. Antidifferentiating, we get $s(t) = -16t^2 + 160t + s_0$. $s_0 = 0$, so $s(t) = -16t^2 + 160t$. At $t = 5$, $s(t) = -400 + 800 = 400$ ft.

18. The equation of motion is $y = -\frac{gt^2}{2} + v_0 t + y_0 = -16t^2 + 128t + 320$. Taking the first derivative, we get $v = -32t + 128$. The second derivative gives us $a = -32$.

    (a) At its highest point, the stone's velocity is zero:
    $v = 0 = -32t + 128$, so $t = 4$.
    (b) At $t = 4$, the height is $y = -16(4)^2 + 128(4) + 320 = 576$ ft
    (c) When the stone hits the beach,
    $$y = 0 = -16t^2 + 128t + 320$$
    $$0 = -t^2 + 8t + 20 = (10 - t)(2 + t).$$
    So $t = 10$ seconds.
    (d) Impact is at $t = 10$. The velocity, v, at this time is $v(10) = -32(10) + 128 = -192$ ft/sec. Upon impact, the stone's velocity is 192 ft/sec downward.

19. (a) $a(t) = 1.6$, so $v(t) = 1.6t + v_0 = 1.6t$, since the initial velocity is 0.
    (b) $s(t) = 0.8t^2 + s_0$, where s_0 is the rock's initial height.

20. (a) $s = v_0 t - 16t^2$, where $v_0 =$ initial velocity, and $v = s' = v_0 - 32t$. At the maximum height, $v = 0$, so $v_0 = 32t_{max}$. Plugging into the distance equation yields $100 = 32t_{max}^2 - 16t_{max}^2 = 16t_{max}^2$, so $t_{max} = \frac{5}{2}$ seconds, from which we get $v_0 = 32\left(\frac{5}{2}\right) = 80$ ft/sec.
    (b) This time $g = 5$ ft/sec^2, so $s = v_0 t - 2.5t^2 = 80t - 2.5t^2$, and $v = s' = 80 - 5t$. At the highest point, $v = 0$, so $t_{max} = \frac{80}{5} = 16$ seconds. Plugging into the distance equation yields $s = 80(16) - 2.5(16)^2 = 640$ ft.

21. The height of an object above the ground which begins at rest and falls for t seconds is

$$s(t) = -16t^2 + K,$$

where K is the initial height. Here the flower pot falls from 200 ft, so $K = 200$. To see when the pot hits the ground, solve $-16t^2 + 200 = 0$. The solution is

$$t = \sqrt{\frac{200}{16}} \approx 3.54 \text{ seconds}.$$

Now, velocity is given by $s'(t) = v(t) = -32t$. So, the velocity when the pot hits the ground is

$$v(3.54) \approx -113.1 \text{ ft/sec},$$

which is approximately 77 mph downwards.

22. The first thing we should do is convert our units. We'll bring everything into feet and seconds. Thus, the initial speed of the car is

$$\frac{70 \text{ miles}}{\text{hour}} \left(\frac{1 \text{ hour}}{3600 \text{ sec}} \right) \left(\frac{5280 \text{ feet}}{1 \text{ mile}} \right) \approx 102.7 \text{ ft/sec}.$$

We assume that the acceleration is constant as the car comes to a stop. A graph of its velocity versus time is given in Figure 6.5. We know that the area under the curve represents the distance that the car travels before it comes to a stop, 157 feet. But this area is a triangle, so it is easy to find t_0, the time the car comes to rest. We solve

$$\frac{1}{2}(102.7)t_0 = 157,$$

which gives

$$t_0 \approx 3.06 \text{ sec}.$$

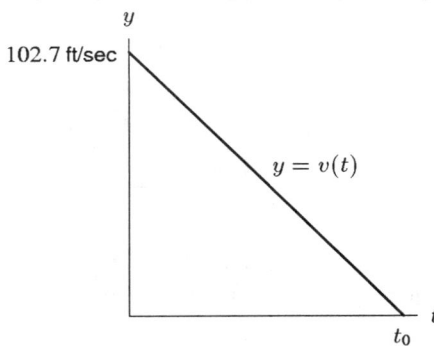

Figure 6.5: Graph of velocity versus time

Since acceleration is the rate of change of velocity, the car's acceleration is given by the slope of the line in Figure 6.5. Thus, the acceleration, k, is given by

$$k = \frac{102.7 - 0}{0 - 3.06} \approx -33.56 \text{ ft/sec}^2.$$

Notice that k is negative because the car is slowing down.

Solutions for Section 6.4

1.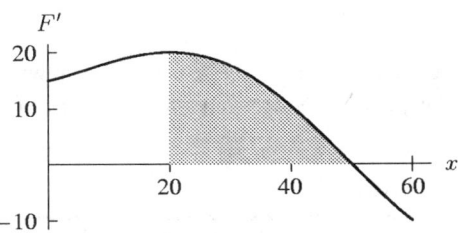

We know that $F(x)$ increases for $x < 50$ because the derivative of F is positive there. See figure above. Similarly, $F(x)$ decreases for $x > 50$. Therefore, the graph of F rises until $x = 50$, and then it begins to fall. Thus, the maximum value attained by F is $F(50)$. To evaluate $F(50)$, we use the Fundamental Theorem:

$$F(50) - F(20) = \int_{20}^{50} F'(x)\, dx,$$

which gives

$$F(50) = F(20) + \int_{20}^{50} F'(x)\, dx = 150 + \int_{20}^{50} F'(x)\, dx.$$

The definite integral equals the area of the shaded region under the graph of F', which is roughly 350. Therefore, the greatest value attained by F is $F(50) \approx 150 + 350 = 500$.

252 CHAPTER SIX /SOLUTIONS

2. Using the Fundamental Theorem, we know that the change in F between $x = 0$ and $x = 0.5$ is given by

$$F(0.5) - F(0) = \int_0^{0.5} \sin t \cos t \, dt \approx 0.115.$$

Since $F(0) = 1.0$, we have $F(0.5) \approx 1.115$. The other values are found similarly, and are given in Table 6.1.

TABLE 6.1

b	0	0.5	1	1.5	2	2.5	3
$F(b)$	1	1.11492	1.35404	1.4975	1.41341	1.17908	1.00996

3.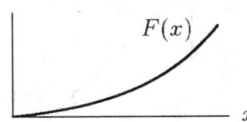

By the Fundamental Theorem, $f(x) = F'(x)$. Since f is positive and increasing, F is increasing and concave up. Since $F(0) = \int_0^0 f(t)dt = 0$, the graph of F must start from the origin.

4.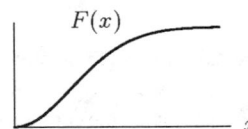

Since f is always positive, F is always increasing. F has an inflection point where $f' = 0$. Since $F(0) = \int_0^0 f(t)dt = 0$, F goes through the origin.

5.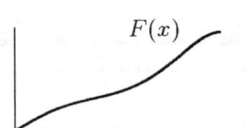

Since f is always non-negative, F is increasing. F is concave up where f is increasing and concave down where f is decreasing; F has inflection points at the critical points of f. Since $F(0) = \int_0^0 f(t)dt = 0$, the graph of F goes through the origin.

6.

TABLE 6.2

x	0	0.5	1	1.5	2
$I(x)$	0	0.50	1.09	2.03	3.65

7. (a) Again using 0.00001 as the lower limit, because the integral is improper, gives $Si(4) = 1.76, Si(5) = 1.55$.
   (b) $Si(x)$ decreases when the integrand is negative, which occurs when $\pi < x < 2\pi$.

8. $\sqrt{3 + \cos(x^2)}$.

9. $(1 + x)^{200}$.

10. $\arctan(x^2)$.

11. $\dfrac{d}{dt}\displaystyle\int_t^{\pi}\cos(z^3)\,dz = \dfrac{d}{dt}\left(-\int_{\pi}^t\cos(z^3)\,dz\right) = -\cos(t^3).$

12. $\dfrac{d}{dx}\displaystyle\int_x^1\ln t\,dt = \dfrac{d}{dx}\left(-\int_1^x\ln t\,dt\right) = -\ln x.$

13. Considering $\mathrm{Si}(x^2)$ as the composition of $\mathrm{Si}(u)$ and $u(x) = x^2$, we may apply the chain rule to obtain

$$\dfrac{d}{dx} = \dfrac{d(\mathrm{Si}(u))}{du}\cdot\dfrac{du}{dx}$$
$$= \dfrac{\sin u}{u}\cdot 2x$$
$$= \dfrac{2\sin(x^2)}{x}.$$

14. (a) Since $\dfrac{d}{dt}(\cos(2t)) = -2\sin(2t)$, we have $F(\pi) = \displaystyle\int_0^{\pi}\sin(2t)\,dt = -\dfrac{1}{2}\cos(2t)\bigg|_0^{\pi} = -\dfrac{1}{2}(1-1) = 0.$

    (b) $F(\pi) =$ (Area above t-axis) $-$ (Area below t-axis) $= 0$. (The two areas are equal.)

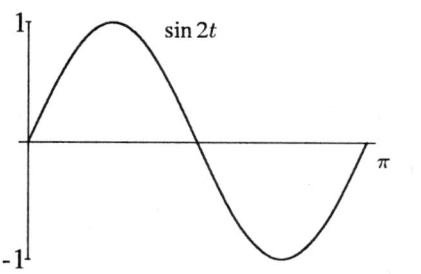

   (c) $F(x) \geq 0$ everywhere. $F(x) = 0$ only at integer multiples of π. This can be seen for $x \geq 0$ by noting $F(x) =$ (area above t-axis) $-$ (area below t-axis), which is always non-negative and only equals zero when x is an integer multiple of π. For $x > 0$

$$F(-x) = \int_0^{-x}\sin 2t\,dt$$
$$= -\int_{-x}^0\sin 2t\,dt$$
$$= \int_0^x\sin 2t\,dt = F(x),$$

   since the area from $-x$ to 0 is the negative of the area from 0 to x. So we have $F(x) \geq 0$ for all x.

15. (a) $F'(x) = \dfrac{1}{\ln x}$ by the Construction Theorem.

    (b) For $x \geq 2$, $F'(x) > 0$, so $F(x)$ is increasing. Since $F''(x) = -\dfrac{1}{x(\ln x)^2} < 0$ for $x \geq 2$, the graph of $F(x)$ is concave down.

    (c)

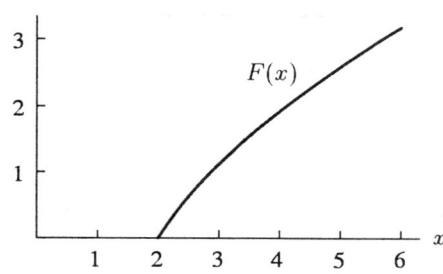

16.
$$\frac{d}{dx}[x\,\text{erf}(x)] = \text{erf}(x)\frac{d}{dx}(x) + x\frac{d}{dx}[\text{erf}(x)]$$
$$= \text{erf}(x) + x\frac{d}{dx}\left(\frac{2}{\sqrt{\pi}}\int_0^x e^{-t^2}\,dt\right)$$
$$= \text{erf}(x) + \frac{2}{\sqrt{\pi}}xe^{-x^2}.$$

17. If we let $f(x) = \text{erf}(x)$ and $g(x) = \sqrt{x}$, then we are looking for $\frac{d}{dx}[f(g(x))]$. By the chain rule, this is the same as $g'(x)f'(g(x))$. Since
$$f'(x) = \frac{d}{dx}\left(\frac{2}{\sqrt{\pi}}\int_0^x e^{-t^2}\,dt\right)$$
$$= \frac{2}{\sqrt{\pi}}e^{-x^2}$$

and $g'(x) = \frac{1}{2\sqrt{x}}$, we have
$$f'(g(x)) = \frac{2}{\sqrt{\pi}}e^{-x},$$

and so
$$\frac{d}{dx}[\text{erf}(\sqrt{x})] = \frac{1}{2\sqrt{x}}\frac{2}{\sqrt{\pi}}e^{-x} = \frac{1}{\sqrt{\pi x}}e^{-x}.$$

18. If we let $f(x) = \text{erf}(x)$ and $g(x) = x^3$, then we use the chain rule because we are looking for $\frac{d}{dx}f(g(x)) = f'(g(x))\cdot g'(x)$. Since $f'(x) = \frac{2}{\sqrt{\pi}}e^{-x^2}$, we have
$$\frac{d}{dx}\left(\frac{2}{\sqrt{\pi}}\int_0^{x^3} e^{-t^2}\,dt\right) = f'(x^3)\cdot 3x^2 = \frac{2}{\sqrt{\pi}}e^{-(x^3)^2}\cdot 3x^2 = \frac{6}{\sqrt{\pi}}x^2 e^{-x^6}.$$

19. We split the integral $\int_x^{x^3} e^{-t^2}\,dt$ into two pieces, say at $t=1$ (though it could be at any other point):
$$\int_x^{x^3} e^{-t^2}\,dt = \int_1^{x^3} e^{-t^2}\,dt + \int_x^1 e^{-t^2}\,dt = \int_1^{x^3} e^{-t^2}\,dt - \int_1^x e^{-t^2}\,dt.$$

We have used the fact that $\int_x^1 e^{-t^2}\,dt = -\int_1^x e^{-t^2}\,dt$. Differentiating gives
$$\frac{d}{dx}\left(\int_x^{x^3} e^{-t^2}\,dt\right) = \frac{d}{dx}\left(\int_1^{x^3} e^{-t^2}\,dt\right) - \frac{d}{dx}\left(\int_1^x e^{-t^2}\,dt\right)$$

For the first integral, we take $g(x) = x^3$ as the inside function, so the final answer is
$$\frac{d}{dx}\left(\int_x^{x^3} e^{-t^2}\,dt\right) = e^{-(x^3)^2}\cdot 3x^2 - e^{-x^2} = 3x^2 e^{-x^6} - e^{-x^2}.$$

Solutions for Chapter 6 Review

1. $\frac{5}{2}x^2 + 7x + C$
2. $3\ln|t| + \frac{2}{t} + C$
3. $e^x + 5x + C$
4. $\frac{2}{5}x^{5/2} + \frac{2}{15}x^{3/2} - 2\ln|x| + C$
5. $\tan x + C$
6. $\frac{1}{\ln 2}2^x + C$, since $\frac{d}{dx}(2^x) = (\ln 2)\cdot 2^x$

7. $\int (x+1)^2 \, dx = \dfrac{(x+1)^3}{3} + C.$

Another way to work the problem is to expand $(x+1)^2$ to $x^2 + 2x + 1$ as follows:

$$\int (x+1)^2 \, dx = \int (x^2 + 2x + 1) \, dx = \dfrac{x^3}{3} + x^2 + x + C.$$

These two answers are the same, since $\dfrac{(x+1)^3}{3} = \dfrac{x^3 + 3x^2 + 3x + 1}{3} = \dfrac{x^3}{3} + x^2 + x + \dfrac{1}{3}$, which is $\dfrac{x^3}{3} + x^2 + x$, plus a constant.

8. $\int (x+1)^3 \, dx = \dfrac{(x+1)^4}{4} + C.$

Another way to work the problem is to expand $(x+1)^3$ to $x^3 + 3x^2 + 3x + 1$:

$$\int (x+1)^3 \, dx = \int (x^3 + 3x^2 + 3x + 1) \, dx = \dfrac{x^4}{4} + x^3 + \dfrac{3}{2}x^2 + x + C.$$

It can be shown that these answers are the same by expanding $\dfrac{(x+1)^4}{4}$.

9. $\frac{1}{10}(x+1)^{10} + C$

10. Since $f(x) = \dfrac{x+1}{x} = 1 + \dfrac{1}{x}$, the indefinite integral is $x + \ln|x| + C$

11. Since $f(x) = x + 1 + \dfrac{1}{x}$, the indefinite integral is $\dfrac{1}{2}x^2 + x + \ln|x| + C$

12. $3\sin\psi + 2\psi^{3/2} + C$

13. $3\sin x + 7\cos x + C$

14. $\dfrac{x^2}{2} + 2\ln|x| - \pi\cos x + C$

15. $2e^x - 8\sin x + C$

16. Antiderivative $F(x) = \dfrac{x^2}{2} + \dfrac{x^6}{6} - \dfrac{x^{-4}}{4} + C$

17. $\ln|x| - \dfrac{1}{x} - \dfrac{1}{2x^2} + C$

18. $F(x) = \dfrac{x^7}{7} - \dfrac{1}{7}\left(\dfrac{x^{-5}}{-5}\right) + C = \dfrac{x^7}{7} + \dfrac{1}{35}x^{-5} + C$

19. $G(t) = 5t + \sin t + C$

20. $-\cos(3\alpha) + C$

21. Antiderivative $H(r) = 2\dfrac{r^{3/2}}{3/2} + \dfrac{1}{2}\dfrac{r^{1/2}}{1/2} + C = \dfrac{4}{3}r^{3/2} + r^{1/2} + C$

22. Antiderivative $G(x) = \dfrac{x^4}{4} + x^3 + \dfrac{3x^2}{2} + x + C = \dfrac{(x+1)^4}{4} + C$

23. Antiderivative $H(t) = t - 2\ln|t| - 1/t + C$

24. $P(y) = \ln|y| + y^2/2 + y + C$

25. $F(z) = e^z + 3z + C$

26. $e^{x^2} + C$

27. $e^x + e^{1+x} + C$

28. $G(\theta) = -\cos\theta - 2\sin\theta + C$

29. $P(r) = \pi r^2 + C$

30. $\dfrac{2^x}{\ln 2} - \dfrac{2^{-x}}{\ln 2} + C$

31. $\dfrac{(1+\sin t)^{30}}{30} + C$

256 CHAPTER SIX /SOLUTIONS

32. $\sin(t^2) + C$

33. $\dfrac{1}{2}\sin(t^2) + C$

34. $\sin(x^3 + 7) + C$

35. $e^2 y + \dfrac{2^y}{\ln 2} + C$

36. $-\cos\theta + \arctan\theta + C$

37. $\dfrac{1}{2}e^{x^2} + C$

38. $4x^{3/2} + \dfrac{1}{x} + 10\ln x + C$

39. The area we want (the shaded area in the figure below) is symmetric about the y-axis and so is given by

$$\text{Area} = 2\int_0^{\pi/3}\left(\cos x - \dfrac{1}{2}\left(\dfrac{3}{\pi}x\right)^2\right)dx$$

$$= 2\int_0^{\pi/3}\cos x\,dx - \int_0^{\pi/3}\dfrac{9}{\pi^2}x^2\,dx$$

$$= 2\sin x\Big|_0^{\pi/3} - \dfrac{9}{\pi^2}\cdot\dfrac{x^3}{3}\Big|_0^{\pi/3}$$

$$= 2\cdot\dfrac{\sqrt{3}}{2} - \dfrac{3}{\pi^2}\cdot\dfrac{\pi^3}{3^3} = \sqrt{3} - \dfrac{\pi}{9}.$$

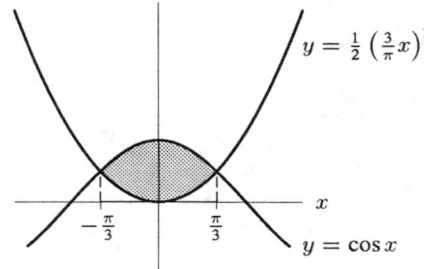

40. Since $y = x^3 - x = x(x-1)(x+1)$, the graph crosses the axis at the three points shown in Figure 6.6. The two regions have the same area (by symmetry). Since the graph is below the axis for $0 < x < 1$, we have

$$\text{Area} = 2\left(-\int_0^1 (x^3 - x)\,dx\right)$$

$$= -2\left[\dfrac{x^4}{4} - \dfrac{x^2}{2}\right]_0^1 = -2\left(\dfrac{1}{4} - \dfrac{1}{2}\right) = \dfrac{1}{2}.$$

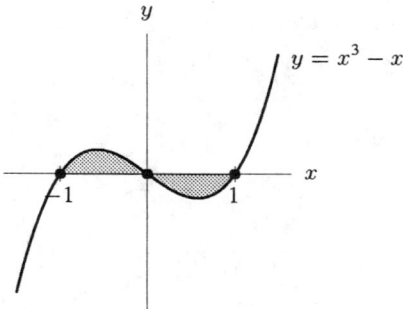

Figure 6.6

41. The graph of $y = c(1-x^2)$ has x-intercepts of $x = \pm 1$. See Figure 6.7. Since it is symmetric about the y-axis, we have

$$\text{Area} = \int_{-1}^{1} c(1-x^2)\,dx = 2c\int_0^1 (1-x^2)\,dx$$

$$= 2c\left(x - \frac{x^3}{3}\right)\bigg|_0^1 = \frac{4c}{3}.$$

We want the area to be 1, so

$$\frac{4c}{3} = 1, \quad \text{giving} \quad c = \frac{3}{4}.$$

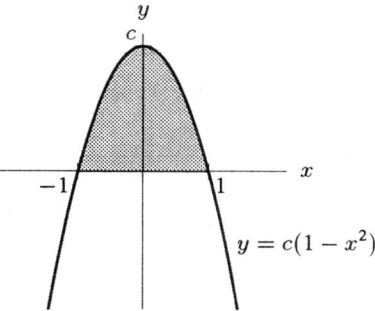

Figure 6.7

42. The curves intersect at $(0, 0)$ and $(\pi, 0)$. At any x-coordinate the "height" between the two curves is $\sin x - x(x - \pi)$.

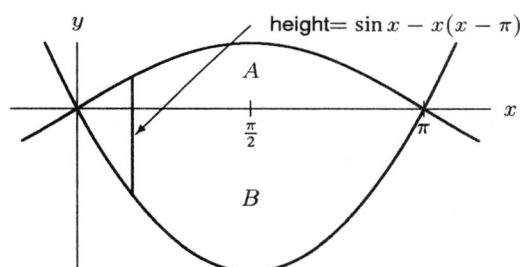

Thus the total area is

$$\int_0^\pi [\sin x - x(x-\pi)]\,dx = \int_0^\pi (\sin x - x^2 + \pi x)\,dx$$

$$= \left(-\cos x - \frac{x^3}{3} + \frac{\pi x^2}{2}\right)\bigg|_0^\pi$$

$$= \left(1 - \frac{\pi^3}{3} + \frac{\pi^3}{2}\right) - (-1)$$

$$= 2 + \frac{\pi^3}{6}.$$

Another approach is to notice that the area between the two curves is (area A) + (area B).

$$\text{Area B} = -\int_0^\pi x(x-\pi)\,dx \text{ since the function is negative on } 0 \le x \le \pi$$

$$= -\left(\frac{x^3}{3} - \frac{\pi x^2}{2}\right)\bigg|_0^\pi = \frac{\pi^3}{2} - \frac{\pi^3}{3} = \frac{\pi^3}{6};$$

$$\text{Area A} = \int_0^\pi \sin x\,dx = -\cos x\bigg|_0^\pi = 2.$$

Thus the area is $2 + \dfrac{\pi^3}{6}$.

258 CHAPTER SIX /SOLUTIONS

43.

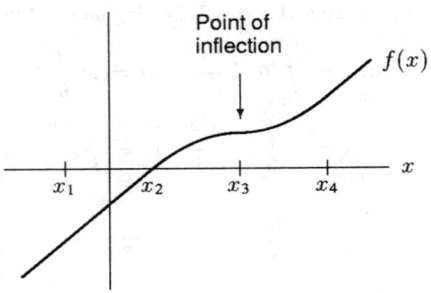

44.

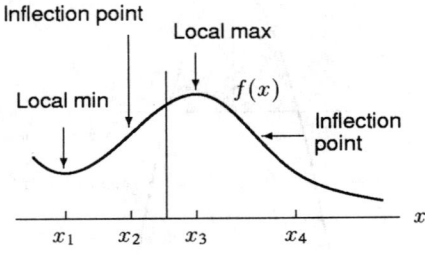

45.

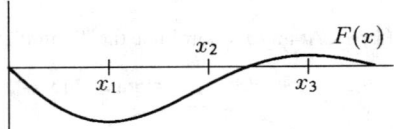

46. The graph of $h(t)$ must slope downwards most steeply when $h'(t)$ has its minimum. The graph of $h(t)$ should have its minimum about two-thirds of the way through the time interval (when the graph of $h'(t)$ intersects the x-axis), and have its final value about half-way between its maximum and minimum values. A possible graph of $h(t)$ is given in Figure 6.8. The placement of the horizontal axis below the graph is arbitrary.

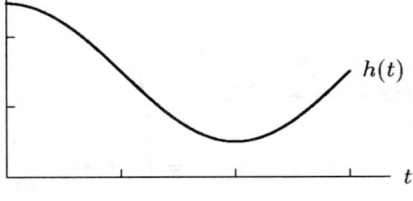

Figure 6.8

47. Let v be the velocity and s be the position of the particle at time t. We know that $a = dv/dt$, so acceleration is the slope of the velocity graph. Similarly, velocity is the slope of the position graph. Graphs of v and s are shown in Figures 6.9 and 6.10, respectively.

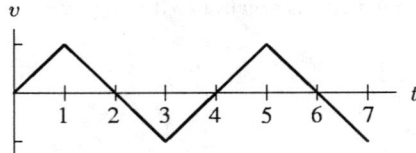

Figure 6.9: Velocity against time

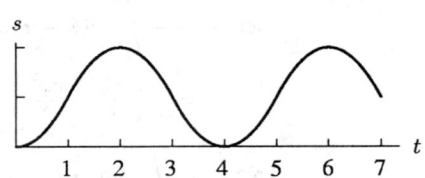

Figure 6.10: Position against time

48. We know the height is given by
$$s = -25t^2 + 72t + 40,$$
so the velocity is given by
$$v = -50t + 72$$
and the acceleraction is given by
$$a = -50.$$
The acceleration due to gravity is -50 ft/sec^2. Since $v(0) = 72$, the object was thrown at 72 ft/sec. Since $s(0) = 40$, the object was thrown from a height of 40 ft.

49. (a) Since 6 sec = 1/10 min,
$$\text{Angular acceleration} = \frac{2500 - 1100}{1/10} = 14{,}000 \text{ revs/min}^2.$$

(b) We know angular acceleration is the derivative of angular velocity. Since
$$\text{Angular acceleration} = 14{,}000,$$
we have
$$\text{Angular velocity} = 14{,}000t + C.$$
Measuring time from the moment at which the angular velocity is 1100 revs/min, we have $C = 1100$. Thus,
$$\text{Angular velocity} = 14{,}000t + 1100.$$
Thus the total number of revolutions performed during the period from $t = 0$ to $t = 1/10$ min is given by
$$\text{Number of revolutions} = \int_0^{1/10} (14000t + 1100)\,dt = 7000t^2 + 1100t \Big|_0^{1/10} = 180 \text{ revolutions}.$$

50. (a) Since the rotor is slowing down at a constant rate,
$$\text{Angular acceleration} = \frac{260 - 350}{1.5} = -60 \text{ revs/min}^2.$$
Units are revolutions per minute per minute, or revs/min^2.

(b) To decrease from 350 to 0 revs/min at a deceleration of 60 revs/min^2,
$$\text{Time needed} = \frac{350}{60} \approx 5.83 \text{ min}.$$

(c) We know angular acceleration is the derivative of angular velocity. Since
$$\text{Angular acceleration} = -60 \text{ revs/min}^2,$$
we have
$$\text{Angular velocity} = -60t + C.$$
Measuring time from the moment when angular velocity is 350 revs/min, we get $C = 350$. Thus
$$\text{Angular velocity} = -60t + 350.$$
So, the total number of revolutions made between the time the angular speed is 350 revs/min and stopping is given by:
$$\text{Number of revolutions} = \int_0^{5.83} (\text{Angular velocity})\,dt$$
$$= \int_0^{5.83} (-60t + 350)\,dt = -30t^2 + 350t \Big|_0^{5.83}$$
$$= 1020.83 \text{ revolutions}.$$

51. (a) Using $g = -32$ ft/sec^2, we have

t (sec)	0	1	2	3	4	5
$v(t)$ (ft/sec)	80	48	16	-16	-48	-80

(b) The object reaches its highest point when $v = 0$, which appears to be at $t = 2.5$ seconds. By symmetry, the object should hit the ground again at $t = 5$ seconds.

(c) Left sum $= 80(1) + 48(1) + 16(\frac{1}{2}) = 136$ ft, which is an overestimate.
Right sum $= 48(1) + 16(1) + (-16)(\frac{1}{2}) = 56$ ft, which is an underestimate.
Note that we used a smaller third rectangle of width $1/2$ to end our sum at $t = 2.5$.

(d) We have $v(t) = 80 - 32t$, so antidifferentiation yields $s(t) = 80t - 16t^2 + s_0$.
But $s_0 = 0$, so $s(t) = 80t - 16t^2$.
At $t = 2.5$, $s(t) = 100$ ft., so 100 ft. is the highest point.

52. The velocity of the car decreases at a constant rate, so we can write: $dv/dt = -a$. Integrating this gives $v = -at + C$. The constant of integration C is the velocity when $t = 0$, so $C = 60$ mph $= 88$ ft/sec, and $v = -at + 88$. From this equation we can see the car comes to rest at time $t = 88/a$.

Integrating the expression for velocity we get $s = -\frac{a}{2}t^2 + 88t + C$, where C is the initial position, so $C = 0$. We can use fact that the car comes to rest at time $t = 88/a$ after traveling 200 feet. Start with

$$s = -\frac{a}{2}t^2 + 88t,$$

and substitute $t = 88/a$ and $s = 200$:

$$200 = -\frac{a}{2}\left(\frac{88}{a}\right)^2 + 88\left(\frac{88}{a}\right) = \frac{88^2}{2a}$$

$$a = \frac{88^2}{2(200)} = 19.36 \text{ ft/sec}^2$$

53. (a) In the beginning, both birth and death rates are small; this is consistent with a very small population. Both rates begin climbing, the birth rate faster than the death rate, which is consistent with a growing population. The birth rate is then high, but it begins to decrease as the population increases.

(b)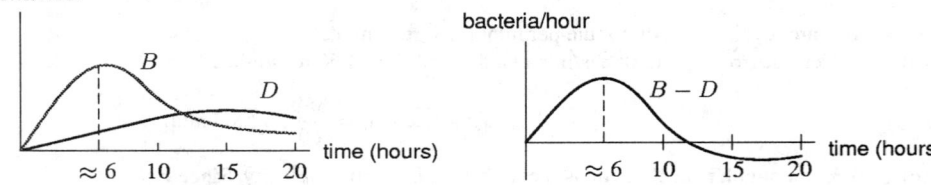

Figure 6.11: Difference between B and D is greatest at $t \approx 6$

The bacteria population is growing most quickly when $B - D$, the rate of change of population, is maximal; that happens when B is farthest above D, which is at a point where the slopes of both graphs are equal. That point is $t \approx 6$ hours.

(c) Total number born by time t is the area under the B graph from $t = 0$ up to time t. See Figure 6.12.

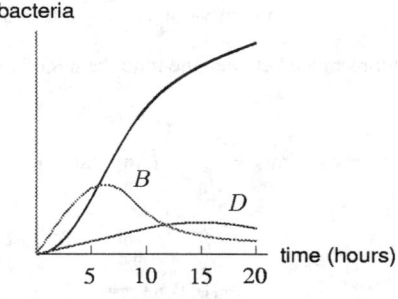

Figure 6.12: Number born by time t is $\int_0^t B(x)\,dx$

Total number alive at time t is the number born minus the number that have died, which is the area under the B graph minus the area under the D graph, up to time t. See Figure 6.13.

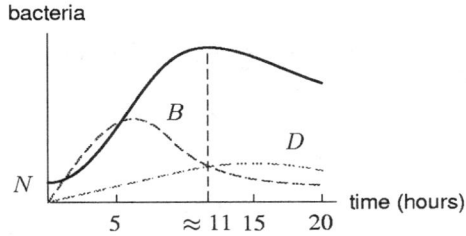

Figure 6.13: Number alive at time t is $\int_0^t (B(x) - D(x))\,dx$

From Figure 6.13, we see that the population is at a maximum when $B = D$, that is, after about 11 hours. This stands to reason, because $B - D$ is the rate of change of population, so population is maximized when $B - D = 0$, that is, when $B = D$.

54.

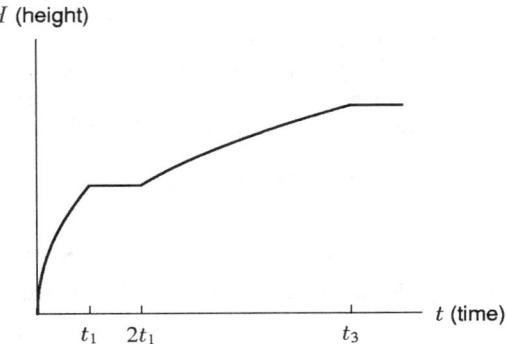

Suppose t_1 is the time to fill the left side to the top of the middle ridge. Since the container gets wider as you go up, the rate dH/dt decreases with time. Therefore, for $0 \le t \le t_1$, graph is concave down.

At $t = t_1$, water starts to spill over to right side and so depth of left side doesn't change. It takes as long for the right side to fill to the ridge as the left side, namely t_1. Thus the graph is horizontal for $t_1 \le t \le 2t_1$.

For $t \ge 2t_1$, water level is above the central ridge. The graph is climbing because the depth is increasing, but at a slower rate than for $t \le t_1$ because the container is wider. The graph is concave down because width is increasing with depth. Time t_3 represents the time when container is full.

55.
- For $[0, t_1]$, the acceleration is constant and positive and the velocity is positive so the displacement is positive. Thus, the work done is positive.
- For $[t_1, t_2]$, the acceleration, and therefore the force, is zero. Therefore, the work done is zero.
- For $[t_2, t_3]$, the acceleration is negative and thus the force is negative. The velocity, and thus the displacement, is positive; therefore the work done is negative.
- For $[t_3, t_4]$, the acceleration (and thus the force) and the velocity (and thus the displacement) are negative. Thus, the work done is positive.
- For $[t_2, t_4]$, the acceleration and thus the force is constant and negative. Velocity both positive and negative; total displacement is 0. Since force is constant, work is 0.

56. $F(x)$ represents the net area between $(\sin t)/t$ and the t-axis from $t = \frac{\pi}{2}$ to $t = x$, with area counted as negative for $(\sin t)/t$ below the t-axis. As long as the integrand is positive $F(x)$ is increasing. Therefore, the global maximum of $F(x)$ occurs at $x = \pi$ and is given by the area

$$A_1 = \int_{\pi/2}^{\pi} \frac{\sin t}{t}\,dt.$$

At $x = \pi/2$, $F(x) = 0$. Figure 6.14 shows that the area A_1 is larger than the area A_2. Thus $F(x) > 0$ for $\frac{\pi}{2} < x \le \frac{3\pi}{2}$. Therefore the global minimum is $F(\frac{\pi}{2}) = 0$.

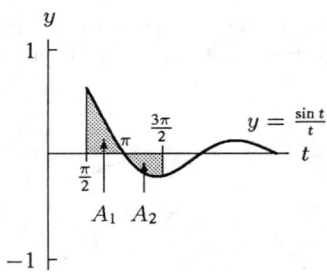

Figure 6.14

Solutions to the Projects and Computer Algebra Investigations

1. (a) If the poorest $p\%$ of the population has exactly $p\%$ of the goods, then $F(x) = x$.
   (b) Any such F is increasing. For example, the poorest 50% of the population includes the poorest 40%, and so the poorest 50% must own more than the poorest 40%. Thus $F(0.4) \leq F(0.5)$, and so, in general, F is increasing. In addition, it is clear that $F(0) = 0$ and $F(1) = 1$.

   The graph of F is concave up by the following argument. Consider $F(0.05) - F(0.04)$. This is the fraction of resources the fifth poorest percent of the population has. Similarly, $F(0.20) - F(0.19)$ is the fraction of resources that the twentieth poorest percent of the population has. Since the twentieth poorest percent owns more than the fifth poorest percent, we have
   $$F(0.05) - F(0.04) \leq F(0.20) - F(0.19).$$
   More generally, we can see that
   $$F(x_1 + \Delta x) - F(x_1) \leq F(x_2 + \Delta x) - F(x_2)$$
   for any x_1 smaller than x_2 and for any increment Δx. Dividing this inequality by Δx and taking the limit as $\Delta x \to 0$, we get
   $$F'(x_1) \leq F'(x_2).$$
   So, the derivative of F is an increasing function, i.e. F is concave up.
   (c) G is twice the shaded area below in the following figure. If the resource is distributed evenly, then G is zero. The larger G is, the more unevenly the resource is distributed. The maximum possible value of G is 1.

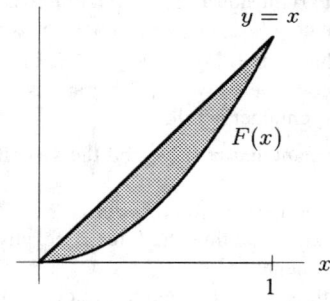

2. (a)

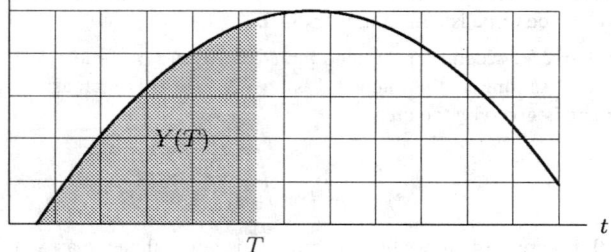

$Y(T)$ is the area of the shaded region in the picture. Thus, $Y(T) = \int_0^T y(t)\,dt$.

(b) Here is a graph of $Y(T)$. Note that the graph of y looked like the graph of a quadratic function. Thus, the graph of Y should look like a cubic, which indeed it does.

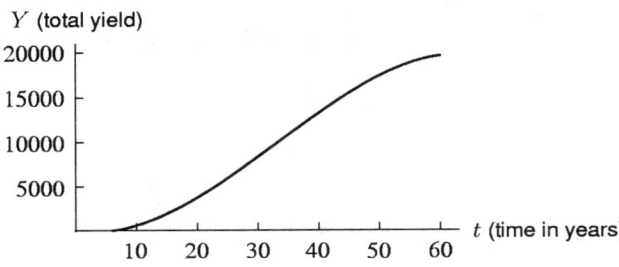

(c) $a(T) = \dfrac{1}{T} Y(T) = \dfrac{1}{T} \displaystyle\int_0^T y(t)\, dt.$

(d) (i) If the function $a(T)$ takes on its maximum at some point T, then $a'(T) = 0$. Since $a(T) = \frac{1}{T} Y(T)$, we may differentiate using the quotient rule:

$$a'(T) = \dfrac{TY'(T) - Y(T)}{T^2} = 0;$$

or, equivalently, $TY'(T) = Y(T)$.

(ii) The expression above may be rewritten in terms of y, giving us

$$T \dfrac{d}{dT} \int_0^T y(t)\, dt = \int_0^T y(t)\, dt.$$

Simplifying, we obtain $Ty(T) = \int_0^T y(t)\, dt$, or, equivalently,

$$y(T) = \dfrac{1}{T} \int_0^T y(t)\, dt = a(T),$$

as a condition on $y(T)$ for maximization of $a(T)$.

To find the value of T which satisfies $Ty(T) = Y(T)$, notice that $Y(T)$ is the area under the curve from 0 to T, and that $Ty(T)$ is the area of a rectangle of height $y(T)$. Thus we want the area under the curve to be equal to the area of the rectangle, or $A = B$ in the figure below. This happens when $T \approx 50$ years. In other words, the orchard should be cut down after about 50 years.

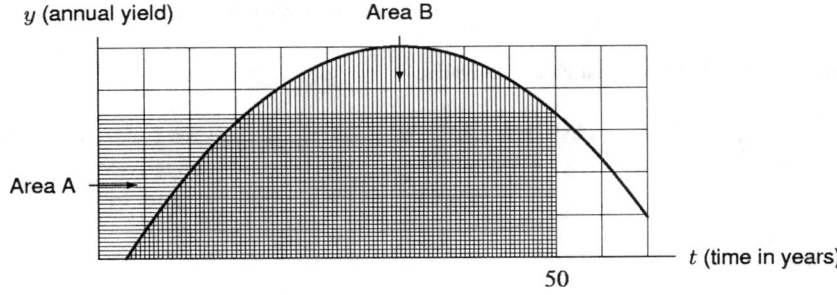

3. (a) A CAS gives

$$\int e^{2x}\, dx = \dfrac{1}{2} e^{2x} \qquad \int e^{3x}\, dx = \dfrac{1}{3} e^{3x} \qquad \int e^{3x+5}\, dx = \dfrac{1}{3} e^{3x+5}.$$

(b) The three integrals in part (a) obey the rule

$$\int e^{ax+b}\, dx = \dfrac{1}{a} e^{ax+b}.$$

(c) Checking the formula by calculating the derivative

$$\frac{d}{dx}\left(\frac{1}{a}e^{ax+b}\right) = \frac{1}{a}\frac{d}{dx}e^{ax+b} \quad \text{by the constant multiple rule}$$

$$= \frac{1}{a}e^{ax+b}\frac{d}{dx}(ax+b) \quad \text{by the chain rule}$$

$$= \frac{1}{a}e^{ax+b} \cdot a = e^{ax+b}.$$

4. (a) A CAS gives

$$\int \sin(3x)\, dx = -\frac{1}{3}\cos(3x) \qquad \int \sin(4x)\, dx = -\frac{1}{4}\cos(4x) \qquad \int \sin(3x-2)\, dx = -\frac{1}{3}\cos(3x-2).$$

(b) The three integrals in part (a) obey the rule

$$\int \sin(ax+b)\, dx = -\frac{1}{a}\cos(ax+b).$$

(c) Checking the formula by calculating the derivative

$$\frac{d}{dx}\left(-\frac{1}{a}\cos(ax+b)\right) = -\frac{1}{a}\frac{d}{dx}\cos(ax+b) \quad \text{by the constant multiple rule}$$

$$= -\frac{1}{a}(-\sin(ax+b))\frac{d}{dx}(ax+b) \quad \text{by the chain rule}$$

$$= -\frac{1}{a}(-\sin(ax+b)) \cdot a = \sin(ax+b).$$

5. (a) A CAS gives

$$\int \frac{x-2}{x-1}\, dx = x + \ln|x-1|$$

$$\int \frac{x-3}{x-1}\, dx = x + 2\ln|x-1|$$

$$\int \frac{x-1}{x-2}\, dx = x - \ln|x-2|$$

Although the absolute values are needed in the answer, some CASs may not include them.

(b) The three integrals in part (a) obey the rule

$$\int \frac{x-a}{x-b}\, dx = x + (b-a)\ln|x-b|.$$

(c) Checking the formula by calculating the derivative

$$\frac{d}{dx}(x + (b-a)\ln|x-b|) = 1 + (b-a)\frac{1}{x-b} \quad \text{by the sum and constant multiple rules}$$

$$= \frac{(x-b)+(b-a)}{x-b} = \frac{x-a}{x-b}$$

6. (a) A CAS gives

$$\int \frac{1}{(x-1)(x-3)}\, dx = \frac{1}{2}(\ln|x-3| - \ln|x-1|)$$

$$\int \frac{1}{(x-1)(x-4)}\, dx = \frac{1}{3}(\ln|x-4| - \ln|x-1|)$$

$$\int \frac{1}{(x-1)(x+3)}\, dx = \frac{1}{4}(\ln|x+3| - \ln|x-1|).$$

Although the absolute values are needed in the answer, some CASs may not include them.

(b) The three integrals in part (a) obey the rule

$$\int \frac{1}{(x-a)(x-b)} \, dx = \frac{1}{b-a} (\ln|x-b| - \ln|x-a|).$$

(c) Checking the formula by calculating the derivative

$$\frac{d}{dx}\left(\frac{1}{b-a}(\ln|x-b| - \ln|x-a|)\right) = \frac{1}{b-a}\left(\frac{1}{x-b} - \frac{1}{x-a}\right)$$

$$= \frac{1}{b-a}\left(\frac{(x-a)-(x-b)}{(x-a)(x-b)}\right)$$

$$= \frac{1}{b-a}\left(\frac{b-a}{(x-a)(x-b)}\right) = \frac{1}{(x-a)(x-b)}.$$

Solutions to Problems on the Equations of Motion

1. The velocity as a function of time is given by: $v = v_0 + at$. Since the object starts from rest, $v_0 = 0$, and the velocity is just the acceleration times time: $v = -32t$. Integrating this, we get position as a function of time: $y = -16t^2 + y_0$, where the last term, y_0, is the initial position at the top of the tower, so $y_0 = 400$ feet. Thus we have a function giving position as a function of time: $y = -16t^2 + 400$.

   To find at what time the object hits the ground, we find t when $y = 0$. We solve $0 = -16t^2 + 400$ for t, getting $t^2 = 400/16 = 25$, so $t = 5$. Therefore the object hits the ground after 5 seconds. At this time it is moving with a velocity $v = -32(5) = -160$ feet/second.

2. In Problem 1 we used the equation $0 = -16t^2 + 400$ to learn that the object hits the ground after 5 seconds. In a more general form this is the equation $y = -\frac{g}{2}t^2 + v_0 t + y_0$, and we know that $v_0 = 0$, $y_0 = 400$ ft. So the moment the object hits the ground is given by $0 = -\frac{g}{2}t^2 + 400$. In Problem 1 we used $g = 32$ ft/sec^2, but in this case we want to find a g that results in the object hitting the ground after only 5/2 seconds. We put in 5/2 for t and solve for g:

$$0 = -\frac{g}{2}\left(\frac{5}{2}\right)^2 + 400, \text{ so } g = \frac{2(400)}{(5/2)^2} = 128 \text{ ft/sec}^2.$$

3. $a(t) = -32$. Since $v(t)$ is the antiderivative of $a(t)$, $v(t) = -32t + v_0$. But $v_0 = 0$, so $v(t) = -32t$. Since $s(t)$ is the antiderivative of $v(t)$, $s(t) = -16t^2 + s_0$, where s_0 is the height of the building. Since the ball hits the ground in 5 seconds, $s(5) = 0 = -400 + s_0$. Hence $s_0 = 400$ feet, so the window is 400 feet high.

4. Let time $t = 0$ be the moment when the astronaut jumps up. If acceleration due to gravity is 5 ft/sec^2 and initial velocity is 10 ft/sec, then the velocity of the astronaut is described by

$$v(t) = 10 - 5t.$$

Suppose $y(t)$ describes his distance from the surface of the moon. By the Fundamental Theorem,

$$y(t) - y(0) = \int_0^t (10 - 5x) \, dx$$

$$y(t) = 10t - \frac{1}{2}5t^2.$$

since $y(0) = 0$ (assuming the astronaut jumps off the surface of the moon).

The astronaut reaches the maximum height when his velocity is 0, i.e. when

$$\frac{dy}{dt} = v(t) = 10 - 5t = 0.$$

Solving for t, we get $t = 2$ sec as the time at which he reaches the maximum height from the surface of the moon. At this time his height is

$$y(2) = 10(2) - \frac{1}{2}5(2)^2 = 10 \text{ ft}.$$

When the astronaut is at height $y = 0$, he either just landed or is about to jump. To find how long it is before he comes back down, we find when he is at height $y = 0$. Set $y(t) = 0$ to get

$$0 = 10t - \frac{1}{2}5t^2$$
$$0 = 20t - 5t^2$$
$$0 = 4t - t^2$$
$$0 = t(t - 4).$$

So we have $t = 0$ sec (when he jumps off) and $t = 4$ sec (when he lands, which gives the time he spent in the air).

5. Let the acceleration due to gravity equal $-k$ meters/sec^2, for some positive constant k, and suppose the object falls from an initial height of $s(0)$ meters. We have $a(t) = dv/dt = -k$, so that

$$v(t) = -kt + v_0.$$

Since the initial velocity is zero, we have

$$v(0) = -k(0) + v_0 = 0,$$

which means $v_0 = 0$. Our formula becomes

$$v(t) = \frac{ds}{dt} = -kt.$$

This means

$$s(t) = \frac{-kt^2}{2} + s_0.$$

Since

$$s(0) = \frac{-k(0)^2}{2} + s_0,$$

we have $s_0 = s(0)$, and our formula becomes

$$s(t) = \frac{-kt^2}{2} + s(0).$$

Suppose that the object falls for t seconds. Assuming it hasn't hit the ground, its height is

$$s(t) = \frac{-kt^2}{2} + s(0),$$

so that the distance traveled is

$$s(0) - s(t) = \frac{kt^2}{2} \text{ meters},$$

which is proportional to t^2.

6. (a) $t = \dfrac{s}{\frac{1}{2}v_{\max}}$, where t is the time it takes for an object to travel the distance s, starting from rest with uniform acceleration a. $v_{\max}$ is the highest velocity the object reaches. Since its initial velocity is 0, the mean of its highest velocity and initial velocity is $\frac{1}{2}v_{\max}$.

(b) By Problem 5, $s = \frac{1}{2}gt^2$, where g is the acceleration due to gravity, so it takes $\sqrt{200/32} = 5/2$ seconds for the body to hit the ground. Since $v = gt$, $v_{\max} = 32(\frac{5}{2}) = 80$ ft/sec. Galileo's statement predicts $(100 \text{ ft})/(40 \text{ ft/sec}) = 5/2$ seconds, and so Galileo's result is verified.

(c) If the acceleration is a constant a, then $s = \frac{1}{2}at^2$, and $v_{\max} = at$. Thus

$$\frac{s}{\frac{1}{2}v_{\max}} = \frac{\frac{1}{2}at^2}{\frac{1}{2}at} = t.$$

7. (a) Since $s(t) = -\frac{1}{2}gt^2$, the distance a body falls in the first second is

$$s(1) = -\frac{1}{2} \cdot g \cdot 1^2 = -\frac{g}{2}.$$

In the second second, the body travels

$$s(2) - s(1) = -\frac{1}{2}\left(g \cdot 2^2 - g \cdot 1^2\right) = -\frac{1}{2}(4g - g) = -\frac{3g}{2}.$$

In the third second, the body travels

$$s(3) - s(2) = -\frac{1}{2}\left(g \cdot 3^2 - g \cdot 2^2\right) = -\frac{1}{2}(9g - 4g) = -\frac{5g}{2},$$

and in the fourth second, the body travels

$$s(4) - s(3) = -\frac{1}{2}\left(g \cdot 4^2 - g \cdot 3^2\right) = -\frac{1}{2}(16g - 9g) = -\frac{7g}{2}.$$

(b) Galileo seems to have been correct. His observation follows from the fact that the differences between consecutive squares are consecutive odd numbers. For, if n is any number, then $n^2 - (n-1)^2 = 2n - 1$, which is the nth odd number (where 1 is the first).

8. If r is the distance from the center of the earth,

$$g = \frac{GM}{r^2},$$

so at 2 meters

$$9.8 = \frac{GM}{(6.4 \times 10^6 + 2)^2}.$$

At 100 meters above the ground,

$$g_{\text{new}} = \frac{GM}{(6.4 \times 10^6 + 100)^2}$$

so

$$\frac{g_{\text{new}}}{9.8} = \frac{GM}{(6.4 \times 10^6 + 100)^2} \bigg/ \frac{GM}{(6.4 \times 10^6 + 2)^2}$$

$$g_{\text{new}} = 9.8 \left(\frac{6,400,002}{6,400,100}\right)^2 = 9.79969\ldots \text{ m/sec}^2.$$

Thus, to the first decimal place, the acceleration due to gravity is still 9.8 m/sec² at 100 m above the ground.

At 100,000 meters above the ground,

$$g_{\text{new}} = 9.8 \left(\frac{6,400,002}{6,500,000}\right)^2 = 9.5 \text{m/sec}^2.$$

CHAPTER SEVEN

Solutions for Section 7.1

1. (a) $\frac{d}{dx}\sin(x^2+1) = 2x\cos(x^2+1)$; $\quad \frac{d}{dx}\sin(x^3+1) = 3x^2\cos(x^3+1)$
    (b) (i) $\frac{1}{2}\sin(x^2+1) + C$ $\quad$ (ii) $\frac{1}{3}\sin(x^3+1) + C$
    (c) (i) $-\frac{1}{2}\cos(x^2+1) + C$ $\quad$ (ii) $-\frac{1}{3}\cos(x^3+1) + C$

2. The general antiderivative is $\int(\pi t^3 + 4t)\,dt = (\pi/4)t^4 + 2t^2 + C$.

3. Make the substitution $w = 3x$, $dw = 3\,dx$. We have
    $$\int \sin 3x\,dx = \frac{1}{3}\int \sin w\,dw = \frac{1}{3}(-\cos w) + C = -\frac{1}{3}\cos 3x + C.$$

4. Make the substitution $w = x^2$, $dw = 2x\,dx$. We have
    $$\int 2x\cos(x^2)\,dx = \int \cos w\,dw = \sin w + C = \sin x^2 + C.$$

5. Make the substitution $w = t^3$, $dw = 3t^2\,dt$. The general antiderivative is $\int 12t^2 \cos(t^3)\,dt = 4\sin(t^3) + C$.

6. Make the substitution $w = 2 - 5x$, then $dw = -5dx$. We have
    $$\int \sin(2-5x)dx = \int \sin w\left(-\frac{1}{5}\right)dw = -\frac{1}{5}(-\cos w) + C = \frac{1}{5}\cos(2-5x) + C.$$

7. Make the substitution $w = \sin x$, $dw = \cos x\,dx$. We have
    $$\int e^{\sin x} \cos x\,dx = \int e^w\,dw = e^w + C = e^{\sin x} + C.$$

8. Make the substitution $w = t^2$, $dw = 2t\,dt$. The general antiderivative is $\int te^{t^2}\,dt = (1/2)e^{t^2} + C$.

9. Make the substitution $w = x^2 + 1$, $dw = 2x\,dx$. We have
    $$\int \frac{x}{x^2+1}\,dx = \frac{1}{2}\int \frac{dw}{w} = \frac{1}{2}\ln|w| + C = \frac{1}{2}\ln(x^2+1) + C.$$
    (Notice that since $x^2 + 1 \geq 0$, $|x^2 + 1| = x^2 + 1$.)

10. Make the substitution $w = 2x$, then $dw = 2dx$. We have
    $$\int \frac{1}{3\cos^2 2x}\,dx = \frac{1}{3}\int \frac{1}{\cos^2 w}\left(\frac{1}{2}\right)dw$$
    $$= \frac{1}{6}\int \frac{1}{\cos^2 w}\,dw = \frac{1}{6}\tan w + C = \frac{1}{6}\tan 2x + C.$$

11. We use the substitution $w = y^2 + 5$, $dw = 2y\,dy$.
    $$\int y(y^2+5)^8\,dy = \frac{1}{2}\int(y^2+5)^8(2y\,dy)$$
    $$= \frac{1}{2}\int w^8\,dw = \frac{1}{2}\frac{w^9}{9} + C$$
    $$= \frac{1}{18}(y^2+5)^9 + C.$$
    Check: $\frac{d}{dy}\left(\frac{1}{18}(y^2+5)^9 + C\right) = \frac{1}{18}[9(y^2+5)^8(2y)] = y(y^2+5)^8$.

12. We use the substitution $w = t^3 - 3$, $dw = 3t^2\,dt$.
    $$\int t^2(t^3-3)^{10}\,dt = \frac{1}{3}\int(t^3-3)^{10}(3t^2dt) = \int w^{10}\left(\frac{1}{3}dw\right)$$
    $$= \frac{1}{3}\frac{w^{11}}{11} + C = \frac{1}{33}(t^3-3)^{11} + C.$$
    Check: $\frac{d}{dt}\left[\frac{1}{33}(t^3-3)^{11} + C\right] = \frac{1}{3}(t^3-3)^{10}(3t^2) = t^2(t^3-3)^{10}$.

13. We use the substitution $w = x^2 - 4$, $dw = 2x\, dx$.

$$\int x(x^2-4)^{\frac{7}{2}}\, dx = \frac{1}{2}\int (x^2-4)^{\frac{7}{2}}(2x\, dx) = \frac{1}{2}\int w^{\frac{7}{2}}\, dw$$
$$= \frac{1}{2}(\frac{2}{9}w^{\frac{9}{2}}) + C = \frac{1}{9}(x^2-4)^{\frac{9}{2}} + C.$$

Check: $\dfrac{d}{dx}[\frac{1}{9}(x^2-4)^{\frac{9}{2}} + C] = \frac{1}{9}\left[\frac{9}{2}(x^2-4)^{\frac{7}{2}}\right]2x = x(x^2-4)^{\frac{7}{2}}.$

14. We use the substitution $w = 4 - x$, $dw = -dx$.

$$\int \frac{1}{\sqrt{4-x}}\, dx = -\int \frac{1}{\sqrt{w}}\, dw = -2\sqrt{w} + C = -2\sqrt{4-x} + C.$$

Check: $\dfrac{d}{dx}(-2\sqrt{4-x} + C) = -2 \cdot \frac{1}{2} \cdot \frac{1}{\sqrt{4-x}} \cdot -1 = \frac{1}{\sqrt{4-x}}.$

15. We use the substitution $w = x^2 + 3$, $dw = 2x\, dx$.

$$\int x(x^2+3)^2\, dx = \int w^2(\frac{1}{2}\, dw) = \frac{1}{2}\frac{w^3}{3} + C = \frac{1}{6}(x^2+3)^3 + C.$$

Check: $\dfrac{d}{dx}\left[\frac{1}{6}(x^2+3)^3 + C\right] = \frac{1}{6}\left[3(x^2+3)^2(2x)\right] = x(x^2+3)^2.$

16. We use the substitution $w = y + 5$, $dw = dy$, to get

$$\int \frac{dy}{y+5} = \int \frac{dw}{w} = \ln|w| + C = \ln|y+5| + C.$$

Check: $\dfrac{d}{dy}(\ln|y+5| + C) = \dfrac{1}{y+5}.$

17. We use the substitution $w = 2t - 7$, $dw = 2\, dt$.

$$\int (2t-7)^{73}\, dt = \frac{1}{2}\int w^{73}\, dw = \frac{1}{(2)(74)}w^{74} + C = \frac{1}{148}(2t-7)^{74} + C.$$

Check: $\dfrac{d}{dt}\left[\frac{1}{148}(2t-7)^{74} + C\right] = \frac{74}{148}(2t-7)^{73}(2) = (2t-7)^{73}.$

18. In this case, it seems easier not to substitute.

$$\int (x^2+3)^2\, dx = \int (x^4 + 6x^2 + 9)\, dx = \frac{x^5}{5} + 2x^3 + 9x + C.$$

Check: $\dfrac{d}{dx}\left[\frac{x^5}{5} + 2x^3 + 9x + C\right] = x^4 + 6x^2 + 9 = (x^2+3)^2.$

19. We use the substitution $w = \cos\theta + 5$, $dw = -\sin\theta\, d\theta$.

$$\int \sin\theta(\cos\theta + 5)^7\, d\theta = -\int w^7\, dw = -\frac{1}{8}w^8 + C$$
$$= -\frac{1}{8}(\cos\theta + 5)^8 + C.$$

Check:

$$\frac{d}{d\theta}\left[-\frac{1}{8}(\cos\theta+5)^8 + C\right] = -\frac{1}{8} \cdot 8(\cos\theta+5)^7 \cdot (-\sin\theta)$$
$$= \sin\theta(\cos\theta+5)^7$$

20. We use the substitution $w = \cos 3t$, $dw = -3\sin 3t\, dt$.

$$\int \sqrt{\cos 3t}\, \sin 3t\, dt = -\frac{1}{3}\int \sqrt{w}\, dw$$
$$= -\frac{1}{3} \cdot \frac{2}{3} w^{\frac{3}{2}} + C = -\frac{2}{9}(\cos 3t)^{\frac{3}{2}} + C.$$

Check:

$$\frac{d}{dt}\left[-\frac{2}{9}(\cos 3t)^{\frac{3}{2}} + C\right] = -\frac{2}{9} \cdot \frac{3}{2}(\cos 3t)^{\frac{1}{2}} \cdot (-\sin 3t) \cdot 3$$
$$= \sqrt{\cos 3t}\, \sin 3t.$$

21. We use the substitution $w = -x^2$, $dw = -2x\, dx$.

$$\int x e^{-x^2}\, dx = -\frac{1}{2}\int e^{-x^2}(-2x\, dx) = -\frac{1}{2}\int e^w\, dw$$
$$= -\frac{1}{2}e^w + C = -\frac{1}{2}e^{-x^2} + C.$$

Check: $\frac{d}{dx}(-\frac{1}{2}e^{-x^2} + C) = (-2x)(-\frac{1}{2}e^{-x^2}) = xe^{-x^2}$.

22. We use the substitution $w = \sin\theta$, $dw = \cos\theta\, d\theta$.

$$\int \sin^6\theta \cos\theta\, d\theta = \int w^6\, dw = \frac{w^7}{7} + C = \frac{\sin^7\theta}{7} + C.$$

Check: $\frac{d}{d\theta}\left[\frac{\sin^7\theta}{7} + C\right] = \sin^6\theta \cos\theta$.

23. We use the substitution $w = \sin 5\theta$, $dw = 5\cos 5\theta\, d\theta$.

$$\int \sin^6 5\theta \cos 5\theta\, d\theta = \frac{1}{5}\int w^6\, dw = \frac{1}{5}\left(\frac{w^7}{7}\right) + C = \frac{1}{35}\sin^7 5\theta + C.$$

Check: $\frac{d}{d\theta}(\frac{1}{35}\sin^7 5\theta + C) = \frac{1}{35}[7\sin^6 5\theta](5\cos 5\theta) = \sin^6 5\theta \cos 5\theta$.

Note that we could also use Problem 22 to solve this problem, substituting $w = 5\theta$ and $dw = 5\, d\theta$ to get:

$$\int \sin^6 5\theta \cos 5\theta\, d\theta = \frac{1}{5}\int \sin^6 w \cos w\, dw$$
$$= \frac{1}{5}\left(\frac{\sin^7 w}{7}\right) + C = \frac{1}{35}\sin^7 5\theta + C.$$

24. We use the substitution $w = x^3 + 1$, $dw = 3x^2\, dx$, to get

$$\int x^2 e^{x^3+1}\, dx = \frac{1}{3}\int e^w\, dw = \frac{1}{3}e^w + C = \frac{1}{3}e^{x^3+1} + C.$$

Check: $\frac{d}{dx}\left(\frac{1}{3}e^{x^3+1} + C\right) = \frac{1}{3}e^{x^3+1} \cdot 3x^2 = x^2 e^{x^3+1}$.

25. We use the substitution $w = \sin\alpha$, $dw = \cos\alpha\, d\alpha$.

$$\int \sin^3\alpha \cos\alpha\, d\alpha = \int w^3\, dw = \frac{w^4}{4} + C = \frac{\sin^4\alpha}{4} + C.$$

Check: $\frac{d}{d\alpha}\left(\frac{\sin^4\alpha}{4} + C\right) = \frac{1}{4} \cdot 4\sin^3\alpha \cdot \cos\alpha = \sin^3\alpha \cos\alpha$.

26. We use the substitution $w = \ln z$, $dw = \frac{1}{z} dz$.

$$\int \frac{(\ln z)^2}{z} dz = \int w^2 \, dw = \frac{w^3}{3} + C = \frac{(\ln z)^3}{3} + C.$$

Check: $\dfrac{d}{dz}\left[\dfrac{(\ln z)^3}{3} + C\right] = 3 \cdot \dfrac{1}{3}(\ln z)^2 \cdot \dfrac{1}{z} = \dfrac{(\ln z)^2}{z}.$

27. We use the substitution $w = e^t + t$, $dw = (e^t + 1) dt$.

$$\int \frac{e^t + 1}{e^t + t} dt = \int \frac{1}{w} dw = \ln|w| + C = \ln|e^t + t| + C.$$

Check: $\dfrac{d}{dt}(\ln|e^t + t| + C) = \dfrac{e^t + 1}{e^t + t}.$

28. We use the substitution $w = y^2 + 4$, $dw = 2y \, dy$.

$$\int \frac{y}{y^2 + 4} dy = \frac{1}{2} \int \frac{dw}{w} = \frac{1}{2} \ln|w| + C = \frac{1}{2} \ln(y^2 + 4) + C.$$

(We can drop the absolute value signs since $y^2 + 4 \geq 0$ for all y.)

Check: $\dfrac{d}{dy}\left[\dfrac{1}{2} \ln(y^2 + 4) + C\right] = \dfrac{1}{2} \cdot \dfrac{1}{y^2 + 4} \cdot 2y = \dfrac{y}{y^2 + 4}.$

29. We use the substitution $w = \cos 2x$, $dw = -2 \sin 2x \, dx$.

$$\int \tan 2x \, dx = \int \frac{\sin 2x}{\cos 2x} dx = -\frac{1}{2} \int \frac{dw}{w}$$

$$= -\frac{1}{2} \ln|w| + C = -\frac{1}{2} \ln|\cos 2x| + C.$$

Check:

$$\frac{d}{dx}\left[-\frac{1}{2} \ln|\cos 2x| + C\right] = -\frac{1}{2} \cdot \frac{1}{\cos 2x} \cdot -2\sin 2x$$

$$= \frac{\sin 2x}{\cos 2x} = \tan 2x.$$

30. We use the substitution $w = \sqrt{x}$, $dw = \frac{1}{2\sqrt{x}} dx$.

$$\int \frac{\cos \sqrt{x}}{\sqrt{x}} dx = \int \cos w (2 \, dw) = 2 \sin w + C = 2 \sin \sqrt{x} + C.$$

Check: $\dfrac{d}{dx}(2 \sin \sqrt{x} + C) = 2 \cos \sqrt{x} \left(\dfrac{1}{2\sqrt{x}}\right) = \dfrac{\cos \sqrt{x}}{\sqrt{x}}.$

31. We use the substitution $w = \sqrt{y}$, $dw = \frac{1}{2\sqrt{y}} dy$.

$$\int \frac{e^{\sqrt{y}}}{\sqrt{y}} dy = 2 \int e^w \, dw = 2e^w + C = 2e^{\sqrt{y}} + C.$$

Check: $\dfrac{d}{dy}(2e^{\sqrt{y}} + C) = 2e^{\sqrt{y}} \cdot \dfrac{1}{2\sqrt{y}} = \dfrac{e^{\sqrt{y}}}{\sqrt{y}}.$

32. We use the substitution $w = x + e^x$, $dw = (1 + e^x) dx$.

$$\int \frac{1 + e^x}{\sqrt{x + e^x}} dx = \int \frac{dw}{\sqrt{w}} = 2\sqrt{w} + C = 2\sqrt{x + e^x} + C.$$

Check: $\dfrac{d}{dx}(2\sqrt{x + e^x} + C) = 2 \cdot \dfrac{1}{2}(x + e^x)^{-\frac{1}{2}} \cdot (1 + e^x) = \dfrac{1 + e^x}{\sqrt{x + e^x}}.$

33. We use the substitution $w = 2 + e^x$, $dw = e^x\, dx$.

$$\int \frac{e^x}{2 + e^x}\, dx = \int \frac{dw}{w} = \ln|w| + C = \ln(2 + e^x) + C.$$

(We can drop the absolute value signs since $2 + e^x \geq 0$ for all x.)

Check: $\dfrac{d}{dx}[\ln(2 + e^x) + C] = \dfrac{1}{2 + e^x} \cdot e^x = \dfrac{e^x}{2 + e^x}.$

34. We use the substitution $w = x^2 + 2x + 19$, $dw = 2(x+1)dx$.

$$\int \frac{(x+1)dx}{x^2 + 2x + 19} = \frac{1}{2}\int \frac{dw}{w} = \frac{1}{2}\ln|w| + C = \frac{1}{2}\ln(x^2 + 2x + 19) + C.$$

(We can drop the absolute value signs, since $x^2 + 2x + 19 = (x+1)^2 + 18 > 0$ for all x.)

Check: $\dfrac{d}{dx}\left[\dfrac{1}{2}\ln(x^2 + 2x + 19)\right] = \dfrac{1}{2}\dfrac{1}{x^2 + 2x + 19}(2x + 2) = \dfrac{x+1}{x^2 + 2x + 19}.$

35. In this case, it seems easier not to substitute.

$$\int y^2(1+y)^2\, dy = \int y^2(y^2 + 2y + 1)\, dy = \int (y^4 + 2y^3 + y^2)\, dy$$
$$= \frac{y^5}{5} + \frac{y^4}{2} + \frac{y^3}{3} + C.$$

Check: $\dfrac{d}{dy}\left(\dfrac{y^5}{5} + \dfrac{y^4}{2} + \dfrac{y^3}{3} + C\right) = y^4 + 2y^3 + y^2 = y^2(y+1)^2.$

36. We use the substitution $w = 1 + 2x^3$, $dw = 6x^2\, dx$.

$$\int x^2(1 + 2x^3)^2\, dx = \int w^2\left(\frac{1}{6}dw\right) = \frac{1}{6}\left(\frac{w^3}{3}\right) + C = \frac{1}{18}(1 + 2x^3)^3 + C.$$

Check: $\dfrac{d}{dx}\left[\dfrac{1}{18}(1 + 2x^2)^3 + C\right] = \dfrac{1}{18}[3(1 + 2x^3)^2(6x^2)] = x^2(1 + 2x^3)^2.$

37. We use the substitution $w = 1 + 3t^2$, $dw = 6t\, dt$.

$$\int \frac{t}{1 + 3t^2}\, dt = \int \frac{1}{w}\left(\frac{1}{6}dw\right) = \frac{1}{6}\ln|w| + C = \frac{1}{6}\ln(1 + 3t^2) + C.$$

(We can drop the absolute value signs since $1 + 3t^2 > 0$ for all t.)

Check: $\dfrac{d}{dt}\left[\dfrac{1}{6}\ln(1 + 3t^2) + C\right] = \dfrac{1}{6}\dfrac{1}{1 + 3t^2}(6t) = \dfrac{t}{1 + 3t^2}.$

38. We use the substitution $w = e^x + e^{-x}$, $dw = (e^x - e^{-x})\, dx$.

$$\int \frac{e^x - e^{-x}}{e^x + e^{-x}}\, dx = \int \frac{dw}{w} = \ln|w| + C = \ln(e^x + e^{-x}) + C.$$

(We can drop the absolute value signs since $e^x + e^{-x} > 0$ for all x).

Check: $\dfrac{d}{dx}[\ln(e^x + e^{-x}) + C] = \dfrac{1}{e^x + e^{-x}}(e^x - e^{-x}).$

39. It seems easier not to substitute.

$$\int \frac{(t+1)^2}{t^2}\, dt = \int \frac{(t^2 + 2t + 1)}{t^2}\, dt$$
$$= \int \left(1 + \frac{2}{t} + \frac{1}{t^2}\right) dt = t + 2\ln|t| - \frac{1}{t} + C.$$

Check: $\dfrac{d}{dt}\left(t + 2\ln|t| - \dfrac{1}{t} + C\right) = 1 + \dfrac{2}{t} + \dfrac{1}{t^2} = \dfrac{(t+1)^2}{t^2}.$

40. We use the substitution $w = \sin(x^2)$, $dw = 2x\cos(x^2)\,dx$.

$$\int \frac{x\cos(x^2)}{\sqrt{\sin(x^2)}}\,dx = \frac{1}{2}\int w^{-\frac{1}{2}}\,dw = \frac{1}{2}(2w^{\frac{1}{2}}) + C = \sqrt{\sin(x^2)} + C.$$

Check: $\dfrac{d}{dx}(\sqrt{\sin(x^2)} + C) = \dfrac{1}{2\sqrt{\sin(x^2)}}[\cos(x^2)]2x = \dfrac{x\cos(x^2)}{\sqrt{\sin(x^2)}}.$

41. We use the substitution $w = x^2 + x$, $dw = (2x+1)\,dx$.

$$\int (2x+1)e^{x^2}e^x\,dx = \int (2x+1)e^{x^2+x}\,dx = \int e^w\,dw$$
$$= e^w + C = e^{x^2+x} + C.$$

Check: $\dfrac{d}{dx}(e^{x^2+x} + C) = e^{x^2+x}\cdot(2x+1) = (2x+1)e^{x^2}e^x.$

42. We use the Pythagorean Identity to change the integrand in the following manner:

$$\sin^3 x = (\sin^2 x)\sin x = (1 - \cos^2 x)\sin x = \sin x - \cos^2 x \sin x.$$

Thus, we have

$$\int \sin^3 x\,dx = \int (\sin x - \cos^2 x \sin x)\,dx$$
$$= \int \sin x\,dx - \int \cos^2 x \sin x\,dx.$$

The first of these new integrals can be easily found. The second can be found using the substitution $w = \cos x$ so $dw = -\sin x\,dx$. The second integral becomes

$$\int \cos^2 x \sin x\,dx = -\int w^2\,dw$$
$$= -\frac{1}{3}w^3 + C$$
$$= -\frac{1}{3}\cos^3 x + C$$

and so our final answer is

$$\int \sin^3 x\,dx = \int \sin x\,dx - \int \cos^2 x \sin x\,dx$$
$$= -\cos x + (1/3)\cos^3 x + C.$$

43. (a) $\int 4x(x^2+1)\,dx = \int (4x^3 + 4x)\,dx = x^4 + 2x^2 + C.$

(b) If $w = x^2 + 1$, then $dw = 2x\,dx$.

$$\int 4x(x^2+1)\,dx = \int 2w\,dw = w^2 + C = (x^2+1)^2 + C.$$

(c) The expressions from parts (a) and (b) look different, but they are both correct. Note that $(x^2+1)^2 + C = x^4 + 2x^2 + 1 + C$. In other words, the expressions from parts (a) and (b) differ only by a constant, so they are both correct antiderivatives.

44. (a) $E(t) = 1.4e^{0.07t}$

(b)
$$\text{Average Yearly Consumption} = \frac{\text{Total Consumption for the Century}}{100 \text{ years}}$$
$$= \frac{1}{100}\int_0^{100} 1.4e^{0.07t}\, dt$$
$$= (0.014)\left[\frac{1}{0.07}e^{0.07t}\Big|_0^{100}\right]$$
$$= (0.014)\left[\frac{1}{0.07}(e^7 - e^0)\right]$$
$$= 0.2(e^7 - 1) \approx 219 \text{ million megawatt-hours.}$$

(c) We are looking for t such that $E(t) \approx 219$:
$$1.4e^{0.07t} \approx 219$$
$$e^{0.07t} = 156.4.$$

Taking natural logs,
$$0.07t = \ln 156.4$$
$$t \approx \frac{5.05}{0.07} \approx 72.18.$$

Thus, consumption was closest to the average during 1972.

(d) Between the years 1900 and 2000 the graph of $E(t)$ looks like

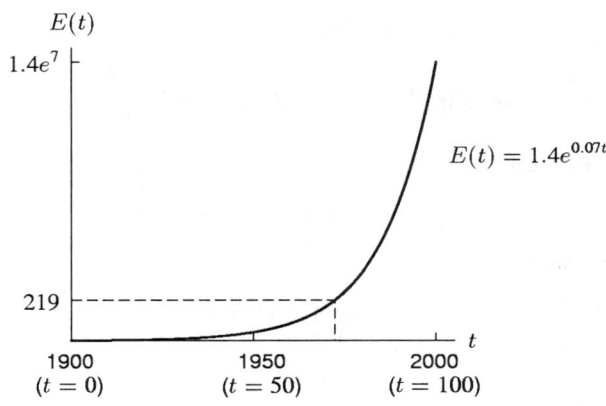

From the graph, we can see the t value such that $E(t) = 219$. It lies to the right of $t = 50$, and is thus in the second half of the century.

45. Since $v = \dfrac{dh}{dt}$, it follows that $h(t) = \displaystyle\int v(t)\, dt$ and $h(0) = h_0$. Since
$$v(t) = \frac{mg}{k}\left(1 - e^{-\frac{k}{m}t}\right) = \frac{mg}{k} - \frac{mg}{k}e^{-\frac{k}{m}t},$$
we have
$$h(t) = \int v(t)\, dt = \frac{mg}{k}\int dt - \frac{mg}{k}\int e^{-\frac{k}{m}t}\, dt.$$

The first integral is simply $\dfrac{mg}{k}t + C$. To evaluate the second integral, make the substitution $w = -\dfrac{k}{m}t$. Then
$$dw = -\frac{k}{m}\, dt,$$

so

$$\int e^{-\frac{k}{m}t}\,dt = \int e^w \left(-\frac{m}{k}\right) dw = -\frac{m}{k}e^w + C = -\frac{m}{k}e^{-\frac{k}{m}t} + C.$$

Thus

$$h(t) = \int v\,dt = \frac{mg}{k}t - \frac{mg}{k}\left(-\frac{m}{k}e^{-\frac{k}{m}t}\right) + C$$

$$= \frac{mg}{k}t + \frac{m^2 g}{k^2}e^{-\frac{k}{m}t} + C.$$

Since $h(0) = h_0$,

$$h_0 = \frac{mg}{k}\cdot 0 + \frac{m^2 g}{k^2}e^0 + C;$$

$$C = h_0 - \frac{m^2 g}{k^2}.$$

Thus

$$h(t) = \frac{mg}{k}t + \frac{m^2 g}{k^2}e^{-\frac{k}{m}t} - \frac{m^2 g}{k^2} + h_0$$

$$h(t) = \frac{mg}{k}t - \frac{m^2 g}{k^2}\left(1 - e^{-\frac{k}{m}t}\right) + h_0.$$

46. (a) In the first case, we are given that $R_0 = 1000$ widgets/year. So we have $R = 1000 e^{0.15t}$. To determine the total number sold, we need to integrate this rate over the time period from 0 to 10. So the total number of widgets sold is

$$\int_0^{10} 1000 e^{0.15t}\,dt = \frac{1000}{0.15}e^{0.15t}\bigg|_0^{10} = 6667(e^{1.5} - 1) \approx 23{,}211 \text{ widgets.}$$

In the second case, the total number of widgets sold is

$$\int_0^{10} 150{,}000{,}000 e^{0.15t}\,dt = 1{,}000{,}000{,}000 e^{0.15t}\bigg|_0^{10} \approx 3.5 \text{ billion widgets.}$$

(b) We want to determine T such that

$$\int_0^T 1000 e^{0.15t}\,dt \approx \frac{23{,}211}{2}.$$

Evaluating both sides, we get

$$6667(e^{0.15T} - 1) = 11{,}606$$
$$6667 e^{0.15T} = 18273$$
$$e^{0.15T} = 2.740$$
$$0.15T = 1.01, \quad \text{so} \quad T = 6.7 \text{ years.}$$

Similarly, in the second case,

$$\int_0^T 150{,}000{,}000 e^{0.15t}\,dt \approx \frac{3{,}500{,}000{,}000}{2}.$$

Evaluating both sides, we get

$$(1 \text{ billion})(e^{0.15T} - 1) = 1.75 \text{ billion}$$
$$e^{0.15T} = 2.75$$
$$T \approx 6.7 \text{ years}$$

So the half way mark is reached at the same time regardless of the initial rate.

(c) Since half the widgets are sold in the last $3\frac{1}{2}$ years of the decade, if each widget is expected to last $3\frac{1}{2}$ years, their claim could easily be true.

Solutions for Section 7.2

1. (a) We substitute $w = 1 + x^2$, $dw = 2x\, dx$.

$$\int_{x=0}^{x=1} \frac{x}{1+x^2}\, dx = \frac{1}{2}\int_{w=1}^{w=2} \frac{1}{w}\, dw = \frac{1}{2}\ln|w|\Big|_1^2 = \frac{1}{2}\ln 2.$$

   (b) We substitute $w = \cos x$, $dw = -\sin x\, dx$.

$$\int_{x=0}^{x=\frac{\pi}{4}} \frac{\sin x}{\cos x}\, dx = -\int_{w=1}^{w=\sqrt{2}/2} \frac{1}{w}\, dw$$

$$= -\ln|w|\Big|_1^{\sqrt{2}/2} = -\ln\frac{\sqrt{2}}{2} = \frac{1}{2}\ln 2.$$

2. $\displaystyle\int_0^\pi \cos(x+\pi)\, dx = \sin(x+\pi)\Big|_0^\pi = \sin(2\pi) - \sin(\pi) = 0 - 0 = 0$

3. We substitute $w = \pi x$. Then $dw = \pi\, dx$.

$$\int_{x=0}^{x=\frac{1}{2}} \cos \pi x\, dx = \int_{w=0}^{w=\pi/2} \cos w \left(\frac{1}{\pi}\, dw\right) = \frac{1}{\pi}(\sin w)\Big|_0^{\pi/2} = \frac{1}{\pi}$$

4. $\displaystyle\int_0^{\pi/2} e^{-\cos\theta} \sin\theta\, d\theta = e^{-\cos\theta}\Big|_0^{\pi/2} = e^{-\cos(\pi/2)} - e^{-\cos(0)} = 1 - \frac{1}{e}$

5. $\displaystyle\int_1^2 2xe^{x^2}\, dx = e^{x^2}\Big|_1^2 = e^{2^2} - e^{1^2} = e^4 - e = e(e^3 - 1)$

6. We substitute $w = \sqrt[3]{x} = x^{\frac{1}{3}}$. Then $dw = \frac{1}{3}x^{-\frac{2}{3}}\, dx = \frac{1}{3\sqrt[3]{x^2}}\, dx$.

$$\int_1^8 \frac{e^{\sqrt[3]{x}}}{\sqrt[3]{x^2}}\, dx = \int_{x=1}^{x=8} e^w (3\, dw) = 3e^w\Big|_{x=1}^{x=8} = 3e^{\sqrt[3]{x}}\Big|_1^8 = 3(e^2 - e).$$

7. We substitute $w = t + 2$, so $dw = dt$.

$$\int_{t=-1}^{t=e-2} \frac{1}{t+2}\, dt = \int_{w=1}^{w=e} \frac{dw}{w} = \ln|w|\Big|_1^e = \ln e - \ln 1 = 1.$$

8. We substitute $w = \sqrt{x}$. Then $dw = \frac{1}{2}x^{-1/2}dx$.

$$\int_{x=1}^{x=4} \frac{\cos\sqrt{x}}{\sqrt{x}}\, dx = \int_{w=1}^{w=2} \cos w (2\, dw)$$

$$= 2(\sin w)\Big|_1^2 = 2(\sin 2 - \sin 1).$$

9. We substitute $w = 1 + x^2$. Then $dw = 2x\, dx$.

$$\int_{x=0}^{x=2} \frac{x}{(1+x^2)^2}\, dx = \int_{w=1}^{w=5} \frac{1}{w^2}\left(\frac{1}{2}\, dw\right) = -\frac{1}{2}\left(\frac{1}{w}\right)\Big|_1^5 = \frac{2}{5}.$$

10. No immediate substitution is apparent, so we must try to modify the integral in some way. First, we put the integral into a more convenient form by using the fact that $\sin^2\theta = 1 - \cos^2\theta$. Thus: $\int_{-\frac{\pi}{4}}^{\frac{\pi}{4}} \cos^2\theta \sin^5\theta \, d\theta = \int_{-\frac{\pi}{4}}^{\frac{\pi}{4}} \cos^2\theta(1-\cos^2\theta)^2 \sin\theta \, d\theta$.

Now, we can make a substitution which helps. We let $w = \cos\theta$, so $dw = -\sin\theta \, d\theta$.

Note that $w = \frac{\sqrt{2}}{2}$ when $\theta = -\frac{\pi}{4}$ and when $\theta = \frac{\pi}{4}$. Thus after our substitution, we get

$$-\int_{w=\frac{\pi}{4}}^{w=\frac{\pi}{4}} w^2(1-w^2)^2 \, dw.$$

Since the upper and lower limits of integration are the same, this definite integral must equal 0. Notice that we could have deduced this fact immediately, since $\cos^2\theta$ is even and $\sin^5\theta$ is odd, so $\cos^2\theta \sin^5\theta$ is odd.

Thus $\int_{-\frac{\pi}{4}}^{0} \cos^2\theta \sin^5\theta \, d\theta = -\int_{0}^{\frac{\pi}{4}} \cos^2\theta \sin^5\theta \, d\theta$, and the given integral must evaluate to 0.

11. $$\int_{-1}^{3}(x^3 + 5x)\,dx = \left.\frac{x^4}{4}\right|_{-1}^{3} + \left.\frac{5x^2}{2}\right|_{-1}^{3} = 40.$$

12. $\int_{-1}^{1} \frac{1}{1+y^2}\,dy = \tan^{-1} y \Big|_{-1}^{1} = \frac{\pi}{2}.$

13. $\int_{1}^{3} \frac{1}{x}\,dx = \ln x \Big|_{1}^{3} = \ln 3.$

14. $\int_{1}^{3} \frac{dt}{(t+7)^2} = \left.\frac{-1}{t+7}\right|_{1}^{3} = \left(-\frac{1}{10}\right) - \left(-\frac{1}{8}\right) = \frac{1}{40}.$

15. $\int_{-1}^{2} \sqrt{x+2}\,dx = \left.\frac{2}{3}(x+2)^{3/2}\right|_{-1}^{2} = \frac{2}{3}\left[(4)^{3/2} - (1)^{3/2}\right] = \frac{2}{3}(7) = \frac{14}{3}.$

16. It turns out that $\frac{\sin x}{x}$ cannot be integrated using elementary methods. However, the function is decreasing on $[1,2]$. One way to see this is to graph the function on a calculator or computer, as has been done below:

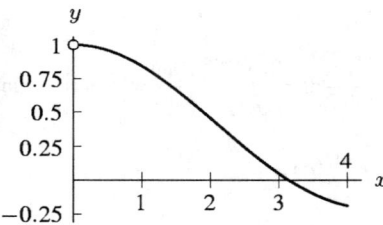

So since our function is monotonic, the error for our left- and right-hand sums is less than or equal to $\left|\frac{\sin 2}{2} - \frac{\sin 1}{1}\right| \Delta t \approx 0.61\Delta t$. So with 13 intervals, our error will be less than 0.05. With $n = 13$, the left sum is about 0.674, and the right sum is about 0.644. For more accurate sums, with $n = 100$ the left sum is about 0.6613 and the right sum is about 0.6574. The actual integral is about 0.6593.

17. Substitute $w = 1 + x^2$, $dw = 2x\,dx$. Then $x\,dx = \frac{1}{2}dw$, and

$$\int_{x=0}^{x=1} x(1+x^2)^{20}\,dx = \frac{1}{2}\int_{w=1}^{w=2} w^{20}\,dw = \left.\frac{w^{21}}{42}\right|_{1}^{2} = \frac{299593}{6} = 49932\frac{1}{6}.$$

18. We substitute $w = \cos\theta + 5$, $dw = -\sin\theta\, d\theta$. Then

$$\int_{\theta=0}^{\theta=\pi} \sin\theta\, d\theta(\cos\theta + 5)^7 = -\int_{w=6}^{w=4} w^7\, dw = \int_{w=4}^{w=6} w^7\, dw = \left.\frac{w^8}{8}\right|_4^6 = 201{,}760.$$

19. Substitute $w = 3\alpha$, $dw = 3\, d\alpha$. Then $d\alpha = \frac{1}{3}\, dw$. We have

$$\int_{\alpha=0}^{\alpha=\frac{\pi}{12}} \sin 3\alpha\, d\alpha = \frac{1}{3}\int_{w=0}^{w=\frac{\pi}{4}} \sin w\, dw$$

$$= -\frac{1}{3}\cos w\Big|_0^{\frac{\pi}{4}}$$

$$= -\frac{1}{3}\left(\frac{\sqrt{2}}{2} - 1\right) = \frac{1}{3}\left(1 - \frac{\sqrt{2}}{2}\right).$$

20. Let $w = 1 + 5x^2$. We have $dw = 10x\, dx$, so $\frac{dw}{10} = x\, dx$. When $x = 0$, $w = 1$. When $x = 1$, $w = 6$.

$$\int \frac{x\, dx}{1 + 5x^2} = \int_1^6 \frac{\frac{1}{10}\, dw}{w} = \frac{1}{10}\int_1^6 \frac{dw}{w} = \frac{1}{10}\ln|w|\Big|_1^6$$

$$= \frac{1}{10}(\ln 6 - \ln 1) = \frac{\ln 6}{10}$$

21.
$$\int_1^2 \frac{x^2 + 1}{x}\, dx = \int_1^2 \left(x + \frac{1}{x}\right) dx = \left(\frac{x^2}{2} + \ln|x|\right)\Big|_1^2 = \frac{3}{2} + \ln 2.$$

22. $\int_0^1 \frac{1}{x^2 + 2x + 1}\, dx = \int_0^1 \frac{1}{(x+1)^2}\, dx.$
We substitute $w = x + 1$, so $dw = dx$. Note that when $x = 1$, we have $w = 2$, and when $x = 0$, we have $w = 1$.

$$\int_{x=0}^{x=1} \frac{1}{(x+1)^2}\, dx = \int_{w=1}^{w=2} \frac{1}{w^2}\, dw = -\frac{1}{w}\Big|_{w=1}^{w=2} = -\frac{1}{2} + 1 = \frac{1}{2}.$$

23. Let $w = x + 2$, giving $dw = dx$. When $x = 0$, $w = 2$, and when $x = 1$, $w = 3$. Thus,

$$\int_0^1 \frac{(x+2)}{(x+2)^2 + 1}\, dx = \int_2^3 \frac{w}{w^2 + 1}\, dw.$$

For the last integral, we make the substitution $u = w^2 + 1$, $du = 2w\, dw$. Then, we have

$$\int_2^3 \frac{w}{w^2+1}\, dw = \frac{1}{2}\ln|w^2 + 1|\Big|_2^3$$

$$= \frac{1}{2}(\ln|10| - \ln|5|)$$

$$= \frac{1}{2}\ln\left(\frac{10}{5}\right) = \frac{1}{2}\ln(2)$$

24. Substitute $w = x^2 + 4$, $dw = 2x\, dx$. Then,

$$\int_{x=4}^{x=1} x\sqrt{x^2+4}\, dx = \frac{1}{2}\int_{w=20}^{w=5} w^{\frac{1}{2}}\, dw = \frac{1}{3}w^{\frac{3}{2}}\Big|_{20}^5$$

$$= \frac{1}{3}\left(5^{\frac{3}{2}} - 8 \cdot 5^{\frac{3}{2}}\right) = -\frac{7}{3} \cdot 5^{3/2} = -\frac{35}{3}\sqrt{5}$$

25. Let $w = x^2$, $dw = 2x\,dx$. When $x = 0$, $w = 0$, and when $x = \frac{1}{\sqrt{2}}$, $w = \frac{1}{2}$. Then

$$\int_0^{\frac{1}{\sqrt{2}}} \frac{x\,dx}{\sqrt{1-x^4}} = \int_0^{\frac{1}{2}} \frac{\frac{1}{2}dw}{\sqrt{1-w^2}} = \frac{1}{2}\arcsin w \Big|_0^{\frac{1}{2}} = \frac{1}{2}(\arcsin\frac{1}{2} - \arcsin 0) = \frac{\pi}{12}.$$

26. $f(t) = \sin\frac{1}{t}$ has no elementary antiderivative, so we will have to use left and right sums. With $n = 100$, left sum $= 0.5462$ and right sum $= 0.5582$. However, since f is not monotonic on $[1/4, 1]$ (see figure), we cannot be sure that the integral is between these values.

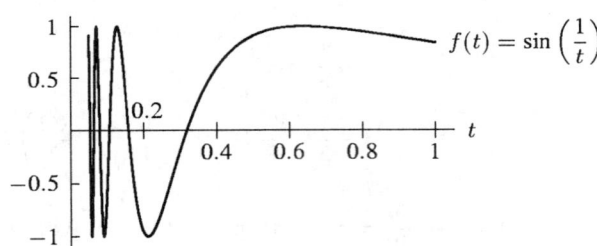

To get an upper and lower bound for this integral, divide the interval $[1/4, 1]$ into subintervals in such a way that f is monotonic on each one. Since $f'(t) = -\frac{1}{t^2}\cos\frac{1}{t} = 0$ when $\frac{1}{t} = \frac{\pi}{2}$ or $t = \frac{2}{\pi}$ (and this is the only point in $[1/4, 1]$ where $f'(t) = 0$), we will write

$$\int_{1/4}^1 \sin\frac{1}{t}\,dt = \int_{1/4}^{2/\pi} \sin\frac{1}{t}\,dt + \int_{2/\pi}^1 \sin\frac{1}{t}\,dt.$$

Now, using $n = 100$, we find $0.209 < \int_{1/4}^{2/\pi} \sin\frac{1}{t}\,dt < 0.216$ (left sum $= 0.209$, right sum $= 0.216$, f is increasing), and $0.339 < \int_{2/\pi}^1 \sin\frac{1}{t}\,dt < 0.340$ (here left sum $= 0.340$, right sum $= 0.339$ because f is decreasing). Thus

$$0.548 < \int_{1/4}^1 \sin\frac{1}{t}\,dt < 0.556.$$

The actual integral is approximately 0.5523.

27. We substitute $w = x^2 + 4x + 5$, so $dw = (2x + 4)\,dx$. Notice that when $x = -2$, $w = 1$, and when $x = 0$, $w = 5$.

$$\int_{x=-2}^{x=0} \frac{2x+4}{x^2+4x+5}\,dx = \int_{w=1}^{w=5} \frac{1}{w}\,dw = \ln|w|\Big|_{w=1}^{w=5} = \ln 5.$$

28. After the substitution $w = e^x$ and $dw = e^x\,dx$, the first integral becomes

$$\frac{1}{2}\int \frac{1}{1+w^2}\,dw,$$

while after the substitution $w = \sin x$ and $dw = \cos x\,dx$, the second integral becomes

$$\int \frac{1}{1+w^2}\,dw.$$

29. The substitution $w = \ln x$, $dw = \frac{1}{x}\,dx$ transforms the first integral into $\int w\,dw$, which is just a respelling of the integral $\int x\,dx$.

30. For the first integral, let $w = \sin x$, $dw = \cos x\,dx$. Then

$$\int e^{\sin x}\cos x\,dx = \int e^w\,dw.$$

For the second integral, let $w = \arcsin x$, $dw = \frac{1}{\sqrt{1-x^2}}dx$. Then

$$\int \frac{e^{\arcsin x}}{\sqrt{1-x^2}}\,dx = \int e^w\,dw.$$

31. For the first integral, let $w = x + 1, dw = dx$. Then

$$\int \sqrt{x+1}\, dx = \int \sqrt{w}\, dw.$$

For the second integral, let $w = 1 + \sqrt{x}, dw = \frac{1}{2}x^{-\frac{1}{2}}\, dx = \frac{1}{2\sqrt{x}}\, dx$. Then, $\frac{dx}{\sqrt{x}} = 2\, dw$, and

$$\int \frac{\sqrt{1+\sqrt{x}}}{\sqrt{x}}\, dx = \int \sqrt{1+\sqrt{x}} \left(\frac{dx}{\sqrt{x}}\right) = 2\int \sqrt{w}\, dw.$$

32. The substitutions $w = \sin x, dw = \cos w$ and $w = x^3 + 1, dw = 3x^2\, dx$ transform the integrals into

$$\int w^3\, dw \quad \text{and} \quad \frac{1}{3}\int w^3\, dw.$$

33.
$$\int \frac{dx}{x^2 + 4x + 5} = \int \frac{dx}{(x+2)^2 + 1}.$$

We make the substitution $\tan \theta = x + 2$. Then $dx = \frac{1}{\cos^2 \theta}\, d\theta$.

$$\int \frac{dx}{(x+2)^2 + 1} = \int \frac{d\theta}{\cos^2 \theta (\tan^2 \theta + 1)}$$

$$= \int \frac{d\theta}{\cos^2 \theta \left(\frac{\sin^2 \theta}{\cos^2 \theta} + 1\right)}$$

$$= \int \frac{d\theta}{\sin^2 \theta + \cos^2 \theta}$$

$$= \int d\theta = \theta + C$$

But since $\tan \theta = x + 2$, $\theta = \arctan(x + 2)$, and so $\theta + C = \arctan(x + 2) + C$.

34. (a) $2\sqrt{x} + C$
    (b) $2\sqrt{x+1} + C$
    (c) To get this last result, we make the substitution $w = \sqrt{x}$. Normally we would like to substitute $dw = \frac{1}{2\sqrt{x}}\, dx$, but in this case we cannot since there are no spare $\frac{1}{\sqrt{x}}$ terms around. Instead, we note $w^2 = x$, so $2w\, dw = dx$. Then

$$\int \frac{1}{\sqrt{x}+1}\, dx = \int \frac{2w}{w+1}\, dw$$

$$= 2\int \frac{(w+1) - 1}{w+1}\, dw$$

$$= 2\int \left(1 - \frac{1}{w+1}\right) dw$$

$$= 2(w - \ln|w+1|) + C$$

$$= 2\sqrt{x} - 2\ln(\sqrt{x} + 1) + C.$$

We also note that we can drop the absolute value signs, since $\sqrt{x} + 1 \geq 0$ for all x.

35. To find the area under the graph of $f(x) = xe^{x^2}$, we need to evaluate the definite integral

$$\int_0^2 xe^{x^2}\, dx.$$

This is done in Example 1, Section 7.2, using the substitution $w = x^2$, the result being

$$\int_0^2 xe^{x^2}\, dx = \frac{1}{2}(e^4 - 1).$$

36.

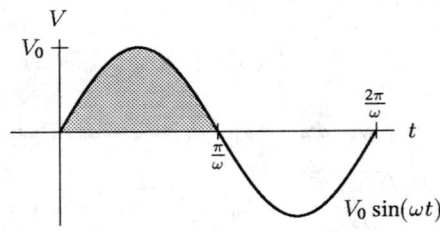

The period of $V = V_0 \sin(\omega t)$ is $2\pi/\omega$, so the area under the first arch is given by

$$\text{Area} = \int_0^{\pi/\omega} V_0 \sin(\omega t)\, dt$$

$$= -\frac{V_0}{\omega} \cos(\omega t)\Big|_0^{\pi/\omega}$$

$$= -\frac{V_0}{\omega} \cos(\pi) + \frac{V_0}{\omega} \cos(0)$$

$$= -\frac{V_0}{\omega}(-1) + \frac{V_0}{\omega}(1) = \frac{2V_0}{\omega}.$$

37. (a) The Fundamental Theorem gives

$$\int_{-\pi}^{\pi} \cos^2\theta \sin\theta\, d\theta = -\frac{\cos^3\theta}{3}\Big|_{-\pi}^{\pi} = \frac{-(-1)^3}{3} - \frac{-(-1)^3}{3} = 0.$$

This agrees with the fact that the function $f(\theta) = \cos^2\theta \sin\theta$ is odd and the interval of integration is centered at $x = 0$, thus we must get 0 for the definite integral.

(b) The area is given by

$$\text{Area} = \int_0^{\pi} \cos^2\theta \sin\theta\, d\theta = -\frac{\cos^3\theta}{3}\Big|_0^{\pi} = \frac{-(-1)^3}{3} - \frac{-(1)^3}{3} = \frac{2}{3}.$$

38. (a) If $w = 2t$, then $dw = 2dt$. When $t = 0$, $w = 0$; when $t = 0.5$, $w = 1$. Thus,

$$\int_0^{0.5} f(2t)\, dt = \int_0^1 f(w)\, \frac{1}{2} dw = \frac{1}{2}\int_0^1 f(w)\, dw = \frac{3}{2}.$$

(b) If $w = 1 - t$, then $dw = -dt$. When $t = 0$, $w = 1$; when $t = 1$, $w = 0$. Thus,

$$\int_0^1 f(1-t)\, dt = \int_1^0 f(w)(-dw) = +\int_0^1 f(w)\, dw = 3.$$

(c) If $w = 3 - 2t$, then $dw = -2dt$. When $t = 1$, $w = 1$; when $t = 1.5$, $w = 0$. Thus,

$$\int_1^{1.5} f(3-2t)\, dt = \int_1^0 f(w)\left(-\frac{1}{2}dw\right) = +\frac{1}{2}\int_0^1 f(w)\, dw = \frac{3}{2}.$$

39. If $f(x) = \dfrac{1}{x+1}$, the average value of f on the interval $0 \le x \le 2$ is defined to be

$$\frac{1}{2-0}\int_0^2 f(x)\, dx = \frac{1}{2}\int_0^2 \frac{dx}{x+1}.$$

We'll integrate by substitution. We let $w = x + 1$ and $dw = dx$, and we have

$$\int_{x=0}^{x=2} \frac{dx}{x+1} = \int_{w=1}^{w=3} \frac{dw}{w} = \ln w\Big|_1^3 = \ln 3 - \ln 1 = \ln 3.$$

Thus, the average value of $f(x)$ on $0 \leq x \leq 2$ is $\frac{1}{2}\ln 3 \approx 0.5493$. See Figure 7.1.

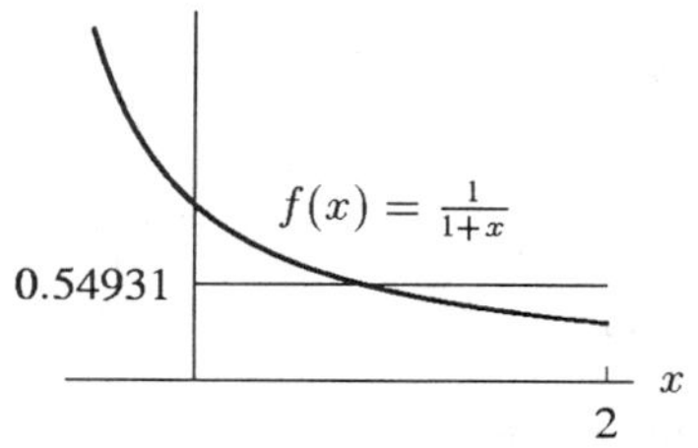

Figure 7.1

40. **(a)** We first try the substitution $w = \sin\theta$, $dw = \cos\theta\, d\theta$. Then
$$\int \sin\theta \cos\theta\, d\theta = \int w\, dw = \frac{w^2}{2} + C = \frac{\sin^2\theta}{2} + C.$$
(b) If we instead try the substitution $w = \cos\theta$, $dw = -\sin\theta\, d\theta$, we get
$$\int \sin\theta \cos\theta\, d\theta = -\int w\, dw = -\frac{w^2}{2} + C = -\frac{\cos^2\theta}{2} + C.$$
(c) Once we note that $\sin 2\theta = 2\sin\theta\cos\theta$, we can also say
$$\int \sin\theta \cos\theta\, d\theta = \frac{1}{2}\int \sin 2\theta\, d\theta.$$
Substituting $w = 2\theta$, $dw = 2\, d\theta$, the above equals
$$\frac{1}{4}\int \sin w\, dw = -\frac{\cos w}{4} + C = -\frac{\cos 2\theta}{4} + C.$$
(d) All these answers are correct. Although they have different forms, they differ from each other only in terms of a constant, and thus they are all acceptable antiderivatives. For example, $1 - \cos^2\theta = \sin^2\theta$, so $\frac{\sin^2\theta}{2} = -\frac{\cos^2\theta}{2} + \frac{1}{2}$. Thus the first two expressions differ only by a constant C.

Similarly, $\cos 2\theta = \cos^2\theta - \sin^2\theta = 2\cos^2\theta - 1$, so $-\frac{\cos 2\theta}{4} = -\frac{\cos^2\theta}{2} + \frac{1}{4}$, and thus the second and third expressions differ only by a constant. Of course, if the first two expressions and the last two expressions differ only in the constant C, then the first and last only differ in the constant as well.

41. **(a)** For a 40,000-word novel,
$$\text{Payment} = \int_0^{40{,}000} f(w)\, dw$$
$$= \int_0^{2000} \left(\frac{1}{2} + \frac{w}{2000}\right) dw + \int_{2000}^{20{,}000} \frac{3}{2}\, dw + \int_{20{,}000}^{40{,}000} \frac{3}{2} e^{20 - \frac{w}{1000}}\, dw$$
$$= \left(\frac{1}{2}w + \frac{w^2}{4000}\right)\bigg|_0^{2000} + \left(\frac{3}{2}w\right)\bigg|_{2000}^{20{,}000} + \left(\frac{-3000}{2} e^{20 - \frac{w}{1000}}\right)\bigg|_{20{,}000}^{40{,}000}$$
$$= 2000 + 27{,}000 + 1500 - 1500e^{-20}$$
$$\approx 30{,}500 \text{ pennies}.$$

(b) Notice from part (a) that the payment for a 40,000-word novel is 2000 pennies (or a penny a word) for the first 2000 pages, 27,000 pennies (or one and a half pennies a word) for the next 18,000 pages, and it is only 1,500 pennies for the final 20,000 words. Thus, the payment is greater for two 15,000-word novels, since the function discourages long novels.

42. **(a)** Amount of water entering tank in a short period of time = rate × time = $r(t)\Delta t$.

(b)

Amount of water entering the tank between $t = 0$ and $t = 5$ $\approx \sum_{i=0}^{n-1} r(t_i)\Delta t$, where $\Delta t = 5/n$.

Amount of water entering the tank between $t = 0$ and $t = 5$ $= \int_0^5 r(t)\, dt$.

(c) If $Q(t)$ is the amount of water in the tank at time t, then $Q'(t) = r(t)$. We want to calculate $Q(5) - Q(0)$. By the Fundamental Theorem,

$$\text{Amount which has entered tank} = Q(5) - Q(0) = \int_0^5 r(t)\, dt = \int_0^5 20e^{0.02t}\, dt = \left. \frac{20}{0.02} e^{0.02t} \right|_0^5$$

$$= 1000(e^{0.02(5)} - 1) \approx 105.17 \text{ gallons.}$$

(d) By the Fundamental Theorem again,

$$\text{Amount which has entered tank} = Q(t) - Q(0) = \int_0^t r(t)\, dt$$

$$Q(t) - 3000 = \int_0^t 20e^{0.02t}\, dt$$

so

$$Q(t) = 3000 + \int_0^t 20e^{0.02t}\, dt = 3000 + \left. \frac{20}{0.02} e^{0.02t} \right|_0^t$$

$$= 3000 + 1000(e^{0.02t} - 1)$$

$$= 1000e^{0.02t} + 2000.$$

43. Since v is given as the velocity of a falling body, the height h is decreasing, so $v = -\frac{dh}{dt}$, and it follows that $h(t) = -\int v(t)\, dt$ and $h(0) = h_0$. Let $w = e^{t\sqrt{gk}} + e^{-t\sqrt{gk}}$. Then

$$dw = \sqrt{gk}\left(e^{t\sqrt{gk}} - e^{-t\sqrt{gk}}\right) dt,$$

so $\dfrac{dw}{\sqrt{gk}} = \left(e^{t\sqrt{gk}} - e^{-t\sqrt{gk}}\right) dt$. Therefore,

$$-\int v(t)\, dt = -\int \sqrt{\frac{g}{k}} \left(\frac{e^{t\sqrt{gk}} - e^{-t\sqrt{gk}}}{e^{t\sqrt{gk}} + e^{-t\sqrt{gk}}} \right) dt$$

$$= -\sqrt{\frac{g}{k}} \int \frac{1}{e^{t\sqrt{gk}} + e^{-t\sqrt{gk}}} \left(e^{t\sqrt{gk}} - e^{-t\sqrt{gk}}\right) dt$$

$$= -\sqrt{\frac{g}{k}} \int \left(\frac{1}{w}\right) \frac{dw}{\sqrt{gk}}$$

$$= -\sqrt{\frac{g}{gk^2}} \ln|w| + C$$

$$= -\frac{1}{k} \ln\left(e^{t\sqrt{gk}} + e^{-t\sqrt{gk}}\right) + C.$$

Since

$$h(0) = -\frac{1}{k} \ln(e^0 + e^0) + C = -\frac{\ln 2}{k} + C = h_0,$$

we have $C = h_0 + \dfrac{\ln 2}{k}$. Thus,

$$h(t) = -\frac{1}{k} \ln\left(e^{t\sqrt{gk}} + e^{-t\sqrt{gk}}\right) + \frac{\ln 2}{k} + h_0 = -\frac{1}{k} \ln\left(\frac{e^{t\sqrt{gk}} + e^{-t\sqrt{gk}}}{2}\right) + h_0.$$

Solutions for Section 7.3

1. Let $u = \arctan x$, $v' = 1$. Then $v = x$ and $u' = \dfrac{1}{1+x^2}$. Integrating by parts, we get:

$$\int 1 \cdot \arctan x \, dx = x \cdot \arctan x - \int x \cdot \frac{1}{1+x^2} \, dx.$$

To compute the second integral use the substitution, $z = 1 + x^2$.

$$\int \frac{x}{1+x^2} \, dx = \frac{1}{2} \int \frac{dz}{z} = \frac{1}{2} \ln |z| + C = \frac{1}{2} \ln(1+x^2) + C.$$

Thus,

$$\int \arctan x \, dx = x \cdot \arctan x - \frac{1}{2} \ln(1+x^2) + C.$$

2. Let $u = t$ and $v' = e^{5t}$, so $u' = 1$ and $v = \frac{1}{5}e^{5t}$.
   Then $\int te^{5t} \, dt = \frac{1}{5}te^{5t} - \int \frac{1}{5}e^{5t} \, dt = \frac{1}{5}te^{5t} - \frac{1}{25}e^{5t} + C$.

3. Let $u = t^2$ and $v' = e^{5t}$, so $u' = 2t$ and $v = \frac{1}{5}e^{5t}$.
   Then $\int t^2 e^{5t} \, dt = \frac{1}{5}t^2 e^{5t} - \frac{2}{5} \int te^{5t} \, dt$.
   Using Problem 2, we have $\int t^2 e^{5t} \, dt = \frac{1}{5}t^2 e^{5t} - \frac{2}{5}(\frac{1}{5}te^{5t} - \frac{1}{25}e^{5t}) + C$
   $= \frac{1}{5}t^2 e^{5t} - \frac{2}{25}te^{5t} + \frac{2}{125}e^{5t} + C$.

4. Let $u = p$ and $v' = e^{(-0.1)p}$, $u' = 1$. Thus, $v = \int e^{(-0.1)p} \, dp = -10e^{(-0.1)p}$. With this choice of u and v, integration by parts gives:

$$\int pe^{(-0.1)p} \, dp = p(-10e^{(-0.1)p}) - \int (-10e^{(-0.1)p}) \, dp$$
$$= -10pe^{(-0.1)p} + 10 \int e^{(-0.1)p} \, dp$$
$$= -10pe^{(-0.1)p} - 100e^{(-0.1)p} + C.$$

5. Let $u = t$, $v' = \sin t$. Thus, $v = -\cos t$ and $u' = 1$. With this choice of u and v, integration by parts gives:

$$\int t \sin t \, dt = -t \cos t - \int (-\cos t) \, dt$$
$$= -t \cos t + \sin t + C.$$

6. Let $u = \ln y$, $v' = y$. Then, $v = \frac{1}{2}y^2$ and $u' = \dfrac{1}{y}$. Integrating by parts, we get:

$$\int y \ln y \, dy = \frac{1}{2}y^2 \ln y - \int \frac{1}{2}y^2 \cdot \frac{1}{y} \, dy$$
$$= \frac{1}{2}y^2 \ln y - \frac{1}{2} \int y \, dy$$
$$= \frac{1}{2}y^2 \ln y - \frac{1}{4}y^2 + C.$$

7. Let $u = \ln x$ and $v' = x^3$, so $u' = \frac{1}{x}$ and $v = \frac{x^4}{4}$.
   Then

$$\int x^3 \ln x \, dx = \frac{x^4}{4} \ln x - \int \frac{x^3}{4} \, dx = \frac{x^4}{4} \ln x - \frac{x^4}{16} + C.$$

8. Let $u = z+1$, $v' = e^{2z}$. Thus, $v = \frac{1}{2}e^{2z}$ and $u' = 1$. Integrating by parts, we get:

$$\int (z+1)e^{2z}\, dz = (z+1) \cdot \frac{1}{2}e^{2z} - \int \frac{1}{2}e^{2z}\, dz$$
$$= \frac{1}{2}(z+1)e^{2z} - \frac{1}{4}e^{2z} + C$$
$$= \frac{1}{4}(2z+1)e^{2z} + C.$$

9. Let $u = z$, $v' = e^{-z}$. Thus $v = -e^{-z}$ and $u' = 1$. Integration by parts gives:

$$\int ze^{-z}\, dz = -ze^{-z} - \int (-e^{-z})\, dz$$
$$= -ze^{-z} - e^{-z} + C$$
$$= -(z+1)e^{-z} + C.$$

10. Let $u = t^2$, $v' = \sin t$ implying $v = -\cos t$ and $u' = 2t$. Integrating by parts, we get:

$$\int t^2 \sin t\, dt = -t^2 \cos t - \int 2t(-\cos t)\, dt.$$

Again, applying integration by parts with $u = t$, $v' = \cos t$, we have:

$$\int t \cos t\, dt = t \sin t + \cos t + C.$$

Thus

$$\int t^2 \sin t\, dt = -t^2 \cos t + 2t \sin t + 2 \cos t + C.$$

11. Let $u = \theta^2$ and $v' = \cos 3\theta$, so $u' = 2\theta$ and $v = \frac{1}{3}\sin 3\theta$.

Then $\int \theta^2 \cos 3\theta\, d\theta = \frac{1}{3}\theta^2 \sin 3\theta - \frac{2}{3} \int \theta \sin 3\theta\, d\theta$. The integral on the right hand side is simpler than our original integral, but to evaluate it we need to again use integration by parts.

To find $\int \theta \sin 3\theta\, d\theta$, let $u = \theta$ and $v' = \sin 3\theta$, so $u' = 1$ and $v = -\frac{1}{3}\cos 3\theta$.

This gives

$$\int \theta \sin 3\theta\, d\theta = -\frac{1}{3}\theta \cos 3\theta + \frac{1}{3}\int \cos 3\theta\, d\theta = -\frac{1}{3}\theta \cos 3\theta + \frac{1}{9}\sin 3\theta + C.$$

Thus,

$$\int \theta^2 \cos 3\theta\, d\theta = \frac{1}{3}\theta^2 \sin 3\theta + \frac{2}{9}\theta \cos 3\theta - \frac{2}{27}\sin 3\theta + C.$$

12. Let $u = \sin \theta$ and $v' = \sin \theta$, so $u' = \cos \theta$ and $v = -\cos \theta$. Then

$$\int \sin^2 \theta\, d\theta = -\sin \theta \cos \theta + \int \cos^2 \theta\, d\theta$$
$$= -\sin \theta \cos \theta + \int (1 - \sin^2 \theta)\, d\theta$$
$$= -\sin \theta \cos \theta + \int 1\, d\theta - \int \sin^2 \theta\, d\theta.$$

By adding $\int \sin^2 \theta\, d\theta$ to both sides of the above equation, we find that $2\int \sin^2 \theta\, d\theta = -\sin \theta \cos \theta + \theta + C$, so $\int \sin^2 \theta\, d\theta = -\frac{1}{2}\sin \theta \cos \theta + \frac{\theta}{2} + C'$.

13. Let $u = \theta + 1$ and $v' = \sin(\theta + 1)$, so $u' = 1$ and $v = -\cos(\theta + 1)$.

$$\int (\theta + 1)\sin(\theta + 1)\, d\theta = -(\theta + 1)\cos(\theta + 1) + \int \cos(\theta + 1)\, d\theta$$
$$= -(\theta + 1)\cos(\theta + 1) + \sin(\theta + 1) + C.$$

14. Let $u = \cos(3\alpha + 1)$ and $v' = \cos(3\alpha + 1)$, so $u' = -3\sin(3\alpha + 1)$, and $v = \frac{1}{3}\sin(3\alpha + 1)$. Then

$$\int \cos^2(3\alpha + 1)\, d\alpha = \int \left(\cos(3\alpha + 1)\right)\cos(3\alpha + 1)\, d\alpha$$
$$= \frac{1}{3}\cos(3\alpha + 1)\sin(3\alpha + 1) + \int \sin^2(3\alpha + 1)\, d\alpha$$
$$= \frac{1}{3}\cos(3\alpha + 1)\sin(3\alpha + 1) + \int \left(1 - \cos^2(3\alpha + 1)\right) d\alpha$$
$$= \frac{1}{3}\cos(3\alpha + 1)\sin(3\alpha + 1) + \alpha - \int \cos^2(3\alpha + 1)\, d\alpha.$$

By adding $\int \cos^2(3\alpha + 1)\, d\alpha$ to both sides of the above equation, we find that

$$2\int \cos^2(3\alpha + 1)\, d\alpha = \frac{1}{3}\cos(3\alpha + 1)\sin(3\alpha + 1) + \alpha + C,$$

which gives

$$\int \cos^2(3\alpha + 1)\, d\alpha = \frac{1}{6}\cos(3\alpha + 1)\sin(3\alpha + 1) + \frac{\alpha}{2} + C.$$

15. Let $u = \ln x$, $v' = x^{-2}$. Then $v = -x^{-1}$ and $u' = x^{-1}$. Integrating by parts, we get:

$$\int x^{-2}\ln x\, dx = -x^{-1}\ln x - \int (-x^{-1})\cdot x^{-1}\, dx$$
$$= -x^{-1}\ln x - x^{-1} + C.$$

16. Let $u = \ln 5q$, $v' = q^5$. Then $v = \frac{1}{6}q^6$ and $u' = \frac{1}{q}$. Integrating by parts, we get:

$$\int q^5 \ln 5q\, dq = \frac{1}{6}q^6 \ln 5q - \int (5\cdot\frac{1}{5q})\cdot\frac{1}{6}q^6\, dq$$
$$= \frac{1}{6}q^6 \ln 5q - \frac{1}{36}q^6 + C.$$

17. Let $u = y$ and $v' = (y + 3)^{1/2}$, so $u' = 1$ and $v = \frac{2}{3}(y + 3)^{3/2}$.
$\int y\sqrt{y + 3}\, dy = \frac{2}{3}y(y + 3)^{3/2} - \int \frac{2}{3}(y + 3)^{3/2}\, dy = \frac{2}{3}y(y + 3)^{3/2} - \frac{4}{15}(y + 3)^{5/2} + C.$

18. Let $u = t + 2$ and $v' = \sqrt{2 + 3t}$, so $u' = 1$ and $v = \frac{2}{9}(2 + 3t)^{3/2}$. Then

$$\int (t + 2)\sqrt{2 + 3t}\, dt = \frac{2}{9}(t + 2)(2 + 3t)^{3/2} - \frac{2}{9}\int (2 + 3t)^{3/2}\, dt$$
$$= \frac{2}{9}(t + 2)(2 + 3t)^{3/2} - \frac{4}{135}(2 + 3t)^{5/2} + C.$$

19. Let $u = y$ and $v' = \frac{1}{\sqrt{5-y}}$, so $u' = 1$ and $v = -2(5 - y)^{1/2}$.

$$\int \frac{y}{\sqrt{5 - y}}\, dy = -2y(5 - y)^{1/2} + 2\int (5 - y)^{1/2}\, dy = -2y(5 - y)^{1/2} - \frac{4}{3}(5 - y)^{3/2} + C.$$

20. $\int \frac{t+7}{\sqrt{5-t}}\, dt = \int \frac{t}{\sqrt{5-t}}\, dt + 7\int (5-t)^{-1/2}\, dt.$

To calculate the first integral, we use integration by parts. Let $u = t$ and $v' = \frac{1}{\sqrt{5-t}}$, so $u' = 1$ and $v = -2(5-t)^{1/2}$. Then

$$\int \frac{t}{\sqrt{5-t}}\, dt = -2t(5-t)^{1/2} + 2\int (5-t)^{1/2}\, dt = -2t(5-t)^{1/2} - \frac{4}{3}(5-t)^{3/2} + C.$$

We can calculate the second integral directly: $7\int (5-t)^{-1/2} = -14(5-t)^{1/2} + C_1$. Thus

$$\int \frac{t+7}{\sqrt{5-t}}\, dt = -2t(5-t)^{1/2} - \frac{4}{3}(5-t)^{3/2} - 14(5-t)^{1/2} + C_2.$$

21. Let $u = (\ln t)^2$ and $v' = 1$, so $u' = \frac{2\ln t}{t}$ and $v = t$. Then

$$\int (\ln t)^2\, dt = t(\ln t)^2 - 2\int \ln t\, dt = t(\ln t)^2 - 2t\ln t + 2t + C.$$

(We use the fact that $\int \ln x\, dx = x\ln x - x + C$, a result which can be derived using integration by parts.)

22. Let $u = (\ln x)^4$ and $v' = x$, so $u' = \frac{4(\ln x)^3}{x}$ and $v = \frac{x^2}{2}$. Then

$$\int x(\ln x)^4\, dx = \frac{x^2(\ln x)^4}{2} - 2\int x(\ln x)^3\, dx.$$

$\int x(\ln x)^3\, dx$ is somewhat less complicated than $\int x(\ln x)^4\, dx$. To calculate it, we again try integration by parts, this time letting $u = (\ln x)^3$ (instead of $(\ln x)^4$) and $v' = x$. We find

$$\int x(\ln x)^3\, dx = \frac{x^2}{2}(\ln x)^3 - \frac{3}{2}\int x(\ln x)^2\, dx.$$

Once again, express the given integral in terms of a less-complicated one. Using integration by parts two more times, we find that

$$\int x(\ln x)^2\, dx = \frac{x^2}{2}(\ln x)^2 - \int x(\ln x)\, dx$$

and that

$$\int x\ln x\, dx = \frac{x^2}{2}\ln x - \frac{x^2}{4} + C.$$

Putting this all together, we have

$$\int x(\ln x)^4\, dx = \frac{x^2}{2}(\ln x)^4 - x^2(\ln x)^3 + \frac{3}{2}x^2(\ln x)^2 - \frac{3}{2}x^2\ln x + \frac{3}{4}x^2 + C.$$

23. Let $u = \arcsin w$ and $v' = 1$, so $u' = \frac{1}{\sqrt{1-w^2}}$ and $v = w$. Then

$$\int \arcsin w\, dw = w\arcsin w - \int \frac{w}{\sqrt{1-w^2}}\, dw = w\arcsin w + \sqrt{1-w^2} + C.$$

24. Let $u = \arctan 7z$ and $v' = 1$, so $u' = \frac{7}{1+49z^2}$ and $v = z$. Now $\int \frac{7z\, dz}{1+49z^2}$ can be evaluated by the substitution $w = 1 + 49z^2$, $dw = 98z\, dz$, so

$$\int \frac{7z\, dz}{1+49z^2} = 7\int \frac{\frac{1}{98}\, dw}{w} = \frac{1}{14}\int \frac{dw}{w} = \frac{1}{14}\ln|w| + C = \frac{1}{14}\ln(1+49z^2) + C$$

So

$$\int \arctan 7z\, dz = z\arctan 7z - \frac{1}{14}\ln(1+49z^2) + C.$$

25. This integral can first be simplified by making the substitution $w = x^2$, $dw = 2x\,dx$. Then

$$\int x \arctan x^2 \, dx = \frac{1}{2} \int \arctan w \, dw.$$

To evaluate $\int \arctan w \, dw$, we'll use integration by parts. Let $u = \arctan w$ and $v' = 1$, so $u' = \frac{1}{1+w^2}$ and $v = w$. Then

$$\int \arctan w \, dw = w \arctan w - \int \frac{w}{1+w^2} \, dw = w \arctan w - \frac{1}{2} \ln|1+w^2| + C.$$

Since $1 + w^2$ is never negative, we can drop the absolute value signs. Thus, we have

$$\int x \arctan x^2 \, dx = \frac{1}{2}\left(x^2 \arctan x^2 - \frac{1}{2}\ln(1+(x^2)^2) + C\right)$$

$$= \frac{1}{2}x^2 \arctan x^2 - \frac{1}{4}\ln(1+x^4) + C.$$

26. Let $u = x^2$ and $v' = xe^{x^2}$, so $u' = 2x$ and $v = \frac{1}{2}e^{x^2}$. Then

$$\int x^3 e^{x^2} \, dx = \frac{1}{2}x^2 e^{x^2} - \int xe^{x^2} \, dx = \frac{1}{2}x^2 e^{x^2} - \frac{1}{2}e^{x^2} + C.$$

Note that we can also do this problem by substitution and integration by parts. If we let $w = x^2$, so $dw = 2x\,dx$, then $\int x^3 e^{x^2} \, dx = \frac{1}{2}\int we^w \, dw$. We could then perform integration by parts on this integral to get the same result.

27. To simplify matters, let us try the substitution $w = x^3$, $dw = 3x^2 \, dx$. Then

$$\int x^5 \cos x^3 \, dx = \frac{1}{3} \int w \cos w \, dw.$$

Now we integrate by parts. Let $u = w$ and $v' = \cos w$, so $u' = 1$ and $v = \sin w$. Then

$$\frac{1}{3}\int w \cos w \, dw = \frac{1}{3}[w \sin w - \int \sin w \, dw]$$

$$= \frac{1}{3}[w \sin w + \cos w] + C$$

$$= \frac{1}{3}x^3 \sin x^3 + \frac{1}{3}\cos x^3 + C$$

28. $\int_1^5 \ln t \, dt = (t \ln t - t)\Big|_1^5 = 5 \ln 5 - 4 \approx 4.047$

29. $\int_3^5 x \cos x \, dx = (\cos x + x \sin x)\Big|_3^5 = \cos 5 + 5 \sin 5 - \cos 3 - 3 \sin 3 \approx -3.944.$

30. We use integration by parts. Let $u = z$ and $v' = e^{-z}$, so $u' = 1$ and $v = -e^{-z}$.

Then $\int_0^{10} ze^{-z} \, dz = -ze^{-z}\Big|_0^{10} + \int_0^{10} e^{-z} \, dz$

$$= -10e^{-10} + (-e^{-z})\Big|_0^{10}$$

$$= -11e^{-10} + 1$$

$$\approx 0.9995.$$

31. $\int_1^3 t \ln t \, dt = \left(\frac{1}{2}t^2 \ln t - \frac{1}{2}t\right)\Big|_1^3 = \frac{9}{2}\ln 3 - 2 \approx 2.944.$

32. We use integration by parts. Let $u = \arctan y$ and $v' = 1$, so $u' = \frac{1}{1+y^2}$ and $v = y$. Thus

$$\int_0^1 \arctan y \, dy = (\arctan y) y \Big|_0^1 - \int_0^1 \frac{y}{1+y^2} \, dy$$

$$= \frac{\pi}{4} - \frac{1}{2}\ln|1+y^2|\Big|_0^1$$

$$= \frac{\pi}{4} - \frac{1}{2}\ln 2 \approx 0.439.$$

33. $\int_0^5 \ln(1+t) \, dt = \left((1+t)\ln(1+t) - (1+t)\right)\Big|_0^5 = 6\ln 6 - 5 \approx 5.751.$

34. First we make the substitution $y = x^2$, so $dy = 2x\,dx$. Thus

$$\int_{x=0}^{x=1} x \arctan x^2 \, dx = \frac{1}{2}\int_{y=0}^{y=1} \arctan y \, dy.$$

From Problem 32, we know that

$$\int_0^1 \arctan y \, dy = \frac{\pi}{4} - \frac{\ln 2}{2}.$$

Thus

$$\int_0^1 x \arctan x^2 \, dx = \frac{1}{2}\left(\frac{\pi}{4} - \frac{1}{2}\ln 2\right) \approx 0.219.$$

35. We use integration by parts. Let $u = \arcsin z$ and $v' = 1$, so $u' = \frac{1}{\sqrt{1-z^2}}$ and $v = z$. Then

$$\int_0^1 \arcsin z \, dz = z \arcsin z \Big|_0^1 - \int_0^1 \frac{z}{\sqrt{1-z^2}} \, dz = \frac{\pi}{2} - \int_0^1 \frac{z}{\sqrt{1-z^2}} \, dz.$$

To find $\int_0^1 \frac{z}{\sqrt{1-z^2}} \, dz$, we substitute $w = 1 - z^2$, so $dw = -2z\,dz$. Then

$$\int_{z=0}^{z=1} \frac{z}{\sqrt{1-z^2}} \, dz = -\frac{1}{2}\int_{w=1}^{w=0} w^{-\frac{1}{2}} \, dw = \frac{1}{2}\int_{w=0}^{w=1} w^{-\frac{1}{2}} \, dw = w^{\frac{1}{2}}\Big|_0^1 = 1.$$

Thus our final answer is $\frac{\pi}{2} - 1 \approx 0.571.$

36. To simplify the integral, we first make the substitution $z = u^2$, so $dz = 2u\,du$. Then

$$\int_{u=0}^{u=1} u \arcsin u^2 \, du = \frac{1}{2}\int_{z=0}^{z=1} \arcsin z \, dz.$$

From Problem 35, we know that $\int_0^1 \arcsin z \, dz = \frac{\pi}{2} - 1$. Thus,

$$\int_0^1 u \arcsin u^2 \, du = \frac{1}{2}\left(\frac{\pi}{2} - 1\right) \approx 0.285.$$

37. From integration by parts in Problem 12, we obtain

$$\int \sin^2 \theta \, d\theta = -\frac{1}{2} \sin \theta \cos \theta + \frac{1}{2}\theta + C.$$

Using the identity given in the book, we have

$$\int \sin^2 \theta \, d\theta = \int \frac{1 - \cos 2\theta}{2} \, d\theta = \frac{1}{2}\theta - \frac{1}{4}\sin 2\theta + C.$$

Although the answers differ in form, they are really the same, since (by one of the standard double angle formulas) $-\frac{1}{4}\sin 2\theta = -\frac{1}{4}(2 \sin \theta \cos \theta) = -\frac{1}{2}\sin \theta \cos \theta.$

38. Integration by parts: let $u = \cos \theta$ and $v' = \cos \theta$, so $u' = -\sin \theta$ and $v = \sin \theta$.

$$\int \cos^2 \theta \, d\theta = \sin \theta \cos \theta - \int (-\sin \theta)(\sin \theta) \, d\theta$$
$$= \sin \theta \cos \theta + \int \sin^2 \theta \, d\theta.$$

Now use $\sin^2 \theta = 1 - \cos^2 \theta$.

$$\int \cos^2 \theta \, d\theta = \sin \theta \cos \theta + \int (1 - \cos^2 \theta) \, d\theta$$
$$= \sin \theta \cos \theta + \int d\theta - \int \cos^2 \theta \, d\theta.$$

Adding $\int \cos^2 \theta \, d\theta$ to both sides, we have

$$2 \int \cos^2 \theta \, d\theta = \sin \theta \cos \theta + \theta + C$$
$$\int \cos^2 \theta \, d\theta = \frac{1}{2} \sin \theta \cos \theta + \frac{1}{2}\theta + C'.$$

Use the identity $\cos^2 \theta = \frac{1+\cos 2\theta}{2}$.

$$\int \cos^2 \theta \, d\theta = \int \frac{1 + \cos 2\theta}{2} \, d\theta = \frac{1}{2}\theta + \frac{1}{4}\sin 2\theta + C.$$

The only difference is in the two terms $\frac{1}{2}\sin \theta \cos \theta$ and $\frac{1}{4}\sin 2\theta$, but since $\sin 2\theta = 2 \sin \theta \cos \theta$, we have $\frac{1}{4}\sin 2\theta = \frac{1}{4}(2\sin \theta \cos \theta) = \frac{1}{2}\sin \theta \cos \theta$, so there is no real difference between the formulas.

39. First, let $u = e^x$ and $v' = \sin x$, so $u' = e^x$ and $v = -\cos x$.
Thus $\int e^x \sin x \, dx = -e^x \cos x + \int e^x \cos x \, dx$. To calculate $\int e^x \cos x \, dx$, we again need to use integration by parts. Let $u = e^x$ and $v' = \cos x$, so $u' = e^x$ and $v = \sin x$.
Thus

$$\int e^x \cos x \, dx = e^x \sin x - \int e^x \sin x \, dx.$$

This gives

$$\int e^x \sin x \, dx = e^x \sin x - e^x \cos x - \int e^x \sin x \, dx.$$

By adding $\int e^x \sin x \, dx$ to both sides, we obtain

$$2 \int e^x \sin x \, dx = e^x(\sin x - \cos x) + C.$$

Thus $\int e^x \sin x \, dx = \frac{1}{2}e^x(\sin x - \cos x) + C.$

This problem could also be done in other ways; for example, we could have started with $u = \sin x$ and $v' = e^x$ as well.

40. Let $u = e^\theta$ and $v' = \cos\theta$, so $u' = e^\theta$ and $v = \sin\theta$. Then $\int e^\theta \cos\theta\, d\theta = e^\theta \sin\theta - \int e^\theta \sin\theta\, d\theta$.

In Problem 39 we found that $\int e^x \sin x\, dx = \frac{1}{2}e^x(\sin x - \cos x) + C$.

$$\int e^\theta \cos\theta\, d\theta = e^\theta \sin\theta - \left[\frac{1}{2}e^\theta(\sin\theta - \cos\theta)\right] + C$$

$$= \frac{1}{2}e^\theta(\sin\theta + \cos\theta) + C.$$

41. We integrate by parts. Since in Problem 39 we found that $\int e^x \sin x\, dx = \frac{1}{2}e^x(\sin x - \cos x)$, we let $u = x$ and $v' = e^x \sin x$, so $u' = 1$ and $v = \frac{1}{2}e^x(\sin x - \cos x)$.

Then
$$\int xe^x \sin x\, dx = \frac{1}{2}xe^x(\sin x - \cos x) - \frac{1}{2}\int e^x(\sin x - \cos x)\, dx$$

$$= \frac{1}{2}xe^x(\sin x - \cos x) - \frac{1}{2}\int e^x \sin x\, dx + \frac{1}{2}\int e^x \cos x\, dx.$$

Using Problems 39 and 40, we see that this equals

$$\frac{1}{2}xe^x(\sin x - \cos x) - \frac{1}{4}e^x(\sin x - \cos x) + \frac{1}{4}e^x(\sin x + \cos x) + C$$

$$= \frac{1}{2}xe^x(\sin x - \cos x) + \frac{1}{2}e^x \cos x + C.$$

42. Again we use Problems 39 and 40. Integrate by parts, letting $u = \theta$ and $v' = e^\theta \cos\theta$, so $u' = 1$ and $v = \frac{1}{2}e^\theta(\sin\theta + \cos\theta)$. Then

$$\int \theta e^\theta \cos\theta\, d\theta = \frac{1}{2}\theta e^\theta(\sin\theta + \cos\theta) - \frac{1}{2}\int e^\theta(\sin\theta + \cos\theta)\, d\theta$$

$$= \frac{1}{2}\theta e^\theta(\sin\theta + \cos\theta) - \frac{1}{2}\int e^\theta \sin\theta\, d\theta - \frac{1}{2}\int e^\theta \cos\theta\, d\theta$$

$$= \frac{1}{2}\theta e^\theta(\sin\theta + \cos\theta) - \frac{1}{4}e^\theta(\sin\theta - \cos\theta) - \frac{1}{4}(\sin\theta + \cos\theta) + C$$

$$= \frac{1}{2}\theta e^\theta(\sin\theta + \cos\theta) - \frac{1}{2}e^\theta \sin\theta + C.$$

43. We integrate by parts. Since we know what the answer is supposed to be, it's easier to choose u and v'. Let $u = x^n$ and $v' = e^x$, so $u' = nx^{n-1}$ and $v = e^x$. Then

$$\int x^n e^x\, dx = x^n e^x - n\int x^{n-1} e^x\, dx.$$

44. We integrate by parts. Let $u = x^n$ and $v' = \cos ax$, so $u' = nx^{n-1}$ and $v = \frac{1}{a}\sin ax$. Then

$$\int x^n \cos ax\, dx = \frac{1}{a}x^n \sin ax - \int (nx^{n-1})(\frac{1}{a}\sin ax)\, dx$$

$$= \frac{1}{a}x^n \sin ax - \frac{n}{a}\int x^{n-1} \sin ax\, dx.$$

45. We integrate by parts. Let $u = x^n$ and $v' = \sin ax$, so $u' = nx^{n-1}$ and $v = -\frac{1}{a}\cos ax$.

Then
$$\int x^n \sin ax\, dx = -\frac{1}{a}x^n \cos ax - \int (nx^{n-1})(-\frac{1}{a}\cos ax)\, dx$$

$$= -\frac{1}{a}x^n \cos ax + \frac{n}{a}\int x^{n-1} \cos ax\, dx.$$

46. We integrate by parts. Since we know what the answer is supposed to be, it's easier to choose u and v'. Let $u = \cos^{n-1} x$ and $v' = \cos x$, so $u' = (n-1)\cos^{n-2} x(-\sin x)$ and $v = \sin x$.
Then

$$\int \cos^n x\, dx = \cos^{n-1} x \sin x + (n-1) \int \cos^{n-2} x \sin^2 x\, dx$$

$$= \cos^{n-1} x \sin x + (n-1) \int \cos^{n-2} x(1 - \cos^2 x)\, dx$$

$$= \cos^{n-1} x \sin x - (n-1) \int \cos^n x\, dx + (n-1) \int \cos^{n-2} x\, dx.$$

Thus, by adding $(n-1) \int \cos^n x\, dx$ to both sides of the equation, we find

$$n \int \cos^n x\, dx = \cos^{n-1} x \sin x + (n-1) \int \cos^{n-2} x\, dx,$$

so $$\int \cos^n dx = \frac{1}{n} \cos^{n-1} x \sin x + \frac{n-1}{n} \int \cos^{n-2} x\, dx.$$

47.

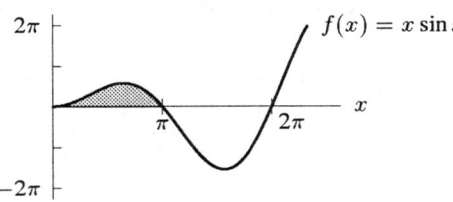

The graph of $f(x) = x \sin x$ is shown above. The first positive zero is at $x = \pi$, so, using integration by parts,

$$\text{Area} = \int_0^\pi x \sin x\, dx$$

$$= -x \cos x \Big|_0^\pi + \int_0^\pi \cos x\, dx$$

$$= -x \cos x \Big|_0^\pi + \sin x \Big|_0^\pi$$

$$= -\pi \cos \pi - (-0 \cos 0) + \sin \pi - \sin 0 = \pi.$$

48. Using integration by parts we have

$$\int_0^1 x f''(x)\, dx = x f'(x) \Big|_0^1 - \int_0^1 f'(x)\, dx$$

$$= 1 \cdot f'(1) - 0 \cdot f'(0) - [f(1) - f(0)]$$

$$= 2 - 0 - 5 + 6 = 3.$$

49. (a) We have

$$F(a) = \int_0^a x^2 e^{-x}\, dx$$

$$= -x^2 e^{-x} \Big|_0^a + \int_0^a 2x e^{-x}\, dx$$

$$= (-x^2 e^{-x} - 2x e^{-x}) \Big|_0^a + 2 \int_0^a e^{-x}\, dx$$

$$= (-x^2 e^{-x} - 2x e^{-x} - 2e^{-x}) \Big|_0^a$$

$$= -a^2 e^{-a} - 2a e^{-a} - 2e^{-a} + 2.$$

(b) $F(a)$ is increasing because $x^2 e^{-x}$ is positive, so as a increases, the area under the curve from 0 to a also increases and thus the integral increases.

(c) We have $F'(a) = a^2 e^{-a}$, so
$$F''(a) = 2ae^{-a} - a^2 e^{-a} = a(2-a)e^{-a}.$$

We see that $F''(a) > 0$ for $0 < a < 2$, so F is concave up on this interval.

50. (a) Increasing V_0 increases the maximum value of V, since this maximum is V_0. Increasing ω or ϕ does not affect the maximum of V.

(b) Since
$$\frac{dV}{dt} = -\omega V_0 \sin(\omega t + \phi),$$
the maximum of dV/dt is ωV_0. Thus, the maximum of dV/dt is increased if V_0 or ω is increased, and is unaffected if ϕ is increased.

(c) The period of $V = V_0 \cos(\omega t + \phi)$ is $2\pi/\omega$, so
$$\text{Average value} = \frac{1}{2\pi/\omega} \int_0^{2\pi/\omega} (V_0 \cos(\omega t + \phi))^2 \, dt.$$

Substituting $x = \omega t + \phi$, we have $dx = \omega dt$. When $t = 0$, $x = \phi$, and when $t = 2\pi/\omega$, $x = 2\pi + \phi$. Thus,
$$\text{Average value} = \frac{\omega}{2\pi} \int_\phi^{2\pi + \phi} V_0^2 (\cos x)^2 \frac{1}{\omega} \, dx$$
$$= \frac{V_0^2}{2\pi} \int_\phi^{2\pi + \phi} (\cos x)^2 \, dx.$$

Using integration by parts and the fact that $\sin^2 x = 1 - \cos^2 x$, we see that
$$\text{Average value} = \frac{V_0^2}{2\pi} \left[\frac{1}{2}(\cos x \sin x + x) \right]_\phi^{2\pi + \phi}$$
$$= \frac{V_0^2}{4\pi} \left[\cos(2\pi + \phi) \sin(2\pi + \phi) + (2\pi + \phi) - \cos\phi \sin\phi - \phi \right]$$
$$= \frac{V_0^2}{4\pi} \cdot 2\pi = \frac{V_0^2}{2}.$$

Thus, increasing V_0 increases the average value; increasing ω or ϕ has no effect.

However, it is not in fact necessary to compute the integral to see that ω does not affect the average value, since all ω's dropped out of the average value expression when we made the substitution $x = \omega t + \phi$.

51. (a) One way to avoid integrating by parts is to take the derivative of the right hand side instead. Since $\int e^{ax} \sin bx \, dx$ is the antiderivative of $e^{ax} \sin bx$,

$$e^{ax} \sin bx = \frac{d}{dx}[e^{ax}(A \sin bx + B \cos bx) + C]$$
$$= ae^{ax}(A \sin bx + B \cos bx) + e^{ax}(Ab \cos bx - Bb \sin bx)$$
$$= e^{ax}[(aA - bB) \sin bx + (aB + bA) \cos bx].$$

Thus $aA - bB = 1$ and $aB + bA = 0$. Solving for A and B in terms of a and b, we get
$$A = \frac{a}{a^2 + b^2}, \quad B = -\frac{b}{a^2 + b^2}.$$

Thus
$$\int e^{ax} \sin bx = e^{ax} \left(\frac{a}{a^2 + b^2} \sin bx - \frac{b}{a^2 + b^2} \cos bx \right) + C.$$

(b) If we go through the same process, we find
$$ae^{ax}[(aA - bB)\sin bx + (aB + bA)\cos bx] = e^{ax}\cos bx.$$
Thus $aA - bB = 0$, and $aB + bA = 1$. In this case, solving for A and B yields
$$A = \frac{b}{a^2 + b^2}, \quad B = \frac{a}{a^2 + b^2}.$$
Thus $\int e^{ax}\cos bx = e^{ax}\left(\frac{b}{a^2+b^2}\sin bx + \frac{a}{a^2+b^2}\cos bx\right) + C.$

52. (a) We know that $\frac{dE}{dt} = r$, so the total energy E used in the first T hours is given by $E = \int_0^T t e^{-at}\, dt$. We use integration by parts. Let $u = t$, $v' = e^{-at}$. Then $u' = 1$, $v = -\frac{1}{a}e^{-at}$.

$$\begin{aligned} E &= \int_0^T t e^{-at}\, dt \\ &= -\frac{t}{a}e^{-at}\Big|_0^T - \int_0^T \left(-\frac{1}{a}e^{-at}\right) dt \\ &= -\frac{1}{a}Te^{-aT} + \frac{1}{a}\int_0^T e^{-at}\, dt \\ &= -\frac{1}{a}Te^{-aT} + \frac{1}{a^2}(1 - e^{-aT}). \end{aligned}$$

(b)
$$\lim_{T\to\infty} E = -\frac{1}{a}\lim_{T\to\infty}\left(\frac{T}{e^{aT}}\right) + \frac{1}{a^2}\left(1 - \lim_{T\to\infty}\frac{1}{e^{aT}}\right).$$

Since $a > 0$, the second limit on the right hand side in the above expression is 0. In the first limit, although both the numerator and the denominator go to infinity, the denominator e^{aT} goes to infinity more quickly than T does. So in the end the denominator e^{aT} is much greater than the numerator T. Hence $\lim_{T\to\infty}\frac{T}{e^{aT}} = 0$. (You can check this by graphing $y = \frac{T}{e^{aT}}$ on a calculator or computer for some values of a.) Thus $\lim_{T\to\infty} E = \frac{1}{a^2}$.

53. (a) We want to compute C_1, with $C_1 > 0$, such that
$$\int_0^1 (\Psi_1(x))^2\, dx = \int_0^1 (C_1 \sin(\pi x))^2\, dx = C_1^2 \int_0^1 \sin^2(\pi x)\, dx = 1.$$

We use integration by parts with $u = v' = \sin(\pi x)$.
So $u' = \pi\cos(\pi x)$ and $v = -\frac{1}{\pi}\cos(\pi x)$. Thus

$$\begin{aligned}\int_0^1 \sin^2(\pi x)\, dx &= -\frac{1}{\pi}\sin(\pi x)\cos(\pi x)\Big|_0^1 + \int_0^1 \cos^2(\pi x)\, dx \\ &= -\frac{1}{\pi}\sin(\pi x)\cos(\pi x)\Big|_0^1 + \int_0^1 (1 - \sin^2(\pi x))\, dx.\end{aligned}$$

Moving $\int_0^1 \sin^2(\pi x)\, dx$ from the right side to the left side of the equation and solving, we get

$$2\int_0^1 \sin^2(\pi x)\, dx = -\frac{1}{\pi}\sin(\pi x)\cos(\pi x)\Big|_0^1 + \int_0^1 1\, dx = 0 + 1 = 1,$$

so
$$\int_0^1 \sin^2(\pi x)\, dx = \frac{1}{2}.$$

Thus, we have
$$\int_0^1 (\Psi_1(x))^2\, dx = C_1^2 \int_0^1 \sin^2(\pi x)\, dx = \frac{C_1^2}{2}.$$

So, to normalize Ψ_1, we take $C_1 > 0$ such that
$$\frac{C_1^2}{2} = 1 \quad \text{so} \quad C_1 = \sqrt{2}.$$

(b) To normalize Ψ_n, we want to compute C_n, with $C_n > 0$, such that

$$\int_0^1 (\Psi_n(x))^2\, dx = C_n^2 \int_0^1 \sin^2(n\pi x)\, dx = 1.$$

The solution to part (a) shows us that

$$\int \sin^2(\pi t)\, dt = -\frac{1}{2\pi} \sin(\pi t) \cos(\pi t) + \frac{1}{2}\int 1\, dt.$$

In the integral for Ψ_n, we make the substitution $t = nx$, so $dx = \frac{1}{n} dt$. Since $t = 0$ when $x = 0$ and $t = n$ when $x = 1$, we have

$$\int_0^1 \sin^2(n\pi x)\, dx = \frac{1}{n} \int_0^n \sin^2(\pi t)\, dt$$

$$= \frac{1}{n}\left(-\frac{1}{2\pi} \sin(\pi t) \cos(\pi t)\Big|_0^n + \frac{1}{2}\int_0^n 1\, dt\right)$$

$$= \frac{1}{n}\left(0 + \frac{n}{2}\right) = \frac{1}{2}.$$

Thus, we have

$$\int_0^1 (\Psi_n(x))^2\, dx = C_n^2 \int_0^1 \sin^2(n\pi x)\, dx = \frac{C_n^2}{2}.$$

So to normalize Ψ_n, we take C_n such that

$$\frac{C_n^2}{2} = 1 \quad \text{so} \quad C_n = \sqrt{2}.$$

Solutions for Section 7.4

1. $\frac{1}{10} e^{(-3\theta)}(-3\cos\theta + \sin\theta) + C.$
   (Let $a = -3, b = 1$ in II-9.)

2. $\frac{1}{6} x^6 \ln x - \frac{1}{36} x^6 + C.$ (Let $n = 5$ in III-13.)

3. $-\frac{1}{5}\cos^5 w + C$
   (Let $x = \cos w$, as suggested in IV-23. Then $-\sin w\, dw = dx$, and $\int \sin w \cos^4 w\, dw = -\int x^4\, dx$.)

4. $-\frac{1}{4}\sin^3 x \cos x - \frac{3}{8}\sin x \cos x + \frac{3}{8}x + C.$
   (Use IV-17.)

5. $\frac{1}{\sqrt{3}}\arctan\frac{y}{\sqrt{3}} + C.$
   (Let $a = \sqrt{3}$ in V-24).

6. Let $m = 3$ in IV-21.

$$\int \frac{1}{\cos^3 x}\, dx = \frac{1}{2}\frac{\sin x}{\cos^2 x} + \frac{1}{2}\int \frac{1}{\cos x}\, dx$$

$$= \frac{1}{2}\frac{\sin x}{\cos^2 x} + \frac{1}{4}\ln\left|\frac{\sin x + 1}{\sin x - 1}\right| + C \text{ by IV-22.}$$

7. $\left(\frac{1}{2}x^3 - \frac{3}{4}x^2 + \frac{3}{4}x - \frac{3}{8}\right)e^{2x} + C.$
   (Let $a = 2, p(x) = x^3$ in III-14.)

8. $\dfrac{5}{16}\sin 3\theta \sin 5\theta + \dfrac{3}{16}\cos 3\theta \cos 5\theta + C.$
   (Let $a = 3, b = 5$ in II-12.)

9. $\dfrac{3}{16}\cos 3\theta \sin 5\theta - \dfrac{5}{16}\sin 3\theta \cos 5\theta + C.$
   (Let $a = 3, b = 5$ in II-10.)

10. $\left(\dfrac{1}{3}x^2 - \dfrac{2}{9}x + \dfrac{2}{27}\right)e^{3x} + C.$
    (Let $a = 3, p(x) = x^2$ in III-14.)

11. $\dfrac{1}{3}e^{x^3} + C.$
    (Substitute $w = x^3$, $dw = 3x^2\, dx$. It isn't necessary to use the table.)

12. $\left(\dfrac{1}{3}x^4 - \dfrac{4}{9}x^3 + \dfrac{4}{9}x^2 - \dfrac{8}{27}x + \dfrac{8}{81}\right)e^{3x} + C.$
    (Let $a = 3, p(x) = x^4$ in III-14.)

13. Substitute $w = 5u$, $dw = 5\, du$. Then

$$\int u^5 \ln(5u)\, du = \dfrac{1}{5^6}\int w^5 \ln w\, dw$$
$$= \dfrac{1}{5^6}\left(\dfrac{1}{6}w^6 \ln w - \dfrac{1}{36}w^6 + C\right)$$
$$= \dfrac{1}{6}u^6 \ln 5u - \dfrac{1}{36}u^6 + C.$$

Or use $\ln 5u = \ln 5 + \ln u$.

$$\int u^5 \ln 5u\, du = \ln 5 \int u^5\, du + \int u^5 \ln u\, du$$
$$= \dfrac{u^6}{6}\ln 5 + \dfrac{1}{6}u^6 \ln u - \dfrac{1}{36}u^6 + C \quad \text{(using III-13)}$$
$$= \dfrac{u^6}{6}\ln 5u - \dfrac{1}{36}u^6 + C.$$

14. Use long division to reorganize the integral:

$$\int \dfrac{t^2 + 1}{t^2 - 1}\, dt = \int \left(1 + \dfrac{2}{t^2 - 1}\right) dt = \int dt + \int \dfrac{2}{(t-1)(t+1)}\, dt.$$

To get this second integral, let $a = 1, b = -1$ in V-26, so

$$\int \dfrac{t^2 + 1}{t^2 - 1}\, dt = t + \ln|t - 1| - \ln|t + 1| + C.$$

15. Substitute $w = x^2, dw = 2x\, dx$. Then $\int x^3 \sin x^2\, dx = \dfrac{1}{2}\int w \sin w\, dw$. By III-15, we have

$$\int w \sin w\, dw = -\dfrac{1}{2}w \cos w + \dfrac{1}{2}\sin w + C = -\dfrac{1}{2}x^2 \cos x^2 + \dfrac{1}{2}\sin x^2 + C.$$

16. $\dfrac{1}{45}(7\cos 2y \sin 7y - 2\sin 2y \cos 7y) + C.$
    (Let $a = 2, b = 7$ in II-11.)

17.
$$\int y^2 \sin 2y\, dy = -\dfrac{1}{2}y^2 \cos 2y + \dfrac{1}{4}(2y)\sin 2y + \dfrac{1}{8}(2)\cos 2y + C$$
$$= -\dfrac{1}{2}y^2 \cos 2y + \dfrac{1}{2}y \sin 2y + \dfrac{1}{4}\cos 2y + C.$$

(Use $a = 2, p(y) = y^2$ in III-15.)

18. $\frac{1}{34}e^{5x}(5\sin 3x - 3\cos 3x) + C.$
(Let $a = 5, b = 3$ in II-8.)

19. Substitute $w = 2\theta$, $dw = 2\,d\theta$. Then use IV-19, letting $m = 2$.

$$\int \frac{1}{\sin^2 2\theta}\,d\theta = \frac{1}{2}\int \frac{1}{\sin^2 w}\,dw = \frac{1}{2}\left(-\frac{\cos w}{\sin w}\right) + C = -\frac{1}{2\tan w} + C = -\frac{1}{2\tan 2\theta} + C.$$

20. Substitute $w = 3\theta$, $dw = 3\,d\theta$. Then use IV-19, letting $m = 3$.

$$\int \frac{1}{\sin^3 3\theta}\,d\theta = \frac{1}{3}\int \frac{1}{\sin^3 w}\,dw = \frac{1}{3}\left[-\frac{1}{2}\frac{\cos w}{\sin^2 w} + \frac{1}{2}\int \frac{1}{\sin w}\,dw\right]$$
$$= -\frac{1}{6}\frac{\cos w}{\sin^2 w} + \frac{1}{6}\left[\frac{1}{2}\ln\left|\frac{\cos(w) - 1}{\cos(w) + 1}\right| + C\right] \text{ by IV-20}$$
$$= -\frac{1}{6}\frac{\cos 3\theta}{\sin^2 3\theta} + \frac{1}{12}\ln\left|\frac{\cos(3\theta) - 1}{\cos(3\theta) + 1}\right| + C.$$

21. Substitute $w = 7x$, $dw = 7\,dx$. Then use IV-21.

$$\int \frac{1}{\cos^4 7x}\,dx = \frac{1}{7}\int \frac{1}{\cos^4 w}\,dw = \frac{1}{7}\left[\frac{1}{3}\frac{\sin w}{\cos^3 w} + \frac{2}{3}\int \frac{1}{\cos^2 w}\,dw\right]$$
$$= \frac{1}{21}\frac{\sin w}{\cos^3 w} + \frac{2}{21}\left[\frac{\sin w}{\cos w} + C\right]$$
$$= \frac{1}{21}\frac{\tan w}{\cos^2 w} + \frac{2}{21}\tan w + C$$
$$= \frac{1}{21}\frac{\tan 7x}{\cos^2 7x} + \frac{2}{21}\tan 7x + C.$$

22.
$$\int \frac{1}{x^2 + 4x + 3}\,dx = \int \frac{1}{(x+1)(x+3)}\,dx = \frac{1}{2}(\ln|x+1| - \ln|x+3|) + C.$$

(Let $a = -1$ and $b = -3$ in V-26).

23. Using the advice in IV-23, since both m and n are even and since n is negative, we convert everything to cosines, since $\cos x$ is in the denominator.

$$\int \tan^4 x\,dx = \int \frac{\sin^4 x}{\cos^4 x}\,dx$$
$$= \int \frac{(1 - \cos^2 x)^2}{\cos^4 x}\,dx$$
$$= \int \frac{1}{\cos^4 x}\,dx - 2\int \frac{1}{\cos^2 x}\,dx + \int 1\,dx.$$

By IV-21

$$\int \frac{1}{\cos^4 x}\,dx = \frac{1}{3}\frac{\sin x}{\cos^3 x} + \frac{2}{3}\int \frac{1}{\cos^2 x}\,dx,$$
$$\int \frac{1}{\cos^2 x}\,dx = \frac{\sin x}{\cos x} + C.$$

Substituting back in, we get

$$\int \tan^4 x\,dx = \frac{1}{3}\frac{\sin x}{\cos^3 x} - \frac{4}{3}\frac{\sin x}{\cos x} + x + C.$$

24.
$$\int \frac{dz}{z(z-3)} = -\frac{1}{3}(\ln|z| - \ln|z-3|) + C.$$
(Let $a = 0, b = 3$ in V-26.)

25.
$$\int \frac{dy}{4-y^2} = -\int \frac{dy}{(y+2)(y-2)} = -\frac{1}{4}(\ln|y-2| - \ln|y+2|) + C.$$
(Let $a = 2, b = -2$ in V-26.)

26. $\arctan(z+2) + C.$
(Substitute $w = z+2$ and use V-24, letting $a = 1$.)

27.
$$\int \frac{1}{y^2 + 4y + 5} dy = \int \frac{1}{1+(y+2)^2} dy = \arctan(y+2) + C.$$
(Substitute $w = y + 2$, and let $a = 1$ in V-24).

28.
$$\int \frac{1}{x^2 + 4x + 4} dx = \int \frac{1}{(x+2)^2} dx = -\frac{1}{x+2} + C.$$
You need not use the table.

29.
$$\int \sin^3 3\theta \cos^2 3\theta \, d\theta = \int (\sin 3\theta)(\cos^2 3\theta)(1 - \cos^2 3\theta) \, d\theta$$
$$= \int \sin 3\theta (\cos^2 3\theta - \cos^4 3\theta) \, d\theta.$$

Using an extension of the tip given in rule IV-23, we let $w = \cos 3\theta$, $dw = -3\sin 3\theta \, d\theta$.

$$\int \sin 3\theta (\cos^2 3\theta - \cos^4 3\theta) \, d\theta = -\frac{1}{3} \int (w^2 - w^4) \, dw$$
$$= -\frac{1}{3}\left(\frac{w^3}{3} - \frac{w^5}{5}\right) + C$$
$$= -\frac{1}{9}(\cos^3 3\theta) + \frac{1}{15}(\cos^5 3\theta) + C.$$

30. If we make the substitution $w = 2z^2$ then $dw = 4z \, dz$, and the integral becomes:
$$\int ze^{2z^2} \cos(2z^2) \, dz = \frac{1}{4} \int e^w \cos w \, dw$$

Now we can use Formula 9 from the table of integrals to get:

$$\frac{1}{4} \int e^w \cos w \, dw = \frac{1}{4} \left[\frac{1}{2} e^w (\cos w + \sin w) + C\right]$$
$$= \frac{1}{8} e^w (\cos w + \sin w) + C$$
$$= \frac{1}{8} e^{2z^2} (\cos 2z^2 + \sin 2z^2) + C$$

31. Using II-10 in the integral table, if $m \neq \pm n$, then
$$\int_{-\pi}^{\pi} \sin m\theta \sin n\theta \, d\theta = \frac{1}{n^2 - m^2} [m \cos m\theta \sin n\theta - n \sin m\theta \cos n\theta] \Big|_{-\pi}^{\pi}$$
$$= \frac{1}{n^2 - m^2}[(m \cos m\pi \sin n\pi - n \sin m\pi \cos n\pi) -$$
$$(m \cos(-m\pi) \sin(-n\pi) - n \sin(-m\pi) \cos(-n\pi))]$$

But $\sin k\pi = 0$ for all integers k, so each term reduces to 0, making the whole integral reduce to 0.

32. Using formula II-11, if $m \neq \pm n$, then

$$\int_{-\pi}^{\pi} \cos m\theta \cos n\theta \, d\theta = \frac{1}{n^2 - m^2}(n \cos m\theta \sin n\theta - m \sin m\theta \cos n\theta)\Big|_{-\pi}^{\pi}.$$

We see that in the evaluation, each term will have a $\sin k\pi$ term, so the expression reduces to 0.

33. (a) $\dfrac{2}{x} + \dfrac{1}{x+3} = \dfrac{2(x+3)}{x(x+3)} + \dfrac{x}{x(x+3)} = \dfrac{3x+6}{x^2+3x}$. Thus

$$\int \frac{3x+6}{x^2+3x} \, dx = \int \left(\frac{2}{x} + \frac{1}{x+3}\right) dx = 2\ln|x| + \ln|x+3| + C.$$

(b) Let $a = 0, b = -3, c = 3$, and $d = 6$ in V-27.

$$\int \frac{3x+6}{x^2+3x} \, dx = \int \frac{3x+6}{x(x+3)} \, dx$$
$$= \frac{1}{3}(6\ln|x| + 3\ln|x+3|) + C = 2\ln|x| + \ln|x+3| + C.$$

34. Since $\dfrac{1}{x^2-1} = \dfrac{1}{(x-1)(x+1)}$, let's imagine that our fraction is the result of adding together two terms, one with a denominator of $x - 1$, the other with a denominator of $x + 1$:

$$\frac{1}{(x-1)(x+1)} = \frac{A}{x-1} + \frac{B}{x+1}.$$

To find A and B, we multiply by the least common multiple of both sides to clear the fractions. This yields

$$1 = A(x+1) + B(x-1)$$
$$= (A+B)x + (A-B).$$

Since the two sides are equal for all values of x in the domain, and there is no x term on the left-hand side, $A + B = 0$. Similarly, since A and $-B$ are constant terms on the right-hand side, and 1 is the constant term on the left-hand side, $A - B = 1$. Therefore, we have the system of equations

$$A + B = 0$$
$$A - B = 1.$$

Solving this gives us $A = 1/2$ and $B = -1/2$, so

$$\frac{1}{x^2-1} = \frac{1}{2(x-1)} - \frac{1}{2(x+1)}.$$

Now, we find the integral

$$\int \frac{1}{x^2-1} \, dx = \int \left(\frac{1}{2(x-1)} - \frac{1}{2(x+1)}\right) dx$$
$$= \frac{1}{2}\ln|x-1| - \frac{1}{2}\ln|x+1| + C.$$

35. (a) We split $\dfrac{1}{x^2 - x} = \dfrac{1}{x(x-1)}$ into partial fractions:

$$\frac{1}{x^2-x} = \frac{A}{x} + \frac{B}{x-1}.$$

Multiplying by $x(x-1)$ gives

$$1 = A(x-1) + Bx = (A+B)x - A,$$

so $-A = 1$ and $A + B = 0$, giving $A = -1, B = 1$. Therefore,

$$\int \frac{1}{x^2-x} \, dx = \int \left(\frac{1}{x-1} - \frac{1}{x}\right) dx = \ln|x-1| - \ln|x| + C.$$

(b) $\displaystyle\int \frac{1}{x^2-x} \, dx = \int \frac{1}{(x-1)(x)} \, dx$. Using $a = 1$ and $b = 0$ in V-26, we get $\ln|x-1| - \ln|x| + C$.

36. Split the integrand into partial fractions, giving

$$\frac{1}{x(L-x)} = \frac{A}{x} + \frac{B}{L-x}$$
$$1 = A(L-x) + Bx = (B-A)x + AL.$$

We have $B - A = 0$ and $AL = 1$, so $A = B = 1/L$. Thus,

$$\int \frac{1}{x(L-x)}\, dx = \int \frac{1}{L}\left(\frac{1}{x} + \frac{1}{L-x}\right) dx = \frac{1}{L}\left(\ln|x| - \ln|L-x|\right) + C.$$

37. We write

$$\frac{1}{3P - 3P^2} = \frac{A}{3P} + \frac{B}{1-P},$$

multiply through by $3P(1-P)$, and then solve for A and B, getting $A = 1$ and $B = 1/3$. So

$$\int \frac{dP}{3P - 3P^2} = \int \left(\frac{1}{3P} + \frac{1}{3(1-P)}\right) dP = \frac{1}{3}\int \frac{dP}{P} + \frac{1}{3}\int \frac{dP}{1-P}$$
$$= \frac{1}{3}\ln|P| - \frac{1}{3}\ln|1-P| + C = \frac{1}{3}\ln\left|\frac{P}{1-P}\right| + C.$$

38. Using the technique of partial fractions, we have:

$$\frac{3x+1}{x^2 - 3x + 2} = \frac{3x+1}{(x-1)(x-2)} = \frac{A}{x-1} + \frac{B}{x-2}.$$

Multiplying by $(x-1)$ and $(x-2)$, this becomes

$$3x + 1 = A(x-2) + B(x-1)$$
$$= (A+B)x - 2A - B$$

which produces the system of equations

$$\begin{cases} A + B = 3 \\ -2A - B = 1. \end{cases}$$

Solving this system yields $A = -4$ and $B = 7$. So,

$$\int \frac{3x+1}{x^2 - 3x + 2}\, dx = \int \left(-\frac{4}{x-1} + \frac{7}{x-2}\right) dx$$
$$= -4\int \frac{dx}{x-1} + 7\int \frac{dx}{x-2}$$
$$= -4\ln|x-1| + 7\ln|x-2| + C.$$

39. (a)

$$\frac{1}{1-0}\int_0^1 V_0 \cos(120\pi t)\, dt = \left.\frac{V_0}{120\pi}\sin(120\pi t)\right|_0^1$$
$$= \frac{V_0}{120\pi}[\sin(120\pi) - \sin(0)]$$
$$= \frac{V_0}{120\pi}[0 - 0] = 0.$$

(b) Let's find the average of V^2 first.

$$\overline{V^2} = \text{Average of } V^2 = \frac{1}{1-0}\int_0^1 V^2 dt$$

$$= \frac{1}{1-0}\int_0^1 (V_0 \cos(120\pi t))^2 dt$$

$$= V_0^2 \int_0^1 \cos^2(120\pi t)dt$$

Now, let $120\pi t = x$, and $dt = \frac{dx}{120\pi}$. So

$$\overline{V^2} = \frac{V_0^2}{120\pi}\int_0^{120\pi} \cos^2 x\, dx.$$

$$= \frac{V_0^2}{120\pi}\left(\frac{1}{2}\cos x \sin x + \frac{1}{2}x\right)\Big|_0^{120\pi} \quad \text{II-18}$$

$$= \frac{V_0^2}{120\pi} 60\pi = \frac{V_0^2}{2}.$$

So, the average of V^2 is $\frac{V_0^2}{2}$ and $\overline{V} = \sqrt{\text{average of } V^2} = \frac{V_0}{\sqrt{2}}$.

(c) $V_0 = \sqrt{2}\cdot\overline{V} = 110\sqrt{2} \approx 156$ volts.

40. (a) Since $R(T)$ is the rate or production, we find the total production by integrating:

$$\int_0^N R(t)\, dt = \int_0^N (A + Be^{-t}\sin(2\pi t))\, dt$$

$$= NA + B\int_0^N e^{-t}\sin(2\pi t)\, dt.$$

Let $a = -1$ and $b = 2\pi$ in II-8.

$$= NA + \frac{B}{1+4\pi^2}e^{-t}(-\sin(2\pi t) - 2\pi \cos(2\pi t))\Big|_0^N.$$

Since N is an integer (so $\sin 2\pi N = 0$ and $\cos 2\pi N = 1$),

$$\int_0^N R(t)\, dt = NA + B\frac{2\pi}{1+4\pi^2}(1-e^{-N}).$$

Thus the total production is $NA + \frac{2\pi B}{1+4\pi^2}(1-e^{-N})$ over the first N years.

(b) The average production over the first N years is

$$\int_0^N \frac{R(t)\, dt}{N} = A + \frac{2\pi B}{1+4\pi^2}\left(\frac{1-e^{-N}}{N}\right).$$

(c) As $N \to \infty$, $A + \frac{2\pi B}{1+4\pi^2}\frac{1-e^{-N}}{N} \to A$, since the second term in the sum goes to 0. This is why A is called the average!

(d) When t gets large, the term $Be^{-t}\sin(2\pi t)$ gets very small. Thus, $R(t) \approx A$ for most t, so it makes sense that the average of $\int_0^N R(t)\, dt$ is A as $N \to \infty$.

(e) This model is not reasonable for long periods of time, since an oil well has finite capacity and will eventually "run dry." Thus, we cannot expect average production to be close to constant over a long period of time.

41. We want to calculate
$$\int_0^1 C_n \sin(n\pi x) \cdot C_m \sin(m\pi x)\, dx.$$
We use II-11 from the table of integrals with $a = n\pi, b = m\pi$. Since $n \neq m$, we see that

$$\int_0^1 \Psi_n(x) \cdot \Psi_m(x)\, dx = C_n C_m \int_0^1 \sin(n\pi x) \sin(m\pi x)\, dx$$
$$= \frac{C_n C_m}{m^2\pi^2 - n^2\pi^2} \left(n\pi \cos(n\pi x) \sin(m\pi x) - m\pi \sin(n\pi x) \cos(m\pi x)\right)\Big|_0^1$$
$$= \frac{C_n C_m}{(m^2 - n^2)\pi^2} \Big(n\pi \cos(n\pi) \sin(m\pi) - m\pi \sin(n\pi) \cos(m\pi)$$
$$- n\pi \cos(0) \sin(0) + m\pi \sin(0) \cos(0)\Big)$$
$$= 0$$

since $\sin(0) = \sin(n\pi) = \sin(m\pi) = 0$.

Solutions for Section 7.5

1. (a)
$$\text{LEFT}(2) = 2 \cdot f(0) + 2 \cdot f(2)$$
$$= 2 \cdot 1 + 2 \cdot 5$$
$$= 12$$
$$\text{RIGHT}(2) = 2 \cdot f(2) + 2 \cdot f(4)$$
$$= 2 \cdot 5 + 2 \cdot 17$$
$$= 44$$

(b)
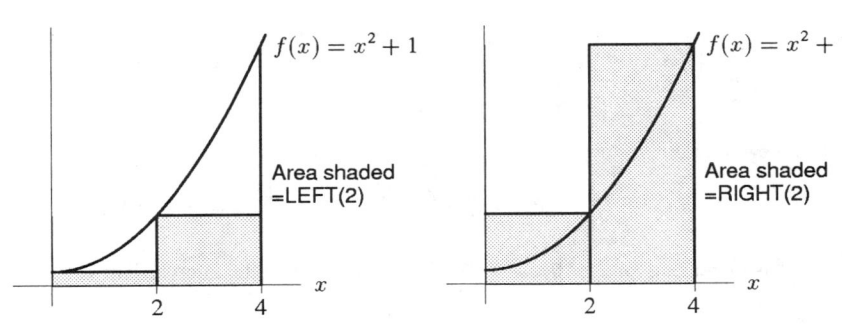

LEFT(2) is an underestimate, while RIGHT(2) is an overestimate.

2. (a)
$$\text{MID}(2) = 2 \cdot f(1) + 2 \cdot f(3)$$
$$= 2 \cdot 2 + 2 \cdot 10$$
$$= 24$$
$$\text{TRAP}(2) = \frac{\text{LEFT}(2) + \text{RIGHT}(2)}{2}$$
$$= \frac{12 + 44}{2} \quad \text{(see Problem 1)}$$
$$= 28$$

(b)

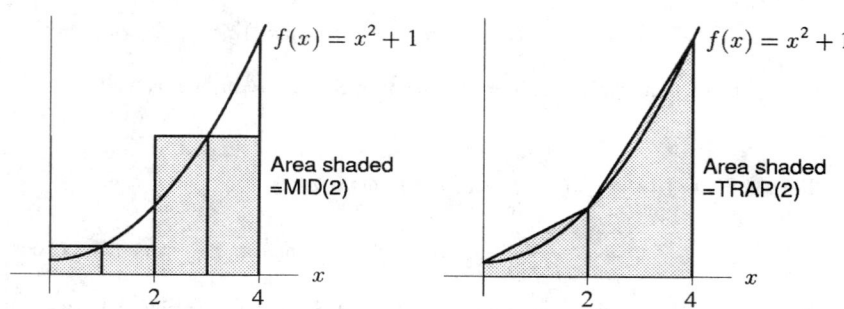

MID(2) is an underestimate, since $f(x) = x^2 + 1$ is concave up and a tangent line will be below the curve. TRAP(2) is an overestimate, since a secant line lies above the curve.

3.

n	1	2	4
LEFT	40	40.7846	41.7116
RIGHT	51.2250	46.3971	44.5179
TRAP	45.6125	43.5909	43.1147
MID	41.5692	42.6386	42.8795

4. (a) (i) Let $f(x) = \frac{1}{1+x^2}$. The left-hand Riemann sum is

$$\frac{1}{8}\left(f(0) + f\left(\frac{1}{8}\right) + f\left(\frac{2}{8}\right) + \cdots + f\left(\frac{7}{8}\right)\right)$$
$$= \frac{1}{8}\left(\frac{64}{64} + \frac{64}{65} + \frac{64}{68} + \frac{64}{73} + \frac{64}{80} + \frac{64}{89} + \frac{64}{100} + \frac{64}{113}\right)$$
$$\approx 8(0.1020) = 0.8160.$$

(ii) Let $f(x) = \frac{1}{1+x^2}$. The right-hand Riemann sum is

$$\frac{1}{8}\left(f\left(\frac{1}{8}\right) + f\left(\frac{2}{8}\right) + f\left(\frac{3}{8}\right) + \cdots + f(1)\right)$$
$$= \frac{1}{8}\left(\frac{64}{65} + \frac{64}{68} + \frac{64}{73} + \frac{64}{80} + \frac{64}{89} + \frac{64}{100} + \frac{64}{113} + \frac{64}{128}\right)$$
$$\approx 0.8160 - \frac{1}{16} = 0.7535.$$

(iii) The trapezoid rule gives us that

$$\text{TRAP}(8) = \frac{\text{LEFT}(8) + \text{RIGHT}(8)}{2} \approx 0.7847.$$

(b) Since $1 + x^2$ is increasing for $x > 0$, so $\frac{1}{1+x^2}$ is decreasing over the interval. Thus

$$\text{RIGHT}(8) < \int_0^1 \frac{1}{1+x^2}\,dx < \text{LEFT}(8)$$
$$0.7535 < \frac{\pi}{4} < 0.8160$$
$$3.014 < \pi < 3.264.$$

5. (a) (i) LEFT(32) = 13.6961, RIGHT(32) = 14.3437, TRAP(32) = 14.0199

Exact value = $(x \ln x - x)\Big|_1^{10} \approx 14.02585093$

(ii) LEFT(32) = 50.3180, RIGHT(32) = 57.0178, TRAP(32) = 53.6679

Exact value = $e^x\Big|_0^4 \approx 53.59815003$

(b) Both $\ln x$ and e^x are increasing, so the left sum underestimates and the right sum overestimates.

   (i) LEFT(32) $\leq$ TRAP(32) $\leq$ Actual value $\leq$ RIGHT(32)

   (ii) LEFT(32) $\leq$ Actual value $\leq$ TRAP(32) $\leq$ RIGHT(32)

   The trapezoid rule is an overestimate if f is concave up, and an underestimate if it is concave down.

   Since $\ln x$ is concave down, the trapezoidal estimate is too small. Since e^x is concave up, the trapezoidal estimate is too large. In each case, however, the trapezoidal estimate should be better than the left- or right-hand sums, since it is the average of the two.

6. Let $s(t)$ be the distance traveled at time t and $v(t)$ be the velocity at time t. Then the distance traveled during the interval $0 \leq t \leq 6$ is

$$s(6) - s(0) = s(t)\Big|_0^6$$
$$= \int_0^6 s'(t)\,dt \quad \text{(by the Fundamental Theorem)}$$
$$= \int_0^6 v(t)\,dt.$$

We estimate the distance by estimating this integral.

From the table, we find: LEFT(6) = 31, RIGHT(6) = 39, TRAP(6) = 35.

7. For a decreasing function whose graph is concave up, the diagrams below show that RIGHT < MID < TRAP < LEFT. Thus,

   (a) $0.664 =$ LEFT, $0.633 =$ TRAP, $0.632 =$ MID, and $0.601 =$ RIGHT.
   (b) $0.632 <$ true value < 0.633.

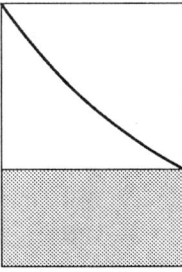
RIGHT = 0.601

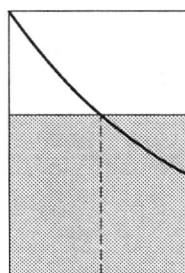
MID = 0.632

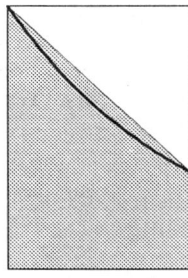
TRAP = 0.633

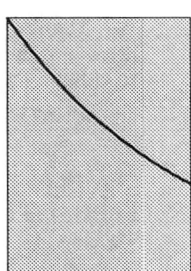
LEFT = 0.664

8. (a) Since $f(x)$ is closer to horizontal (that is, $|f'| < |g'|$), LEFT and RIGHT will be more accurate with $f(x)$.
   (b) Since $g(x)$ has more curvature, MID and TRAP will be more accurate with $f(x)$.

9. (a) TRAP(4) gives probably the best estimate of the integral. We cannot calculate MID(4).

$$\text{LEFT}(4) = 3 \cdot 100 + 3 \cdot 97 + 3 \cdot 90 + 3 \cdot 78 = 1095$$
$$\text{RIGHT}(4) = 3 \cdot 97 + 3 \cdot 90 + 3 \cdot 78 + 3 \cdot 55 = 960$$
$$\text{TRAP}(4) = \frac{1095 + 960}{2} = 1027.5.$$

   (b) Because there are no points of inflection, the graph is either concave down or concave up. By plotting points, we see that it is concave down. So TRAP(4) is an underestimate.

10. $f(x)$ is increasing, so RIGHT gives an overestimate and LEFT gives an underestimate.

11. $f(x)$ is concave down, so MID gives an overestimate and TRAP gives an underestimate.

12. $f(x)$ is decreasing and concave up, so LEFT and TRAP give overestimates and RIGHT and MID give underestimates.

13. $f(x)$ is concave up, so TRAP gives an overestimate and MID gives an underestimate.

14. (a) $\int_0^{2\pi} \sin\theta\, d\theta = -\cos\theta \Big|_0^{2\pi} = 0.$

    (b) MID(1) is 0 since the midpoint of 0 and 2π is π, and $\sin\pi = 0$. Thus MID(1) $= 2\pi(\sin\pi) = 0$. The midpoints we use for MID(2) are $\pi/2$ and $3\pi/2$, and $\sin(\pi/2) = -\sin(3\pi/2)$. Thus MID(2) $= \pi\sin(\pi/2) + \pi\sin(3\pi/2) = 0$.

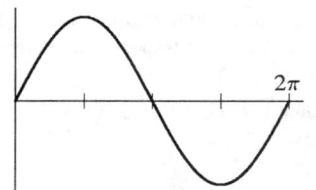

   (c) MID(3) = 0.

   In general, MID(n) = 0 for all n, even though your calculator (because of round-off error) might not return it as such. The reason is that $\sin(x) = -\sin(2\pi - x)$. If we use MID($n$), we will always take sums where we are adding pairs of the form $\sin(x)$ and $\sin(2\pi - x)$, so the sum will cancel to 0. (If n is odd, we will get a $\sin\pi$ in the sum which doesn't pair up with anything — but $\sin\pi$ is already 0.)

15. (a)

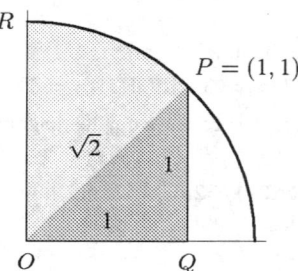

   The graph of $y = \sqrt{2 - x^2}$ is the upper half of a circle of radius $\sqrt{2}$ centered at the origin. The integral represents the area under this curve between the lines $x = 0$ and $x = 1$. From the picture, we see that this area can be split into 2 parts, A_1 and A_2. Notice since $OQ = QP = 1$, $\triangle OQP$ is isoceles. Thus $\angle POQ = \angle ROP = \frac{\pi}{4}$, and A_1 is exactly $\frac{1}{8}$ of the entire circle. Thus the total area is

   $$\text{Area} = A_1 + A_2 = \frac{1}{8}\pi(\sqrt{2})^2 + \frac{1\cdot 1}{2} = \frac{\pi}{4} + \frac{1}{2}.$$

   (b) LEFT(5) ≈ 1.32350, RIGHT(5) ≈ 1.24066, T
   TRAP(5) ≈ 1.28208, MID(5) ≈ 1.28705

   Exact value ≈ 1.285398163

   Left-hand error ≈ -0.03810, Right-hand error ≈ 0.04474,
   Trapezoidal error ≈ 0.00332, Midpoint error ≈ -0.001656

   Thus right-hand error > trapezoidal error > 0 > midpoint error > left-hand error, and |midpt error| < |trap error| < |left-error| < |right-error|.

16. One such graph is shown below.

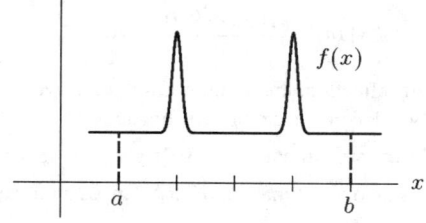

17. We approximate the area of the playing field by using Riemann sums. From the data provided,

$$\text{LEFT}(10) = \text{RIGHT}(10) = \text{TRAP}(10) = 89{,}000 \text{ square feet}.$$

Thus approximately

$$\frac{89{,}000 \text{ sq. ft.}}{200 \text{ sq. ft./lb.}} = 445 \text{ lbs. of fertilizer}$$

should be necessary.

18. (a)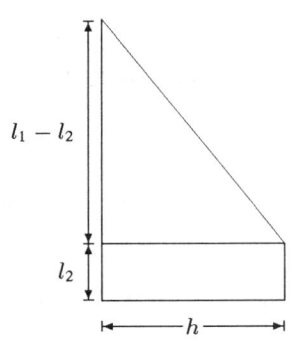

The area can be obtained by breaking the trapezoid into a rectangle of height l_2, base h and a triangle of height $(l_1 - l_2)$, base h. So

$$\text{Area} = \text{area of rectangle} + \text{area of triangle}$$
$$= h \cdot l_2 + \frac{1}{2} \cdot h \cdot (l_1 - l_2)$$
$$= h \cdot \frac{l_1 + l_2}{2}$$

The formula suggests that the area of a trapezoid equals the base times the average of l_1 and l_2.

(b)

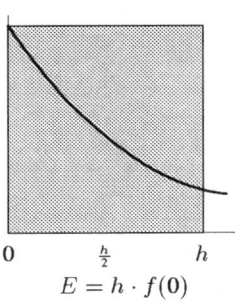

$E = h \cdot f(0)$

$F = h \cdot f(h)$

$R = h \cdot f(h/2)$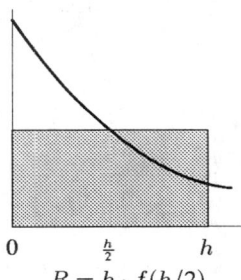

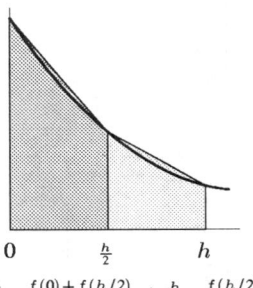

$C = h \cdot \frac{f(0) + f(h)}{2}$

$N = \frac{h}{2} \cdot \frac{f(0) + f(h/2)}{2} + \frac{h}{2} \cdot \frac{f(h/2) + f(h)}{2}$

(c) Since the region F is inside the region A, $F < A$. (We abuse the notation by labeling the regions with the same letters that denote their areas.) Similarly, $A < N < C < E$ by the inclusion relations among the corresponding regions. Since $N = \frac{R+C}{2}$ is the average of R and C, it lies halfway between R and C. Thus, $R < N < C < E$. To determine the relationship of R and A, notice that the following statements are equivalent:
- $R < A$
- $h \cdot f(\frac{h}{2}) < \int_0^h f(t)\,dt$
- $f(\frac{h}{2}) < \frac{1}{h} \int_0^h f(t)\,dt$
- $f(\frac{h}{2}) <$ Average value of f in the interval $[0, h]$

If we construct the tangent line to the graph of f at the point $(\frac{1}{2}h, f(\frac{1}{2}h))$ and call it g, we see that the tangent line g lies below the graph of f, because f is evidently concave up.

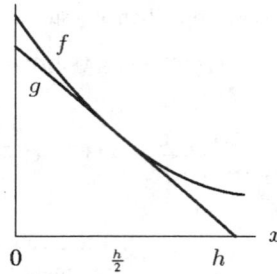

Thus the average value of f is greater than the average value of g. Moreover, $f(h/2)$ is clearly the average value of g in the interval $[0, h]$, so

$$f(h/2) = \text{(average value of } g) < \text{(average value of } f),$$

and so $R < A$. The region F is included in the region R, so we have $F < R$. Putting everything together:

$$F < R < A < N < C < E.$$

(d) Visually, F is better. Since $C = \frac{E+F}{2}$, and $A < C$, the average of E and F, A must be closer to the *smaller* of E and F, so F is the better estimate.

(e) If we mark off R and C on a number line, N marks the midpoint of $\overline{RC}$ because N is the average of R and C. Since $R < A < N$, A lies closer to R than it does to C, so R is the better estimate.

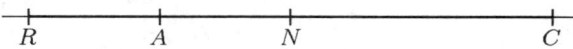

19.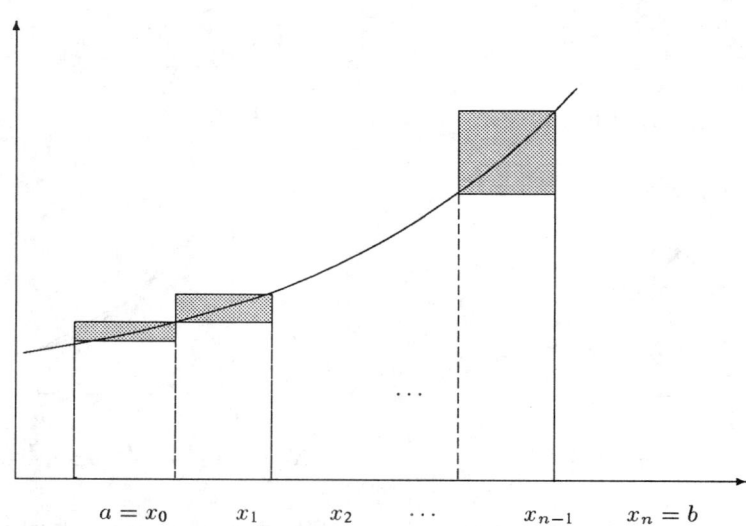

From the diagram, the difference between RIGHT(n) and LEFT(n) is the area of the shaded rectangles.

RIGHT(n) $= f(x_1)\Delta x + f(x_2)\Delta x + \cdots + f(x_n)\Delta x$
LEFT(n) $= f(x_0)\Delta x + f(x_1)\Delta x + \cdots + f(x_{n-1})\Delta x$

Notice that the terms in these two sums are the same, except that RIGHT(n) contains $f(x_n)\Delta x$ ($= f(b)\Delta x$), and LEFT(n) contains $f(x_0)\Delta x$ ($= f(a)\Delta x$). Thus

$$\begin{aligned}\text{RIGHT}(n) &= \text{LEFT}(n) + f(x_n)\Delta x - f(x_0)\Delta x \\ &= \text{LEFT}(n) + f(b)\Delta x - f(a)\Delta x\end{aligned}$$

20.
$$\text{TRAP}(n) = \frac{\text{LEFT}(n) + \text{RIGHT}(n)}{2}$$
$$= \frac{\text{LEFT}(n) + \text{LEFT}(n) + f(b)\Delta x - f(a)\Delta x}{2}$$
$$= \text{LEFT}(n) + \frac{1}{2}(f(b) - f(a))\Delta x$$

21.

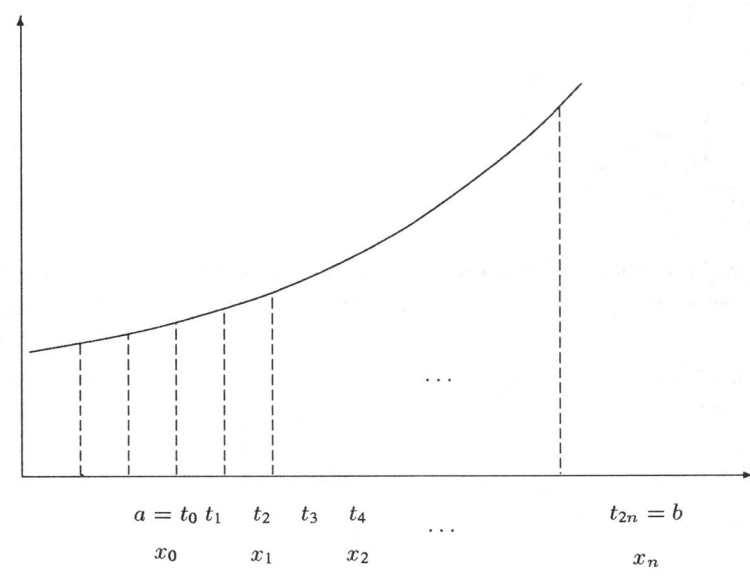

Divide the interval $[a, b]$ into n pieces, by $x_0, x_1, x_2, \ldots, x_n$, and also into $2n$ pieces, by $t_0, t_1, t_2, \ldots, t_{2n}$. Then the x's coincide with the even t's, so $x_0 = t_0$, $x_1 = t_2$, $x_2 = t_4$, $\ldots$, $x_n = t_{2n}$ and $\Delta t = \frac{1}{2}\Delta x$.

$$\text{LEFT}(n) = f(x_0)\Delta x + f(x_1)\Delta x + \cdots + f(x_{n-1})\Delta x$$

Since $\text{MID}(n)$ is obtained by evaluating f at the midpoints $t_1, t_3, t_5, \ldots$ of the x intervals, we get

$$\text{MID}(n) = f(t_1)\Delta x + f(t_3)\Delta x + \cdots + f(t_{2n-1})\Delta x$$

Now
$$\text{LEFT}(2n) = f(t_0)\Delta t + f(t_1)\Delta t + f(t_2)\Delta t + \cdots + f(t_{2n-1})\Delta t.$$

Regroup terms, putting all the even t's first, the odd t's last:

$$\text{LEFT}(2n) = f(t_0)\Delta t + f(t_2)\Delta t + \cdots + f(t_{2n-2})\Delta t + f(t_1)\Delta t + f(t_3)\Delta t + \cdots + f(t_{2n-1})\Delta t$$

$$= \underbrace{f(x_0)\frac{\Delta x}{2} + f(x_1)\frac{\Delta x}{2} + \cdots + f(x_{n-1})\frac{\Delta x}{2}}_{\text{LEFT}(n)/2} + \underbrace{f(t_1)\frac{\Delta x}{2} + f(t_3)\frac{\Delta x}{2} + \cdots + f(t_{2n-1})\frac{\Delta x}{2}}_{\text{MID}(n)/2}$$

So
$$\text{LEFT}(2n) = \frac{1}{2}(\text{LEFT}(n) + \text{MID}(n))$$

22. When $n = 10$, we have $a = 1; b = 2; \Delta x = \frac{1}{10}; f(a) = 1; f(b) = \frac{1}{2}$.
$\text{LEFT}(10) \approx 0.71877, \text{RIGHT}(10) \approx 0.66877, \text{TRAP}(10) \approx 0.69377$
We have
$\text{RIGHT}(10) = \text{LEFT}(10) + f(b)\Delta x - f(a)\Delta x = 0.71877 + \frac{1}{10}(\frac{1}{2}) - \frac{1}{10}(1) = 0.66877$, and $\text{TRAP}(10) = \text{LEFT}(10) + \frac{\Delta x}{2}(f(b) - f(a)) = 0.71877 + \frac{1}{10}\frac{1}{2}(\frac{1}{2} - 1) = 0.69377$,
so the equations are verified.

310 CHAPTER SEVEN /SOLUTIONS

23. First, we compute:

$$(f(b) - f(a))\Delta x = (f(b) - f(a))\left(\frac{b-a}{n}\right)$$

$$= (f(5) - f(2))\left(\frac{3}{n}\right)$$

$$= (21 - 13)\left(\frac{3}{n}\right)$$

$$= \frac{24}{n}$$

RIGHT(10) = LEFT(10) + 24 = 3.156 + 2.4 = 5.556.
TRAP(10) = LEFT(10) + $\frac{1}{2}$(2.4) = 3.156 + 1.2 = 4.356.
LEFT(20) = $\frac{1}{2}$(LEFT(10) + MID(10)) = $\frac{1}{2}$(3.156 + 3.242) = 3.199.
RIGHT(20) = LEFT(20) + 2.4 = 3.199 + 1.2 = 4.399.
TRAP(20) = LEFT(20) + $\frac{1}{2}$(1.2) = 3.199 + 0.6 = 3.799.

Solutions for Section 7.6

1. (a) From Problem 2 on page 304, for $\int_0^4 (x^2 + 1)\, dx$, we have MID(2)= 24 and TRAP(2)= 28. Thus,

$$\text{SIMP}(2) = \frac{2\text{MID}(2) + \text{TRAP}(2)}{3}$$

$$= \frac{2(24) + 28}{3}$$

$$= \frac{76}{3}.$$

   (b)
$$\int_0^4 (x^2 + 1)\, dx = \left(\frac{x^3}{3} + x\right)\Bigg|_0^4 = \left(\frac{64}{3} + 4\right) - (0 + 0) = \frac{76}{3}$$

   (c) Error= 0. Simpson's Rule gives the exact answer.

2. SIMP(10) $\approx$ 0.23182. When we do $n = 10$ intervals and $n = 20$ intervals, these digits match, so they are probably correct digits since Simpson's rule will improve the number of digits of accuracy when we double the number of intervals. This reason also holds true for Problems 3-7.

3. 4.2365. ($n = 10$ intervals, or more).

4. 1.4301. ($n = 10$ intervals, or more)

5. 1.0894. ($n = 10$ intervals, or more)

6. 29.09346. ($n = 20$ intervals, or more)

7. 0.904524. ($n = 10$ intervals, or more)

8. (a) **TABLE 7.1** *Errors for the left and right rule approximations to* $\int_1^2 \frac{1}{x}\, dx = 0.6931471806\ldots$

n	LEFT(n)	Left error	RIGHT(n)	Right error
2	0.833333	−0.14019	0.583333	0.10981
4	0.759524	−0.06638	0.634524	0.05862
8	0.725372	−0.03222	0.662872	0.03028
16	0.709016	−0.01587	0.677766	0.01538
32	0.701021	−0.00787	0.685396	0.00775
64	0.697069	−0.00392	0.689256	0.00389
128	0.695104	−0.00196	0.691198	0.00195

(b) The left errors are negative and the right errors are positive. This occurs because $f(x) = 1/x$ is decreasing, meaning that the left sums are overestimates and the right sums are underestimates. Doubling n approximately halves the error.

(c) **TABLE 7.2** *Errors for the trapezoid and midpoint rule approximations to* $\int_1^2 \frac{1}{x} dx = 0.6931471806\ldots$

n	TRAP(n)	Trap error	MID(n)	Mid error
2	0.708333	−0.01518	0.685714	0.00743
4	0.697024	−0.00387	0.691220	0.00193
8	0.694122	−0.00097	0.692661	0.00049
16	0.6933912	−0.000244	0.6930252	0.000122
32	0.6932082	−0.000061	0.6931166	0.000031
64	0.6931624	−0.000015	0.6931396	0.000008
128	0.6931510	−0.000004	0.6931453	0.000002

(d) The trapezoid errors are negative because $f(x) = 1/x$ is concave up, and thus, the trapezoids overestimate. The midpoint errors are positive. Doubling n approximately quarters the error.

(e) **TABLE 7.3** *Errors for Simpson's rule for* $\int_1^2 \frac{1}{x} dx = 0.6931471806\ldots$

n	SIMP(n)	error
2	0.69325396825	−0.000106788
4	0.69315453065	−0.000007350
8	0.69314765282	−0.000000472
16	0.69314721029	−0.000000030
32	0.69314718242	−0.000000002

The error is multiplied by approximately $1/16$ when n is doubled.

9. (a) $\int_0^4 e^x \, dx = e^x \Big|_0^4 = e^4 - e^0 \approx 53.598\ldots$

(b) Computing the sums directly, since $\Delta x = 2$, we have
LEFT(2)= $2 \cdot e^0 + 2 \cdot e^2 \approx 2(1) + 2(7.389) = 16.778;$ error $= 36.820$.
RIGHT(2)= $2 \cdot e^2 + 2 \cdot e^4 \approx 2(7.389) + 2(54.598) = 123.974;$ error $= -70.376$.
TRAP(2)= $\frac{16.778 + 123.974}{2} = 70.376;$ error $= 16.778$.
MID(2)= $2 \cdot e^1 + 2 \cdot e^3 \approx 2(2.718) + 2(20.086) = 45.608;$ error $= 7.990$.
SIMP(2)= $\frac{2(45.608) + 70.376}{3} = 53.864;$ error $= -0.266$.

(c) Similarly, since $\Delta x = 1$, we have LEFT(4)= 31.193; error $= 22.405$
RIGHT(4)= 84.791; error $= -31.193$
TRAP(4)= 57.992; error $= -4.394$
MID(4)= 51.428; error $= 2.170$
SIMP(4)= 53.616; error $= -0.018$

(d) For LEFT and RIGHT, we expect the error to go down by $1/2$, and this is very roughly what we see. For MID and TRAP, we expect the error to go down by $1/4$, and this is approximately what we see. For SIMP, we expect the error to go down by $1/2^4 = 1/16$, and this is approximately what we see.

10. (a) $\int_0^2 (x^3 + 3x^2)\,dx = \left(\dfrac{x^4}{4} + x^3\right)\Big|_0^2 = 12.$

(b) SIMP(2) = 12.
SIMP(4) = 12.
SIMP(100) = 12.
SIMP(n) = 12 for all n. Simpson's rule always gives the exact answer if the integrand is a polynomial of degree less than 4.

11. Here, the error in the approximation using $n = 10$ is $4 - 2.346 = 1.654$.

(a) Since the error in the LEFT approximation is proportional to $1/n$, when we triple n from 10 to 30 the error is divided by 3, so the error here is $1.654/3 = 0.551333$, giving LEFT(30) = $4 - 0.551333 \approx 3.449$.

(b) The procedure here is identical to part (a), except that the TRAP error is proportional to $1/n^2$, so the error in TRAP(30) will be $1.654/3^2 = 0.183778$, giving TRAP(30) = $4 - 0.183778 \approx 3.816$.

(c) For SIMP, the error will be $1.654/3^4 = 0.0204198$, giving SIMP(30) = $4 - 0.0204198 \approx 3.980$.

12. (a) For the left-hand rule, error is approximately proportional to $\frac{1}{n}$. If we let n_p be the number of subdivisions needed for accuracy to p places, then there is a constant k such that

$$5 \times 10^{-5} = \frac{1}{2} \times 10^{-4} \approx \frac{k}{n_4}$$

$$5 \times 10^{-9} = \frac{1}{2} \times 10^{-8} \approx \frac{k}{n_8}$$

$$5 \times 10^{-13} = \frac{1}{2} \times 10^{-12} \approx \frac{k}{n_{12}}$$

$$5 \times 10^{-21} = \frac{1}{2} \times 10^{-20} \approx \frac{k}{n_{20}}$$

Thus the ratios $n_4 : n_8 : n_{12} : n_{20} \approx 1 : 10^4 : 10^8 : 10^{16}$, and assuming the computer time necessary is proportional to n_p, the computer times are approximately

4 places:	2 seconds	
8 places:	2×10^4 seconds	$\approx$ 6 hours
12 places:	2×10^8 seconds	$\approx$ 6 years
20 places:	2×10^{16} seconds	$\approx$ 600 million years

(b) For the trapezoidal rule, error is approximately proportional to $\frac{1}{n^2}$. If we let N_p be the number of subdivisions needed for accuracy to p places, then there is a constant C such that

$$5 \times 10^{-5} = \frac{1}{2} \times 10^{-4} \approx \frac{C}{N_4^2}$$

$$5 \times 10^{-9} = \frac{1}{2} \times 10^{-8} \approx \frac{C}{N_8^2}$$

$$5 \times 10^{-13} = \frac{1}{2} \times 10^{-12} \approx \frac{C}{N_{12}^2}$$

$$5 \times 10^{-21} = \frac{1}{2} \times 10^{-20} \approx \frac{C}{N_{20}^2}$$

Thus the ratios $N_4^2 : N_8^2 : N_{12}^2 : N_{20}^2 \approx 1 : 10^4 : 10^8 : 10^{16}$, and the ratios $N_4 : N_8 : N_{12} : N_{20} \approx 1 : 10^2 : 10^4 : 10^8$. So the computer times are approximately

4 places:	2 seconds	
8 places:	2×10^2 seconds	$\approx$ 3 minutes
12 places:	2×10^4 seconds	$\approx$ 6 hours
20 places:	2×10^8 seconds	$\approx$ 6 years

13. (a) Since the time for a left-hand approximation is proportional to its accuracy, getting 8 more digits of accuracy will take a factor of 10^8 more time, or, in other words, will take 3 seconds $\times$ 10^8, or 9.5 years.
    (b) Similarly to part (a), the midpoint rule increases in time with the square root of accuracy, so it will take 3 seconds $\times$ 10^4, or 8.33 hours.
    (c) Simpson's rule increases in time with the fourth root of accuracy, so it will only take 3 seconds $\times$ 10^2, or 5 minutes, to get our desired accuracy.

14. We assume that the error is of the same sign for both LEFT(10) and LEFT(20); that is, they are both underestimates or overestimates. Since LEFT(20) < LEFT(10), and LEFT(20) is more accurate, they must both be overestimates.

We assume that LEFT(10) is twice as far from the actual value as LEFT(20). Thus

$$\text{Actual} - \text{LEFT}(20) = \text{LEFT}(20) - \text{LEFT}(10)$$
$$\text{Actual} = 2\,\text{LEFT}(20) - \text{LEFT}(10)$$
$$= 0.34289.$$

Thus the error for LEFT(10) is 0.04186.

15. Since the midpoint rule is sensitive to f'', the simplifying assumption should be that f'' does not change sign in the interval of integration. Thus MID(10) and MID(20) will both be overestimates or will both be underestimates. Since the larger number, MID(10) is less accurate than the smaller number, they must both be overestimates. Then the information that ERROR(10) = 4 $\times$ ERROR(20) means that the the value of the integral and the two sums are arranged as follows:

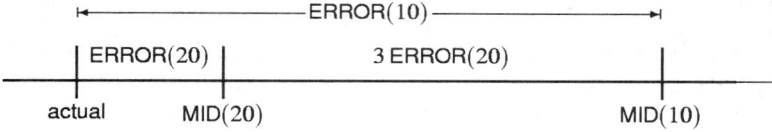

Thus

$$3 \times \text{ERROR}(20) = \text{MID}(10) - \text{MID}(20) = 35.619 - 35.415 = 0.204,$$

so ERROR(20) = 0.068 and ERROR(10) = 4 $\times$ ERROR(20) = 0.272.

16. Since TRAP(n) seems to be decreasing as n increases, we can assume that TRAP(10) and TRAP(30) are both overestimates. We know that the error in the trapezoid rule is approximately proportional to $1/n^2$. In going from $n = 10$ to $n = 30$, n is multiplied by 3 and so we expect the error to go down roughly by a factor of $1/3^2$, or $1/9$. Therefore, if we let $d = |\text{error}(30)|$, then we have $9d \approx |\text{error}(10)|$.

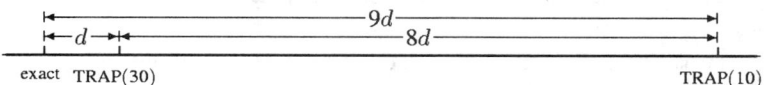

We see from the figure above that the difference between TRAP(10) and TRAP(30) is $8d$, so

$$8d = \text{TRAP(10)} - \text{TRAP(30)}$$
$$8d = 12.676 - 10.420$$
$$d = 0.282.$$

Since d is the magnitude of the error for TRAP(30), and since the exact value is less than TRAP(30), we have

$$\text{Exact} \approx \text{TRAP(30)} - d$$
$$= 10.420 - 0.282$$
$$= 10.138.$$

The exact value[1] of the integral is about 10.138.

---

[1]This method of improving numerical estimates is essentially equivalent to Richardson's h^2 extrapolation, also called extrapolation to the limit. See, for instance, *Survey of Numerical Analysis*, ed. John Todd, (New York: McGraw-Hill, 1962).

17. (a) Since MID(n) seems to be decreasing as n increases, we can assume that MID(10) and MID(20) are both overestimates. We know that the error in the midpoint rule is approximately proportional to $1/n^2$. In going from $n = 10$ to $n = 20$, n is multiplied by 2 and so we expect the error to go down roughly by a factor of $1/2^2$, or $1/4$. Therefore, if we let $d = |\text{error}(20)|$, then we have $4d \approx |\text{error}(10)|$.

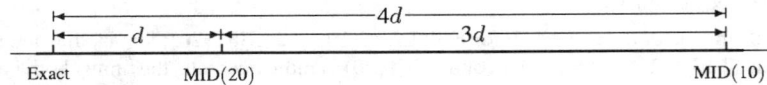

As we see from the figure above, $3d$ is equal to the difference between MID(10) and MID(20), so

$$3d = \text{MID}(10) - \text{MID}(20)$$
$$3d = 5.364 - 4.926$$
$$d = 0.146.$$

Since d is the magnitude of the error for MID(20), and since the exact value is less than MID(20), we have

$$\text{Exact} \approx \text{MID}(20) - d$$
$$= 4.926 - 0.146 = 4.780.$$

The exact value of the integral is about 4.780.

(b) In part (a), we estimated that the error for MID(20) ≈ 0.146. As n goes from 20 to 60, n is multiplied by 3 and so we expect the error to go down by $1/9$. The error for MID(60) is about $(1/9)0.146 = 0.0162$. Since we estimated the exact value at 4.780, we estimate that MID(60) is about $4.780 + 0.0162 = 4.7962$.

18. Here, the error in the left-hand approximation with $n = 2$ is just Exact $-$ LEFT(2) $= 0.69315 - 0.83333 = -0.14018$. We know that the accuracy of the LEFT rule is proportional to $1/n$, so when the number of subintervals multiplies by 10 to $n = 20$, the error decreases by a factor of 10 to about 0.014018, yielding LEFT(20) $=$ Exact $-$ Error $\approx 0.69315 + 0.014018 = 0.707168$. We can repeat an analogous procedure for all the rules in the table, with the exception that for TRAP and MID the error is proportional to $1/n^2$, so the error decreases by a factor of 100 when n goes to 20, and for SIMP the error is proportional to $1/n^4$, so the error decreases by a factor of 10,000 when $n = 20$. The result of such calculations is shown below.

	Approximation $n = 2$	Error $n = 2$	Error $n = 20$
LEFT	0.83333	-0.14018	-0.014018
RIGHT	0.58333	0.10982	0.010982
TRAP	0.70833	-0.01518	-0.0001518
MID	0.68571	0.00744	0.0000744
SIMP	0.69325	-0.0001	-0.00000001

19. (a) Since LEFT is too small and RIGHT is too big, the function appears to be increasing. Since TRAP is too small and MID is too big, the function appears to be concave down.

(b) Error = Exact $-$ Approximation = $7.621372 -$ Approximation. See column (b) below.

(c) Since n goes from 3 to 30, it is multiplied by 10. Thus, the errors for the LEFT and RIGHT go down by a factor of $1/10$, the errors for TRAP and MID go down by a factor of $1/10^2 = 1/100$, and the error for SIMP goes down by a factor of $1/10^4 = 1/10000$. See column (c) below.

	approximation for $n = 3$	(b) error for $n = 3$	(c) error for $n = 30$
LEFT	5.416101	2.205271	0.2205721
RIGHT	9.307921	-1.686549	-0.1686549
TRAP	7.362011	0.259361	0.0025936
MID	7.742402	-0.12103	-0.0012103
SIMP	7.615605	0.005767	0.0000006

20. (a) If $f(x) = 1$, then
$$\int_a^b f(x)\, dx = (b-a).$$
Also,
$$\frac{h}{3}\left(\frac{f(a)}{2} + 2f(m) + \frac{f(b)}{2}\right) = \frac{b-a}{3}\left(\frac{1}{2} + 2 + \frac{1}{2}\right) = (b-a).$$
So the equation holds for $f(x) = 1$.

If $f(x) = x$, then
$$\int_a^b f(x)\, dx = \left.\frac{x^2}{2}\right|_a^b = \frac{b^2 - a^2}{2}.$$
Also,
$$\frac{h}{3}\left(\frac{f(a)}{2} + 2f(m) + \frac{f(b)}{2}\right) = \frac{b-a}{3}\left(\frac{a}{2} + 2\frac{a+b}{2} + \frac{b}{2}\right)$$
$$= \frac{b-a}{3}\left(\frac{a}{2} + a + b + \frac{b}{2}\right)$$
$$= \frac{b-a}{3}\left(\frac{3}{2}b + \frac{3}{2}a\right)$$
$$= \frac{(b-a)(b+a)}{2}$$
$$= \frac{b^2 - a^2}{2}.$$
So the equation holds for $f(x) = x$.

If $f(x) = x^2$, then $\int_a^b f(x)\, dx = \left.\frac{x^3}{3}\right|_a^b = \frac{b^3 - a^3}{3}$. Also,
$$\frac{h}{3}\left(\frac{f(a)}{2} + 2f(m) + \frac{f(b)}{2}\right) = \frac{b-a}{3}\left(\frac{a^2}{2} + 2\left(\frac{a+b}{2}\right)^2 + \frac{b^2}{2}\right)$$
$$= \frac{b-a}{3}\left(\frac{a^2}{2} + \frac{a^2 + 2ab + b^2}{2} + \frac{b^2}{2}\right)$$
$$= \frac{b-a}{3}\left(\frac{2a^2 + 2ab + 2b^2}{2}\right)$$
$$= \frac{b-a}{3}\left(a^2 + ab + b^2\right)$$
$$= \frac{b^3 - a^3}{3}.$$
So the equation holds for $f(x) = x^2$.

(b) For any quadratic function, $f(x) = Ax^2 + Bx + C$, the "Facts about Sums and Constant Multiples of Integrands" give us:
$$\int_a^b f(x)\, dx = \int_a^b (Ax^2 + Bx + C)\, dx = A\int_a^b x^2\, dx + B\int_a^b x\, dx + C\int_a^b 1\, dx.$$
Now we use the results of part (a) to get:
$$\int_a^b f(x)\, dx = A\frac{h}{3}\left(\frac{a^2}{2} + 2m^2 + \frac{b^2}{2}\right) + B\frac{h}{3}\left(\frac{a}{2} + 2m + \frac{b}{2}\right) + C\frac{h}{3}\left(\frac{1}{2} + 2\cdot 1 + \frac{1}{2}\right)$$
$$= \frac{h}{3}\left(\frac{Aa^2 + Ba + C}{2} + 2(Am^2 + Bm + C) + \frac{Ab^2 + Bb + C}{2}\right)$$
$$= \frac{h}{3}\left(\frac{f(a)}{2} + 2f(m) + \frac{f(b)}{2}\right)$$

21. (a) Suppose $q_i(x)$ is the quadratic function approximating $f(x)$ on the subinterval $[x_i, x_{i+1}]$, and m_i is the midpoint of the interval, $m_i = (x_i + x_{i+1})/2$. Then, using the equation in Problem 20, with $a = x_i$ and $b = x_{i+1}$ and $h = \Delta x = x_{i+1} - x_i$:

$$\int_{x_i}^{x_{i+1}} f(x)\,dx \approx \int_{x_i}^{x_{i+1}} q_i(x)\,dx = \frac{\Delta x}{3}\left(\frac{q_i(x_i)}{2} + 2q_i(m_i) + \frac{q_i(x_{i+1})}{2}\right).$$

(b) Summing over all subintervals gives

$$\int_a^b f(x)\,dx \approx \sum_{i=0}^{n-1} \int_{x_i}^{x_{i+1}} q_i(x)\,dx = \sum_{i=0}^{n-1} \frac{\Delta x}{3}\left(\frac{q_i(x_i)}{2} + 2q_i(m_i) + \frac{q_i(x_{i+1})}{2}\right).$$

Splitting the sum into two parts:

$$= \frac{2}{3}\sum_{i=0}^{n-1} q_i(m_i)\Delta x + \frac{1}{3}\sum_{i=0}^{n-1} \frac{q_i(x_i) + q_i(x_{i+1})}{2}\Delta x$$

$$= \frac{2}{3}\text{MID}(n) + \frac{1}{3}\text{TRAP}(n)$$

$$= \text{SIMP}(n).$$

Solutions for Section 7.7

1.
$$\int_1^\infty e^{-2x}\,dx = \lim_{b\to\infty}\int_1^b e^{-2x}\,dx = \lim_{b\to\infty} -\frac{e^{-2x}}{2}\bigg|_1^b$$
$$= \lim_{b\to\infty}(-e^{-2b}/2 + e^{-2}/2) = 0 + e^{-2}/2 = e^{-2}/2,$$

where the first limit is 0 because $\lim_{x\to\infty} e^{-x} = 0$.

2. Using integration by parts with $u = x$ and $v' = e^{-x}$, we find that

$$\int xe^{-x}\,dx = -xe^{-x} - \int -e^{-x}\,dx = -(1+x)e^{-x}$$

so

$$\int_0^\infty \frac{x}{e^x}\,dx = \lim_{b\to\infty}\int_0^b \frac{x}{e^x}\,dx$$
$$= \lim_{b\to\infty} -1(1+x)e^{-x}\bigg|_0^b$$
$$= \lim_{b\to\infty}\left[1 - (1+b)e^{-b}\right]$$
$$= 1.$$

3.
$$\int_1^\infty \frac{x}{4+x^2} = \lim_{b\to\infty}\int_1^b \frac{x}{4+x^2}\,dx = \lim_{b\to\infty}\frac{1}{2}\ln|4+x^2|\bigg|_1^b = \lim_{b\to\infty}\frac{1}{2}\ln|4+b^2| - \frac{1}{2}\ln 5.$$

As $b \to \infty$, $\ln|4+b^2| \to \infty$, so the limit diverges.

4.
$$\int_{-\infty}^{0} \frac{e^x}{1+e^x}\,dx = \lim_{b\to -\infty}\int_{b}^{0} \frac{e^x}{1+e^x}\,dx$$
$$= \lim_{b\to -\infty} \ln|1+e^x|\Big|_{b}^{0}$$
$$= \lim_{b\to -\infty} [\ln|1+e^0| - \ln|1+e^b|]$$
$$= \ln(1+1) - \ln(1+0) = \ln 2.$$

5. First, we note that $1/(z^2+25)$ is an even function. Therefore,
$$\int_{-\infty}^{\infty} \frac{dz}{z^2+25} = \int_{-\infty}^{0} \frac{dz}{z^2+25} + \int_{0}^{\infty} \frac{dz}{z^2+25} = 2\int_{0}^{\infty} \frac{dz}{z^2+25}.$$

We'll now evaluate this improper integral by using a limit:
$$\int_{0}^{\infty} \frac{dz}{z^2+25} = \lim_{b\to\infty}\left(\frac{1}{5}\arctan(b/5) - \frac{1}{5}\arctan(0)\right) = \frac{1}{5}\cdot\frac{\pi}{2} = \frac{\pi}{10}.$$

So the original integral is twice that, namely $\pi/5$.

6. This is an improper integral because $\sqrt{16-x^2} = 0$ at $x=4$. So
$$\int_{0}^{4} \frac{dx}{\sqrt{16-x^2}} = \lim_{b\to 4^-}\int_{0}^{b}\frac{dx}{\sqrt{16-x^2}}$$
$$= \lim_{b\to 4^-} (\arcsin x/4)\Big|_{0}^{b}$$
$$= \lim_{b\to 4^-} [\arcsin(b/4) - \arcsin(0)] = \pi/2 - 0 = \pi/2.$$

7.
$$\int_{\pi/4}^{\pi/2} \frac{\sin x}{\sqrt{\cos x}}\,dx = \lim_{b\to \pi/2^-}\int_{\pi/4}^{b}\frac{\sin x}{\sqrt{\cos x}}\,dx$$
$$= \lim_{b\to \pi/2^-} -\int_{\pi/4}^{b} (\cos x)^{-1/2}(-\sin x)\,dx$$
$$= \lim_{b\to \pi/2^-} -2(\cos x)^{1/2}\Big|_{\pi/4}^{b}$$
$$= \lim_{b\to \pi/2^-} [-2(\cos b)^{1/2} + 2(\cos \pi/4)^{1/2}]$$
$$= 2\left(\frac{\sqrt{2}}{2}\right)^{\frac{1}{2}} = 2^{\frac{3}{4}}.$$

8. This integral is improper because $1/v$ blows up at $v=0$. To evaluate it, we must split the region of integration up into two pieces, from 0 to 1 and from -1 to 0. But notice,
$$\int_{0}^{1}\frac{1}{v}\,dv = \lim_{b\to 0^+}\int_{b}^{1}\frac{1}{v}\,dv = \lim_{b\to 0^+}\left(\ln v\Big|_{b}^{1}\right) = -\ln b.$$

As $b\to 0^+$, this goes to infinity and the integral diverges, so our original integral also diverges.

9.
$$\lim_{a\to 0^+}\int_{a}^{1}\frac{x^4+1}{x}\,dx = \lim_{a\to 0^+}\left(\frac{x^4}{4}+\ln x\right)\Big|_{a}^{1} = \lim_{a\to 0^+}[1/4 - (a^4/4 + \ln a)],$$

which diverges as $a\to 0$, since $\ln a \to -\infty$.

318 CHAPTER SEVEN /SOLUTIONS

10.
$$\int_1^\infty \frac{1}{x^2+1}\,dx = \lim_{b\to\infty} \int_1^b \frac{1}{x^2+1}\,dx$$
$$= \lim_{b\to\infty} \arctan(x)\Big|_1^b$$
$$= \lim_{b\to\infty} [\arctan(b) - \arctan(1)]$$
$$= \pi/2 - \pi/4 = \pi/4.$$

11.
$$\int_1^\infty \frac{1}{\sqrt{x^2+1}}\,dx = \lim_{b\to\infty} \int_1^b \frac{1}{\sqrt{x^2+1}}\,dx$$
$$= \lim_{b\to\infty} \ln|x + \sqrt{x^2+1}|\Big|_1^b$$
$$= \lim_{b\to\infty} \ln(b + \sqrt{b^2+1}) - \ln(1+\sqrt{2}).$$

As $b \to \infty$, this limit does not exist, so the integral diverges.

12. We use V-26 with $a = 4$ and $b = -4$:
$$\int_0^4 \frac{1}{u^2-16}\,du = \lim_{b\to 4^-} \int_0^b \frac{1}{u^2-16}\,du$$
$$= \lim_{b\to 4^-} \int_0^b \frac{1}{(u-4)(u+4)}\,du$$
$$= \lim_{b\to 4^-} \frac{(\ln|u-4| - \ln|u+4|)}{8}\Big|_0^b$$
$$= \lim_{b\to 4^-} \frac{1}{8}\left(\ln|b-4| + \ln 4 - \ln|b+4| - \ln 4\right).$$

As $b \to 4^-$, $\ln|b-4| \to -\infty$, so the limit does not exist and the integral diverges.

13.
$$\int_1^\infty \frac{y}{y^4+1}\,dy = \lim_{b\to\infty} \frac{1}{2}\int_1^b \frac{2y}{(y^2)^2+1}\,dy$$
$$= \lim_{b\to\infty} \frac{1}{2}\arctan(y^2)\Big|_1^b$$
$$= \lim_{b\to\infty} \frac{1}{2}[\arctan(b^2) - \arctan 1]$$
$$= (1/2)[\pi/2 - \pi/4] = \pi/8.$$

14. With the substitution $w = \ln x$, $dw = \frac{1}{x}dx$,
$$\int \frac{dx}{x\ln x} = \int \frac{1}{w}\,dw = \ln|w| + C = \ln|\ln x| + C$$

so
$$\int_2^\infty \frac{dx}{x\ln x} = \lim_{b\to\infty} \int_2^b \frac{dx}{x\ln x}$$
$$= \lim_{b\to\infty} \ln|\ln x|\Big|_2^b$$
$$= \lim_{b\to\infty} [\ln|\ln b| - \ln|\ln 2|].$$

As $b \to \infty$, the limit goes to ∞ and hence the integral diverges.

15. With the substitution $w = \ln x$, $dw = \frac{1}{x}dx$,

$$\int \frac{\ln x}{x}\, dx = \int w\, dw = \frac{1}{2}w^2 + C = \frac{1}{2}(\ln x)^2 + C$$

so

$$\int_0^1 \frac{\ln x}{x}\, dx = \lim_{a \to 0^+} \int_a^1 \frac{\ln x}{x}\, dx = \lim_{a \to 0^+} \frac{1}{2}[\ln(x)]^2 \bigg|_a^1 = \lim_{a \to 0^+} -\frac{1}{2}[\ln(a)]^2.$$

As $a \to 0^+$, $\ln a \to -\infty$, so the integral diverges.

16. This is a proper integral; use V-26 in the integral table with $a = 4$ and $b = -4$.

$$\int_{16}^{20} \frac{1}{y^2 - 16}\, dy = \int_{16}^{20} \frac{1}{(y-4)(y+4)}\, dy$$
$$= \frac{\ln|y-4| - \ln|y+4|}{8} \bigg|_{16}^{20}$$
$$= \frac{\ln 16 - \ln 24 - (\ln 12 - \ln 20)}{8}$$
$$= \frac{\ln 320 - \ln 288}{8} = \frac{1}{8}\ln(10/9) = 0.01317.$$

17. As in Problem 14, $\int \frac{dx}{x \ln x} = \ln|\ln x| + C$, so

$$\int_1^2 \frac{dx}{x \ln x} = \lim_{b \to 1^+} \int_b^2 \frac{dx}{x \ln x}$$
$$= \lim_{b \to 1^+} \ln|\ln x| \bigg|_b^2$$
$$= \lim_{b \to 1^+} \ln(\ln 2) - \ln(\ln b).$$

As $b \to 1^+$, $\ln(\ln b) \to -\infty$, so the integral diverges.

18. Using the substitution $w = -x^{\frac{1}{2}}$, $-2dw = x^{-\frac{1}{2}}\, dx$,

$$\int e^{-x^{\frac{1}{2}}} x^{-\frac{1}{2}}\, dx = -2\int e^w\, dw = -2e^{-x^{\frac{1}{2}}} + C.$$

So

$$\int_0^\pi \frac{1}{\sqrt{x}} e^{-\sqrt{x}}\, dx = \lim_{b \to 0^+} \int_b^\pi \frac{1}{\sqrt{x}} e^{-\sqrt{x}}\, dx$$
$$= \lim_{b \to 0^+} -2e^{-\sqrt{x}} \bigg|_b^\pi$$
$$= 2 - 2e^{-\sqrt{\pi}}.$$

19. Letting $w = \ln x$, $dw = \frac{1}{x}dx$,

$$\int \frac{dx}{x(\ln x)^2} = \int w^{-2}\, dw = -w^{-1} + C = -\frac{1}{\ln x} + C,$$

so

$$\int_3^\infty \frac{dx}{x(\ln x)^2} = \lim_{b \to \infty} \int_3^b \frac{dx}{x(\ln x)^2}$$
$$= \lim_{b \to \infty} \left(-\frac{1}{\ln b} + \frac{1}{\ln 3} \right)$$
$$= \frac{1}{\ln 3}.$$

320 CHAPTER SEVEN /SOLUTIONS

20.
$$\int_0^2 \frac{1}{\sqrt{4-x^2}} \, dx = \lim_{b \to 2^-} \int_0^b \frac{1}{\sqrt{4-x^2}} \, dx$$
$$= \lim_{b \to 2^-} \arcsin \frac{x}{2} \Big|_0^b$$
$$= \lim_{b \to 2^-} \arcsin \frac{b}{2} = \arcsin 1 = \frac{\pi}{2}.$$

21. $\displaystyle\int_4^\infty \frac{dx}{(x-1)^2} = \lim_{b \to \infty} \int_4^b \frac{dx}{(x-1)^2} = \lim_{b \to \infty} -\frac{1}{(x-1)} \Big|_4^b = \lim_{b \to \infty} \left[-\frac{1}{b-1} + \frac{1}{3}\right] = \frac{1}{3}.$

22. $\displaystyle\int \frac{dx}{x^2-1} = \int \frac{dx}{(x-1)(x+1)} = \frac{1}{2}(\ln|x-1| - \ln|x+1|) + C = \frac{1}{2}\left(\ln \frac{|x-1|}{|x+1|}\right) + C$, so

$$\int_4^\infty \frac{dx}{x^2-1} = \lim_{b \to \infty} \int_4^b \frac{dx}{x^2-1}$$
$$= \lim_{b \to \infty} \frac{1}{2} \left(\ln \frac{|x-1|}{|x+1|}\right) \Big|_4^b$$
$$= \lim_{b \to \infty} \left[\frac{1}{2} \ln \left(\frac{b-1}{b+1}\right) - \frac{1}{2} \ln \frac{3}{5}\right]$$
$$= -\frac{1}{2} \ln \frac{3}{5} = \frac{1}{2} \ln \frac{5}{3}.$$

23.
$$\int_7^\infty \frac{dy}{\sqrt{y-5}} = \lim_{b \to \infty} \int_7^b \frac{dy}{\sqrt{y-5}}$$
$$= \lim_{b \to \infty} 2\sqrt{y-5} \Big|_7^b$$
$$= \lim_{b \to \infty} (2\sqrt{b-5} - 2\sqrt{2}).$$

As $b \to \infty$, this limit goes to ∞, so the integral diverges.

24.
$$\int_\pi^\infty \sin y \, dy = \lim_{b \to \infty} \int_\pi^b \sin y \, dy$$
$$= \lim_{b \to \infty} (-\cos y) \Big|_\pi^b$$
$$= \lim_{b \to \infty} [-\cos b - (-\cos \pi)].$$

As $b \to \infty$, $-\cos b$ fluctuates between -1 and 1, so the limit fails to exist: the integral diverges. (This doesn't follow right from the fact that $\sin y$ fluctuates between -1 and 1!)

25. Since the graph is above the x-axis for $x \geq 0$, we have

$$\text{Area} = \int_0^\infty xe^{-x} \, dx = \lim_{b \to \infty} \int_0^b xe^{-x} \, dx$$
$$= \lim_{b \to \infty} \left(-xe^{-x} \Big|_0^b + \int_0^b e^{-x} \, dx\right)$$
$$= \lim_{b \to \infty} \left(-be^{-b} - e^{-x} \Big|_0^b\right)$$
$$= \lim_{b \to \infty} (-be^{-b} - e^{-b} + e^0) = 1.$$

26. The curve has an asymptote at $t = \frac{\pi}{2}$, and so the area integral is improper there.

$$\text{Area} = \int_0^{\frac{\pi}{2}} \frac{dt}{\cos^2 t} = \lim_{b \to \frac{\pi}{2}} \int_0^b \frac{dt}{\cos^2 t} = \lim_{b \to \frac{\pi}{2}} \tan t \Big|_0^b,$$

which diverges. Therefore the area is infinite.

27. (a)

$$\int_0^\infty h(x)\,dx = \int_0^4 h(x)\,dx + \int_4^\infty h(x)\,dx$$

$$\int_4^\infty h(x)\,dx = \int_4^\infty x^{-2}\,dx$$

$$= \lim_{b \to \infty} \int_4^b x^{-2}\,dx$$

$$= \lim_{b \to \infty} [-x^{-1}]\Big|_4^b = \lim_{b \to \infty} [\tfrac{1}{4} - \tfrac{1}{b}] = \tfrac{1}{4}.$$

$$\int_0^4 h(x)\,dx = \int_0^4 \left(x^{-\frac{3}{2}} - \tfrac{1}{16}\right) dx$$

$$= \int_0^4 x^{-\frac{3}{2}}\,dx - \int_0^4 \tfrac{1}{16}\,dx$$

$$= \lim_{b \to 0^+} \int_b^4 x^{-\frac{3}{2}}\,dx - \tfrac{1}{4}$$

$$= \lim_{b \to 0^+} -2x^{-\frac{1}{2}}\Big|_b^4 - \tfrac{1}{4}$$

$$= \lim_{b \to 0^+} \left(\tfrac{2}{\sqrt{b}} - 1\right) - \tfrac{1}{4}.$$

As $b \to 0^+$, this limit diverges, so the entire integral must diverge.

(b) Since h is differentiable for $0 < x < 4$ and for $x > 4$, the question is whether the slope to the left of $x = 4$ is the same as the slope to the right of $x = 4$. If so, h is differentiable. For $0 < x < 4$, $h'(x) = -\frac{3}{2}x^{-5/2}$ so $\lim_{x \to 4^-} h'(x) = -\frac{3}{2}(4)^{-5/2} = -\frac{3}{64}$. For $x > 4$, $h'(x) = -\frac{2}{x^3}$ so $\lim_{x \to 4^+} h'(x) = -\frac{2}{4^3} = -\frac{1}{32}$. This shows that, to the left of $x = 4$, the slope is $-\frac{3}{64}$ and to the right of $x = 4$, the slope is $-\frac{1}{32}$. Since the slope is not the same on either side of $x = 4$, the curve is not smooth there. If magnified you would see a "corner" at this point, so h is not differentiable there.

28. We let $t = (x - a)/\sqrt{b}$. This means that $dt = dx/\sqrt{b}$, and that $t = \pm\infty$ when $x = \pm\infty$. We have

$$\int_{-\infty}^\infty e^{-(x-a)^2/b}\,dx = \int_{-\infty}^\infty e^{-t^2}(\sqrt{b}\,dt) = \sqrt{b}\int_{-\infty}^\infty e^{-t^2}\,dt = \sqrt{b}\sqrt{\pi} = \sqrt{b\pi}.$$

29. (a)

$$\Gamma(1) = \int_0^\infty e^{-t}\,dt$$

$$= \lim_{b \to \infty} \int_0^b e^{-t}\,dt$$

$$= \lim_{b \to \infty} -e^{-t}\Big|_0^b$$

$$= \lim_{b \to \infty} [1 - e^{-b}] = 1.$$

Using Problem 2,

$$\Gamma(2) = \int_0^\infty te^{-t}\,dt = 1.$$

(b) We integrate by parts. Let $u = t^n$, $v' = e^{-t}$. Then $u' = nt^{n-1}$ and $v = -e^{-t}$, so
$$\int t^n e^{-t}\, dt = -t^n e^{-t} + n \int t^{n-1} e^{-t}\, dt.$$

So
$$\begin{aligned}
\Gamma(n+1) &= \int_0^\infty t^n e^{-t}\, dt \\
&= \lim_{b\to\infty} \int_0^b t^n e^{-t}\, dt \\
&= \lim_{b\to\infty} \left[-t^n e^{-t} \Big|_0^b + n \int_0^b t^{n-1} e^{-t}\, dt \right] \\
&= \lim_{b\to\infty} -b^n e^{-b} + \lim_{b\to\infty} n \int_0^b t^{n-1} e^{-t}\, dt \\
&= 0 + n \int_0^\infty t^{n-1} e^{-t}\, dt \\
&= n\Gamma(n).
\end{aligned}$$

(c) We already have $\Gamma(1) = 1$ and $\Gamma(2) = 1$. Using $\Gamma(n+1) = n\Gamma(n)$ we can get
$$\begin{aligned}
\Gamma(3) &= 2\Gamma(2) = 2 \\
\Gamma(4) &= 3\Gamma(3) = 3\cdot 2 \\
\Gamma(5) &= 4\Gamma(4) = 4\cdot 3\cdot 2.
\end{aligned}$$

So it appears that $\Gamma(n)$ is just the first $n-1$ numbers multiplied together, so $\Gamma(n) = (n-1)!$.

30. (a) Using a calculator or a computer, the graph is:

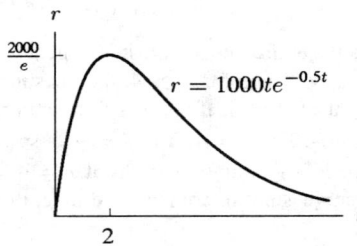

(b) People are getting sick fastest when the rate of infection is highest, i.e. when r is at its maximum. Since
$$\begin{aligned}
r' &= 1000e^{-0.5t} - 1000(0.5)te^{-0.5t} \\
&= 500e^{-0.5t}(2 - t)
\end{aligned}$$
this must occur at $t = 2$.

(c) The total number of sick people $= \int_0^\infty 1000te^{-0.5t}\, dt$.

Using integration by parts, with $u = t$, $v' = e^{-0.5t}$:
$$\begin{aligned}
\text{Total} &= \lim_{b\to\infty} 1000 \left(\frac{-t}{0.5} e^{-0.5t} \Big|_0^b - \int_0^b \frac{-1}{0.5} e^{-0.5t}\, dt \right) \\
&= \lim_{b\to\infty} 1000 \left(-2be^{-0.5b} - \frac{2}{0.5} e^{-0.5b} \right) \Big|_0^b \\
&= \lim_{b\to\infty} 1000 \left(-2be^{-0.5b} - 4e^{-0.5b} + 4 \right) \\
&= 4000 \text{ people.}
\end{aligned}$$

31. The energy required is

$$E = \int_1^\infty \frac{kq_1q_2}{r^2}\,dr = kq_1q_2 \lim_{b\to\infty} \left.-\frac{1}{r}\right|_1^b$$
$$= (9\times 10^9)(1)(1)(1) = 9\times 10^9 \text{ joules}$$

32. The factor $\ln x$ grows slowly enough not to change the convergence or divergence of the integral, although it will change what it converges or diverges to.

Integrating by parts or using the table of integrals, we get

$$\int_e^\infty x^p \ln x\,dx = \lim_{b\to\infty} \int_e^b x^p \ln x\,dx$$
$$= \lim_{b\to\infty} \left[\frac{1}{p+1}x^{p+1}\ln x - \frac{1}{(p+1)^2}x^{p+1}\right]\Big|_e^b$$
$$= \lim_{b\to\infty}\left[\left(\frac{1}{p+1}b^{p+1}\ln b - \frac{1}{(p+1)^2}b^{p+1}\right)\right.$$
$$\left. - \left(\frac{1}{p+1}e^{p+1} - \frac{1}{(p+1)^2}e^{p+1}\right)\right].$$

If $p > -1$, then $(p+1)$ is positive and the limit does not exist since b^{p+1} and $\ln b$ both approach ∞ as b does.

If $p < -1$, then $(p+1)$ is negative and both b^{p+1} and $b^{p+1}\ln b$ approach 0 as $b\to\infty$. (This follows by looking at graphs of $x^{p+1}\ln x$ (for different values of p), or by noting that $\ln x$ grows more slowly than x^{p+1} tends to 0.) So the value of the integral is $-pe^{p+1}/(p+1)^2$.

The case $p = -1$ has to be handled separately. For $p = -1$,

$$\int_e^\infty \frac{\ln x}{x}\,dx = \lim_{b\to\infty}\int_e^b \frac{\ln x}{x}\,dx = \lim_{b\to\infty} \frac{(\ln x)^2}{2}\Big|_e^b = \lim_{b\to\infty}\left(\frac{(\ln b)^2 - 1}{2}\right).$$

As $b\to\infty$, this limit does not exist, so the integral diverges if $p = -1$.

To summarize, $\int_e^\infty x^p \ln x\,dx$ converges for $p < -1$ to the value $-pe^{p+1}/(p+1)^2$.

33. The factor $\ln x$ grows slowly enough (as $x\to 0^+$) not to change the convergence or divergence of the integral, although it will change what it converges or diverges to.

The integral is always improper, because $\ln x$ is not defined for $x = 0$. Integrating by parts (or, alternatively, the integral table) yields

$$\int_0^e x^p \ln x\,dx = \lim_{a\to 0^+} \int_a^e x^p \ln x\,dx$$
$$= \lim_{a\to 0^+} \left(\frac{1}{p+1}x^{p+1}\ln x - \frac{1}{(p+1)^2}x^{p+1}\right)\Big|_a^e$$
$$= \lim_{a\to 0^+}\left[\left(\frac{1}{p+1}e^{p+1} - \frac{1}{(p+1)^2}e^{p+1}\right)\right.$$
$$\left.-\left(\frac{1}{p+1}a^{p+1}\ln a - \frac{1}{(p+1)^2}a^{p+1}\right)\right].$$

If $p < -1$, then $(p+1)$ is negative, so as $a\to 0^+$, $a^{p+1}\to\infty$ and $\ln a\to -\infty$, and therefore the limit does not exist.

If $p > -1$, then $(p+1)$ is positive and it's easy to see that $a^{p+1}\to 0$ as $a\to 0$. Looking at graphs of $x^{p+1}\ln x$ (for different values of p) shows that $a^{p+1}\ln a \to 0$ as $a\to 0$. This isn't so easy to see analytically. It's true because if we let $t = \frac{1}{a}$ then

$$\lim_{a\to 0^+} a^{p+1}\ln a = \lim_{t\to\infty}\left(\frac{1}{t}\right)^{p+1}\ln\left(\frac{1}{t}\right) = \lim_{t\to\infty} -\frac{\ln t}{t^{p+1}}.$$

This last limit is zero because $\ln t$ grows very slowly, much more slowly than t^{p+1}. So if $p > -1$, the integral converges and equals $e^{p+1}[1/(p+1) - 1/(p+1)^2] = pe^{p+1}/(p+1)^2$.

What happens if $p = -1$? Then we get

$$\int_0^e \frac{\ln x}{x} dx = \lim_{a \to 0^+} \int_a^e \frac{\ln x}{x} dx$$

$$= \lim_{a \to 0^+} \frac{(\ln x)^2}{2}\bigg|_a^e$$

$$= \lim_{a \to 0^+} \left(\frac{1 - (\ln a)^2}{2}\right).$$

Since $\ln a \to -\infty$ as $a \to 0^+$, this limit does not exist.

To summarize, $\int_0^e x^p \ln x$ converges for $p > -1$ to the value $pe^{p+1}/(p+1)^2$.

Solutions for Section 7.8

1. (a) The area is infinite. The area under $1/x$ is infinite and the area under $1/x^2$ is 1. So the area between the two has to be infinite also.
   (b) Since $f(x)$ is bounded between 0 and $1/x^2$, and the area under $1/x^2$ is finite, $f(x)$ will have finite area by the comparison test. Similarly, $h(x)$ lies above $1/x$, whose area is infinite, so $h(x)$ must have infinite area as well. We can tell nothing about the area of $g(x)$, because the comparison test tells us nothing about a function larger than a function with finite area but smaller than one with infinite area. Finally, $k(x)$ will certainly have infinite area, because it has a lower bound m, for some $m > 0$. Thus, $\int_0^a k(x)\,dx \geq ma$, and since the latter does not converge as $a \to \infty$, neither can the former.

2. It converges:

$$\int_{50}^{\infty} \frac{dz}{z^3} = \lim_{b \to \infty} \int_{50}^{b} \frac{dz}{z^3} = \lim_{b \to \infty} \left(-\frac{1}{2}z^{-2}\bigg|_{50}^{b}\right) = \frac{1}{2} \lim_{b \to \infty} \left(\frac{1}{50^2} - \frac{1}{b^2}\right) = \frac{1}{5000}$$

3. The integral converges.

$$\int_{0.5}^{1} \frac{1}{x^{(19/20)}} dx = 20(x^{1/20})\bigg|_{0.5}^{1}$$

$$= 20\left(1 - (0.5)^{1/20}\right) \approx 0.681.$$

4. Since $\frac{1}{1+x} \geq \frac{1}{2x}$ and $\frac{1}{2}\int_0^{\infty} \frac{1}{x} dx$ diverges, we have that $\int_1^{\infty} \frac{dx}{1+x}$ diverges.

5. If $x \geq 1$, we know that $\frac{1}{x^3+1} \leq \frac{1}{x^3}$, and since $\int_1^{\infty} \frac{dx}{x^3}$ converges, the improper integral $\int_1^{\infty} \frac{dx}{x^3+1}$ converges.

6. For $\theta \geq 2$, we have $\frac{1}{\sqrt{\theta^3+1}} \leq \frac{1}{\sqrt{\theta^3}} = \frac{1}{\theta^{\frac{3}{2}}}$, and $\int_2^{\infty} \frac{d\theta}{\theta^{3/2}}$ converges (check by integration), so $\int_2^{\infty} \frac{d\theta}{\sqrt{\theta^3+1}}$ converges.

7. This integral diverges. To see this, substitute $t + 1 = w$, $dt = dw$. So,

$$\int_{t=-1}^{t=5} \frac{dt}{(t+1)^2} = \int_{w=0}^{w=6} \frac{dw}{w^2},$$

which diverges.

8. Since $\frac{1}{1+e^y} \leq \frac{1}{e^y} = e^{-y}$ and $\int_0^{\infty} e^{-y} dy$ converges, the integral $\int_0^{\infty} \frac{dy}{1+e^y}$ converges.

9. This integral is convergent because, for $\phi \geq 1$,
$$\frac{2+\cos\phi}{\phi^2} \leq \frac{3}{\phi^2},$$
and $\int_1^\infty \frac{3}{\phi^2}d\phi = 3\int_1^\infty \frac{1}{\phi^2}d\phi$ converges.

10. Since we know the antiderivative of $\frac{1}{1+u^2}$, we can use the Fundamental Theorem of Calculus to evaluate the integral. Since the integrand is even, we write
$$\int_{-\infty}^\infty \frac{du}{1+u^2} = 2\int_0^\infty \frac{du}{1+u^2} = 2\lim_{b\to\infty}\int_0^b \frac{du}{1+u^2}$$
$$= 2\lim_{b\to\infty}\arctan b = 2\left(\frac{\pi}{2}\right) = \pi.$$
Thus, the integral converges to π.

11. Since $\frac{1}{u+u^2} < \frac{1}{u^2}$ for $u \geq 1$, and since $\int_1^\infty \frac{du}{u^2}$ converges, $\int_1^\infty \frac{du}{u+u^2}$ converges.

12. This improper integral diverges. We expect this because, for large θ, $\frac{1}{\sqrt{\theta^2+1}} \approx \frac{1}{\sqrt{\theta^2}} = \frac{1}{\theta}$ and $\int_1^\infty \frac{d\theta}{\theta}$ diverges. More precisely, for $\theta \geq 1$
$$\frac{1}{\sqrt{\theta^2+1}} \geq \frac{1}{\sqrt{\theta^2+\theta^2}} = \frac{1}{\sqrt{2}\sqrt{\theta^2}} = \frac{1}{\sqrt{2}}\cdot\frac{1}{\theta}$$
and $\int_1^\infty \frac{d\theta}{\theta}$ diverges. (The factor $\frac{1}{\sqrt{2}}$ doesn't affect the divergence.)

13. This integral is improper at $\theta = 0$. For $0 \leq \theta \leq 1$, we have $\frac{1}{\sqrt{\theta^3+\theta}} \leq \frac{1}{\sqrt{\theta}}$, and since $\int_0^1 \frac{1}{\sqrt{\theta}}d\theta$ converges, $\int_0^1 \frac{d\theta}{\sqrt{\theta^3+\theta}}$ converges.

14. Since $\frac{1}{e^z+2^z} < \frac{1}{e^z} = e^{-z}$ for $z \geq 0$, and $\int_0^\infty e^{-z}dz$ converges, $\int_0^\infty \frac{dz}{e^z+2^z}$ converges.

15. Since $\frac{1}{\phi^2} \leq \frac{2-\sin\phi}{\phi^2}$ for $0 < \phi \leq \pi$, and since $\int_0^\pi \frac{1}{\phi^2}d\phi$ diverges, $\int_0^\pi \frac{2-\sin\phi}{\phi^2}d\phi$ must diverge.

16. Since $\frac{3+\sin\alpha}{\alpha} \geq \frac{2}{\alpha}$ for $\alpha \geq 4$, and since $\int_4^\infty \frac{2}{\alpha}d\alpha$ diverges, then $\int_4^\infty \frac{3+\sin\alpha}{\alpha}d\alpha$ diverges.

17. If we integrate e^{-x^2} from 1 to 10, we get 0.139. This answer doesn't change noticeably if you extend the region of integration to from 1 to 11, say, or even up to 1000. There's a reason for this; and the reason is that the tail, $\int_{10}^\infty e^{-x^2}dx$, is very small indeed. In fact
$$\int_{10}^\infty e^{-x^2}dx \leq \int_{10}^\infty e^{-x}dx = e^{-10},$$
which is very small. (In fact, the tail integral is less than $e^{-100}/10$. Can you prove that? [Hint: $e^{-x^2} \leq e^{-10x}$ for $x \geq 10$.])

18. Approximating the integral by $\int_0^{10} e^{-x^2}\cos^2 x\, dx$ yields 0.606 to two decimal places. This is a good approximation to the improper integral because the "tail" is small:
$$\int_{10}^\infty e^{-x^2}\cos^2 x\, dx \leq \int_{10}^\infty e^{-x}dx = e^{-10},$$
which is very small.

19. To find a, we first calculate $\int_0^{10} e^{-\frac{x^2}{2}} dx$. Since $\frac{x^2}{2} \geq x$ for $x \geq 10$, this will differ from $\int_0^\infty e^{-\frac{x^2}{2}} dx$ by at most

$$\int_{10}^\infty e^{-\frac{x^2}{2}} dx \leq \int_{10}^\infty e^{-x} dx = e^{-10},$$

which is very small. Using Simpson's rule with 100 intervals (well more than necessary), we find $\int_0^{10} e^{-\frac{x^2}{2}} dx \approx 1.253314137$. Thus, since $e^{-\frac{x^2}{2}}$ is even, $\int_{-10}^{10} e^{-\frac{x^2}{2}} dx \approx 2.506628274$, and this is extremely close to $\int_{-\infty}^\infty e^{-\frac{x^2}{2}} dx$.

To find a, we need $\int_{-\infty}^\infty a e^{-\frac{x^2}{2}} dx = 1$.

$$a = \frac{1}{\int_{-\infty}^\infty e^{-\frac{x^2}{2}} dx} \approx 0.399 \text{ to three decimal places.}$$

20. (a) If we substitute $w = x - k$ and $dw = dx$, we find

$$\int_{-\infty}^\infty a e^{-\frac{(x-k)^2}{2}} dx = \int_{-\infty}^\infty a e^{-\frac{w^2}{2}} dw.$$

This integral is the same as the integral in Problem 19, so the value of a will be the same, namely 0.399.

(b) The answer is the same because $g(x)$ is the same as $f(x)$ in Problem 19 except that it is shifted by k to the right. Since we are integrating from $-\infty$ to ∞, however, this shift doesn't mean anything for the integral.

21. (a) Since $e^{-x^2} \leq e^{-3x}$ for $x \geq 3$,

$$\int_3^\infty e^{-x^2} dx \leq \int_3^\infty e^{-3x} dx$$

Now

$$\int_3^\infty e^{-3x} dx = \lim_{b \to \infty} \int_3^b e^{-3x} dx = \lim_{b \to \infty} -\frac{1}{3} e^{-3x} \Big|_3^b$$

$$= \lim_{b \to \infty} \frac{e^{-9}}{3} - \frac{e^{-3b}}{3} = \frac{e^{-9}}{3}.$$

Thus

$$\int_3^\infty e^{-x^2} dx \leq \frac{e^{-9}}{3}.$$

(b) By reasoning similar to part (a),

$$\int_n^\infty e^{-x^2} dx \leq \int_n^\infty e^{-nx} dx,$$

and

$$\int_n^\infty e^{-nx} dx = \frac{1}{n} e^{-n^2},$$

so

$$\int_n^\infty e^{-x^2} dx \leq \frac{1}{n} e^{-n^2}.$$

22. First let's calculate the indefinite integral $\int \frac{dx}{x (\ln x)^p}$. Let $\ln x = w$, then $\frac{dx}{x} = dw$. So

$$\int \frac{dx}{x (\ln x)^p} = \int \frac{dw}{w^p}$$

$$= \begin{cases} \ln|w| + C, & \text{if } p = 1 \\ \frac{1}{1-p} w^{1-p} + C, & \text{if } p \neq 1 \end{cases}$$

$$= \begin{cases} \ln|\ln x| + C, & \text{if } p = 1 \\ \frac{1}{1-p} (\ln x)^{1-p} + C, & \text{if } p \neq 1. \end{cases}$$

Notice that $\lim_{x \to \infty} \ln x = +\infty$.

(a) $p = 1$:
$$\int_2^\infty \frac{dx}{x \ln x} = \lim_{b \to \infty} \left(\ln|\ln b| - \ln|\ln 2| \right) = +\infty.$$

(b) $p < 1$:
$$\int_2^\infty \frac{dx}{x(\ln x)^p} = \frac{1}{1-p} \left(\lim_{b \to \infty} (\ln b)^{1-p} - (\ln 2)^{1-p} \right) = +\infty.$$

(c) $p > 1$:
$$\int_2^\infty \frac{dx}{x(\ln x)^p} = \frac{1}{1-p} \left(\lim_{b \to \infty} (\ln b)^{1-p} - (\ln 2)^{1-p} \right)$$
$$= \frac{1}{1-p} \left(\lim_{b \to \infty} \frac{1}{(\ln b)^{p-1}} - (\ln 2)^{1-p} \right)$$
$$= -\frac{1}{1-p} (\ln 2)^{1-p}.$$

Thus, $\int_2^\infty \frac{dx}{x(\ln x)^p}$ is convergent for $p > 1$, divergent for $p \leq 1$.

23. The indefinite integral $\int \frac{dx}{x(\ln x)^p}$ is computed in Problem 22. Let $\ln x = w$, then $\frac{dx}{x} = dw$. Notice that $\lim_{x \to 1} \ln x = 0$, and $\lim_{x \to 0^+} \ln x = -\infty$.

For this integral notice that $\ln 1 = 0$, so the integrand blows up at $x = 1$.

(a) $p = 1$:
$$\int_1^2 \frac{dx}{x \ln x} = \lim_{a \to 1^+} \left(\ln|\ln 2| - \ln|\ln a| \right)$$

Since $\ln a \to 0$ as $a \to 1$, $\ln|\ln a| \to -\infty$ as $b \to 1$. So the integral is divergent.

(b) $p < 1$:
$$\int_1^2 \frac{dx}{x(\ln x)^p} = \frac{1}{1-p} \lim_{a \to 1^+} \left((\ln 2)^{1-p} - (\ln a)^{1-p} \right)$$
$$= \frac{1}{1-p} (\ln 2)^{1-p}.$$

(c) $p > 1$:
$$\int_1^2 \frac{dx}{x(\ln x)^p} = \frac{1}{1-p} \lim_{a \to 1^+} \left((\ln 2)^{1-p} - (\ln a)^{1-p} \right)$$

As $\lim_{a \to 1^+} (\ln a)^{1-p} = \lim_{a \to 1^+} \frac{1}{(\ln a)^{p-1}} = +\infty$, the integral diverges.

Thus, $\int_1^2 \frac{dx}{x(\ln x)^p}$ is convergent for $p < 1$, divergent for $p \geq 1$.

24. (a) For large x,
$$\frac{2x^2 + 1}{4x^4 + 4x^2 - 2} \approx \frac{2x^2}{4x^4} = \frac{1}{2x^2},$$

and since $\int_1^\infty \frac{dx}{2x^2}$ converges, we expect the original integral to converge also. More precisely, we can say that for $x \geq 1$, $2x^2 + 1 \leq 3x^2$ and $4x^4 + 4x^2 - 2 \geq 4x^4$, so
$$\int_1^\infty \frac{2x^2 + 1}{4x^4 + 4x^2 - 2} dx \leq \int_1^\infty \frac{3x^2}{4x^4} dx = \frac{3}{4}$$

(b) For large x,

$$\left(\frac{2x^4+1}{4x^4+4x^2-2}\right)^{\frac{1}{4}} \approx \left(\frac{2x^2}{4x^4}\right)^{\frac{1}{4}} = \frac{1}{2^{\frac{1}{4}}x^{\frac{1}{2}}},$$

and since $\int_1^\infty \frac{dx}{2^{\frac{1}{4}}x^{\frac{1}{2}}}$ diverges, we expect the original integral will diverge. To show this, notice that for $x \geq 1$, $2x^2 + 1 \geq 2x^2$ and $4x^4 + 4x^2 - 2 \leq 4x^4 + 4x^4 = 8x^4$, so

$$\int_1^\infty \left(\frac{2x^2+1}{4x^4+4x^2-2}\right)^{\frac{1}{4}} dx \geq \int_1^\infty \left(\frac{2x^2}{8x^4}\right)^{\frac{1}{4}} dx = \frac{1}{\sqrt{2}}\int_1^\infty \frac{dx}{\sqrt{x}}$$

So the original integral diverges.

25. (a) The tangent line to e^t has slope $(e^t)' = e^t$. Thus at $t = 0$, the slope is $e^0 = 1$. The line passes through $(0, e^0) = (0, 1)$. Thus the equation of the tangent line is $y = 1 + t$. Since e^t is everywhere concave up, its graph is always above the graph of any of its tangent lines; in particular, e^t is always above the line $y = 1 + t$. This is tantamount to saying

$$1 + t \leq e^t,$$

with equality holding only at the point of tangency, $t = 0$.

(b) If $t = \frac{1}{x}$, then the above inequality becomes

$$1 + \frac{1}{x} \leq e^{1/x}, \text{ or } e^{1/x} - 1 \geq \frac{1}{x}.$$

Since $t = \frac{1}{x}$, t is never zero. Therefore, the inequality is strict, and we write

$$e^{1/x} - 1 > \frac{1}{x}.$$

(c) Since $e^{1/x} - 1 > \frac{1}{x}$,

$$\frac{1}{x^5\left(e^{1/x}-1\right)} < \frac{1}{x^5\left(\frac{1}{x}\right)} = \frac{1}{x^4}.$$

Since $\int_1^\infty \frac{dx}{x^4}$ converges, $\int_1^\infty \frac{dx}{x^5\left(e^{1/x}-1\right)}$ converges.

Solutions for Chapter 7 Review

1. The limits of integration are 0 and b, and the rectangle represents the region under the curve $f(x) = h$ between these limits. Thus,

$$\text{Area of rectangle} = \int_0^b h\, dx = hx\Big|_0^b = hb.$$

2. Name the slanted line $y = f(x)$. Then the triangle is the region under the line $y = f(x)$ and between the lines $y = 0$ and $x = b$. Thus,

$$\text{Area of triangle} = \int_0^b f(x)\, dx.$$

Since $f(x)$ is a line of slope h/b which passes through the origin, its equation is $f(x) = hx/b$. Thus,

$$\text{Area of triangle} = \int_0^b \frac{hx}{b}\, dx = \frac{hx^2}{2b}\Big|_0^b = \frac{hb^2}{2b} = \frac{hb}{2}.$$

3. The circle $x^2 + y^2 = r^2$ cannot be expressed as a function $y = f(x)$, since for every x with $-r < x < r$, there are two corresponding y values on the circle. However, if we consider the top half of the circle only, as shown below, we have $x^2 + y^2 = r^2$, or $y^2 = r^2 - x^2$, and taking the positive square root, we have that $y = \sqrt{r^2 - x^2}$ is the equation of the top semicircle.

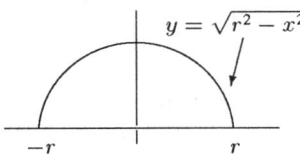

Then
$$\text{Area of Circle} = 2(\text{Area of semicircle}) = 2\int_{-r}^{r} \sqrt{r^2 - x^2}\, dx$$

We evaluate this using integral table formula 30.

$$2\int_{x=-r}^{x=r} \sqrt{r^2 - x^2}\, dx = 2\left[\frac{1}{2}\left(x\sqrt{r^2 - x^2} + r^2 \arcsin \frac{x}{r}\right)\right]\Big|_{-r}^{r}$$
$$= r^2(\arcsin 1 - \arcsin(-1))$$
$$= r^2\left(\frac{\pi}{2} - \left(-\frac{\pi}{2}\right)\right) = \pi r^2.$$

4. Since the definition of f is different on $0 \leq t \leq 1$ than it is on $1 \leq t \leq 2$, break the definite integral at $t = 1$.

$$\int_0^2 f(t)\, dt = \int_0^1 f(t)\, dt + \int_1^2 f(t)\, dt$$
$$= \int_0^1 t^2\, dt + \int_1^2 (2 - t)\, dt$$
$$= \frac{t^3}{3}\Big|_0^1 + \left(2t - \frac{t^2}{2}\right)\Big|_1^2$$
$$= 1/3 + 1/2 = 5/6 \approx 0.833$$

5. (a) i. 0   ii. $\frac{2}{\pi}$ iii. $\frac{1}{2}$

   (b) Average value of $f(t)$ < Average value of $k(t)$ < Average value of $g(t)$

   We can look at the three functions in the range $-\frac{\pi}{2} \leq x \leq \frac{3\pi}{2}$, since they all have periods of 2π ($|\cos t|$ and $(\cos t)^2$ also have a period of π, but that doesn't hurt our calculation). It is clear from the graphs of the three functions below that the average value for $\cos t$ is 0 (since the area above the x-axis is equal to the area below it), while the average values for the other two are positive (since they are everywhere positive, except where they are 0).

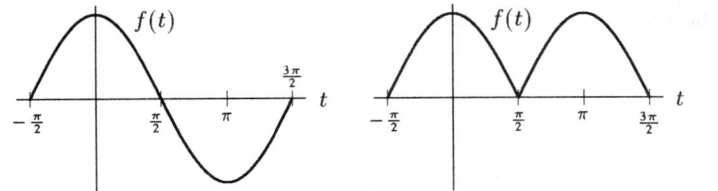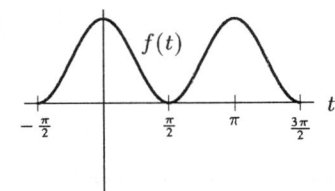

It is also fairly clear from the graphs that the average value of $g(t)$ is greater than the average value of $k(t)$; it is also possible to see this algebraically, since

$$(\cos t)^2 = |\cos t|^2 \leq |\cos t|$$

because $|\cos t| \leq 1$ (and both of these $\leq$'s are $<$'s at all the points where the functions are not 0 or 1).

6. (a) $\ln(1 + e^x) + C$. (Let $w = 1 + e^x$.)
   (b) $\arctan(e^x) + C$. (Let $w = e^x$, getting $\int \dfrac{dw}{1 + w^2}$, and use formula V-24.)
   (c) $x - \ln(1 + e^x) + C$.
       [Note that $\dfrac{1}{1 + e^x} = 1 - \dfrac{e^x}{1 + e^x}$ and use part (a).]

7. (a) Let $w = 2x$, so $w^2 = 4x^2$ and $dw = 2\,dx$.

$$\int \frac{dx}{\sqrt{1 - 4x^2}} = \frac{1}{2}\int \frac{dw}{\sqrt{1 - w^2}} = \frac{1}{2}\arcsin w + C \text{ by VI-28}$$
$$= \frac{1}{2}\arcsin 2x + C.$$

   (b)
$$\int_0^{\frac{\pi}{8}} \frac{dx}{\sqrt{1 - 4x^2}} = \frac{1}{2}\arcsin 2x \Big|_0^{\frac{\pi}{8}} \approx 0.45167.$$

   (c) Simpson's rule with 100 intervals also yields ≈ 0.45167.

8. (a) Recall that $x = e^{\ln x}$. Thus $x^x = (e^{\ln x})^x = e^{x \ln x}$.
   (b)
$$\frac{d}{dx}(x^x) = \frac{d}{dx}(e^{x \ln x}) = e^{x \ln x} \frac{d}{dx}(x \ln x) \text{ by the chain rule}$$
$$= e^{x \ln x}(\ln x + 1)$$
$$= x^x(\ln x + 1).$$

   (c) By the Fundamental Theorem of Calculus and part (b),

$$\int x^x(1 + \ln x)\,dx = x^x + C.$$

   (d) By the Fundamental Theorem of Calculus,

$$\int_1^2 x^x(1 + \ln x)\,dx = x^x \Big|_1^2$$
$$= 2^2 - 1^1$$
$$= 3.$$

   Using a calculator, we can check our answer numerically. With 50 subdivisions, the left-hand sum ≈ 2.943 and the right-hand sum ≈ 3.058. With 100 subdivisions, the left-hand sum ≈ 2.971 and the right-hand sum ≈ 3.029.

9. After the substitution $w = x + 2$, the first integral becomes

$$\int w^{-2}\,dw.$$

   After the substitution $w = x^2 + 1$, the second integral becomes

$$\frac{1}{2}\int w^{-2}\,dw.$$

10. After the substitution $w = x^2$, the second integral becomes

$$\frac{1}{2}\int \frac{dw}{\sqrt{1 - w^2}}.$$

11. After the substitution $w = 1 - x^2$, the first integral becomes

$$-\frac{1}{2}\int w^{-1}\,dw.$$

After the substitution $w = \ln x$, the second integral becomes

$$\int w^{-1}\,dw.$$

12. *First solution*: After the substitution $w = x + 1$, the first integral becomes

$$\int \frac{w-1}{w}\,dw = w - \int w^{-1}\,dw.$$

With this same substitution, the second integral becomes

$$\int w^{-1}\,dw.$$

Second solution: We note that the sum of the integrands is 1, so the sum of the integrals is x. Thus

$$\int \frac{x}{x+1}\,dx = x - \int \frac{1}{x+1}\,dx.$$

13. $\int_4^\infty \frac{dt}{t^{3/2}}$ should converge, since $\int_1^\infty \frac{dt}{t^n}$ converges for $n > 1$.
We calculate its value.

$$\int_4^\infty \frac{dt}{t^{3/2}} = \lim_{b\to\infty}\int_4^b t^{-3/2}\,dt = \lim_{b\to\infty} -2t^{-1/2}\bigg|_4^b = \lim_{b\to\infty}\left(1 - \frac{2}{\sqrt{b}}\right) = 1.$$

14. $\int \frac{dx}{x\ln x} = \ln|\ln x| + C$. (Substitute $w = \ln x$, $dw = \frac{1}{x}\,dx$).
Thus

$$\int_{10}^\infty \frac{dx}{x\ln x} = \lim_{b\to\infty}\int_{10}^b \frac{dx}{x\ln x} = \lim_{b\to\infty} \ln|\ln x|\bigg|_{10}^b = \lim_{b\to\infty} \ln(\ln b) - \ln(\ln 10).$$

As $b \to \infty$, $\ln(\ln b) \to \infty$, so this diverges.

15. To find $\int we^{-w}\,dw$, integrate by parts, with $u = w$ and $v' = e^{-w}$. Then $u' = 1$ and $v = -e^{-w}$.
Then

$$\int we^{-w}\,dw = -we^{-w} + \int e^{-w}\,dw = -we^{-w} - e^{-w} + C.$$

Thus

$$\int_0^\infty we^{-w}\,dw = \lim_{b\to\infty}\int_0^b we^{-w}\,dw = \lim_{b\to\infty} \left(-we^{-w} - e^{-w}\right)\bigg|_0^b = 1.$$

16. The trouble spot is at $x = 0$, so we write

$$\int_{-1}^1 \frac{1}{x^4}\,dx = \int_{-1}^0 \frac{1}{x^4}\,dx + \int_0^1 \frac{1}{x^4}\,dx.$$

However, both these integrals diverge. For example,

$$\int_0^1 \frac{1}{x^4}\,dx = \lim_{a\to 0^+}\int_a^1 \frac{1}{x^4}\,dx = \lim_{a\to 0^+} -\frac{x^{-3}}{3}\bigg|_a^1 = \lim_{a\to 0^+}\left(\frac{1}{3a^3} - \frac{1}{3}\right).$$

Since this limit does not exist, $\int_0^1 \frac{1}{x^4}\,dx$ diverges and so the original integral diverges.

17. Since the value of $\tan\theta$ is between -1 and 1 on the interval $-\pi/4 \leq \theta \leq \pi/4$, our integral is not improper and so converges. Moreover, since $\tan\theta$ is an odd function, we have

$$\int_{-\frac{\pi}{4}}^{\frac{\pi}{4}} \tan\theta\, d\theta = \int_{-\frac{\pi}{4}}^{0} \tan\theta\, d\theta + \int_{0}^{\frac{\pi}{4}} \tan\theta\, d\theta$$

$$= -\int_{-\frac{\pi}{4}}^{0} \tan(-\theta)\, d\theta + \int_{0}^{\frac{\pi}{4}} \tan\theta\, d\theta$$

$$= -\int_{0}^{\frac{\pi}{4}} \tan\theta\, d\theta + \int_{0}^{\frac{\pi}{4}} \tan\theta\, d\theta = 0.$$

18. It is easy to see that this integral converges:

$$\frac{1}{4+z^2} < \frac{1}{z^2}, \quad \text{and so} \quad \int_{2}^{\infty} \frac{1}{4+z^2}\, dz < \int_{2}^{\infty} \frac{1}{z^2}\, dz = \frac{1}{2}.$$

We can also find its exact value.

$$\int_{2}^{\infty} \frac{1}{4+z^2}\, dz = \lim_{b\to\infty} \int_{2}^{b} \frac{1}{4+z^2}\, dz$$

$$= \lim_{b\to\infty} \left(\frac{1}{2}\arctan\frac{z}{2}\bigg|_{2}^{b}\right)$$

$$= \lim_{b\to\infty} \left(\frac{1}{2}\arctan\frac{b}{2} - \frac{1}{2}\arctan 1\right)$$

$$= \frac{1}{2}\frac{\pi}{2} - \frac{1}{2}\frac{\pi}{4} = \frac{\pi}{8}.$$

Note that $\frac{\pi}{8} < \frac{1}{2}$.

19. We find the exact value:

$$\int_{10}^{\infty} \frac{1}{z^2-4}\, dz = \int_{10}^{\infty} \frac{1}{(z+2)(z-2)}\, dz$$

$$= \lim_{b\to\infty} \int_{10}^{b} \frac{1}{(z+2)(z-2)}\, dz$$

$$= \lim_{b\to\infty} \frac{1}{4}\left(\ln|z-2| - \ln|z+2|\right)\bigg|_{10}^{b}$$

$$= \frac{1}{4}\lim_{b\to\infty}\left[(\ln|b-2| - \ln|b+2|) - (\ln 8 - \ln 12)\right]$$

$$= \frac{1}{4}\lim_{b\to\infty}\left[\left(\ln\frac{b-2}{b+2}\right) + \ln\frac{3}{2}\right]$$

$$= \frac{1}{4}(\ln 1 + \ln 3/2) = \frac{\ln 3/2}{4}.$$

20. Substituting $w = t+5$, we see that our integral is just $\int_{0}^{15} \frac{dw}{\sqrt{w}}$. This will converge, since $\int_{0}^{b} \frac{dw}{w^p}$ converges for $0 < p < 1$. We find its exact value:

$$\int_{0}^{15} \frac{dw}{\sqrt{w}} = \lim_{a\to 0^+} \int_{a}^{15} \frac{dw}{\sqrt{w}} = \lim_{a\to 0^+} 2w^{\frac{1}{2}}\bigg|_{a}^{15} = 2\sqrt{15}.$$

21. Since $\sin\phi < \phi$ for $\phi > 0$,
$$\int_0^{\frac{\pi}{2}} \frac{1}{\sin\phi}\, d\phi > \int_0^{\frac{\pi}{2}} \frac{1}{\phi}\, d\phi,$$
The integral on the right diverges, so the integral on the left must also. Alternatively, we use IV-20 in the integral table to get
$$\int_0^{\frac{\pi}{2}} \frac{1}{\sin\phi}\, d\phi = \lim_{b\to 0^+} \int_b^{\frac{\pi}{2}} \frac{1}{\sin\phi}\, d\phi$$
$$= \lim_{b\to 0^+} \frac{1}{2} \ln\left|\frac{\cos\phi - 1}{\cos\phi + 1}\right|\Bigg|_b^{\frac{\pi}{2}}$$
$$= -\frac{1}{2} \lim_{b\to 0^+} \ln\left|\frac{\cos b - 1}{\cos b + 1}\right|.$$

As $b \to 0^+$, $\cos b - 1 \to 0$ and $\cos b + 1 \to 2$, so $\ln\left|\frac{\cos b - 1}{\cos b + 1}\right| \to -\infty$. Thus the integral diverges.

22. Let $\phi = 2\theta$. Then $d\phi = 2\, d\theta$, and
$$\int_0^{\pi/4} \tan 2\theta\, d\theta = \int_0^{\pi/2} \frac{1}{2} \tan\phi\, d\phi = \int_0^{\pi/2} \frac{1}{2} \frac{\sin\phi}{\cos\phi}\, d\phi$$
$$= \lim_{b\to (\pi/2)^-} \int_0^{b} \frac{1}{2} \frac{\sin\phi}{\cos\phi}\, d\phi = \lim_{b\to (\pi/2)^-} -\frac{1}{2} \ln|\cos\phi|\Bigg|_0^b.$$

As $b \to \pi/2$, $\cos\phi \to 0$, so $\ln|\cos\phi| \to -\infty$. Thus the integral diverges.

One could also see this by noting that $\cos x \approx \pi/2 - x$ and $\sin x \approx 1$ for x close to $\pi/2$: therefore, $\tan x \approx 1/(\frac{\pi}{2} - x)$, the integral of which diverges.

23. The integrand $\frac{x}{x+1} \to 1$ as $x \to \infty$, so there's no way $\int_1^\infty \frac{x}{x+1}\, dx$ can converge.

24. This function is difficult to integrate, so instead we try to compare it with some other function. Since $\frac{\sin^2\theta}{\theta^2 + 1} \geq 0$, we see that $\int_0^\infty \frac{\sin^2\theta}{\theta^2+1}\, d\theta \geq 0$. Also, since $\sin^2\theta \leq 1$,
$$\int_0^\infty \frac{\sin^2\theta}{\theta^2 + 1}\, d\theta \leq \int_0^\infty \frac{1}{\theta^2 + 1}\, d\theta = \lim_{b\to\infty} \arctan\theta\Bigg|_0^b = \frac{\pi}{2}.$$

Thus $\int_0^\infty \frac{\sin^2\theta}{\theta^2+1}\, d\theta$ converges, and its value is between 0 and $\frac{\pi}{2}$.

25. $\int_0^\pi \tan^2\theta\, d\theta = \tan\theta - \theta + C$, by formula IV-23. The integrand blows up at $\theta = \frac{\pi}{2}$, so
$$\int_0^\pi \tan^2\theta\, d\theta = \int_0^{\frac{\pi}{2}} \tan^2\theta\, d\theta + \int_{\frac{\pi}{2}}^\pi \tan^2\theta\, d\theta = \lim_{b\to\frac{\pi}{2}} [\tan\theta - \theta]_0^b + \lim_{a\to\frac{\pi}{2}} [\tan\theta - \theta]_a^\pi$$
which is undefined.

26. Since $0 \leq \sin x < 1$ for $0 \leq x \leq 1$, we have
$$(\sin x)^{\frac{3}{2}} < (\sin x)$$
so $\frac{1}{(\sin x)^{\frac{3}{2}}} > \frac{1}{(\sin x)}$
or $(\sin x)^{-\frac{3}{2}} > (\sin x)^{-1}$

Thus $\int_0^1 (\sin x)^{-1} dx = \lim_{a\to 0} \ln\left|\frac{1}{\sin x} - \frac{1}{\tan x}\right|_a^1$, which is infinite.

Hence, $\int_0^1 (\sin x)^{-\frac{3}{2}} dx$ is infinite.

27. The point of intersection of the two curves $y = x^2$ and $y = 6 - x$ is at $(2,4)$. The average height of the shaded area is the average value of the difference between the functions:
$$\frac{1}{(2-0)} \int_0^2 ((6-x) - x^2)\, dx = \left(3x - \frac{x^2}{4} - \frac{x^3}{6}\right)\Bigg|_0^2 = \frac{11}{3}.$$

334 CHAPTER SEVEN /SOLUTIONS

28. The average width of the shaded area in the figure below is the average value of the horizontal distance between the two functions. If we call this horizontal distance $h(y)$, then the average width is

$$\frac{1}{(6-0)} \int_0^6 h(y)\, dy.$$

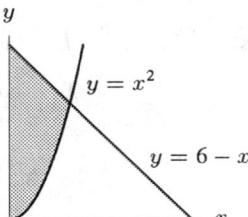

We could compute this integral if we wanted to, but we don't need to. We can simply note that the integral (without the $\frac{1}{6}$ term) is just the area of the shaded region; similarly, the integral in Problem 27 is *also* just the area of the shaded region. So they are the same. Now we know that our average width is just $\frac{1}{3}$ as much as the average height, since we divide by 6 instead of 2. So the answer is $\frac{11}{9}$.

29.

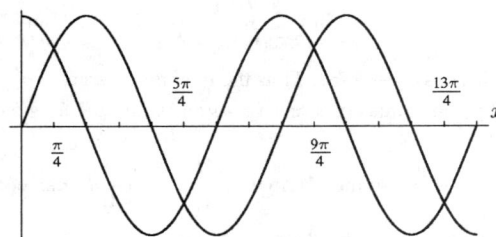

As is evident from the accompanying figure of the graphs of $y = \sin x$ and $y = \cos x$, the crossings occur at $x = \frac{\pi}{4}, \frac{5\pi}{4}, \frac{9\pi}{4}, \ldots$, and the regions bounded by any two consecutive crossings have the same area. So picking two consecutive crossings, we get an area of

$$\text{Area} = \int_{\frac{\pi}{4}}^{\frac{5\pi}{4}} (\sin x - \cos x)\, dx$$
$$= 2\sqrt{2}.$$

(Note that we integrated $\sin x - \cos x$ here because for $\frac{\pi}{4} \leq x \leq \frac{5\pi}{4}$, $\sin x \geq \cos x$.)

30. True. $y^2 - 1$ is concave up, and the midpoint rule always underestimates for a function that is concave up.

31. False. If the function $f(x)$ is a line, then the trapezoid rule gives the exact answer to $\int_a^b f(x)\, dx$.

32. False. Suppose f is the following:

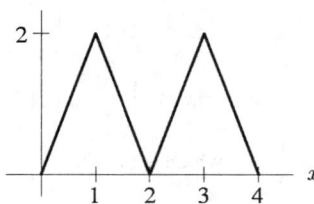

Then LEFT(2) = 0, LEFT(4) = 4, and

$$\int_a^b f(x)\, dx = 4.$$

33. True. Since f' and g' are greater than 0, all left rectangles give underestimates. The bigger the derivative, the bigger the underestimate, so the bigger the error. (Note: if we didn't have $0 < f' < g'$, but instead just had $f' < g'$, the statement wouldn't necessarily be true. This is because some left rectangles could be overestimates and some could be underestimates–so, for example, it could be that the error in approximating g is 0! If $0 < f' < g'$, however, this can't happen.

34. Since $e^x \geq 1$ for x between 0 and 1, it follows that
$$\frac{e^x}{x} \geq \frac{1}{x}$$
and so
$$\int_0^1 \frac{e^x}{x} dx \geq \int_0^1 \frac{1}{x} dx.$$
We know that the improper integral $\int_0^1 \frac{1}{x} dx$ diverges. Therefore, since $\int_0^1 \frac{e^x}{x} dx$ is larger, it also diverges.

35. (a) $f(x) = 1 + e^{-x}$ is concave up for $0 \leq x \leq 0.5$, so trapezoids will overestimate $\int_0^{0.5} f(x)dx$, and the midpoint rule will underestimate.

    (b) $f(x) = e^{-x^2}$ is concave down for $0 \leq x \leq 0.5$, so trapezoids will underestimate $\int_0^{0.5} f(x)dx$ and midpoint will overestimate the integral.

    (c) Both the trapezoid rule and the midpoint rule will give the exact value of the integral. Note that upper and lower sums will not, unless the line is horizontal.

36. Let's assume that TRAP(10) and TRAP(50) are either both overestimates or both underestimates. Since TRAP(50) is more accurate, and it is bigger than TRAP(10), both are underestimates. Since TRAP(50) is 25 times more accurate, we have
$$I - \text{TRAP}(10) = 25(I - \text{TRAP}(50)),$$
where I is the value of the integral. Solving for I, we have
$$I \approx \frac{25 \text{ TRAP}(50) - \text{TRAP}(10)}{24} \approx 4.6969$$
Thus the error for TRAP(10) is approximately 0.0078.

37. Let us assume that SIMP(5) and (10) are both overestimates or both underestimates. Then since SIMP(10) is more accurate and bigger than SIMP(5), they are both underestimates. Since SIMP(10) is 16 times more accurate,
$$I - \text{SIMP}(5) = 16(I - \text{SIMP}(10)).$$
Solving for I, we have
$$I = \frac{16\text{SIMP}(10) - \text{SIMP}(5)}{15} \approx 7.4175.$$

38. If $I(t)$ is average per capita income t years after 1987, then $I'(t) = r(t)$.

    (a) Since $t = 8$ in 1995, by the Fundamental Theorem,
$$I(8) - I(0) = \int_0^8 r(t)\, dt = \int_0^8 480(1.024)^t\, dt$$
$$= \frac{480(1.024)^t}{\ln(1.024)} \bigg|_0^8 = 4228$$
so $I(8) = 26{,}000 + 4228 = 30{,}228$.

(b)
$$I(t) - I(0) = \int_0^t r(t)\,dt = \int_0^t 480(1.024)^t\,dt$$
$$= \frac{480(1.024)^t}{\ln(1.024)}\bigg|_0^t$$
$$= \frac{480}{\ln(1.024)}\left((1.024)^t - 1\right)$$
$$= 20{,}239\left((1.024)^t - 1\right)$$

Thus, since $I(0) = 26{,}000$,
$$I(t) = 26{,}000 + 20{,}239(1.024^t - 1) = 20{,}239(1.024)^t + 5761.$$

39. Let $C(y)$ be the consumption of petroleum from 1991 through the year $1991 + y$. Let $a = 1.02$ and $K = 1.4 \times 10^{20}$. We are told that in the year $1990 + t$, the annual rate of consumption will be Ka^t joules/year. Thus

$$C(y) = \sum_{t=1}^{y} Ka^t.$$

Since $a > 1$, the function $u = Ka^t$ is increasing and $C(y)$ can be viewed as a right-hand Riemann sum overestimate for a definite integral

$$C(y) \geq \int_0^y Ka^t\,dt = \frac{K}{\ln a}(a^y - 1).$$

Thus we seek y such that
$$\frac{K}{\ln a}(a^y - 1) = 10^{22},$$
or
$$a^y = \frac{10^{22} \ln a}{K} + 1 \approx 2.414.$$

Taking logarithms, we get $y \ln a \approx \ln 2.414$, which gives $y \approx 45$. So in about 45 years, we will run out of petroleum!

Solutions to the Projects and Computer Algebra Investigations

1. (a) If $e^t \geq 1 + t$, then
$$e^x = 1 + \int_0^x e^t\,dt$$
$$\geq 1 + \int_0^x (1+t)\,dt = 1 + x + \frac{1}{2}x^2.$$
We can keep going with this idea. Since $e^t \geq 1 + t + \frac{1}{2}t^2$,
$$e^x = 1 + \int_0^x e^t\,dt$$
$$\geq 1 + \int_0^x \left(1 + t + \frac{1}{2}t^2\right)dt = 1 + x + \frac{1}{2}x^2 + \frac{1}{6}x^3.$$
We notice that each term in our summation is of the form $\frac{x^n}{n!}$. Furthermore, we see that if we have a sum $1 + x + \frac{x^2}{2} + \cdots + \frac{x^n}{n!}$ such that
$$e^x \geq 1 + x + \frac{x^2}{2} + \cdots + \frac{x^n}{n!},$$

then
$$e^x = 1 + \int_0^x e^t\, dt$$
$$\geq 1 + \int_0^x \left(1 + t + \frac{t^2}{2} + \cdots + \frac{t^n}{n!}\right) dt$$
$$= 1 + x + \frac{x^2}{2} + \frac{x^3}{6} + \cdots + \frac{x^{n+1}}{(n+1)!}.$$

Thus we can continue this process as far as we want, so
$$e^x \geq 1 + x + \frac{1}{2}x^2 + \cdots + \frac{1}{n!}x^n = \sum_{j=0}^{n} \frac{x^j}{j!} \text{ for any } n.$$

(In fact, it turns out that if you let n get larger and larger and keep adding up terms, your values approach exactly e^x.)

(b) We note that $\sin x = \int_0^x \cos t\, dt$ and $\cos x = 1 - \int_0^x \sin t\, dt$. Thus, since $\cos t \leq 1$, we have
$$\sin x \leq \int_0^x 1\, dt = x.$$

Now using $\sin t \leq t$, we have
$$\cos x \leq 1 - \int_0^x t\, dt = 1 - \frac{1}{2}x^2.$$

Then we just keep going:
$$\sin x \leq \int_0^x \left(1 - \frac{1}{2}t^2\right) dt = x - \frac{1}{6}x^3.$$

Therefore
$$\cos x \leq 1 - \int_0^x \left(t - \frac{1}{6}t^3\right) dt = 1 - \frac{1}{2}x^2 + \frac{1}{24}x^4.$$

2. (a) (i)

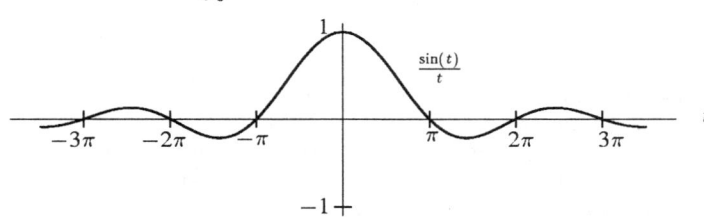

(ii) Si(x) neither always decreases nor always increases, since its derivative, $x^{-1}\sin x$, has both positive and negative values for $x > 0$. For positive x, Si(x) is the area under the curve $\frac{\sin t}{t}$ and above the t-axis from $t = 0$ to $t = x$, minus the area above the curve and below the t-axis. Looking at the graph above, one can see that this difference of areas is going to always be positive.

(iii)

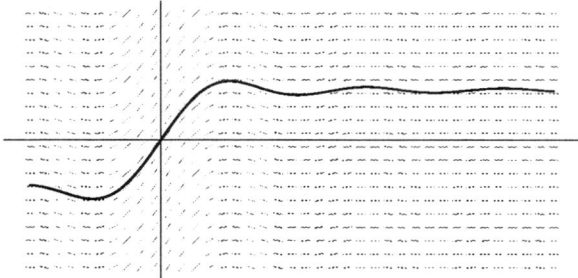

It seems that the limit exists: the curve drawn in the slope field,
$$y = Si(x) = \int_0^x \frac{\sin t}{t}\, dt,$$

seems to approach some limiting height as $x \to \infty$. (In fact, the limiting height is $\pi/2$.)

(b)

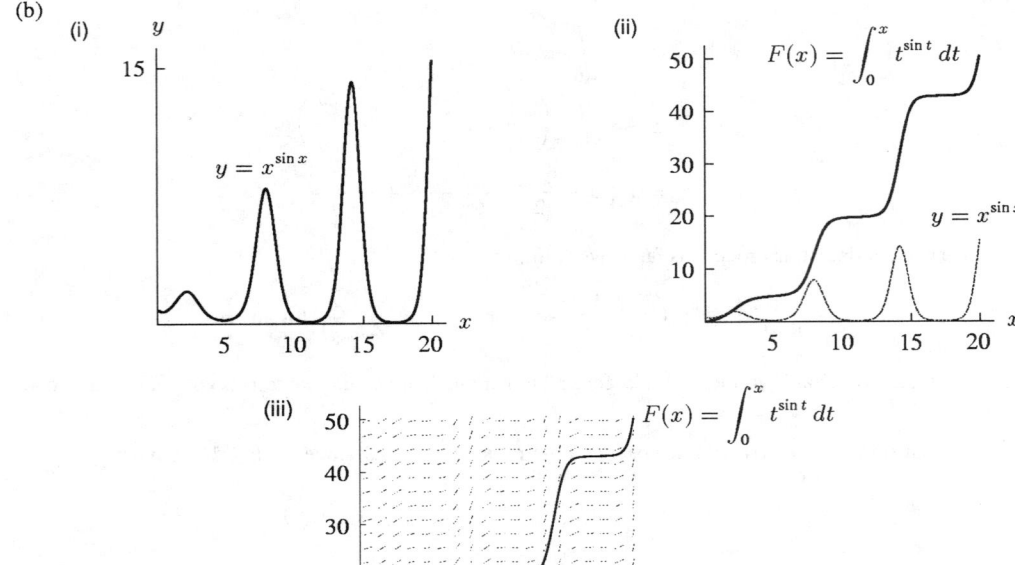

(c) (i) The most obvious feature of the graph of $y = \sin(x^2)$ is its symmetry about the y-axis. This means the function $g(x) = \sin(x^2)$ is an even function, i.e. for all x, we have $g(x) = g(-x)$. Since $\sin(x^2)$ is even, its antiderivative F must be odd, that is $F(-x) = -F(-x)$. To see this, set $F(t) = \int_0^t \sin(x^2)\,dx$, then

$$F(-t) = \int_0^{-t} \sin(x^2)\,dx = -\int_{-t}^0 \sin(x^2)\,dx = -\int_0^t \sin(x^2)\,dx = -F(t),$$

since the area from $-t$ to 0 is the same as the area from 0 to t. Thus $F(t) = -F(-t)$ and F is odd.

The second obvious feature of the graph of $y = \sin(x^2)$ is that it oscillates between -1 and 1 with a "period" which goes to zero as $|x|$ increases. This implies that $F'(x)$ alternates between intervals where it is positive or negative, and increasing or decreasing, with frequency growing arbitrarily large as $|x|$ increases. Thus $F(x)$ itself similarly alternates between intervals where it is increasing or decreasing, and concave up or concave down.

Finally, since $y = \sin(x^2) = F'(x)$ passes through $(0,0)$, and $F(0) = 0$, F is tangent to the x-axis at the origin.

(ii)

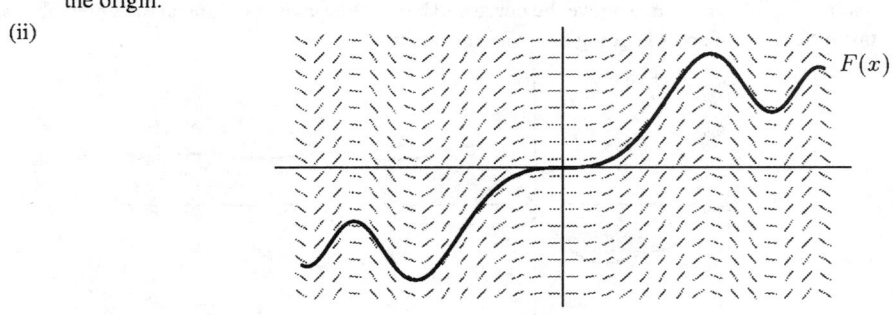

Figure 7.2

F never crosses the x-axis in the region $x > 0$, and $\lim\limits_{x \to \infty} F(x)$ exists. One way to see these facts is to note that by the Construction Theorem,

$$F(x) = F(x) - F(0) = \int_0^x F'(t)\,dt.$$

So $F(x)$ is just the area between the curve $y = \sin(t^2)$ and the t-axis for $0 \leq t \leq x$ (with area above the t-axis counting positively, and area below the t-axis counting negatively). Now looking at the graph of curve, we see that this area will include alternating pieces above and below the t-axis. We can also see that the area of these pieces is approaching 0 as we go further out. So we add a piece, take a piece away, add another piece, take another piece away, and so on.

It turns out that this means that the sums of the pieces converge. To see this, think of walking from point A to point B. If you walk almost to B, then go a smaller distance toward A, then a yet smaller distance back toward B, and so on, you will eventually approach some point between A and B. So we can see that $\lim_{x \to \infty} F(x)$ exists. Also, since we always subtract a smaller piece than we just added, and the first piece is added instead of subtracted, we see that we never get a negative sum; thus $F(x)$ is never negative in the region $x > 0$, so $F(x)$ never crosses the x-axis there.

3. (a) A CAS gives

$$\int \frac{\ln x}{x}\, dx = \frac{(\ln x)^2}{2}$$

$$\int \frac{(\ln x)^2}{x}\, dx = \frac{(\ln x)^3}{3}$$

$$\int \frac{(\ln x)^3}{x}\, dx = \frac{(\ln x)^4}{4}$$

(b) Looking at the answers to part (a),

$$\int \frac{(\ln x)^n}{x}\, dx = \frac{(\ln x)^{n+1}}{n+1}.$$

(c) Let $w = \ln x$. Then $dw = (1/x)dx$, and

$$\int \frac{(\ln x)^n}{x}\, dx = \int w^n\, dw = \frac{w^{n+1}}{n+1} = \frac{(\ln x)^{n+1}}{n}.$$

4. (a) A CAS gives

$$\int \ln x\, dx = -x + x\ln x$$

$$\int (\ln x)^2\, dx = 2x - 2x\ln x + x(\ln x)^2$$

$$\int (\ln x)^3\, dx = -6x + 6x\ln x - 3x(\ln x)^2 + x(\ln x)^3$$

$$\int (\ln x)^4\, dx = 24x - 24x\ln x + 12x(\ln x)^2 - 4x(\ln x)^3 + x(\ln x)^4$$

(b) In each of the cases in part (a), the expression for the integral $\int (\ln x)^n\, dx$ has two parts. The first part is simply a multiple of the expression for $\int (\ln x)^{n-1}\, dx$. For example, $\int (\ln x)^2\, dx$ starts out with $2x - 2x\ln x = -2\int \ln x\, dx$. Similarly, $\int (\ln x)^3\, dx$ starts out with $-6x + 6x\ln x - 3(\ln x)^2 = -3\int (\ln x)^2\, dx$, and $\int (\ln x)^4\, dx$ starts out with $-4\int (\ln x)^3\, dx$. The remaining part of each antiderivative is a single term: it's $x(\ln x)^2$ in the case $n = 2$, it's $x(\ln x)^3$ for $n = 3$, and it's $x(\ln x)^4$ for $n = 4$. The general pattern is

$$\int (\ln x)^n\, dx = -n\int (\ln x)^{n-1}\, dx + x(\ln x)^n.$$

To check this formula, we use integration by parts. Let $u = (\ln x)^n$ so $u' = n(\ln x)^{n-1}/x$ and $v' = 1$ so $v = x$. Then

$$\int (\ln x)^n\, dx = x(\ln x)^n - \int n\frac{(\ln x)^{n-1}}{x} \cdot x\, dx$$

$$\int (\ln x)^n\, dx = x(\ln x)^n - n\int (\ln x)^{n-1}\, dx.$$

This is the result we obtained before.

Alternatively, we can check out result by differentiation:

$$\frac{d}{dx}\left(-n\int (\ln x)^{n-1}\,dx + x(\ln x)^n\right) = -n(\ln x)^{n-1} + \frac{d}{dx}(x(\ln x)^n)$$

$$= -n(\ln x)^{n-1} + (\ln x)^n + x \cdot n(\ln x)^{n-1}\frac{1}{x}$$

$$= -n(\ln x)^{n-1} + (\ln x)^n + n(\ln x)^{n-1} = (\ln x)^n$$

$$= \frac{d}{dx}\left(\int (\ln x)^n\,dx\right).$$

5. (a) A possible answer from the CAS is

$$\int \sin^3 x\,dx = \frac{-9\cos(x) + \cos(3x)}{12}.$$

(b) Differentiating

$$\frac{d}{dx}\left(\frac{-9\cos(x) + \cos(3x)}{12}\right) = \frac{9\sin(x) - 3\sin(3x)}{12} = \frac{3\sin x - \sin(3x)}{4}.$$

(c) Using the identities

$$\sin(3x) = 3\cos^2 x \sin x - \sin^3 x \qquad \sin^2 x + \cos^2 x = 1,$$

we get

$$\sin(3x) = 3\sin x - 4\sin^3 x,$$

so

$$3\sin x - \sin(3x) = 3\sin x - (3\sin x - 4\sin^3 x) = 4\sin^3 x.$$

Therefore

$$\frac{3\sin x - \sin(3x)}{4} = \sin^3 x.$$

6. (a) A possible answer is

$$\int \sin x \cos x \cos(2x)\,dx = -\frac{\cos(4x)}{16}.$$

Different systems may give the answer in a different form.

(b)

$$\frac{d}{dx}\left(-\frac{\cos(4x)}{16}\right) = \frac{\sin(4x)}{4}.$$

(c) Using the double angle formula $\sin 2A = 2\sin A \cos A$ twice, we get

$$\frac{\sin(4x)}{4} = \frac{2\sin(2x)\cos(2x)}{4} = \frac{2\cdot 2\sin x \cos x \cos(2x)}{4} = \sin x \cos x \cos(2x).$$

7. (a) A possible answer from the CAS is

$$\int \frac{x^4}{(1+x^2)^2}\,dx = x + \frac{x}{2(1+x^2)} - \frac{3}{2}\arctan(x).$$

Different systems may give the answer in different form.

(b) Differentiating gives

$$\frac{d}{dx}\left(x + \frac{x}{2(1+x^2)} - \frac{3}{2}\arctan(x)\right) = 1 - \frac{x^2}{(1+x^2)^2} - \frac{1}{1+x^2}.$$

(c) Putting the result of part (b) over a common denominator, we get

$$1 - \frac{x^2}{(1+x^2)^2} - \frac{1}{1+x^2} = \frac{(1+x^2)^2 - x^2 - (1+x^2)}{(1+x^2)^2}$$

$$= \frac{1 + 2x^2 + x^4 - x^2 - 1 - x^2}{(1+x^2)^2} = \frac{x^4}{(1+x^2)^2}.$$

Solutions to Practice Problems on Integration

1. Since $\frac{d}{dt}\cos t = -\sin t$, we have
$$\int \sin t\, dt = -\cos t + C, \text{ where } C \text{ is a constant.}$$

2. Let $2t = w$, then $2dt = dw$, so $dt = \frac{1}{2}dw$, so
$$\int \cos 2t\, dt = \int \frac{1}{2}\cos w\, dw = \frac{1}{2}\sin w + C = \frac{1}{2}\sin 2t + C,$$
where C is a constant.

3. Let $5z = w$, then $5dz = dw$, which means $dz = \frac{1}{5}dw$, so
$$\int e^{5z}\, dz = \int e^w \cdot \frac{1}{5}dw = \frac{1}{5}\int e^w\, dw = \frac{1}{5}e^w + C = \frac{1}{5}e^{5z} + C,$$
where C is a constant.

4. Using the power rule gives $\frac{3}{2}w^2 + 7w + C$.

5. Since $\int \sin w\, d\theta = -\cos w + C$, the substitution $w = 2\theta$, $dw = 2\, d\theta$ gives $\int \sin 2\theta\, d\theta = -\frac{1}{2}\cos 2\theta + C$.

6. Let $w = x^3 - 1$, then $dw = 3x^2 dx$ so that
$$\int (x^3-1)^4 x^2\, dx = \frac{1}{3}\int w^4\, dw = \frac{1}{15}w^5 + C = \frac{1}{15}(x^3-1)^5 + C.$$

7. The power rule gives $\frac{2}{5}x^{5/2} + \frac{3}{5}x^{5/3} + C$

8. From the rule for antidifferentiation of exponentials, we get
$$\int \left(e^x + 3^x\right)\, dx = e^x + \frac{1}{\ln 3}\cdot 3^x + C.$$

9. Either expand $(r+1)^3$ or use the substitution $w = r+1$. If $w = r+1$, then $dw = dr$ and
$$\int (r+1)^3\, dr = \int w^3\, dw = \frac{1}{4}w^4 + C = \frac{1}{4}(r+1)^4 + C.$$

10. Rewrite the integrand as
$$\int \left(\frac{4}{x^2} - \frac{3}{x^3}\right)\, dx = 4\int x^{-2}\, dx - 3\int x^{-3}\, dx = -4x^{-1} + \frac{3}{2}x^{-2} + C.$$

11. Dividing by x^2 gives
$$\int \left(\frac{x^3+x+1}{x^2}\right)\, dx = \int \left(x + \frac{1}{x} + \frac{1}{x^2}\right)\, dx = \frac{1}{2}x^2 + \ln|x| - \frac{1}{x} + C.$$

12. Let $w = 1 + \ln x$, then $dw = dx/x$ so that

$$\int \frac{(1+\ln x)^2}{x} dx = \int w^2 dw = \frac{1}{3}w^3 + C = \frac{1}{3}(1+\ln x)^3 + C.$$

13. Substitute $w = t^2$, so $dw = 2t\, dt$.

$$\int te^{t^2} dt = \frac{1}{2}\int e^{t^2} 2t\, dt = \frac{1}{2}\int e^w \, dw = \frac{1}{2}e^w + C = \frac{1}{2}e^{t^2} + C.$$

Check:

$$\frac{d}{dt}\left(\frac{1}{2}e^{t^2} + C\right) = 2t\left(\frac{1}{2}e^{t^2}\right) = te^{t^2}.$$

14. Integration by parts with $u = x$, $v' = \cos x$ gives

$$\int x\cos x\, dx = x\sin x - \int \sin x\, dx + C = x\sin x + \cos x + C.$$

Or use III-16 with $p(x) = x$ and $a = 1$ in the integral table.

15. Let $w = 2 + 3\cos x$, so $dw = -3\sin x\, dx$, giving $-\frac{1}{3} dw = \sin x\, dx$. Then

$$\int \sin x \left(\sqrt{2+3\cos x}\right) dx = \int \sqrt{w}\left(-\frac{1}{3}\right) dw = -\frac{1}{3}\int \sqrt{w}\, dw$$

$$= \left(-\frac{1}{3}\right)\frac{w^{\frac{3}{2}}}{\frac{3}{2}} + C = -\frac{2}{9}(2+3\cos x)^{\frac{3}{2}} + C.$$

16. Integration by parts twice gives

$$\int x^2 e^{2x} dx = \frac{x^2 e^{2x}}{2} - \int 2xe^{2x} dx = \frac{x^2}{2}e^{2x} - \frac{x}{2}e^{2x} + \frac{1}{4}e^{2x} + C$$

$$= \left(\frac{1}{2}x^2 - \frac{1}{2}x + \frac{1}{4}\right)e^{2x} + C.$$

Or use the integral table, III-14 with $p(x) = x^2$ and $a = 1$.

17. Using substitution with $w = 1 - x$ and $dw = -dx$, we get

$$\int x\sqrt{1-x}\, dx = -\int (1-w)\sqrt{w}\, dw = \frac{2}{5}w^{5/2} - \frac{2}{3}w^{3/2} + C = \frac{2}{5}(1-x)^{5/2} - \frac{2}{3}(1-x)^{3/2} + C.$$

18. Integration by parts with $u = \ln x$, $v' = x$ gives

$$\int x\ln x\, dx = \frac{x^2}{2}\ln x - \int \frac{1}{2}x\, dx = \frac{1}{2}x^2 \ln x - \frac{1}{4}x^2 + C.$$

Or use the integral table, III-13, with $n = 1$.

19. We integrate by parts, with $u = y$, $v' = \sin y$. We have $u' = 1$, $v = -\cos y$, and

$$\int y\sin y\, dy = -y\cos y - \int (-\cos y)\, dy = -y\cos y + \sin y + C.$$

Check:

$$\frac{d}{dy}(-y\cos y + \sin y + C) = -\cos y + y\sin y + \cos y = y\sin y.$$

20. We integrate by parts, using $u = (\ln x)^2$ and $v' = 1$. Then $u' = 2\frac{\ln x}{x}$ and $v = x$, so

$$\int (\ln x)^2 \, dx = x(\ln x)^2 - 2 \int \ln x \, dx.$$

But, integrating by parts or using the integral table, $\int \ln x \, dx = x \ln x - x + C$. Therefore,

$$\int (\ln x)^2 \, dx = x(\ln x)^2 - 2x \ln x + 2x + C.$$

Check:
$$\frac{d}{dx}\left[x(\ln x)^2 - 2x \ln x + 2x + C\right] = (\ln x)^2 + x\frac{2\ln x}{x} - 2\ln x - 2x\frac{1}{x} + 2 = (\ln x)^2.$$

21. Remember that $\ln(x^2) = 2 \ln x$. Therefore,

$$\int \ln(x^2) \, dx = 2 \int \ln x \, dx = 2x \ln x - 2x + C.$$

Check:
$$\frac{d}{dx}(2x \ln x - 2x + C) = 2\ln x + \frac{2x}{x} - 2 = 2\ln x = \ln(x^2).$$

22. Using the exponent rules and the chain rule, we have

$$\int e^{0.5-0.3t} \, dt = e^{0.5} \int e^{-0.3t} \, dt = -\frac{e^{0.5}}{0.3}e^{-0.3t} + C = -\frac{e^{0.5-0.3t}}{0.3} + C.$$

23. Let $\sin \theta = w$, then $\cos \theta \, d\theta = dw$, so

$$\int \sin^2 \theta \cos \theta \, d\theta = \int w^2 \, dw = \frac{1}{3}w^3 + C = \frac{1}{3}\sin^3 \theta + C,$$

where C is a constant.

24. Substitute $w = 4 - x^2$, $dw = -2x \, dx$:

$$\int x\sqrt{4-x^2} \, dx = -\frac{1}{2}\int \sqrt{w} \, dw = -\frac{1}{3}w^{3/2} + C = -\frac{1}{3}(4-x^2)^{3/2} + C.$$

Check
$$\frac{d}{dx}\left[-\frac{1}{3}(4-x^2)^{3/2} + C\right] = -\frac{1}{3}\left[\frac{3}{2}(4-x^2)^{1/2}(-2x)\right] = x\sqrt{4-x^2}.$$

25. Expanding the numerator and dividing, we have

$$\int \frac{(u+1)^3}{u^2} \, du = \int \frac{(u^3+3u^2+3u+1)}{u^2} \, du = \int \left(u+3+\frac{3}{u}+\frac{1}{u^2}\right) du$$
$$= \frac{u^2}{2} + 3u + 3\ln|u| - \frac{1}{u} + C.$$

Check:
$$\frac{d}{du}\left(\frac{u^2}{2} + 3u + 3\ln|u| - \frac{1}{u} + C\right) = u + 3 + 3/u + 1/u^2 = \frac{(u+1)^3}{u^2}.$$

26. Substitute $w = \sqrt{y}$, $dw = 1/(2\sqrt{y})\,dy$. Then
$$\int \frac{\cos\sqrt{y}}{\sqrt{y}}\,dy = 2\int \cos w\,dw = 2\sin w + C = 2\sin\sqrt{y} + C.$$
Check:
$$\frac{d}{dy} 2\sin\sqrt{y} + C = \frac{2\cos\sqrt{y}}{2\sqrt{y}} = \frac{\cos\sqrt{y}}{\sqrt{y}}.$$

27. Since $\frac{d}{dz}(\tan z) = \frac{1}{\cos^2 z}$, we have
$$\int \frac{1}{\cos^2 z}\,dz = \tan z + C.$$
Check:
$$\frac{d}{dz}(\tan z + C) = \frac{d}{dz}\frac{\sin z}{\cos z} = \frac{(\cos z)(\cos z) - (\sin z)(-\sin z)}{\cos^2 z} = \frac{1}{\cos^2 z}.$$

28. Denote $\int \cos^2\theta\,d\theta$ by A. Let $u = \cos\theta$, $v' = \cos\theta$. Then, $v = \sin\theta$ and $u' = -\sin\theta$. Integrating by parts, we get:
$$A = \cos\theta\sin\theta - \int(-\sin\theta)\sin\theta\,d\theta.$$
Employing the identity $\sin^2\theta = 1 - \cos^2\theta$, the equation above becomes:
$$A = \cos\theta\sin\theta + \int d\theta - \int \cos^2\theta\,d\theta$$
$$= \cos\theta\sin\theta + \theta - A + C.$$
Solving this equation for A, and using the identity $\sin 2\theta = 2\cos\theta\sin\theta$ we get:
$$A = \int \cos^2\theta\,d\theta = \frac{1}{4}\sin 2\theta + \frac{1}{2}\theta + C.$$
[Note: An alternate solution would have been to use the identity $\cos^2\theta = \frac{1}{2}\cos 2\theta + \frac{1}{2}$.]

29. Multiplying out and integrating term by term:
$$\int t^{10}(t-10)\,dt = \int (t^{11} - 10t^{10})\,dt = \int t^{11}\,dt - 10\int t^{10}\,dt = \frac{1}{12}t^{12} - \frac{10}{11}t^{11} + C.$$

30. Substitute $w = 2x - 6$. Then $dw = 2\,dx$ and
$$\int \tan(2x-6)\,dx = \frac{1}{2}\int \tan w\,dw = \frac{1}{2}\int \frac{\sin w}{\cos w}\,dw$$
$$= -\frac{1}{2}\ln|\cos w| + C \text{ by substitution or by I-7 of the integral table.}$$
$$= -\frac{1}{2}\ln|\cos(2x-6)| + C.$$

31. Using integration by parts, we have
$$\int_1^3 \ln(x^3)\,dx = 3\int_1^3 \ln x\,dx = 3(x\ln x - x)\Big|_1^3 = 9\ln 3 - 6 \approx 3.8875.$$
This matches the approximation given by Simpson's rule with 10 intervals.

32. In Problem 20, we found that

$$\int (\ln x)^2 \, dx = x(\ln x)^2 - 2x \ln x + 2x + C.$$

Thus

$$\int_1^e (\ln x)^2 \, dx = \left[x(\ln x)^2 - 2x \ln x + 2x \right]\Big|_1^e = e - 2 \approx 0.71828.$$

This matches the approximation given by Simpson's rule with 10 intervals.

33. Integrating by parts, we take $u = e^{2x}$, $u' = 2e^{2x}$, $v' = \sin 2x$, and $v = -\frac{1}{2}\cos 2x$, so

$$\int e^{2x} \sin 2x \, dx = -\frac{e^{2x}}{2} \cos 2x + \int e^{2x} \cos 2x \, dx.$$

Integrating by parts again, with $u = e^{2x}$, $u' = 2e^{2x}$, $v' = \cos 2x$, and $v = \frac{1}{2}\sin 2x$, we get

$$\int e^{2x} \cos 2x \, dx = \frac{e^{2x}}{2} \sin 2x - \int e^{2x} \sin 2x \, dx.$$

Substituting into the previous equation, we obtain

$$\int e^{2x} \sin 2x \, dx = -\frac{e^{2x}}{2} \cos 2x + \frac{e^{2x}}{2} \sin 2x - \int e^{2x} \sin 2x \, dx.$$

Solving for $\int e^{2x} \sin 2x \, dx$ gives

$$\int e^{2x} \sin 2x \, dx = \frac{1}{4} e^{2x} (\sin 2x - \cos 2x) + C.$$

This result can also be obtained using II-8 in the integral table. Thus

$$\int_{-\pi}^{\pi} e^{2x} \sin 2x = \left[\frac{1}{4} e^{2x}(\sin 2x - \cos 2x) \right]\Big|_{-\pi}^{\pi} = \frac{1}{4}(e^{-2\pi} - e^{2\pi}) \approx -133.8724.$$

We get -133.37 using Simpson's rule with 10 intervals. With 100 intervals, we get -133.8724. Thus our answer matches the approximation of Simpson's rule.

34.
$$\int_0^{10} z e^{-z} \, dz = \left[-z e^{-z} \right]\Big|_0^{10} - \int_0^{10} -e^{-z} \, dz \quad (\text{let } z = u, e^{-z} = v', -e^{-z} = v)$$
$$= -10 e^{-10} - \left[e^{-z} \right]\Big|_0^{10}$$
$$= -10 e^{-10} - e^{-10} + 1$$
$$= -11 e^{-10} + 1.$$

35. Let $\sin \theta = w$, $\cos \theta \, d\theta = dw$. So, if $\theta = -\frac{\pi}{3}$, then $w = -\frac{\sqrt{3}}{2}$, and if $\theta = \frac{\pi}{4}$, then $w = \frac{\sqrt{2}}{2}$. So we have

$$\int_{-\pi/3}^{\pi/4} \sin^3 \theta \cos \theta \, d\theta = \int_{-\sqrt{3}/2}^{\sqrt{2}/2} w^3 \, dw = \frac{1}{4} w^4 \Big|_{-\sqrt{3}/2}^{\sqrt{2}/2} = \frac{1}{4}\left[\left(\frac{\sqrt{2}}{2}\right)^4 - \left(\frac{-\sqrt{3}}{2}\right)^4 \right] = -\frac{5}{64}.$$

36. This integral is 0 because the function $x^3 \cos(x^2)$ is odd (meaning $f(-x) = -f(x)$), and so the negative contribution to the integral from $-\frac{\pi}{4} < x < 0$ exactly cancels the positive contribution from $0 < x < \frac{\pi}{4}$. See figure below.

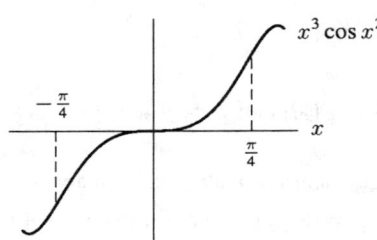

37. Let $\sqrt{x} = w$, $\frac{1}{2}x^{-\frac{1}{2}}dx = dw$, $\frac{dx}{\sqrt{x}} = 2\,dw$. If $x = 1$ then $w = 1$, and if $x = 4$ so $w = 2$. So we have

$$\int_1^4 \frac{e^{\sqrt{x}}}{\sqrt{x}}\,dx = \int_1^2 e^w \cdot 2\,dw = 2e^w \Big|_1^2 = 2(e^2 - e) \approx 9.34.$$

38.
$$\int_0^1 \frac{dx}{x^2 + 1} = \tan^{-1} x \Big|_0^1 = \tan^{-1} 1 - \tan^{-1} 0 = \frac{\pi}{4} - 0 = \frac{\pi}{4}.$$

39. Let $\ln x = w$, then $\frac{1}{x}\,dx = dw$, so

$$\int \frac{(\ln x)^2}{x}\,dx = \int w^2\,dw = \frac{1}{3}w^3 + C = \frac{1}{3}(\ln x)^3 + C, \text{ where } C \text{ is a constant.}$$

40. Multiplying out, dividing, and then integrating yields

$$\int \frac{(t+2)^2}{t^3}\,dt = \int \frac{t^2 + 4t + 4}{t^3}\,dt = \int \frac{1}{t}\,dt + \int \frac{4}{t^2}\,dt + \int \frac{4}{t^3}\,dt = \ln|t| - \frac{4}{t} - \frac{2}{t^2} + C,$$

where C is a constant.

41. Integrating term by term:

$$\int \left(x^2 + 2x + \frac{1}{x}\right) dx = \frac{1}{3}x^3 + x^2 + \ln|x| + C,$$

where C is a constant.

42. Dividing and then integrating, we obtain

$$\int \frac{t+1}{t^2}\,dt = \int \frac{1}{t}\,dt + \int \frac{1}{t^2}\,dt = \ln|t| - \frac{1}{t} + C, \text{ where } C \text{ is a constant.}$$

43. Let $t^2 + 1 = w$, then $2t\,dt = dw$, $t\,dt = \frac{1}{2}dw$, so

$$\int t e^{t^2+1}\,dt = \int e^w \cdot \frac{1}{2}\,dw = \frac{1}{2}\int e^w\,dw = \frac{1}{2}e^w + C = \frac{1}{2}e^{t^2+1} + C,$$

where C is a constant.

44. Let $\cos\theta = w$, then $-\sin\theta\,d\theta = dw$, so

$$\int \tan\theta\,d\theta = \int \frac{\sin\theta}{\cos\theta}\,d\theta = \int \frac{-1}{w}\,dw$$
$$= -\ln|w| + C = -\ln|\cos\theta| + C,$$

where C is a constant.

45. If $u = \sin(5\theta)$, $du = \cos(5\theta) \cdot 5\,d\theta$, so

$$\int \sin(5\theta)\cos(5\theta)d\theta = \frac{1}{5}\int \sin(5\theta) \cdot 5\cos(5\theta)d\theta = \frac{1}{5}\int u\,du$$

$$= \frac{1}{5}\left(\frac{u^2}{2}\right) + C = \frac{1}{10}\sin^2(5\theta) + C$$

or

$$\int \sin(5\theta)\cos(5\theta)d\theta = \frac{1}{2}\int 2\sin(5\theta)\cos(5\theta)d\theta = \frac{1}{2}\int \sin(10\theta)d\theta \quad \text{(using } \sin(2x) = 2\sin x \cos x\text{)}$$

$$= \frac{-1}{20}\cos(10\theta) + C.$$

46. Using substitution,

$$\int \frac{x}{x^2+1}\,dx = \int \frac{1/2}{w}\,dw \quad (x^2+1 = w,\, 2x\,dx = dw,\, x\,dx = \frac{1}{2}dw)$$

$$= \frac{1}{2}\int \frac{1}{w}\,dw = \frac{1}{2}\ln|w| + C = \frac{1}{2}\ln|x^2+1| + C,$$

where C is a constant.

47. Since $\frac{d}{dz}(\arctan z) = \frac{1}{1+z^2}$, we have

$$\int \frac{dz}{1+z^2} = \arctan z + C, \quad \text{where } C \text{ is a constant.}$$

48. Let $w = 2z$, so $dw = 2dz$. Then, since $\frac{d}{dw}\arctan w = \frac{1}{1+w^2}$, we have

$$\int \frac{dz}{1+4z^2} = \int \frac{\frac{1}{2}dw}{1+w^2} = \frac{1}{2}\arctan w + C = \frac{1}{2}\arctan 2z + C.$$

49. Let $w = \cos 2\theta$. Then $dw = -2\sin 2\theta\,d\theta$, hence

$$\int \cos^3 2\theta \sin 2\theta\,d\theta = -\frac{1}{2}\int w^3\,dw = -\frac{w^4}{8} + C = -\frac{\cos^4 2\theta}{8} + C.$$

Check:
$$\frac{d}{d\theta}\left(-\frac{\cos^4 2\theta}{8}\right) = -\frac{(4\cos^3 2\theta)(-\sin 2\theta)(2)}{8} = \cos^3 2\theta \sin 2\theta.$$

50. Let $\cos 5\theta = w$, then $-5\sin 5\theta\,d\theta = dw$, $\sin 5\theta\,d\theta = -\frac{1}{5}dw$. So

$$\int \sin 5\theta \cos^3 5\theta\,d\theta = \int w^3 \cdot \left(-\frac{1}{5}\right)dw = -\frac{1}{5}\int w^3\,dw = -\frac{1}{20}w^4 + C$$

$$= -\frac{1}{20}\cos^4 5\theta + C,$$

where C is a constant.

51.
$$\int \sin^3 z \cos^3 z \, dz = \int \sin z (1 - \cos^2 z) \cos^3 z \, dz$$
$$= \int \sin z \cos^3 z \, dz - \int \sin z \cos^5 z \, dz$$
$$= \int w^3 (-dw) - \int w^5 (-dw) \quad (\text{let } \cos z = w, \text{ so } -\sin z \, dz = dw)$$
$$= -\int w^3 \, dw + \int w^5 \, dw$$
$$= -\frac{1}{4}w^4 + \frac{1}{6}w^6 + C$$
$$= -\frac{1}{4}\cos^4 z + \frac{1}{6}\cos^6 z + C,$$

where C is a constant.

52. If $u = t - 10, t = u + 10$ and $dt = 1 \, du$, so substituting we get
$$\int (u+10) u^{10} du = \int (u^{11} + 10 u^{10}) \, du = \frac{1}{12} u^{12} + \frac{10}{11} u^{11} + C$$
$$= \frac{1}{12}(t-10)^{12} + \frac{10}{11}(t-10)^{11} + C.$$

53. Let $\sin \theta = w$, then $\cos \theta \, d\theta = dw$, so
$$\int \cos \theta \sqrt{1 + \sin \theta} \, d\theta = \int \sqrt{1 + w} \, dw$$
$$= \frac{(1+w)^{3/2}}{3/2} + C = \frac{2}{3}(1 + \sin \theta)^{3/2} + C,$$

where C is a constant.

54.
$$\int x e^x \, dx = x e^x - \int e^x \, dx \quad (\text{let } x = u, e^x = v', e^x = v)$$
$$= x e^x - e^x + C,$$

where C is a constant.

55.
$$\int t^3 e^t \, dt = t^3 e^t - \int 3 t^2 e^t \, dt \quad (\text{let } t^3 = u, e^t = v', 3t^2 = u', e^t = v)$$
$$= t^3 e^t - 3 \int t^2 e^t \, dt \quad (\text{let } t^2 = u, e^t = v')$$
$$= t^3 e^t - 3(t^2 e^t - \int 2t e^t \, dt)$$
$$= t^3 e^t - 3t^2 e^t + 6 \int t e^t \, dt \quad (\text{let } t = u, e^t = v')$$
$$= t^3 e^t - 3t^2 e^t + 6(t e^t - \int e^t \, dt)$$
$$= t^3 e^t - 3t^2 e^t + 6t e^t - 6 e^t + C,$$

where C is a constant.

SOLUTIONS TO PRACTICE PROBLEMS ON INTEGRATION **349**

56. Let $x^2 = w$, then $2x\,dx = dw$, $x = 1 \Rightarrow w = 1$, $x = 3 \Rightarrow w = 9$. Thus,

$$\int_1^3 x(x^2+1)^{70}\,dx = \int_1^9 (w+1)^{70}\frac{1}{2}\,dw$$
$$= \frac{1}{2}\cdot\frac{1}{71}(w+1)^{71}\Big|_1^9$$
$$= \frac{1}{142}(10^{71} - 2^{71}).$$

57. Let $w = 3z + 5$ and $dw = 3\,dz$. Then

$$\int (3z+5)^3\,dz = \frac{1}{3}\int w^3\,dw = \frac{1}{12}w^4 + C = \frac{1}{12}(3z+5)^4 + C.$$

58. Rewrite $9 + u^2$ as $9[1 + (u/3)^2]$ and let $w = u/3$, then $dw = du/3$ so that

$$\int \frac{du}{9+u^2} = \frac{1}{3}\int \frac{dw}{1+w^2} = \frac{1}{3}\arctan w + C = \frac{1}{3}\arctan\left(\frac{u}{3}\right) + C.$$

59. Let $u = \sin w$, then $du = \cos w\,dw$ so that

$$\int \frac{\cos w}{1+\sin^2 w}\,dw = \int \frac{du}{1+u^2} = \arctan u + C = \arctan(\sin w) + C.$$

60. Let $w = \ln x$, then $dw = (1/x)dx$ which gives

$$\int \frac{1}{x}\tan(\ln x)\,dx = \int \tan w\,dw = \int \frac{\sin w}{\cos w}\,dw = -\ln(|\cos w|) + C = -\ln(|\cos(\ln x)|) + C.$$

61. Let $w = \ln x$, then $dw = (1/x)dx$ so that

$$\int \frac{1}{x}\sin(\ln x)\,dx = \int \sin w\,dw = -\cos w + C = -\cos(\ln x) + C.$$

62. Let $u = 2x$, then $du = 2\,dx$ so that

$$\int \frac{dx}{\sqrt{1-4x^2}} = \frac{1}{2}\int \frac{du}{\sqrt{1-u^2}} = \frac{1}{2}\arcsin u + C = \frac{1}{2}\arcsin(2x) + C.$$

63. Let $u = 16 - w^2$, then $du = -2w\,dw$ so that

$$\int \frac{w\,dw}{\sqrt{16-w^2}} = -\frac{1}{2}\int \frac{du}{\sqrt{u}} = -\sqrt{u} + C = -\sqrt{16-w^2} + C.$$

64. Dividing and then integrating term by term, we get

$$\int \frac{e^{2y}+1}{e^{2y}}\,dy = \int \left(\frac{e^{2y}}{e^{2y}} + \frac{1}{e^{2y}}\right)dy = \int (1+e^{-2y})\,dy = \int dy + \left(-\frac{1}{2}\right)\int e^{-2y}(-2)\,dy$$
$$= y - \frac{1}{2}e^{-2y} + C.$$

65. Let $u = 1 - \cos w$, then $du = \sin w\,dw$ which gives

$$\int \frac{\sin w\,dw}{\sqrt{1-\cos w}} = \int \frac{du}{\sqrt{u}} = 2\sqrt{u} + C = 2\sqrt{1-\cos w} + C.$$

66. Let $w = \ln x$. Then $dw = (1/x)dx$ which gives
$$\int \frac{dx}{x \ln x} = \int \frac{dw}{w} = \ln|w| + C = \ln|\ln x| + C.$$

67. Let $w = 3u + 8$, then $dw = 3du$ and
$$\int \frac{du}{3u + 8} = \int \frac{dw}{3w} = \frac{1}{3}\ln|3u + 8| + C.$$

68. Let $w = \sqrt{x^2 + 1}$, then $dw = \dfrac{x\,dx}{\sqrt{x^2 + 1}}$ so that
$$\int \frac{x}{\sqrt{x^2 + 1}} \cos\sqrt{x^2 + 1}\,dx = \int \cos w\,dw = \sin w + C = \sin\sqrt{x^2 + 1} + C.$$

69. Integrating by parts using $u = t^2$ and $dv = \dfrac{t\,dt}{\sqrt{1+t^2}}$ gives $du = 2t\,dt$ and $v = \sqrt{1 + t^2}$. Now
$$\int \frac{t^3}{\sqrt{1+t^2}}\,dt = t^2\sqrt{1+t^2} - \int 2t\sqrt{1+t^2}\,dt$$
$$= t^2\sqrt{1+t^2} - \frac{2}{3}(1+t^2)^{3/2} + C$$
$$= \sqrt{1+t^2}(t^2 - \frac{2}{3}(1+t^2)) + C$$
$$= \sqrt{1+t^2}\frac{(t^2 - 2)}{3} + C.$$

70. Using integration by parts, let $r = u$ and $dt = e^{ku}du$, so $dr = du$ and $t = (1/k)e^{ku}$. Thus
$$\int ue^{ku}\,du = \frac{u}{k}e^{ku} - \frac{1}{k}\int e^{ku}\,du = \frac{u}{k}e^{ku} - \frac{1}{k^2}e^{ku} + C.$$

71. Let $u = w + 5$, then $du = dw$ and noting that $w = u - 5$ we obtain
$$\int (w+5)^4 w\,dw = \int u^4(u - 5)\,du$$
$$= \int (u^5 - 5u^4)\,du$$
$$= \frac{1}{6}u^6 - u^5 + C$$
$$= \frac{1}{6}(w+5)^6 - (w+5)^5 + C.$$

72. $\int e^{\sqrt{2}x+3}\,dx = \dfrac{1}{\sqrt{2}}\int e^{\sqrt{2}x+3}\sqrt{2}\,dx$. If $u = \sqrt{2}x + 3$, $du = \sqrt{2}\,dx$, so
$$\frac{1}{\sqrt{2}}\int e^u\,du = \frac{1}{\sqrt{2}}e^u + C = \frac{1}{\sqrt{2}}e^{\sqrt{2}x+3} + C.$$

73. Integrate by parts letting $u = (\ln r)^2$ and $dv = r\,dr$, then $du = (2/r)\ln r\,dr$ and $v = r^2/2$. We get
$$\int r(\ln r)^2\,dr = \frac{1}{2}r^2(\ln r)^2 - \int r\ln r\,dr.$$

Then using integration by parts again with $u = \ln r$ and $dv = r\,dr$, so $du = dr/r$ and $v = r^2/2$, we get
$$\int r\ln^2 r\,dr = \frac{1}{2}r^2(\ln r)^2 - \left[\frac{1}{2}r^2\ln r - \frac{1}{2}\int r\,dr\right] = \frac{1}{2}r^2(\ln r)^2 - \frac{1}{2}r^2\ln r + \frac{1}{4}r^2 + C.$$

74. $\int (e^x + x)^2 dx = \int (e^{2x} + 2xe^x + x^2) dy$. Separating into three integrals, we have

$$\int e^{2x} dx = \frac{1}{2} \int e^{2x} 2\, dx = \frac{1}{2} e^{2x} + C_1,$$

$$\int 2xe^x dx = 2 \int xe^x dx = 2xe^x - 2e^x + C_2$$

from Formula II-13 of the integral table or integration by parts, and

$$\int x^2 dx = \frac{x^3}{3} + C_3.$$

Combining the results and writing $C = C_1 + C_2 + C_3$, we get

$$\frac{1}{2} e^{2x} + 2xe^x - 2e^x + \frac{x^3}{3} + C.$$

75. Integrate by parts, $r = \ln u$ and $dt = u^2\, du$, so $dr = (1/u)\, du$ and $t = (1/3) u^3$. We have

$$\int u^2 \ln u\, du = \frac{1}{3} u^3 \ln u - \frac{1}{3} \int u^2\, du = \frac{1}{3} u^3 \ln u - \frac{1}{9} u^3 + C.$$

76. The integral table yields

$$\int \frac{5x + 6}{x^2 + 4} dx = \frac{5}{2} \ln |x^2 + 4| + \frac{6}{2} \arctan \frac{x}{2} + C$$

$$= \frac{5}{2} \ln |x^2 + 4| + 3 \arctan \frac{x}{2} + C.$$

Check:

$$\frac{d}{dx} \left(\frac{5}{2} \ln |x^2 + 4| + \frac{6}{2} \arctan \frac{x}{2} + C \right) = \frac{5}{2} \left(\frac{1}{x^2 + 4} (2x) + 3 \frac{1}{1 + (x/2)^2} \frac{1}{2} \right)$$

$$= \frac{5x}{x^2 + 4} + \frac{6}{x^2 + 4} = \frac{5x + 6}{x^2 + 4}.$$

77. Using Table IV-19, let $m = 3$, $w = 2x$, and $dw = 2dx$. Then

$$\int \frac{1}{\sin^3(2x)} dx = \frac{1}{2} \int \frac{1}{\sin^3 w} dw$$

$$= \frac{1}{2} \left[\frac{-1}{(3-1)} \frac{\cos w}{\sin^2 w} \right] + \frac{1}{4} \int \frac{1}{\sin w} dw,$$

and using Table IV-20, we have

$$\int \frac{1}{\sin w} dw = \frac{1}{2} \ln \left| \frac{\cos w - 1}{\cos w + 1} \right| + C.$$

Thus,

$$\int \frac{1}{\sin^3(2x)} dx = -\frac{\cos 2x}{4 \sin^2 2x} + \frac{1}{8} \ln \left| \frac{\cos 2x - 1}{\cos 2x + 1} \right| + C.$$

78. We can factor $r^2 - 100 = (r - 10)(r + 10)$ so we can use Table V-26 (with $a = 10$ and $b = -10$) to get

$$\int \frac{dr}{r^2 - 100} = \frac{1}{20} \left[\ln |r - 10| + \ln |r + 10| \right] + C.$$

79. Integration by parts will be used twice here. First let $u = y^2$ and $dv = \sin(cy)\,dy$, then $du = 2y\,dy$ and $v = -(1/c)\cos(cy)$. Thus

$$\int y^2 \sin(cy)\,dy = -\frac{y^2}{c}\cos(cy) + \frac{2}{c}\int y\cos(cy)\,dy.$$

Now use integration by parts to evaluate the integral in the right hand expression. Here let $u = y$ and $dv = \cos(cy)dy$ which gives $du = dy$ and $v = (1/c)\sin(cy)$. Then we have

$$\int y^2 \sin(cy)\,dy = -\frac{y^2}{c}\cos(cy) + \frac{2}{c}\left(\frac{y}{c}\sin(cy) - \frac{1}{c}\int \sin(cy)\,dy\right)$$

$$= -\frac{y^2}{c}\cos(cy) + \frac{2y}{c^2}\sin(cy) + \frac{2}{c^3}\cos(cy) + C.$$

80. Integration by parts will be used twice. First let $u = e^{-ct}$ and $dv = \sin(kt)dt$, then $du = -ce^{-ct}dt$ and $v = (-1/k)\cos kt$. Then

$$\int e^{-ct}\sin kt\,dt = -\frac{1}{k}e^{-ct}\cos kt - \frac{c}{k}\int e^{-ct}\cos kt\,dt$$

$$= -\frac{1}{k}e^{-ct}\cos kt - \frac{c}{k}\left(\frac{1}{k}e^{-ct}\sin kt + \frac{c}{k}\int e^{-ct}\sin kt\,dt\right)$$

$$= -\frac{1}{k}e^{-ct}\cos kt - \frac{c}{k^2}e^{-ct}\sin kt - \frac{c^2}{k^2}\int e^{-ct}\sin kt\,dt$$

Solving for $\int e^{-ct}\sin kt\,dt$ gives

$$\frac{k^2+c^2}{k^2}\int e^{-ct}\sin kt\,dt = -\frac{e^{-ct}}{k^2}(k\cos kt + c\sin kt),$$

so

$$\int e^{-ct}\sin kt\,dt = -\frac{e^{-ct}}{k^2+c^2}(k\cos kt + c\sin kt) + C.$$

81. Using II-9 from the integral table, with $a = 5$ and $b = 3$, we have

$$\int e^{5x}\cos(3x)\,dx = \frac{1}{25+9}e^{5x}\left[5\cos(3x) + 3\sin(3x)\right] + C$$

$$= \frac{1}{34}e^{5x}\left[5\cos(3x) + 3\sin(3x)\right] + C.$$

82. $\int (x^{\sqrt{k}} + \sqrt{k}^x)\,dx = \int x^{\sqrt{k}}\,dx + \int \sqrt{k}^x\,dx$. For the first integral, use Formula I-1 with $n = \sqrt{k}$. For the second integral, use Formula I-3 with $a = \sqrt{k}$.

83. Factor $\sqrt{3}$ out of the integrand and use VI-30 of the integral table with $u = 2x$ and $du = 2dx$ to get

$$\int \sqrt{3+12x^2}\,dx = \int \sqrt{3}\sqrt{1+4x^2}\,dx$$

$$= \frac{\sqrt{3}}{2}\int \sqrt{1+u^2}\,du$$

$$= \frac{\sqrt{3}}{4}\left(u\sqrt{1+u^2} + \int \frac{1}{\sqrt{1+u^2}}\,du\right).$$

Then from VI-29, simplify the integral on the right to get

$$\int \sqrt{3+12x^2}\,dx = \frac{\sqrt{3}}{4}\left(u\sqrt{1+u^2} + \ln|u + \sqrt{1+u^2}|\right) + C$$

$$= \frac{\sqrt{3}}{4}\left(2x\sqrt{1+(2x)^2} + \ln|2x + \sqrt{1+(2x)^2}|\right) + C.$$

84. Using Table III-14, with $a = -4$ we have
$$\int (x^2 - 3x + 2)e^{-4x}\, dx = -\frac{1}{4}(x^2 - 3x + 2)e^{-4x}$$
$$-\frac{1}{16}(2x - 3)e^{-4x} - \frac{1}{64}(2)e^{-4x} + C.$$
$$= \frac{1}{32}e^{-4x}(-11 + 20x - 8x^2) + C.$$

85. We know $x^2 + 5x + 4 = (x+1)(x+4)$, so we can use V-26 of the integral table with $a = -1$ and $b = -4$ to write
$$\int \frac{dx}{x^2 + 5x + 4} = \frac{1}{3}(\ln|x+1| - \ln|x+4|) + C.$$

86. By completing the square, we get
$$x^2 - 3x + 2 = (x^2 - 3x + (-\tfrac{3}{2})^2) + 2 - \frac{9}{4} = (x - \tfrac{3}{2})^2 - \frac{1}{4}.$$
Then
$$\int \frac{1}{\sqrt{x^2 - 3x + 2}}\, dx = \int \frac{1}{\sqrt{(x - \tfrac{3}{2})^2 - \tfrac{1}{4}}}\, dx.$$
Let $w = (x - (3/2))$, then $dw = dx$ and $a^2 = 1/4$. Then we have
$$\int \frac{1}{\sqrt{x^2 - 3x + 2}}\, dx = \int \frac{1}{\sqrt{w^2 - a^2}}\, dw$$
and from VI-29 of the integral table we have
$$\int \frac{1}{\sqrt{w^2 - a^2}}\, dw = \ln\left|w + \sqrt{w^2 - a^2}\right| + C$$
$$= \ln\left|\left(x - \tfrac{3}{2}\right) + \sqrt{\left(x - \tfrac{3}{2}\right)^2 - \tfrac{1}{4}}\right| + C$$
$$= \ln\left|\left(x - \tfrac{3}{2}\right) + \sqrt{x^2 - 3x + 2}\right| + C.$$

87. First divide $x^2 + 3x + 2$ into x^3 to obtain
$$\frac{x^3}{x^2 + 3x + 2} = x - 3 + \frac{7x + 6}{x^2 + 3x + 2}.$$
Since $x^2 + 3x + 2 = (x+1)(x+2)$, we can use V-27 of the integral table (with $c = 7$, $d = 6$, $a = -1$, and $b = -2$) to get
$$\int \frac{7x + 6}{x^2 + 3x + 2}\, dx = -\ln|x+1| + 8\ln|x+2| + C.$$
Including the terms $x - 3$ from the long division and integrating them gives
$$\int \frac{x^3}{x^2 + 3x + 2}\, dx = \int \left(x - 3 + \frac{7x + 6}{x^2 + 3x + 6}\right) dx = \frac{1}{2}x^2 - 3x - \ln|x+1| + 8\ln|x+2| + C.$$

88. First divide $x^2 + 1$ by $x^2 - 3x + 2$ to obtain
$$\frac{x^2 + 1}{x^2 - 3x + 2} = 1 + \frac{3x - 1}{x^2 - 3x + 2}.$$
Factoring $x^2 - 3x + 2 = (x-2)(x-1)$ we can use V-27 (with $c = 3$, $d = -1$, $a = 2$ and $b = 1$) to write
$$\int \frac{3x - 1}{x^2 - 3x + 2}\, dx = 5\ln|x-2| - 2\ln|x-1| + C.$$
Remembering to include the extra term of $+1$ we got when dividing, we get
$$\int \frac{x^2 + 1}{x^2 - 3x + 2}\, dx = \int \left(1 + \frac{3x - 1}{x^2 - 3x + 2}\right) dx = x + 5\ln|x-2| - 2\ln|x-1| + C.$$

354   CHAPTER SEVEN /SOLUTIONS

89. We can factor the denominator into $ax(x + \frac{b}{a})$, so

$$\int \frac{dx}{ax^2 + bx} = \frac{1}{a} \int \frac{1}{x(x + \frac{b}{a})}$$

Now we can use V-26 (with $A = 0$ and $B = -\frac{b}{a}$ to give

$$\frac{1}{a} \int \frac{1}{x(x + \frac{b}{a})} = \frac{1}{a} \cdot \frac{a}{b} \left(\ln|x| - \ln\left|x + \frac{b}{a}\right| \right) + C = \frac{1}{b} \left(\ln|x| - \ln\left|x + \frac{b}{a}\right| \right) + C.$$

90. Let $w = ax^2 + 2bx + c$, then $dw = (2ax + 2b)dx$ so that

$$\int \frac{ax + b}{ax^2 + 2bx + c} dx = \frac{1}{2} \int \frac{dw}{w} = \frac{1}{2} \ln|w| + C = \frac{1}{2} \ln|ax^2 + 2bx + c| + C.$$

91.
$$\int \frac{dz}{z^2 + z} = \int \frac{dz}{z(z+1)} = \int \left(\frac{1}{z} - \frac{1}{z+1} \right) dz = \ln|z| - \ln|z + 1| + C.$$

(This is formula V–26 in the integral table.)
Check:

$$\frac{d}{dz} \left(\ln|z| - \ln|z + 1| + C \right) = \frac{1}{z} - \frac{1}{z+1} = \frac{1}{z^2 + z}.$$

92. Multiplying out and integrating term by term,

$$\int \left(\frac{x}{3} + \frac{3}{x} \right)^2 dx = \int \left(\frac{x^2}{9} + 2 + \frac{9}{x^2} \right) dx = \frac{1}{9} \left(\frac{x^3}{3} \right) + 2x + 9 \left(\frac{x^{-1}}{-1} \right) + C = \frac{x^3}{27} + 2x - \frac{9}{x} + C.$$

93. If $u = 2^t + 1$, $du = 2^t (\ln 2) \, dt$, so

$$\int \frac{2^t}{2^t + 1} dt = \frac{1}{\ln 2} \int \frac{2^t \ln 2}{2^t + 1} dt = \frac{1}{\ln 2} \int \frac{1}{u} = \frac{1}{\ln 2} \ln|u| + C = \frac{1}{\ln 2} \ln|2^t + 1| + C.$$

94. If $u = 1 - x$, $du = -1 \, dx$, so

$$\int 10^{1-x} dx = -1 \int 10^{1-x}(-1 \, dx) = -1 \int 10^u \, du = -1 \frac{10^u}{\ln 10} + C = -\frac{1}{\ln 10} 10^{1-x} + C.$$

95. Multiplying out and integrating term by term gives

$$\int (x^2 + 5)^3 dx = \int (x^6 + 15x^4 + 75x^2 + 125) dx = \frac{1}{7}x^7 + 15\frac{x^5}{5} + 75\frac{x^3}{3} + 125x + C$$
$$= \frac{1}{7}x^7 + 3x^5 + 25x^3 + 125x + C.$$

96. Integrate by parts letting $r = v$ and $dt = \arcsin v \, dv$ then $dr = dv$ and to find t we integrate $\arcsin v \, dv$ by parts letting $x = \arcsin v$ and $dy = dv$. This gives

$$t = v \arcsin v - \int (1/\sqrt{1 - v^2}) v \, dv = v \arcsin v + \sqrt{1 - v^2}.$$

Now, back to the original integration by parts, and we have

$$\int v \arcsin v \, dv = v^2 \arcsin v + v\sqrt{1 - v^2} - \int \left[v \arcsin v + \sqrt{1 - v^2} \right] dv.$$

Adding $\int v \arcsin v \, dv$ to both sides of the above line we obtain

$$2 \int v \arcsin v \, dv = v^2 \arcsin v + v\sqrt{1-v^2} - \int \sqrt{1-v^2} \, dv$$

$$= v^2 \arcsin v + v\sqrt{1-v^2} - \frac{1}{2}v\sqrt{1-v^2} - \frac{1}{2}\arcsin v + C.$$

Dividing by 2 gives

$$\int v \arcsin v \, dv = \left(\frac{v^2}{2} - \frac{1}{4}\right) \arcsin v + \frac{1}{4}v\sqrt{1-v^2} + K,$$

where $K = C/2$.

97. Let $x = 2\theta$, then $dx = 2d\theta$. Thus

$$\int \sin^2(2\theta) \cos^3(2\theta) \, d\theta = \frac{1}{2} \int \sin^2 x \cos^3 x \, dx.$$

We let $w = \sin x$ and $dw = \cos x \, dx$. Then

$$\frac{1}{2} \int \sin^2 x \cos^3 x \, dx = \frac{1}{2} \int \sin^2 x \cos^2 x \cos x \, dx$$

$$= \frac{1}{2} \int \sin^2 x (1 - \sin^2 x) \cos x \, dx$$

$$= \frac{1}{2} \int w^2 (1 - w^2) \, dw = \frac{1}{2} \int (w^2 - w^4) \, dw$$

$$= \frac{1}{2}\left(\frac{w^3}{3} - \frac{w^5}{5}\right) + C = \frac{1}{6}\sin^3 x - \frac{1}{10}\sin^5 x + C$$

$$= \frac{1}{6}\sin^3(2\theta) - \frac{1}{10}\sin^5(2\theta) + C.$$

98. If $u = 2\sin x$, then $du = 2\cos x \, dx$, so

$$\int \cos(2\sin x) \cos x \, dx = \frac{1}{2} \int \cos(2\sin x) 2\cos x \, dx = \frac{1}{2} \int \cos u \, du$$

$$= \frac{1}{2} \sin u + C = \frac{1}{2} \sin(2\sin x) + C.$$

99. By VI-30 in the table of integrals, we have

$$\int \sqrt{4-x^2} \, dx = \frac{x\sqrt{4-x^2}}{2} + 2 \int \frac{1}{\sqrt{4-x^2}} \, dx.$$

The same table informs us in formula VI-28 that

$$\int \frac{1}{\sqrt{4-x^2}} \, dx = \arcsin \frac{x}{2} + C.$$

Thus

$$\int \sqrt{4-x^2} \, dx = \frac{x\sqrt{4-x^2}}{2} + 2 \arcsin \frac{x}{2} + C.$$

100. By long division, $\dfrac{z^3}{z-5} = z^2 + 5z + 25 + \dfrac{125}{z-5}$, so

$$\int \frac{z^3}{z-5} \, dz = \int \left(z^2 + 5z + 25 + \frac{125}{z-5}\right) dz = \frac{z^3}{3} + \frac{5z^2}{2} + 25z + 125 \int \frac{1}{z-5} \, dz$$

$$= \frac{z^3}{3} + \frac{5}{2}z^2 + 25z + 125 \ln|z-5| + C.$$

356   CHAPTER SEVEN /SOLUTIONS

101. If $u = 1 + \cos^2 w$, $du = 2(\cos w)^1(-\sin w)\,dw$, so

$$\int \frac{\sin w \cos w}{1 + \cos^2 w}\,dw = -\frac{1}{2}\int \frac{-2\sin w \cos w}{1 + \cos^2 w}\,dw = -\frac{1}{2}\int \frac{1}{u}\,du = -\frac{1}{2}\ln|u| + C$$
$$= -\frac{1}{2}\ln|1 + \cos^2 w| + C.$$

102. $\int \frac{1}{\tan(3\theta)}\,d\theta = \int \frac{1}{\left(\frac{\sin(3\theta)}{\cos(3\theta)}\right)}\,d\theta = \int \frac{\cos(3\theta)}{\sin(3\theta)}\,d\theta$. If $u = \sin(3\theta)$, $du = \cos(3\theta) \cdot 3d\theta$, so

$$\int \frac{\cos(3\theta)}{\sin(3\theta)}\,d\theta = \frac{1}{3}\int \frac{3\cos(3\theta)}{\sin(3\theta)}\,d\theta = \frac{1}{3}\int \frac{1}{u}\,du = \frac{1}{3}\ln|u| + C = \frac{1}{3}\ln|\sin(3\theta)| + C.$$

103. $\int \frac{x}{\cos^2 x}\,dx = \int x \frac{1}{\cos^2 x}\,dx$. Using integration by parts with $u = x$, $du = dx$ and $dv = \frac{1}{\cos^2 x}\,dx$, $v = \tan x$, we have

$$\int x\left(\frac{1}{\cos^2 x}\,dx\right) = x\tan x - \int \tan x\,dx.$$

Formula I-7 gives the final result of $x\tan x - (-\ln|\cos x|) + C = x\tan x + \ln|\cos x| + C$.

104. Dividing and integrating term by term gives

$$\int \frac{x+1}{\sqrt{x}}\,dx = \int \left(\frac{x}{\sqrt{x}} + \frac{1}{\sqrt{x}}\right)dx = \int (x^{1/2} + x^{-1/2})\,dx = \frac{x^{3/2}}{\frac{3}{2}} + \frac{x^{1/2}}{\frac{1}{2}} + C = \frac{2}{3}x^{3/2} + 2\sqrt{x} + C.$$

105. If $u = \sqrt{x+1}$, $u^2 = x + 1$ with $x = u^2 - 1$ and $dx = 2u\,du$. Substituting, we get

$$\int \frac{x}{\sqrt{x+1}}\,dx = \int \frac{(u^2-1)2u\,du}{u} = \int (u^2 - 1)2\,du = 2\int (u^2 - 1)\,du$$
$$= \frac{2u^3}{3} - 2u + C = \frac{2(\sqrt{x+1})^3}{3} - 2\sqrt{x+1} + C.$$

106. $\int \frac{\sqrt{\sqrt{x}+1}}{\sqrt{x}} = \int (\sqrt{x}+1)^{1/2}\frac{1}{\sqrt{x}}\,dx$; if $u = \sqrt{x} + 1$, $du = \frac{1}{2\sqrt{x}}\,dx$, so we have

$$2\int (\sqrt{x}+1)^{1/2}\frac{1}{2\sqrt{x}}\,dx = 2\int u^{1/2}\,du = 2\left(\frac{u^{3/2}}{\frac{3}{2}}\right) + C = \frac{4}{3}u^{3/2} + C = \frac{4}{3}(\sqrt{x}+1)^{3/2} + C.$$

107. If $u = e^{2y} + 1$, then $du = e^{2y}2\,dy$, so

$$\int \frac{e^{2y}}{e^{2y}+1}\,dy = \frac{1}{2}\int \frac{2e^{2y}}{e^{2y}+1}\,dy = \frac{1}{2}\int \frac{1}{u}\,du = \frac{1}{2}\ln|u| + C = \frac{1}{2}\ln|e^{2y}+1| + C.$$

108. If $u = z^2 - 5$, $du = 2z\,dz$, then

$$\int \frac{z}{(z^2-5)^3}\,dz = \int (z^2-5)^{-3}z\,dz = \frac{1}{2}\int (z^2-5)^{-3}2z\,dz = \frac{1}{2}\int u^{-3}\,du = \frac{1}{2}\left(\frac{u^{-2}}{-2}\right) + C$$
$$= \frac{1}{-4(z^2-5)^2} + C.$$

109. Letting $u = z - 5$, $z = u + 5$, $dz = du$, and substituting, we have

$$\int \frac{z}{(z-5)^3}\,dz = \int \frac{u+5}{u^3}\,du = \int (u^{-2} + 5u^{-3})\,du = \frac{u^{-1}}{-1} + 5\left(\frac{u^{-2}}{-2}\right) + C$$
$$= \frac{-1}{(z-5)} + \frac{-5}{2(z-5)^2} + C.$$

110. If $u = 1 + \tan x$ then $du = \dfrac{1}{\cos^2 x} dx$, and so

$$\int \frac{(1+\tan x)^3}{\cos^2 x}\, dx = \int (1+\tan x)^3 \frac{1}{\cos^2 x}\, dx = \int u^3\, du = \frac{u^4}{4} + C = \frac{(1+\tan x)^4}{4} + C.$$

111. $\displaystyle\int \frac{(2x-1)e^{x^2}}{e^x}\, dx = \int e^{x^2-x}(2x-1)\, dx.$ If $u = x^2 - x$, $du = (2x-1)dx$, so

$$\int e^{x^2-x}(2x-1)\, dx = \int e^u\, du$$
$$= e^u + c$$
$$= e^{x^2-x} + C.$$

112. Let $w = x + \sin x$, then $dw = (1 + \cos x)\, dx$ which gives

$$\int (x + \sin x)^3 (1 + \cos x)\, dx = \int w^3\, dw = \frac{1}{4}w^4 + C = \frac{1}{4}(x + \sin x)^4 + C.$$

113. Using Table III-16,

$$\int (2x^3 + 3x + 4)\cos(2x)\, dx = \frac{1}{2}(2x^3 + 3x + 4)\sin(2x)$$
$$+ \frac{1}{4}(6x^2 + 3)\cos(2x)$$
$$- \frac{1}{8}(12x)\sin(2x) - \frac{3}{4}\cos(2x) + C.$$
$$= 2\sin(2x) + x^3 \sin(2x) + \frac{3x^2}{2}\cos(2x) + C.$$

CHAPTER EIGHT

Solutions for Section 8.1

1. Vertical slices are circular. Horizontal slices would be similar to ellipses in cross-section, or at least ovals (a word derived from *ovum*, the Latin word for egg).

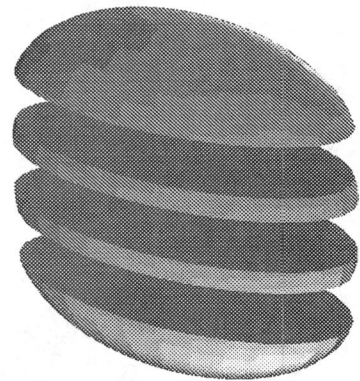

Figure 8.1

2.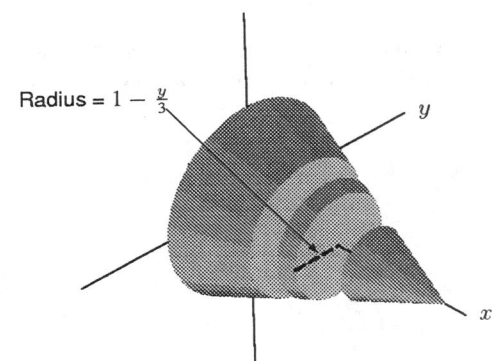

   Slice parallel to the base of the cone, or, equivalently, rotate the line $x = (3-y)/3$ about the y–axis. (One can also slice the other way.) The volume V is given by

   $$V = \int_{y=0}^{y=3} \pi x^2 \, dy = \int_0^3 \pi \left(\frac{3-y}{3}\right)^2 dy$$
   $$= \pi \int_0^3 \left(1 - \frac{2y}{3} + \frac{y^2}{9}\right) dy$$
   $$= \pi \left(y - \frac{y^2}{3} + \frac{y^3}{27}\right) \Big|_0^3 = \pi.$$

3.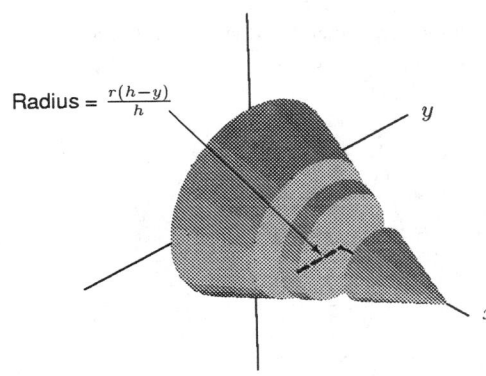

   This cone is what you get when you rotate the line $x = r(h-y)/h$ about the y–axis. So slicing perpendicular to the y–axis yields

   $$V = \int_{y=0}^{y=h} \pi x^2 \, dy = \pi \int_0^h \left(\frac{(h-y)r}{h}\right)^2 dy$$
   $$= \pi \frac{r^2}{h^2} \int_0^h (h^2 - 2hy + y^2) \, dy$$
   $$= \frac{\pi r^2}{h^2}\left[h^2 y - hy^2 + \frac{y^3}{3}\right]\Big|_0^h = \frac{\pi r^2 h}{3}.$$

360 CHAPTER EIGHT /SOLUTIONS

4.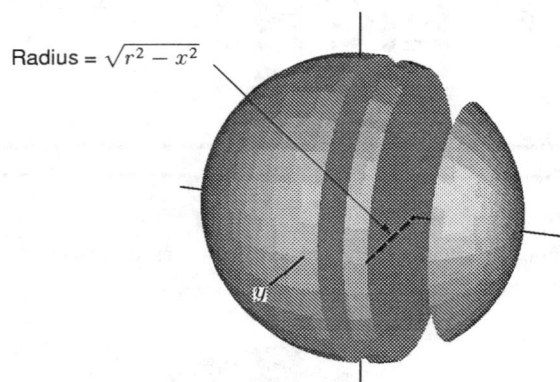

Radius = $\sqrt{r^2 - x^2}$

We slice up the sphere in planes perpendicular to the x-axis. Each slice is a circle, with radius $y = \sqrt{r^2 - x^2}$; that's the radius because $x^2 + y^2 = r^2$ when $z = 0$. Then the volume is

$$V \approx \sum \pi(y^2)\Delta x = \sum \pi(r^2 - x^2)\Delta x.$$

Therefore, as Δx tends to zero, we get

$$V = \int_{x=-r}^{x=r} \pi(r^2 - x^2)\,dx$$

$$= 2\int_{x=0}^{x=r} \pi(r^2 - x^2)\,dx$$

$$= 2\left(\pi r^2 x - \frac{\pi x^3}{3}\right)\bigg|_0^r$$

$$= \frac{4\pi r^3}{3}.$$

5.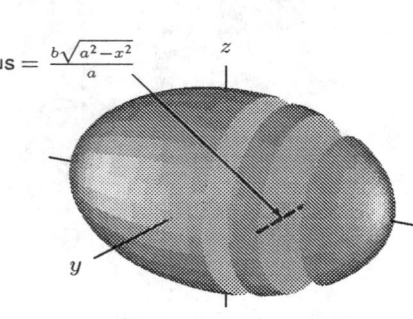

Radius = $\frac{b\sqrt{a^2 - x^2}}{a}$

$y^2 = b^2\left(1 - \frac{x^2}{a^2}\right).$

$$V = \int_{-a}^{a} \pi y^2\,dx = \pi\int_{-a}^{a} b^2\left(1 - \frac{x^2}{a^2}\right)dx$$

$$= 2\pi b^2 \int_0^a \left(1 - \frac{x^2}{a^2}\right)dx = 2\pi b^2 \left[x - \frac{x^3}{3a^2}\right]_0^a$$

$$= 2\pi b^2 \left(a - \frac{a^3}{3a^2}\right) = 2\pi b^2\left(a - \frac{1}{3}a\right)$$

$$= \frac{4}{3}\pi ab^2.$$

6.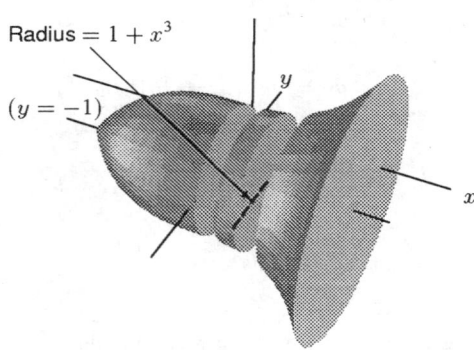

Radius = $1 + x^3$

$(y = -1)$

We slice the region perpendicular to the x–axis. The Riemann sum we get is $\sum \pi(x^3 + 1)^2 \Delta x$. So the volume V is the integral

$$V = \int_{-1}^{1} \pi(x^3 + 1)^2\,dx$$

$$= \pi \int_{-1}^{1}(x^6 + 2x^3 + 1)\,dx$$

$$= \pi\left(\frac{x^7}{7} + \frac{x^4}{2} + x\right)\bigg|_{-1}^{1}$$

$$= (16/7)\pi \approx 7.18.$$

7.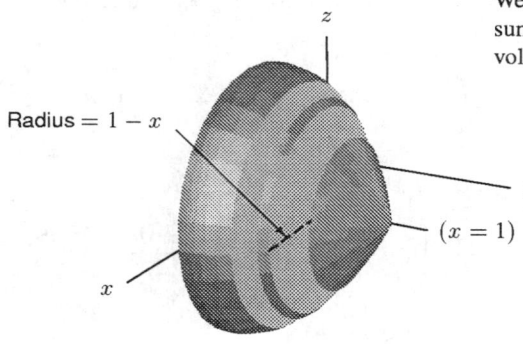

Radius = $1 - x$

$(x = 1)$

We slice the region perpendicular to the y–axis. The Riemann sum we get is $\sum \pi(1 - x)^2\Delta y = \sum \pi(1 - y^2)^2 \Delta y$. So the volume V is the integral

$$V = \int_0^1 \pi(1 - y^2)^2\,dy$$

$$= \pi \int_0^1 (1 - 2y^2 + y^4)\,dy$$

$$= \pi\left(y - \frac{2y^3}{3} + \frac{y^5}{5}\right)\bigg|_0^1$$

$$= (8/15)\pi \approx 1.68.$$

8.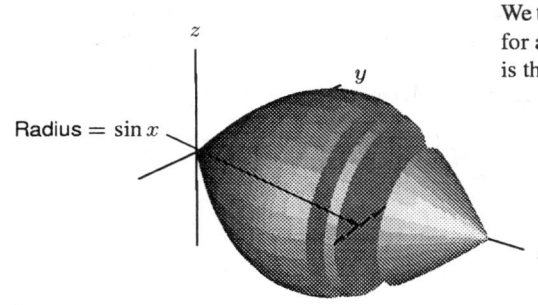

We take slices perpendicular to the x-axis. The Riemann sum for approximating the volume is $\sum \pi \sin^2 x \Delta x$. The volume is the integral corresponding to that sum, namely

$$V = \int_0^\pi \pi \sin^2 x \, dx$$
$$= \pi \left[-\frac{1}{2} \sin x \cos x + \frac{1}{2} x \right] \Big|_0^\pi = \frac{\pi^2}{2} \approx 4.935.$$

9.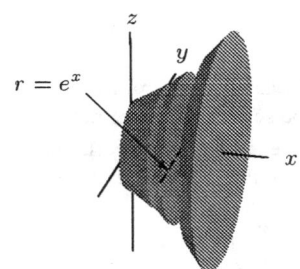

This is the volume of revolution gotten from the rotating the curve $y = e^x$. Take slices perpendicular to the x-axis. They will be circles with radius e^x, so

$$V = \int_{x=0}^{x=1} \pi y^2 \, dx = \pi \int_0^1 e^{2x} \, dx$$
$$= \frac{\pi e^{2x}}{2} \Big|_0^1 = \frac{\pi(e^2 - 1)}{2} \approx 10.036.$$

10.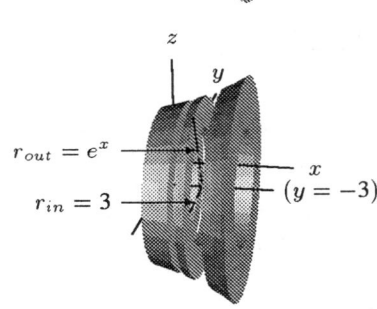

We slice the volume with planes perpendicular to the line $y = -3$. This divides the curve into thin washers, as in Example 5 on page 342 of the text, whose volumes are

$$\pi r_{out}^2 dx - \pi r_{in}^2 dx = \pi(3+y)^2 dx - \pi 3^2 dx.$$

So the integral we get from adding all these washers up is

$$V = \int_{x=0}^{x=1} [\pi(3+y)^2 - \pi 3^2] \, dx$$
$$= \pi \int_0^1 [(3+e^x)^2 - 9] \, dx$$
$$= \pi \int_0^1 [e^{2x} + 6e^x] \, dx = \pi \left[\frac{e^{2x}}{2} + 6e^x \right] \Big|_0^1$$
$$= \pi[(e^2/2 + 6e) - (1/2 + 6)] \approx 42.42.$$

11.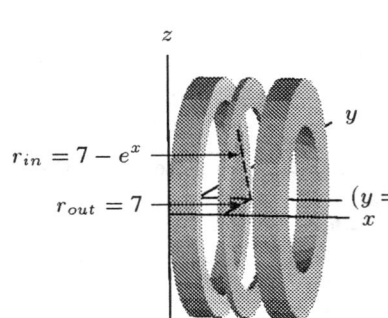

This problem can be done by slicing the volume into washers with planes perpendicular to the axis of rotation, $y = 7$, just like in Example 5. This time the outside radius of a washer is 7, and the inside radius is $7 - e^x$. Therefore, the volume V is

$$V = \int_{x=0}^{x=1} [\pi 7^2 - \pi(7-e^x)^2] \, dx = \pi \int_0^1 (14e^x - e^{2x}) \, dx$$
$$= \pi \left[14e^x - \frac{1}{2} e^{2x} \right] \Big|_0^1 = \pi \left[14e - \frac{1}{2} e^2 - \left(14 - \frac{1}{2} \right) \right]$$
$$\approx 65.54.$$

12.

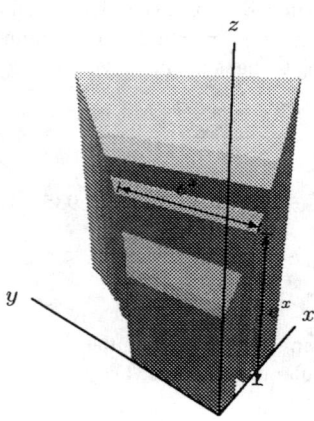

We now slice perpendicular to the x-axis. As stated in the problem, the cross-sections obtained thereby will be squares, with base length e^x. The volume of one square slice is $(e^x)^2\,dx$. (Look at the picture.) Adding up the volumes of the slices yields

$$\text{Volume} = \int_{x=0}^{x=1} y^2\,dx = \int_0^1 e^{2x}\,dx$$
$$= \left.\frac{e^{2x}}{2}\right|_0^1 = \frac{e^2 - 1}{2} \approx 3.195.$$

13.

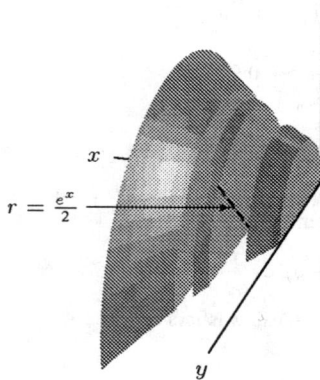

We slice perpendicular to the x-axis. As stated in the problem, the cross-sections obtained thereby will be semicircles, with radius $\frac{e^x}{2}$. The volume of one semicircular slice is $\frac{1}{2}\pi\left(\frac{e^x}{2}\right)^2 dx$. (Look at the picture.) Adding up the volumes of the slices yields

$$\text{Volume} = \int_{x=0}^{x=1} \pi\frac{y^2}{2}\,dx = \frac{\pi}{8}\int_0^1 e^{2x}\,dx$$
$$= \left.\frac{\pi e^{2x}}{16}\right|_0^1 = \frac{\pi(e^2 - 1)}{16} \approx 1.25.$$

14.

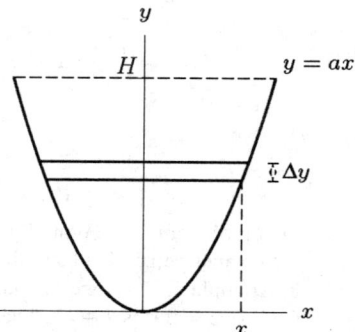

We divide the interior of the boat into flat slabs of thickness Δy and width $2x = 2\sqrt{y/a}$. (See above.) We have

$$\text{Volume of slab} \approx 2xL\Delta y = 2L\sqrt{\frac{y}{a}}\Delta y.$$

We are interested in the total volume of the region $0 \le y \le H$, so

$$\text{Total volume} = \lim_{\Delta y \to 0} \sum 2L\left(\frac{y}{a}\right)^{(1/2)}\Delta y = \int_0^H 2L\left(\frac{y}{a}\right)^{(1/2)} dy$$
$$= \frac{2L}{\sqrt{a}}\int_0^H y^{(1/2)}\,dy = \frac{4LH^{(3/2)}}{3\sqrt{a}}.$$

If L and H are in meters,

$$\text{Buoyancy force} = \frac{40{,}000LH^{(3/2)}}{3\sqrt{a}}\text{ newtons.}$$

15. (a) We can begin by slicing the pie into horizontal slabs of thickness Δh located at height h. To find the radius of each slice, we note that radius increases linearly with height. Since $r = 4.5$ when $h = 3$ and $r = 3.5$ when $h = 0$, we should have $r = 3.5 + h/3$. Then the volume of each slab will be $\pi r^2 \, \Delta h = \pi(3.5 + h/3)^2 \, \Delta h$. To find the total volume of the pie, we integrate this from $h = 0$ to $h = 3$:

$$V = \pi \int_0^3 \left(3.5 + \frac{h}{3}\right)^2 dh$$

$$= \pi \left[\frac{h^3}{27} + \frac{7h^2}{6} + \frac{49h}{4}\right]\Big|_0^3$$

$$= \pi \left[\frac{3^3}{27} + \frac{7(3^2)}{6} + \frac{49(3)}{4}\right] \approx 152 \text{ in}^3.$$

(b) We use 1.5 in as a rough estimate of the radius of an apple. This gives us a volume of $(4/3)\pi(1.5)^3 \approx 10 \text{ in}^3$. Since $152/10 \approx 15$, we would need about 15 apples to make a pie.

16. Although we could work this problem by using the formula for the volume of a right pyramid with a square base, we'll find the volume of the dump by using slices instead. The slices will be squares, and we'll start slicing at the base of the pyramid. The side of a square slice at the base is 100 yards; for every yard above the base, the side of the square slice decreases by 1 yard. Therefore, the side of a slice y yards above the base is $(100 - y)$, and the volume of the slice is $(100 - y)^2 \Delta y$.

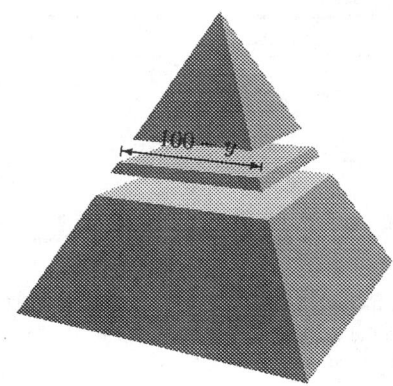

Thus, the volume of the dump is

$$V = \int_0^{20} (100 - y)^2 \, dy$$

$$= \int_0^{20} (100^2 - 200y + y^2) \, dy$$

$$= \left(10{,}000y - 100y^2 + \frac{y^3}{3}\right)\Big|_0^{20} \approx 162{,}666.67 \text{ cubic yards.}$$

If 65 cubic yards arrive at the dump every day, then 365(65) cubic yards arrive each year. This means it will take approximately

$$\frac{162{,}667}{365(65)} \approx 6.87 \text{ years}$$

for the dump to fill up.

17. We can find the volume of the tree by slicing it into a series of thin horizontal cylinders of height dh and circumference C. The volume of each cylindrical disk will then be

$$V = \pi r^2 \, dh = \pi \left(\frac{C}{2\pi}\right)^2 dh = \frac{C^2 \, dh}{4\pi}.$$

Summing all such cylinders, we have the total volume of the tree as

$$\text{Total volume} = \frac{1}{4\pi} \int_0^{120} C^2 \, dh.$$

We can estimate this volume using a trapezoidal approximation to the integral with $\Delta h = 20$:

$$\text{LEFT estimate} = \frac{1}{4\pi}[20(31^2 + 28^2 + 21^2 + 17^2 + 12^2 + 8^2)] = \frac{1}{4\pi}(53660).$$

$$\text{RIGHT estimate} = \frac{1}{4\pi}[20(28^2 + 21^2 + 17^2 + 12^2 + 8^2 + 2^2)] = \frac{1}{4\pi}(34520).$$

$$\text{TRAP} = \frac{1}{4\pi}(44090) \approx 3509 \text{ cubic inches.}$$

18. The problem appears complicated, because we are now working in three dimensions. However, if we take one dimension at a time, we will see that the solution is not too difficult. For example, let's just work at a constant depth, say 0. We apply the trapezoid rule to find the approximate area along the length of the boat. For example, by the trapezoid rule the approximate area at depth 0 from the front of the boat to 10 feet toward the back is $\frac{(2+8)\cdot 10}{2} = 50$. Overall, at depth 0 we have that the area for each length span is as follows:

TABLE 8.1

length span:	0–10	10–20	20–30	30–40	40–50	50–60
depth 0	50	105	145	165	165	130

We can fill in the whole chart the same way:

TABLE 8.2

	length span:	0–10	10–20	20–30	30–40	40–50	50–60
	0	50	105	145	165	165	130
	2	25	60	90	105	105	90
depth	4	15	35	50	65	65	50
	6	5	15	25	35	35	25
	8	0	5	10	10	10	10

Now, to find the volume, we just apply the trapezoid rule to the depths and areas. For example, according to the trapezoid rule the approximate volume as the depth goes from 0 to 2 and the length goes from 0 to 10 is $\frac{(50+25)\cdot 2}{2} = 75$. Again, we fill in a chart:

TABLE 8.3

	length span:	0–10	10–20	20–30	30–40	40–50	50–60
	0–2	75	165	235	270	270	220
depth	2–4	40	95	140	170	170	140
span	4–6	20	50	75	100	100	75
	6–8	5	20	35	45	45	35

Adding all this up, we find the volume is approximately 2595 cubic feet.

You might wonder what would have happened if we had done our trapezoids along the depth axis first instead of along the length axis. If you try this, you'd find that you come up with the same answers in the volume chart! For the trapezoid rule, it doesn't matter which axis you choose first.

19. This is a one-quarter of the circumference of a circle of radius 2. That circumference is $2 \cdot 2\pi = 4\pi$, so the length is $\frac{4\pi}{4} = \pi$.

20. Note that this function is actually $x^{3/2}$ in disguise. So

$$L = \int_0^2 \sqrt{1 + \left[\frac{3}{2}x^{\frac{1}{2}}\right]^2}\, dx = \int_{x=0}^{x=2} \sqrt{1 + \frac{9}{4}x}\, dx$$

$$= \frac{4}{9}\int_{w=1}^{w=\frac{11}{2}} w^{\frac{1}{2}}\, dw$$

$$= \frac{8}{27} w^{\frac{3}{2}}\Big|_1^{\frac{11}{2}} = \frac{8}{27}\left(\left(\frac{11}{2}\right)^{\frac{3}{2}} - 1\right) \approx 3.526,$$

where we set $w = 1 + \frac{9}{4}x$, so $dx = \frac{4}{9}dw$.

21. (a) The equation of a circle of radius r around the origin is $x^2 + y^2 = r^2$. This means that $y^2 = r^2 - x^2$, so $2y(dy/dx) = -2x$, and $dy/dx = -x/y$. Since the circle is symmetric about both axes, its arc length is 4 times the arc length in the first quadrant, namely

$$4\int_0^r \sqrt{1+\left(\frac{dy}{dx}\right)^2}\,dx = 4\int_0^r \sqrt{1+\left(-\frac{x}{y}\right)^2}\,dx.$$

(b) Evaluating this integral yields

$$4\int_0^r \sqrt{1+\left(-\frac{x}{y}\right)^2}\,dx = 4\int_0^r \sqrt{1+\frac{x^2}{r^2-x^2}}\,dx = 4\int_0^r \sqrt{\frac{r^2}{r^2-x^2}}\,dx$$

$$= 4r\int_0^r \sqrt{\frac{1}{r^2-x^2}}\,dx = 4r(\arcsin(x/r))\Big|_0^r = 2\pi r.$$

This is the expected answer.

22.

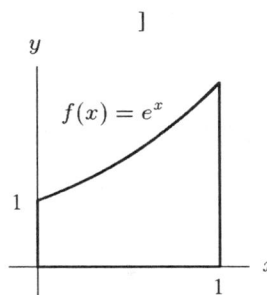

As can be seen above, the region has three straight sides and one curved one. The lengths of the straight sides are 1, 1, and e. The curved side is given by the equation $y = f(x) = e^x$. We can find its length by the formula

$$\int_0^1 \sqrt{1+f'(x)^2}\,dx = \int_0^1 \sqrt{1+(e^x)^2}\,dx$$

$$= \int_0^1 \sqrt{1+e^{2x}}\,dx.$$

But this is hard to integrate, so we approximate. Using Simpson's rule with $n = 10$ and $n = 20$ intervals, we find the integral ≈ 2.003. We could also approximate the integral by noting the the second derivative of $g(x) = \sqrt{1+e^{2x}}$ is $g''(x) = \dfrac{e^{4x}+2e^{2x}}{(1+e^{2x})^{3/2}}$. Thus our integrand is concave up, so the trapezoid rule gives an overestimate and the midpoint rule gives an underestimate. TRAP(20) = 2.0038... and MID(10) = 2.0033..., so 2.003 is correct within 0.001. The total length, therefore, is about $1+1+e+2.003 \approx 6.7212$. (Note that we use e to 4 decimal places to ensure that the total roundoff error is less than 0.001.)

23. Since $y = (e^x + e^{-x})/2$, $y' = (e^x - e^{-x})/2$. The length of the catenary is

$$\int_{-1}^1 \sqrt{1+(y')^2}\,dx = \int_{-1}^1 \sqrt{1+\left[\frac{e^x-e^{-x}}{2}\right]^2}\,dx = \int_{-1}^1 \sqrt{1+\frac{e^{2x}}{4}-\frac{1}{2}+\frac{e^{-2x}}{4}}\,dx$$

$$= \int_{-1}^1 \sqrt{\left[\frac{e^x+e^{-x}}{2}\right]^2}\,dx = \int_{-1}^1 \frac{e^x+e^{-x}}{2}\,dx$$

$$= \left[\frac{e^x-e^{-x}}{2}\right]\Big|_{-1}^1 = e-e^{-1} \approx 2.35.$$

24. Since the ellipse is symmetric about its axes, we can just find its arc length in the first quadrant and multiply that by 4. To determine the arc length of this section, we first solve for y in terms of x: since $x^2/4 + y^2 = 1$ is the equation for the ellipse, we have $y^2 = 1 - x^2/4$, so $y = \sqrt{1 - x^2/4}$. We also need to find dy/dx; we can do this by differentiating $y^2 = 1 - x^2/4$ implicitly, obtaining $2y\,dy/dx = -x/2$, whence $dy/dx = -x/(4y)$. We now set up the integral:

$$\begin{aligned}\text{Circumference of ellipse} \\ \text{in first quadrant}\end{aligned} = \int_0^2 \sqrt{1 + \left(-\frac{x}{4y}\right)^2}\,dx = \int_0^2 \sqrt{1 + \frac{x^2}{16y^2}}\,dx$$

$$= \int_0^2 \sqrt{1 + \frac{x^2}{16 - 4x^2}}\,dx = \int_0^2 \sqrt{\frac{16 - 3x^2}{16 - 4x^2}}\,dx.$$

This is an improper integral, since $16 - 4x^2 = 0$ for $x = 2$. Hence, integrating it numerically is somewhat tricky. However, we can integrate numerically from 0 to 1.999, and then use a vertical line to approximate the last section. The upper point of the line is $(1.999, 0.016)$; the lower point is $(2, 0)$. The length of the line connecting these two points is $\sqrt{(2 - 1.999)^2 + (0 - 0.016)^2} \approx 0.016$. Approximating the integral from 0 to 1.999 gives 2.391; hence the total arc length of the first quadrant is approximately $2.391 + 0.016 = 2.407$. So the arc length of the whole ellipse is about $4 \cdot 2.407 \approx 9.63$.

25. Here are many functions which "work."

- Any linear function $y = mx + b$ "works." This follows because $\frac{dy}{dx} = m$ is constant for such functions. So

$$\int_a^b \sqrt{1 + \left(\frac{dy}{dx}\right)^2}\,dx = \int_a^b \sqrt{1 + m^2}\,dx = (b - a)\sqrt{1 + m^2}.$$

- The function $y = \frac{x^4}{8} + \frac{1}{4x^2}$ "works": $\frac{dy}{dx} = \frac{1}{2}(x^3 - 1/x^3)$, and

$$\int \sqrt{1 + \left(\frac{dy}{dx}\right)^2}\,dx = \int \sqrt{1 + \frac{(x^3 - \frac{1}{x^3})^2}{4}}\,dx = \int \sqrt{1 + \frac{x^6}{4} - \frac{1}{2} + \frac{1}{4x^6}}\,dx$$

$$= \int \sqrt{\frac{1}{4}\left(x^3 + \frac{1}{x^3}\right)^2}\,dx = \int \frac{1}{2}\left(x^3 + \frac{1}{x^3}\right)dx$$

$$= \left[\frac{x^4}{8} - \frac{1}{4x^2}\right] + C.$$

- One more function that "works" is $y = \ln(\cos x)$; we have $\frac{dy}{dx} = -\sin x/\cos x$. Hence

$$\int \sqrt{1 + \left(\frac{dy}{dx}\right)^2}\,dx = \int \sqrt{1 + \left(\frac{-\sin x}{\cos x}\right)^2}\,dx = \int \sqrt{1 + \frac{\sin^2 x}{\cos^2 x}}\,dx$$

$$= \int \sqrt{\frac{\sin^2 x + \cos^2 x}{\cos^2 x}}\,dx = \int \sqrt{\frac{1}{\cos^2 x}}\,dx$$

$$= \int \frac{1}{\cos x}\,dx = \frac{1}{2}\ln\left|\frac{\sin x + 1}{\sin x - 1}\right| + C,$$

where the last integral comes from IV-22 of the integral tables.

26. (a) If $f(x) = \int_0^x \sqrt{g'(t)^2 - 1}\,dt$, then, by the Fundamental Theorem of Calculus, $f'(x) = \sqrt{g'(x)^2 - 1}$. So the arc length of f from 0 to x is

$$\int_0^x \sqrt{1 + (f'(t))^2}\,dt = \int_0^x \sqrt{1 + (\sqrt{g'(t)^2 - 1})^2}\,dt$$

$$= \int_0^x \sqrt{1 + g'(t)^2 - 1}\,dt$$

$$= \int_0^x g'(t)\,dt = g(x) - g(0) = g(x).$$

(b) If g is the arc length of any function f, then by the Fundamental Theorem of Calculus, $g'(x) = \sqrt{1 + f'(x)^2} \geq 1$. So if $g'(x) < 1$, g cannot be the arc length of a function.

(c) We find a function f whose arc length from 0 to x is $g(x) = 2x$. Using part (a), we see that

$$f(x) = \int_0^x \sqrt{(g'(t))^2 - 1}\, dt = \int_0^x \sqrt{2^2 - 1}\, dt = \sqrt{3}x.$$

This is the equation of a line. Does it make sense to you that the arc length of a line segment depends linearly on its right endpoint?

Solutions for Section 8.2

1. (a) Suppose we choose an x, $0 \leq x \leq 2$. If Δx is a small fraction of a meter, then the density of the rod is approximately $\rho(x)$ anywhere from x to $x + \Delta x$ meters from the left end of the rod (see below). The mass of the rod from x to $x + \Delta x$ meters is therefore approximately $\rho(x)\Delta x = (2 + 6x)\Delta x$. If we slice the rod into N pieces, then a Riemann sum is $\sum_{i=1}^{N}(2 + 6x_i)\Delta x$.

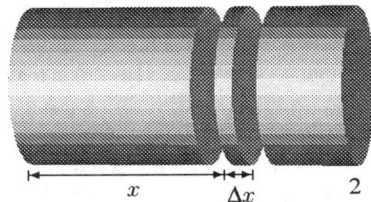

(b) The definite integral is

$$M = \int_0^2 \rho(x)\, dx = \int_0^2 (2 + 6x)\, dx = (2x + 3x^2)\Big|_0^2 = 16 \text{ grams}.$$

2. (a)

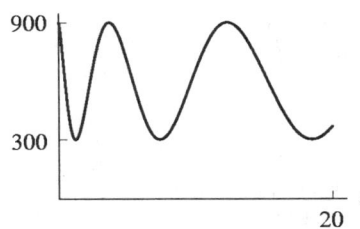

(b) Suppose we choose an x, $0 \leq x \leq 20$. We approximate the density of the number of the cars between x and $x + \Delta x$ miles as $p(x)$ cars per mile. Therefore, the number of cars between x and $x + \Delta x$ is approximately $p(x)\Delta x$. If we slice the 20 mile strip into N slices, we get that the total number of cars is

$$C \approx \sum_{i=1}^{N} p(x_i)\Delta x = \sum_{i=1}^{N}\left[600 + 300\sin(4\sqrt{x_i + 0.15})\right]\Delta x,$$

where $\Delta x = 20/N$. (This is a right-hand approximation; the corresponding left-hand approximation is $\sum_{i=0}^{N-1} p(x_i)\Delta x$.)

(c) As $N \to \infty$, the Riemann sum above approaches the integral

$$C = \int_0^{20} (600 + 300\sin 4\sqrt{x + 0.15})\, dx.$$

If we approximate the integral using one of our approximation methods (like Simpson's rule), we find $C \approx 11513$. We can also find the integral exactly as follows:

$$C = \int_0^{20} (600 + 300\sin 4\sqrt{x + 0.15})\, dx$$

$$= \int_0^{20} 600\, dx + \int_0^{20} 300\sin 4\sqrt{x + 0.15}\, dx$$

$$= 12000 + 300 \int_0^{20} \sin 4\sqrt{x + 0.15}\, dx.$$

Let $w = \sqrt{x + 0.15}$, so $x = w^2 - 0.15$ and $dx = 2w\, dw$. Then

$$\int_{x=0}^{x=20} \sin 4\sqrt{x + 0.15}\, dx = 2\int_{w=\sqrt{0.15}}^{w=\sqrt{20.15}} w \sin 4w\, dw, \text{ (using integral table III-15)}$$

$$= 2\left[-\frac{1}{4}w\cos 4w + \frac{1}{16}\sin 4w\right]\Bigg|_{\sqrt{0.15}}^{\sqrt{20.15}}$$

$$\approx -1.624.$$

Using this, we have $C \approx 12000 + 300(-1.624) \approx 11513$, which matches our numerical approximation.

3. (a) Orient the rectangle in the coordinate plane in such a way that the side referred to in the problem—call it S—lies on the y-axis from $y = 0$ to $y = 5$, as shown in the figure to the right. We may subdivide the rectangle into strips of width Δx and length 5. If the left side of a given strip is a distance x away from S (i.e., the y-axis), its density 2 is $1/(1 + x^4)$. If Δx is small enough, the density of the strip is approximately constant—i.e., the density of the whole strip is about $1/(1 + x^4)$. The mass of the strip is just its density times its area, or $5\Delta x/(1 + x^4)$. Thus the mass of the whole rectangle is approximated by the left Riemann sum

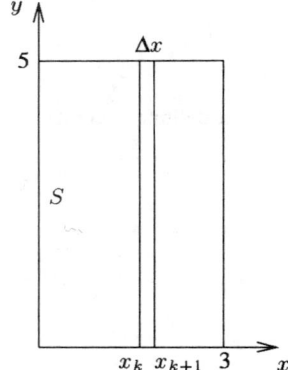

$$\sum_{k=0}^{n-1} \frac{5\Delta x}{1 + x_k^4},$$

where $0 = x_0 < x_1 < x_2 < \cdots < x_{n-1} < x_n = 3$ and $\Delta x = x_k - x_{k-1}$. (Since the density function $1/(1 + x^4)$ is strictly decreasing on the entire interval, the left Riemann sum overestimates the mass.) If we had used the right hand side of each strip in approximating its density, we would have obtained the right Riemann sum

$$\sum_{k=1}^{n} \frac{5\Delta x}{1 + x_k^4},$$

which is an underestimate of the mass.

(b) The exact mass of the rectangle is obtained by letting $\Delta x \to 0$ in the Riemann sums above, giving us the integral

$$\int_0^3 \frac{5\, dx}{1 + x^4}.$$

Since it is not easy to find an antiderivative of $5/(1 + x^4)$, we evaluate this integral numerically. Because the integrand is decreasing, we know that the value of the integral is between the left and right hand sums. For $n = 256$, the right sum is 5.46 and the left sum is 5.52; since both of these quantities are 5.5 to one decimal place, then the exact mass must be 5.5 to one decimal place as well.

4. (a) Partition $[0, 10,000]$ into N subintervals of width Δr. The area in the ith subinterval is $\approx 2\pi r_i \Delta r$. So the total mass in the slick $= M \approx \sum_{i=1}^{N} 2\pi r_i \left(\frac{50}{1+r_i}\right) \Delta r$.

(b) $M = \int_0^{10,000} 100\pi \frac{r}{1+r} dr$. We may rewrite $\frac{r}{1+r}$ as $\frac{1+r}{1+r} - \frac{1}{1+r} = 1 - \frac{1}{1+r}$, so that

$$M = \int_0^{10,000} 100\pi \left(1 - \frac{1}{1+r}\right) dr = 100\pi \left(r - \ln|1+r| \Big|_0^{10,000} \right)$$

$$= 100\pi(10,000 - \ln(10,001)) \approx 3.14 \times 10^6 \text{ kg}.$$

(c) We wish to find an R such that

$$\int_0^R 100\pi \frac{r}{1+r} dr = \frac{1}{2} \int_0^{10,000} 100\pi \frac{r}{1+r} dr \approx 1.57 \times 10^6.$$

So $100\pi(R - \ln|R+1|) \approx 1.57 \times 10^6$; $R - \ln|R+1| \approx 5000$. By trial and error, we find $R \approx 5009$ meters.

5. (a) We form a Riemann sum by slicing the region into concentric rings of radius r and width Δr. Then the volume deposited on one ring will be the height $H(r)$ multiplied by the area of the ring. A ring of width Δr will have an area given by

$$\text{Area} = \pi(r + \Delta r)^2 - \pi(r^2)$$
$$= \pi(r^2 + 2r\Delta r + (\Delta r)^2 - r^2)$$
$$= \pi(2r\Delta r + (\Delta r)^2).$$

Since Δr is approaching zero, we can approximate

$$\text{Area of ring} \approx \pi(2r\Delta r + 0) = 2\pi r \Delta r.$$

From this, we have

$$\Delta V \approx H(r) \cdot 2\pi r \Delta r.$$

Thus, summing the contributions from all rings we have

$$V \approx \sum H(r) \cdot 2\pi r \Delta r.$$

Taking the limit as $\Delta r \to 0$, we get

$$V = \int_0^5 2\pi r \left(0.115 e^{-2r}\right) dr.$$

(b) We use integration by parts:

$$V = 0.23\pi \int_0^5 \left(r e^{-2r}\right) dr$$

$$= 0.23\pi \left(\frac{r e^{-2r}}{-2} - \frac{e^{-2r}}{4} \right) \Big|_0^5$$

$$\approx 0.181 (\text{millimeters}) \cdot (\text{kilometers})^2 = 0.181 \cdot 10^{-3} \cdot 10^6 \text{ meters}^3 = 181 \text{ cubic meters}.$$

6. We have

$$\text{Moment} = \int_0^2 x\rho(x)\, dx = \int_0^2 x(2 + 6x)\, dx$$

$$= \int_0^2 (6x^2 + 2x)\, dx = (2x^3 + x^2) \Big|_0^2 = 20 \text{ gram-meters}.$$

Now, using this and Problem 1 (b), we have

$$\text{Center of Mass} = \frac{\text{moment}}{\text{total mass}} = \frac{20 \text{ gram-meters}}{16 \text{ grams}} = \frac{5}{4} \text{ meters (from its left end)}.$$

370 CHAPTER EIGHT /SOLUTIONS

7. We have
$$\text{Total mass of the rod} = \int_0^3 (1+x^2)\,dx = \left[x + \frac{x^3}{3}\right]\Big|_0^3 = 12 \text{ grams}.$$

In addition,
$$\text{Moment} = \int_0^3 x(1+x^2)\,dx = \left[\frac{x^2}{2} + \frac{x^4}{4}\right]\Big|_0^3 = \frac{99}{4} \text{ gram-meters}.$$

Thus, the center of mass is at the position $\bar{x} = \frac{99/4}{12} = 2.06$ meters.

8. (a) We find that
$$\text{Moment} = \int_0^1 x(1+kx^2)\,dx = \left(\frac{x^2}{2} + \frac{kx^4}{4}\right)\Big|_0^1 = \frac{1}{2} + \frac{k}{4} \text{ gram-meters},$$

and that
$$\text{Total mass} = \int_0^1 (1+kx^2)\,dx = \left(x + \frac{kx^3}{3}\right)\Big|_0^1 = 1 + \frac{k}{3} \text{ grams}.$$

Thus, the center of mass is
$$\bar{x} = \frac{\frac{1}{2} + \frac{k}{4}}{1 + \frac{k}{3}} = \frac{3}{4}\left(\frac{2+k}{3+k}\right) \text{ meters}.$$

(b) Let $f(k) = \frac{3}{4}\left(\frac{2+k}{3+k}\right)$. Then $f'(k) = \frac{3}{4}\left(\frac{1}{(3+k)^2}\right)$, which is always positive, so f is an increasing function of k. Since $f(0) = 0.5$, this is the smallest value of f. As $k \to \infty$, $f(k) \to 3/4 = 0.75$. So $f(k)$ is always between 0.5 and 0.75.

9. (a) The density is minimum at $x = -1$ and increases as x increases, so more of the mass of the rod is in the right half of the rod. We thus expect the balancing point to be to the right of the origin.

(b) We need to compute
$$\int_{-1}^1 x(3 - e^{-x})\,dx = \left(\frac{3}{2}x^2 + xe^{-x} + e^{-x}\right)\Big|_{-1}^1 \quad \text{(using integration by parts)}$$
$$= \frac{3}{2} + e^{-1} + e^{-1} - \left(\frac{3}{2} - e^1 + e^1\right) = \frac{2}{e}.$$

We must divide this result by the total mass, which is given by
$$\int_{-1}^1 (3 - e^{-x})\,dx = (3x + e^{-x})\Big|_{-1}^1 = 6 - e + \frac{1}{e}.$$

We therefore have
$$\bar{x} = \frac{2/e}{6 - e + (1/e)} = \frac{2}{1 + 6e - e^2} \approx 0.2.$$

10. The area of the disk is $\pi/2$ m^2, so it has density $1/(\pi/2) = 2/\pi$ kg/m^2. We find the mass of a small strip of width Δx located at x_i (see Figure 8.2). The height of the strip is $\sqrt{1 - x_i^2}$, so

$$\text{Area of the small strip} \approx 2 \cdot \sqrt{1 - x_i^2}\Delta x \text{ m}^2.$$

When multiplied by the density $2/\pi$, we get

$$\text{Mass of the strip} \approx \frac{4}{\pi} \cdot \sqrt{1 - x_i^2}\Delta x \text{ kg}.$$

We then sum the product of these masses with x_i, and take the limit as $\Delta x \to 0$ to get

$$\bar{x} = \int_0^1 \frac{4}{\pi} x\sqrt{1 - x^2}\,dx = -\frac{4}{3\pi}(1-x^2)^{3/2}\Big|_0^1 = \frac{4}{3\pi} \text{ meter}.$$

Finally, we divide by the total mass 1 kg to get the result $\bar{x} = 4/(3\pi)$ meters.

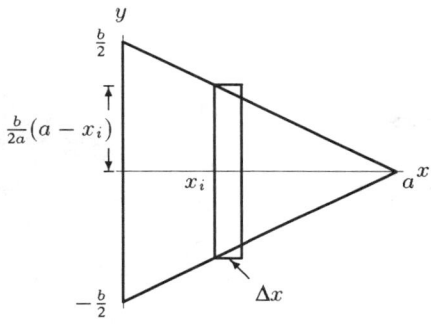

Figure 8.2: Area of a small strip

11. The area of the triangle is $ab/2$, so it has density $2m/(ab)$ where m is the total mass of the triangle. We need to find the mass of a small strip of width Δx located at x_i (see below).

$$\text{Area of the small strip} \approx 2 \cdot \frac{b(a-x_i)}{2a}\Delta x.$$

Multiplying by the density $2m/(ab)$ gives

$$\text{Mass of the strip} \approx 2m\frac{(a-x_i)}{a^2}\Delta x.$$

We then sum the product of these masses with x_i, and take the limit as $\Delta x \to 0$ to get

$$\int_0^a \frac{2mx(a-x)}{a^2}\,dx = \frac{2m}{a^2}\left(\frac{ax^2}{2} - \frac{x^3}{3}\right)\bigg|_0^a = \frac{2m}{a^2}\left(\frac{a^3}{2} - \frac{a^3}{3}\right) = \frac{ma}{3}.$$

Finally, we divide by the total mass m to get the desired result $\bar{x} = a/3$, which is independent of the length of the base b.

12. Partition $a \leq x \leq b$ into N subintervals of width $\Delta x = \frac{(b-a)}{N}$; $a = x_0 < x_1 < \cdots < x_N = b$. The mass of the strip on the ith subinterval is approximately $m_i = \rho(x_i)[f(x_i) - g(x_i)]\Delta x$. If we use a right-hand Riemann sum, the approximation for the total mass is
$\sum_{i=1}^{N} \rho(x_i)[f(x_i) - g(x_i)]\Delta x$, and the exact mass is $M = \int_a^b \rho(x)[f(x) - g(x)]dx$.

372 CHAPTER EIGHT /SOLUTIONS

13. (a) Use the formula for the volume of a cylinder:
$$\text{Volume} = \pi r^2 l.$$

Since it is only a half cylinder
$$\text{Volume of shed} = \frac{1}{2}\pi r^2 l.$$

(b) Set up the axes as shown in Figure 8.3. The density can be defined as
$$\text{Density} = ky.$$

Now slice the sawdust horizontally into slabs of thickness Δy as shown in Figure 8.4, and calculate
$$\text{Volume of slab} \approx 2xl\Delta y = 2l(\sqrt{r^2 - y^2})\Delta y.$$
$$\text{Mass of slab} = \text{Density} \cdot \text{Volume} \approx 2kly\sqrt{r^2 - y^2}\Delta y.$$

Finally, we compute the total mass of sawdust:
$$\text{Total mass of sawdust} = \int_0^r 2kly\sqrt{r^2 - y^2}\, dy = -\frac{2}{3}kl(r^2 - y^2)^{3/2}\Big|_0^r = \frac{2klr^3}{3}.$$

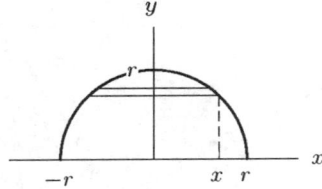
Figure 8.3

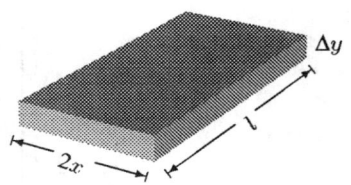
Figure 8.4

14. First we rewrite the chart, listing the density with the corresponding distance from the center of the earth (x km below the surface is equivalent to $6370 - x$ km from the center):

This gives us spherical shells whose volumes are $\frac{4}{3}\pi(r_i^3 - r_{i+1}^3)$ for any two consecutive distances from the origin. We will assume that the density of the earth is increasing with depth. Therefore, the average density of the ith shell is between D_i and D_{i+1}, the densities at top and bottom of shell i. So $\frac{4}{3}\pi D_{i+1}(r_i^3 - r_{i+1}^3)$ and $\frac{4}{3}\pi D_i(r_i^3 - r_{i+1}^3)$ are upper and lower bounds for the mass of the shell.

TABLE 8.4

i	x_i	$r_i = 6370 - x_i$	D_i
0	0	6370	3.3
1	1000	5370	4.5
2	2000	4370	5.1
3	2900	3470	5.6
4	3000	3370	10.1
5	4000	2370	11.4
6	5000	1370	12.6
7	6000	370	13.0
8	6370	0	13.0

To get a rough approximation of the mass of the earth, we don't need to use all the data. Let's just use the densities at $x = 0, 2900, 5000$ and 6370 km. Calculating an upper bound on the mass,

$$M_U = \frac{4}{3}\pi[13.0(1370^3 - 0^3) + 12.6(3470^3 - 1370^3) + 5.6(6370^3 - 3470^3)] \cdot 10^{15} \approx 7.29 \times 10^{27} \text{ g}.$$

The factor of 10^{15} may appear unusual. Remember the radius is given in kilometers and the density is given in g/cm^3, so we must convert kilometers to centimeters: 1 km $= 10^5$ cm , so 1 km$^3 = 10^{15}$ cm^3.

The lower bound is

$$M_L = \frac{4}{3}\pi[12.6(1370^3 - 0^3) + 5.6(3470^3 - 1370^3) + 3.3(6370^3 - 3470^3)] \cdot 10^{15} \approx 4.05 \times 10^{27} \text{ g}.$$

Here, our upper bound is just under 2 times our lower bound.

Using all our data, we can find a more accurate estimate. The upper and lower bounds are

$$M_U = \frac{4}{3}\pi \sum_{i=0}^{7} D_{i+1}(r_i^3 - r_{i+1}^3) \cdot 10^{15} \text{ g}$$

and

$$M_L = \frac{4}{3}\pi \sum_{i=0}^{7} D_i(r_i^3 - r_{i+1}^3) \cdot 10^{15} \text{ g}.$$

We have

$$M_U = \frac{4}{3}\pi[4.5(6370^3 - 5370^3) + 5.1(5370^3 - 4370^3) + 5.6(4370^3 - 3470^3)$$
$$+ 10.1(3470^3 - 3370^3) + 11.4(3370^3 - 2370^3) + 12.6(2370^3 - 1370^3)$$
$$+ 13.0(1370^3 - 370^3) + 13.0(370^3 - 0^3)] \cdot 10^{15} \text{ g}$$
$$\approx 6.50 \times 10^{27} \text{ g}$$

and

$$M_L = \frac{4}{3}\pi[3.3(6370^3 - 5370^3) + 4.5(5370^3 - 4370^3) + 5.1(4370^3 - 3470^3)$$
$$+ 5.6(3470^3 - 3370^3) + 10.1(3370^3 - 2370^3) + 11.4(2370^3 - 1370^3)$$
$$+ 12.6(1370^3 - 370^3) + 13.0(370^3 - 0^3)] \cdot 10^{15} \text{ g}$$
$$\approx 5.46 \times 10^{27} \text{ g}.$$

15. (a) Partition $0 \le h \le 100$ into N subintervals of width $\Delta h = \frac{100}{N}$ (see right). The density is taken to be approximately $\rho(h_i)$ on the ith spherical shell, and the volume is approximately the surface area of a sphere of radius $r_e + h_i$ meters times Δh, where $r_e = 6.37 \times 10^6$ meters is the radius of the earth. If the volume of the ith shell is V_i, then $V_i \approx 4\pi(r_e + h_i)^2 \Delta h$, and a left-hand Riemann sum for the total mass is

$$M \approx \sum_{i=0}^{N-1} 4\pi(r_e + h_i)^2 \times 1.28 e^{-0.000124 h_i} \Delta h.$$

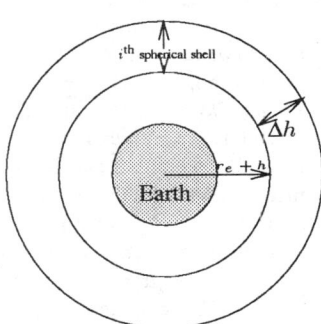

(b) This Riemann sum becomes the integral

$$M = 4\pi \int_0^{100} (r_e + h)^2 \times 1.28 e^{-0.000124 h} \, dh$$
$$= 4\pi \int_0^{100} (6.37 \times 10^6 + h)^2 \times 1.28 e^{-0.000124 h} \, dh.$$

Evaluating the integral using numerical methods gives $M = 6.48 \times 10^{16}$ kg.

16. We want to take a cross-section of the pipe and cut it up in such a way that the speed of the water is nearly uniform on each slice.
We will use thin rings around the pipe's center; if a given ring is narrow enough, all points on it will be roughly equidistant from the center. Since the water speed is a function of the distance from the center, it will be nearly constant on the entire ring.

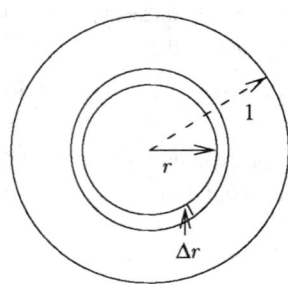

Let r be the distance from the center to the inner boundary of the ring, and let Δr be the width of the ring, as in the figure above. By straightening the ring into a thin rectangle, we find that its area is approximately given by the quantity $2\pi r \Delta r$. The speed across any part of the ring is roughly equal to the speed across the inner boundary, $10(1 - r^2)$ inches per second. The flow is defined as the speed times the area; thus on any given ring we have

$$\text{Flow} \approx 10(1 - r^2) \cdot 2\pi r \Delta r.$$

The total flow across the pipe cross-section is approximated by a Riemann sum incorporating all of the rings:

$$\text{Total Flow} \approx 20\pi \sum (1 - r^2) r \Delta r,$$

where r is in between 0 and 1. Letting $\Delta r \to 0$, we obtain the exact solution:

$$\text{Total Flow} = 20\pi \int_0^1 (1 - r^2) r \, dr = 20\pi \left(\frac{r^2}{2} - \frac{r^4}{4} \right) \Big|_0^1 = 5\pi \text{ cubic inches/second}.$$

Solutions for Section 8.3

1. (a) Looking at the graph, it appears that the graph of B is above $F = 10$ between $t = 2.3$ and $t = 4.2$, or for about 1.9 seconds.
   (b) Although the total impulse is defined as the integral from 0 to ∞, the thrust is 0 after a certain time, so the integral is actually not improper. From $t = 0$ to $t = 2$, the graph of A looks like a triangle with base 2 and height 12, for an area of 12. From $t = 2$ to $t = 4$, the graph of A looks a trapezoid with base 2 and heights 13 and 6, for an area of 19. From $t = 4$ to $t = 16$, A is approximately a rectangle with height 5.8 and width 12, for an area of 69.6. Finally, from $t = 16$ to $t = 17$, A looks like a triangle with base 1 and height 5.8, for an area of 2.9. So, the total area under the curve of A's thrust, which is A's total impulse, is about 103.5 newton-seconds.
   (c) Note that when we calculated the impulse in part (b), we multiplied height, measured in newtons, by width, measured in seconds. So the units of impulse are newton-seconds.
   (d) The graph of B's thrust looks like a triangle with base 6 and height 22, for a total impulse of about 66 newton-seconds. So rocket A, with total impulse 103.5 newton-seconds, has a larger total impulse than rocket B.
   (e) As we can see from the graph, rocket B reaches a maximum thrust of 22, whereas A only reaches a maximum thrust of 13. So rocket B has the largest maximum thrust.

2. Let x be the distance from ground to the bucket of cement. At height x, if the bucket is lifted by Δx, the work done is $500\Delta x + 5(75 - x)\Delta x$. The $500\Delta x$ term is due to the bucket of cement; the $5(75 - x)\Delta x$ term is due to the remaining cable. So the total work required to lift the bucket is

$$W = \int_0^{30} 500 \, dx + \int_0^{30} 5(75 - x) \, dx$$
$$= (500)(30) + 5(75(30) - \frac{1}{2}30^2)$$
$$= 24000 \text{ ft-lb}.$$

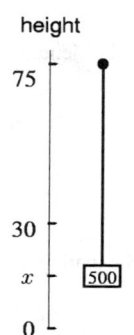

3. To lift the weight an additional height Δh off the ground from a height of h, we must do work on the weight and the amount of rope not yet pulled onto the roof. Since the roof is 30 ft off the ground, there will be $30 - h$ feet remaining of rope, for a weight of $4(30 - h)$. So the work required to raise the weight and the rope a height Δh will be $\Delta h(1000 + 4(30 - h))$. To find the total work, we integrate this quantity from $h = 0$ to $h = 10$:

$$\text{Work} = \int_0^{10} (1000 + 4(30 - h))\, dh$$
$$= \int_0^{10} (1120 - 4h)\, dh$$
$$= (1120h - 2h^2)\Big|_0^{10}$$
$$= 11{,}200 - 200$$
$$= 11{,}000 \text{ ft-lbs.}$$

4. Consider lifting a rectangular slab of water h feet from the top up to the top. The area of such a slab is $(10)(20) = 200$ square feet; if the thickness is dh, then the volume of such a slab is $200\, dh$ cubic feet. This much water weighs 62.4 pounds per ft^3, so the weight of such a slab is $(200\, dh)(62.4) = 12480\, dh$ pounds. To lift that much water h feet requires $12480 h\, dh$ foot-pounds of work. To lift the whole tank, we lift one plate at a time; integrating over the slabs yields

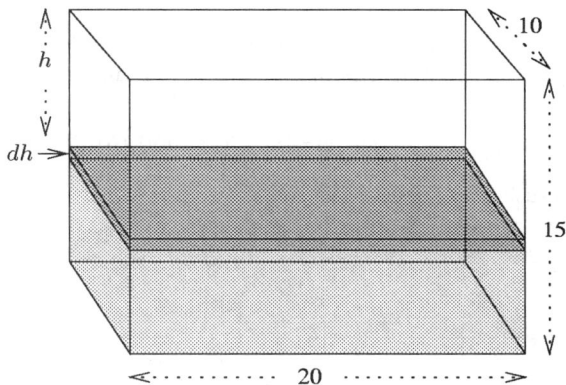

$$\int_0^{15} 12480 h\, dh = \frac{12480 h^2}{2}\Big|_0^{15} = \frac{12480 \cdot 15^2}{2} = 1{,}404{,}000 \text{ foot-pounds.}$$

5. Let x be the distance measured from the bottom the tank. To pump a layer of water of thickness Δx at x feet from the bottom, the work needed is

$$(62.4)\pi 6^2 (20 - x)\Delta x.$$

Therefore, the total work is

$$W = \int_0^{10} 36 \cdot (62.4)\pi (20 - x)\, dx$$
$$= 36 \cdot (62.4)\pi \left(20x - \frac{1}{2}x^2\right)\Big|_0^{10}$$
$$= 36 \cdot (62.4)\pi (200 - 50)$$
$$\approx 1{,}058{,}591.1 \text{ ft-lb.}$$

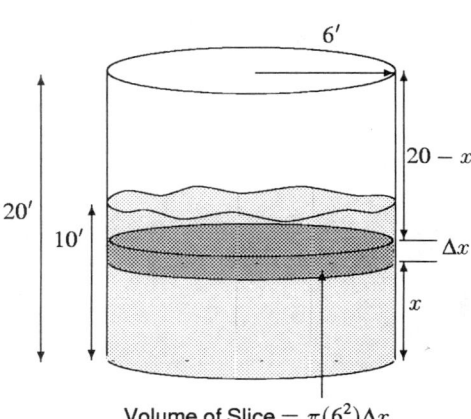

Volume of Slice $= \pi(6^2)\Delta x$

376 CHAPTER EIGHT /SOLUTIONS

6. Let x be the distance from the bottom of the tank. To pump a layer of water of thickness Δx at x feet from the bottom to 10 feet above the tank, the work done is $(62.4)\pi 6^2(30 - x)\Delta x$. Thus the total work is

$$\int_0^{20} 36 \cdot (62.4)\pi(30 - x)\, dx$$

$$= 36 \cdot (62.4)\pi \left(30x - \frac{1}{2}x^2\right)\bigg|_0^{20}$$

$$= 36 \cdot (62.4)\pi(30(20) - \frac{1}{2}20^2)$$

$$\approx 2{,}822{,}909.50 \text{ ft-lb}.$$

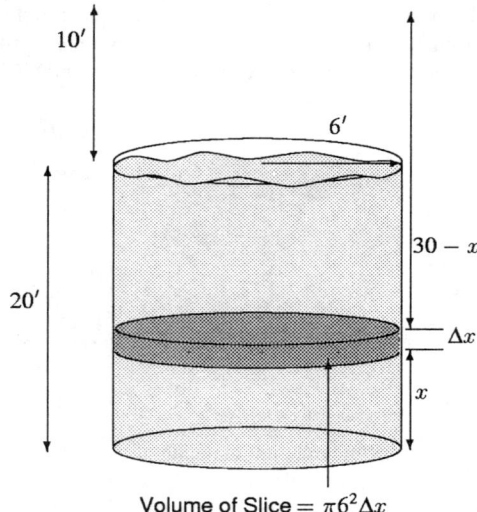

Volume of Slice $= \pi 6^2 \Delta x$

7. We begin by slicing the oil into slabs at a distance h below the surface with thickness Δh. We can then calculate the volume of the slab and the work needed to raise this slab to the surface, a distance of h.

$$\text{Volume of } \Delta h \text{ disk} = \pi r^2 \Delta h = 25\pi\Delta h$$
$$\text{Weight of } \Delta h \text{ disk} = (25\pi)(50)\Delta h$$
$$\text{Distance to raise} = h$$
$$\text{Work to raise} = (25\pi)(50)(h)\Delta h.$$

Integrating the work over all such slabs, we have

$$\text{Work} = \int_{19}^{25} (50)(25\pi)(h)\, dh$$

$$= 625\pi h^2 \bigg|_{19}^{25}$$

$$= 390{,}625\pi - 225{,}625\pi$$

$$\approx 518{,}363 \text{ ft-lbs}.$$

A diagram of this tank is shown in Figure 8.5.

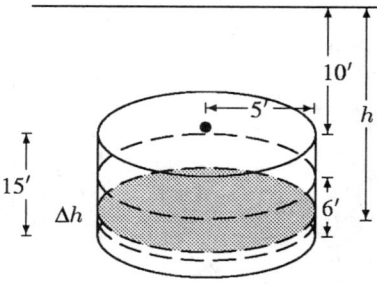

Figure 8.5

8.

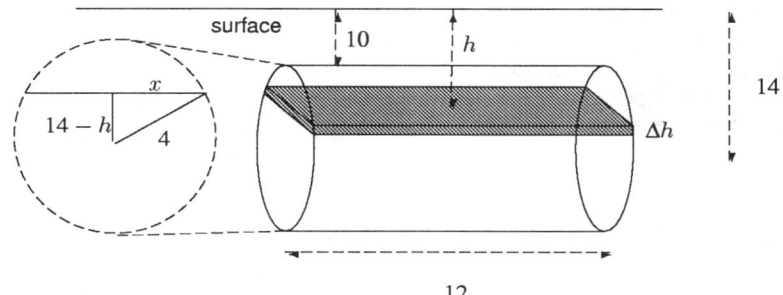

Let h represent distance below the surface in feet. We slice the tank up into horizontal slabs of thickness Δh. From looking at the figure, we can see that the slabs will be rectangular. The length of any slab is 12 feet. The width w of a slab h units below the ground will equal $2x$, where $(14-h)^2 + x^2 = 16$, so $w = 2\sqrt{4^2 - (14-h)^2}$. The volume of such a slab is therefore $12w\,\Delta h = 24\sqrt{16 - (14-h)^2}\,\Delta h$ cubic feet; the slab weighs $42 \cdot 24\sqrt{16-(14-h)^2}\,\Delta h = 1008\sqrt{16 - (14-h)^2}\,\Delta h$ pounds. So the total work done in pumping out all the gasoline is

$$\int_{10}^{18} 1008h\sqrt{16 - (14-h)^2}\,dh = 1008\int_{10}^{18} h\sqrt{16-(14-h)^2}\,dh.$$

Substitute $s = 14 - h$, $ds = -dh$. We get

$$1008\int_{10}^{18} h\sqrt{16-(14-h)^2}\,dh = -1008\int_{4}^{-4}(14-s)\sqrt{16-s^2}\,ds$$

$$= 1008 \cdot 14 \int_{-4}^{4}\sqrt{16-s^2}\,ds - 1008\int_{-4}^{4} s\sqrt{16-s^2}\,ds.$$

The first integral represents the area of a semicircle of radius 4, which is 8π. The second is the integral of an odd function, over the interval $-4 \leq s \leq 4$, and is therefore 0. Hence, the total work is $1008 \cdot 14 \cdot 8\pi \approx 354{,}673$ foot-pounds.

9. The force exerted on the satellite by the earth (and vice versa!) is GMm/r^2, where r is the distance from the center of the earth to the center of the satellite, m is the mass of the satellite, M is the mass of the earth, and G is the gravitational constant. So the total work done is

$$\int_{6.4\times 10^6}^{8.4\times 10^6} \frac{GMm}{r^2}\,dr = \left(\frac{-GMm}{r}\right)\Big|_{6.4\times 10^6}^{8.4\times 10^6} \approx 1.489 \times 10^{10} \text{ joules.}$$

10. Setting the initial kinetic energy and escape work equal to each other gives

$$\frac{1}{2}mv^2 = \frac{GMm}{R}, \text{ or } v^2 = \frac{2GM}{R}.$$

Since the planet is assumed to be a sphere of radius R and density ρ, we have $M = \rho(\frac{4}{3}\pi)R^3$. Hence

$$v^2 = \frac{2G\rho(\frac{4}{3}\pi)R^3}{R}$$

and therefore
$$v = k\sqrt{\rho}R$$

where $k = \sqrt{\frac{8\pi G}{3}}$. That is, the escape velocity is proportional to R and $\sqrt{\rho}$.

11. On page 355 of the text it is stated that:

$$v = \sqrt{\frac{2GM}{R}}$$

where M is the mass of the planet we're trying to leave—in this case, the Moon. Since the force on a mass m on the surface of the moon is $F = mg = GMm/R^2$, we have $GM/R = gR$. Therefore,

$$v = \sqrt{\frac{2GM}{R}} = \sqrt{2gR} = \sqrt{2(1.6)(1740 \cdot 10^3)} \approx 2360 \text{ m/sec.}$$

378 CHAPTER EIGHT /SOLUTIONS

12. (a) Divide the wall into N horizontal strips, each of which is of height Δh. The area of each strip is $1000\Delta h$, and the pressure at depth h_i is $62.4 h_i$, so we approximate the force on the strip as $1000(62.4 h_i)\Delta h$.

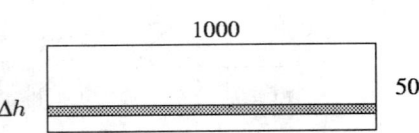

Therefore,
$$\text{Force on the Dam} \approx \sum_{i=0}^{N-1} 1000(62.4 h_i)\Delta h.$$

(b) As $N \to \infty$, the Riemann sum becomes the integral, so the force on the dam is
$$\int_0^{50} (1000)(62.4h)\,dh = 62400 \frac{h^2}{2}\bigg|_0^{50} = 78{,}000{,}000 \text{ pounds}.$$

13.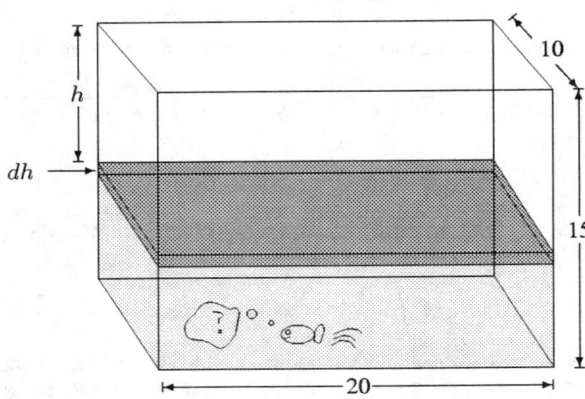

- Bottom: The bottom of the tank is at constant depth 15 feet, and therefore is under constant pressure, $15 \cdot 62.4 = 936$ lb/ft^2. The area of the base is 200 ft^2 and so the total force is $200 \text{ ft}^2 \cdot 936 \text{ lb/ft}^2 = 187200$ lb.
- 15×10 side: The area of a horizontal strip of width dh is $10\,dh$ square feet, and the pressure at height h is $62.4h$ pounds per square foot. Therefore, the force on such a strip is $62.4h(10\,dh)$ pounds. Hence, the total force on this side is
$$\int_0^{15} (62.4h)(10)\,dh = 624\frac{h^2}{2}\bigg|_0^{15} = 70200 \text{ lbs}.$$
- 15×20 side: Similarly, the total force on this side is
$$\int_0^{15} (62.4h)(20)\,dh = 1248\frac{h^2}{2}\bigg|_0^{15} = 140400 \text{ lbs}.$$

14. Bottom:
$$\text{Water force} = 62.4(2)(12) = 1497.6 \text{ lbs}.$$

Front and back:
$$\text{Water force} = (62.4)(4)\int_0^2 (2-x)\,dx = (62.4)(4)\left(2x - \frac{1}{2}x^2\right)\bigg|_0^2$$
$$= (62.4)(4)(2) = 499.2 \text{ lbs}.$$

Both sides:
$$\text{Water force} = (62.4)(3)\int_0^2 (2-x)\,dx = (62.4)(3)(2) = 374.4 \text{ lbs}.$$

15. The density of the rod is $10 \text{ kg}/6 \text{ m} = \frac{5}{3}\frac{\text{kg}}{\text{m}}$. A little piece, dx m, of the rod thus has mass $5/3\, dx$ kg. If this piece has an angular velocity of 2 rad/sec, then its actual velocity is $2|x|$ m/sec. This is because a radian angle sweeps out an arc length equal to the radius of the circle, and in this case the little piece moves in circles about the origin of radius $|x|$. The kinetic energy of the little piece is $mv^2/2 = (5/3\, dx)(2|x|)^2/2 = \frac{10}{3}x^2\, dx$.

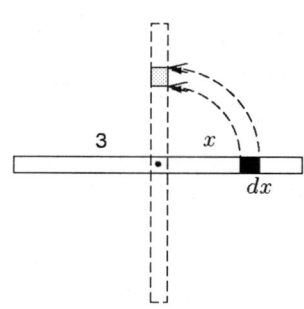

Therefore,

$$\text{Total Kinetic Energy} = \int_{-3}^{3} \frac{10x^2}{3}\, dx = \frac{20}{3}\left[\frac{x^3}{3}\right]\bigg|_{0}^{3} = 60 \text{ kg} \cdot \text{m}^2/\text{sec}^2 = 60 \text{ joules}.$$

16. We slice the record into rings in such a way that every point has approximately the same speed: use concentric circles around the hole. We assume the record is a flat disk of uniform density: since its mass is 50 grams and its area is $\pi(10\text{cm})^2 = 100\pi$ cm^2, the record has density $\frac{50}{100\pi} = \frac{1}{2\pi}\frac{\text{gram}}{\text{cm}^2}$. So a ring of width dr, having area about $2\pi r\, dr$ cm^2, has mass approximately $(2\pi r\, dr)(1/2\pi) = r\, dr$ gm. At radius r, the velocity of the ring is

$$33\tfrac{1}{3}\frac{\text{rev}}{\text{min}} \cdot \frac{1 \text{ min}}{60 \text{ sec}} \cdot \frac{2\pi r \text{ cm}}{1 \text{ rev}} = \frac{10\pi r}{9}\frac{\text{cm}}{\text{sec}}.$$

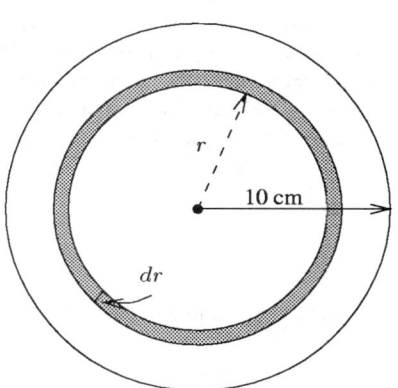

The kinetic energy of the ring is

$$\frac{1}{2}mv^2 = \frac{1}{2}(r\, dr \text{ grams})\left(\frac{10\pi r}{9}\frac{\text{cm}}{\text{sec}}\right)^2 = \frac{50\pi^2 r^3\, dr}{81}\frac{\text{gram} \cdot \text{cm}^2}{\text{sec}^2}.$$

So the kinetic energy of the record, summing the energies of all these rings, is

$$\int_{0}^{10} \frac{50\pi^2 r^3\, dr}{81} = \frac{25\pi^2 r^4}{162}\bigg|_{0}^{10} \approx 15231 \frac{\text{gram} \cdot \text{cm}^2}{\text{sec}^2} = 15231 \text{ ergs}.$$

17. We need to divide the disk up into circular rings of charge and integrate their contributions to the potential (at P) from 0 to a. These rings, however, are not uniformly distant from the point P. A ring of radius z is $\sqrt{R^2 + z^2}$ away from point P (see below).

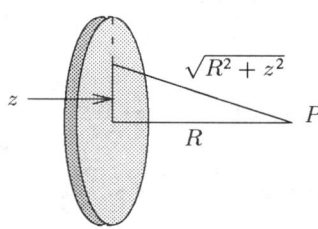

The ring has area $2\pi z\, \Delta z$, and charge $2\pi z \sigma\, \Delta z$. The potential of the ring is then $\dfrac{2\pi z\sigma\, \Delta z}{\sqrt{R^2+z^2}}$ and the total potential at point P is

$$\int_0^a \frac{2\pi z\sigma\, dz}{\sqrt{R^2+z^2}} = \pi\sigma \int_0^a \frac{2z\, dz}{\sqrt{R^2+z^2}}.$$

We make the substitution $u = z^2$. Then $du = 2z\, dz$. We obtain

$$\pi\sigma \int_0^a \frac{2z\, dz}{\sqrt{R^2+z^2}} = \pi\sigma \int_0^{a^2} \frac{du}{\sqrt{R^2+u}} = \pi\sigma(2\sqrt{R^2+u})\Big|_0^{a^2}$$

$$= \pi\sigma(2\sqrt{R^2+z^2})\Big|_0^a = 2\pi\sigma(\sqrt{R^2+a^2} - R).$$

(The substitution $u = R^2 + z^2$ or $\sqrt{R^2+z^2}$ works also.)

18. Cut the wine into horizontal pieces. Let y denote height above the top of the stem of the wine glass. In this case, each layer is a circular disk of thickness Δy and base area πx^2, where $x = y/2$ by the similar triangles in Figure 8.6. The volume of the disc is $\pi x^2\, \Delta y$, so the mass of the disc of wine is $1.2\pi x^2\, \Delta y$ grams. This means that the force due to gravity acting on this mass (i.e., its weight) is $980(1.2\pi x^2 \Delta y)$ dynes.

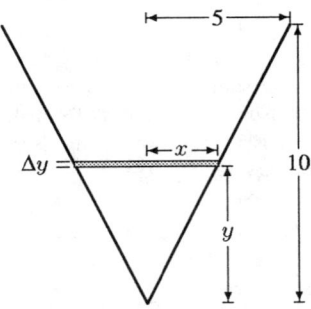

Figure 8.6: Note: Δy is the thickness of the disk

To get that layer of liquid to the top of the straw, we must raise it a height of $15 - y$. Thus the total work is

$$\int_0^8 980(1.2\pi x^2)(15-y)\, dy = 980(1.2\pi) \int_0^8 \frac{y^2}{4}(15-y)\, dy$$

$$= 294\pi \int_0^8 (15y^2 - y^3)\, dy = 294\pi \left(5y^3 - \frac{y^4}{4}\right)\Big|_0^8$$

$$\approx 1{,}418{,}693 \text{ ergs (about } 0.142 \text{ joules)}.$$

19. Divide the disk into rings of radius r, width Δr, as shown in Figure 8.7.

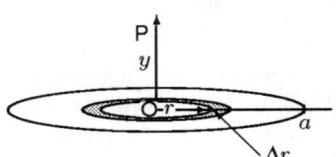

Figure 8.7

Then

$$\text{Area of ring} \approx 2\pi r \Delta r.$$

Since total area of disk is πa^2,

$$\text{Mass of ring} \approx \frac{2\pi r \Delta r}{\pi a^2} M = \frac{2rM}{a^2} \Delta r.$$

Thus, calculating the gravitational force due to the ring, we have

$$\text{Gravitational force on } m \text{ due to ring} = G\left(\frac{2rM}{a^2}\Delta r\right) \frac{my}{(r^2+y^2)^{3/2}} = \frac{2GMmyr}{a^2(r^2+y^2)^{3/2}} \Delta r.$$

Summing over all rings, we get

$$\text{Total gravitational force on } m \text{ due to disk} \approx \sum \frac{2GMmyr}{a^2(r^2+y^2)^{3/2}} \Delta r.$$

As $\Delta r \to 0$, we get

$$\text{Gravitational force on } m \text{ due to disk} = \int_0^a \frac{2GMmyr}{a^2(r^2+y^2)^{3/2}} dr = \frac{2GMmy}{a^2} \cdot \frac{-1}{(r^2+y^2)^{1/2}}\bigg|_0^a$$

$$= \frac{2GMmy}{a^2}\left(\frac{1}{y} - \frac{1}{(a^2+y^2)^{1/2}}\right).$$

20.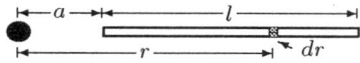

The density of the rod, in mass per unit length, is M/l (see above). So a slice of size dr has mass $\frac{M\,dr}{l}$. It pulls the small mass m with force $Gm\frac{M\,dr}{l}/r^2 = \frac{GmM\,dr}{lr^2}$. So the total gravitational attraction between the rod and point is

$$\int_a^{a+l} \frac{GmM\,dr}{lr^2} = \frac{GmM}{l}\left(-\frac{1}{r}\right)\bigg|_a^{a+l}$$

$$= \frac{GmM}{l}\left(\frac{1}{a} - \frac{1}{a+l}\right)$$

$$= \frac{GmM}{l}\cdot\frac{l}{a(a+l)} = \frac{GmM}{a(a+l)}.$$

21.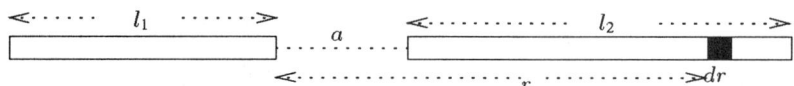

This time, let's split the second rod into small slices of length dr. Each slice is of mass $\frac{M_2}{l_2}dr$, since the density of the second rod is $\frac{M_2}{l_2}$. Since the slice is small, we can treat it as a particle at distance r away from the end of the first rod, as in Problem 20. By that problem, the force of attraction between the first rod and particle is

$$\frac{GM_1\frac{M_2}{l_2}dr}{(r)(r+l_1)}.$$

So the total force of attraction between the rods is

$$\int_a^{a+l_2} \frac{GM_1\frac{M_2}{l_2}dr}{(r)(r+l_1)} = \frac{GM_1M_2}{l_2}\int_a^{a+l_2}\frac{dr}{(r)(r+l_1)}$$

$$= \frac{GM_1M_2}{l_2}\int_a^{a+l_2}\frac{1}{l_1}\left(\frac{1}{r} - \frac{1}{r+l_1}\right)dr.$$

$$= \frac{GM_1M_2}{l_1l_2}\left(\ln|r| - \ln|r+l_1|\right)\bigg|_a^{a+l_2}$$

$$= \frac{GM_1M_2}{l_1l_2}\left[\ln|a+l_2| - \ln|a+l_1+l_2| - \ln|a| + \ln|a+l_1|\right]$$

$$= \frac{GM_1M_2}{l_1l_2}\ln\left[\frac{(a+l_1)(a+l_2)}{a(a+l_1+l_2)}\right].$$

This result is symmetric: if you switch l_1 and l_2 or M_1 and M_2, you get the same answer. That means it's not important which rod is "first," and which is "second."

Solutions for Section 8.4

1.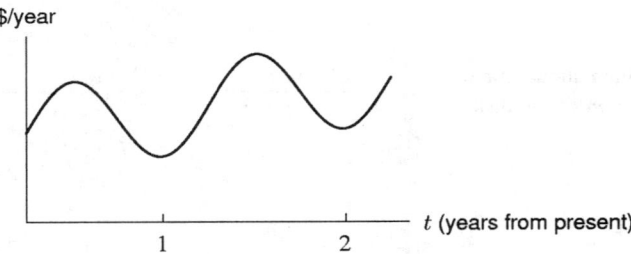

   The graph reaches a peak each summer, and a trough each winter. The graph shows sunscreen sales increasing from cycle to cycle. This gradual increase may be due in part to inflation and to population growth.

2. You should choose the payment which gives you the highest present value. The immediate lump-sum payment of $2800 obviously has a present value of exactly $2800, since you are getting it now. We can calculate the present value of the installment plan as:

$$PV = 1000e^{-0.06(0)} + 1000e^{-0.06(1)} + 1000e^{-0.06(2)}$$
$$\approx \$2828.68.$$

   Since the installment payments offer a (slightly) higher present value, you should accept this option.

3. (a) We calculate the future values of the two options:

$$FV_1 = 6e^{0.1(3)} + 2e^{0.1(2)} + 2e^{0.1(1)} + 2e^{0.1(0)}$$
$$\approx 8.099 + 2.443 + 2.210 + 2$$
$$= \$14.752 \text{ million.}$$

$$FV_2 = e^{0.1(3)} + 2e^{0.1(2)} + 4e^{0.1(1)} + 6e^{0.1(0)}$$
$$\approx 1.350 + 2.443 + 4.421 + 6$$
$$= \$14.214 \text{ million.}$$

   As we can see, the first option gives a higher future value, so he should choose Option 1.

   (b) From the future value we can easily derive the present value using the formula $PV = FVe^{-rt}$. So the present value is

$$\text{Option 1: } PV = 14.752e^{0.1(-3)} \approx \$10.929 \text{ million.}$$
$$\text{Option 2: } PV = 14.214e^{0.1(-3)} \approx \$10.530 \text{ million.}$$

4.
$$\text{Future Value} = \int_0^{15} 3000e^{0.06(15-t)}\, dt = 3000e^{0.9} \int_0^{15} e^{-0.06t}\, dt$$
$$= 3000e^{0.9} \left(\frac{1}{-0.06}e^{-0.06t}\right)\Big|_0^{15} = 3000e^{0.9}\left(\frac{1}{-0.06}e^{-0.9} + \frac{1}{0.06}e^0\right)$$
$$\approx \$72{,}980.16$$

$$\text{Present Value} = \int_0^{15} 3000e^{-0.06t}\, dt = 3000\left(-\frac{1}{0.06}\right)e^{-0.06t}\Big|_0^{15}$$
$$\approx \$29{,}671.52.$$

There's a quicker way to calculate the present value of the income stream, since the future value of the income stream is (as we've shown) $72,980.16, the present value of the income stream must be the present value of $72,980.16. Thus,

$$\text{Present Value} = \$72{,}980.16(e^{-.06 \cdot 15})$$
$$\approx \$29{,}671.52,$$

which is what we got before.

5. (a) Solve for $P(t) = P$.

$$100000 = \int_0^{10} Pe^{0.10(10-t)}\, dt = Pe \int_0^{10} e^{-0.10t}\, dt$$

$$= \frac{Pe}{-0.10} e^{-0.10t} \Big|_0^{10} = Pe(-3.678 + 10)$$

$$= P \cdot 17.183.$$

So, $P \approx \$5820$ per year.

(b) To answer this, we'll calculate the present value of $\$100,000$:

$$100000 = Pe^{0.10(10)}$$

$$P \approx \$36,787.94.$$

6. (a) Let L be the number of years for the balance to reach $\$10,000$. Since our income stream is $\$1000$ per year, the future value of this income stream should equal (in L years) $\$10,000$. Thus

$$10000 = \int_0^L 1000 e^{0.05(L-t)}\, dt = 1000 e^{0.05L} \int_0^L e^{-0.05t}\, dt$$

$$= 1000 e^{0.05L} \left(-\frac{1}{0.05}\right) e^{-0.05t} \Big|_0^L = 20000 e^{0.05L} \left(1 - e^{-0.05L}\right)$$

$$= 20000 e^{0.05L} - 20000$$

so $e^{0.05L} = \frac{3}{2}$

$$L = 20 \ln\left(\frac{3}{2}\right) \approx 8.11 \text{ years.}$$

(b) We want

$$10000 = 2000 e^{0.05L} + \int_0^L 1000 e^{0.05(L-t)}\, dt.$$

The first term on the right hand side is the future value of our initial balance. The second term is the future value of our income stream. We want this sum to equal $\$10,000$ in L years. We solve for L:

$$10000 = 2000 e^{0.05L} + 1000 e^{0.05L} \int_0^L e^{-0.05t}\, dt$$

$$= 2000 e^{0.05L} + 1000 e^{0.05L} \left(\frac{1}{-0.05}\right) e^{-0.05t} \Big|_0^L$$

$$= 2000 e^{0.05L} + 20000 e^{0.05L} \left(1 - e^{-0.05L}\right)$$

$$= 2000 e^{0.05L} + 20000 e^{0.05L} - 20000.$$

So,

$$22000 e^{0.05L} = 30000$$

$$e^{0.05L} = \frac{30000}{22000}$$

$$L = 20 \ln \frac{15}{11} \approx 6.203 \text{ years.}$$

7. At any time t, the company receives income of $s(t)$ per year. It will then invest this money for a length of $2-t$ years at 6% interest, giving it future value of $s(t)e^{(0.06)(2-t)}$ from this income. If we sum all such incomes over the two-year period, we can find the total value of the sales:

$$\text{Value} = \int_0^2 s(t) e^{(0.06)(2-t)}\, dt$$

$$= \int_0^2 \left[50 e^{-t} e^{(0.06)(2-t)}\right] dt$$

$$= \int_0^2 \left[50 e^{0.12 - 1.06t}\right] dt = \left(\frac{-53.1838}{e^{1.06t}}\right)\Big|_0^2 = \$46,800.$$

8. Price in future $= P(1 + 20\sqrt{t})$.
   The present value V of price satisfies $V = P(1 + 20\sqrt{t})e^{-0.05t}$.
   We want to maximize V. To do so, we find the critical points of $V(t)$ for $t \geq 0$. (Recall that $\sqrt{t}$ is nondifferentiable at $t = 0$.)

$$\frac{dV}{dt} = P\left[\frac{20}{2\sqrt{t}}e^{-0.05t} + (1 + 20\sqrt{t})(-0.05e^{-0.05t})\right]$$

$$= Pe^{-0.05t}\left[\frac{10}{\sqrt{t}} - 0.05 - \sqrt{t}\right].$$

Setting $\frac{dV}{dt} = 0$ gives $\frac{10}{\sqrt{t}} - 0.05 - \sqrt{t} = 0$. Using a calculator, we find $t \approx 10$ years. Since $V'(t) > 0$ for $0 < t < 10$ and $V'(t) < 0$ for $t > 10$, we confirm that this is a maximum. Thus, the best time to sell the wine is in 10 years.

9. (a)

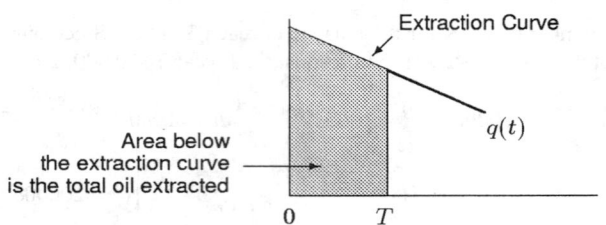

Suppose the oil extracted over the time period $[0, T]$ is S. (See above.) Since $q(t)$ is the rate of oil extraction, we have:

$$S = \int_0^T q(t)dt = \int_0^T (a - bt)dt = \int_0^T (10 - 0.1t)\,dt.$$

To calculate the time at which the oil is exhausted, set $S = 100$ and try different values of T. We find $T = 10.6$ gives

$$\int_0^{10.6} (10 - 0.1t)\,dt = 100,$$

so the oil is exhausted in 10.6 years.

(b) Suppose p is the oil price, C is the extraction cost per barrel, and r is the interest rate. We have the present value of the profit as

$$\text{Present value of profit} = \int_0^T (p - C)q(t)e^{-rt}dt$$

$$= \int_0^{10.6} (20 - 10)(10 - 0.1t)e^{-0.1t}\,dt$$

$$= 624.9 \text{ million dollars.}$$

10. One good way to approach the problem is in terms of present values. In 1980, the present value of Germany's loan was 20 billion DM. Now let's figure out the rate that the Soviet Union would have to give money to Germany to pay off 10% interest on the loan by using the formula for the present value of a continuous stream. Since the Soviet Union sends gas at a constant rate, the rate of deposit, $P(t)$, is a constant c. Since they don't start sending the gas until after 5 years have passed, the present value of the loan is given by:

$$\text{Present Value} = \int_5^\infty P(t)e^{-rt}\,dt.$$

We want to find c so that

$$20{,}000{,}000{,}000 = \int_5^\infty ce^{-rt}\,dt = c\int_5^\infty e^{-rt}\,dt$$

$$= c \lim_{b\to\infty} (-10e^{-0.10t})\Big|_5^b = ce^{-0.10(5)}$$

$$\approx 6.065c.$$

Dividing, we see that c should be about 3.3 billion DM per year. At 0.10 DM per m^3 of natural gas, the Soviet Union must deliver gas at the constant, continuous rate of about 33 billion m^3 per year.

11.

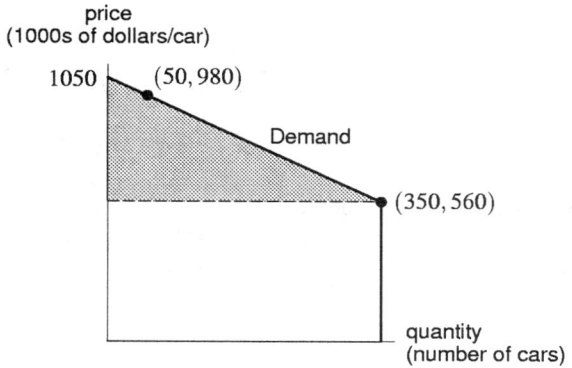

Measuring money in thousands of dollars, the equation of the line representing the demand curve passes through (50, 980) and (350, 560). So the equation is $y - 560 = \frac{420}{-300}(x - 350)$, i.e. $y - 560 = -\frac{7}{5}x + 490$.
The consumer surplus is thus

$$\int_0^{350} \left(-\frac{7}{5}x + 1050\right) dx - (350)(560) = -\frac{7}{10}x^2 + 1050x \Big|_0^{350} - 196000$$
$$= 85750.$$

(Note that $85750 = \frac{1}{2} \cdot 490 \cdot 350$, the area of the triangle in the diagram. We thus could have avoided the formula for consumer surplus in solving the problem.)

Recalling that our unit measure for the price axis is $1000/car, the consumer surplus is $85,750,000.

12.

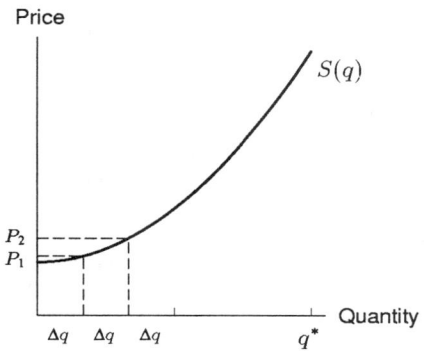

The supply curve, $S(q)$, represents the minimum price p per unit that the suppliers will be willing to supply some quantity q of the good for. If the suppliers have q^* of the good and q^* is divided into subintervals of size Δq, then if the consumers could offer the suppliers for each Δq a price increase just sufficient to induce the suppliers to sell an additional Δq of the good, the consumers' total expenditure on q^* goods would be

$$p_1 \Delta q + p_2 \Delta q + \cdots = \sum p_i \Delta q.$$

As $\Delta q \to 0$ the Riemann sum becomes the integral $\int_0^{q^*} S(q)\, dq$. Thus $\int_0^{q^*} S(q)\, dq$ is the amount the consumers would pay if suppliers could be forced to sell at the lowest price they would be willing to accept.

386 CHAPTER EIGHT /SOLUTIONS

13.
$$\int_0^{q^*} (p^* - S(q))\, dq = \int_0^{q^*} p^*\, dq - \int_0^{q^*} S(q)\, dq$$
$$= p^* q^* - \int_0^{q^*} S(q)\, dq.$$

Using Problem 12, this integral is the extra amount consumers pay (i.e., suppliers earn over and above the minimum they would be willing to accept for supplying the good). It results from charging the equilibrium price.

14. (a) $p^* q^*$ = the total amount paid for q^* of the good at equilibrium.

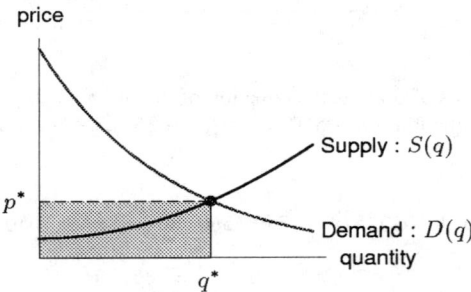

(b) $\int_0^{q^*} D(q)\, dq$ = the maximum consumers would be willing to pay if they had to pay the highest price acceptable to them for each additional unit of the good.

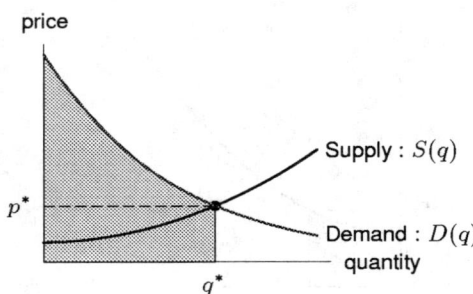

(c) $\int_0^{q^*} S(q)\, dq$ = the minimum suppliers would be willing to accept if they were paid the minimum price acceptable to them for each additional unit of the good.

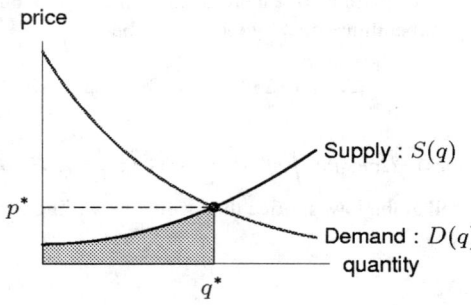

(d) $\int_0^{q^*} D(q)\,dq - p^*q^* =$ consumer surplus.

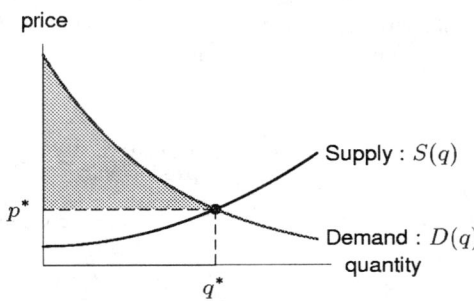

(e) $p^*q^* - \int_0^{q^*} S(q)\,dq =$ producer surplus.

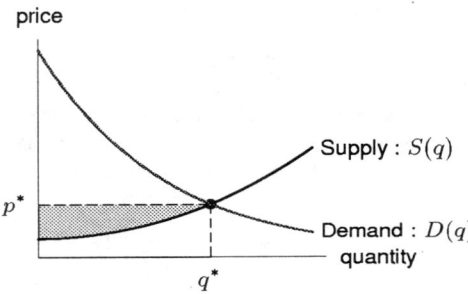

(f) $\int_0^{q^*} (D(q) - S(q))\,dq =$ producer surplus and consumer surplus.

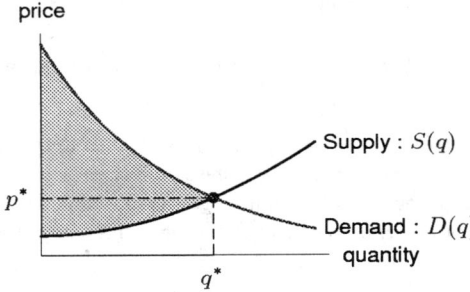

15.

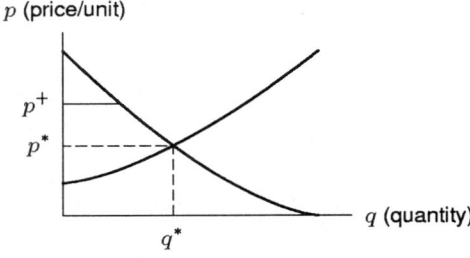

Figure 8.8: What effect does the artificially high price, p^+, have?

(a) A graph of possible demand and supply curves for the milk industry is given in Figure 8.8, with the equilibrium price and quantity labeled p^* and q^* respectively. Suppose that the price is fixed at the artificially high price labeled p^+ in Figure 8.8. Recall that the consumer surplus is the difference between the amount the consumers did pay (p^+) and the amount they would have been willing to pay (given on the demand curve). This is the area shaded in Figure 8.9(i). Notice that this consumer surplus is clearly less than the consumer surplus at the equilibrium price, shown in Figure 8.9(ii).

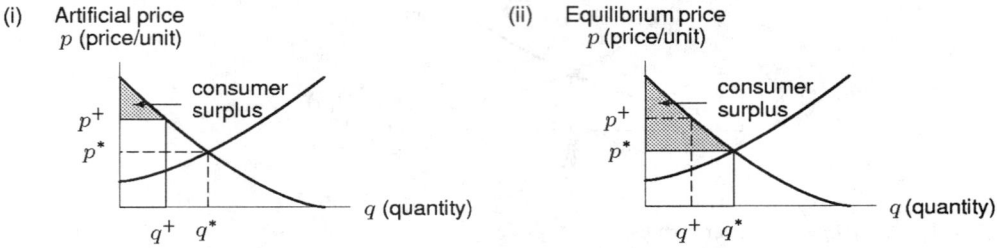

Figure 8.9: Consumer surplus for the milk industry

(b) At a price of p^+, the quantity sold, q^+, is less than it would have been at the equilibrium price. The producer surplus is the area between p^+ and the supply curve *at this reduced demand*. This area is shaded in Figure 8.10(i). Compare this producer surplus (at the artificially high price) to the producer surplus in Figure 8.10(ii) (at the equilibrium price). It appears that in this case, producer surplus is greater at the artificial price than at the equilibrium price. (Different supply and demand curves might have led to a different answer.)

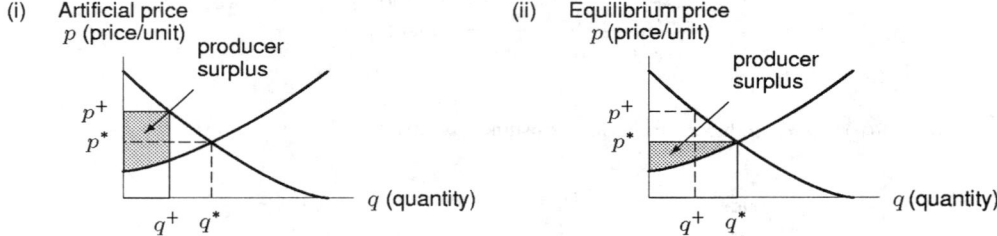

Figure 8.10: Producer surplus for the milk industry

(c) The total gains from trade (Consumer surplus + Producer surplus) at the artificially high price of p^+ is the area shaded in Figure 8.11(i). The total gains from trade at the equilibrium price of p^* is the area shaded in Figure 8.11(ii). It is clear that, under artificial price conditions, total gains from trade go down. The total financial effect of the artificially high price on all producers and consumers combined is a negative one.

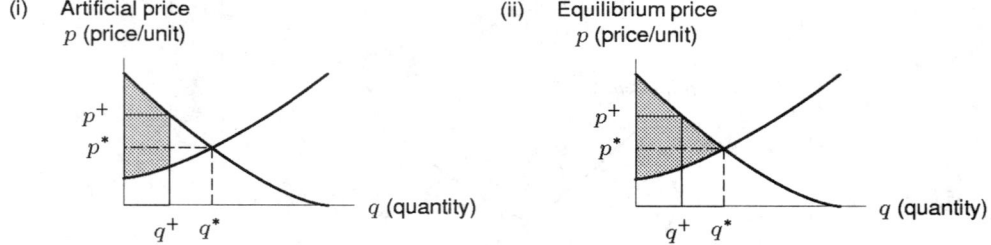

Figure 8.11: Total gains from trade

16.

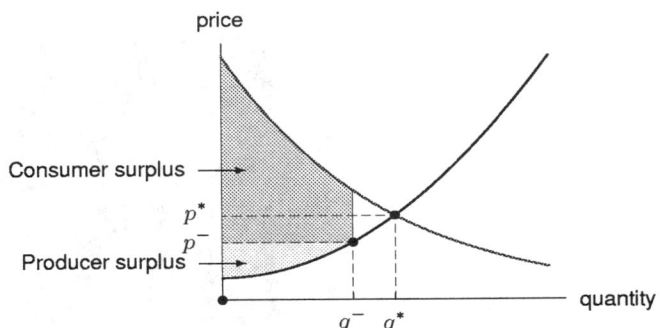

(a) The producer surplus is the area on the graph between p^- and the supply function. Lowering the price also lowers the producer surplus.

(b) Note that the consumer surplus—the area between the line p^- and the supply curve—increases or decreases depending on the functions describing the supply and demand and on the lowered price. (For example, the consumer surplus seems to be increased in the graph above, but if the price were brought down to $0 then the consumer surplus would be zero, and hence clearly less than the consumer surplus at equilibrium.)

(c) The graph above shows that the total gains from the trade are decreased.

Solutions for Chapter 8 Review

1. (a)

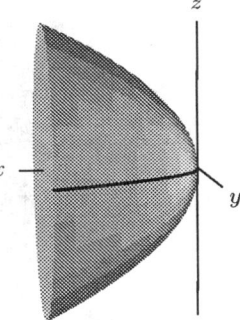

(b) Divide [0,1] into N subintervals of width $\Delta x = \frac{1}{N}$. The volume of the ith disc is $\pi(\sqrt{x_i})^2 \Delta x = \pi x_i \Delta x$. So, $V \approx \sum_{i=1}^{N} \pi x_i \Delta x$.

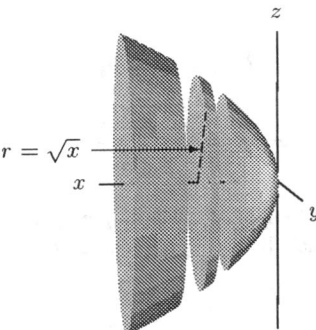

(c)

$$\text{Volume} = \int_0^1 \pi x \, dx = \frac{\pi}{2} x^2 \Big|_0^1 = \frac{\pi}{2} \approx 1.57.$$

2. (a)

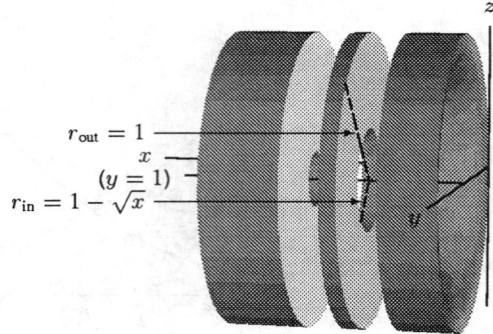

Slice the figure perpendicular to the x–axis. One gets washers of inner radius $1 - \sqrt{x}$ and outer radius 1. Therefore,

$$V = \int_0^1 \left(\pi 1^2 - \pi(1-\sqrt{x})^2\right) dx$$

$$= \pi \int_0^1 (1 - [1 - 2\sqrt{x} + x]) dx$$

$$= \pi \left[\frac{4}{3}x^{\frac{3}{2}} - \frac{1}{2}x^2\right]_0^1 = \frac{5\pi}{6} \approx 2.62.$$

(b)

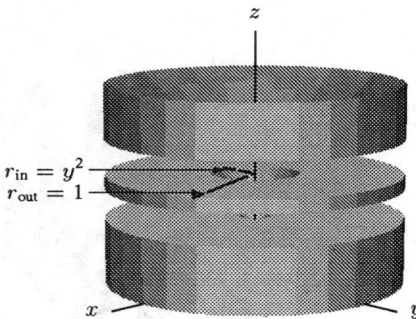

Note that $x = y^2$. We now integrate over y instead of x, slicing perpendicular to the y–axis. This gives us washers of inner radius x and outer radius 1. So

$$V = \int_{y=0}^{y=1} (\pi 1^2 - \pi x^2) dy$$

$$= \int_0^1 \pi(1-y^4) dy$$

$$= \left(\pi y - \frac{\pi}{5}y^5\right)\bigg|_0^1 = \pi - \frac{\pi}{5} = \frac{4\pi}{5} \approx 2.51.$$

3. (a) Slice the headlight into N disks of height Δx by cutting perpendicular to the x–axis. The radius of each disk is y; the height is Δx. The volume of each disk is $\pi y^2 \Delta x$. Therefore, the Riemann sum approximating the volume of the headlight is

$$\sum_{i=1}^N \pi y_i^2 \Delta x = \sum_{i=1}^N \pi \frac{9x_i}{4} \Delta x.$$

(b)

$$\pi \int_0^4 \frac{9x}{4} dx = \pi \frac{9}{8} x^2 \bigg|_0^4 = 18\pi.$$

4. (a) The line $y = ax$ must pass through (l, b). Hence $b = al$, so $a = b/l$.
   (b) Cut the cone into N slices, slicing perpendicular to the x–axis. Each piece is almost a cylinder. The radius of the ith cylinder is $r(x_i) = \dfrac{bx_i}{l}$, so the volume
   $$V \approx \sum_{i=1}^{N} \pi \left(\frac{bx_i}{l}\right)^2 \Delta x.$$
   Therefore, as $N \to \infty$, we get
   $$V = \int_0^l \pi b^2 l^{-2} x^2 dx$$
   $$= \pi \frac{b^2}{l^2}\left[\frac{x^3}{3}\right]_0^l = \left(\pi\frac{b^2}{l^2}\right)\left(\frac{l^3}{3}\right) = \frac{1}{3}\pi b^2 l.$$

5. (a) If you slice the apple perpendicular to the core, you expect that the cross section will be approximately a circle.

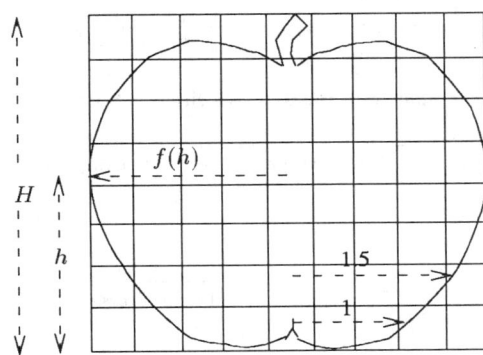

If $f(h)$ is the radius of the apple at height h above the bottom, and H is the height of the apple, then
$$\text{Volume} = \int_0^H \pi f(h)^2 \, dh.$$

Ignoring the stem, $H \approx 3.5$. Although we do not have a formula for $f(h)$, we can estimate it at various points. (Remember, we measure here from the bottom of the *apple*, which is not quite the bottom of the graph.)

h	0	0.5	1	1.5	2	2.5	3	3.5
$f(h)$	1	1.5	2	2.1	2.3	2.2	1.8	1.2

Now let $g(h) = \pi f(h)^2$, the area of the cross-section at height h. From our approximations above, we get the following table.

h	0	0.5	1	1.5	2	2.5	3	3.5
$g(h)$	3.14	7.07	12.57	13.85	16.62	13.85	10.18	4.52

We can now take left- and right-hand sum approximations. Note that $\Delta h = 0.5$ inches. Thus
$$\text{LEFT}(9) = (3.14 + 7.07 + 12.57 + 13.85 + 16.62 + 13.85 + 10.18)(0.5) = 38.64.$$
$$\text{RIGHT}(9) = (7.07 + 12.57 + 13.85 + 16.62 + 13.85 + 10.18 + 4.52)(0.5) = 39.33.$$

Thus the volume of the apple is ≈ 39 cu.in.

(b) The apple weighs $0.03 \times 39 \approx 1.17$ pounds, so it costs about 94¢. (Expensive apple!)

6.

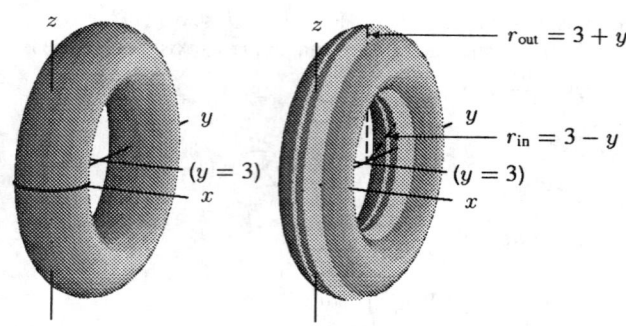

Figure 8.12: The Torus **Figure 8.13:** Slice of Torus

As shown in Figure 8.13, we slice the torus perpendicular to the line $y = 3$. We obtain washers with width dx, inner radius $r_{in} = 3-y$, and outer radius $r_{out} = 3+y$. Therefore, the area of the washer is $\pi r_{out}^2 - \pi r_{in}^2 = \pi[(3+y)^2-(3-y)^2] = 12\pi y$. Since $y = \sqrt{1-x^2}$, the volume is gotten by summing up the volumes of the washers: we get

$$\int_{-1}^{1} 12\pi\sqrt{1-x^2}\,dx = 12\pi \int_{-1}^{1} \sqrt{1-x^2}\,dx.$$

But $\int_{-1}^{1} \sqrt{1-x^2}\,dx$ is the area of a semicircle of radius 1, which is $\frac{\pi}{2}$. So we get $12\pi \cdot \frac{\pi}{2} = 6\pi^2 \approx 59.22$. (Or, you could use

$$\int \sqrt{1-x^2}\,dx = \left[x\sqrt{1-x^2} + \arcsin(x)\right],$$

by VI-30 and VI-28.)

7. Since $f(x) = \sin x$, $f'(x) = \cos(x)$, so

$$\text{Arc Length} = \int_0^{\pi} \sqrt{1+\cos^2 x}\,dx.$$

8. We'll find the arc length of the top half of the ellipse, and multiply that by 2. In the top half of the ellipse, the equation $(x^2/a^2) + (y^2/b^2) = 1$ implies

$$y = +b\sqrt{1-\frac{x^2}{a^2}}.$$

Differentiating $(x^2/a^2) + (y^2/b^2) = 1$ implicitly with respect to x gives us

$$\frac{2x}{a^2} + \frac{2y}{b^2}\frac{dy}{dx} = 0,$$

so

$$\frac{dy}{dx} = \frac{\frac{-2x}{a^2}}{\frac{2y}{b^2}} = -\frac{b^2 x}{a^2 y}.$$

Substituting this into the arc length formula, we get

$$\text{Arc Length} = \int_{-a}^{a} \sqrt{1+\left(-\frac{b^2 x}{a^2 y}\right)^2}\,dx$$

$$= \int_{-a}^{a} \sqrt{1+\left(\frac{b^4 x^2}{a^4(b^2)(1-\frac{x^2}{a^2})}\right)}\,dx$$

$$= \int_{-a}^{a} \sqrt{1+\left(\frac{b^2 x^2}{a^2(a^2-x^2)}\right)}\,dx.$$

Hence the arc length of the entire ellipse is

$$2\int_{-a}^{a} \sqrt{1+\left(\frac{b^2 x^2}{a^2(a^2-x^2)}\right)}\,dx.$$

9. We'll divide up time between 1971 and 1992 into intervals of length dt, and figure out how much of the strontium-90 produced during that time interval is still around.

First, strontium-90 decays exponentially, so if a quantity S_0 was produced t years ago, and S is the quantity around today, $S = S_0 e^{-kt}$. Since the half-life is 28 years, $\frac{1}{2} = e^{-k(28)}$, giving $k = \frac{-\ln(\frac{1}{2})}{28} \approx 0.025$.

Suppose we measure t in years from 1971, so that 1992 is $t = 21$.

```
   1971 (t = 0)    dt              1992 (t = 21)
   |───────────────┤┤──────────────────────────|
   - - - - - - - ->│<- - - - - - - - - - - - ->
         t                 (21 − t)
```

Since strontium-90 is produced at a rate of 1 kg/year, during the interval dt we know that a quantity $1\,dt$ kg was produced. Since this was $(21 - t)$ years ago, the quantity remaining now is $1\,dt\,e^{-0.025(21-t)}$. Summing over all such intervals gives

$$\text{Strontium in 1992} \approx \int_0^{21} e^{-0.025(21-t)}\,dt$$

$$= \left.\frac{e^{-0.025(21-t)}}{0.025}\right|_0^{21} = 16.34 \text{ kg}.$$

[Note: This is exactly like a future value problem from economics, with a negative interest rate.]

10. Let x be the height from ground to the weight. It follows that $0 \leq x \leq 20$. At height x, to lift the weight Δx more, the work needed is $200\Delta x + 2(20-x)\Delta x = (240 - 2x)\Delta x$. So the total work is

$$W = \int_0^{20} (240 - 2x)\,dx$$

$$= \left.(240x - x^2)\right|_0^{20}$$

$$= 240(20) - 20^2 = 4400 \text{ ft-lb}.$$

11. Let x be the distance from the bucket to the surface of the water. It follows that $0 \leq x \leq 40$. At x feet, the bucket weighs $(30 - \frac{1}{4}x)$, where the $\frac{1}{4}x$ term is due to the leak. When the bucket is x feet from the surface of the water, the work done by raising it Δx feet is $(30 - \frac{1}{4}x)\Delta x$. So the total work required to raise the bucket to the top is

$$W = \int_0^{40} (30 - \frac{1}{4}x)\,dx$$

$$= \left.\left(30x - \frac{1}{8}x^2\right)\right|_0^{40}$$

$$= 30(40) - \frac{1}{8}40^2 = 1000 \text{ ft-lb}.$$

12.

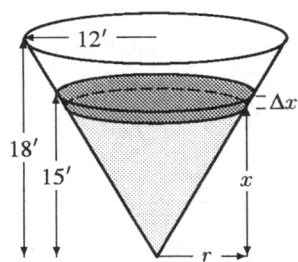

Figure 8.14

Let x be the depth of the water measured from the bottom of the tank. See Figure 8.14. It follows that $0 \le x \le 15$. Let r be the radius of the section of the cone with height x. By similar triangles, $\frac{r}{x} = \frac{12}{18}$, so $r = \frac{2}{3}x$. Then the work required to pump a layer of water with thickness of Δx at depth x over the top of the tank is $62.4\pi \left(\frac{2}{3}x\right)^2 \Delta x (18-x)$. So the total work done by pumping the water over the top of the tank is

$$W = \int_0^{15} 62.4\pi \left(\frac{2}{3}x\right)^2 (18-x)\,dx$$

$$= \frac{4}{9} 62.4\pi \int_0^{15} x^2(18-x)\,dx$$

$$= \frac{4}{9} 62.4\pi \left(6x^3 - \frac{1}{4}x^4\right)\Big|_0^{15}$$

$$= \frac{4}{9} 62.4\pi (7593.75) \approx 661{,}619.41 \text{ ft-lb}.$$

13. Let h be height above the bottom of the dam. Then

$$\text{Water force} = \int_0^{25} (62.4)(25-h)(60)\,dh$$

$$= (62.4)(60)\left(25h - \frac{h^2}{2}\right)\Big|_0^{25}$$

$$= (62.4)(60)(625 - 312.5)$$

$$= (62.4)(60)(312.5)$$

$$= 1{,}170{,}000 \text{ lbs}.$$

14. (a)

$$\text{Future Value} = \int_0^{20} 100 e^{0.10(20-t)}\,dt$$

$$= 100 \int_0^{20} e^2 e^{-0.10t}\,dt$$

$$= \frac{100 e^2}{-0.10} e^{-0.10t}\Big|_0^{20}$$

$$= \frac{100 e^2}{0.10}\left(1 - e^{-0.10(20)}\right) \approx \$6389.06.$$

The present value of the income stream is

$$\int_0^{20} 100 e^{-0.10t}\,dt = 100\left(\frac{1}{-0.10}\right) e^{-0.10t}\Big|_0^{20}$$

$$= 1000\left(1 - e^{-2}\right) = \$864.66.$$

Note that this is also the present value of the sum $6389.06.

(b) Let T be the number of years for the balance to reach $5000. Then

$$5000 = \int_0^T 100 e^{0.10(T-t)}\,dt$$

$$50 = e^{0.10T} \int_0^T e^{-0.10t}\,dt$$

$$= \frac{e^{0.10T}}{-0.10} e^{-0.10t}\Big|_0^T$$

$$= 10 e^{0.10T}\left(1 - e^{-0.10T}\right) = 10 e^{0.10}T - 10.$$

So, $60 = 10 e^{0.10T}$, and $T = 10 \ln 6 \approx 17.92$ years.

15. (a) Let's split the time interval into n parts, each of length Δt.

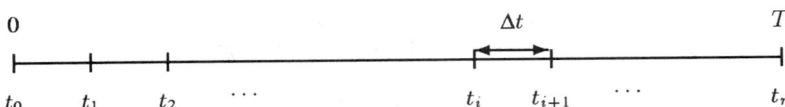

During the interval from t_i to t_{i+1}, profit is earned at a rate of approximately $(2 - 0.1t_i)$ thousand dollars per year, or $(2000 - 100t_i)$ dollars per year. Thus during this period, a total profit of $(2000 - 100t_i)\Delta t$ dollars is earned. Since this profit is earned t_i years in the future, its present value is $(2000 - 100t_i)\Delta t e^{-0.1t_i}$ dollars. Thus

$$\text{Total Present Value} \approx \sum_{i=0}^{n-1}(2000 - 100t_i)e^{-0.1t_i}\Delta t.$$

(b) The Riemann sum corresponds to the integral:

$$\text{Present value} = \int_0^T e^{-0.10t}(2000 - 100t)\,dt.$$

(c) To find where the present value is maximized, we take the derivative of

$$P(T) = \int_0^T e^{-0.10t}(2000 - 100t)\,dt,$$

and obtain

$$P'(T) = e^{-0.10T}(2000 - 100T).$$

This is 0 exactly when $2000 - 100T = 0$, that is, when $T = 20$ years. The value $T = 20$ maximizes $P(T)$, since $P'(T) > 0$ for $T < 20$, and $P'(T) < 0$ for $T > 20$. To determine what the maximum is, we evaluate the integral representation for $P(T)$ by III-14 in the integral table:

$$P(20) = \int_0^{20} e^{-0.10t}(2000 - 100t)\,dt$$

$$= \left[\frac{(2000 - 100t)}{-0.10}e^{-0.10t} + 10000e^{-0.10t}\right]\Big|_0^{20} \approx \$11353.35.$$

16. Pick a small interval of time Δt which takes place at time t. Fuel is consumed at a rate of $(25 + 0.1v)^{-1}$ gallons per mile. In the time Δt, the car moves $v\,\Delta t$ miles, so it consumes $v\,\Delta t/(25 + 0.1v)$ gallons during the instant Δt. Since $v = 50\frac{t}{t+1}$, the car consumes

$$\frac{v\,\Delta t}{25 + 0.1v} = \frac{50\frac{t}{t+1}\,\Delta t}{25 + 0.1\left(50\frac{t}{t+1}\right)} = \frac{50t\,\Delta t}{25(t+1) + 5t} = \frac{10t\,\Delta t}{6t + 5}$$

gallons of gas, in terms of the time t at which the instant occurs. To find the total gas consumed, sum up the instants in an integral:

$$\text{Gas consumed} = \int_2^3 \frac{10t}{6t + 5}\,dt \approx 1.25 \text{ gallons}.$$

17. (a) Slice the mountain horizontally into N cylinders of height Δh. The sum of the volumes of the cylinders will be

$$\sum_{i=1}^{N}\pi r^2\Delta h = \sum_{i=1}^{N}\pi\left(\frac{3.5\cdot 10^5}{\sqrt{h + 600}}\right)^2\Delta h.$$

(b)
$$\text{Volume} = \int_{400}^{14400} \pi \left(\frac{3.5 \cdot 10^5}{\sqrt{h + 600}} \right)^2 dh$$
$$= 1.23 \cdot 10^{11} \pi \int_{400}^{14400} \frac{1}{(h + 600)} dh$$
$$= 1.23 \cdot 10^{11} \pi \ln(h + 600) \Big|_{400}^{14400} dh$$
$$= 1.23 \cdot 10^{11} \pi [\ln 15000 - \ln 1000]$$
$$= 1.23 \cdot 10^{11} \pi \ln(15000/1000)$$
$$= 1.23 \cdot 10^{11} \pi \ln 15 \approx 1.05 \cdot 10^{12} \text{ cubic feet.}$$

18. Look at the disc-shaped slab of water at height y and of thickness dy. The rate at which water is flowing out when it is at depth y is $k\sqrt{y}$ (Torricelli's Law, with k constant). Then, if $x = g(y)$, we have

$$dt = \begin{pmatrix} \text{Time for water to} \\ \text{drop by this amount} \end{pmatrix} = \frac{\text{Volume}}{\text{Rate}} = \frac{\pi (g(y))^2 dy}{k\sqrt{y}}.$$

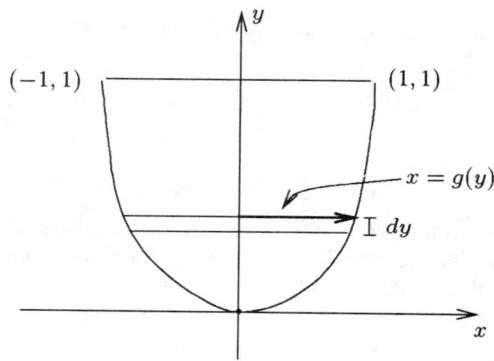

If the rate at which the depth of the water is dropping is constant, then $\frac{dy}{dt}$ is constant, so we want

$$\frac{\pi (g(y))^2}{k\sqrt{y}} = \text{constant},$$

or $\frac{(g(y))^2}{\sqrt{y}} = $ constant, so $g(y) = c\sqrt[4]{y}$, for some constant c. Since $x = 1$ when $y = 1$, we have $c = 1$ and so $x = \sqrt[4]{y}$, or $y = x^4$.

19. First we find the volume of the body up to the horizontal line through Q.

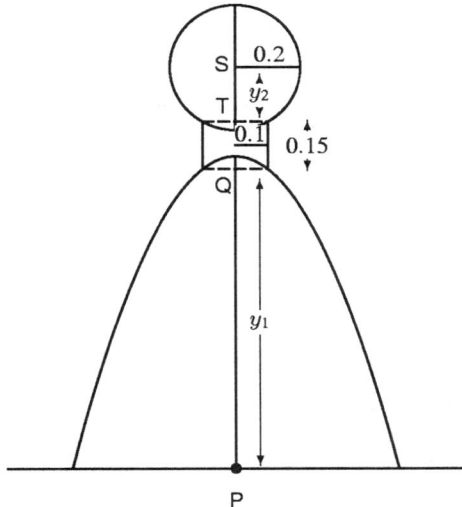

We put the origin at P, the x-axis horizontal and the y-axis pointing upward, and compute the volume obtained by rotating the curve $y = 1 - 4x^2$ around the y-axis up to Q. At Q, we have $x = 0.1$, so
$$y_1 = 1 - 4(0.1^2) = 0.96.$$
Slicing the body horizontally into disks of radius x, thickness Δy, we have
$$\text{Volume of disk in body} \approx \pi x^2 \Delta y = \frac{\pi}{4}(1-y)\Delta y.$$
Thus,
$$\text{Volume of body up to Q} = \int_0^{0.96} \frac{\pi}{4}(1-y)dy = \frac{\pi}{4}\left(y - \frac{y^2}{2}\right)\Big|_0^{0.96} = 0.3921.$$
To find the volume of the head, it is easiest to consider the origin at S, the x-axis horizontal, and the y-axis pointed upward. Then think of the head as the volume obtained by rotating the circle $x^2 + y^2 = (0.2)^2$ about the y-axis. We compute the volume of the head down to the horizontal line through T, at which point $x = 0.1$. Thus
$$(0.1)^2 + y_2^2 = (0.2)^2.$$
So
$$y_2 = -\sqrt{0.03} = -0.1732.$$
Slicing the head into circular disks, we have
$$\text{Volume of disk in head} \approx \pi x^2 \Delta y = \pi(0.2^2 - y^2)\Delta y.$$
Thus,
$$\text{Volume of head down to T} = \int_{-0.1732}^{0.2} \pi(0.2^2 - y^2)dy = \pi\left(0.2^2 y - \frac{y^3}{3}\right)\Big|_{-0.1732}^{0.2}$$
$$= 0.0331.$$
The neck is exactly cylindrical, with
$$\text{Volume of neck} = \pi(0.1^2)0.15 = 0.0047.$$
Thus,
$$\text{Total volume} = \text{Vol body} + \text{Vol head} + \text{Vol neck}$$
$$= 0.3921 + 0.0331 + 0.0047$$
$$= 0.4299 \approx 0.43 \text{m}^3.$$

20. (a) Divide the cross-section of the blood into rings of radius r, width Δr. See Figure 8.15.

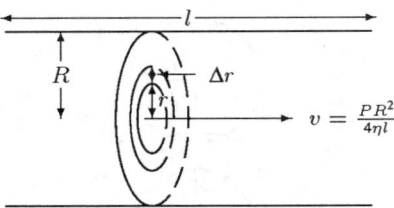

Figure 8.15

Then
$$\text{Area of ring} \approx 2\pi r \Delta r.$$

The velocity of the blood is approximately constant throughout the ring, so

$$\text{Rate blood flows through ring} \approx \text{Velocity} \cdot \text{Area}$$
$$= \frac{P}{4\eta l}(R^2 - r^2) \cdot 2\pi r \Delta r.$$

Thus, summing over all rings, we find the total blood flow:

$$\text{Rate blood flowing through blood vessel} \approx \sum \frac{P}{4\eta l}(R^2 - r^2) 2\pi r \Delta r.$$

Taking the limit as $\Delta r \to 0$, we get

$$\text{Rate blood flowing through blood vessel} = \int_0^R \frac{\pi P}{2\eta l}(R^2 r - r^3)\, dr$$
$$= \frac{\pi P}{2\eta l}\left(\frac{R^2 r^2}{2} - \frac{r^4}{4}\right)\Bigg|_0^R = \frac{\pi P R^4}{8\eta l}.$$

(b) Since
$$\text{Rate of blood flow} = \frac{\pi P R^4}{8\eta l},$$
if we take $k = \pi P/(8\eta l)$, then we have
$$\text{Rate of blood flow} = kR^4,$$
that is, rate of blood flow is proportional to R^4, in accordance with Poiseuille's Law.

21. (a) Slicing horizontally, as shown in Figure 8.16, we see that the volume of one disk-shaped slab is
$$\Delta V \approx \pi x^2 \Delta y = \frac{\pi y}{a}\Delta y.$$

Thus, the volume of the water is given by
$$V = \int_0^h \frac{\pi}{a} y\, dy = \frac{\pi}{a}\frac{y^2}{2}\Bigg|_0^h = \frac{\pi h^2}{2a}.$$

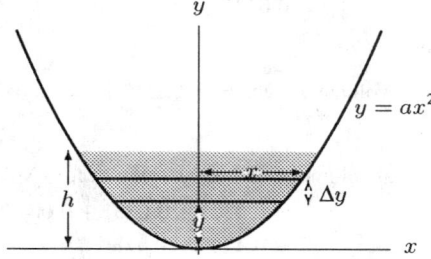

Figure 8.16

(b) The surface of the water is a circle of radius x. Since at the surface, $y = h$, we have $h = ax^2$. Thus, at the surface, $x = \sqrt{(h/a)}$. Therefore the area of the surface of water is given by

$$A = \pi x^2 = \frac{\pi h}{a}.$$

(c) If the rate at which water is evaporating is proportional to the surface area, we have

$$\frac{dV}{dt} = -kA.$$

(The negative sign is included because the volume is decreasing.) By the chain rule, $\frac{dV}{dt} = \frac{dV}{dh} \cdot \frac{dh}{dt}$. We know $\frac{dV}{dh} = \frac{\pi h}{a}$ and $A = \frac{\pi h}{a}$ so

$$\frac{\pi h}{a}\frac{dh}{dt} = -k\frac{\pi h}{a} \quad \text{giving} \quad \frac{dh}{dt} = -k.$$

(d) Integrating gives

$$h = -kt + h_0.$$

Solving for t when $h = 0$ gives

$$t = \frac{h_0}{k}.$$

22. (a) The volume of water in the centrifuge is $\pi(1^2) \cdot 1 = \pi$ cubic meters. The centrifuge has total volume 2π cubic meters, so the volume of the air in the centrifuge is π cubic meters. Now suppose the equation of the parabola is $y = h + bx^2$. We know that the volume of air in the centrifuge is the volume of the top part (a cylinder) plus the volume of the middle part (shaped like a bowl). See Figure 8.17.

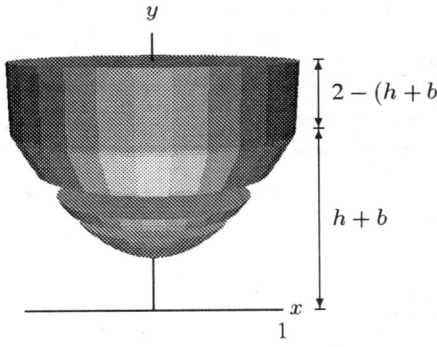

Figure 8.17: The Volume of Air

To find the volume of the cylinder of air, we find the maximum water depth. If $x = 1$, then $y = h + b$. Therefore the height of the water at the edge of the bowl, 1 meter away from the center, is $h + b$. The volume of the cylinder of air is therefore $[2 - (h + b)] \cdot \pi \cdot (1)^2 = [2 - h - b]\pi$.

To find the volume of the bowl of air, we note that the bowl is a volume of rotation with radius x at height y, where $y = h + bx^2$. Solving for x^2 gives $x^2 = (y - h)/b$. Hence, slicing horizontally as shown in the picture:

$$\text{Bowl Volume} = \int_h^{h+b} \pi x^2 \, dy = \int_h^{h+b} \pi \frac{y-h}{b} \, dy = \left.\frac{\pi(y-h)^2}{2b}\right|_h^{h+b} = \frac{b\pi}{2}.$$

So the volume of both pieces together is $[2 - h - b]\pi + b\pi/2 = (2 - h - b/2)\pi$. But we know the volume of air should be π, so $(2 - h - b/2)\pi = \pi$, hence $h + b/2 = 1$ and $b = 2 - 2h$. Therefore, the equation of the parabolic cross-section is $y = h + (2 - 2h)x^2$.

(b) The water spills out the top when $h + b = h + (2 - 2h) = 2$, or when $h = 0$. The bottom is exposed when $h = 0$. Therefore, the two events happen simultaneously.

23. Any small piece of mass ΔM on either of the two spheres has kinetic energy $\frac{1}{2}v^2\Delta M$. Since the angular velocity of the two spheres is the same, the actual velocity of the piece ΔM will depend on how far away it is from the axis of revolution. The further away a piece is from the axis, the faster it must be moving and the larger its velocity v. This is because if ΔM is at a distance r from the axis, in one revolution it must trace out a circular path of length $2\pi r$ about the axis. Since every piece in either sphere takes 1 minute to make 1 revolution, pieces farther from the axis must move faster, as they travel a greater distance.

   Thus, since the thin spherical shell has more of its mass concentrated farther from the axis of rotation than does the solid sphere, the bulk of it is traveling faster than the bulk of the solid sphere. So, it has the higher kinetic energy.

24. Any small piece of mass ΔM on either of the two hoops has kinetic energy $\frac{1}{2}v^2\Delta M$. Since the angular velocity of the two hoops is the same, the actual velocity of the piece ΔM will depend on how far away it is from the axis of revolution. The further away a piece is from the axis, the faster it must be moving and the larger its velocity v. This is because if ΔM is at a distance r from the axis, in one revolution it must trace out a circular path of length $2\pi r$ about the axis. Since every piece in either hoop takes 1 minute to make 1 revolution, pieces farther from the axis must move faster, as they travel a greater distance.

   The hoop rotating about the cylindrical axis has all of its mass at a distance R from the axis, whereas the other hoop has a good bit of its mass close (or on) the axis of rotation. So, since the bulk of the hoop rotating about the cylindrical axis is traveling faster than the bulk of the other hoop, it must have the higher kinetic energy.

Solutions to the Projects and Computer Algebra Investigations

1. Let us make coordinate axes with the origin at the center of the box. The x and y axes will lie along the central axes of the cylinders, and the (height) axis will extend vertically to the top of the box. If one slices the cylinders horizontally, one gets a cross. The cross is what you get if you cut out four corner squares from a square of side length 2. If h is the height of the cross above (or below) the xy plane, the equation of a cylinder is $h^2 + y^2 = 1$ (or $h^2 + x^2 = 1$). Thus the "armpits" of the cross occur where $y^2 - 1 = -h^2 = x^2 - 1$ for some fixed height h—that is, out $\sqrt{1-h^2}$ units from the center, or $1 - \sqrt{1-h^2}$ units away from the edge. Each corner square has area $(1 - \sqrt{1-h^2})^2 = 2 - h^2 - 2\sqrt{1-h^2}$. The whole big square has area 4. Therefore, the area of the cross is

$$4 - 4(2 - h^2 - 2\sqrt{1-h^2}) = -4 + 4h^2 + 8\sqrt{1-h^2}.$$

We integrate this from $h = -1$ to $h = 1$, and obtain the volume, V:

$$V = \int_{-1}^{1} -4 + 4h^2 + 8\sqrt{1-h^2}\, dh$$

$$= \left[-4h + \frac{4h^3}{3} + 8 \cdot \frac{1}{2}\left(h\sqrt{1-h^2} + \arcsin h\right)\right]\Bigg|_{-1}^{1}$$

$$= -8 + \frac{8}{3} + 4\pi = 4\pi - \frac{16}{3} \approx 7.23.$$

This is a reasonable answer, as the volume of the cube is 8, and the volume of one cylinder alone is $2\pi \approx 6.28$.

2. (a) Let y represent height, and let x represent horizontal distance from the lowest point of the cable. Then the stretched cable is a parabola of the form $y = kx^2$ passing through the point $(1280/2, 143) = (640, 143)$. Therefore, $143 = k(640)^2$ so $k \approx 3.491 \times 10^{-4}$. To find the arc length of the parabola, we take twice the arc length of the part to the right of the lowest point. Since $dy/dx = 2kx$,

$$\text{Arc Length} = 2\int_0^{640} \sqrt{1 + (2kx)^2}\, dx = 2\int_0^{640} \sqrt{1 + 4k^2x^2}\, dx.$$

The easiest way to find this integral is to substitute the value of k and find the integral's value numerically, giving

$$\text{Arc Length} \approx 1321.4 \text{ meters.}$$

Alternatively, we can make the substitution $w = 2kx$:

$$\text{Arc Length} = \frac{2}{2k}\int_0^{1280k} \sqrt{1 + w^2}\, dw$$

$$= \frac{1}{k}\int_0^{1280k} \sqrt{1 + w^2}\, dw$$

$$= \frac{1}{2k}\left(w\sqrt{1+w^2}\Big|_0^{1280k}\right) + \frac{1}{2k}\left(\int_0^{1280k} \frac{1}{\sqrt{1+w^2}}\, dw\right)$$

[Using the integral table, Formula VI-29]

$$= \frac{1}{2k}\left(1280k\sqrt{1 + (1280k)^2}\right) + \frac{1}{2k}\left(\ln\left|x + \sqrt{1 + x^2}\right|\Big|_0^{1280k}\right)$$

$$= \frac{1}{2k}\left(1280k\sqrt{1 + (1280k)^2}\right) + \frac{1}{2k}\left(\ln\left|1280k + \sqrt{1 + (1280k)^2}\right|\right)$$

$$\approx 1321.4 \text{ meters.}$$

(b) Adding 0.05% to the length of the cable gives a cable length of $(1321.4)(1.0005) = 1322.1$. We now want to calculate the new shape of the parabola; that is, we want to find a new k so that the arc length is 1322.1. Since

$$\text{Arc Length} = 2\int_0^{640} \sqrt{1 + 4k^2x^2}\, dx$$

we can find k numerically by trial and error. Trying values close to our original value of k, we find $k \approx 3.52 \times 10^{-4}$. To find the sag for this new k, we find the height $y = kx^2$ for which the cable hangs from the towers. This is

$$y = k(640)^2 \approx 144.2.$$

Thus the cable sag is 144.2 meters, over a meter more than on a cold winter day. Notice, though, that although the length increases by 0.05%, the sag increases by more: $144.2/143 \approx 1.0084$, an increase of 0.84%.

3. (a) We need to check that the point with the given coordinates is on the curve, i.e., that

$$x = a\sin^2 t, \quad y = \frac{a\sin^3 t}{\cos t}$$

satisfies the equation

$$y = \sqrt{\frac{x^3}{a - x}}.$$

This can be done by substituting into the computer algebra system and asking it to simplify the difference between the two sides, or by hand calculation:

$$\text{Right-hand side} = \sqrt{\frac{(a\sin^2 t)^3}{a - a\sin^2 t}} = \sqrt{\frac{a^3 \sin^6 t}{a(1 - \sin^2 t)}}$$

$$= \sqrt{\frac{a^3 \sin^6 t}{a\cos^2 t}} = \sqrt{\frac{a^2 \sin^6 t}{\cos^2 t}}$$

$$= \frac{a\sin^3 t}{\cos t} = y = \text{Left-hand side.}$$

We chose the positive square root because both $\sin t$ and $\cos t$ are positive for $0 \le t \le \pi/2$. Thus the point always lies on the curve. In addition, when $t = 0$, $x = 0$ and $y = 0$, so the point starts at the point where $x = 0$. As t approaches $\pi/2$, the value of $x = a \sin^2 t$ approachs a and the value of $y = a \sin^3 t / \cos t$ increases without bound (or approaches ∞), so the point on the curve approaches the vertical asymptote at $x = a$.

(b) We calculate the volume using horizontal slices. See the graph of $y = \sqrt{x^3/(a-x)}$ in Figure 8.18.

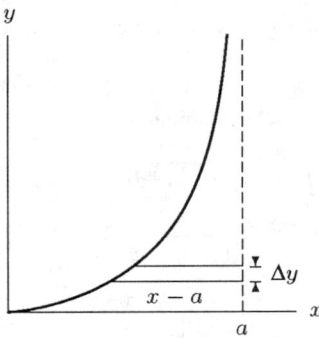

Figure 8.18

The slice at y is a disk of thickness Δy and radius $x - a$, hence it has volume $\pi(x-a)^2 \Delta y$. So the volume is given by the improper integral

$$\text{Volume} = \int_0^\infty \pi(x-a)^2 \, dy.$$

(c) We substitute

$$x = a \sin^2 t, \quad y = \frac{a \sin^3 t}{\cos t}$$

and

$$dy = \frac{d}{dt}\left(\frac{a \sin^3 t}{\cos t}\right) dt = a\left(3 \sin^2 t + \frac{\sin^4 t}{\cos^2 t}\right) dt.$$

Since $t = 0$ where $y = 0$ and $t = \pi/2$ at the asymptote where $y \to \infty$, we get

$$\text{Volume} = \int_0^{\pi/2} \pi(a \sin^2 t - a)^2 a \left(3 \sin^2 t + \frac{\sin^4 t}{\cos^2 t}\right) dt$$

$$= \pi a^3 \int_0^{\pi/2} (3 \sin^2 t \cos^4 t + \sin^4 t \cos^2 t) \, dt = \frac{\pi^2 a^3}{8}.$$

You can use a CAS to calculate this integral; it can also be done using trigonometric identities.

4. (a) The expression for arclength in terms of a definite integral gives

$$A(t) = \int_0^t \sqrt{1 + 4x^2} \, dx = \frac{2t\sqrt{1+4t^2} + \operatorname{arcsinh}(2t)}{4}.$$

The integral was evaluated using a computer algebra system; different systems may give the answer in different forms. Here arcsinh is the inverse function of the hyperbolic sine function.

(b) Figure 8.19 shows that the graphs of $A(t)$ and t^2 look very similar. This suggests that $A(t) \approx t^2$.

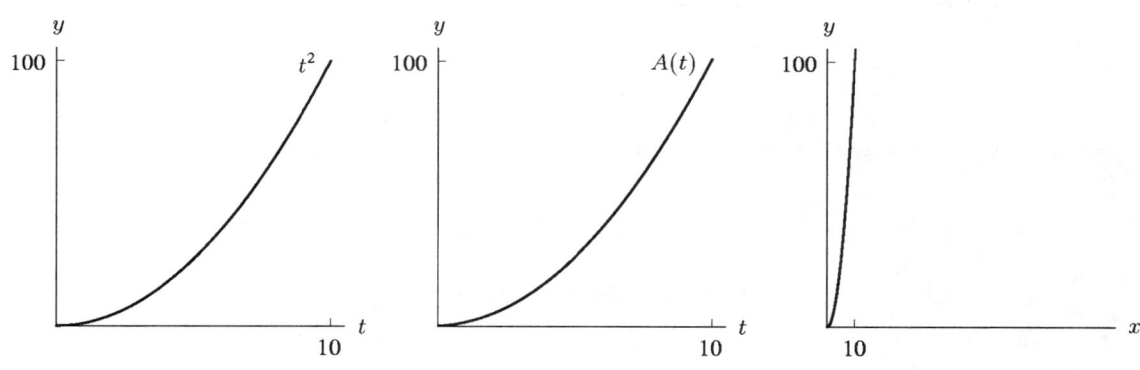

Figure 8.19　　　　　　　　　　　　　　　　　　**Figure 8.20**

(c) The graph in Figure 8.20 is approximately vertical and close to the y axis. Thus, if we measure the arclength up to a certain y-value, the answer is approximately the same as if we had measured the length straight up the y-axis. Hence
$$A(t) \approx y = f(t) = t^2.$$
So
$$A(t) \approx t^2.$$

5. (a) The expression for arclength in terms of a definite integral gives
$$A(t) = \int_0^t \sqrt{1 + \left(\frac{1}{2\sqrt{x}}\right)^2}\, dx = \frac{2\sqrt{t}\sqrt{1+4t} + \operatorname{arcsinh}(2\sqrt{t})}{4}.$$

The integral was evaluated using a computer algebra system; different systems may give the answer in different forms. Some may involve ln instead of arcsinh, which is the inverse function of the hyperbolic sine function.

(b) Figure 8.22 shows that the graphs of $A(t)$ and the graph of $y = t$ look very similar. This suggests that $A(t) \approx t$.

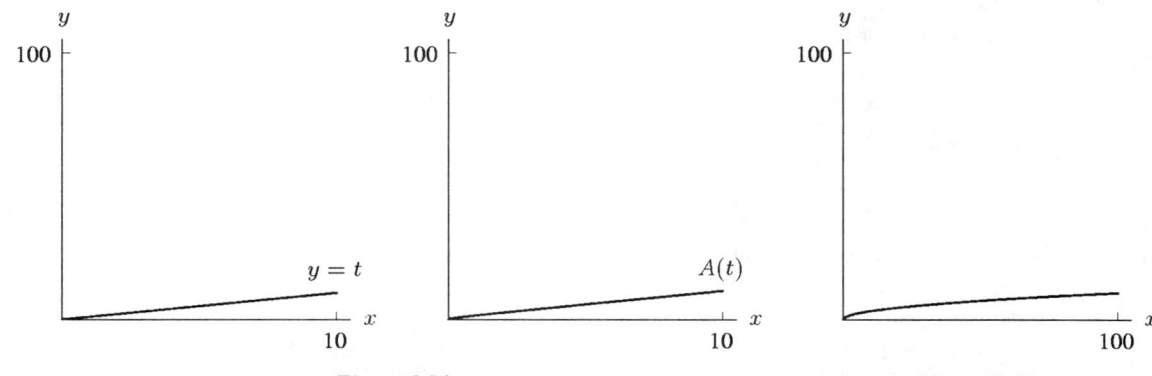

Figure 8.21　　　　　　　　　　　　　　　　　　**Figure 8.22**

(c) The graph in Figure 8.22 is approximately horizontal and close to the x-axis. Thus, if we measure the arclength up to a certain x-value, the answer is approximately the same as if we had measured the length straight along the x-axis. Hence
$$A(t) \approx x = t.$$
So
$$A(t) \approx t.$$

404 CHAPTER EIGHT /SOLUTIONS

6. (a) Slice the sphere at right angles to the axis of the cylinder. The cross-section is an annulus with internal radius a and external radius $\sqrt{(r^2 - x^2)}$ so that its area is

$$\pi \left(\sqrt{r^2 - x^2}\right)^2 - \pi a^2 = \pi(r^2 - x^2 - a^2).$$

The left and right endpoints of the cylinder are where the points on the wall of cylinder also lie on the surface of the sphere, i.e., where $x^2 + a^2 = r^2$, or $x = \pm\sqrt{r^2 - a^2}$. Thus

$$\text{Volume of bead} = \int_{-\sqrt{(r^2-a^2)}}^{\sqrt{(r^2-a^2)}} \pi(r^2 - x^2 - a^2)\,dx.$$

(b) Using a computer algebra system to evaluate the integral,

$$\text{Volume of bead} = \frac{4\pi}{3}\left(r^2 - a^2\right)^{\frac{3}{2}}.$$

Solutions to Problems on Distribution Functions

1.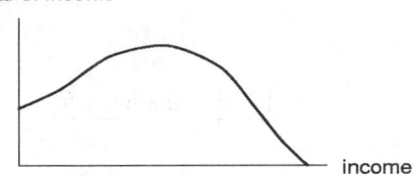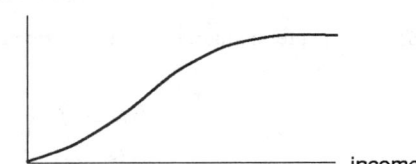

   Figure 8.23: Density function   Figure 8.24: Cumulative distribution function

2.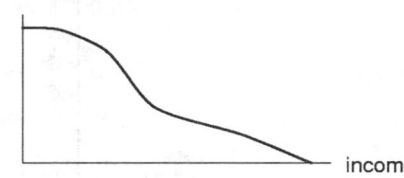

   Figure 8.25: Density function   Figure 8.26: Cumulative distribution function

3.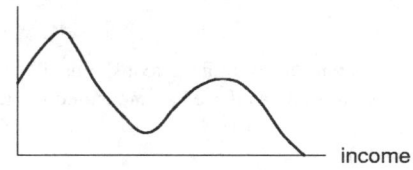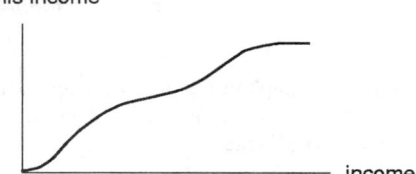

   Figure 8.27: Density function   Figure 8.28: Cumulative distribution function

4. No. Though the density function has its maximum value at 50, this does not mean that a large fraction of the population receives scores near 50. The value $p(50)$ can not be interpreted as a probability. Probability corresponds to *area* under the graph of a density function. Most of the area in this case is in the broad hump covering the range $0 \leq x \leq 40$, very little in the peak around $x = 50$. Most people score in the range $0 \leq x \leq 40$.

5. (a) Let $P(x)$ be the cumulative distribution function of the heights of the unfertilized plants. As do all cumulative distribution functions, $P(x)$ rises from 0 to 1 as x increases. The greatest number of plants will have heights in the range where $P(x)$ rises the most. The steepest rise appears to occur at about $x = 1$ m. Reading from the graph we see that $P(0.9) \approx 0.2$ and $P(1.1) \approx 0.8$, so that approximately $P(1.1) - P(0.9) = 0.8 - 0.2 = 0.6 = 60\%$ of the unfertilized plants grow to heights between 0.9 m and 1.1 m. Most of the plants grow to heights in the range 0.9 m to 1.1 m.

   (b) Let $P_A(x)$ be the cumulative distribution function of the plants that were fertilized with A. Since $P_A(x)$ rises the most in the range $0.7 \text{ m} \leq x \leq 0.9 \text{ m}$, many of the plants fertilized with A will have heights in the range 0.7 m to 0.9 m. Reading from the graph of P_A, we find that $P_A(0.7) \approx 0.2$ and $P_A(0.9) \approx 0.8$, so $P_A(0.9) - P_A(0.7) \approx 0.8 - 0.2 = 0.6 = 60\%$ of the plants fertilized with A have heights between 0.7 m and 0.9 m. Fertilizer A had the effect of stunting the growth of the plants.

   On the other hand, the cumulative distribution function $P_B(x)$ of the heights of the plants fertilized with B rises the most in the range $1.1 \text{ m} \leq x \leq 1.3 \text{ m}$, so most of these plants have heights in the range 1.1 m to 1.3 m. Fertilizer B caused the plants to grow about 0.2 m taller than they would have with no fertilizer.

6. (a) $F(7) = 0.6$ tells us that 60% of the trees in the forest have height 7 meters or less.
   (b) $F(7) > F(6)$. There are more trees of height less than 7 meters than trees of height less than 6 meters because every tree of height ≤ 6 meters also has height ≤ 7 meters.

7. For a small interval Δx around 68, the fraction of the population of American men with heights in this interval is about $(0.2)\Delta x$. For example, taking $\Delta x = 0.1$, we can say that approximately $(0.2)(0.1) = 0.02 = 2\%$ of American men have heights between 68 and 68.1 inches.

8. We want to find the cumulative distribution function for the age density function. We see that $P(10)$ is equal to 0.15 since the table shows that 15% of the population is between 0 and 10 years of age. Also,

$$P(20) = \begin{array}{c}\text{Fraction of the population}\\ \text{between 0 and 20 years old}\end{array} = 0.15 + 0.14 = 0.29$$

and

$$P(30) = 0.15 + 0.14 + 0.14 = 0.43$$

Continuing in this way, we obtain the values for $P(t)$ shown in Table 8.5.

TABLE 8.5 *Cumulative distribution function of ages in the US*

t	0	10	20	30	40	50	60	70	80	90	100
$P(t)$	0	0.15	0.29	0.43	0.60	0.74	0.84	0.92	0.97	0.99	1.00

9. (a) The two functions are shown below. The choice is based on the fact that the cumulative distribution does not decrease.
   (b) The cumulative distribution levels off to 1, so the top mark on the vertical scale must be 1.

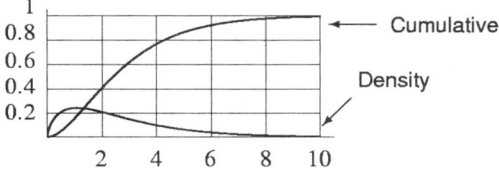

The total area under the density function must be 1. Since the area under the density function is about 2.5 boxes, each box must have area $1/2.5 = 0.4$. Since each box has a height of 0.2, the base must be 2.

10. (a) The area under the graph of the height density function $p(x)$ is concentrated in two humps centered at 0.5 m and 1.1 m. The plants can therefore be separated into two groups, those with heights in the range 0.3 m to 0.7 m, corresponding to the first hump, and those with heights in the range 0.9 m to 1.3 m, corresponding to the second hump. This grouping of the grasses according to height is probably close to the species grouping. Since the second hump contains more area than the first, there are more plants of the tall grass species in the meadow.

(b) As do all cumulative distribution functions, the cumulative distribution function $P(x)$ of grass heights rises from 0 to 1 as x increases. Most of this rise is achieved in two spurts, the first as x goes from 0.3 m to 0.7 m, and the second as x goes from 0.9 m to 1.3 m. The plants can therefore be separated into two groups, those with heights in the range 0.3 m to 0.7 m, corresponding to the first spurt, and those with heights in the range 0.9 m to 1.3 m, corresponding to the second spurt. This grouping of the grasses according to height is the same as the grouping we made in part (a), and is probably close to the species grouping.

(c) The fraction of grasses with height less than 0.7 m equals $P(0.7) = 0.25 = 25\%$. The remaining 75% are the tall grasses.

11. (a) The percentage of calls lasting from 1 to 2 minutes is given by the integral

$$\int_1^2 p(x)\,dx = \int_1^2 0.4e^{-0.4x}\,dx = e^{-0.4} - e^{-0.8} \approx 22.1\%.$$

(b) A similar calculation (changing the limits of integration) gives the percentage of calls lasting 1 minute or less as

$$\int_0^1 p(x)\,dx = \int_0^1 0.4e^{-0.4x}\,dx = 1 - e^{-0.4} \approx 33.0\%.$$

(c) The percentage of calls lasting 3 minutes or more is given by the improper integral

$$\int_3^\infty p(x)\,dx = \lim_{b \to \infty} \int_3^b 0.4e^{-0.4x}\,dx = \lim_{b \to \infty}(e^{-1.2} - e^{-0.4b}) = e^{-1.2} \approx 30.1\%.$$

(d) The cumulative distribution function is the integral of the probability density; thus,

$$C(h) = \int_0^h p(x)\,dx = \int_0^h 0.4e^{-0.4x}\,dx = 1 - e^{-0.4h}.$$

12. (a) We must have $\int_0^\infty f(t)\,dt = 1$, for even though it is possible that any given person survives the disease, everyone eventually dies. Therefore,

$$\int_0^\infty cte^{-kt}\,dt = 1.$$

Integrating by parts gives

$$\int_0^b cte^{-kt}\,dt = -\frac{c}{k}te^{-kt}\Big|_0^b + \int_0^b \frac{c}{k}e^{-kt}\,dt$$

$$= \left(-\frac{c}{k}te^{-kt} - \frac{c}{k^2}e^{-kt}\right)\Big|_0^b$$

$$= \frac{c}{k^2} - \frac{c}{k}be^{-kb} - \frac{c}{k^2}e^{-kb}.$$

As $b \to \infty$, we see

$$\int_0^\infty cte^{-kt}\,dt = \frac{c}{k^2} = 1 \quad \text{so} \quad c = k^2.$$

(b) We are told that $\int_0^5 f(t)\,dt = 0.4$, so using the fact that $c = k^2$ and the antiderivatives from part (a), we have

$$\int_0^5 k^2 te^{-kt}\,dt = \left(-\frac{k^2}{k}te^{-kt} - \frac{k^2}{k^2}e^{-kt}\right)\Big|_0^5$$

$$= 1 - 5ke^{-5k} - e^{-5k} = 0.4$$

so
$$5ke^{-5k} + e^{-5k} = 0.6.$$

Since this equation cannot be solved exactly, we use a calculator or computer to find $k = 0.275$. Since $c = k^2$, we have $c = (0.275)^2 = 0.076$.

(c) The cumulative death distribution function, $C(t)$, represents the fraction of the population that have died up to time t. Thus,
$$C(t) = \int_0^t k^2 x e^{-kx} dx = \left(-kxe^{-kx} - e^{-kx}\right)\Big|_0^t$$
$$= 1 - kte^{-kt} - e^{-kt}.$$

13. (a) The fraction of maintenance checks completed in 15 minutes is $P(15) = 0.21$, or 21%.
    (b) Since $P(30) = 0.98$, we see that 98% of maintenance checks take 30 minutes or less. Therefore, only 2% take more than 30 minutes.
    (c) Since 8% take less than or equal to 10 minutes and 21% take less than or equal to 15 minutes, the fraction taking between 10 and 15 minutes must be $0.21 - 0.08 = 0.13$, or 13%.
    (d) We begin by making a table showing how the times are distributed. Reading from the table given in the problem, we see that the fraction of jobs completed between 0 and 5 minutes is 0.03, and the fraction completed between 5 and 10 minutes is 0.05. See Table 8.6.

TABLE 8.6 *Distribution of times for routine maintenance*

time period (minutes)	0-5	5-10	10-15	15-20	20-25	25-30	> 30
fraction of jobs	0.03	0.05	0.13	0.17	0.42	0.18	0.02

We now draw the histogram, arranging the vertical scale in such a way that the area of each bar in the histogram equals the fraction of jobs completed in the corresponding time period. For instance, since the first bar is to have area 0.03 and width 5 minutes, its height must be $0.03/5 = 0.006$. See Figure 8.29.

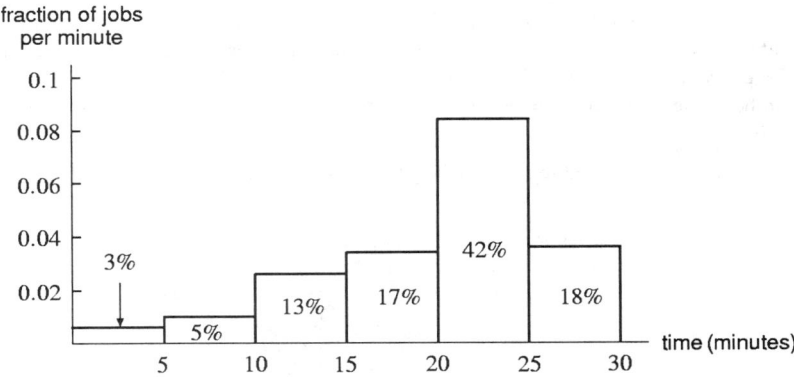

Figure 8.29: Histogram of maintenance times

(e) From Figure 8.29, we see that more of the jobs take between 20 and 25 minutes to complete, so this is the most likely length of time.
(f) The density function is a smoothed version of the histogram in Figure 8.29. Without more detailed information, we cannot know exactly how to draw it. A reasonable sketch is given in Figure 8.30.
(g) A graph is given in Figure 8.31. Since P is a cumulative distribution function, we know that $P(t)$ is approaching 1 as t gets large, but is never larger than 1.

408 CHAPTER EIGHT /SOLUTIONS

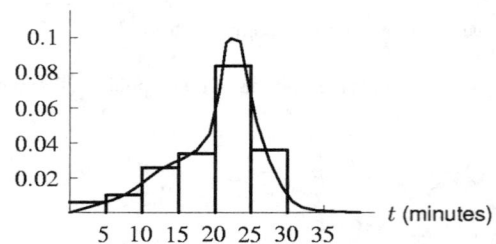

Figure 8.30: Density function for routine maintenance checks

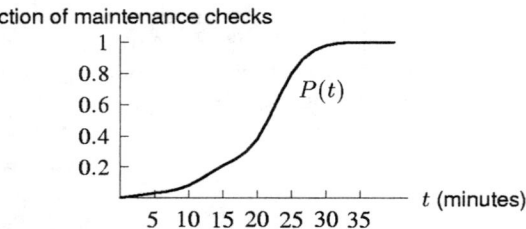

Figure 8.31: Cumulative distribution function for routine maintenance checks

14. (a) The fraction of students passing is given by the area under the curve from 2 to 4 divided by the total area under the curve. This appears to be about $\frac{2}{3}$.
    (b) The fraction with honor grades corresponds to the area under the curve from 3 to 4 divided by the total area. This is about $\frac{1}{3}$.
    (c) The peak around 2 probably exists because many students work to get just a passing grade.
    (d)

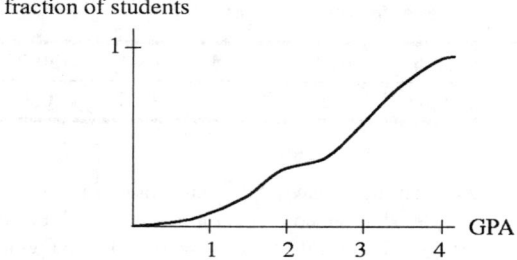

15. (a) Most of the earth's surface is below sea level. Much of the earth's surface is either around 3 miles below sea level or exactly at sea level. It appears that essentially all of the surface is between 4 miles below sea level and 2 miles above sea level. Very little of the surface is around 1 mile below sea level.
    (b) The fraction below sea level corresponds to the area under the curve from −4 to 0 divided by the total area under the curve. This appears to be about $\frac{3}{4}$.

16. (a) The shaded region in Figure 8.32 represents the probability that the bus will be from 2 to 4 minutes late.

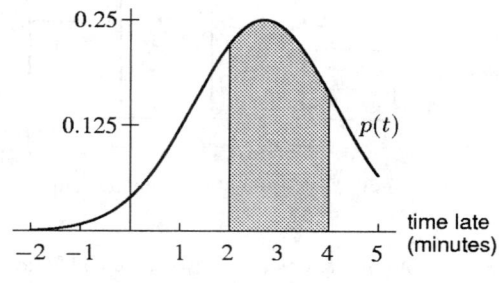

Figure 8.32

(b)

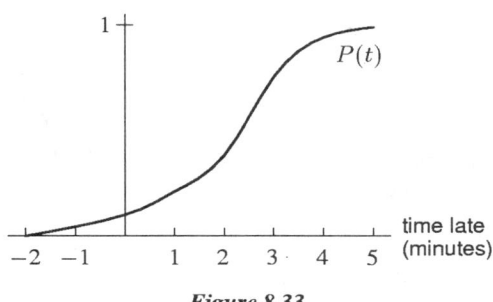

Figure 8.33

The probability that the bus will be 2 to 4 minutes late (the area shaded in Figure 8.32) is $P(4) - P(2)$. The inflection point on the graph of $P(t)$ in Figure 8.33 corresponds to where $p(t)$ is a maximum. To the left of the inflection point, P is increasing at an increasing rate, while to the right of the inflection point P is increasing at a decreasing rate. Thus, the inflection point marks where the rate at which P is increasing is a maximum (i.e., where the derivative of P, which is p, is a maximum).

17. (a) The density function $f(r)$ will be zero outside the range $0 < r < 5$ and equal to a nonzero constant k inside this range. The area of the region under the density curve equals $5k$, which must equal 1, so $k = 0.2$. We have

$$f(r) = \begin{cases} 0 & \text{if } r \leq 0 \\ 0.2 & \text{if } 0 < r < 5 \\ 0 & \text{if } 5 \leq r. \end{cases}$$

The graph of $f(r)$ is given in Figure 8.34.

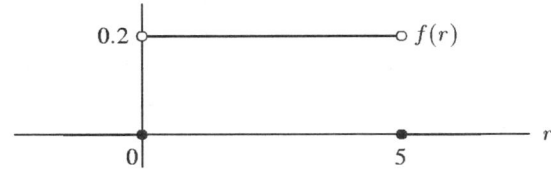

Figure 8.34

(b) The cumulative distribution function $F(r)$ equals the area of the region under the density function up to r. From the graph in Figure 8.34 we see that the area is zero if $r < 0$; for $0 \leq r \leq 5$ the region is rectangular of height 0.2, width r, and area $0.2r$; and for $r > 5$ the area is 1. Thus

$$F(r) = \begin{cases} 0 & \text{if } r < 0 \\ 0.2r & \text{if } 0 \leq r \leq 5 \\ 1 & \text{if } 5 < r. \end{cases}$$

(c) The cumulative distribution function $G(v)$ is the fraction of raindrops of volume less than or equal to v. Since volume $v = 4\pi r^3/3$, or equivalently $r = (3v/(4\pi))^{1/3}$, we see that $G(v)$ is the same as the fraction of raindrops of radius less than or equal to $(3v/(4\pi))^{1/3} = 0.62v^{1/3}$. In other words,

$$G(v) = F(0.62v^{1/3}) = \begin{cases} 0 & \text{if } 0.62v^{1/3} < 0 \\ (0.2)(0.62)v^{1/3} & \text{if } 0 \leq 0.62v^{1/3} \leq 5 \\ 1 & \text{if } 5 < 0.62v^{1/3}. \end{cases}$$

The final answer is thus

$$G(v) = \begin{cases} 0 & \text{if } v < 0 \\ 0.124v^{1/3} & \text{if } 0 \leq v \leq 523.6 \\ 1 & \text{if } 523.6 < v. \end{cases}$$

Note that the volume of a raindrop of radius 5 is $v = 4\pi 5^3/3 = 523.6$.

410 CHAPTER EIGHT /SOLUTIONS

(d) The density function $g(v)$ equals the derivative of the cumulative distribution function $G(v)$. We have

$$g(v) = \begin{cases} 0 & \text{if } v \leq 0 \\ 0.0413v^{-2/3} & \text{if } 0 < v < 523.6 \\ 0 & \text{if } 523.6 \leq v. \end{cases}$$

The density function is graphed in Figure 8.35. Notice that the density functions $f(r)$ for the radii of the raindrops and $g(v)$ for the volumes are quite different.

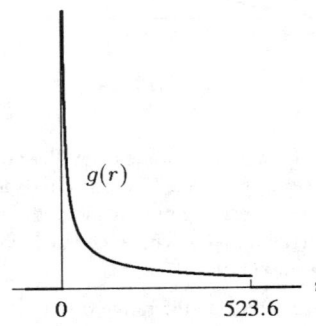

Figure 8.35

18. (a)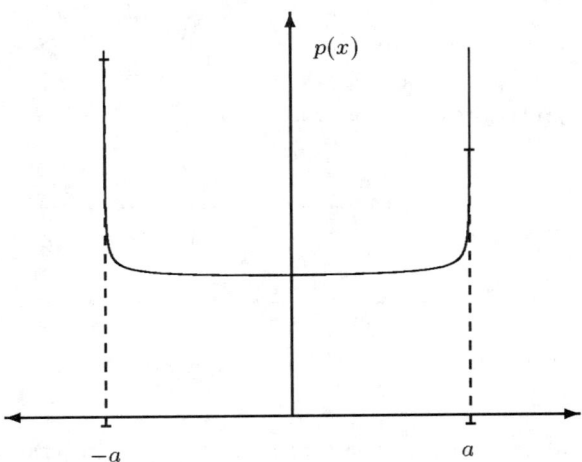

(b) The graphs should look similar.

(c) We expect $\displaystyle\int_{-a}^{a} \frac{dx}{\pi\sqrt{a^2 - x^2}} = 1$, since

$$f(x) = \begin{cases} \frac{1}{\pi\sqrt{a^2-x^2}} & -a < x < a; \\ 0 & |x| \geq a, \end{cases}$$

and thus $\displaystyle\int_{-a}^{a} \frac{dx}{\pi\sqrt{a^2 - x^2}} = 1$ by the definition of a probability density function. Indeed,

$$\int_{-a}^{a} \frac{dx}{\pi\sqrt{a^2 - x^2}} = \frac{1}{\pi} \arcsin\frac{x}{a}\Big|_{-a}^{a}$$
$$= \frac{1}{\pi}(\arcsin 1 - \arcsin(-1))$$
$$= 1.$$

(d) It does make sense, physically speaking. The fact that $f(x) \to \infty$ as $x \to a$ does not mean that the ball spends an infinite amount of time at a, but just that the ratio of the time spent near $-a$ and a to the time spent elsewhere goes to ∞. This makes sense—if we watch a pendulum, we note that more time is spent near the ends of its path (where its velocity is small) than in the middle of the path (where its velocity is largest).

Solutions to Problems on Probability and More on Distributions

1.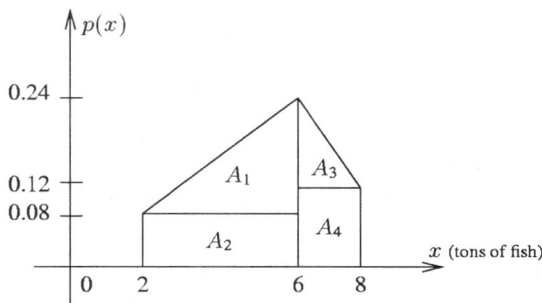

   Splitting the figure into four pieces, we see that

   $$\text{Area under the curve} = A_1 + A_2 + A_3 + A_4$$
   $$= \frac{1}{2}(0.16)4 + 4(0.08) + \frac{1}{2}(0.12)2 + 2(0.12)$$
   $$= 1.$$

   We expect the area to be 1, since $\int_{-\infty}^{\infty} p(x)\,dx = 1$ for any probability density function, and $p(x)$ is 0 except when $2 \le x \le 8$.

2. Recall that the mean is $\int_{-\infty}^{\infty} xp(x)\,dx$. In the fishing example, $p(x) = 0$ except when $2 \le x \le 8$, so the mean is

   $$\int_2^8 xp(x)\,dx.$$

   Using the equation for $p(x)$ from the graph,

   $$\int_2^8 xp(x)\,dx = \int_2^6 xp(x)\,dx + \int_6^8 xp(x)\,dx$$
   $$= \int_2^6 x(0.04x)\,dx + \int_6^8 x(-0.06x + 0.6)\,dx$$
   $$= \left.\frac{0.04x^3}{3}\right|_2^6 + \left.\left(-0.02x^3 + 0.3x^2\right)\right|_6^8$$
   $$\approx 5.253 \text{ tons}.$$

3. (a) Since $d(e^{-ct})/dt = ce^{-ct}$, we have

   $$c\int_0^6 e^{-ct}\,dt = \left.-e^{-ct}\right|_0^6 = 1 - e^{-6c} = 0.1,$$

   so

   $$c = -\frac{1}{6}\ln 0.9 \approx 0.0176.$$

   (b) Similarly, with $c = 0.0176$, we have

   $$c\int_6^{12} e^{-ct}\,dt = \left.-e^{-ct}\right|_6^{12}$$
   $$= e^{-6c} - e^{-12c} = 0.9 - 0.81 = 0.09,$$

   so the probability is 9%.

4. (a) We can find the proportion of students by integrating the density $p(x)$ between $x = 1.5$ and $x = 2$:

$$P(2) - P(1.5) = \int_{1.5}^{2} \frac{x^3}{4} \, dx$$

$$= \frac{x^4}{16}\bigg|_{1.5}^{2}$$

$$= \frac{(2)^4}{16} - \frac{(1.5)^4}{16} = 0.684,$$

so that the proportion is 0.684 : 1 or 68.4%.

(b) We find the mean by integrating x times the density over the relevant range:

$$\text{Mean} = \int_{0}^{2} x \left(\frac{x^3}{4}\right) dx$$

$$= \int_{0}^{2} \frac{x^4}{4} \, dx$$

$$= \frac{x^5}{20}\bigg|_{0}^{2}$$

$$= \frac{2^5}{20} = 1.6 \text{ hours}.$$

(c) The median will be the time T such that exactly half of the students are finished by time T, or in other words

$$\frac{1}{2} = \int_{0}^{T} \frac{x^3}{4} \, dx$$

$$\frac{1}{2} = \frac{x^4}{16}\bigg|_{0}^{T}$$

$$\frac{1}{2} = \frac{T^4}{16}$$

$$T = \sqrt[4]{8} = 1.682 \text{ hours}.$$

5. (a) Since $\int_{0}^{\infty} p(x) \, dx = 1$, we have

$$1 = \int_{0}^{\infty} ae^{-0.122x} \, dx$$

$$= \frac{a}{-0.122} e^{-0.122x}\bigg|_{0}^{\infty} = \frac{a}{0.122}.$$

So $a = 0.122$.

(b)

$$P(x) = \int_{0}^{x} p(t) \, dt$$

$$= \int_{0}^{x} 0.122 e^{-0.122t} \, dt$$

$$= -e^{0.122t}\bigg|_{0}^{x} = 1 - e^{-0.122x}.$$

(c) Median is the x such that

$$P(x) = 1 - e^{-0.122x} = 0.5.$$

So $e^{-0.122x} = 0.5$. Thus,

$$x = -\frac{\ln 0.5}{0.122} \approx 5.68 \text{ seconds}$$

and

$$\text{Mean} = \int_0^\infty x(0.122)e^{-0.122x}\,dx = -\int_0^\infty x\left(-0.122e^{-0.122x}\right)\,dx.$$

We now use integration by parts. Let $u = -x$ and $v' = -0.122e^{-0.122x}$. Then $u' = -1$, and $v = e^{-0.122x}$. Therefore,

$$\text{Mean} = -xe^{-0.122x}\Big|_0^\infty + \int_0^\infty e^{-0.122x}\,dx = \frac{1}{0.122} \approx 8.20 \text{ seconds}.$$

(d)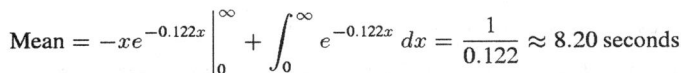

6. (a) Since $\mu = 100$ and $\sigma = 15$:

$$p(x) = \frac{1}{15\sqrt{2\pi}}e^{-\frac{1}{2}\left(\frac{x-100}{15}\right)^2}.$$

(b) The fraction of the population with IQ scores between 115 and 120 is (integrating numerically)

$$\int_{115}^{120} p(x)\,dx = \int_{115}^{120} \frac{1}{15\sqrt{2\pi}}e^{-\frac{(x-100)^2}{450}}\,dx$$

$$= \frac{1}{15\sqrt{2\pi}}\int_{115}^{120} e^{-\frac{(x-100)^2}{450}}\,dx$$

$$\approx 0.067 = 6.7\% \text{ of the population}.$$

7. The fraction of the population within one standard deviation of the mean is given by

$$\text{Fraction within } \sigma \text{ of mean} = \int_{-\sigma}^{\sigma} \frac{1}{\sqrt{2\pi}\sigma}e^{-x^2/(2\sigma^2)}\,dx.$$

Let us substitute $w = \dfrac{x}{\sigma}$ so that $dw = \dfrac{1}{\sigma}dx$, and when $x = \pm\sigma$, $w = \pm 1$. Then we have

$$\text{Fraction} = \int_{-\sigma}^{\sigma}\frac{1}{\sqrt{2\pi}\sigma}e^{-x^2/(2\sigma^2)}\,dx = \int_{-1}^{1}\frac{1}{\sqrt{2\pi}\sigma}e^{-w^2/2}\cdot \sigma\,dw = \int_{-1}^{1}\frac{1}{\sqrt{2\pi}}e^{-w^2/2}\,dw.$$

This integral is independent of σ. Evaluating the integral numerically gives 0.68, showing that about 68% of the population lies within one standard deviation of the mean.

8. (a)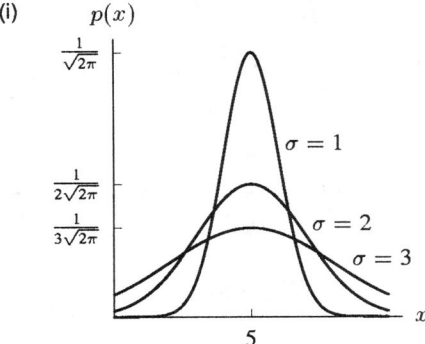

(b) Recall that the mean is the "balancing point." In other words, if the area under the curve was made of cardboard, we'd expect it to balance at the mean. All of the graphs are symmetric across the line $x = \mu$, so μ is the "balancing point" and hence the mean.

As the graphs also show, increasing σ flattens out the graph, in effect lessening the concentration of the data near the mean. Thus, the smaller the σ value, the more data is clustered around the mean.

9. (a) First, we find the critical points of $p(x)$:

$$\frac{d}{dx}p(x) = \frac{1}{\sigma\sqrt{2\pi}}\left[\frac{-2(x-\mu)}{2\sigma^2}\right]e^{-\frac{(x-\mu)^2}{2\sigma^2}}$$

$$= -\frac{(x-\mu)}{\sigma^3\sqrt{2\pi}}e^{-\frac{(x-\mu)^2}{2\sigma^2}}.$$

This implies $x = \mu$ is the only critical point of $p(x)$.

To confirm that $p(x)$ is maximized at $x = \mu$, we rely on the first derivative test. As $-\frac{1}{\sigma^3\sqrt{2\pi}}e^{-\frac{(x-\mu)^2}{2\sigma^2}}$ is always negative, the sign of $p'(x)$ is the opposite of the sign of $(x-\mu)$; thus $p'(x) > 0$ when $x < \mu$, and $p'(x) < 0$ when $x > \mu$.

(b) To find the inflection points, we need to find where $p''(x)$ changes sign; that will happen only when $p''(x) = 0$. As

$$\frac{d^2}{dx^2}p(x) = -\frac{1}{\sigma^3\sqrt{2\pi}}e^{-\frac{(x-\mu)^2}{2\sigma^2}}\left[-\frac{(x-\mu)^2}{\sigma^2}+1\right],$$

$p''(x)$ changes sign when $\left[-\frac{(x-\mu)^2}{\sigma^2}+1\right]$ does, since the sign of the other factor is always negative. This occurs when

$$-\frac{(x-\mu)^2}{\sigma^2}+1 = 0,$$
$$-(x-\mu)^2 = -\sigma^2,$$
$$x-\mu = \pm\sigma.$$

Thus, $x = \mu + \sigma$ or $x = \mu - \sigma$. Since $p''(x) > 0$ for $x < \mu - \sigma$ and $x > \mu + \sigma$ and $p''(x) < 0$ for $\mu - \sigma \leq x \leq \mu + \sigma$, these are in fact points of inflection.

(c) μ represents the mean of the distribution, while σ is the standard deviation. In other words, σ gives a measure of the "spread" of the distribution, i.e., how tightly the observations are clustered about the mean. A small σ tells us that most of the data are close to the mean; a large σ tells us that the data is spread out.

10. (a) The cumulative distribution function

$$P(t) = \int_0^t p(x)dx = \text{Area under graph of density function } p(x) \text{ for } 0 \leq x \leq t$$

$$= \text{Fraction of population who survive } t \text{ years or less after treatment}$$

$$= \text{Fraction of population who survive up to } t \text{ years after treatment.}$$

(b) The probability that a randomly selected person survives for at least t years is the probability that he lives t years or longer, so

$$S(t) = \int_t^\infty p(x)\,dx = \lim_{b\to\infty}\int_t^b Ce^{-Ct}\,dx$$

$$= \lim_{b\to\infty} -e^{-Ct}\Big|_t^b = \lim_{b\to\infty} -e^{-Cb} - (-e^{-Ct}) = e^{-Ct},$$

or equivalently,

$$S(t) = 1 - \int_0^t p(x)\,dx = 1 - \int_0^t Ce^{-Ct}\,dx = 1 + e^{-Ct}\Big|_0^t = 1 + (e^{-Ct} - 1) = e^{-Ct}.$$

(c) The probability of surviving at least two years is

$$S(2) = e^{-C(2)} = 0.70$$

so

$$\ln e^{-C(2)} = \ln 0.70$$
$$-2C = \ln 0.7$$
$$C = -\frac{1}{2}\ln 0.7 \approx 0.178.$$

11. (a) The probability you dropped the glove within a kilometer of home is given by
$$\int_0^1 2e^{-2x}\,dx = -e^{-2x}\Big|_0^1 = -e^{-2} + 1 \approx 0.865.$$

(b) Since the probability that the glove was dropped within y km $= \int_0^y p(x)\,dx = 1 - e^{-2y}$, we solve
$$1 - e^{-2y} = 0.95$$
$$e^{-2y} = 0.05$$
$$y = \frac{\ln 0.05}{-2} \approx 1.5 \text{ km}.$$

12. It is not (a) since a probability density must be a non-negative function; not (c) since the total integral of a probability density must be 1; (b) and (d) are probability density functions, but (d) is not a good model. According to (d), the probability that the next customer comes after 4 minutes is 0. In real life there should be a positive probability of not having a customer in the next 4 minutes. So (b) is the best answer.

13. (a) P is the cumulative distribution function, so the percentage of the population that made between \$20,000 and \$50,000 is
$$P(50) - P(20) = 99\% - 75\% = 24\%.$$
Therefore $\frac{6}{25}$ of the population made between \$20,000 and \$50,000.

(b) The median income is the income such that half the people made less than this amount. Looking at the chart, we see that $P(12.6) = 50\%$, so the median must be \$12,600.

(c) The cumulative distribution function looks something like this:

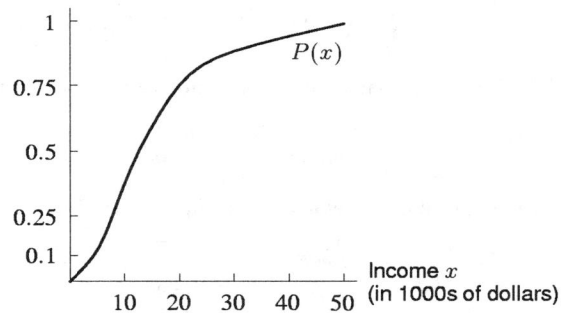

The density function is the derivative of the cumulative distribution. Qualitatively it looks like:

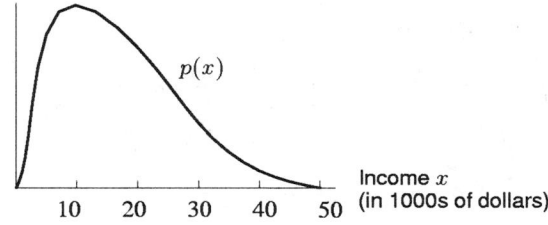

The density function has a maximum at about \$8000. This means that more people have incomes around \$8000 than around any other amount. On the density function, this is the highest point. On the cumulative distribution, this is the point of steepest slope (because $P' = p$), which is also the point of inflection.

14. (a) We want to find a such that $\int_0^\infty p(v)\,dv = \lim_{r\to\infty} a \int_0^r v^2 e^{-mv^2/2kT}\,dv = 1$. Therefore,
$$\frac{1}{a} = \lim_{r\to\infty} \int_0^r v^2 e^{-mv^2/2kT}\,dv.$$

416 CHAPTER EIGHT /SOLUTIONS

To evaluate the integral, use integration by parts with the substitutions $u = v$ and $w' = ve^{-mv^2/2kT}$:

$$\int_0^r \underbrace{v}_{u} \underbrace{ve^{-mv^2/2kT}}_{w'} dv = \underbrace{v}_{u} \underbrace{\frac{e^{-mv^2/2kT}}{-m/kT}}_{w}\bigg|_0^r - \int_0^r \underbrace{1}_{u'} \underbrace{\frac{e^{-mv^2/2kT}}{-m/kT}}_{w} dv$$

$$= -\frac{kTr}{m}e^{-mr^2/2kT} + \frac{kT}{m}\int_0^r e^{-mv^2/2kT}\, dv.$$

From the normal distribution we know that $\int_0^\infty \frac{1}{\sqrt{2\pi}}e^{-x^2/2}\, dx = \frac{1}{2}$, so

$$\int_0^\infty e^{-x^2/2}\, dx = \frac{\sqrt{2\pi}}{2}.$$

Therefore in the above integral, make the substitution $x = \sqrt{\frac{m}{kT}}v$, so that $dx = \sqrt{\frac{m}{kT}}\, dv$, or $dv = \sqrt{\frac{kT}{m}}\, dx$. Then

$$\frac{kT}{m}\int_0^r e^{-mv^2/2kT}\, dv = \left(\frac{kT}{m}\right)^{3/2}\int_0^{\sqrt{\frac{m}{kT}}r} e^{-x^2/2}\, dx.$$

Substituting this into Equation 14a we get

$$\frac{1}{a} = \lim_{r \to \infty}\left(-\frac{kTr}{m}e^{-mr^2/2kT} + \left(\frac{kT}{m}\right)^{3/2}\int_0^{\sqrt{\frac{m}{kT}}r} e^{-x^2/2}\, dx\right) = 0 + \left(\frac{kT}{m}\right)^{3/2} \cdot \frac{\sqrt{2\pi}}{2}.$$

Therefore, $a = \frac{2}{\sqrt{2\pi}}(\frac{m}{kT})^{3/2}$. Substituting the values for k, T, and m gives $a \approx 3.4 \times 10^{-8}$.

(b) To find the median, we wish to find the speed x such that

$$\int_0^x p(v)\, dv = \int_0^x av^2 e^{-\frac{mv^2}{2kT}}\, dv = \frac{1}{2},$$

where $a = \frac{2}{\sqrt{2\pi}}(\frac{m}{kT})^{3/2}$. Using a calculator, by trial and error we get $x \approx 441$ m/sec.
To find the mean, we find

$$\int_0^\infty vp(v)\, dv = \int_0^\infty av^3 e^{-\frac{mv^2}{2kT}}\, dv.$$

This integral can be done by substitution. Let $u = v^2$, so $du = 2v\, dv$. Then

$$\int_0^\infty av^3 e^{-\frac{mv^2}{2kT}}\, dv = \frac{a}{2}\int_{v=0}^{v=\infty} v^2 e^{-\frac{mv^2}{2kT}} 2v\, dv$$

$$= \frac{a}{2}\int_{u=0}^{u=\infty} ue^{-\frac{mu}{2kT}}\, du$$

$$= \lim_{r \to \infty}\frac{a}{2}\int_0^r ue^{-\frac{mu}{2kT}}\, du.$$

Now, using the integral table, we have

$$\int_0^\infty av^3 e^{-\frac{mv^2}{2kT}}\, dv = \lim_{r \to \infty}\frac{a}{2}\left[-\frac{2kT}{m}ue^{-\frac{mu}{2kT}} - \left(-\frac{2kT}{m}\right)^2 e^{-\frac{mu}{2kT}}\right]\bigg|_0^r$$

$$= \frac{a}{2}\left(-\frac{2kT}{m}\right)^2$$

$$\approx 457.7 \text{ m/sec}.$$

The maximum for $p(v)$ will be at a point where $p'(v) = 0$.

$$p'(v) = a(2v)e^{-\frac{mv^2}{2kT}} + av^2\left(-\frac{2mv}{2kT}\right)e^{-\frac{mv^2}{2kT}}$$

$$= ae^{-\frac{mv^2}{2kT}}\left(2v - v^3\frac{m}{kT}\right).$$

Thus $p'(v) = 0$ at $v = 0$ and at $v = \sqrt{\frac{2kT}{m}} \approx 405$. It's obvious that $p(0) = 0$, and that $p \to 0$ as $v \to \infty$. So $v = 405$ gives us a maximum: $p(405) \approx 0.002$.

(c) The mean, as we found in part (b), is $\frac{a}{2} \frac{4k^2T^2}{m^2} = \frac{4}{\sqrt{2\pi}} \frac{k^{1/2}T^{1/2}}{m^{1/2}}$. It is clear, then, that as T increases so does the mean. We found in part (b) that $p(v)$ reached its maximum at $v = \sqrt{\frac{2kT}{m}}$. Thus

$$\text{The maximum value of } p(v) = \frac{2}{\sqrt{2\pi}} \left(\frac{m}{kT}\right)^{3/2} \frac{2kT}{m} e^{-1}$$

$$= \frac{4}{e\sqrt{2\pi}} \frac{m^{1/2}}{kT^{1/2}}.$$

Thus as T increases, the maximum value decreases.

15. (a) Let the $p(r)$ be the density function. Then $P(r) = \int_0^r p(x)\,dx$, and from the Fundamental Theorem of Calculus, $p(r) = \frac{d}{dr}P(r) = \frac{d}{dr}(1 - (2r^2 + 2r + 1)e^{-2r}) = -(4r+2)e^{-2r} + 2(2r^2 + 2r + 1)e^{-2r}$, or $p(r) = 4r^2e^{-2r}$.
We have that $p'(r) = 8r(e^{-2r}) - 8r^2e^{-2r} = e^{-2r} \cdot 8r(1-r)$, which is zero when $r = 0$ or $r = 1$, negative when $r > 1$, and positive when $r < 1$. Thus $p(1) = 4e^{-2} \approx 0.54$ is a relative maximum.
Here are sketches of $p(r)$ and the cumulative position $P(r)$:

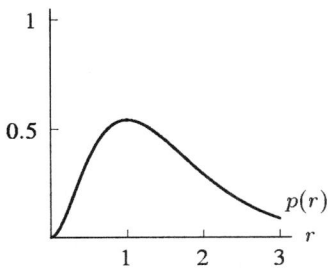
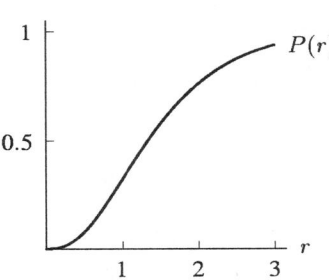

(b) The median distance is the distance r such that $P(r) = 1 - (2r^2 + 2r + 1)e^{-2r} = 0.5$, or equivalently, $(2r^2 + 2r + 1)e^{-2r} = 0.5$.
By experimentation with a calculator, we find that $r \approx 1.33$ Bohr radii is the median distance.
The mean distance is equal to the value of the integral $\int_0^\infty rp(r)\,dr = \lim_{x \to \infty} \int_0^x rp(r)\,dr$. We have that $\int_0^x rp(r)\,dr = \int_0^x 4r^3 e^{-2r}\,dr$. Using the integral table, we get

$$\int_0^x 4r^3 e^{-2r}\,dr = \left[\left(-\frac{1}{2}\right)4r^3 - \frac{1}{4}(12r^2) - \frac{1}{8}(24r) - \frac{1}{16}(24)\right] e^{-2x} \Big|_0^x$$

$$= \frac{3}{2} - \left[2x^3 + 3x^2 + 3x + \frac{3}{2}\right] e^{-2x}.$$

Taking the limit of this expression as $x \to \infty$, we see that all terms involving (powers of x or constants) $\cdot e^{-2x}$ have limit 0, and thus the mean distance is 1.5 Bohr radii.
The most likely distance is obtained by maximizing $p(r) = 4r^2 e^{-2r}$; as we have already seen this corresponds to $r = 1$ Bohr unit.

(c) Because it is the most likely distance of the electron from the nucleus.

CHAPTER NINE

Solutions for Section 9.1

1. Let $\dfrac{1}{1+x} = (1+x)^{-1}$. Then $f(0) = 1$.

$$f'(x) = -1!(1+x)^{-2} \qquad f'(0) = -1,$$
$$f''(x) = 2!(1+x)^{-3} \qquad f''(0) = 2!,$$
$$f'''(x) = -3!(1+x)^{-4} \qquad f'''(0) = -3!,$$
$$f^{(4)}(x) = 4!(1+x)^{-5} \qquad f^{(4)}(0) = 4!,$$
$$f^{(5)}(x) = -5!(1+x)^{-6} \qquad f^{(5)}(0) = -5!,$$
$$f^{(6)}(x) = 6!(1+x)^{-7} \qquad f^{(6)}(0) = 6!,$$
$$f^{(7)}(x) = -7!(1+x)^{-8} \qquad f^{(7)}(0) = -7!,$$
$$f^{(8)}(x) = 8!(1+x)^{-9} \qquad f^{(8)}(0) = 8!.$$

$$P_4(x) = 1 - x + x^2 - x^3 + x^4,$$
$$P_6(x) = 1 - x + x^2 - x^3 + x^4 - x^5 + x^6,$$
$$P_8(x) = 1 - x + x^2 - x^3 + x^4 - x^5 + x^6 - x^7 + x^8.$$

2. Let $f(x) = \dfrac{1}{1-x} = (1-x)^{-1}$. Then $f(0) = 1$.

$$f'(x) = 1!(1-x)^{-2} \qquad f'(0) = 1!,$$
$$f''(x) = 2!(1-x)^{-3} \qquad f''(0) = 2!,$$
$$f'''(x) = 3!(1-x)^{-4} \qquad f'''(0) = 3!,$$
$$f^{(4)}(x) = 4!(1-x)^{-5} \qquad f^{(4)}(0) = 4!,$$
$$f^{(5)}(x) = 5!(1-x)^{-6} \qquad f^{(5)}(0) = 5!,$$
$$f^{(6)}(x) = 6!(1-x)^{-7} \qquad f^{(6)}(0) = 6!,$$
$$f^{(7)}(x) = 7!(1-x)^{-8} \qquad f^{(7)}(0) = 7!.$$

$$P_3(x) = 1 + x + x^2 + x^3,$$
$$P_5(x) = 1 + x + x^2 + x^3 + x^4 + x^5,$$
$$P_7(x) = 1 + x + x^2 + x^3 + x^4 + x^5 + x^6 + x^7.$$

3. Let $f(x) = \sqrt{1+x} = (1+x)^{1/2}$. Then $f(0) = 1$, and

$$f'(x) = \tfrac{1}{2}(1+x)^{-1/2} \qquad f'(0) = \tfrac{1}{2},$$
$$f''(x) = -\tfrac{1}{4}(1+x)^{-3/2} \qquad f''(0) = -\tfrac{1}{4},$$
$$f'''(x) = \tfrac{3}{8}(1+x)^{-5/2} \qquad f'''(0) = \tfrac{3}{8},$$
$$f^{(4)}(x) = -\tfrac{15}{16}(1+x)^{-7/2} \qquad f^{(4)}(0) = -\tfrac{15}{16}.$$

Thus,

$$P_2(x) = 1 + \tfrac{1}{2}x - \tfrac{1}{8}x^2,$$
$$P_3(x) = 1 + \tfrac{1}{2}x - \tfrac{1}{8}x^2 + \tfrac{1}{16}x^3,$$
$$P_4(x) = 1 + \tfrac{1}{2}x - \tfrac{1}{8}x^2 + \tfrac{1}{16}x^3 - \tfrac{5}{128}x^4.$$

4. Let $f(x) = \cos x$. Then $f(0) = \cos(0) = 1$, and

$$f'(x) = -\sin x \qquad f'(0) = 0,$$
$$f''(x) = -\cos x \qquad f''(0) = -1,$$
$$f'''(x) = \sin x \qquad f'''(0) = 0,$$
$$f^{(4)}(x) = \cos x \qquad f^{(4)}(0) = 1,$$
$$f^{(5)}(x) = -\sin x \qquad f^{(5)}(0) = 0,$$
$$f^{(6)}(x) = -\cos x \qquad f^{(6)}(0) = -1.$$

Thus,

$$P_2(x) = 1 - \frac{x^2}{2!},$$
$$P_4(x) = 1 - \frac{x^2}{2!} + \frac{x^4}{4!},$$
$$P_6(x) = 1 - \frac{x^2}{2!} + \frac{x^4}{4!} - \frac{x^6}{6!}.$$

5. Let $f(x) = \arctan x$. Then $f(0) = \arctan 0 = 0$, and

$$f'(x) = 1/(1+x^2) = (1+x^2)^{-1} \qquad f'(0) = 1,$$
$$f''(x) = (-1)(1+x^2)^{-2} 2x \qquad f''(0) = 0,$$
$$f'''(x) = 2!(1+x^2)^{-3} 2^2 x^2 + (-1)(1+x^2)^{-2} 2 \qquad f'''(0) = -2,$$
$$f^{(4)}(x) = -3!(1+x^2)^{-4} 2^3 x^3 + 2!(1+x^2)^{-3} 2^3 x$$
$$\qquad\qquad + 2!(1+x^2)^{-3} 2^2 x \qquad f^{(4)}(0) = 0.$$

Therefore,

$$P_3(x) = P_4(x) = x - \frac{1}{3}x^3.$$

6. Let $f(x) = \tan x$. So $f(0) = \tan 0 = 0$, and

$$f'(x) = 1/\cos^2 x \qquad f'(0) = 1,$$
$$f''(x) = 2\sin x/\cos^3 x \qquad f''(0) = 0,$$
$$f'''(x) = (2/\cos^2 x) + (6\sin^2 x/\cos^4 x) \qquad f'''(0) = 2,$$
$$f^{(4)}(x) = (16\sin x/\cos^3 x) + (24\sin^3 x/\cos^5 x) \qquad f^{(4)}(0) = 0.$$

Thus,

$$P_3(x) = P_4(x) = x + \frac{x^3}{3}.$$

7. Let $f(x) = \sqrt[3]{1-x} = (1-x)^{1/3}$. Then $f(0) = 1$, and

$$f'(x) = -\tfrac{1}{3}(1-x)^{-2/3} \qquad f'(0) = -\tfrac{1}{3},$$
$$f''(x) = -\tfrac{2}{3^2}(1-x)^{-5/3} \qquad f''(0) = -\tfrac{2}{3^2},$$
$$f'''(x) = -\tfrac{10}{3^3}(1-x)^{-8/3} \qquad f'''(0) = -\tfrac{10}{3^3},$$
$$f^{(4)}(x) = -\tfrac{80}{3^4}(1-x)^{-11/3} \qquad f^{(4)}(0) = -\tfrac{80}{3^4}.$$

Then,

$$P_2(x) = 1 - \frac{1}{3}x - \frac{1}{2!}\frac{2}{3^2}x^2 = 1 - \frac{1}{3}x - \frac{1}{9}x^2,$$
$$P_3(x) = P_2(x) - \frac{1}{3!}\left(\frac{10}{3^3}\right)x^3 = 1 - \frac{1}{3}x - \frac{1}{9}x^2 - \frac{5}{81}x^3,$$
$$P_4(x) = P_3(x) - \frac{1}{4!}\frac{80}{3^4}x^4 = 1 - \frac{1}{3}x - \frac{1}{9}x^2 - \frac{5}{81}x^3 - \frac{10}{243}x^4.$$

8. Let $f(x) = \ln(1+x)$. Then $f(0) = \ln 1 = 0$, and

$$f'(x) = (1+x)^{-1} \qquad f'(0) = 1,$$
$$f''(x) = (-1)(1+x)^{-2} \qquad f''(0) = -1,$$
$$f'''(x) = 2(1+x)^{-3} \qquad f'''(0) = 2,$$
$$f^{(4)}(x) = -3!(1+x)^{-4} \qquad f^{(4)}(0) = -3!,$$
$$f^{(5)}(x) = 4!(1+x)^{-5} \qquad f^{(5)}(0) = 4!,$$
$$f^{(6)}(x) = -5!(1+x)^{-6} \qquad f^{(6)}(0) = -5!,$$
$$f^{(7)}(x) = 6!(1+x)^{-7} \qquad f^{(7)}(0) = 6!,$$
$$f^{(8)}(x) = -7!(1+x)^{-8} \qquad f^{(8)}(0) = -7!,$$
$$f^{(9)}(x) = 8!(1+x)^{-9} \qquad f^{(9)}(0) = 8!.$$

So,

$$P_5(x) = x - \frac{x^2}{2} + \frac{x^3}{3} - \frac{x^4}{4} + \frac{x^5}{5},$$
$$P_7(x) = x - \frac{x^2}{2} + \frac{x^3}{3} - \frac{x^4}{4} + \frac{x^5}{5} - \frac{x^6}{6} + \frac{x^7}{7},$$
$$P_9(x) = x - \frac{x^2}{2} + \frac{x^3}{3} - \frac{x^4}{4} + \frac{x^5}{5} - \frac{x^6}{6} + \frac{x^7}{7} - \frac{x^8}{8} + \frac{x^9}{9}.$$

9. Let $f(x) = \dfrac{1}{\sqrt{1+x}} = (1+x)^{-1/2}$. Then $f(0) = 1$.

$$f'(x) = -\tfrac{1}{2}(1+x)^{-3/2} \qquad f'(0) = -\tfrac{1}{2},$$
$$f''(x) = \tfrac{3}{2^2}(1+x)^{-5/2} \qquad f''(0) = \tfrac{3}{2^2},$$
$$f'''(x) = -\tfrac{3 \cdot 5}{2^3}(1+x)^{-7/2} \qquad f'''(0) = -\tfrac{3 \cdot 5}{2^3},$$
$$f^{(4)}(x) = \tfrac{3 \cdot 5 \cdot 7}{2^4}(1+x)^{-9/2} \qquad f^{(4)}(0) = \tfrac{3 \cdot 5 \cdot 7}{2^4}$$

Then,

$$P_2(x) = 1 - \frac{1}{2}x + \frac{1}{2!}\frac{3}{2^2}x^2 = 1 - \frac{1}{2}x + \frac{3}{8}x^2,$$
$$P_3(x) = P_2(x) - \frac{1}{3!}\frac{3 \cdot 5}{2^3}x^3 = 1 - \frac{1}{2}x + \frac{3}{8}x^2 - \frac{5}{16}x^3,$$
$$P_4(x) = P_3(x) + \frac{1}{4!}\frac{3 \cdot 5 \cdot 7}{2^4}x^4 = 1 - \frac{1}{2}x + \frac{3}{8}x^2 - \frac{5}{16}x^3 + \frac{35}{128}x^4.$$

10. Let $f(x) = (1+x)^p$.
    (a) Suppose that $p = 0$. Then $f(x) = 1$ and $f^{(k)}(x) = 0$ for any $k \geq 1$. Thus $P_2(x) = P_3(x) = P_4(x) = 1$.
    (b) If $p = 1$ then $f(x) = 1 + x$, so

    $$f(0) = 1,$$
    $$f'(x) = 1,$$
    $$f^{(k)}(x) = 0 \qquad k \geq 2.$$

    Thus $P_2(x) = P_3(x) = P_4(x) = 1 + x$.
    (c) In general:

    $$f(x) = (1+x)^p,$$
    $$f'(x) = p(1+x)^{p-1},$$
    $$f''(x) = p(p-1)(1+x)^{p-2},$$
    $$f'''(x) = p(p-1)(p-2)(1+x)^{p-3},$$
    $$f^{(4)}(x) = p(p-1)(p-2)(p-3)(1+x)^{p-4}.$$

$$f(0) = 1,$$
$$f'(0) = p,$$
$$f''(0) = p(p-1),$$
$$f'''(0) = p(p-1)(p-2),$$
$$f^{(4)}(0) = p(p-1)(p-2)(p-3).$$

$$P_2(x) = 1 + px + \frac{p(p-1)}{2}x^2,$$
$$P_3(x) = 1 + px + \frac{p(p-1)}{2}x^2 + \frac{p(p-1)(p-2)}{6}x^3,$$
$$P_4(x) = 1 + px + \frac{p(p-1)}{2}x^2 + \frac{p(p-1)(p-2)}{6}x^3$$
$$+ \frac{p(p-1)(p-2)(p-3)}{24}x^4.$$

11. Using the fact that

$$f(x) \approx P_6(x) = f(0) + f'(0)x + \frac{f''(0)}{2!}x^2 + \frac{f'''(0)}{3!}x^3 + \frac{f^{(4)}(0)}{4!}x^4 + \frac{f^{(5)}(0)}{5!}x^5 + \frac{f^{(6)}(0)}{6!}x^6$$

and identifying coefficients with those given for $P_6(x)$, we obtain the following:

(a) $f(0) =$ constant term which equals 0, so $f(0) = 0$.
(b) $f'(0) =$ coefficient of x which equals 3, so $f'(0) = 3$.
(c) $\frac{f'''(0)}{3!} =$ coefficient of x^3 which equals -4, so $f'''(0) = -24$.
(d) $\frac{f^{(5)}(0)}{5!} =$ coefficient of x^5 which equals 0, so $f^{(5)}(0) = 0$.
(e) $\frac{f^{(6)}(0)}{6!} =$ coefficient of x^6 which equals 5, so $f^{(6)}(0) = 5(6!) = 3600$.

12. (a) We have

$$g(x) = g(5) + g'(5)(x-5) + \frac{g''(5)}{2!}(x-5)^2 + \frac{g'''(5)}{3!}(x-5)^3 + \ldots$$

Substituting gives

$$g(x) = 3 - 2(x-5) + \frac{1}{2!}(x-5)^2 - \frac{3}{3!}(x-5)^3 + \ldots$$

The degree 2 Taylor polynomial, $P_2(x)$, is obtained by truncating after the $(x-5)^2$ term:

$$P_2(x) = 3 - 2(x-5) + \frac{1}{2}(x-5)^2.$$

The degree 3 Taylor polynomial, $P_3(x)$, is obtained by truncating after the $(x-5)^3$ term:

$$P_3(x) = 3 - 2(x-5) + \frac{1}{2}(x-5)^2 - \frac{1}{2}(x-5)^3.$$

(b) Substitute $x = 4.9$ into the Taylor polynomial of degree 2:

$$P_2(4.9) = 3 - 2(4.9-5) + \frac{1}{2}(4.9-5)^2 = 3.205.$$

From the Taylor polynomial of degree 3, we obtain

$$P_3(4.9) = 3 - 2(4.9-5) + \frac{1}{2}(4.9-5)^2 - \frac{1}{2}(4.9-5)^3 = 3.2055.$$

13. Let $f(x) = \sin x$. $f(\frac{\pi}{2}) = 1$.

$$f'(x) = \cos x \qquad f'(\tfrac{\pi}{2}) = 0,$$
$$f''(x) = -\sin x \qquad f''(\tfrac{\pi}{2}) = -1,$$
$$f'''(x) = -\cos x \qquad f'''(\tfrac{\pi}{2}) = 0,$$
$$f^{(4)}(x) = \sin x \qquad f^{(4)}(\tfrac{\pi}{2}) = 1.$$

So,
$$P_4(x) = 1 + 0 - \frac{1}{2!}\left(x - \frac{\pi}{2}\right)^2 + 0 + \frac{1}{4!}\left(x - \frac{\pi}{2}\right)^4$$
$$= 1 - \frac{1}{2!}\left(x - \frac{\pi}{2}\right)^2 + \frac{1}{4!}\left(x - \frac{\pi}{2}\right)^4.$$

14. Let $f(x) = \cos x$. Then $\cos\frac{\pi}{4} = \sin\frac{\pi}{4} = \frac{\sqrt{2}}{2}$.
Then $f'(x) = -\sin x$, $f''(x) = -\cos x$, and $f'''(x) = \sin x$, so the Taylor polynomial for $\cos x$ of degree three about $x = \pi/4$ is

$$P_3(x) = \cos\frac{\pi}{4} + \left(-\sin\frac{\pi}{4}\right)\left(x - \frac{\pi}{4}\right) + \frac{-\cos\frac{\pi}{4}}{2!}\left(x - \frac{\pi}{4}\right)^2 + \frac{\sin\frac{\pi}{4}}{3!}\left(x - \frac{\pi}{4}\right)^3$$
$$= \frac{\sqrt{2}}{2}\left(1 - \left(x - \frac{\pi}{4}\right) - \frac{1}{2}\left(x - \frac{\pi}{4}\right)^2 + \frac{1}{6}\left(x - \frac{\pi}{4}\right)^3\right).$$

15. Let $f(x) = e^x$. Since $f^{(k)}(x) = e^x = f(x)$ for all $k \geq 1$, the Taylor polynomial of degree 4 for $f(x) = e^x$ about $x = 1$ is

$$P_4(x) = e^1 + e^1(x - 1) + \frac{e^1}{2!}(x - 1)^2 + \frac{e^1}{3!}(x - 1)^3 + \frac{e^1}{4!}(x - 1)^4$$
$$= e\left[1 + (x - 1) + \frac{1}{2}(x - 1)^2 + \frac{1}{6}(x - 1)^3 + \frac{1}{24}(x - 1)^4\right].$$

16. Let $f(x) = \sqrt{1 + x} = (1 + x)^{1/2}$.
Then $f'(x) = \frac{1}{2}(1 + x)^{-1/2}$, $f''(x) = -\frac{1}{4}(1 + x)^{-3/2}$, and $f'''(x) = \frac{3}{8}(1 + x)^{-5/2}$. The Taylor polynomial of degree three about $x = 1$ is thus

$$P_3(x) = (1 + 1)^{1/2} + \frac{1}{2}(1 + 1)^{-1/2}(x - 1) + \frac{-\frac{1}{4}(1 + 1)^{-3/2}}{2!}(x - 1)^2$$
$$+ \frac{\frac{3}{8}(1 + 1)^{-5/2}}{3!}(x - 1)^3$$
$$= \sqrt{2}\left(1 + \frac{x - 1}{4} - \frac{(x - 1)^2}{32} + \frac{(x - 1)^3}{128}\right).$$

17. Since $P_2(x)$ is the second degree Taylor polynomial for $f(x)$ about $x = 0$, $P_2(0) = f(0)$, which says $a = f(0)$. Since

$$\left.\frac{d}{dx}P_2(x)\right|_{x=0} = f'(0),$$

$b = f'(0)$; and since

$$\left.\frac{d^2}{dx^2}P_2(x)\right|_{x=0} = f''(0),$$

$2c = f''(0)$. In other words, a is the y-intercept of $f(x)$, b is the slope of the tangent line to $f(x)$ at $x = 0$ and c tells us the concavity of $f(x)$ near $x = 0$. So $c < 0$ since f is concave down; $b > 0$ since f is increasing; $a > 0$ since $f(0) > 0$.

18. As we can see from Problem 17, a is the y-intercept of $f(x)$, b is the slope of the tangent line to $f(x)$ at $x = 0$ and c tells us the concavity of $f(x)$ near $x = 0$.
So $a > 0, b < 0$ and $c < 0$.

424 CHAPTER NINE /SOLUTIONS

19. As we can see from Problem 17, a is the y-intercept of $f(x)$, b is the slope of the tangent line to $f(x)$ at $x = 0$ and c tells us the concavity of $f(x)$ near $x = 0$.
So $a < 0, b > 0$ and $c > 0$.

20. As we can see from Problem 17, a is the y-intercept of $f(x)$, b is the slope of the tangent line to $f(x)$ at $x = 0$ and c tells us the concavity of $f(x)$ near $x = 0$.
So $a < 0, b < 0$ and $c > 0$.

21.
$$f(x) = \sin x \qquad f(\tfrac{\pi}{4}) = \tfrac{\sqrt{2}}{2},$$
$$f'(x) = \cos x \qquad f'(\tfrac{\pi}{4}) = \tfrac{\sqrt{2}}{2},$$
$$f''(x) = -\sin x \qquad f''(\tfrac{\pi}{4}) = -\tfrac{\sqrt{2}}{2},$$
$$f'''(x) = -\cos x \qquad f'''(\tfrac{\pi}{4}) = -\tfrac{\sqrt{2}}{2}.$$

$$\sin x = \frac{\sqrt{2}}{2} + \frac{\sqrt{2}}{2}\left(x - \frac{\pi}{4}\right) - \frac{\sqrt{2}}{2}\frac{(x - \frac{\pi}{4})^2}{2!} - \frac{\sqrt{2}}{2}\frac{(x - \frac{\pi}{4})^3}{3!} - \cdots$$
$$= \frac{\sqrt{2}}{2} + \frac{\sqrt{2}}{2}\left(x - \frac{\pi}{4}\right) - \frac{\sqrt{2}}{4}\left(x - \frac{\pi}{4}\right)^2 - \frac{\sqrt{2}}{12}\left(x - \frac{\pi}{4}\right)^3 - \cdots$$

22.
$$f(\theta) = \cos \theta \qquad f(\tfrac{\pi}{4}) = \tfrac{\sqrt{2}}{2},$$
$$f'(\theta) = -\sin \theta \qquad f'(\tfrac{\pi}{4}) = -\tfrac{\sqrt{2}}{2},$$
$$f''(\theta) = -\cos \theta \qquad f''(\tfrac{\pi}{4}) = -\tfrac{\sqrt{2}}{2},$$
$$f'''(\theta) = \sin \theta \qquad f'''(\tfrac{\pi}{4}) = \tfrac{\sqrt{2}}{2}.$$

$$\cos \theta = \frac{\sqrt{2}}{2} - \frac{\sqrt{2}}{2}\left(\theta - \frac{\pi}{4}\right) - \frac{\sqrt{2}}{2}\frac{(\theta - \frac{\pi}{4})^2}{2!} + \frac{\sqrt{2}}{2}\frac{(\theta - \frac{\pi}{4})^3}{3!} - \cdots$$
$$= \frac{\sqrt{2}}{2} - \frac{\sqrt{2}}{2}\left(\theta - \frac{\pi}{4}\right) - \frac{\sqrt{2}}{4}\left(\theta - \frac{\pi}{4}\right)^2 + \frac{\sqrt{2}}{12}\left(\theta - \frac{\pi}{4}\right)^3 - \cdots$$

23.
$$f(\theta) = \sin \theta \qquad f(-\tfrac{\pi}{4}) = -\tfrac{\sqrt{2}}{2},$$
$$f'(\theta) = \cos \theta \qquad f'(-\tfrac{\pi}{4}) = \tfrac{\sqrt{2}}{2},$$
$$f''(\theta) = -\sin \theta \qquad f''(-\tfrac{\pi}{4}) = \tfrac{\sqrt{2}}{2},$$
$$f'''(\theta) = -\cos \theta \qquad f'''(-\tfrac{\pi}{4}) = -\tfrac{\sqrt{2}}{2}.$$

$$\sin \theta = -\frac{\sqrt{2}}{2} + \frac{\sqrt{2}}{2}\left(\theta + \frac{\pi}{4}\right) + \frac{\sqrt{2}}{2}\frac{(\theta + \frac{\pi}{4})^2}{2!} - \frac{\sqrt{2}}{2}\frac{(\theta + \frac{\pi}{4})^3}{3!} + \cdots$$
$$= -\frac{\sqrt{2}}{2} + \frac{\sqrt{2}}{2}\left(\theta + \frac{\pi}{4}\right) + \frac{\sqrt{2}}{4}\left(\theta + \frac{\pi}{4}\right)^2 - \frac{\sqrt{2}}{12}\left(\theta + \frac{\pi}{4}\right)^3 + \cdots.$$

24.
$$f(x) = \tan x \qquad f(\tfrac{\pi}{4}) = 1,$$
$$f'(x) = \tfrac{1}{\cos^2 x} \qquad f'(\tfrac{\pi}{4}) = 2,$$
$$f''(x) = \tfrac{-2(-\sin x)}{\cos^3 x} = \tfrac{2\sin x}{\cos^3 x} \qquad f''(\tfrac{\pi}{4}) = 4,$$
$$f'''(x) = \tfrac{-6\sin x(-\sin x)}{\cos^4 x} + \tfrac{2}{\cos^2 x} \qquad f'''(\tfrac{\pi}{4}) = 16.$$

$$\tan x = 1 + 2\left(x - \frac{\pi}{4}\right) + 4\frac{(x - \frac{\pi}{4})^2}{2!} + 16\frac{(x - \frac{\pi}{4})^3}{3!} + \cdots$$
$$= 1 + 2\left(x - \frac{\pi}{4}\right) + 2\left(x - \frac{\pi}{4}\right)^2 + \frac{8}{3}\left(x - \frac{\pi}{4}\right)^3 + \cdots$$

25. Let C_n be the coefficient of the nth term in the series. Note that

$$0 = C_1 = \frac{d}{dx}(x^2 e^{x^2})\bigg|_{x=0},$$

and since

$$\frac{1}{2} = C_6 = \frac{\frac{d^6}{dx^6}(x^2 e^{x^2})\bigg|_{x=0}}{6!},$$

we have

$$\frac{d^6}{dx^6}(x^2 e^{x^2})\bigg|_{x=0} = \frac{6!}{2} = 360.$$

26. Let C_n be the coefficient of the nth term in the series. $C_1 = f'(0)/1!$, so $f'(0) = 1!C_1 = 1 \cdot 1 = 1$.
Similarly, $f''(0) = 2!C_2 = 2! \cdot \frac{1}{2} = 1$;
$f'''(0) = 3!C_3 = 3! \cdot \frac{1}{3} = 2! = 2$;
$f^{(10)}(0) = 10!C_{10} = 10! \cdot \frac{1}{10} = \frac{10!}{10} = 9! = 362880.$

27.
$$f(x) = 4x^2 - 7x + 2 \quad f(0) = 2$$
$$f'(x) = 8x - 7 \quad f'(0) = -7$$
$$f''(x) = 8 \quad f''(0) = 8,$$

so $P_2(x) = 2 + (-7)x + \frac{8}{2}x^2 = 4x^2 - 7x + 2$. We notice that $f(x) = P_2(x)$ in this case.

28. $f'(x) = 3x^2 + 14x - 5$, $f''(x) = 6x + 14$, $f'''(x) = 6$. Thus, about $a = 0$,

$$P_3(x) = 1 + \frac{-5}{1!}x + \frac{14}{2!}x^2 + \frac{6}{3!}x^3$$
$$= 1 - 5x + 7x^2 + x^3$$
$$= f(x).$$

29. (a) We'll make the following conjecture:
"If $f(x)$ is a polynomial of degree n, i.e.

$$f(x) = a_0 + a_1 x + a_2 x^2 + \cdots + a_{n-1} x^{n-1} + a_n x^n,$$

then $P_n(x)$, the nth degree Taylor polynomial for $f(x)$ about $x = 0$, is $f(x)$ itself."

(b) All we need to do is to calculate $P_n(x)$, the nth degree Taylor polynomial for f about $x = 0$ and see if it is the same as $f(x)$.

$$f(0) = a_0;$$
$$f'(0) = (a_1 + 2a_2 x + \cdots + na_n x^{n-1})\big|_{x=0}$$
$$= a_1;$$
$$f''(0) = (2a_2 + 3 \cdot 2a_3 x + \cdots + n(n-1)a_n x^{n-2})\big|_{x=0}$$
$$= 2!a_2.$$

If we continue doing this, we'll see in general

$$f^{(k)}(0) = k!a_k, \quad k = 1, 2, 3, \cdots, n.$$

Therefore,

$$P_n(x) = f(0) + \frac{f'(0)}{1!}x + \frac{f''(0)}{2!}x^2 + \cdots + \frac{f^{(n)}(0)}{n!}x^n$$
$$= a_0 + a_1 x + a_2 x^2 + \cdots + a_n x^n$$
$$= f(x).$$

30.
$$\lim_{x\to 0}\frac{\sin x}{x}=\lim_{x\to 0}\frac{x-\frac{x^3}{3!}}{x}=\lim_{x\to 0}\left(1-\frac{x^2}{3!}\right)=1.$$

31.
$$\lim_{x\to 0}\frac{1-\cos x}{x^2}=\lim_{x\to 0}\frac{1-\left(1-\frac{x^2}{2!}+\frac{x^4}{4!}\right)}{x^2}=\lim_{x\to 0}\left(\frac{1}{2}-\frac{x^2}{4!}\right)=\frac{1}{2}.$$

32. For $f(h)=e^h$, $P_4(h)=1+h+\frac{h^2}{2}+\frac{h^3}{3!}+\frac{h^4}{4!}$. So

(a)
$$\lim_{h\to 0}\frac{e^h-1-h}{h^2}=\lim_{h\to 0}\frac{P_4(h)-1-h}{h^2}$$
$$=\lim_{h\to 0}\frac{\frac{h^2}{2}+\frac{h^3}{3!}+\frac{h^4}{4!}}{h^2}$$
$$=\lim_{h\to 0}\left(\frac{1}{2}+\frac{h}{3!}+\frac{h^2}{4!}\right)$$
$$=\frac{1}{2}.$$

(b)
$$\lim_{h\to 0}\frac{e^h-1-h-\frac{h^2}{2}}{h^3}=\lim_{h\to 0}\frac{P_4(h)-1-h-\frac{h^2}{2}}{h^3}$$
$$=\lim_{h\to 0}\frac{\frac{h^3}{3!}+\frac{h^4}{4!}}{h^3}=\lim_{h\to 0}\left(\frac{1}{3!}+\frac{h}{4!}\right)$$
$$=\frac{1}{3!}=\frac{1}{6}.$$

Using Taylor polynomials of higher degree would not have changed the results since the terms with higher powers of h all go to zero as $h\to 0$.

33. Let $f(x)$ be a function that has derivatives up to order n at $x=a$. Let
$$P_n(x)=C_0+C_1(x-a)+\cdots+C_n(x-a)^n$$
be the polynomial of degree n that approximates $f(x)$ about $x=a$. We require that $P_n(x)$ and all of its first n derivatives agree with those of the function $f(x)$ at $x=a$, i.e., we want
$$f(a)=P_n(a),$$
$$f'(a)=P_n'(a),$$
$$f''(a)=P_n''(a),$$
$$\vdots$$
$$f^{(n)}(a)=P_n^{(n)}(a).$$

When we substitute $x=a$ in $P_n(x)$, all the terms except the first drop out, so
$$f(a)=C_0.$$
Now differentiate $P_n(x)$:
$$P_n'(x)=C_1+2C_2(x-a)+3C_3(x-a)^2+\cdots+nC_n(x-a)^{n-1}.$$
Substitute $x=a$ again, which yields
$$f'(a)=P_n'(a)=C_1.$$

Differentiate $P'_n(x)$:

$$P'''_n(x) = 2C_2 + 3 \cdot 2C_3(x-a) + \cdots + n(n-1)C_n(x-a)^{n-2}$$

and substitute $x = a$ again:

$$f''(a) = P''_n(a) = 2C_2.$$

Differentiating and substituting again gives

$$f'''(a) = P'''_n(a) = 3 \cdot 2C_3.$$

Similarly,

$$f^{(k)}(a) = P_n^{(k)}(a) = k!C_k.$$

So, $C_0 = f(a)$, $C_1 = f'(a)$, $C_2 = \frac{f''(a)}{2!}$, $C_3 = \frac{f'''(a)}{3!}$, and so on.

If we adopt the convention that $f^{(0)}(a) = f(a)$ and $0! = 1$, then

$$C_k = \frac{f^{(k)}(a)}{k!}, \quad k = 0, 1, 2, \cdots, n.$$

Therefore,

$$f(x) \approx P_n(x) = C_0 + C_1(x-a) + C_2(x-a)^2 \cdots + C_n(x-a)^n$$

$$= f(a) + f'(a)(x-a) + \frac{f''(a)}{2!}(x-a)^2 + \cdots + \frac{f^{(n)}(a)}{n!}(x-a)^n.$$

34. (a) $f(x) = e^{x^2}$.

$f'(x) = 2xe^{x^2}$, $f''(x) = 2(1 + 2x^2)e^{x^2}$, $f'''(x) = 4(3x + 2x^3)e^{x^2}$,

$f^{(4)}(x) = 4(3 + 6x^2)e^{x^2} + 4(3x + 2x^3)2xe^{x^2}$.

The Taylor polynomial about $x = 0$ is

$$P_4(x) = 1 + \frac{0}{1!}x + \frac{2}{2!}x^2 + \frac{0}{3!}x^3 + \frac{12}{4!}x^4$$

$$= 1 + x^2 + \frac{1}{2}x^4.$$

(b) $f(x) = e^x$. The Taylor polynomial of degree 2 is

$$Q_2(x) = 1 + \frac{x}{1!} + \frac{x^2}{2!} = 1 + x + \frac{1}{2}x^2.$$

If we substitute x^2 for x in the Taylor polynomial for e^x of degree 2, we will get $P_4(x)$, the Taylor polynomial for e^{x^2} of degree 4:

$$Q_2(x^2) = 1 + x^2 + \frac{1}{2}(x^2)^2$$

$$= 1 + x^2 + \frac{1}{2}x^4$$

$$= P_4(x).$$

(c) Let $Q_{10}(x) = 1 + \frac{x}{1!} + \frac{x^2}{2!} + \cdots + \frac{x^{10}}{10!}$ be the Taylor polynomial of degree 10 for e^x about $x = 0$. Then

$$P_{20}(x) = Q_{10}(x^2)$$

$$= 1 + \frac{x^2}{1!} + \frac{(x^2)^2}{2!} + \cdots + \frac{(x^2)^{10}}{10!}$$

$$= 1 + \frac{x^2}{1!} + \frac{x^4}{2!} + \cdots + \frac{x^{20}}{10!}.$$

(d) Let $e^x \approx Q_5(x) = 1 + \frac{x}{1!} + \cdots + \frac{x^5}{5!}$. Then

$$e^{-2x} \approx Q_5(-2x)$$

$$= 1 + \frac{-2x}{1!} + \frac{(-2x)^2}{2!} + \frac{(-2x)^3}{3!} + \frac{(-2x)^4}{4!} + \frac{(-2x)^5}{5!}$$

$$= 1 - 2x + 2x^2 - \frac{4}{3}x^3 + \frac{2}{3}x^4 - \frac{4}{15}x^5.$$

35. (a)
$$f(x) = \sin x^2$$
$$f'(x) = (\cos x^2)2x$$
$$f''(x) = (-\sin x^2)4x^2 + (\cos x^2)2$$
$$f'''(x) = (-\cos x^2)8x^3 + (-\sin x^2)8x + (-\sin x^2)4x$$
$$= (-\cos x^2)8x^3 + (-\sin x^2)12x$$
$$f^{(4)}(x) = (\sin x^2)16x^4 + (-\cos x^2)24x^2 + (-\cos x^2)24x^2 + (-\sin x^2)12$$
$$= (\sin x^2)16x^4 + (-\cos x^2)48x^2 + (-\sin x^2)12$$
$$f^{(5)}(x) = (\cos x^2)32x^5 + (\sin x^2)64x^3 + (\sin x^2)96x^3$$
$$+(-\cos x^2)96x + (-\cos x^2)24x$$
$$= (\cos x^2)32x^5 + (\sin x^2)160x^3 + (-\cos x^2)120x$$
$$f^{(6)}(x) = (-\sin x^2)64x^6 + (\cos x^2)160x^4 + (\cos x^2)320x^4 + (\sin x^2)480x^2$$
$$+(\sin x^2)240x^2 + (-\cos x^2)120$$
$$= (-\sin x^2)64x^6 + (\cos x^2)480x^4 + (\sin x^2)720x^2 + (-\cos x^2)120$$

So,
$$f(0) = 0 \qquad f^{(4)}(0) = 0,$$
$$f'(0) = 0 \qquad f^{(5)}(0) = 0,$$
$$f''(0) = 2 \qquad f^{(6)}(0) = -120,$$
$$f'''(0) = 0.$$

Thus
$$f(x) = \sin x^2 = \frac{2}{2!}x^2 - \frac{120}{6!}x^6 + \cdots$$
$$= x^2 - \frac{1}{3!}x^6 + \cdots.$$

As we can see, the amount of calculation in order to find the higher derivatives of $\sin x^2$ increases very rapidly. In fact, the next non-zero term in the Taylor expansion of $\sin x^2$ is the 10th derivative term, which really requires a lot of work to get.

(b)
$$\sin x = x - \frac{1}{3!}x^3 + \frac{1}{5!}x^5 - \cdots$$

The first couple of coefficients of the above expansion are the same as those in part (a). If we substitute x^2 for x in the Taylor expansion of $\sin x$, we should get the Taylor expansion of $\sin x^2$.

$$\sin x^2 = x^2 - \frac{1}{3!}(x^2)^3 + \frac{1}{5!}(x^2)^5 - \cdots$$
$$= x^2 - \frac{1}{3!}x^6 + \frac{1}{5!}x^{10} - \cdots.$$

36. (a) $\dfrac{\sin t}{t} \approx \dfrac{t - \frac{t^3}{3!}}{t} = 1 - \dfrac{t^2}{6}$

$$\int_0^1 \frac{\sin t}{t}\, dt \approx \int_0^1 \left(1 - \frac{t^2}{6}\right) dt = t - \frac{t^3}{18}\Big|_0^1 = 0.94444\cdots$$

(b) $\dfrac{\sin t}{t} \approx \dfrac{t - \frac{t^3}{3!} + \frac{t^5}{5!}}{t} = 1 - \dfrac{t^2}{6} + \dfrac{t^4}{120}$

$$\int_0^1 \frac{\sin t}{t}\, dt \approx \int_0^1 \left(1 - \frac{t^2}{6} + \frac{t^4}{120}\right) dt = t - \frac{t^3}{18} + \frac{t^5}{600}\Big|_0^1 = 0.94611\cdots$$

Solutions for Section 9.2

1. Yes.
2. No, because it contains negative powers of x.
3. No, each term is a power of a different quantity.
4. Yes. It's a Taylor polynomial or a series with all coefficients beyond the 7th being zero.
5.
$$f(x) = \tfrac{1}{1-x} = (1-x)^{-1} \qquad f(0) = 1,$$
$$f'(x) = -(1-x)^{-2}(-1) = (1-x)^{-2} \qquad f'(0) = 1,$$
$$f''(x) = -2(1-x)^{-3}(-1) = 2(1-x)^{-3} \qquad f''(0) = 2,$$
$$f'''(x) = -6(1-x)^{-4}(-1) = 6(1-x)^{-4} \qquad f'''(0) = 6.$$

$$f(x) = \frac{1}{1-x} = 1 + 1 \cdot x + \frac{2x^2}{2!} + \frac{6x^3}{3!} + \cdots$$
$$= 1 + x + x^2 + x^3 + \cdots$$

6.
$$f(x) = \sqrt{1+x} = (1+x)^{\frac{1}{2}} \qquad f(0) = 1$$
$$f'(x) = \tfrac{1}{2}(1+x)^{-\frac{1}{2}} \qquad f'(0) = \tfrac{1}{2}$$
$$f''(x) = -\tfrac{1}{4}(1+x)^{-\frac{3}{2}} \qquad f''(0) = -\tfrac{1}{4}$$
$$f'''(x) = \tfrac{3}{8}(1+x)^{-\frac{5}{2}} \qquad f'''(0) = \tfrac{3}{8}$$

$$f(x) = \sqrt{1+x} = 1 + \frac{1}{2}x + \frac{(-\tfrac{1}{4})x^2}{2!} + \frac{(\tfrac{3}{8})x^3}{3!} + \cdots$$
$$= 1 + \frac{x}{2} - \frac{x^2}{8} + \frac{x^3}{16} + \cdots$$

7.
$$f(x) = \tfrac{1}{\sqrt{1+x}} = (1+x)^{-\frac{1}{2}} \qquad f(0) = 1$$
$$f'(x) = -\tfrac{1}{2}(1+x)^{-\frac{3}{2}} \qquad f'(0) = -\tfrac{1}{2}$$
$$f''(x) = \tfrac{3}{4}(1+x)^{-\frac{5}{2}} \qquad f''(0) = \tfrac{3}{4}$$
$$f'''(x) = -\tfrac{15}{8}(1+x)^{-\frac{7}{2}} \qquad f'''(0) = -\tfrac{15}{8}$$

$$f(x) = \frac{1}{\sqrt{1+x}} = 1 + \left(-\frac{1}{2}\right)x + \frac{(\tfrac{3}{4})x^2}{2!} + \frac{(-\tfrac{15}{8})x^3}{3!} + \cdots$$
$$= 1 - \frac{x}{2} + \frac{3x^2}{8} - \frac{5x^3}{16} + \cdots$$

8.
$$f(y) = \sqrt[3]{1-y} = (1-y)^{\frac{1}{3}} \qquad f(0) = 1$$
$$f'(y) = \tfrac{1}{3}(1-y)^{-\frac{2}{3}}(-1) = -\tfrac{1}{3}(1-y)^{-\frac{2}{3}} \qquad f'(0) = -\tfrac{1}{3}$$
$$f''(y) = \tfrac{2}{9}(1-y)^{-\frac{5}{3}}(-1) = -\tfrac{2}{9}(1-y)^{-\frac{5}{3}} \qquad f''(0) = \tfrac{2}{9}$$
$$f'''(y) = \tfrac{10}{27}(1-y)^{-\frac{8}{3}}(-1) = -\tfrac{10}{27}(1-y)^{-\frac{8}{3}} \qquad f'''(0) = -\tfrac{10}{27}$$

$$f(y) = \sqrt[3]{1-y} = 1 + \left(-\frac{1}{3}\right)y + \frac{(-\tfrac{2}{9})y^2}{2!} + \frac{(-\tfrac{10}{27})y^3}{3!} + \cdots$$
$$= 1 - \frac{y}{3} - \frac{y^2}{9} - \frac{5y^3}{81} - \cdots$$

9.
$$f(x) = \tfrac{1}{x} \qquad f(1) = 1$$
$$f'(x) = -\tfrac{1}{x^2} \qquad f'(1) = -1$$
$$f''(x) = \tfrac{2}{x^3} \qquad f''(1) = 2$$
$$f'''(x) = -\tfrac{6}{x^4} \qquad f'''(1) = -6$$

$$\frac{1}{x} = 1 - (x-1) + \frac{2(x-1)^2}{2!} - \frac{6(x-1)^3}{3!} + \cdots$$
$$= 1 - (x-1) + (x-1)^2 - (x-1)^3 + \cdots.$$

10. Again using the derivatives found in Problem 9, we have

$$f(2) = \tfrac{1}{2}, \qquad f'(2) = -\tfrac{1}{4}, \qquad f''(2) = \tfrac{1}{4}, \qquad f'''(2) = -\tfrac{3}{8}.$$

$$\frac{1}{x} = \frac{1}{2} - \frac{x-2}{4} + \frac{(x-2)^2}{4 \cdot 2!} - \frac{3(x-2)^3}{8 \cdot 3!} + \cdots$$
$$= \frac{1}{2} - \frac{(x-2)}{4} + \frac{(x-2)^2}{8} - \frac{(x-2)^3}{16} + \cdots.$$

11. Using the derivatives from Problem 9, we have

$$f(-1) = -1, \quad f'(-1) = -1, \quad f''(-1) = -2, \quad f'''(-1) = -6.$$

Hence,

$$\frac{1}{x} = -1 - (x+1) - \frac{2(x+1)^2}{2!} - \frac{6(x+1)^3}{3!} - \cdots$$
$$= -1 - (x+1) - (x+1)^2 - (x+1)^3 - \cdots.$$

12. (a)
$$f(x) = \ln(1+2x) \qquad f(0) = 0$$
$$f'(x) = \tfrac{2}{1+2x} \qquad f'(0) = 2$$
$$f''(x) = -\tfrac{4}{(1+2x)^2} \qquad f''(0) = -4$$
$$f'''(x) = \tfrac{16}{(1+2x)^3} \qquad f'''(0) = 16$$

$$\ln(1+2x) = 2x - 2x^2 + \tfrac{8}{3}x^3 + \cdots$$

(b) To get the expression for $\ln(1+2x)$ from the series for $\ln(1+x)$, substitute $2x$ for x in the series

$$\ln(1+x) = x - \frac{x^2}{2} + \frac{x^3}{3} - \frac{x^4}{4} + \cdots$$

to get

$$\ln(1+2x) = 2x - \frac{(2x)^2}{2} + \frac{(2x)^3}{3} - \frac{(2x)^4}{4} + \cdots$$
$$= 2x - 2x^2 + \frac{8x^3}{3} - 4x^4 + \cdots$$

(c) Since the interval of convergence for $\ln(1+x)$ is $-1 < x < 1$, substituting $2x$ for x suggests the interval of convergence of $\ln(1+2x)$ is $-1 < 2x < 1$, or $-\tfrac{1}{2} < x < \tfrac{1}{2}$.

13.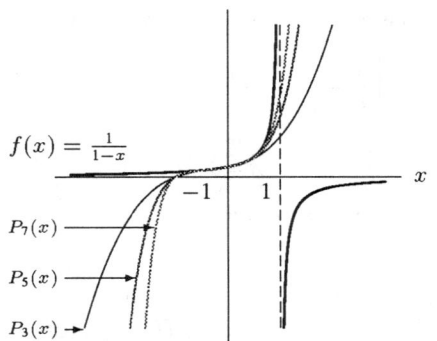

The graph suggests that the Taylor polynomials converge to $f(x) = \dfrac{1}{1-x}$ on the interval $-1 < x < 1$.

14. By looking at the graph we can see that the Taylor polynomials are reasonable approximations for the function $f(x) = \sqrt{1+x}$ between $x = -1$ and $x = 1$. Thus a good guess is that the interval of convergence is $-1 < x < 1$. ($P_n(x)$ represents the nth Taylor approximation of $f(x) = \sqrt{1+x}$.)

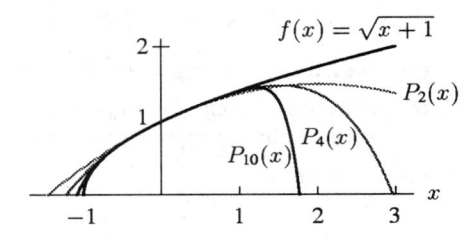

15. By looking at the graph we can see that the Taylor polynomials are reasonable approximations for the function $f(x) = \dfrac{1}{\sqrt{1+x}}$ between $x = -1$ and $x = 1$. Thus a good guess is that the interval of convergence is $-1 < x < 1$. ($P_n(x)$ represents the nth Taylor approximation of $f(x) = \dfrac{1}{\sqrt{1+x}}$.)

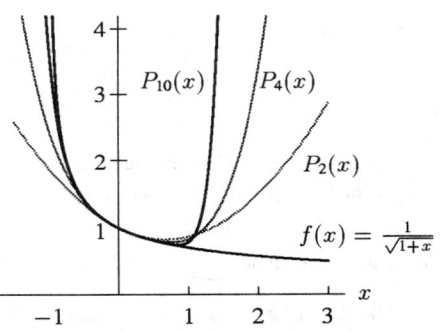

16. To find R, we consider the following limit, where the coefficient of the nth term is given by $a_n = n^2$.

$$\lim_{n \to \infty} \frac{|a_n|}{|a_{n+1}|} = \lim_{n \to \infty} \frac{n^2}{(n+1)^2}$$
$$= \lim_{n \to \infty} \frac{n^2}{n^2 + 2n + 1}$$
$$= \lim_{n \to \infty} \left(\frac{1}{1 + \left(\frac{2}{n}\right) + \left(\frac{1}{n^2}\right)} \right) = 1.$$

Thus, the radius of convergence is $R = 1$.

17. The coefficient of the nth term is $a_n = (-1)^{n+1}/n^2$. Now consider the ratio

$$\left| \frac{a_n}{a_{n+1}} \right| = \frac{(n+1)^2}{n^2} \to 1 = R \quad \text{as} \quad n \to \infty.$$

Thus, the radius of convergence is $R = 1$.

18. Here the coefficient of the nth term a_n is $(2^n/n!)$. Now we have

$$\left| \frac{a_n}{a_{n+1}} \right| = \frac{2^n/n!}{2^{n+1}/(n+1)!} = \frac{(n+1)}{2} \to \infty \text{ as } n \to \infty.$$

Thus, the radius of convergence is $R = \infty$, and the series converges for all x.

19. Here the coefficient of the nth term is $a_n = n/(2n+1)$. Now we have

$$\left|\frac{a_n}{a_{n+1}}\right| = \frac{n/(2n+1)}{(n+1)/(2n+3)}$$
$$= \frac{n(2n+3)}{(n+1)(2n+1)} \to 1 \text{ as } n \to \infty.$$

Thus, by the ratio test, the radius of convergence is $R = 1$.

20. Here $a_n = (2n)!/(n!)^2$. We have:

$$\left|\frac{a_n}{a_{n+1}}\right| = \frac{(2n)!/(n!)^2}{(2(n+1))!/((n+1)!)^2}$$
$$= \frac{(2n)!}{(2(n+1))!} \cdot \frac{(n+1)!(n+1)!}{(n!)(n!)}$$
$$= \frac{(n+1)^2}{(2n+2)(2n+1)} \to \frac{1}{4} \text{ as } n \to \infty.$$

Thus, the radius of convergence is $R = 1/4$.

21. (a) Show that the sum of each group of fractions is more than $1/2$.
    (b) Explain why this shows that the harmonic series does not converge.

    (a) Notice that

$$\frac{1}{3} + \frac{1}{4} > \frac{1}{4} + \frac{1}{4} = \frac{2}{4} = \frac{1}{2}$$
$$\frac{1}{5} + \frac{1}{6} + \frac{1}{7} + \frac{1}{8} > \frac{1}{8} + \frac{1}{8} + \frac{1}{8} + \frac{1}{8} = \frac{4}{8} = \frac{1}{2}$$
$$\frac{1}{9} + \frac{1}{10} + \cdots + \frac{1}{16} > \frac{1}{16} + \frac{1}{16} + \cdots + \frac{1}{16} = \frac{8}{16} = \frac{1}{2}.$$

In the same way, we can see that the sum of the fractions in each grouping is greater than $1/2$.

(b) Since the sum of the first n groups is greater than $n/2$, it follows that the harmonic series does not converge.

22. We want to estimate $\sum_{k=1}^{100{,}000} \frac{1}{k}$ using left and right Riemann sum approximations to $f(x) = 1/x$ on the interval $1 \leq x \leq 100{,}000$. Figure 9.1 shows a left Riemann sum approximation with 99,999 terms. Since $f(x)$ is decreasing, the left Riemann sum overestimates the area under the curve. Figure 9.1 shows that the first term in the sum is $f(1) \cdot 1$ and the last is $f(99{,}999) \cdot 1$, so we have

$$\int_1^{100{,}000} \frac{1}{x}\,dx < \text{LHS} = f(1) \cdot 1 + f(2) \cdot 1 + \cdots + f(99{,}999) \cdot 1.$$

Since $f(x) = 1/x$, the left Riemann sum is

$$\text{LHS} = \frac{1}{1} \cdot 1 + \frac{1}{2} \cdot 1 + \cdots + \frac{1}{99{,}999} \cdot 1 = \sum_{k=1}^{99{,}999} \frac{1}{k},$$

so

$$\int_1^{100{,}000} \frac{1}{x}\,dx < \sum_{k=1}^{99{,}999} \frac{1}{k}.$$

Since we want the sum to go $k = 100{,}000$ rather than $k = 99{,}999$, we add $1/100{,}000$ to both sides:

$$\int_1^{100{,}000} \frac{1}{x}\,dx + \frac{1}{100{,}000} < \sum_{k=1}^{99{,}999} \frac{1}{k} + \frac{1}{100{,}000} = \sum_{k=1}^{100{,}000} \frac{1}{k}.$$

The left Riemann sum has therefore given us an underestimate for our sum. We now use the right Riemann sum in Figure 9.2 to get an overestimate for our sum.

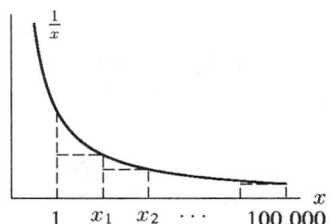

Figure 9.1   **Figure 9.2**

The right Riemann sum again has 99,999 terms, but this time the sum underestimates the area under the curve. Figure 9.2 shows that the first rectangle has area $f(2) \cdot 1$ and the last $f(100,000) \cdot 1$, so we have

$$\text{RHS} = f(2) \cdot 1 + f(3) \cdot 1 + \cdots + f(100,000) \cdot 1 < \int_1^{100,000} \frac{1}{x}\,dx.$$

Since $f(x) = 1/x$, the right Riemann sum is

$$\text{RHS} = \frac{1}{2} \cdot 1 + \frac{1}{3} \cdot 1 + \cdots + \frac{1}{100,000} \cdot 1 = \sum_{k=2}^{100,000} \frac{1}{k}.$$

So

$$\sum_{k=2}^{100,000} \frac{1}{k} < \int_1^{100,000} \frac{1}{x}\,dx.$$

Since we want the sum to start at $k = 1$, we add 1 to both sides:

$$\sum_{k=1}^{100,000} \frac{1}{k} = \frac{1}{1} + \sum_{k=2}^{100,000} \frac{1}{k} < 1 + \int_1^{100,000} \frac{1}{x}\,dx.$$

Putting these under- and overestimates together, we have

$$\int_1^{100,000} \frac{1}{x}\,dx + \frac{1}{100,000} < \sum_{k=1}^{100,000} \frac{1}{k} < 1 + \int_1^{100,000} \frac{1}{x}\,dx.$$

Since $\int_1^{100,000} \frac{1}{x}\,dx = \ln 100,000 - \ln 1 = 11.513$, we have

$$11.513 < \sum_{k=1}^{100,000} \frac{1}{k} < 12.513.$$

Therefore we have $\sum_{k=1}^{100,000} \frac{1}{k} \approx 12$.

23. Using a right-hand sum, we have

$$\frac{1}{2} + \frac{1}{3} + \frac{1}{4} + \cdots + \frac{1}{n} < \int_1^n \frac{dx}{x} = \ln n.$$

If a computer could add a million terms in one second, then it could add

$$60\,\frac{\text{sec}}{\text{min}} \cdot 60\,\frac{\text{min}}{\text{hour}} \cdot 24\,\frac{\text{hour}}{\text{day}} \cdot 365\,\frac{\text{days}}{\text{year}} \cdot 1\text{ million }\frac{\text{terms}}{\text{sec}}$$

terms per year. Thus,

$$1 + \frac{1}{2} + \frac{1}{3} + \cdots + \frac{1}{n} < 1 + \ln n = 1 + \ln(60 \cdot 60 \cdot 24 \cdot 365 \cdot 10^6) \approx 32.082 < 33.$$

So the sum after one year is about 32.

434   CHAPTER NINE /SOLUTIONS

24. The partial sums look like: $S_1 = 1$, $S_2 = 0.9$, $S_3 = 0.91$, $S_4 = 0.909$, $S_5 = 0.9091$, $S_6 = 0.90909$. The series appears to be converging to $0.909090\ldots$ or $10/11$.

25. The partial sums are $S_1 = 1$, $S_2 = -1$, $S_3 = 2$, $S_{10} = -5$, $S_{11} = 6$, $S_{100} = -50$, $S_{101} = 51$, $S_{1000} = -500$, $S_{1001} = 501$, which appear to be oscillating further and further from 0. This series does not converge.

26. The partial sums look like: $S_1 = 1$, $S_2 = 0$, $S_3 = 0.5$, $S_4 = 0.3333$, $S_5 = 0.375$, $S_{10} = 0.3679$, $S_{20} = 0.3679$, and higher partial sums agree with these first 6 decimal places. The series appears to be converging to about 0.3679.

27. This is the series for e^x with x replaced by 2, so the series converges to e^2.

28. This is the series for $\sin x$ with x replaced by 1, so the series converges to $\sin 1$.

29. This is the series for $1/(1-x)$ with x replaced by $1/4$, so the series converges to $1/(1-(1/4)) = 4/3$.

30. This is the series for $\cos x$ with x replaced by 10, so the series converges to $\cos 10$.

31. This is the series for $\ln(1+x)$ with x replaced by $1/2$, so the series converges to $\ln(3/2)$.

Solutions for Section 9.3

1. We'll use
$$\sqrt{1+y} = (1+y)^{\frac{1}{2}} = 1 + \left(\frac{1}{2}\right)y + \left(\frac{1}{2}\right)\left(\frac{-1}{2}\right)\frac{y^2}{2!}$$
$$+ \left(\frac{1}{2}\right)\left(\frac{-1}{2}\right)\left(\frac{-3}{2}\right)\frac{y^3}{3!} + \cdots$$
$$= 1 + \frac{y}{2} - \frac{y^2}{8} + \frac{y^3}{16} - \cdots.$$

Substitute $y = -2x$.
$$\sqrt{1-2x} = 1 + \frac{(-2x)}{2} - \frac{(-2x)^2}{8} + \frac{(-2x)^3}{16} - \cdots$$
$$= 1 - x - \frac{x^2}{2} - \frac{x^3}{2} - \cdots$$

2. Substitute $x = \theta^2$ into series for $\cos x$:
$$\cos(\theta^2) = 1 - \frac{(\theta^2)^2}{2!} + \frac{(\theta^2)^4}{4!} - \frac{(\theta^2)^6}{6!} + \cdots$$
$$= 1 - \frac{\theta^4}{2!} + \frac{\theta^8}{4!} - \frac{\theta^{12}}{6!} + \cdots.$$

3. Substitute $y = -x$ into $e^y = 1 + y + \frac{y^2}{2!} + \frac{y^3}{3!} + \cdots$. We get
$$e^{-x} = 1 + (-x) + \frac{(-x)^2}{2!} + \frac{(-x)^3}{3!} + \cdots$$
$$= 1 - x + \frac{x^2}{2!} - \frac{x^3}{3!} + \cdots.$$

4.
$$\frac{t}{1+t} = t(1+t)^{-1} = t\left(1 + (-1)t + \frac{(-1)(-2)}{2!}t^2 + \frac{(-1)(-2)(-3)}{3!}t^3 + \cdots\right)$$
$$= t - t^2 + t^3 - t^4 + \cdots$$

5. Substituting $x = -2y$ into $\ln(1+x) = x - \frac{x^2}{2} + \frac{x^3}{3} - \frac{x^4}{4} + \cdots$ gives

$$\ln(1-2y) = (-2y) - \frac{(-2y)^2}{2} + \frac{(-2y)^3}{3} - \frac{(-2y)^4}{4} + \cdots$$
$$= -2y - 2y^2 - \frac{8}{3}y^3 - 4y^4 - \cdots.$$

6. Since $\frac{d}{dx}(\arcsin x) = \frac{1}{\sqrt{1-x^2}} = 1 + \frac{1}{2}x^2 + \frac{3}{8}x^4 + \frac{5}{16}x^6 + \cdots$, integrating gives

$$\arcsin x = c + x + \frac{1}{6}x^3 + \frac{3}{40}x^5 + \frac{5}{112}x^7 + \cdots.$$

Since $\arcsin 0 = 0$, $c = 0$.

7. Substituting $x = -z^2$ into $\frac{1}{\sqrt{1+x}} = (1+x)^{-\frac{1}{2}} = 1 - \frac{1}{2}x + \frac{3}{8}x^2 - \frac{5}{16}x^3 + \cdots$ gives

$$\frac{1}{\sqrt{1-z^2}} = 1 - \frac{(-z^2)}{2} + \frac{3(-z^2)^2}{8} - \frac{5(-z^2)^3}{16} + \cdots$$
$$= 1 + \frac{1}{2}z^2 + \frac{3}{8}z^4 + \frac{5}{16}z^6 + \cdots.$$

8.
$$\phi^3 \cos(\phi^2) = \phi^3 \left(1 - \frac{(\phi^2)^2}{2!} + \frac{(\phi^2)^4}{4!} - \frac{(\phi^2)^6}{6!} + \cdots\right)$$
$$= \phi^3 - \frac{\phi^7}{2!} + \frac{\phi^{11}}{4!} - \frac{\phi^{15}}{6!} + \cdots$$

9.
$$\frac{z}{e^{z^2}} = ze^{-z^2} = z\left(1 + (-z^2) + \frac{(-z^2)^2}{2!} + \frac{(-z^2)^3}{3!} + \cdots\right)$$
$$= z - z^3 + \frac{z^5}{2!} - \frac{z^7}{3!} + \cdots$$

10.
$$\sqrt{(1+t)}\sin t = \left(1 + \frac{t}{2} - \frac{t^2}{8} + \frac{t^3}{16} - \cdots\right)\left(t - \frac{t^3}{3!} + \frac{t^5}{5!} - \cdots\right)$$

Multiplying and collecting terms yields

$$\sqrt{(1+t)}\sin t = t + \frac{t^2}{2} - \left(\frac{t^3}{3!} + \frac{t^3}{8}\right) + \left(\frac{t^4}{16} - \frac{t^4}{12}\right) + \cdots$$
$$= t + \frac{1}{2}t^2 - \frac{7}{24}t^3 - \frac{1}{48}t^4 + \cdots.$$

11.
$$e^t \cos t = \left(1 + t + \frac{t^2}{2!} + \frac{t^3}{3!} + \frac{t^4}{4!} + \cdots\right)\left(1 - \frac{t^2}{2!} + \frac{t^4}{4!} - \frac{t^6}{6!} + \cdots\right)$$

Multiplying out and collecting terms gives

$$e^t \cos t = 1 + t + \left(\frac{t^2}{2!} - \frac{t^2}{2!}\right) + \left(\frac{t^3}{3!} - \frac{t^3}{2!}\right) + \left(\frac{t^4}{4!} + \frac{t^4}{4!} - \frac{t^4}{(2!)^2}\right) + \cdots$$
$$= 1 + t - \frac{t^3}{3} - \frac{t^4}{6} + \cdots.$$

12. Substituting the series for $\sin\theta = \theta - \frac{\theta^3}{3!} + \frac{\theta^5}{5!} - \cdots$ into

$$\sqrt{1+y} = 1 + \frac{1}{2}y - \frac{1}{8}y^2 + \frac{1}{16}y^3 - \cdots$$

gives

$$\sqrt{1+\sin\theta} = 1 + \left(\theta - \frac{\theta^3}{3!} + \frac{\theta^5}{5!} - \cdots\right) - \frac{1}{8}\left(\theta - \frac{\theta^3}{3!} + \frac{\theta^5}{5!} - \cdots\right)^2$$
$$+ \frac{1}{16}\left(\theta - \frac{\theta^3}{3!} + \frac{\theta^5}{5!} - \cdots\right)^3 - \cdots$$
$$= 1 + \theta - \frac{\theta^2}{8} + \left(\frac{\theta^3}{16} - \frac{\theta^3}{3!}\right) + \cdots$$
$$= 1 + \theta - \frac{1}{8}\theta^2 - \frac{5}{48}\theta^3 + \cdots$$

13. The Taylor expansion about $\theta = 0$ for $\sin\theta$ is

$$\theta - \frac{\theta^3}{3!} + \frac{\theta^5}{5!} - \frac{\theta^7}{7!} + \cdots.$$

So

$$1 + \sin\theta = 1 + \theta - \frac{\theta^3}{3!} + \frac{\theta^5}{5!} - \frac{\theta^7}{7!} + \cdots.$$

The Taylor expansion about $\theta = 0$ for $\cos\theta$ is

$$\cos\theta = 1 - \frac{\theta^2}{2!} + \frac{\theta^4}{4!} - \frac{\theta^6}{6!} + \cdots.$$

The Taylor expansion for $\frac{1}{1+\theta}$ about $\theta = 0$ is

$$\frac{1}{1+\theta} = 1 - \theta + \theta^2 - \theta^3 + \theta^4 - \cdots.$$

So, substituting $-\theta^2$ for θ:

$$\frac{1}{1-\theta^2} = 1 - (-\theta^2) + (-\theta^2)^2 - (-\theta^2)^3 + (-\theta^2)^4 + \cdots$$
$$= 1 + \theta^2 + \theta^4 + \theta^6 + \theta^8 + \cdots.$$

For small θ, we can neglect the terms above quadratic in these expansions, giving:

$$1 + \sin\theta \approx 1 + \theta$$
$$\cos\theta \approx 1 - \frac{\theta^2}{2}$$
$$\frac{1}{1-\theta^2} \approx 1 + \theta^2.$$

For all $\theta \neq 0$, we have

$$1 - \frac{\theta^2}{2} < 1 + \theta^2.$$

Also, since $\theta^2 < \theta$ for $0 < \theta < 1$, we have

$$1 - \frac{\theta^2}{2} < 1 + \theta^2 < 1 + \theta.$$

So, for small positive θ, we have

$$\cos\theta < \frac{1}{1-\theta^2} < 1 + \sin\theta.$$

14. From the series for $\ln(1+y)$,
$$\ln(1+y) = y - \frac{y^2}{2} + \frac{y^3}{3} - \frac{y^4}{4} + \cdots,$$
we get
$$\ln(1+y^2) = y^2 - \frac{y^4}{2} + \frac{y^6}{3} - \frac{y^8}{4} + \cdots$$

The Taylor series for $\sin y$ is
$$\sin y = y - \frac{y^3}{3!} + \frac{y^5}{5!} - \frac{y^7}{7!} + \cdots$$

So
$$\sin y^2 = y^2 - \frac{y^6}{3!} + \frac{y^{10}}{5!} - \frac{y^{14}}{7!} + \cdots$$

The Taylor series for $\cos y$ is
$$\cos y = 1 - \frac{y^2}{2!} + \frac{y^4}{4!} - \frac{y^6}{6!} + \cdots$$

So
$$1 - \cos y = \frac{y^2}{2!} - \frac{y^4}{4!} + \frac{y^6}{6!} + \cdots$$

Near $y = 0$, we can drop terms beyond the fourth degree in each expression:
$$\ln(1+y^2) \approx y^2 - \frac{y^4}{2}$$
$$\sin y^2 \approx y^2$$
$$1 - \cos y \approx \frac{y^2}{2!} - \frac{y^4}{4!}.$$

(Note: These functions are all even, so what holds for negative y will hold for positive y.)
Clearly $1 - \cos y$ is smallest, because the y^2 term has a factor of $\frac{1}{2}$. Thus, for small y,
$$\frac{y^2}{2!} - \frac{y^4}{4!} < y^2 - \frac{y^4}{2} < y^2$$
so
$$1 - \cos y < \ln(1+y^2) < \sin(y^2).$$

15. The Taylor series about 0 for $y = \frac{1}{1-x^2}$ is
$$y = 1 + x^2 + x^4 + x^6 + \cdots.$$

The series for $y = (1+x)^{1/4}$ is, using the binomial expansion,
$$y = 1 + \frac{1}{4}x + \frac{1}{4}\left(-\frac{3}{4}\right)\frac{x^2}{2!} + \frac{1}{4}\left(-\frac{3}{4}\right)\left(-\frac{7}{4}\right)\frac{x^3}{3!} + \cdots.$$

The series for $y = \sqrt{1+\frac{x}{2}} = (1+\frac{x}{2})^{1/2}$ is, again using the binomial expansion,
$$y = 1 + \frac{1}{2}\cdot\frac{x}{2} + \frac{1}{2}\left(-\frac{1}{2}\right)\cdot\frac{x^2}{8} + \frac{1}{2}\left(-\frac{1}{2}\right)\left(-\frac{3}{2}\right)\cdot\frac{x^3}{48} + \cdots.$$

Similarly for $y = \frac{1}{\sqrt{1-x}} = (1-x)^{-(1/2)}$,
$$y = 1 + \left(-\frac{1}{2}\right)(-x) + \left(-\frac{1}{2}\right)\left(-\frac{3}{2}\right)\cdot\frac{x^2}{2!} + \left(-\frac{1}{2}\right)\left(-\frac{3}{2}\right)\left(-\frac{5}{2}\right)\cdot\frac{-x^3}{3!} + \cdots.$$

438 CHAPTER NINE /SOLUTIONS

Near 0, let's truncate these series after their x^2 terms:
$$\frac{1}{1-x^2} \approx 1+x^2,$$
$$(1+x)^{1/4} \approx 1+\frac{1}{4}x-\frac{3}{32}x^2,$$
$$\sqrt{1+\frac{x}{2}} \approx 1+\frac{1}{4}x-\frac{1}{32}x^2,$$
$$\frac{1}{\sqrt{1-x}} \approx 1+\frac{1}{2}x+\frac{3}{8}x^2.$$

Thus $\frac{1}{1-x^2}$ looks like a parabola opening upward near the origin, with y-axis as the axis of symmetry, so (a) = I.

Now $\frac{1}{\sqrt{1-x}}$ has the largest positive slope ($\frac{1}{2}$), and is concave up (because the coefficient of x^2 is positive). So (d) = II.

The last two both have positive slope ($\frac{1}{4}$) and are concave down. Since $(1+x)^{\frac{1}{4}}$ has the smallest second derivative (i.e., the most negative coefficient of x^2), (b) = IV and therefore (c) = III.

16.

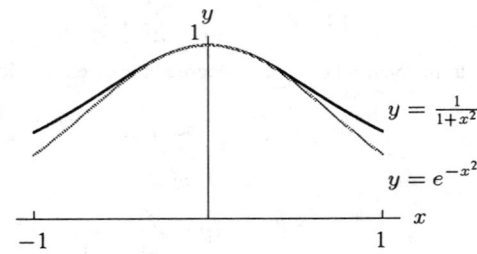

(a)
$$e^{-x^2} = 1-x^2+\frac{x^4}{2!}-\frac{x^6}{3!}+\cdots$$
$$\frac{1}{1+x^2} = 1-x^2+x^4-x^6+\cdots$$

Notice that the first two terms are the same in both series.

(b) $\frac{1}{1+x^2}$ is greater.

(c) Even, because the only terms involved are of even degree.

(d) The coefficients for e^{-x^2} become extremely small for higher powers of x, and we can "counteract" the effect of these powers for large values of x. The series for $\frac{1}{1+x^2}$ has no such coefficients.

17.
$$\frac{1}{2+x} = \frac{1}{2(1+\frac{x}{2})} = \frac{1}{2}\left(1+\frac{x}{2}\right)^{-1}$$
$$= \frac{1}{2}\left(1-\frac{x}{2}+\left(\frac{x}{2}\right)^2-\left(\frac{x}{2}\right)^3+\cdots\right)$$

18.
$$\frac{a}{\sqrt{a^2+x^2}} = \frac{a}{a(1+\frac{x^2}{a^2})^{\frac{1}{2}}} = \left(1+\frac{x^2}{a^2}\right)^{-\frac{1}{2}}$$
$$= 1+\left(-\frac{1}{2}\right)\frac{x^2}{a^2}+\frac{1}{2!}\left(-\frac{1}{2}\right)\left(-\frac{3}{2}\right)\left(\frac{x^2}{a^2}\right)^2$$
$$+\frac{1}{3!}\left(-\frac{1}{2}\right)\left(-\frac{3}{2}\right)\left(-\frac{5}{2}\right)\left(\frac{x^2}{a^2}\right)^3+\cdots$$
$$= 1-\frac{1}{2}\left(\frac{x}{a}\right)^2+\frac{3}{8}\left(\frac{x}{a}\right)^4-\frac{5}{16}\left(\frac{x}{a}\right)^6+\cdots$$

19. (a) $f(x) = (1+ax)(1+bx)^{-1} = (1+ax)\left(1 - bx + (bx)^2 - (bx)^3 + \cdots\right)$
$= 1 + (a-b)x + (b^2 - ab)x^2 + \cdots$

(b) $e^x = 1 + x + \frac{x^2}{2} + \cdots$
Equating coefficients:
$$a - b = 1,$$
$$b^2 - ab = \frac{1}{2}.$$

Solving gives $a = \frac{1}{2}, b = -\frac{1}{2}$.

20.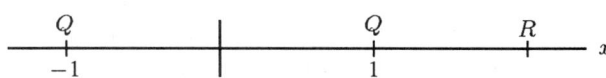

$$E = kQ\left(\frac{1}{(R-1)^2} - \frac{1}{(R+1)^2}\right)$$
$$= \frac{kQ}{R^2}\left(\frac{1}{(1-\frac{1}{R})^2} - \frac{1}{(1+\frac{1}{R})^2}\right)$$

Since $|\frac{1}{R}| < 1$, we can expand the two terms using the binomial expansion:

$$\frac{1}{(1-\frac{1}{R})^2} = \left(1 - \frac{1}{R}\right)^{-2}$$
$$= 1 - 2\left(-\frac{1}{R}\right) + (-2)(-3)\frac{(-\frac{1}{R})^2}{2!} + (-2)(-3)(-4)\frac{(-\frac{1}{R})^3}{3!} + \cdots$$

$$\frac{1}{(1+\frac{1}{R})^2} = \left(1 + \frac{1}{R}\right)^{-2}$$
$$= 1 - 2\left(\frac{1}{R}\right) + (-2)(-3)\frac{(\frac{1}{R})^2}{2!} + (-2)(-3)(-4)\frac{(\frac{1}{R})^3}{3!} + \cdots$$

Substituting, we get:
$$E = \frac{kQ}{R^2}\left[1 + \frac{2}{R} + \frac{3}{R^2} + \frac{4}{R^3} + \cdots - \left(1 - \frac{2}{R} + \frac{3}{R^2} - \frac{4}{R^3} + \cdots\right)\right] \approx \frac{kQ}{R^2}\left(\frac{4}{R} + \frac{8}{R^3}\right),$$

using only the first two non-zero terms.

21. Using the binomial expansion we have
$$\sqrt{a^2 + x^2} = a\left(1 + \frac{x^2}{a^2}\right)^{1/2}$$
$$= a\left(1 + \frac{1}{2}\frac{x^2}{a^2} + \frac{(1/2)(-1/2)}{2!}\frac{x^4}{a^4} + \frac{(1/2)(-1/2)(-3/2)}{3!}\frac{x^6}{a^6} + \cdots\right)$$
$$= a\left(1 + \frac{1}{2}\frac{x^2}{a^2} - \frac{1}{8}\frac{x^4}{a^4} + \frac{1}{16}\frac{x^6}{a^6} + \cdots\right).$$

Similarly, we have
$$\sqrt{a^2 - x^2} = a\left(1 - \frac{1}{2}\frac{x^2}{a^2} - \frac{1}{8}\frac{x^4}{a^4} - \frac{1}{16}\frac{x^6}{a^6} - \cdots\right).$$

Combining gives
$$z = \sqrt{a^2 + x^2} - \sqrt{a^2 - x^2} = a\left(2 \cdot \frac{1}{2}\frac{x^2}{a^2} + 2 \cdot \frac{1}{16}\frac{x^6}{a^6} + \cdots\right) = \frac{x^2}{a} + \frac{1}{8}\frac{x^6}{a^5} + \cdots.$$

22. This time we are interested in how a function behaves at large values in its domain. Therefore, we don't want to expand $V = 2\pi\sigma(\sqrt{R^2 + a^2} - R)$ about $R = 0$. We want to find a variable which becomes small as R gets large. Since $R > a$, it is helpful to write

$$V = R2\pi\sigma \left(\sqrt{1 + \frac{a^2}{R^2}} - 1 \right).$$

We can now expand a series in terms of $(\frac{a}{R})^2$. This may seem strange, but suspend your disbelief. The Taylor series for $\sqrt{1 + \frac{a^2}{R^2}}$ is

$$1 + \frac{1}{2}\frac{a^2}{R^2} + \frac{(1/2)(-1/2)}{2}\left(\frac{a^2}{R^2}\right)^2 + \cdots$$

So $V = R2\pi\sigma \left(1 + \frac{1}{2}\frac{a^2}{R^2} - \frac{1}{8}\left(\frac{a^2}{R^2}\right)^2 + \cdots - 1 \right)$. For large R, we can drop the $-\frac{1}{8}\frac{a^4}{R^4}$ term and terms of higher order, so

$$V \approx \frac{\pi\sigma a^2}{R}.$$

Notice that what we really did by expanding around $(\frac{a}{R})^2 = 0$ was expanding around $R = \infty$. We then get a series that converges for large R.

23. (a) If $\phi = 0$,
$$\text{left side} = b(1 + 1 + 1) = 3b \approx 0$$
so the equation is almost satisfied and there could be a solution near $\phi = 0$.

(b) We have
$$\sin\phi = \phi - \frac{\phi^3}{3!} + \frac{\phi^5}{5!} - \cdots$$
$$\cos\phi = 1 - \frac{\phi^2}{2!} + \frac{\phi^4}{4!} - \cdots$$

So
$$\cos^2\phi = \left(1 - \frac{\phi^2}{2!} + \frac{\phi^4}{4!} - \cdots\right)\left(1 - \frac{\phi^2}{2!} + \frac{\phi^4}{4!} - \cdots\right).$$

Neglecting terms of order ϕ^2 and higher, we get
$$\sin\phi \approx \phi$$
$$\cos\phi \approx 1$$
$$\cos^2\phi \approx 1.$$

So $\phi + b(1 + 1 + 1) \approx 0$, whence $\phi \approx -3b$.

24. (a) Factoring the expression for $t_1 - t_2$, we get
$$\Delta t = t_1 - t_2 = \frac{2l_2}{c(1 - v^2/c^2)} - \frac{2l_1}{c\sqrt{1 - v^2/c^2}} - \frac{2l_2}{c\sqrt{1 - v^2/c^2}} + \frac{2l_1}{c(1 - v^2/c^2)}$$
$$= \frac{2(l_1 + l_2)}{c(1 - v^2/c^2)} - \frac{2(l_1 + l_2)}{c\sqrt{1 - v^2/c^2}}$$
$$= \frac{2(l_1 + l_2)}{c}\left(\frac{1}{1 - v^2/c^2} - \frac{1}{\sqrt{1 - v^2/c^2}}\right).$$

Expanding the two terms within the parentheses in terms of v^2/c^2 gives

$$\left(1 - \frac{v^2}{c^2}\right)^{-1} = 1 + \frac{v^2}{c^2} + \frac{(-1)(-2)}{2!}\left(\frac{-v^2}{c^2}\right)^2 + \frac{(-1)(-2)(-3)}{3!}\left(\frac{-v^2}{c^2}\right)^3 + \cdots$$
$$= 1 + \frac{v^2}{c^2} + \frac{v^4}{c^4} + \frac{v^6}{c^6} + \cdots$$

$$\left(1-\frac{v^2}{c^2}\right)^{-1/2} = 1 + \frac{1}{2}\frac{v^2}{c^2} + \frac{\left(\frac{-1}{2}\right)\left(\frac{-3}{2}\right)}{2!}\left(\frac{-v^2}{c^2}\right)^2 + \frac{\left(\frac{-1}{2}\right)\left(\frac{-3}{2}\right)\left(\frac{-5}{2}\right)}{3!}\left(\frac{-v^2}{c^2}\right)^3 + \cdots$$

$$= 1 + \frac{1}{2}\frac{v^2}{c^2} + \frac{3}{8}\frac{v^4}{c^4} + \frac{5}{16}\frac{v^6}{c^6} + \cdots$$

Thus, we have

$$\Delta t = \frac{2(l_1+l_2)}{c}\left(1 + \frac{v^2}{c^2} + \frac{v^4}{c^4} + \frac{v^6}{c^6} + \cdots - 1 - \frac{1}{2}\frac{v^2}{c^2} - \frac{3}{8}\frac{v^4}{c^4} - \frac{5}{16}\frac{v^6}{c^6} - \cdots\right)$$

$$= \frac{2(l_1+l_2)}{c}\left(\frac{1}{2}\frac{v^2}{c^2} + \frac{5}{8}\frac{v^4}{c^4} + \frac{11}{16}\frac{v^6}{c^6} + \cdots\right)$$

$$\Delta t \approx \frac{(l_1+l_2)}{c}\left(\frac{v^2}{c^2} + \frac{5}{4}\frac{v^4}{c^4}\right).$$

(b) For small v. we can neglect all but the first nonzero term, so

$$\Delta t \approx \frac{(l_1+l_2)}{c}\cdot\frac{v^2}{c^2} = \frac{(l_1+l_2)}{c^3}v^2.$$

Thus, Δt is proportional to v^2 with constant of proportionality $(l_1+l_2)/c^3$.

25. (a) $\mu = \frac{mM}{m+M}$.

If $M \gg m$, then the denominator $m + M \approx M$, so $\mu \approx \frac{mM}{M} = m$.

(b)
$$\mu = m\left(\frac{M}{m+M}\right) = m\left(\frac{\frac{1}{M}M}{\frac{m}{M}+\frac{M}{M}}\right) = m\left(\frac{1}{1+\frac{m}{M}}\right)$$

We can use the binomial expansion since $\frac{m}{M} < 1$.

$$\mu = m\left[1 - \frac{m}{M} + \left(\frac{m}{M}\right)^2 - \left(\frac{m}{M}\right)^3 + \cdots\right]$$

(c) If $m \approx \frac{1}{1836}M$, then $\frac{m}{M} \approx \frac{1}{1836} \approx 0.000545$.

So a first order approximation to μ would give $\mu = m(1 - 0.000545)$. The percentage difference from $\mu = m$ is -0.0545%.

26. (a) For $a/h < 1$, we have

$$\frac{1}{(a^2+h^2)^{1/2}} = \frac{1}{h(1+a^2/h^2)^{1/2}} = \frac{1}{h}\left(1 - \frac{1}{2}\frac{a^2}{h^2} + \frac{3}{8}\frac{a^4}{h^4} - \cdots\right).$$

Thus

$$F = \frac{2GMmh}{a^2}\left(\frac{1}{h} - \frac{1}{h}\left(1 - \frac{1}{2}\frac{a^2}{h^2} + \frac{3}{8}\frac{a^4}{h^4} - \cdots\right)\right)$$

$$= \frac{2GMmh}{a^2h}\left(1 - 1 + \frac{1}{2}\frac{a^2}{h^2} - \frac{3}{8}\frac{a^4}{h^4} - \cdots\right)$$

$$= \frac{2GMm}{a^2}\frac{1}{2}\frac{a^2}{h^2}\left(1 - \frac{3}{4}\frac{a^2}{h^2}\cdots\right) = \frac{GMm}{h^2}\left(1 - \frac{3}{4}\frac{a^2}{h^2} - \cdots\right).$$

(b) Taking only the first nonzero term gives
$$F \approx \frac{GMm}{h^2}.$$

Notice that this approximation to F is independent of a.

(c) If $a/h = 0.02$, then $a^2/h^2 = 0.0004$, so

$$F \approx \frac{GMm}{h^2}(1 - \frac{3}{4}(0.0004)) = \frac{GMm}{h^2}(1 - 0.0003).$$

Thus, the approximations differ by $0.0003 = 0.03\%$.

442 CHAPTER NINE /SOLUTIONS

27. (a) If h is much smaller than R, we can say that $(R + h) \approx R$, giving the approximation
$$F = \frac{mgR^2}{(R+h)^2} \approx \frac{mgR^2}{R^2} = mg.$$

(b)
$$F = \frac{mgR^2}{(R+h)^2} = \frac{mg}{(1+h/R)^2} = mg(1+h/R)^{-2}$$
$$= mg\left(1 + \frac{(-2)}{1!}\left(\frac{h}{R}\right) + \frac{(-2)(-3)}{2!}\left(\frac{h}{R}\right)^2 + \frac{(-2)(-3)(-4)}{3!}\left(\frac{h}{R}\right)^3 + \cdots\right)$$
$$= mg\left(1 - \frac{2h}{R} + \frac{3h^2}{R^2} - \frac{4h^3}{R^3} + \cdots\right)$$

(c) The first order correction comes from term $-2h/R$. The approximation for F is then given by
$$F \approx mg\left(1 - \frac{2h}{R}\right).$$
If the first order correction alters the estimate for F by 10%, we have
$$\frac{2h}{R} = 0.10 \quad \text{so} \quad h = 0.05R \approx 0.05(6400) = 320 \text{ km}.$$
The approximation $F \approx mg$ is good to within 10% — that is, up to about 300 km.

28. (a) We take the left-hand Riemann sum with the formula
$$\text{Left-hand sum} = (1 + 0.9608 + 0.8521 + 0.6977 + 0.5273)(0.2) = 0.8076.$$
Similarly,
$$\text{Right-hand sum} = (0.9608 + 0.8521 + 0.6977 + 0.5273 + 0.3679)(0.2) = 0.6812.$$

(b) Since
$$e^x = 1 + x + \frac{x^2}{2!} + \frac{x^3}{3!} + \cdots,$$
$$e^{-x^2} \approx 1 + (-x^2) + \frac{(-x^2)^2}{2!} + \frac{(-x^2)^3}{3!}$$
$$= 1 - x^2 + \frac{x^4}{2} - \frac{x^6}{6}.$$

(c)
$$\int_0^1 e^{-x^2} dx \approx \int_0^1 \left(1 - x^2 + \frac{x^4}{2} - \frac{x^6}{6}\right) dx$$
$$= \left(x - \frac{x^3}{3} + \frac{x^5}{10} - \frac{x^7}{42}\right)\Big|_0^1 = 0.74286.$$

(d) We can improve the left and right sum values by averaging them to get 0.74439 or by increasing the number of subdivisions. We can improve on the estimate using the Taylor approximation by taking more terms.

29. (a) The Taylor series for $1/(1-x) = 1 + x + x^2 + x^3 + \ldots$, so
$$\frac{1}{0.98} = \frac{1}{1-0.02} = 1 + (0.02) + (0.02)^2 + (0.02)^3 + \cdots$$
$$= 1.020408\ldots$$

(b) Since $d/dx(1/(1-x)) = (1/(1-x))^2$, the Taylor series for $1/(1-x)^2$ is
$$\frac{d}{dx}(1 + x + x^2 + x^3 + \ldots) = 1 + 2x + 3x^2 + 4x^3 + \cdots$$
Thus
$$\frac{1}{(0.99)^2} = \frac{1}{(1-0.01)^2} = 1 + 2(0.01) + 3(0.0001) + 4(0.000001) + \cdots$$
$$= 1.0203040506\ldots$$

30. (a) From a calculator, $4\tan^{-1}(1/5) - \tan^{-1}(1/239) = 0.7853981634$, which agrees with $\pi/4$ to ten decimal places. Notice that you cannot verify that Machin's formula is *exactly* true numerically (because any calculator has only a finite number of digits.) Showing that the formula is exactly true requires a theoretical argument.

(b) The Taylor polynomial of degree 5 approximating $\arctan x$ is

$$\arctan x \approx x - \frac{x^3}{3} + \frac{x^5}{5}.$$

Thus,

$$\pi = 4\left(4\arctan\left(\frac{1}{5}\right) - \arctan\left(\frac{1}{239}\right)\right)$$
$$\approx 4\left(4\left(\frac{1}{5} - \frac{1}{3}\left(\frac{1}{5}\right)^3 + \frac{1}{5}\left(\frac{1}{5}\right)^5\right) - \left(\frac{1}{239} - \frac{1}{3}\left(\frac{1}{239}\right)^3 + \frac{1}{5}\left(\frac{1}{239}\right)^5\right)\right)$$
$$\approx 3.141621029.$$

(c) Because the values of x, namely $x = 1/5$ and $x = 1/239$, are much smaller than 1, the terms in the series get smaller much faster.

Solutions for Section 9.4

1. Yes, $a = 1$, ratio $= -1/2$.
2. No. Ratio between successive terms is not constant: $\frac{1/3}{1/2} = 0.66\ldots$, while $\frac{1/4}{1/3} = 0.75$.
3. Yes, $a = 5$, ratio $= -2$.
4. Yes, $a = 2$, ratio $= 1/2$.
5. No. Ratio between successive terms is not constant: $\frac{2x^2}{x} = 2x$, while $\frac{3x^3}{2x^2} = \frac{3}{2}x$.
6. Yes, $a = y^2$, ratio $= y$.
7. Yes, $a = 1$, ratio $= -x$.
8. Yes, $a = 1$, ratio $= -y^2$.
9. No. Ratio between successive terms is not constant: $\frac{6z^2}{3z} = 2z$, while $\frac{9z^3}{6z^2} = \frac{3}{2}z$.
10. Yes, $a = 1$, ratio $= 2z$.
11. Sum $= \dfrac{y^2}{1-y}$, $|y| < 1$
12. Sum $= \dfrac{1}{1-(-x)} = \dfrac{1}{1+x}$, $|x| < 1$
13. Sum $= \dfrac{1}{1-(-y^2)} = \dfrac{1}{1+y^2}$, $|y| < 1$.
14. Sum $= \dfrac{1}{1-2z}$, $|z| < 1/2$
15. (a) Factoring out $7(1.02)^3$ and using the formula for the sum of a finite geometric series with $a = 7(1.02)^3$ and $r = 1/1.02$, we see

$$\text{Sum} = 7(1.02)^3 + 7(1.02)^2 + 7(1.02) + 7 + \frac{7}{(1.02)} + \frac{7}{(1.02)^3} + \cdots + \frac{7}{(1.02)^{100}}$$
$$= 7(1.02)^3\left(1 + \frac{1}{(1.02)} + \frac{1}{(1.02)^2} + \cdots + \frac{1}{(1.02)^{103}}\right)$$
$$= 7(1.02)^3 \frac{\left(1 - \frac{1}{(1.02)^{104}}\right)}{1 - \frac{1}{1.02}}$$

$$= 7(1.02)^3 \left(\frac{(1.02)^{104} - 1}{(1.02)^{104}} \frac{1.02}{0.02} \right)$$

$$= \frac{7(1.02^{104} - 1)}{0.02(1.02)^{100}}.$$

(b) Using the Taylor expansion for e^x with $x = (0.1)^2$, we see

$$\text{Sum} = 7 + 7(0.1)^2 + \frac{7(0.1)^4}{2!} + \frac{7(0.1)^6}{3!} + \cdots$$

$$= 7 \left(1 + (0.1)^2 + \frac{(0.1)^4}{2!} + \frac{(0.1)^6}{3!} + \cdots \right)$$

$$= 7e^{(0.1)^2}$$

$$= 7e^{0.01}.$$

16. $-2 + 1 - \frac{1}{2} + \frac{1}{4} - \frac{1}{8} + \frac{1}{16} - \cdots = \sum_{n=0}^{\infty} (-2) \left(-\frac{1}{2} \right)^n$, a geometric series.

Let $a = -2$ and $x = -\frac{1}{2}$. Then

$$\sum_{n=0}^{\infty} (-2) \left(-\frac{1}{2} \right)^n = \frac{a}{1-x} = \frac{-2}{1-(-\frac{1}{2})} = -\frac{4}{3}.$$

17. $3 + \frac{3}{2} + \frac{3}{4} + \frac{3}{8} \cdots + \frac{3}{2^{10}} = 3 \left(1 + \frac{1}{2} + \cdots + \frac{1}{2^{10}} \right) = \frac{3\left(1 - \frac{1}{2^{11}}\right)}{1 - \frac{1}{2}} = \frac{3\left(2^{11} - 1\right)}{2^{10}}$

18.
$$\sum_{n=4}^{\infty} \left(\frac{1}{3} \right)^n = \left(\frac{1}{3} \right)^4 + \left(\frac{1}{3} \right)^5 + \cdots = \left(\frac{1}{3} \right)^4 \left(1 + \frac{1}{3} + \left(\frac{1}{3} \right)^2 + \cdots \right) = \frac{(\frac{1}{3})^4}{1 - \frac{1}{3}} = \frac{1}{54}$$

19. $\sum_{n=0}^{\infty} \frac{3^n + 5}{4^n} = \sum_{n=0}^{\infty} \left(\frac{3}{4} \right)^n + \sum_{n=0}^{\infty} \frac{5}{4^n}$, a sum of two geometric series.

$$\sum_{n=0}^{\infty} \left(\frac{3}{4} \right)^n = \frac{1}{1 - \frac{3}{4}} = 4$$

$$\sum_{n=0}^{\infty} \frac{5}{4^n} = \frac{5}{1 - \frac{1}{4}} = \frac{20}{3}$$

so $\sum_{n=0}^{\infty} \frac{3^n + 5}{4^n} = 4 + \frac{20}{3} = \frac{32}{3}.$

20. (a) The amount of atenolol in the blood is given by $Q(t) = Q_0 e^{-kt}$, where $Q_0 = Q(0)$ and k is a constant. Since the half-life is 6.3 hours,

$$\frac{1}{2} = e^{-6.3k}, \quad k = -\frac{1}{6.3} \ln \frac{1}{2} \approx 0.11.$$

After 24 hours

$$Q = Q_0 e^{-k(24)} \approx Q_0 e^{-0.11(24)} \approx Q_0(0.07).$$

Thus, the percentage of the atenolol that remains after 24 hours $\approx 7\%$.

(b)
$Q_0 = 50$
$Q_1 = 50 + 50(0.07)$
$Q_2 = 50 + 50(0.07) + 50(0.07)^2$
$Q_3 = 50 + 50(0.07) + 50(0.07)^2 + 50(0.07)^3$
$\vdots$
$Q_n = 50 + 50(0.07) + 50(0.07)^2 + \cdots + 50(0.07)^n = \frac{50(1 - (0.07)^{n+1})}{1 - 0.07}$

(c)

$P_1 = 50(0.07)$

$P_2 = 50(0.07) + 50(0.07)^2$

$P_3 = 50(0.07) + 50(0.07)^2 + 50(0.07)^3$

$P_4 = 50(0.07) + 50(0.07)^2 + 50(0.07)^3 + 50(0.07)^4$

$\vdots$

$P_n = 50(0.07) + 50(0.07)^2 + 50(0.07)^3 + \cdots + 50(0.07)^n$

$= 50(0.07)\left(1 + (0.07) + (0.07)^2 + \cdots + (0.07)^{n-1}\right) = \dfrac{0.07(50)(1 - (0.07)^n)}{1 - 0.07}$

21. (a)

$P_1 = 0$

$P_2 = 250(0.04)$

$P_3 = 250(0.04) + 250(0.04)^2$

$P_4 = 250(0.04) + 250(0.04)^2 + 250(0.04)^3$

$\vdots$

$P_n = 250(0.04) + 250(0.04)^2 + 250(0.04)^3 + \cdots + 250(0.04)^{n-1}$

(b) $P_n = 250(0.04)\left(1 + (0.04) + (0.04)^2 + (0.04)^3 + \cdots + (0.04)^{n-2}\right) = 250\dfrac{0.04(1 - (0.04)^{n-1})}{1 - 0.04}$

(c)

$P = \lim_{n \to \infty} P_n$

$= \lim_{n \to \infty} 250\dfrac{0.04(1 - (0.04)^{n-1})}{1 - 0.04}$

$= \dfrac{(250)(0.04)}{0.96} = 0.04Q \approx 10.42$

Thus, $\lim_{n \to \infty} P_n = 10.42$ and $\lim_{n \to \infty} Q_n = 260.42$. We would expect these limits to differ because one is right before taking a tablet, one is right after. We would expect the difference between them to be 250 mg, the amount of ampicillin in one tablet.

22.

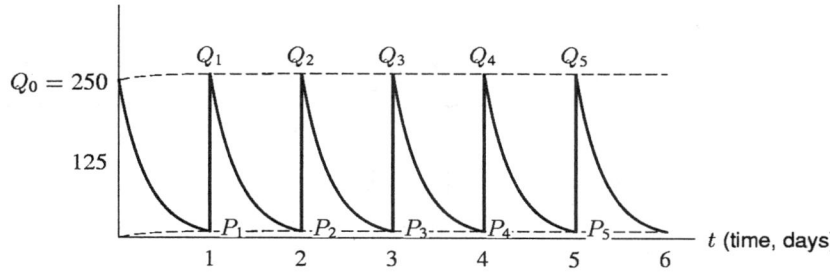

23. (a) Let h_n be the height of the nth bounce after the ball hits the floor for the nth time. Then from Figure 9.3,

$h_0 = \text{height before first bounce} = 10 \text{ feet,}$

$h_1 = \text{height after first bounce} = 10\left(\dfrac{3}{4}\right) \text{ feet,}$

$h_2 = \text{height after second bounce} = 10\left(\dfrac{3}{4}\right)^2 \text{ feet.}$

Generalizing gives

$$h_n = 10\left(\dfrac{3}{4}\right)^n.$$

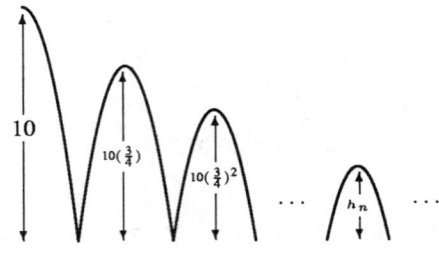

Figure 9.3

(b) When the ball hits the floor for the first time, the total distance it has traveled is just $D_1 = 10$ feet. (Notice that this is the same as $h_0 = 10$.) Then the ball bounces back to a height of $h_1 = 10\left(\frac{3}{4}\right)$, comes down and hits the floor for the second time. See Figure 9.3. The total distance it has traveled is

$$D_2 = h_0 + 2h_1 = 10 + 2 \cdot 10\left(\frac{3}{4}\right) = 25 \text{ feet.}$$

Then the ball bounces back to a height of $h_2 = 10\left(\frac{3}{4}\right)^2$, comes down and hits the floor for the third time. It has traveled

$$D_3 = h_0 + 2h_1 + 2h_2 = 10 + 2 \cdot 10\left(\frac{3}{4}\right) + 2 \cdot 10\left(\frac{3}{4}\right)^2 = 25 + 2 \cdot 10\left(\frac{3}{4}\right)^2 = 36.25 \text{ feet.}$$

Similarly,

$$\begin{aligned} D_4 &= h_0 + 2h_1 + 2h_2 + 2h_3 \\ &= 10 + 2 \cdot 10\left(\frac{3}{4}\right) + 2 \cdot 10\left(\frac{3}{4}\right)^2 + 2 \cdot 10\left(\frac{3}{4}\right)^3 \\ &= 36.25 + 2 \cdot 10\left(\frac{3}{4}\right)^3 \\ &\approx 44.69 \text{ feet.} \end{aligned}$$

(c) When the ball hits the floor for the nth time, its last bounce was of height h_{n-1}. Thus, by the method used in part (b), we get

$$\begin{aligned} D_n &= h_0 + 2h_1 + 2h_2 + 2h_3 + \cdots + 2h_{n-1} \\ &= 10 + \underbrace{2 \cdot 10\left(\frac{3}{4}\right) + 2 \cdot 10\left(\frac{3}{4}\right)^2 + 2 \cdot 10\left(\frac{3}{4}\right)^3 + \cdots + 2 \cdot 10\left(\frac{3}{4}\right)^{n-1}}_{\text{finite geometric series}} \\ &= 10 + 2 \cdot 10 \cdot \left(\frac{3}{4}\right)\left(1 + \left(\frac{3}{4}\right) + \left(\frac{3}{4}\right)^2 + \cdots + \left(\frac{3}{4}\right)^{n-2}\right) \\ &= 10 + 15\left(\frac{1 - \left(\frac{3}{4}\right)^{n-1}}{1 - \left(\frac{3}{4}\right)}\right) \\ &= 10 + 60\left(1 - \left(\frac{3}{4}\right)^{n-1}\right). \end{aligned}$$

24. (a) The acceleration of gravity is 32 ft/sec^2 so acceleration $= 32$ and velocity $v = 32t + C$. Since the ball is dropped, its initial velocity is 0 so $v = 32t$. Thus the position is $s = 16t^2 + C$. Calling the initial position $s = 0$, we have $s = 6t$. The distance traveled is h so $h = 16t$. Solving for t we get $t = \frac{1}{4}\sqrt{h}$.

(b) The first drop from 10 feet takes $\frac{1}{4}\sqrt{10}$ seconds. The first full bounce (to $10 \cdot \left(\frac{3}{4}\right)$ feet) takes $\frac{1}{4}\sqrt{10 \cdot \left(\frac{3}{4}\right)}$ seconds to rise, therefore the same time to come down. Thus, the full bounce, up and down, takes $2\left(\frac{1}{4}\right)\sqrt{10 \cdot \left(\frac{3}{4}\right)}$ seconds.

The next full bounce takes $2(\frac{1}{4})10 \cdot (\frac{3}{4})^2 = 2(\frac{1}{4})\sqrt{10}\left(\sqrt{\frac{3}{4}}\right)^2$ seconds. The nth bounce takes $2(\frac{1}{4})\sqrt{10}\left(\sqrt{\frac{3}{4}}\right)^n$ seconds. Therefore the

Total amount of time

$$= \frac{1}{4}\sqrt{10} + \underbrace{\frac{2}{4}\sqrt{10}\sqrt{\frac{3}{4}} + \frac{2}{4}\sqrt{10}\left(\sqrt{\frac{3}{4}}\right)^2 + \frac{2}{4}\sqrt{10}\left(\sqrt{\frac{3}{4}}\right)^3 + \cdots}_{\text{Geometric series with } a = \frac{2}{4}\sqrt{10}\sqrt{\frac{3}{4}} = \frac{1}{2}\sqrt{10}\sqrt{\frac{3}{4}} \text{ and } x = \sqrt{\frac{3}{4}}}$$

$$= \frac{1}{4}\sqrt{10} + \frac{1}{2}\sqrt{10}\sqrt{\frac{3}{4}}\left(\frac{1}{1-\sqrt{3/4}}\right) \text{ seconds.}$$

25. (a)

$$\text{Total amount of money deposited} = 100 + 92 + 84.64 + \cdots$$
$$= 100 + 100(0.92) + 100(0.92)^2 + \cdots$$
$$= \frac{100}{1-0.92} = 1250 \text{ dollars}$$

(b) Credit multiplier $= 1250/100 = 12.50$
The 12.50 is the factor by which the bank has increased its deposits, from \$100 to \$1250.

26. The amount of additional income generated directly by people spending their extra money is $\$100(0.8) = \$80$ million. This additional money in turn is spent, generating another $(\$100(0.8))(0.8) = \$100(0.8)^2$ million. This continues indefinitely, resulting in

$$\text{Total additional income} = 100(0.8) + 100(0.8)^2 + 100(0.8)^3 + \cdots = \frac{100(0.8)}{1-0.8} = \$400 \text{ million}$$

27. The total of the spending and respending of the additional income is given by the series: Total additional income $= 100(0.9) + 100(0.9)^2 + 100(0.9)^3 + \cdots = \frac{100(0.9)}{1-0.9} = \$900$ million.
Notice the large effect of changing the assumption about the fraction of money spent has: the additional spending more than doubles.

28. (a) Since the fly is traveling at 20 km/hr and the trains at 10 km/hr, each time the fly goes to meet the other train, it will cover 2/3 of the distance between them, while the train covers 1/3 of the distance.
Thus, the first time the fly turns around, it will have covered a distance of

$$d_1 = 30\left(\frac{2}{3}\right) = 20 \text{ km.}$$

By this time each train has gone 10 km, so the distance between the trains is now reduced to one-third of what it was originally, namely $30(1/3) = 10$ km. Thus, by the next turn around the fly has gone an additional $10(2/3) = 30(1/3)(2/3)$ km. By the second turn-around, the fly has covered a total distance of

$$d_2 = 30\left(\frac{2}{3}\right) + 30\left(\frac{1}{3}\right)\left(\frac{2}{3}\right) \text{ km.}$$

At the time of the second turn-around the distance between the trains has been reduced to one third of what it was the time before, or $10(1/3) = 30(1/3)^2$ km. The third turn around adds a distance of $30(1/3)^2(2/3)$:

$$d_3 = 30\left(\frac{2}{3}\right) + 30\left(\frac{1}{3}\right)\left(\frac{2}{3}\right) + 30\left(\frac{1}{3}\right)^2\left(\frac{2}{3}\right) \text{ km.}$$

By the third turn-around, the distance between the trains is $30(1/3)^3$ km, so, at the fourth turn-around

$$d_4 = 30\left(\frac{2}{3}\right) + 30\left(\frac{1}{3}\right)\left(\frac{2}{3}\right) + 30\left(\frac{1}{3}\right)^2\left(\frac{2}{3}\right) + 30\left(\frac{1}{3}\right)^3\left(\frac{2}{3}\right) \text{ km.}$$

(b) Generalizing of the answers to part (a) gives

$$d_n = 30\left(\frac{2}{3}\right) + 30\left(\frac{1}{3}\right)\left(\frac{2}{3}\right) + 30\left(\frac{1}{3}\right)^2\left(\frac{2}{3}\right) + \cdots + 30\left(\frac{1}{3}\right)^{n-1}\left(\frac{2}{3}\right)$$

$$= 30\left(\frac{2}{3}\right)\left(1 + \left(\frac{1}{3}\right) + \left(\frac{1}{3}\right)^2 + \cdots + \left(\frac{1}{3}\right)^{n-1}\right)$$

$$= \frac{20\left(1 - \left(\frac{1}{3}\right)^n\right)}{1 - \frac{1}{3}}.$$

(c) By the time the trains meet and squash the fly, the poor thing will have flown

$$d = \lim_{n \to \infty} d_n$$

$$= \lim_{n \to \infty} \frac{20\left(1 - \left(\frac{1}{3}\right)^n\right)}{1 - \frac{1}{3}}$$

$$= \frac{20}{1 - \frac{1}{3}}$$

$$= 30 \text{ km}.$$

(d) It will take $30/20 = 3/2$ hours for the trains to meet, since they are 30 km apart and the distance between them is decreasing at a speed of 20 km/hr. So the fly will have traveled $20(3/2) = 30$ km by the time trains meet.

There is a joke that accompanies this problem. The mathematician, John von Neumann (called the father of the computer) was asked this question, and instantaneously gave the correct answer. The person who had posed the problem complimented him on seeing the trick. Von Neumann responded "What trick? I just summed the geometric series."

Solutions for Section 9.5

1. No, a Fourier series has terms of the form $\cos nx$, not $\cos^n x$.
2. Not a Fourier series because terms are not of the form $\sin nx$.
3. Yes. Terms are of the form $\sin nx$ and $\cos nx$.
4. Yes. This is a Fourier series where the $\cos nx$ terms all have coefficients of zero.
5.

$$a_0 = \frac{1}{2\pi}\int_{-\pi}^{\pi} f(x)\,dx = \frac{1}{2\pi}\left[\int_{-\pi}^{0} -1\,dx + \int_{0}^{\pi} 1\,dx\right] = 0$$

$$a_1 = \frac{1}{\pi}\int_{-\pi}^{\pi} f(x)\cos x\,dx = \frac{1}{\pi}\left[\int_{-\pi}^{0} -\cos x\,dx + \int_{0}^{\pi} \cos x\,dx\right]$$

$$= \frac{1}{\pi}\left[-\sin x\Big|_{-\pi}^{0} + \sin x\Big|_{0}^{\pi}\right] = 0.$$

Similarly, a_2 and a_3 are both 0.
(In fact, notice $f(x)\cos nx$ is an odd function, so $\int_{-\pi}^{\pi} f(x)\cos nx = 0$.)

$$b_1 = \frac{1}{\pi}\int_{-\pi}^{\pi} f(x)\sin x\,dx = \frac{1}{\pi}\left[\int_{-\pi}^{0} -\sin x\,dx + \int_{0}^{\pi} \sin x\,dx\right]$$

$$= \frac{1}{\pi}\left[\cos x\Big|_{-\pi}^{0} + (-\cos x)\Big|_{0}^{\pi}\right] = \frac{4}{\pi}.$$

$$b_2 = \frac{1}{\pi}\int_{-\pi}^{\pi} f(x)\sin 2x\,dx = \frac{1}{\pi}\left[\int_{-\pi}^{0} -\sin 2x\,dx + \int_{0}^{\pi} \sin 2x\,dx\right]$$

$$= \frac{1}{\pi}\left[\frac{1}{2}\cos 2x\Big|_{-\pi}^{0} + \left(-\frac{1}{2}\cos 2x\right)\Big|_{0}^{\pi}\right] = 0.$$

$$b_3 = \frac{1}{\pi}\int_{-\pi}^{\pi} f(x)\sin 3x\,dx = \frac{1}{\pi}\left[\int_{-\pi}^{0} -\sin 3x\,dx + \int_{0}^{\pi} \sin 3x\,dx\right]$$

$$= \frac{1}{\pi}\left[\frac{1}{3}\cos 3x\Big|_{-\pi}^{0} + \left(-\frac{1}{3}\cos 3x\right)\Big|_{0}^{\pi}\right] = \frac{4}{3\pi}.$$

Thus, $F_1(x) = F_2(x) = \frac{4}{\pi}\sin x$ and $F_3(x) = \frac{4}{\pi}\sin x + \frac{4}{3\pi}\sin 3x$.

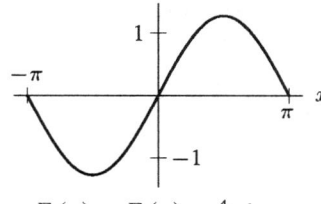

$F_1(x) = F_2(x) = \frac{4}{\pi}\sin x$

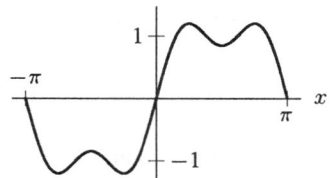

$F_3(x) = \frac{4}{\pi}\sin x + \frac{4}{3\pi}\sin 3x$

6. First,

$$a_0 = \frac{1}{2\pi}\int_{-\pi}^{\pi} f(x)\,dx = \frac{1}{2\pi}\left[\int_{-\pi}^{0} -x\,dx + \int_{0}^{\pi} x\,dx\right] = \frac{1}{2\pi}\left[-\frac{x^2}{2}\Big|_{-\pi}^{0} + \frac{x^2}{2}\Big|_{0}^{\pi}\right] = \frac{\pi}{2}.$$

To find the a_i's, we use the integral table. For $n \geq 1$,

$$a_n = \frac{1}{\pi}\int_{-\pi}^{\pi} f(x)\cos(nx)\,dx = \frac{1}{\pi}\left[\int_{-\pi}^{0} -x\cos(nx)\,dx + \int_{0}^{\pi} x\cos(nx)\,dx\right]$$

$$= \frac{1}{\pi}\left[\left(-\frac{x}{n}\sin(nx) - \frac{1}{n^2}\cos(nx)\right)\Big|_{-\pi}^{0}\right.$$

$$\left. + \left(\frac{x}{n}\sin(nx) + \frac{1}{n^2}\cos(nx)\right)\Big|_{0}^{\pi}\right]$$

$$= \frac{1}{\pi}\left(-\frac{1}{n^2} + \frac{1}{n^2}\cos(-n\pi) + \frac{1}{n^2}\cos(n\pi) - \frac{1}{n^2}\right)$$

$$= \frac{2}{\pi n^2}(\cos n\pi - 1)$$

Thus, $a_1 = -\frac{4}{\pi}, a_2 = 0$, and $a_3 = -\frac{4}{9\pi}$. To find the b_i's, note that $f(x)$ is even, so for $n \geq 1$, $f(x)\sin(nx)$ is odd. Thus, $\int_{-\pi}^{\pi} f(x)\sin(nx) = 0$, so all the b_i's are 0. $F_1 = F_2 = \frac{\pi}{2} - \frac{4}{\pi}\cos x$, $F_3 = \frac{\pi}{2} - \frac{4}{\pi}\cos x - \frac{4}{9\pi}\cos 3x$.

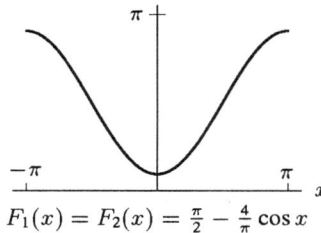

$F_1(x) = F_2(x) = \frac{\pi}{2} - \frac{4}{\pi}\cos x$

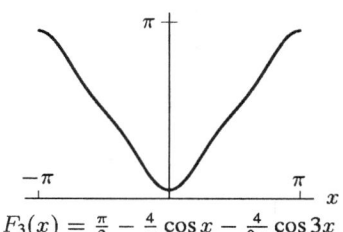

$F_3(x) = \frac{\pi}{2} - \frac{4}{\pi}\cos x - \frac{4}{9\pi}\cos 3x$

450 CHAPTER NINE /SOLUTIONS

7. The energy of the function $f(x)$ is

$$E = \frac{1}{\pi}\int_{-\pi}^{\pi} (f(x))^2\, dx = \frac{1}{\pi}\int_{-\pi}^{\pi} x^2\, dx = \frac{1}{3\pi} x^3 \Big|_{-\pi}^{\pi}$$

$$= \frac{1}{3\pi}(\pi^3 - (-\pi^3)) = \frac{2\pi^3}{3\pi} = \frac{2}{3}\pi^2 = 6.57974.$$

From Problem 6, we know all the b_i's are 0 and $a_0 = \frac{\pi}{2}$, $a_1 = -\frac{4}{\pi}$, $a_2 = 0$, $a_3 = -\frac{4}{9\pi}$. Therefore the energy in the constant term and first three harmonics is

$$A_0^2 + A_1^2 + A_2^2 + A_3^2 = 2a_0^2 + a_1^2 + a_2^2 + a_3^2$$

$$= 2\left(\frac{\pi^2}{4}\right) + \frac{16}{\pi^2} + 0 + \frac{16}{81\pi^2} = 6.57596$$

which means that they contain $\dfrac{6.57596}{6.57974} = 0.99942 \approx 99.942\%$ of the total energy.

8. First, we find a_0.

$$a_0 = \frac{1}{2\pi}\int_{-\pi}^{\pi} x^2\, dx = \frac{1}{2\pi}\left(\frac{x^3}{3}\Big|_{-\pi}^{\pi}\right) = \frac{\pi^2}{3}$$

To find a_n, $n \geq 1$, we use the integral table (III-15 and III-16).

$$a_n = \frac{1}{\pi}\int_{-\pi}^{\pi} x^2 \cos nx\, dx = \frac{1}{\pi}\left[\frac{x^2}{n}\sin(nx) + \frac{2x}{n^2}\cos(nx) - \frac{2}{n^3}\sin(nx)\right]\Big|_{-\pi}^{\pi}$$

$$= \frac{1}{\pi}\left[\frac{2\pi}{n^2}\cos(n\pi) + \frac{2\pi}{n^2}\cos(-n\pi)\right]$$

$$= \frac{4}{n^2}\cos(n\pi)$$

Again, $\cos(n\pi) = (-1)^n$ for all integers n, so $a_n = (-1)^n \frac{4}{n^2}$. Note that

$$b_n = \frac{1}{\pi}\int_{-\pi}^{\pi} x^2 \sin nx\, dx.$$

x^2 is an even function, and $\sin nx$ is odd, so $x^2 \sin nx$ is odd. Thus $\int_{-\pi}^{\pi} x^2 \sin nx\, dx = 0$, and $b_n = 0$ for all n. We deduce that the nth Fourier polynomial for f (where $n \geq 1$) is

$$F_n(x) = \frac{\pi^2}{3} + \sum_{i=1}^{n}(-1)^i \frac{4}{i^2}\cos(ix).$$

In particular, we have the following graphs:

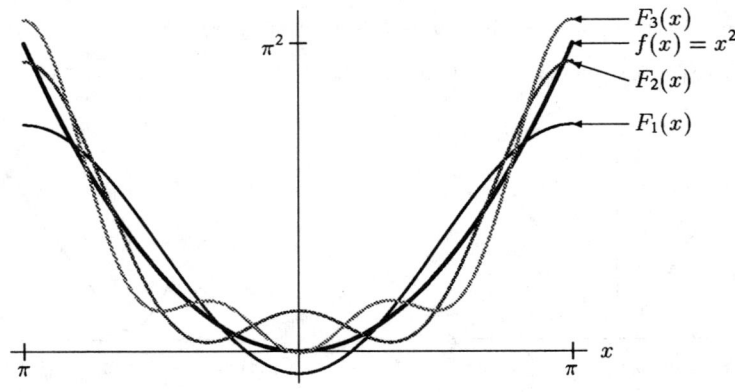

9.

$$a_0 = \frac{1}{2\pi}\int_{-\pi}^{\pi} h(x)\,dx = \frac{1}{2\pi}\int_0^{\pi} x\,dx = \frac{\pi}{4}$$

As in Problem 10, we use the integral table (III-15 and III-16) to find formulas for a_n and b_n.

$$a_n = \frac{1}{\pi}\int_{-\pi}^{\pi} h(x)\cos(nx)\,dx = \frac{1}{\pi}\int_0^{\pi} x\cos nx\,dx = \frac{1}{\pi}\left(\frac{x}{n}\sin(nx) + \frac{1}{n^2}\cos(nx)\right)\Big|_0^{\pi}$$

$$= \frac{1}{\pi}\left(\frac{1}{n^2}\cos(n\pi) - \frac{1}{n^2}\right)$$

$$= \frac{1}{n^2\pi}\left(\cos(n\pi) - 1\right).$$

Note that since $\cos(n\pi) = (-1)^n$, $a_n = 0$ if n is even and $a_n = -\frac{2}{n^2\pi}$ if n is odd.

$$b_n = \frac{1}{\pi}\int_{-\pi}^{\pi} h(x)\cos(nx)\,dx = \frac{1}{\pi}\int_0^{\pi} x\sin x\,dx$$

$$= \frac{1}{\pi}\left(-\frac{x}{n}\cos(nx) + \frac{1}{n^2}\sin(nx)\right)\Big|_0^{\pi}$$

$$= \frac{1}{\pi}\left(-\frac{\pi}{n}\cos(n\pi)\right)$$

$$= -\frac{1}{n}\cos(n\pi)$$

$$= \frac{1}{n}(-1)^{n+1} \quad \text{if } n \geq 1$$

We have that the nth Fourier polynomial for h (for $n \geq 1$) is

$$H_n(x) = \frac{\pi}{4} + \sum_{i=1}^{n}\left(\frac{1}{i^2\pi}\left(\cos(i\pi) - 1\right)\cdot\cos(ix) + \frac{(-1)^{i+1}\sin(ix)}{i}\right).$$

This can also be written as

$$H_n(x) = \frac{\pi}{4} + \sum_{i=1}^{n}\frac{(-1)^{i+1}\sin(ix)}{i} + \sum_{i=1}^{\left[\frac{n}{2}\right]}\frac{-2}{(2i-1)^2\pi}\cos((2i-1)x)$$

where $\left[\frac{n}{2}\right]$ denotes the biggest integer smaller than or equal to $\frac{n}{2}$. In particular, we have the following graphs:

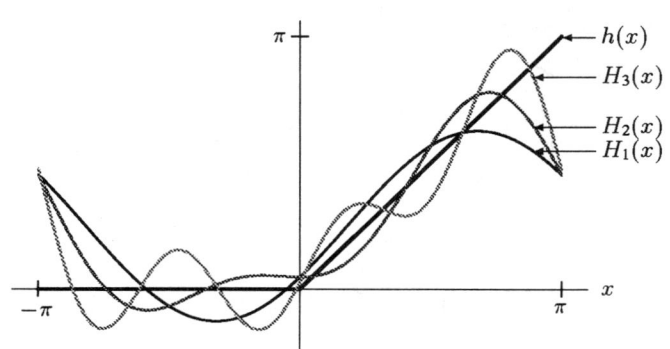

10. To find the nth Fourier polynomial, we must come up with a general formula for a_n and b_n. First, we find a_0.

$$a_0 = \frac{1}{2\pi}\int_{-\pi}^{\pi} g(x)\,dx = \frac{1}{2\pi}\int_{-\pi}^{\pi} x\,dx = \frac{1}{2\pi}\left[\frac{x^2}{2}\right]_{-\pi}^{\pi} = 0$$

Now we use the integral table (III-15 and III-16) to find a_n and b_n for $n \geq 1$.

$$a_n = \frac{1}{\pi}\int_{-\pi}^{\pi} x\cos nx\,dx = \frac{1}{\pi}\left(\frac{x}{n}\sin(nx) + \frac{1}{n^2}\cos(nx)\right)\bigg|_{-\pi}^{\pi}$$

$$= \frac{1}{\pi}\left(\frac{1}{n^2}\cos(n\pi) - \frac{1}{n^2}\cos(-n\pi)\right) = 0$$

(Note that since $x\cos nx$ is odd, we could have deduced that $\int_{-\pi}^{\pi} x\cos nx = 0$.)

$$b_n = \frac{1}{\pi}\int_{-\pi}^{\pi} x\sin nx\,dx = \frac{1}{\pi}\left(-\frac{x}{n}\cos(nx) + \frac{1}{n^2}\sin(nx)\right)\bigg|_{-\pi}^{\pi}$$

$$= \frac{1}{\pi}\left(-\frac{\pi}{n}\cos(n\pi) - \frac{\pi}{n}\cos(-n\pi)\right)$$

$$= -\frac{2}{n}\cos(n\pi)$$

Notice that $\cos(n\pi) = (-1)^n$ for all integers n, so $b_n = (-1)^{n+1}\left(\frac{2}{n}\right)$.
Thus the nth Fourier polynomial for g is

$$G_n(x) = \sum_{i=1}^{n}(-1)^{i+1}\frac{2}{i}\sin(ix).$$

In particular, we have the following graphs:

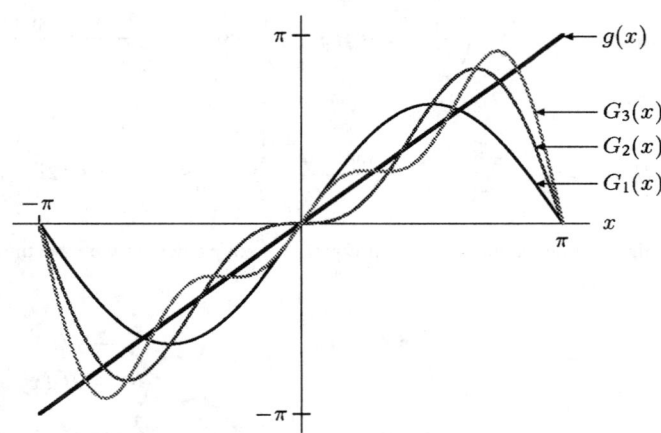

11. (a) The graph of $g(x)$ is

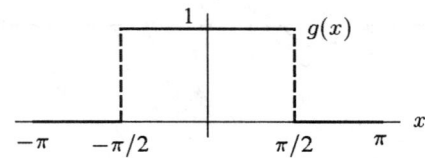

First find the Fourier coefficients: a_0 is the average value of g on $[-\pi, \pi]$ so from the graph, it is clear that
$$a_0 = \frac{1}{2\pi}(\pi \times 1) = \frac{1}{2},$$
or analytically,
$$a_0 = \frac{1}{2\pi}\int_{-\pi}^{\pi} g(x)\,dx = \frac{1}{2\pi}\int_{-\pi/2}^{\pi/2} 1\,dx = \frac{1}{2\pi}x\Big|_{-\pi/2}^{\pi/2} = \frac{1}{2\pi}\left(\frac{\pi}{2} - \left(-\frac{\pi}{2}\right)\right)$$
$$= \frac{1}{2\pi}(\pi) = \frac{1}{2},$$

$$a_k = \frac{1}{\pi}\int_{-\pi}^{\pi} g(x)\cos kx\,dx = \frac{1}{\pi}\int_{-\pi/2}^{\pi/2} \cos kx\,dx = \frac{1}{k\pi}\sin kx\Big|_{-\pi/2}^{\pi/2}$$
$$= \frac{1}{k\pi}\left(\sin\frac{k\pi}{2} - \sin\left(-\frac{k\pi}{2}\right)\right) = \frac{1}{k\pi}\left(2\sin\frac{k\pi}{2}\right),$$

$$b_k = \frac{1}{\pi}\int_{-\pi}^{\pi} g(x)\sin kx\,dx = \frac{1}{\pi}\int_{-\pi/2}^{\pi/2} \sin kx\,dx = -\frac{1}{k\pi}\cos kx\Big|_{-\pi/2}^{\pi/2}$$
$$= -\frac{1}{k\pi}\left(\cos\frac{k\pi}{2} - \cos\left(-\frac{k\pi}{2}\right)\right) = -\frac{1}{k\pi}(0) = 0$$

So,
$$a_1 = \frac{1}{\pi}\left(2\sin\frac{\pi}{2}\right) = \frac{2}{\pi},$$
$$a_2 = \frac{1}{2\pi}\left(2\sin\frac{2\pi}{2}\right) = 0,$$
$$a_3 = \frac{1}{3\pi}\left(2\sin\frac{3\pi}{2}\right) = -\frac{2}{3\pi},$$

which gives
$$F_3(x) = \frac{1}{2} + \frac{2}{\pi}\cos x - \frac{2}{3\pi}\cos 3x.$$

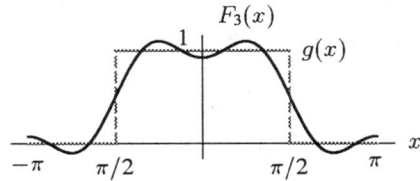

(b) There are cosines instead of sines (but the energy spectrum remains the same).

12. We have $f(x) = x$, $0 \le x < 1$. Let $t = 2\pi x - \pi$. Notice that as x varies from 0 to 1, t varies from $-\pi$ to π. Thus if we rewrite the function in terms of t, we can find the Fourier series in terms of t in the usual way. To do this, let $g(t) = f(x) = x = \frac{t+\pi}{2\pi}$ on $-\pi \le t < \pi$. We now find the fourth degree Fourier polynomial for g.

$$a_o = \frac{1}{2\pi}\int_{-\pi}^{\pi} g(t)\,dt = \frac{1}{2\pi}\int_{-\pi}^{\pi} \frac{t+\pi}{2\pi}\,dt = \frac{1}{(2\pi)^2}\left(\frac{t^2}{2} + \pi t\right)\Big|_{-\pi}^{\pi} = \frac{1}{2}$$

Notice, a_0 is the average value of both f and g. For $n \ge 1$,

$$a_n = \frac{1}{\pi}\int_{-\pi}^{\pi} \frac{t+\pi}{2\pi}\cos(nt)\,dt = \frac{1}{2\pi^2}\int_{-\pi}^{\pi} (t\cos(nt) + \pi\cos(nt))\,dt$$
$$= \frac{1}{2\pi^2}\left[\frac{t}{n}\sin(nt) + \frac{1}{n^2}\cos(nt) + \frac{\pi}{n}\sin(nt)\right]\Big|_{-\pi}^{\pi}$$
$$= 0.$$

$$b_n = \frac{1}{\pi}\int_{-\pi}^{\pi}\frac{t+\pi}{2\pi}\sin(nt)\,dt = \frac{1}{2\pi^2}\int_{-\pi}^{\pi}(t\sin(nt) + \pi\sin(nt))\,dt$$

$$= \frac{1}{2\pi^2}\left[-\frac{t}{n}\cos(nt) + \frac{1}{n^2}\sin(nt) - \frac{\pi}{n}\cos(nt)\right]\bigg|_{-\pi}^{\pi}$$

$$= \frac{1}{2\pi^2}\left(-\frac{4\pi}{n}\cos(\pi n)\right) = -\frac{2}{\pi n}\cos(\pi n) = \frac{2}{\pi n}(-1)^{n+1}.$$

We get the integrals for a_n and b_n using the integral table (formulas III-15 and III-16).

Thus, the Fourier polynomial of degree 4 for g is:

$$G_4(t) = \frac{1}{2} + \frac{2}{\pi}\sin t - \frac{1}{\pi}\sin 2t + \frac{2}{3\pi}\sin 3t - \frac{1}{2\pi}\sin 4t.$$

Now, since $g(t) = f(x)$, the Fourier polynomial of degree 4 for f can be found by replacing t in terms of x again. Thus,

$$F_4(x) = \frac{1}{2} + \frac{2}{\pi}\sin(2\pi x - \pi) - \frac{1}{\pi}\sin(4\pi x - 2\pi) + \frac{2}{3\pi}\sin(6\pi x - 3\pi) - \frac{1}{2\pi}\sin(8\pi x - 4\pi).$$

Now, using the fact that $\sin(x - \pi) = -\sin x$ and $\sin(x - 2\pi) = \sin x$, etc., we have:

$$F_4(x) = \frac{1}{2} - \frac{2}{\pi}\sin(2\pi x) - \frac{1}{\pi}\sin(4\pi x) - \frac{2}{3\pi}\sin(6\pi x) - \frac{1}{2\pi}\sin(8\pi x).$$

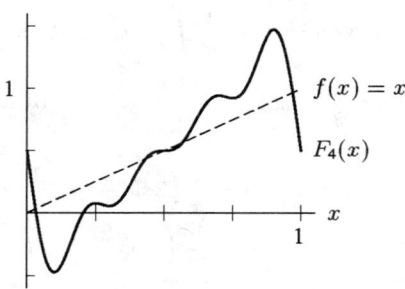

13. Since the period is 2, we make the substitution $t = \pi x - \pi$. Thus, $x = \frac{t+\pi}{\pi}$. We find the Fourier coefficients. Notice that all of the integrals are the same as in Problem 12 except for an extra factor of 2. Thus, $a_0 = 1$, $a_n = 0$, and $b_n = \frac{4}{\pi n}(-1)^{n+1}$, so:

$$G_4(t) = 1 + \frac{4}{\pi}\sin t - \frac{2}{\pi}\sin 2t + \frac{4}{3\pi}\sin 3t - \frac{1}{\pi}\sin 4t.$$

Again, we substitute back in to get a Fourier polynomial in terms of x:

$$F_4(x) = 1 + \frac{4}{\pi}\sin(\pi x - \pi) - \frac{2}{\pi}\sin(2\pi x - 2\pi)$$

$$+ \frac{4}{3\pi}\sin(3\pi x - 3\pi) - \frac{1}{\pi}\sin(4\pi x - 4\pi)$$

$$= 1 - \frac{4}{\pi}\sin(\pi x) - \frac{2}{\pi}\sin(2\pi x) - \frac{4}{3\pi}\sin(3\pi x) - \frac{1}{\pi}\sin(4\pi x).$$

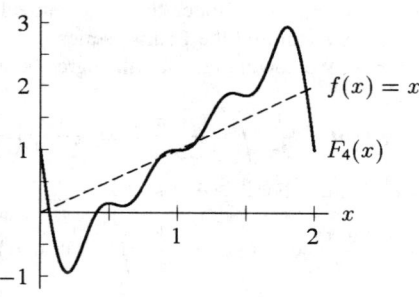

Notice in this case, the terms in our series are $\sin(n\pi x)$, not $\sin(2\pi n x)$, as in Problem 12. In general, the terms will be $\sin(n\frac{2\pi}{b}x)$, where b is the period.

14. The signal received on earth is in the form of a periodic function $h(t)$, which can be expanded in a Fourier series

$$h(t) = a_0 + a_1 \cos t + a_2 \cos 2t + a_3 \cos 3t + \cdots$$
$$+ b_1 \sin t + b_2 \sin 2t + b_3 \sin 3t + \cdots$$

If the periodic noise consists of *only* the second and higher harmonics of the Fourier series, then the original signal contributed the fundamental harmonic plus the constant term, i.e.,

$$\underbrace{a_0}_{\text{constant term}} + \underbrace{a_1 \cos t + b_1 \sin t}_{\text{fundamental harmonic}} = \underbrace{A \cos t}_{\text{original signal}}$$

In order to find A, we need to find a_0, a_1, and b_1. Looking at the graph of $h(t)$, we see

$$a_0 = \text{average value of } h(t) = \frac{1}{2\pi}(\text{Area above the } x\text{-axis} - \text{Area below the } x\text{-axis})$$

$$= \frac{1}{2\pi}\left[80\left(\frac{\pi}{2}\right) - \left(50\left(\frac{\pi}{4}\right) + 30\left(\frac{\pi}{4}\right) + 30\left(\frac{\pi}{4}\right) + 50\left(\frac{\pi}{4}\right)\right)\right]$$

$$= \frac{1}{2\pi}\left[80\left(\frac{\pi}{2}\right) - 80\left(\frac{\pi}{2}\right)\right] = \frac{1}{2\pi} \cdot 0, = 0$$

$$a_1 = \frac{1}{\pi}\int_{-\pi}^{\pi} h(t) \cos t \, dt$$

$$= \frac{1}{\pi}\left[\int_{-\pi}^{-3\pi/4} -50 \cos t \, dt + \int_{-3\pi/4}^{-\pi/2} 0 \cos t \, dt + \int_{-\pi/2}^{-\pi/4} -30 \cos t \, dt \right.$$
$$\left. + \int_{-\pi/4}^{\pi/4} 80 \cos t \, dt + \int_{\pi/4}^{\pi/2} -30 \cos t \, dt + \int_{\pi/2}^{3\pi/4} 0 \cos t \, dt + \int_{3\pi/4}^{\pi} -50 \cos t \, dt\right]$$

$$= \frac{1}{\pi}\left[-50 \sin t \Big|_{-\pi}^{-3\pi/4} - 30 \sin t \Big|_{-\pi/2}^{-\pi/4} \right.$$
$$\left. + 80 \sin t \Big|_{-\pi/4}^{\pi/4} - 30 \sin t \Big|_{\pi/4}^{\pi/2} - 50 \sin t \Big|_{3\pi/4}^{\pi}\right]$$

$$= \frac{1}{\pi}\left[-50\left(-\frac{\sqrt{2}}{2} - 0\right) - 30\left(-\frac{\sqrt{2}}{2} - (-1)\right) + 80\left(\frac{\sqrt{2}}{2} - \left(-\frac{\sqrt{2}}{2}\right)\right)\right.$$
$$\left. - 30\left(1 - \frac{\sqrt{2}}{2}\right) - 50\left(0 - \frac{\sqrt{2}}{2}\right)\right]$$

$$= \frac{1}{\pi}[25\sqrt{2} + 15\sqrt{2} - 30 + 40\sqrt{2} + 40\sqrt{2} - 30 + 15\sqrt{2} + 25\sqrt{2}]$$

$$= \frac{1}{\pi}[160\sqrt{2} - 60] = 52.93,$$

$$b_1 = \frac{1}{\pi}\int_{-\pi}^{\pi} h(t) \sin t \, dt$$

$$= \frac{1}{\pi}\left[\int_{-\pi}^{-3\pi/4} -50 \sin t \, dt + \int_{-3\pi/4}^{-\pi/2} 0 \sin t \, dt + \int_{-\pi/2}^{-\pi/4} -30 \sin t \, dt \right.$$
$$\left. + \int_{-\pi/4}^{\pi/4} 80 \sin t \, dt + \int_{\pi/4}^{\pi/2} -30 \sin t \, dt + \int_{\pi/2}^{3\pi/4} 0 \sin t \, dt + \int_{3\pi/4}^{\pi} -50 \sin t \, dt\right]$$

$$= \frac{1}{\pi}\left[50 \cos t \Big|_{-\pi}^{-3\pi/4} + 30 \cos t \Big|_{-\pi/2}^{-\pi/4} - 80 \cos t \Big|_{-\pi/4}^{\pi/4} + 30 \cos t \Big|_{\pi/4}^{\pi/2} + 50 \cos t \Big|_{3\pi/4}^{\pi}\right]$$

$$= \frac{1}{\pi}\left[50\left(-\frac{\sqrt{2}}{2} - (-1)\right) + 30\left(\frac{\sqrt{2}}{2} - 0\right) - 80\left(\frac{\sqrt{2}}{2} - \frac{\sqrt{2}}{2}\right)\right.$$

$$+30\left(0-\frac{\sqrt{2}}{2}\right)+50\left(-1-\left(-\frac{\sqrt{2}}{2}\right)\right)\Bigg]$$
$$=\frac{1}{\pi}\left[-25\sqrt{2}+50+15\sqrt{2}-0-15\sqrt{2}-50+25\sqrt{2}\right]=\frac{1}{\pi}(0)=0.$$

Also, we could have just noted that $b_1 = \frac{1}{\pi}\int_{-\pi}^{\pi} h(t)\sin t\, dt = 0$ because $h(t)\sin t$ is an odd function. Substituting in, we get

$$a_0 + a_1\cos t + b_1\sin t = 0 + 52.93\cos t + 0 = A\cos t.$$

So $A = 52.93$.

15. The energy spectrum of the flute shows that the first two harmonics have equal energies and contribute the most energy by far. The higher harmonics contribute relatively little energy. In contrast, the energy spectrum of the bassoon shows the comparative weakness of the first two harmonics to the third harmonic which is the strongest component.

16. Let $f(x) = a_k\cos kx + b_k\sin kx$. Then the energy of f is given by

$$\frac{1}{\pi}\int_{-\pi}^{\pi}(f(x))^2\,dx = \frac{1}{\pi}\int_{-\pi}^{\pi}(a_k\cos kx + b_k\sin kx)^2\,dx$$
$$= \frac{1}{\pi}\int_{-\pi}^{\pi}(a_k^2\cos^2 kx - 2a_kb_k\cos kx\sin kx + b_k^2\sin^2 kx)\,dx$$
$$= \frac{1}{\pi}\left[a_k^2\int_{-\pi}^{\pi}\cos^2 kx\,dx - 2a_kb_k\int_{-\pi}^{\pi}\cos kx\sin kx\,dx + b_k^2\int_{-\pi}^{\pi}\sin^2 kx\,dx\right]$$
$$= \frac{1}{\pi}\left[a_k^2\pi - 2a_kb_k\cdot 0 + b_k^2\pi\right] = a_k^2 + b_k^2.$$

17. Since each square in the graph has area $\left(\frac{\pi}{4}\right)\cdot(0.2)$,

$$a_0 = \frac{1}{2\pi}\int_{-\pi}^{\pi} f(x)\,dx$$
$$= \frac{1}{2\pi}\cdot\left(\frac{\pi}{4}\right)\cdot(0.2)\left[\text{Number of squares under graph above } x\text{-axis}\right.$$
$$\left.- \text{Number of squares above graph below } x \text{ axis}\right]$$
$$\approx \frac{1}{2\pi}\cdot\left(\frac{\pi}{4}\right)\cdot(0.2)\cdot[13+11-14]=0.25.$$

Approximate the Fourier coefficients using Riemann sums.

$$a_1 = \frac{1}{\pi}\int_{-\pi}^{\pi} f(x)\cos x\,dx$$
$$\approx \frac{1}{\pi}\left[f(-\pi)\cos(-\pi)+f\left(-\frac{\pi}{2}\right)\cos\left(-\frac{\pi}{2}\right)+f(0)\cos(0)+f\left(\frac{\pi}{2}\right)\cos\left(\frac{\pi}{2}\right)\right]\cdot\frac{\pi}{2}$$
$$= \frac{1}{\pi}\left[(0.92)(-1)+(1)(0)+(-1.7)(1)+(0.7)(0)\right]\cdot\frac{\pi}{2}$$
$$= -1.31$$

Similarly for b_1:

$$b_1 = \frac{1}{\pi}\int_{-\pi}^{\pi} f(x)\sin x\,dx$$
$$\approx \frac{1}{\pi}\left[f(-\pi)\sin(-\pi)+f\left(-\frac{\pi}{2}\right)\sin\left(-\frac{\pi}{2}\right)+f(0)\sin(0)+f\left(\frac{\pi}{2}\right)\sin\left(\frac{\pi}{2}\right)\right]\cdot\frac{\pi}{2}$$
$$= \frac{1}{\pi}\left[(0.92)(0)+(1)(-1)+(-1.7)(0)+(0.7)(1)\right]\cdot\frac{\pi}{2}$$
$$= -0.15.$$

So our first Fourier approximation is

$$F_1(x) = 0.25 - 1.31\cos x - 0.15\sin x.$$

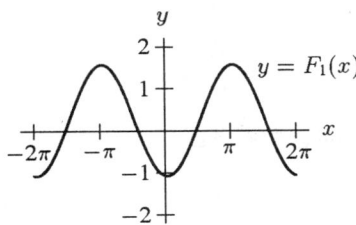

Similarly for a_2:

$$a_2 = \frac{1}{\pi}\int_{-\pi}^{\pi} f(x)\cos 2x\, dx$$

$$\approx \frac{1}{\pi}\left[f(-\pi)\cos(-2\pi) + f\left(-\frac{\pi}{2}\right)\cos(-\pi) + f(0)\cos(0) + f\left(\frac{\pi}{2}\right)\cos(-\pi)\right] \cdot \frac{\pi}{2}$$

$$= \frac{1}{\pi}\left[(0.92)(1) + (1)(-1) + (-1.7)(1) + (0.7)(-1)\right] \cdot \frac{\pi}{2}$$

$$= -1.24$$

Similarly for b_2:

$$b_2 = \frac{1}{\pi}\int_{-\pi}^{\pi} f(x)\sin 2x\, dx$$

$$\approx \frac{1}{\pi}\left[f(-\pi)\sin(-2\pi) + f\left(-\frac{\pi}{2}\right)\sin(-\pi) + f(0)\sin(0) + f\left(\frac{\pi}{2}\right)\sin(-\pi)\right] \cdot \frac{\pi}{2}$$

$$= \frac{1}{\pi}\left[(0.92)(0) + (1)(0) + (-1.7)(0) + (0.7)(0)\right] \cdot \frac{\pi}{2}$$

$$= 0.$$

So our second Fourier approximation is

$$F_2(x) = 0.25 - 1.31\cos x - 0.15\sin x - 1.24\cos 2x.$$

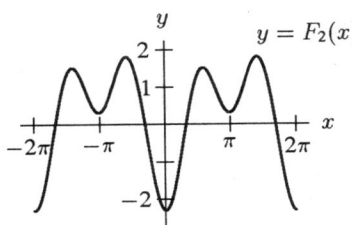

As you can see from comparing our graphs of F_1 and F_2 to the original, our estimates of the Fourier coefficients are not very accurate.

There are other methods of estimating the Fourier coefficients such as taking other Riemann sums, using Simpson's rule, and using the trapezoid rule. With each method, the greater the number of subdivisions, the more accurate the estimates of the Fourier coefficients.

The actual function graphed in the problem was

$$y = \frac{1}{4} - 1.3\cos x - \frac{\sin(\frac{3}{5})}{\pi}\sin x - \frac{2}{\pi}\cos 2x - \frac{\cos 1}{3\pi}\sin 2x$$

$$= 0.25 - 1.3\cos x - 0.18\sin x - 0.63\cos 2x - 0.057\sin 2x.$$

18. The Fourier series for f is
$$f(x) = a_0 + \sum_{k=1}^{\infty} a_k \cos kx + \sum_{k=1}^{\infty} b_k \sin kx.$$
Pick any positive integer m. Then multiply through by $\sin mx$, to get
$$f(x) \sin mx = a_0 \sin mx + \sum_{k=1}^{\infty} a_k \cos kx \sin mx + \sum_{k=1}^{\infty} b_k \sin kx \sin mx.$$
Now, integrate term-by-term on the interval $[-\pi, \pi]$ to get
$$\int_{-\pi}^{\pi} f(x) \sin mx \, dx = \int_{-\pi}^{\pi} \left(a_0 \sin mx + \sum_{k=1}^{\infty} a_k \cos kx \sin mx + \sum_{k=1}^{\infty} b_k \sin kx \sin mx \right) dx$$
$$= a_0 \int_{-\pi}^{\pi} \sin mx \, dx + \sum_{k=1}^{\infty} \left(a_k \int_{-\pi}^{\pi} \cos kx \sin mx \, dx \right)$$
$$+ \sum_{k=1}^{\infty} \left(b_k \int_{-\pi}^{\pi} \sin kx \sin mx \, dx \right).$$
Since m is a positive integer, we know that the first term of the above expression is zero (because $\int_{-\pi}^{\pi} \sin mx \, dx = 0$). Since $\int_{-\pi}^{\pi} \cos kx \sin mx \, dx = 0$, we know that everything in the first infinite sum is zero. Since $\int_{-\pi}^{\pi} \sin kx \sin mx \, dx = 0$ where $k \neq m$, the second infinite sum reduces down to the case where $k = m$ so
$$\int_{-\pi}^{\pi} f(x) \sin mx \, dx = b_m \int_{-\pi}^{\pi} \sin mx \sin mx \, dx = b_m \pi.$$
Divide by π to get
$$b_m = \frac{1}{\pi} \int_{-\pi}^{\pi} f(x) \sin mx \, dx.$$

19. (a)

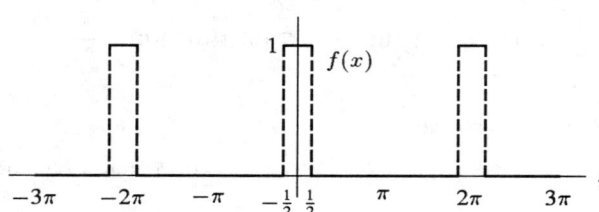

The energy of the pulse train f is
$$E = \frac{1}{\pi} \int_{-\pi}^{\pi} (f(x))^2 \, dx = \frac{1}{\pi} \int_{-1/2}^{1/2} 1^2 \, dx = \frac{1}{\pi} \left(\frac{1}{2} - \left(-\frac{1}{2}\right) \right) = \frac{1}{\pi}.$$

Next, find the Fourier coefficients:
$$a_0 = \text{average value of } f \text{ on } [-\pi, \pi] = \frac{1}{2\pi}(\text{Area}) = \frac{1}{2\pi}(1) = \frac{1}{2\pi},$$

$$a_k = \frac{1}{\pi} \int_{-\pi}^{\pi} f(x) \cos kx \, dx = \frac{1}{\pi} \int_{-1/2}^{1/2} \cos kx \, dx = \frac{1}{k\pi} \sin kx \Big|_{-1/2}^{1/2}$$
$$= \frac{1}{k\pi} \left(\sin\left(\frac{k}{2}\right) - \sin\left(-\frac{k}{2}\right) \right) = \frac{1}{k\pi} \left(2 \sin\left(\frac{k}{2}\right) \right),$$

$$b_k = \frac{1}{\pi} \int_{-\pi}^{\pi} f(x) \sin kx \, dx = \frac{1}{\pi} \int_{-1/2}^{1/2} \sin kx \, dx = -\frac{1}{k\pi} \cos kx \Big|_{-1/2}^{1/2}$$
$$= -\frac{1}{k\pi} \left(\cos\left(\frac{k}{2}\right) - \cos\left(-\frac{k}{2}\right) \right) = \frac{1}{k\pi}(0) = 0.$$

The energy of f contained in the constant term is

$$A_0^2 = 2a_0^2 = 2\left(\frac{1}{2\pi}\right)^2 = \frac{1}{2\pi^2}$$

which is

$$\frac{A_0^2}{E} = \frac{1/2\pi^2}{1/\pi} = \frac{1}{2\pi} \approx 0.159155 = 15.9155\% \quad \text{of the total.}$$

The fraction of energy contained in the first harmonic is

$$\frac{A_1^2}{E} = \frac{a_1^2}{E} = \frac{\left(\frac{2\sin\frac{1}{2}}{\pi}\right)^2}{\frac{1}{\pi}} \approx 0.292653.$$

The fraction of energy contained in both the constant term and the first harmonic together is

$$\frac{A_0^2}{E} + \frac{A_1^2}{E} \approx 0.159155 + 0.292653 = 0.451808\%.$$

(b) The formula for the energy of the kth harmonic is

$$A_k^2 = a_k^2 + b_k^2 = \left(\frac{2\sin\frac{k}{2}}{k\pi}\right)^2 + 0^2 = \frac{4\sin^2\frac{k}{2}}{k^2\pi^2}.$$

By graphing it as a continuous function for $k \geq 1$, we see its overall behavior as k gets larger. See Figure 9.4. The energy spectrum for the first five terms is graphed below as well in Figure 9.5.

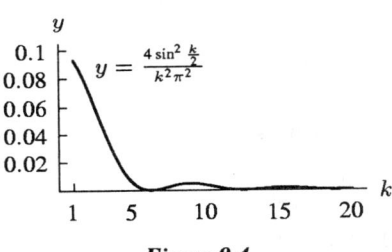

Figure 9.4

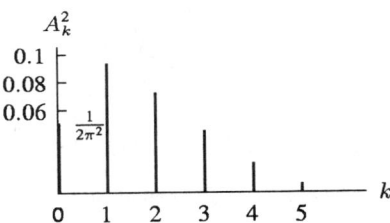

Figure 9.5

(c) The constant term and the first five harmonics are needed to capture 90% of the energy of f. This was determined by adding the fractions of energy of f contained in each harmonic until the sum reached at least 90% of the total energy of f:

$$\frac{A_0^2}{E} + \frac{A_1^2}{E} + \frac{A_2^2}{E} + \frac{A_3^2}{E} + \frac{A_4^2}{E} + \frac{A_5^2}{E} \approx 90.1995\%.$$

(d) $F_5(x) = \frac{1}{2\pi} + \frac{2\sin(\frac{1}{2})}{\pi}\cos x + \frac{\sin 1}{\pi}\cos 2x + \frac{2\sin(\frac{3}{2})}{3\pi}\cos 3x + \frac{\sin 2}{2\pi}\cos 4x + \frac{2\sin(\frac{5}{2})}{5\pi}\cos 5x$

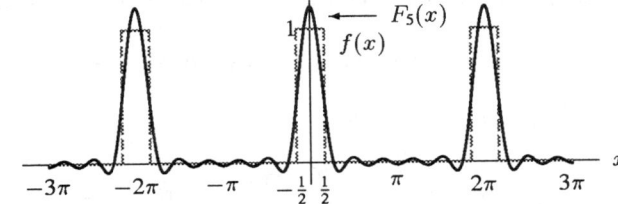

460 CHAPTER NINE /SOLUTIONS

20. (a)

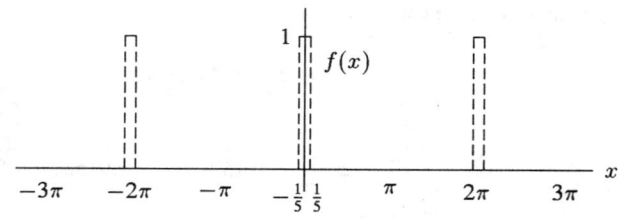

The energy of the pulse train f is

$$E = \frac{1}{\pi}\int_{-\pi}^{\pi}(f(x))^2\,dx = \frac{1}{\pi}\int_{-1/5}^{1/5}1^2\,dx = \frac{1}{\pi}\left(\frac{1}{5}-\left(-\frac{1}{5}\right)\right) = \frac{2}{5\pi}.$$

Next, find the Fourier coefficients:

$$a_0 = \text{average value of } f \text{ on } [-\pi,\pi] = \frac{1}{2\pi}(\text{Area}) = \frac{1}{2\pi}\left(\frac{2}{5}\right) = \frac{1}{5\pi},$$

$$a_k = \frac{1}{\pi}\int_{-\pi}^{\pi} f(x)\cos kx\,dx = \frac{1}{\pi}\int_{-1/5}^{1/5}\cos kx\,dx = \frac{1}{k\pi}\sin kx\Big|_{-1/5}^{1/5}$$
$$= \frac{1}{k\pi}\left(\sin\left(\frac{k}{5}\right)-\sin\left(-\frac{k}{5}\right)\right) = \frac{1}{k\pi}\left(2\sin\left(\frac{k}{5}\right)\right),$$

$$b_k = \frac{1}{\pi}\int_{-\pi}^{\pi} f(x)\sin kx\,dx = \frac{1}{\pi}\int_{-1/5}^{1/5}\sin kx\,dx = -\frac{1}{k\pi}\cos kx\Big|_{-1/5}^{1/5}$$
$$= -\frac{1}{k\pi}\left(\cos\left(\frac{k}{5}\right)-\cos\left(-\frac{k}{5}\right)\right) = \frac{1}{k\pi}(0) = 0.$$

The energy of f contained in the constant term is

$$A_0^2 = 2a_0^2 = 2\left(\frac{1}{5\pi}\right)^2 = \frac{2}{25\pi^2}$$

which is

$$\frac{A_0^2}{E} = \frac{2/25\pi^2}{2/5\pi} = \frac{1}{5\pi} \approx 0.063662 = 6.3662\% \quad \text{of the total.}$$

The fraction of energy contained in the first harmonic is

$$\frac{A_1^2}{E} = \frac{a_1^2}{E} = \frac{\left(\frac{2\sin\frac{1}{5}}{\pi}\right)^2}{\frac{2}{5\pi}} \approx 0.12563.$$

The fraction of energy contained in both the constant term and the first harmonic together is

$$\frac{A_0^2}{E} + \frac{A_1^2}{E} \approx 0.06366 + 0.12563 = 0.18929 = 18.929\%.$$

(b) The formula for the energy of the kth harmonic is

$$A_k^2 = a_k^2 + b_k^2 = \left(\frac{2\sin\frac{k}{5}}{k\pi}\right)^2 + 0^2 = \frac{4\sin^2\frac{k}{5}}{k^2\pi^2}.$$

By graphing this formula as a continuous function for $k \geq 1$, we see its overall behavior as k gets larger in Figure 9.6. The energy spectrum for the first five terms is shown in Figure 9.7.

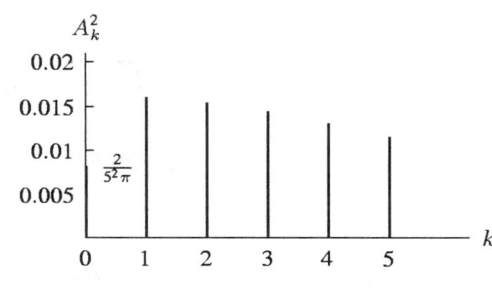

Figure 9.6   **Figure 9.7**

(c) The constant term and the first five harmonics contain
$$\frac{A_0^2}{E} + \frac{A_1^2}{E} + \frac{A_2^2}{E} + \frac{A_3^2}{E} + \frac{A_4^2}{E} + \frac{A_5^2}{E} \approx 61.5255\%$$
of the total energy of f.

(d) The fifth Fourier approximation to f is
$$F_5(x) = \tfrac{1}{5\pi} + \tfrac{2\sin(\tfrac{1}{5})}{\pi}\cos x + \tfrac{\sin(\tfrac{2}{5})}{\pi}\cos 2x + \tfrac{2\sin(\tfrac{3}{5})}{3\pi}\cos 3x + \tfrac{\sin(\tfrac{4}{5})}{2\pi}\cos 4x + \tfrac{2\sin 1}{5\pi}\cos 5x.$$

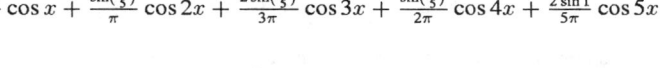

For comparison, below is the thirteenth Fourier approximation to f.

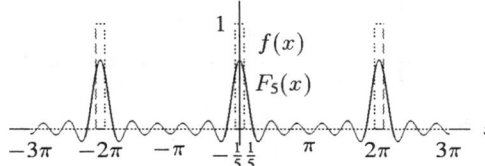

21. (a)

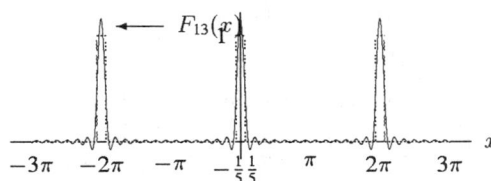

The energy of the pulse train f is
$$E = \frac{1}{\pi}\int_{-\pi}^{\pi}(f(x))^2\,dx = \frac{1}{\pi}\int_{-1}^{1}1^2 = \frac{1}{\pi}(1-(-1)) = \frac{2}{\pi}.$$

Next, find the Fourier coefficients:
$$a_0 = \text{average value of } f \text{ on } [-\pi,\pi] = \frac{1}{2\pi}(\text{Area}) = \frac{1}{2\pi}(2) = \frac{1}{\pi},$$

$$a_k = \frac{1}{\pi}\int_{-\pi}^{\pi} f(x)\cos kx\, dx = \frac{1}{\pi}\int_{-1}^{1} \cos kx\, dx = \frac{1}{k\pi}\sin kx\bigg|_{-1}^{1}$$
$$= \frac{1}{k\pi}(\sin k - \sin(-k)) = \frac{1}{k\pi}(2\sin k),$$

$$b_k = \frac{1}{\pi}\int_{-\pi}^{\pi} f(x)\sin kx\, dx = \frac{1}{\pi}\int_{-1}^{1} \sin kx\, dx = -\frac{1}{k\pi}\cos kx\bigg|_{-1}^{1}$$
$$= -\frac{1}{k\pi}(\cos k - \cos(-k)) = \frac{1}{k\pi}(0) = 0.$$

The energy of f contained in the constant term is
$$A_0^2 = 2a_0^2 = 2\left(\frac{1}{\pi}\right)^2 = \frac{2}{\pi^2}$$

which is
$$\frac{A_0^2}{E} = \frac{2/\pi^2}{2/\pi} = \frac{1}{\pi} \approx 0.3183 = 31.83\% \quad \text{of the total.}$$

The fraction of energy contained in the first harmonic is
$$\frac{A_1^2}{E} = \frac{a_1^2}{E} = \frac{\left(\frac{2\sin 1}{\pi}\right)^2}{\frac{2}{\pi}} \approx 0.4508 = 45.08\%.$$

The fraction of energy contained in both the constant term and the first harmonic together is
$$\frac{A_0^2}{E} + \frac{A_1^2}{E} \approx 0.7691 = 76.91\%.$$

(b) The fraction of energy contained in the second harmonic is
$$\frac{A_2^2}{E} = \frac{a_2^2}{E} = \frac{\left(\frac{\sin 2}{\pi}\right)^2}{\frac{2}{\pi}} \approx 0.1316 = 13.16\%$$

so the fraction of energy contained in the constant term and first two harmonics is
$$\frac{A_0^2}{E} + \frac{A_1^2}{E} + \frac{A_2^2}{E} \approx 0.7691 + 0.1316 = 0.9007 = 90.07\%.$$

Therefore, the constant term and the first two harmonics are needed to capture 90% of the energy of f.

(c)
$$F_3(x) = \frac{1}{\pi} + \frac{2\sin 1}{\pi}\cos x + \frac{\sin 2}{\pi}\cos 2x + \frac{2\sin 3}{3\pi}\cos 3x$$

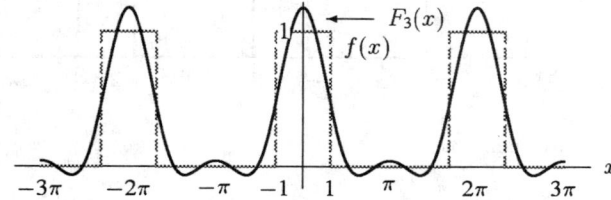

22. As c gets closer and closer to 0, the energy of the pulse train will also approach 0, since

$$E = \frac{1}{\pi}\int_{-\pi}^{\pi}(f(x))^2\,dx = \frac{1}{\pi}\int_{-c/2}^{c/2}1^2\,dx = \frac{1}{\pi}\left(\frac{c}{2}-\left(-\frac{c}{2}\right)\right) = \frac{c}{\pi}.$$

The energy spectrum shows the *relative* distribution of the energy of f among its harmonics. The fraction of energy carried by each harmonic gets smaller as c gets closer to 0, as shown by comparing the kth terms of the Fourier series for pulse trains with $c = 2, 1, 0.4$. For instance, notice that the *fraction* or *percentage* of energy carried by the constant term gets smaller as c gets smaller; the same is true for the energy carried by the first harmonic.

If each harmonic contributes less energy, then more harmonics are needed to capture a fixed percentage of energy. For example, if $c = 2$, only the constant term and the first two harmonics are needed to capture 90% of the total energy of that pulse train. If $c = 1$, the constant term and the first five harmonics are needed to get 90% of the energy of that pulse train. If $c = 0.4$, the constant term and the first thirteen harmonics are needed to get 90% of the energy of that pulse train. This means that more harmonics, or more terms in the series, are needed to get an accurate approximation. Compare the graphs of the fifth and thirteenth Fourier approximations of f in Problem 20.

23. By formula II-11 of the integral table,

$$\int_{-\pi}^{\pi}\cos kx \cos mx\,dx = \frac{1}{m^2-k^2}\left(m\cos(kx)\sin(mx)-k\sin(kx)\cos(mx)\right)\bigg|_{-\pi}^{\pi}.$$

Again, since $\sin(n\pi) = 0$ for any integer n, it is easy to see that this expression is simply 0.

24. We make the substitution $u = mx$, $dx = \frac{1}{m}du$. Then

$$\int_{-\pi}^{\pi}\cos^2 mx\,dx = \frac{1}{m}\int_{u=-m\pi}^{u=m\pi}\cos^2 u\,du.$$

By Formula IV-18 of the integral table, this equals

$$\frac{1}{m}\left[\frac{1}{2}\cos u \sin u\right]\bigg|_{-m\pi}^{m\pi} + \frac{1}{m}\frac{1}{2}\int_{-m\pi}^{m\pi}1\,du = 0 + \frac{1}{2m}u\bigg|_{-m\pi}^{m\pi} = \frac{1}{2m}u\bigg|_{-m\pi}^{m\pi}$$

$$= \frac{1}{2m}(2m\pi) = \pi.$$

25. The easiest way to do this is to use Problem 24.

$$\int_{-\pi}^{\pi}\sin^2 mx\,dx = \int_{-\pi}^{\pi}(1-\cos^2 mx)\,dx = \int_{-\pi}^{\pi}dx - \int_{-\pi}^{\pi}\cos^2 mx\,dx$$

$$= 2\pi - \pi \quad \text{using Problem 24}$$

$$= \pi.$$

26. By formula II-12 of the integral table,

$$\int_{-\pi}^{\pi}\sin kx \cos mx\,dx$$

$$= \frac{1}{m^2-k^2}\left(m\sin(kx)\sin(mx)+k\cos(kx)\cos(mx)\right)\bigg|_{-\pi}^{\pi}$$

$$= \frac{1}{m^2-k^2}\bigg[m\sin(k\pi)\sin(m\pi)+k\cos(k\pi)\cos(m\pi)$$

$$-m\sin(-k\pi)\sin(-m\pi)-k\cos(-k\pi)\cos(-m\pi)\bigg].$$

Since k and m are positive integers, $\sin(k\pi) = \sin(m\pi) = \sin(-k\pi) = \sin(-m\pi) = 0$. Also, $\cos(k\pi) = \cos(-k\pi)$ since $\cos x$ is even. Thus this expression reduces to 0. [Note: since $\sin kx \cos mx$ is odd, so $\int_{-\pi}^{\pi}\sin kx \cos mx\,dx$ must be 0.]

464 CHAPTER NINE /SOLUTIONS

27. Using formula II-10 in the integral table,

$$\int_{-\pi}^{\pi} \sin kx \sin mx \, dx = \frac{1}{m^2 - k^2} \left[k \cos(kx) \sin(mx) - m \sin(kx) \cos(mx) \right] \Big|_{-\pi}^{\pi}.$$

Again, since $\sin(n\pi) = 0$ for all integers n, this expression reduces to 0.

Solutions for Chapter 9 Review

1. $e^x \approx 1 + e(x-1) + \frac{e}{2}(x-1)^2$

2. $\ln x \approx \ln 2 + \frac{1}{2}(x-2) - \frac{1}{8}(x-2)^2$

3. $\sin x \approx -\frac{1}{\sqrt{2}} + \frac{1}{\sqrt{2}}\left(x + \frac{\pi}{4}\right) + \frac{1}{2\sqrt{2}}\left(x + \frac{\pi}{4}\right)^2$

4. $f'(x) = 3x^2 + 14x - 5$, $f''(x) = 6x + 14$, $f'''(x) = 6$. The Taylor polynomial about $x = 1$ is

$$P_3(x) = 4 + \frac{12}{1!}(x-1) + \frac{20}{2!}(x-1)^2 + \frac{6}{3!}(x-1)^3$$
$$= 4 + 12(x-1) + 10(x-1)^2 + (x-1)^3.$$

Notice that if you multiply out and collect terms in $P_3(x)$, you will get $f(x)$ back.

5.
$$\theta^2 \cos \theta^2 = \theta^2 \left(1 - \frac{(\theta^2)^2}{2!} + \frac{(\theta^2)^4}{4!} - \frac{(\theta^2)^6}{6!} + \cdots \right)$$
$$= \theta^2 - \frac{\theta^6}{2!} + \frac{\theta^{10}}{4!} - \frac{\theta^{14}}{6!} + \cdots$$

6. Substituting $y = t^2$ in $\sin y = y - \frac{y^3}{3!} + \frac{y^5}{5!} - \frac{y^7}{7!} + \cdots$ gives

$$\sin t^2 = t^2 - \frac{t^6}{3!} + \frac{t^{10}}{5!} - \frac{t^{14}}{7!} + \cdots$$

7.
$$\frac{1}{\sqrt{4-x}} = \frac{1}{2\sqrt{1 - \frac{x}{2}}} = \frac{1}{2}\left(1 - \frac{x}{2}\right)^{-\frac{1}{2}}$$
$$= \frac{1}{2}\left(1 - \left(-\frac{1}{2}\right)\left(\frac{x}{2}\right) + \frac{1}{2!}\left(-\frac{1}{2}\right)\left(-\frac{3}{2}\right)\left(\frac{x}{2}\right)^2 \right.$$
$$\left. - \frac{1}{3!}\left(-\frac{1}{2}\right)\left(-\frac{3}{2}\right)\left(-\frac{5}{2}\right)\left(\frac{x}{2}\right)^3 + \cdots \right)$$
$$= \frac{1}{2} + \frac{1}{8}x + \frac{3}{64}x^2 + \frac{5}{256}x^3 + \cdots$$

8. Substituting $y = -4z^2$ into $\frac{1}{1+y} = 1 - y + y^2 - y^3 + \cdots$ gives

$$\frac{1}{1 - 4z^2} = 1 + 4z^2 + 16z^4 + 64z^6 + \cdots$$

9.
$$\frac{a}{a+b} = \frac{a}{a(1+\frac{b}{a})} = \left(1+\frac{b}{a}\right)^{-1} = 1 - \frac{b}{a} + \left(\frac{b}{a}\right)^2 - \left(\frac{b}{a}\right)^3 + \cdots$$

10.
$$\sqrt{R-r} = \sqrt{R}\left(1 - \frac{r}{R}\right)^{\frac{1}{2}}$$
$$= \sqrt{R}\left(1 + \frac{1}{2}\left(-\frac{r}{R}\right) + \frac{1}{2!}\left(\frac{1}{2}\right)\left(-\frac{1}{2}\right)\left(-\frac{r}{R}\right)^2 \right.$$
$$\left. + \frac{1}{3!}\left(\frac{1}{2}\right)\left(-\frac{1}{2}\right)\left(-\frac{3}{2}\right)\left(-\frac{r}{R}\right)^3 + \cdots\right)$$
$$= \sqrt{R}\left(1 - \frac{1}{2}\frac{r}{R} - \frac{1}{8}\frac{r^2}{R^2} - \frac{1}{16}\frac{r^3}{R^3} - \cdots\right)$$

11. Infinite geometric series with $a = 1$, $x = -1/3$, so
$$\text{Sum} = \frac{1}{1-(-1/3)} = \frac{3}{4}.$$

12. Finite geometric series which can be rewritten as
$$8\left(1 + \frac{1}{2} + \frac{1}{4} + \frac{1}{8} + \cdots + \frac{1}{2^{13}}\right) = 8\left(\frac{1 - 1/2^{14}}{1 - 1/2}\right) = 16\left(1 - \frac{1}{2^{14}}\right).$$

13. Factoring out a 3, we see
$$3\left(1 + 1 + \frac{1}{2!} + \frac{1}{3!} + \frac{1}{4!} + \frac{1}{5!} + \cdots\right) = 3e^1 = 3e.$$

14. Factoring out a 0.1, we see
$$0.1\left(0.1 - \frac{(0.1)^3}{3!} + \frac{(0.1)^5}{5!} - \frac{(0.1)^7}{7!} + \cdots\right) = 0.1\sin(0.1).$$

15. Write out series expansions about $x = 0$, and compare the first few terms:
$$\sin x = x - \frac{x^3}{3!} + \frac{x^5}{5!} + \cdots$$
$$\ln(1+x) = x - \frac{x^2}{2} + \frac{x^3}{3} - \cdots$$
$$1 - \cos x = 1 - \left(1 - \frac{x^2}{2!} + \frac{x^4}{4!} - \cdots\right) = \frac{x^2}{2!} - \frac{x^4}{4!} + \cdots$$
$$e^x - 1 = x + \frac{x^2}{2!} + \frac{x^3}{3!} + \cdots$$
$$\arctan x = \int \frac{dx}{1+x^2} = \int (1 - x^2 + x^4 - \cdots)\,dx$$
$$= x - \frac{x^3}{3} + \frac{x^5}{5} + \cdots \quad \text{(note that the arbitrary constant is 0)}$$
$$x\sqrt{1-x} = x(1-x)^{1/2} = x\left(1 - \frac{1}{2}x + \frac{(1/2)(-1/2)}{2}x^2 + \cdots\right)$$
$$= x - \frac{x^2}{2} + \frac{x^3}{8} + \cdots$$

So, considering just the first term or two (since we are interested in small x)
$$1 - \cos x < x\sqrt{1-x} < \ln(1+x) < \arctan x < \sin x < x < e^x - 1.$$

16.

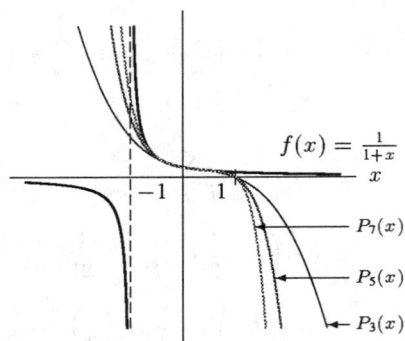

The graph suggests that the Taylor polynomials converge to $f(x) = \dfrac{1}{1+x}$ on the interval $(-1, 1)$.

17. First we use the Taylor series expansion for $\ln(1+t)$,

$$\ln(1+t) = t - \frac{1}{2}t^2 + \frac{1}{3}t^3 - \frac{1}{4}t^4 + \cdots$$

to find the Taylor series expansion of $\ln(1 + x + x^2)$ by putting $t = x + x^2$. We get

$$\ln(1 + x + x^2) = x + \frac{1}{2}x^2 - \frac{2}{3}x^3 + \frac{1}{4}x^4 + \cdots.$$

Next we use the Taylor series for $\sin x$ to get

$$\sin^2 x = (\sin x)^2 = (x - \frac{1}{6}x^3 + \frac{1}{120}x^5 - \cdots)^2 = x^2 - \frac{1}{3}x^4 + \cdots.$$

Finally,

$$\frac{\ln(1 + x + x^2) - x}{\sin^2 x} = \frac{\frac{1}{2}x^2 - \frac{2}{3}x^3 + \frac{1}{4}x^4 + \cdots}{x^2 - \frac{1}{3}x^4 + \cdots} \to \frac{1}{2}, \quad \text{as} \quad x \to 0.$$

18. (a) The series for $\frac{\sin 2\theta}{\theta}$ is

$$\frac{\sin 2\theta}{\theta} = \frac{1}{\theta}\left(2\theta - \frac{(2\theta)^3}{3!} + \frac{(2\theta)^5}{5!} - \cdots\right) = 2 - \frac{4\theta^2}{3} + \frac{4\theta^4}{15} - \cdots$$

so $\lim_{\theta \to 0} \dfrac{\sin 2\theta}{\theta} = 2$.

(b) Near $\theta = 0$, we make the approximation

$$\frac{\sin 2\theta}{\theta} \approx 2 - \frac{4}{3}\theta^2$$

so the parabola is $y = 2 - \frac{4}{3}\theta^2$.

19. (a) $f(t) = te^t$.

Use the Taylor expansion for e^t:

$$f(t) = t\left(1 + t + \frac{t^2}{2!} + \frac{t^3}{3!} + \cdots\right)$$

$$= t + t^2 + \frac{t^3}{2!} + \frac{t^4}{3!} + \cdots$$

(b)

$$\int_0^x f(t)\,dt = \int_0^x te^t\,dt = \int_0^x \left(t + t^2 + \frac{t^3}{2!} + \frac{t^4}{3!} + \cdots\right)dt$$

$$= \left.\frac{t^2}{2} + \frac{t^3}{3} + \frac{t^4}{4 \cdot 2!} + \frac{t^5}{5 \cdot 3!} + \cdots\right|_0^x$$

$$= \frac{x^2}{2} + \frac{x^3}{3} + \frac{x^4}{4 \cdot 2!} + \frac{x^5}{5 \cdot 3!} + \cdots$$

(c) Substitute $x = 1$:
$$\int_0^1 te^t\, dt = \frac{1}{2} + \frac{1}{3} + \frac{1}{4\cdot 2!} + \frac{1}{5\cdot 3!} + \cdots$$

In the integral above, to integrate by parts, let $u = t$, $dv = e^t\, dt$, so $du = dt$, $v = e^t$.
$$\int_0^1 te^t\, dt = te^t\Big|_0^1 - \int_0^1 e^t\, dt = e - (e-1) = 1$$

Hence
$$\frac{1}{2} + \frac{1}{3} + \frac{1}{4\cdot 2!} + \frac{1}{5\cdot 3!} + \cdots = 1.$$

20. (a) Since the expression under the square root sign, $1 - \frac{v^2}{c^2}$ must be positive in order to give a real value of m, we have
$$1 - \frac{v^2}{c^2} > 0$$
$$\frac{v^2}{c^2} < 1$$
$$v^2 < c^2,$$
so $-c < v < c.$

In other words, the object can never travel faster that the speed of light.

(b)

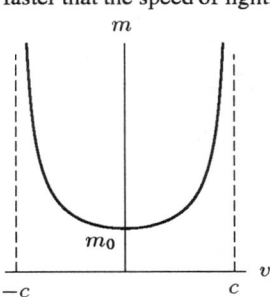

(c) Notice that $m = m_0\left(1 - \frac{v^2}{c^2}\right)^{-1/2}$. If we substitute $u = -\frac{v^2}{c^2}$, we get $m = m_0(1+u)^{-1/2}$ and we can use the binomial expansion to get:
$$m = m_0\left(1 - \frac{1}{2}u + \frac{(-1/2)(-3/2)}{2!}u^2 + \cdots\right)$$
$$= m_0\left(1 + \frac{1}{2}\frac{v^2}{c^2} + \frac{3}{8}\frac{v^4}{c^4} + \cdots\right).$$

(d) We would expect this series to converge only for values of the original function that exist, namely when $|v| < c$.

21. (a) To find when V takes on its minimum values, set $\frac{dV}{dr} = 0$. So
$$-V_0 \frac{d}{dr}\left(2\left(\frac{r_0}{r}\right)^6 - \left(\frac{r_0}{r}\right)^{12}\right) = 0$$
$$-V_0\left(-12r_0^6 r^{-7} + 12r_0^{12} r^{-13}\right) = 0$$
$$12r_0^6 r^{-7} = 12r_0^{12} r^{-13}$$
$$r_0^6 = r^6$$
$$r = r_0.$$

Rewriting $V'(r)$ as $\frac{12r_0^6 V_0}{r^7}\left(1 - \left(\frac{r_0}{r}\right)^6\right)$, we see that $V'(r) > 0$ for $r > r_0$ and $V'(r) < 0$ for $r < r_0$. Thus, $V = -V_0(2(1)^6 - (1)^{12}) = -V_0$ is a minimum.
(Note: We discard the negative root $-r_0$ since the distance r must be positive.)

468 CHAPTER NINE /SOLUTIONS

(b)
$$V(r) = -V_0 \left(2\left(\frac{r_0}{r}\right)^6 - \left(\frac{r_0}{r}\right)^{12}\right)$$
$$V'(r) = -V_0(-12r_0^6 r^{-7} + 12r_0^{12} r^{-13})$$
$$V''(r) = -V_0(84r_0^6 r^{-8} - 156r_0^{12} r^{-14})$$

$$V(r_0) = -V_0$$
$$V'(r_0) = 0$$
$$V''(r_0) = 72V_0 r_0^{-2}$$

The Taylor series is thus:
$$V(r) = -V_0 + 72V_0 r_0^{-2} \cdot (r - r_0)^2 \cdot \frac{1}{2} + \cdots$$

(c) The difference between V and its minimum value $-V_0$ is
$$V - (-V_0) = 36V_0 \frac{(r - r_0)^2}{r_0^2} + \cdots$$
which is approximately proportional to $(r - r_0)^2$ since terms containing higher powers of $(r - r_0)$ have relatively small values for r near r_0.

(d) From part (a) we know that $dV/dr = 0$ when $r = r_0$, hence $F = 0$ when $r = r_0$. Since, if we discard powers of $(r - r_0)$ higher than the second,
$$V(r) \approx -V_0 \left(1 - 36\frac{(r - r_0)^2}{r_0^2}\right)$$
giving
$$F = -\frac{dV}{dr} \approx 72 \cdot \frac{r - r_0}{r_0^2}(-V_0) = -72V_0 \frac{r - r_0}{r_0^2}.$$
So F is approximately proportional to $(r - r_0)$.

22. (a) $F = \frac{GM}{R^2} + \frac{Gm}{(R+r)^2}$

(b) $F = \frac{GM}{R^2} + \frac{Gm}{R^2} \frac{1}{(1+\frac{r}{R})^2}$

Since $\frac{r}{R} < 1$, use the binomial expansion:
$$\frac{1}{(1+\frac{r}{R})^2} = \left(1+\frac{r}{R}\right)^{-2} = 1 - 2\left(\frac{r}{R}\right) + (-2)(-3)\frac{(\frac{r}{R})^2}{2!} + \cdots$$
$$F = \frac{GM}{R^2} + \frac{Gm}{R^2}\left[1 - 2\left(\frac{r}{R}\right) + 3\left(\frac{r}{R}\right)^2 - \cdots\right].$$

(c) Discarding higher power terms, we get
$$F \approx \frac{GM}{R^2} + \frac{Gm}{R^2} - \frac{2Gmr}{R^3}$$
$$= \frac{G(M+m)}{R^2} - \frac{2Gmr}{R^3}.$$

Looking at the expression, we see that the term $\frac{G(M+m)}{R^2}$ is the field strength at a distance R from a single particle of mass $M + m$. The correction term, $-\frac{2Gmr}{R^3}$, is negative because the field strength exerted by a particle of mass $(M + m)$ at a distance R would clearly be larger than the field strength at P in the question.

23. (a) $0.232323\ldots = 0.23 + 0.23(0.01) + 0.23(0.01)^2 + \cdots$ which is a geometric series with $a = 0.23$ and $x = 0.01$.

(b) The sum is $\frac{0.23}{1 - 0.01} = \frac{0.23}{0.99} = \frac{23}{99}$.

24. The amount of cephalexin in the body is given by $Q(t) = Q_0 e^{-kt}$, where $Q_0 = Q(0)$ and k is a constant. Since the half-life is 0.9 hours,
$$\frac{1}{2} = e^{-0.9k}, \quad k = -\frac{1}{0.9}\ln\frac{1}{2} \approx 0.8.$$

(a) After 6 hours
$$Q = Q_0 e^{-k(6)} \approx Q_0 e^{-0.8(6)} = Q_0(0.01).$$
Thus, the percentage of the cephalexin that remains after 6 hours $\approx 1\%$.

(b)
$$Q_1 = 250$$
$$Q_2 = 250 + 250(0.01)$$
$$Q_3 = 250 + 250(0.01) + 250(0.01)^2$$
$$Q_4 = 250 + 250(0.01) + 250(0.01)^2 + 250(0.01)^3$$

(c)
$$Q_3 = \frac{250(1 - (0.01)^3)}{1 - 0.01}$$
$$\approx 252.5$$
$$Q_4 = \frac{250(1 - (0.01)^4)}{1 - 0.01}$$
$$\approx 252.5$$

Thus, by the time a patient has taken three cephalexin tablets, the quantity of drug in the body has leveled off to 252.5 mg.

(d) Looking at the answers to part (b) shows that
$$Q_n = 250 + 250(0.01) + 250(0.01)^2 + \cdots + 250(0.01)^{n-1}$$
$$= \frac{250(1 - (0.01)^n)}{1 - 0.01}.$$

(e) In the long run, $n \to \infty$. So,
$$Q = \lim_{n \to \infty} Q_n = \frac{250}{1 - 0.01} = 252.5.$$

25. Since expanding $f(x + h)$ and $g(x + h)$ in Taylor series gives
$$f(x + h) = f(x) + f'(x)h + \frac{f''(x)}{2!}h^2 + \cdots,$$
$$g(x + h) = g(x) + g'(x)h + \frac{f''(x)}{2!}h^2 + \cdots,$$

we substitute to get

$$\frac{f(x+h)g(x+h) - f(x)g(x)}{h}$$
$$= \frac{(f(x) + f'(x)h + \frac{1}{2}f''(x)h^2 + \ldots)(g(x) + g'(x)h + \frac{1}{2}g''(x)h^2 + \ldots) - f(x)g(x)}{h}$$
$$= \frac{f(x)g(x) + (f'(x)g(x) + f(x)g'(x))h + \text{Terms in } h^2 \text{ and higher powers} - f(x)g(x)}{h}$$
$$= \frac{h(f'(x)g(x) + f(x)g'(x) + \text{Terms in } h \text{ and higher powers})}{h}$$
$$= f'(x)g(x) + f(x)g'(x) + \text{Terms in } h \text{ and higher powers}.$$

Thus, taking the limit as $h \to 0$, we get
$$\frac{d}{dx}(f(x)g(x)) = \lim_{h \to 0} \frac{f(x+h)g(x+h) - f(x)g(x)}{h}$$
$$= f'(x)g(x) + f(x)g'(x).$$

26. Expanding $f(y + k)$ and $g(x + h)$ in Taylor series gives
$$f(y + k) = f(y) + f'(y)k + \frac{f''(y)}{2!}k^2 + \cdots,$$
$$g(x + h) = g(x) + g'(x)h + \frac{g''(x)}{2!}h^2 + \cdots.$$

Now let $y = g(x)$ and $y + k = g(x + h)$. Then $k = g(x + h) - g(x)$ so

$$k = g'(x)h + \frac{g''(x)}{2!}h^2 + \cdots.$$

Substituting $g(x + h) = y + k$ and $y = g(x)$ in the series for $f(y + k)$ gives

$$f(g(x + h)) = f(g(x)) + f'(g(x))k + \frac{f''(g(x))}{2!}k^2 + \cdots.$$

Now, substituting for k, we get

$$f(g(x + h)) = f(g(x)) + f'(g(x)) \cdot (g'(x)h + \frac{g''(x)}{2!}h^2 + \cdots) + \frac{f''(g(x))}{2!}(g'(x)h + \ldots)^2 + \cdots$$
$$= f(g(x)) + (f'(g(x))) \cdot g'(x)h + \text{Terms in } h^2 \text{ and higher powers}.$$

So, substituting for $f(g(x + h))$ and dividing by h, we get

$$\frac{f(g(x + h)) - f(g(x))}{h} = f'(g(x)) \cdot g'(x) + \text{Terms in } h \text{ and higher powers},$$

and thus, taking the limit as $h \to 0$,

$$\frac{d}{dx}f(g(x)) = \lim_{h \to 0} \frac{f(g(x + h)) - f(g(x))}{h}$$
$$= f'(g(x)) \cdot g'(x).$$

27. (a) Notice $g'(0) = 0$ because g has a critical point at $x = 0$. So, for $n \geq 2$,

$$g(x) \approx P_n(x) = g(0) + \frac{g''(0)}{2!}x^2 + \frac{g'''(0)}{3!}x^3 + \cdots + \frac{g^{(n)}(0)}{n!}x^n.$$

(b) The Second Derivative test says that if $g''(0) > 0$, then 0 is a local minimum and if $g''(0) < 0$, 0 is a local maximum.

(c) Let $n = 2$. Then $P_2(x) = g(0) + \frac{g''(0)}{2!}x^2$. So, for x near 0,

$$g(x) - g(0) \approx \frac{g''(0)}{2!}x^2.$$

If $g''(0) > 0$, then $g(x) - g(0) \geq 0$, as long as x stays near 0. In other words, there exists a small interval around $x = 0$ such that for any x in this interval $g(x) \geq g(0)$. So $g(0)$ is a local minimum.
The case when $g''(0) < 0$ is treated similarly; then $g(0)$ is a local maximum.

28. The situation is more complicated. Let's first consider the case when $g'''(0) \neq 0$. To be specific let $g'''(0) > 0$. Then

$$g(x) \approx P_3(x) = g(0) + \frac{g'''(0)}{3!}x^3.$$

So, $g(x) - g(0) \approx \frac{g'''(0)}{3!}x^3$. (Notice that $\frac{g'''(0)}{3!} > 0$ is a constant.) Now, no matter how small an open interval I around $x = 0$ is, there are always some x_1 and x_2 in I such that $x_1 < 0$ and $x_2 > 0$, which means that $\frac{g'''(0)}{3!}x_1^3 < 0$ and $\frac{g'''(0)}{3!}x_2^3 > 0$, i.e. $g(x_1) - g(0) < 0$ and $g(x_2) - g(0) > 0$. Thus, $g(0)$ is neither a local minimum nor a local maximum. (If $g'''(0) < 0$, the same conclusion still holds. Try it! The reasoning is similar.)

Now let's consider the case when $g'''(0) = 0$. If $g^{(4)}(0) > 0$, then by the fourth degree Taylor polynomial approximation to g at $x = 0$, we have

$$g(x) - g(0) \approx \frac{g^{(4)}(0)}{4!}x^4 > 0$$

for x in a small open interval around $x = 0$. So $g(0)$ is a local minimum. (If $g^{(4)}(0) < 0$, then $g(0)$ is a local maximum.)

In general, suppose that $g^{(k)}(0) \neq 0$, $k \geq 2$, and all the derivatives of g with order less than k are 0. In this case g looks like cx^k near $x = 0$, which determines its behavior there. Then $g(0)$ is neither a local minimum nor a local maximum if k is odd. For k even, $g(0)$ is a local minimum if $g^{(k)}(0) > 0$, and $g(0)$ is a local maximum if $g^{(k)}(0) < 0$.

29. Let us begin by finding the Fourier coefficients for $f(x)$. Since f is odd, $\int_{-\pi}^{\pi} f(x)\,dx = 0$ and $\int_{-\pi}^{\pi} f(x)\cos nx\,dx = 0$. Thus $a_i = 0$ for all $i \geq 0$. On the other hand,

$$b_i = \frac{1}{\pi}\int_{-\pi}^{\pi} f(x)\sin nx\,dx = \frac{1}{\pi}\left[\int_{-\pi}^{0} -\sin(nx)\,dx + \int_{0}^{\pi}\sin(nx)\,dx\right]$$

$$= \frac{1}{\pi}\left[\frac{1}{n}\cos(nx)\bigg|_{-\pi}^{0} - \frac{1}{n}\cos(nx)\bigg|_{0}^{\pi}\right]$$

$$= \frac{1}{n\pi}\left[\cos 0 - \cos(-n\pi) - \cos(n\pi) + \cos 0\right]$$

$$= \frac{2}{n\pi}\left(1 - \cos(n\pi)\right).$$

Since $\cos(n\pi) = (-1)^n$, this is 0 if n is even, and $\frac{4}{n\pi}$ if n is odd. Thus the nth Fourier polynomial (where n is odd) is

$$F_n(x) = \frac{4}{\pi}\sin x + \frac{4}{3\pi}\sin 3x + \cdots + \frac{4}{n\pi}\sin(nx).$$

As $n \to \infty$, the nth Fourier polynomial must approach $f(x)$ on the interval $(-\pi, \pi)$, except at the point $x = 0$ (where f is not continuous). In particular, if $x = \frac{\pi}{2}$,

$$F_n(1) = \frac{4}{\pi}\sin\frac{\pi}{2} + \frac{4}{3\pi}\sin\frac{3\pi}{2} + \frac{4}{5\pi}\sin\frac{5\pi}{2} + \frac{4}{7\pi}\sin\frac{7\pi}{2} + \cdots + \frac{4}{n\pi}\sin\frac{n\pi}{2}$$

$$= \frac{4}{\pi}\left(1 - \frac{1}{3} + \frac{1}{5} - \frac{1}{7} + \cdots + (-1)^{2n+1}\frac{1}{2n+1}\right).$$

But $F_n(1)$ approaches $f(\frac{\pi}{2}) = 1$ as $n \to \infty$, so

$$\frac{\pi}{4}F_n(1) = 1 - \frac{1}{3} + \frac{1}{5} - \frac{1}{7} + \cdots + (-1)^{2n+1}\frac{1}{2n+1} \to \frac{\pi}{4} \cdot 1 = \frac{\pi}{4}.$$

30. Let $t = 2\pi x - \pi$. Then, $g(t) = f(x) = e^{2\pi x} = e^{t+\pi}$. Notice that as x varies from 0 to 1, t varies from $-\pi$ to π. Thus, we can find the Fourier coefficients for $g(t)$:

$$a_o = \frac{1}{2\pi}\int_{-\pi}^{\pi} g(t)\,dt = \frac{1}{2\pi}\int_{-\pi}^{\pi} e^{t+\pi}\,dt = \frac{1}{2\pi}e^{t+\pi}\bigg|_{-\pi}^{\pi} = \frac{e^{2\pi} - 1}{2\pi},$$

$$a_n = \frac{1}{\pi}\int_{-\pi}^{\pi} e^{t+\pi}\cos(nt)\,dt = \frac{e^{\pi}}{\pi}\int_{-\pi}^{\pi} e^{t}\cos(nt)\,dt.$$

Using the integral table, Formula II-8, yields:

$$= \frac{e^{\pi}}{\pi}\frac{1}{n^2 + 1}e^{t}(\cos(nt) + n\sin(nt))\bigg|_{-\pi}^{\pi}$$

$$= \frac{e^{\pi}}{\pi}\frac{1}{n^2 + 1}(e^{\pi} - e^{-\pi})(\cos(n\pi))$$

$$= \frac{(e^{2\pi} - 1)}{\pi}\frac{(-1)^n}{n^2 + 1}$$

$$b_n = \frac{1}{\pi}\int_{-\pi}^{\pi} e^{t+\pi}\sin(nt)\,dt = \frac{e^{\pi}}{\pi}\int_{-\pi}^{\pi} e^{t}\sin(nt)\,dt.$$

Again, using the integral table, Formula II-9, yields:

$$= \frac{e^{\pi}}{\pi}\frac{1}{n^2 + 1}e^{t}(\sin(nt) - n\cos(nt))\bigg|_{-\pi}^{\pi}$$

$$= -\frac{e^{\pi}}{\pi}\frac{n}{n^2 + 1}(e^{\pi} - e^{-\pi})\cos(n\pi)$$

$$= \frac{(e^{2\pi} - 1)}{\pi}\frac{(-1)^{n+1}n}{n^2 + 1}.$$

Thus, after factoring a bit, we get:

$$G_3(t) = \frac{e^{2\pi}-1}{\pi}\left(\frac{1}{2} - \frac{1}{2}\cos t + \frac{1}{2}\sin t + \frac{1}{5}\cos 2t - \frac{2}{5}\sin 2t - \frac{1}{10}\cos 3t + \frac{3}{10}\sin 3t\right).$$

Now, we substitute x back in for t:

$$F_3(x) = \frac{e^{2\pi}-1}{\pi}(\frac{1}{2} - \frac{1}{2}\cos(2\pi x - \pi) + \frac{1}{2}\sin(2\pi x - \pi) + \frac{1}{5}\cos(4\pi x - 2\pi)$$
$$-\frac{2}{5}\sin(4\pi x - 2\pi) - \frac{1}{10}\cos(6\pi x - 3\pi) + \frac{3}{10}\sin(6\pi x - 3\pi)).$$

Recalling that $\cos(x-\pi) = -\cos x, \sin(x-\pi) = -\sin x, \cos(x-2\pi) = \cos x$, and $\sin(x-2\pi) = \sin x$, we have:

$$F_3(x) = \frac{e^{2\pi}-1}{\pi}\left(\frac{1}{2} + \frac{1}{2}\cos 2\pi x - \frac{1}{2}\sin 2\pi x + \frac{1}{5}\cos 4\pi x - \frac{2}{5}\sin 4\pi x\right.$$
$$\left. + \frac{1}{10}\cos 6\pi x - \frac{3}{10}\sin 6\pi x\right).$$

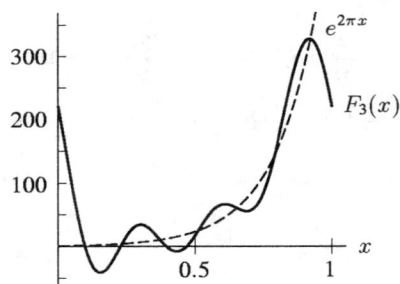

31. (a) Expand $f(x)$ into its Fourier series:

$$f(x) = a_0 + a_1 \cos x + a_2 \cos 2x + a_3 \cos 3x + \cdots + a_k \cos kx + \cdots$$
$$+ b_1 \sin x + b_2 \sin 2x + b_3 \sin 3x + \cdots + b_k \sin kx + \cdots$$

Then differentiate term-by-term:

$$f'(x) = -a_1 \sin x - 2a_2 \sin 2x - 3a_3 \sin 3x - \cdots - ka_k \sin kx - \cdots$$
$$+ b_1 \cos x + 2b_2 \cos 2x + 3b_3 \cos 3x + \cdots + kb_k \cos kx + \cdots$$

Regroup terms:

$$f'(x) = +b_1 \cos x + 2b_2 \cos 2x + 3b_3 \cos 3x + \cdots + kb_k \cos kx + \cdots$$
$$-a_1 \sin x - 2a_2 \sin 2x - 3a_3 \sin 3x - \cdots - ka_k \sin kx - \cdots$$

which forms a Fourier series for the derivative $f'(x)$. The Fourier coefficient of $\cos kx$ is kb_k and the Fourier coefficient of $\sin kx$ is $-ka_k$. Note that there is no constant term as you would expect from the formula ka_k with $k = 0$. Note also that if the kth harmonic f is absent, so is that of f'.

(b) If the amplitude of the kth harmonic of f is

$$A_k = \sqrt{a_k^2 + b_k^2}, \quad k \geq 1,$$

then the amplitude of the kth harmonic of f' is

$$\sqrt{(kb_k)^2 + (-ka_k)^2} = \sqrt{k^2(b_k^2 + a_k^2)} = k\sqrt{a_k^2 + b_k^2} = kA_k.$$

(c) The energy of the kth harmonic of f' is k^2 times the energy of the kth harmonic of f.

32. Let r_k and s_k be the Fourier coefficients of $Af + Bg$. Then

$$r_0 = \frac{1}{2\pi}\int_{-\pi}^{\pi}\left[Af(x) + Bg(x)\right]dx$$

$$= A\left[\frac{1}{2\pi}\int_{-\pi}^{\pi}f(x)\,dx\right] + B\left[\frac{1}{2\pi}\int_{-\pi}^{\pi}g(x)\,dx\right]$$

$$= Aa_0 + Bc_0.$$

Similarly,

$$r_k = \frac{1}{\pi}\int_{-\pi}^{\pi}\left[Af(x) + Bg(x)\right]\cos(kx)\,dx$$

$$= A\left[\frac{1}{\pi}\int_{-\pi}^{\pi}f(x)\cos(kx)\,dx\right] + B\left[\frac{1}{\pi}\int_{-\pi}^{\pi}g(x)\cos(kx)\,dx\right]$$

$$= Aa_k + Bc_k.$$

And finally,

$$s_k = \frac{1}{\pi}\int_{-\pi}^{\pi}\left[Af(x) + Bg(x)\right]\sin(kx)\,dx$$

$$= A\left[\frac{1}{\pi}\int_{-\pi}^{\pi}f(x)\sin(kx)\,dx\right] + B\left[\frac{1}{\pi}\int_{-\pi}^{\pi}g(x)\sin(kx)\,dx\right]$$

$$= Ac_k + Bd_k.$$

33. Since $g(x) = f(x+c)$, we have that $[g(x)]^2 = [f(x+c)]^2$, so g^2 is f^2 shifted horizontally by c. Since f has period 2π, so does f^2 and g^2. If you think of the definite integral as an area, then because of the periodicity, integrals of f^2 over any interval of length 2π have the same value. So

$$\text{Energy of } f = \int_{-\pi}^{\pi}(f(x))^2\,dx = \int_{-\pi+c}^{\pi+c}(f(x))^2\,dx.$$

Now we know that

$$\text{Energy of } g = \frac{1}{\pi}\int_{-\pi}^{\pi}(g(x))^2\,dx$$

$$= \frac{1}{\pi}\int_{-\pi}^{\pi}(f(x+c))^2\,dx.$$

Using the substitution $t = x + c$, we see that the two energies are equal.

Solutions to the Projects and Computer Algebra Investigations

1. (a) (i) Using a Taylor series expansion, we have

$$f(x_0 - h) = f(x_0) - f'(x_0)h + \frac{f''(x_0)}{2}h^2 - \frac{f'''(x_0)}{3!}h^3 + \cdots.$$

So we have

$$\frac{f(x_0) - f(x_0 - h)}{h} - f'(x_0) \approx \frac{f''(x_0)}{2}h + \cdots.$$

This suggests the following bound for small h:

$$\left|\frac{f(x_0) - f(x_0 - h)}{h} - f'(x_0)\right| \leq \frac{Mh}{2},$$

where $|f''(x)| \leq M$ for $|x - x_0| < |h|$.

474 CHAPTER NINE /SOLUTIONS

(ii) We use Taylor series expansions:

$$f(x_0 + h) = f(x_0) + f'(x_0)h + \frac{f''(x_0)}{2}h^2 + \frac{f'''(x_0)}{3!}h^3 + \cdots$$

$$f(x_0 - h) = f(x_0) - f'(x_0)h + \frac{f''(x_0)}{2}h^2 - \frac{f'''(x_0)}{3!}h^3 + \cdots.$$

Subtracting gives

$$f(x_0 + h) - f(x_0 - h) = 2f'(x_0)h + \frac{2f'''(x_0)}{3!}h^3 + \cdots$$

$$= 2f'(x_0)h + \frac{1}{3}f'''(x_0)h^3 + \cdots.$$

So

$$\frac{f(x_0 + h) - f(x_0 - h)}{2h} = f'(x_0) + \frac{f'''(x_0)}{6}h^2 + \cdots.$$

This suggests the following bound for small h:

$$\left| \frac{f(x_0 + h) - f(x_0 - h)}{2h} - f'(x_0) \right| \leq \frac{Mh^2}{6},$$

where $|f'''(x)| \leq M$ for $|x - x_0| < |h|$.

(iii) Expanding each term in the numerator is a Taylor series, we have

$$f(x_0 + 2h) = f(x_0) + 2f'(x_0)h + 2f''(x_0)h^2 + \frac{4}{3}f'''(x_0)h^3$$

$$+ \frac{2}{3}f^{(4)}(x_0)h^4 + \frac{4}{15}f^{(5)}(x_0)h^5 + \cdots$$

$$f(x_0 + h) = f(x_0) + f'(x_0)h + \frac{f''(x_0)}{2}h^2 + \frac{f'''(x_0)}{3!}h^3$$

$$+ \frac{f^{(4)}(x_0)}{4!}h^4 + \frac{f^{(5)}(x_0)}{5!}h^5 + \cdots,$$

$$f(x_0 - h) = f(x_0) - f'(x_0)h + \frac{f''(x_0)}{2}h^2 - \frac{f'''(x_0)}{3!}h^3$$

$$+ \frac{f^{(4)}(x_0)}{4!}h^4 - \frac{f^{(5)}(x_0)}{5!}h^5 + \cdots,$$

$$f(x_0 - 2h) = f(x_0) - 2f'(x_0)h + 2f''(x_0)h^2 - \frac{4}{3}f'''(x_0)h^3$$

$$+ \frac{2}{3}f^{(4)}(x_0)h^4 - \frac{4}{15}f^{(5)}(x_0)h^5 + \cdots.$$

Combining the expansions in pairs, we have

$$8f(x_0 + h) - 8f(x_0 - h) = 16f'(x_0)h + \frac{8}{3}f'''(x_0)h^3 + \frac{2}{15}f^{(5)}(x_0)h^5 + \cdots$$

$$f(x_0 + 2h) - f(x_0 - 2h) = 4f'(x_0)h + \frac{8}{3}f'''(x_0)h^3 + \frac{8}{15}f^{(5)}(x_0)h^5 + \cdots.$$

Thus,

$$-f(x_0 + 2h) + 8f(x_0 + h) - 8f(x_0 - h) + f(x_0 - 2h) = 12f'(x_0)h - \frac{6}{15}f^{(5)}(x_0)h^5 + \cdots$$

so

$$\frac{-f(x_0 + 2h) + 8f(x_0 + h) - 8f(x_0 - h) + f(x_0 - 2h)}{12h} = f'(x_0) - \frac{f^{(5)}(x_0)}{30}h^4 + \cdots.$$

This suggests the following bound for small h,

$$\left| \frac{-f(x_0 + 2h) + 8f(x_0 + h) - 8f(x_0 - h) + f(x_0 - 2h)}{12h} - f'(x_0) \right| \leq \frac{Mh^4}{30},$$

where $|f^{(5)}(x)| \leq M$ for $|x - x_0| \leq |h|$.

(b) (i)

h	$(f(x_0) - f(x_0 - h))/h$	Error
10^{-1}	0.951626	4.837×10^{-2}
10^{-2}	0.995017	4.983×10^{-3}
10^{-3}	0.9995	4.998×10^{-4}
10^{-4}	0.99995	5×10^{-5}

The errors are roughly proportional to h, agreeing with part (a).

(ii)

h	$(f(x_0 + h) - f(x_0 - h))/(2h)$	Error
10^{-1}	1.00167	1.668×10^{-3}
10^{-2}	1.00001667	1.667×10^{-5}
10^{-3}	1.0000001667	1.667×10^{-7}
10^{-4}	1.000000001667	1.667×10^{-9}

The errors are roughly proportional to h^2, agreeing with part (a).

(iii)

h	$(-f(x_0 + 2h) + 8f(x_0 + h) - 8f(x_0 - h) + f(x_0 - 2h))/(12h)$	Error
10^{-1}	0.99999667	3.337×10^{-6}
10^{-2}	0.9999999999667	3.333×10^{-10}
10^{-3}	0.99999999999999667	3.333×10^{-14}
10^{-4}	0.999999999999999999667	3.333×10^{-18}

The errors are roughly proportional to h^4, agreeing with part (a). This is the most accurate formula.

(c) (i)

h	$(f(x_0) - f(x_0 - h))/h$	Error
10^{-1}	1.0001×10^6	1.00×10^{10}
10^{-2}	1.0001×10^7	1.00×10^{10}
10^{-3}	1.0101×10^8	1.01×10^{10}
10^{-4}	1.11111×10^9	1.11×10^{10}
10^{-5}	Undefined	Undefined
10^{-6}	-1.11111×10^{10}	-1.11×10^9
10^{-7}	-1.0101×10^{10}	-1.01×10^8
10^{-8}	-1.001×10^{10}	-1.00×10^7
10^{-9}	-1.0001×10^{10}	-1.00×10^6

(ii)

h	$(f(x_0 + h) - f(x_0 - h))/(2h)$	Error
10^{-1}	1×10^2	1×10^{10}
10^{-2}	1×10^4	1×10^{10}
10^{-3}	1.0001×10^6	1.0001×10^{10}
10^{-4}	1.0101×10^8	1.0101×10^{10}
10^{-5}	Undefined	Undefined
10^{-6}	-1.0101×10^{10}	-1.01×10^8
10^{-7}	-1.0001×10^{10}	-1.00×10^6
10^{-8}	-1.000001×10^{10}	-1.00×10^4
10^{-9}	$-1.00000001 \times 10^{10}$	-1.00×10^2

(iii)

h	$(-f(x_0+2h)+8f(x_0+h)-8f(x_0-h)+f(x_0-2h))/(12h)$	Error
10^{-1}	1.25×10^2	1.00×10^{10}
10^{-2}	1.25×10^4	1.00×10^{10}
10^{-3}	1.25013×10^6	1.00×10^{10}
10^{-4}	1.26326×10^8	1.01×10^{10}
10^{-5}	Undefined	Undefined
10^{-6}	-9.99579×10^9	4.21×10^6
10^{-7}	$-9.9999995998 \times 10^{10}$	4.00×10^2
10^{-8}	$-9.99999999996 \times 10^{10}$	4.00×10^{-2}
10^{-9}	$-9.999999999999996 \times 10^{10}$	4.00×10^{-6}

For relatively large values of h, these approximation formulas fail miserably. The main reason is that $f(x) = 1/x$ changes very quickly at $x_0 = 10^{-5}$. In fact, $f(x) \to \pm\infty$ as $x \to 0$. So we must use very small values for h when estimating a limit (involving f and $x_0 = 10^{-5}$) as $h \to 0$. Here, $h > 10^{-5}$ is too big, since the values of $x_0 - h$ cross over the discontinuity at $x = 0$. For smaller values of h, that make sure we stay on the good side of the abyss, these formulas work quite well. Already by $h = 10^{-6}$, formula (c) is the best approximation.

2. (a) (i) p^2

(ii) There are two ways to do this. One way is to compute your opponent's probability of winning two in a row, which is $(1-p)^2$. Then the probability that neither of you win the next points is:

$$1 - (\text{Probability you win next two} + \text{Probability opponent wins next two})$$
$$= 1 - (p^2 + (1-p)^2)$$
$$= 1 - (p^2 + 1 - 2p + p^2)$$
$$= 2p^2 - 2p$$
$$= 2p(1-p).$$

The other way to compute this is to observe either you win the first point and lose the second or vice versa. Both have probability $p(1-p)$, so the probability you split the points is $2p(1-p)$.

(iii)

$$\text{Probability} = (\text{Probability of splitting next two}) \cdot (\text{Probability of winning two after that})$$
$$= 2p(1-p)p^2$$

(iv)

$$\text{Probability} = (\text{Probability of winning next two}) + (\text{Probability of splitting next two,}$$
$$\text{winning two after that})$$
$$= p^2 + 2p(1-p)p^2$$

(v) The probability is:

$w = (\text{Probability of winning first two})$
$+ (\text{Probability of splitting first two}) \cdot (\text{Probability of winning next two})$
$+ (\text{Prob. of split. first two}) \cdot (\text{Prob. of split. next two}) \cdot (\text{Prob. of winning next two})$
$+ \cdots$
$= p^2 + 2p(1-p)p^2 + (2p(1-p))^2 p^2 + \cdots.$

This is an infinite geometric series with a first term of p^2 and a ratio of $2p(1-p)$. Therefore the probability of winning is

$$w = \frac{p^2}{1 - 2p(1-p)}.$$

(vi) For $p = 0.5$, $w = \frac{(0.5)^2}{1-2(0.5)(1-(0.5))} = 0.5$. This is what we would expect. If you and your opponent are equally likely to score the next point, you and your opponent are equally likely to win the next game.

For $p = 0.6$, $w = \frac{(0.6)^2}{1-2(0.6)(0.4)} = 0.69$. Here your probability of winning the next point has been magnified to a probability 0.69 of winning the game. Thus it gives the better player an advantage to have to win by two points, rather than the "sudden death" of winning by just one point. This makes sense: when you have to win by two, the stronger player always gets a second chance to overcome the weaker player's winning the first point on a "fluke."

For $p = 0.7$, $w = \frac{(0.7)^2}{1-2(0.7)(0.3)} = 0.84$. Again, the stronger player's probability of winning is magnified.

For $p = 0.4$, $w = \frac{(0.4)^2}{1-2(0.4)(0.6)} = 0.31$. We already computed that for $p = 0.6$, $w = 0.69$. Thus the value for w when $p = 0.4$, should be the same as the probability of your opponent winning for $p = 0.6$, namely $1 - 0.69 = 0.31$.

(b) (i)
$S = $ (Prob. you score first point)
$\quad$ +(Prob. you lose first point, your opponent loses the next,
$\quad\quad$ you win the next)
$\quad$ +(Prob. you lose a point, opponent loses, you lose,
$\quad\quad$ opponent loses, you win)
$\quad + \cdots$

$= $ (Prob. you score first point)
$\quad$ +(Prob. you lose)·(Prob. opponent loses)·(Prob. you win)
$\quad$ +(Prob. you lose)·(Prob. opponent loses)·(Prob. you lose)
$\quad\quad$ ·(Prob. opponent loses)·(Prob. you win)$+ \cdots$

$= p + (1-p)(1-q)p + \big((1-p)(1-q)\big)^2 p + \cdots$

$= \dfrac{p}{1-(1-p)(1-q)}$

(ii) Since S is your probability of winning the next point, we can use the formula computed in part (v) of (a) for winning two points in a row, thereby winning the game:
$$w = \frac{S^2}{1 - 2S(1-S)}.$$

- When $p = 0.5$ and $q = 0.5$,
$$S = \frac{0.5}{1-(0.5)(0.5)} = 0.67.$$
Therefore
$$w = \frac{S^2}{1-2S(1-S)} = \frac{(0.67)^2}{1-2(0.67)(1-0.67)} = 0.80.$$

- When $p = 0.6$ and $q = 0.5$,
$$S = \frac{0.6}{1-(0.4)(0.5)} = 0.75 \quad \text{and} \quad w = \frac{(0.75)^2}{1-2(0.75)(1-0.75)} = 0.9.$$

3. (a) The Taylor polynomials of degree 10 are

For $\sin^2 x$, $\quad P_{10}(x) = x^2 - \dfrac{x^4}{3} + \dfrac{2x^6}{45} - \dfrac{x^8}{315} + \dfrac{2x^{10}}{14175}$

For $\cos^2 x$, $\quad Q_{10}(x) = 1 - x^2 + \dfrac{x^4}{3} - \dfrac{2x^6}{45} + \dfrac{x^8}{315} - \dfrac{2x^{10}}{14175}$

(b) The coefficients in $P_{10}(x)$ are the negatives of the corresponding coefficients of $Q_{10}(x)$. The constant term of $P_{10}(x)$ is 0 and the contant term of $Q_{10}(x)$ is 1. Thus, $P_{10}(x)$ and $Q_{10}(x)$ satisfy
$$Q_{10}(x) = 1 - P_{10}(x).$$
This makes sense because $\cos^2 x$ and $\sin^2 x$ satisfy the identity
$$\cos^2 x = 1 - \sin^2 x.$$

478 CHAPTER NINE /SOLUTIONS

4. (a) The Taylor polynomials of degree 7 are

$$\text{For } \sin x, \quad P_7(x) = x - \frac{x^3}{6} + \frac{x^5}{120} - \frac{x^7}{5040}$$

$$\text{For } \sin x \cos x, \quad Q_7(x) = x - \frac{2x^3}{3} + \frac{2x^5}{15} - \frac{4x^7}{315}$$

(b) The coefficient of x^3 in $Q_7(x)$ is $-2/3$, and the coefficient of x^3 in $P_7(x)$ is $-1/6$, so the ratio is

$$\frac{-2/3}{-1/6} = 4.$$

The corresponding ratios for x^5 and x^7 are

$$\frac{2/15}{1/120} = 16 \quad \text{and} \quad \frac{-4/315}{-1/5040} = 64.$$

(c) It appears that the ratio is always a power of 2. For x^3, it is $4 = 2^2$; for x^5, it is $16 = 2^4$; for x^7, it is $64 = 2^6$. This suggests that in general, for the coefficient of x^n, it is 2^{n-1}.

(d) From the identity $\sin(2x) = 2\sin x \cos x$, we expect that $P_7(2x) = 2Q_7(x)$. So, if a_n is the coefficient of x^n in $P_7(x)$, and if b_n is the coefficient of x^n in $Q_7(x)$, then, since the x^n terms $P_7(2x)$ and $2Q_7(x)$ must be equal, we have

$$a_n(2x)^n = 2b_n x^n.$$

Dividing both sides by x^n and combining the powers of 2, this gives the pattern we observed:

$$\frac{b_n}{a_n} = 2^{n-1}.$$

5. (a) The Taylor polynomial is

$$P_{10}(x) = 1 + \frac{x^2}{12} - \frac{x^4}{720} + \frac{x^6}{30240} - \frac{x^8}{1209600} + \frac{x^{10}}{47900160}$$

(b) All the terms have even degree. A polynomial with only terms of even degree is an even function. This suggests that f might be an even function.

(c) To show that f is even, we must show that $f(-x) = f(x)$.

$$f(-x) = \frac{-x}{e^{-x} - 1} + \frac{-x}{2} = \frac{x}{1 - \frac{1}{e^x}} - \frac{x}{2} = \frac{xe^x}{e^x - 1} - \frac{x}{2}$$

$$= \frac{xe^x - \frac{1}{2}x(e^x - 1)}{e^x - 1}$$

$$= \frac{xe^x - \frac{1}{2}xe^x + \frac{1}{2}x}{e^x - 1} = \frac{\frac{1}{2}xe^x + \frac{1}{2}x}{e^x - 1} = \frac{\frac{1}{2}x(e^x - 1) + x}{e^x - 1}$$

$$= \frac{1}{2}x + \frac{x}{e^x - 1} = \frac{x}{e^x - 1} + \frac{x}{2} = f(x)$$

6. (a) The Taylor polynomial is

$$P_{11}(x) = \frac{x^3}{3} - \frac{x^7}{42} + \frac{x^{11}}{1320}.$$

(b) Evaluating, we get

$$P_{11}(1) = \frac{1^3}{3} - \frac{1^7}{42} + \frac{1^{11}}{1320} = 0.310281$$

$$S(1) = \int_0^1 \sin(t^2)\, dt = 0.310268.$$

We need to take about 6 decimal places in the answer as this allows us to see the error. (The values of $P_{11}(1)$ and $S(1)$ start to differ in the fifth decimal place.) Thus, the percentage error is $(0.310281 - 0.310268)/0.310268 = 0.000013/0.310268 = 0.000042 = 0.0042\%$. On the other hand,

$$P_{11}(2) = \frac{2^3}{3} - \frac{2^7}{42} + \frac{2^{11}}{1320} = 1.17056$$

$$S(2) = \int_0^2 \sin(t^2)\, dt = 0.804776.$$

The percentage error in this case is $(1.17056 - 0.804776)/0.804776 = 0.365786/0.804776 = 0.454519$, or about 45%.

Solutions to Problems on Convergence Theorems

1. Let S_n be the nth partial sum for $\sum a_n$ and let T_n be the nth partial sum for $\sum b_n$. Then the nth partial sums for $\sum(a_n + b_n)$, $\sum(a_n - b_n)$, and $\sum ka_n$ are $S_n + T_n$, $S_n - T_n$, and kS_n, respectively. To show that these series converge, we have to show that the limits of their partial sums exist. By the properties of limits,

$$\lim_{n \to \infty} (S_n + T_n) = \lim_{n \to \infty} S_n + \lim_{n \to \infty} T_n$$
$$\lim_{n \to \infty} (S_n - T_n) = \lim_{n \to \infty} S_n - \lim_{n \to \infty} T_n$$
$$\lim_{n \to \infty} kS_n = k \lim_{n \to \infty} S_n.$$

This proves that the limits of the partial sums exist, so the series converge.

2. Let S_n be the n-th partial sum for $\sum a_n$ and let T_n be the n-th partial sum for $\sum b_n$. Suppose that $S_N = T_N + k$. Since $a_n = b_n$ for $n \geq N$, we have $S_n = T_n + k$ for $n \geq N$. Hence if S_n converges to a limit, so does T_n, and vice versa.

3. We have $a_n = S_n - S_{n-1}$. If $\sum a_n$ converges, then $S = \lim_{n \to \infty} S_n$ exists. Hence $\lim_{n \to \infty} S_{n-1}$ exists and is equal to S also. Thus

$$\lim_{n \to \infty} a_n = \lim_{n \to \infty} (S_n - S_{n-1}) = \lim_{n \to \infty} S_n - \lim_{n \to \infty} S_{n-1} = S - S = 0.$$

4. Since $n^3 \geq n^2$, we have $1/n^3 \leq 1/n^2$. Hence the series converges by comparison with $1/n^2$, which we showed converges on page 440 of the text.

5. Since $\ln n \leq n$ for $n \geq 2$, we have $1/\ln n \geq 1/n$, so the series diverges by comparison with the harmonic series, $\sum 1/n$.

6. Using left hand approximating sums or upper sums for the integral of $f(x) = 1/(4x - 3)$ over the interval $1 \leq x \leq n+1$ with uniform subdivisions of length 1 gives:

$$1 + \frac{1}{5} + \frac{1}{9} + \cdots + \frac{1}{4n - 3} > \int_1^{n+1} \frac{dx}{4x - 3} = \frac{1}{4} \ln(4x - 3) \Big|_1^{n+1} = \frac{1}{4} \ln(4n + 1).$$

Since $\ln(4n+1)$ increases without bound as $n \to \infty$, it follows that the partial sums of the series are also unbounded. Thus, this is not a convergent series.

7. Using left hand approximating sums or upper sums for the integral of $f(x) = x/(x^2 + 1)$ over the interval $1 \leq x \leq n+1$ with uniform subdivisions of length 1 gives:

$$\frac{1}{2} + \frac{2}{5} + \frac{3}{10} + \cdots + \frac{n}{n^2 + 1} > \int_1^{n+1} \frac{x}{x^2 + 1} dx = \frac{1}{2} \ln(x^2 + 1) \Big|_1^{n+1} = \frac{1}{2} \left[\ln((n+1)^2 + 1) - \ln 2 \right].$$

Since $\ln((n+1)^2 + 1)$ increases without bound as $n \to \infty$, it follows that the partial sums of the series are also unbounded. Thus, this is not a convergent series.

8. Using right hand approximating sums or lower sums for the integral of $f(x) = x^{(-3/2)}$ over the interval $1 \leq x \leq n$ with uniform subdivisions of width 1 gives:

$$\frac{1}{2^{3/2}} + \cdots + \frac{1}{n^{3/2}} \leq \int_1^n x^{(-3/2)} dx = -2(n^{-1/2} - 1).$$

Thus, as $n \to \infty$, the sequence of partial sums is bounded. Hence the series converges.

9. Using right hand approximating sums or lower sums for the integral of $f(x) = x^{-p}$ over the interval $1 \leq x \leq n$ with uniform subdivisions of width 1 gives:

$$\frac{1}{2^p} + \cdots + \frac{1}{n^p} \leq \int_1^n x^{-p} dx = \frac{1}{1-p} x^{1-p} \Big|_1^n = \frac{1}{1-p} (n^{1-p} - 1).$$

Since $p > 1$, the exponent $1 - p$ is negative, so as $n \to \infty$, the sequence of partial sums converges. Hence the series converges.

10. (a) Since
$$|a_n| = a_n \quad \text{if } a_n \geq 0$$
$$|a_n| = -a_n \quad \text{if } a_n < 0,$$

we have
$$a_n + |a_n| = 2|a_n| \quad \text{if } a_n > 0$$
$$a_n + |a_n| = 0 \quad \text{if } a_n < 0.$$

Thus, for all n,
$$0 \leq a_n + |a_n| \leq 2|a_n|.$$

(b) If $\sum |a_n|$ converges, then $\sum 2|a_n|$ is convergent, so, by comparison, $\sum (a_n + |a_n|)$ is convergent. Then
$$\sum \left((a_n + |a_n|) - |a_n|\right) = \sum a_n$$
is convergent, as it is the difference of two convergent series.

11. The kth coefficient in the series $\sum kC_k x^k$ is $D_k = k \cdot C_k$. We are given that the series $\sum C_k x^k$ has radius of convergence R by the radius of convergence test. Thus, applying the radius of convergence test to the new series, we have
$$\left|\frac{D_k}{D_{k+1}}\right| = \left|\frac{kC_k}{(k+1)C_{k+1}}\right| \to 1 \cdot R \quad \text{as} \quad k \to \infty.$$

Hence the new series has radius of convergence R.

Solutions to Problems on The Error in Taylor Polynomials

1. (a) The Taylor polynomial of degree 0 about $t = 0$ for $f(t) = e^t$ is simply $P_0(x) = 1$. Since $e^t \geq 1$ on $[0, 0.5]$, the approximation is an underestimate.
   (b) Using the zero degree error bound, if $|f'(t)| \leq M$ for $0 \leq t \leq 0.5$, then
   $$|E_0| \leq M \cdot |t| \leq M(0.5).$$
   Since $|f'(t)| = |e^t| = e^t$ is increasing on $[0, 0.5]$,
   $$|f'(t)| \leq e^{0.5} < \sqrt{4} = 2.$$
   Therefore
   $$|E_0| \leq (2)(0.5) = 1.$$
   (Note: By looking at a graph of $f(t)$ and its 0th degree approximation, it is easy to see that the greatest error occurs when $t = 0.5$, and the error is $e^{0.5} - 1 \approx 0.65 < 1$. So our error bound works.)

2. (a) The second-degree Taylor polynomial for $f(t) = e^t$ is $P_2(t) = 1 + t + t^2/2$. Since the full expansion of $e^t = 1 + t + t^2/2 + t^3/6 + t^4/24 + \cdots$ is clearly larger than $P_2(t)$ for $t > 0$, $P_2(t)$ is an underestimate on $[0, 0.5]$.
   (b) Using the second-degree error bound, if $|f^{(3)}(t)| \leq M$ for $0 \leq t \leq 0.5$, then
   $$|E_2| \leq \frac{M}{3!} \cdot |t|^3 \leq \frac{M(0.5)^3}{6}.$$
   Since $|f^{(3)}(t)| = e^t$, and e^t is increasing on $[0, 0.5]$,
   $$f^{(3)}(t) \leq e^{0.5} < \sqrt{4} = 2.$$
   So
   $$|E_2| \leq \frac{(2)(0.5)^3}{6} < 0.047.$$

3. (a) θ is the first degree approximation of $f(\theta) = \sin\theta$; it is also the second degree approximation, since the next term in the Taylor expansion is 0.

    $P_1(\theta) = \theta$ is an overestimate for $0 < \theta \leq 1$, and is an underestimate for $-1 \leq \theta < 0$. (This can be seen easily from a graph.)

    (b) Using the second degree error bound, if $|f^{(3)}(\theta)| \leq M$ for $-1 \leq \theta \leq 1$, then

    $$|E_2| \leq \frac{M \cdot |\theta|^3}{3!} \leq \frac{M}{6}.$$

    For what value of M is $|f^{(3)}(\theta)| \leq M$ for $-1 \leq \theta \leq 1$? Well, $|f^{(3)}(\theta)| = |-\cos\theta| \leq 1$. So $|E_2| \leq \frac{1}{6} = 0.17$.

4. (a) $\theta - \frac{\theta^3}{3!}$ is the third degree Taylor approximation of $f(\theta) = \sin\theta$; it is also the fourth degree approximation, since the next term in the Taylor expansion is 0.

    $P_3(\theta)$ is an underestimate for $0 < \theta \leq 1$, and is an overestimate for $-1 \leq \theta < 0$. (This can be checked with a calculator.)

    (b) Using the fourth degree error bound, if $|f^{(5)}(\theta)| \leq M$ for $-1 \leq \theta \leq 1$, then

    $$|E_4| \leq \frac{M \cdot |\theta|^5}{5!} \leq \frac{M}{120}.$$

    For what value of M is $|f^{(5)}(\theta)| \leq M$ for $-1 \leq \theta \leq 1$? Since $f^{(5)}(\theta) = \cos\theta$ and $|\cos\theta| \leq 1$, we have

    $$|E_4| \leq \frac{1}{120} \leq 0.0084.$$

5. Let $f(x) = \tan x$. The error bound for the Taylor approximation of degree three for $f(1) = \tan 1$ about $x = 0$ is:

    $$|E_3| = |f(1) - P_3(x)| \leq \frac{M \cdot |1 - 0|^4}{4!} = \frac{M}{24}$$

    where $|f^{(4)}(x)| \leq M$ for $0 \leq x \leq 1$. Now, $f^{(4)}(x) = \frac{16 \sin x}{\cos^3 x} + \frac{24 \sin^3 x}{\cos^5 x}$. From a graph of $f^{(4)}(x)$, we see that $f^{(4)}(x)$ is increasing for x between 0 and 1. Thus,

    $$|f^{(4)}(x)| \leq |f^{(4)}(1)| \approx 396,$$

    so

    $$|E_3| \leq \frac{396}{24} = 16.5.$$

    This is not a very helpful error bound! The reason the error bound is so huge is that $x = 1$ is getting near the vertical asymptote of the tangent graph, and the fourth derivative is enormous there.

6. Let $f(x) = (1-x)^{1/3}$, so $f(0.5) = (0.5)^{1/3}$. The error bound in the Taylor approximation of degree 3 for $f(0.5) = 0.5^{\frac{1}{3}}$ about $x = 0$ is:

    $$|E_3| = |f(0.5) - P_3(0.5)| \leq \frac{M \cdot |0.5 - 0|^4}{4!} = \frac{M(0.5)^4}{24},$$

    where $|f^{(4)}(x)| \leq M$ for $0 \leq x \leq 0.5$. Now, $f^{(4)}(x) = -\frac{80}{81}(1-x)^{-(11/3)}$. By looking at the graph of $(1-x)^{-(11/3)}$, we see that $|f^{(4)}(x)|$ is maximized for x between 0 and 0.5 when $x = 0.5$. Thus,

    $$|f^{(4)}| \leq \frac{80}{81}\left(\frac{1}{2}\right)^{-(11/3)} = \frac{80}{81} \cdot 2^{11/3},$$

    so

    $$|E_3| \leq \frac{80 \cdot 2^{11/3} \cdot (0.5)^4}{81 \cdot 24} \approx 0.033.$$

7. Let $f(x) = \ln(1+x)$. The error bound in the Taylor approximation of degree 3 about $x = 0$ is:

    $$|E_4| = |f(0.5) - P_3(0.5)| \leq \frac{M \cdot |0.5 - 0|^4}{4!} = \frac{M(0.5)^4}{24},$$

    where $|f^{(4)}(x)| \leq M$ for $0 \leq x \leq 0.5$. Since $f^{(4)}(x) = \frac{3!}{(1+x)^4}$ and the denominator attains its minimum when $x = 0$, we have $|f^{(4)}(x)| \leq 3!$, so

    $$|E_4| \leq \frac{3!(0.5)^4}{24} \approx 0.016.$$

8. Let $f(x) = (1+x)^{-\frac{1}{2}} = \frac{1}{\sqrt{1+x}}$. The error bound for the Taylor approximation of degree three for $f(2) = \frac{1}{\sqrt{3}}$ about $x = 0$ is:

$$|E_3| = |f(2) - P_3(2)| \le \frac{M \cdot |2-0|^4}{4!} = \frac{M \cdot 2^4}{24},$$

where $|f^{(4)}| \le M$ for $0 \le x \le 2$. Since $f^{(4)}(x) = \frac{105}{16}(1+x)^{-(9/2)}$, we see that if x is between 0 and 2, $|f^{(4)}x)| \le \frac{105}{16}$. Thus,

$$|E_3| \le \frac{105}{16} \cdot \frac{2^4}{24} = \frac{105}{24} = 4.375.$$

Again, this is not a very helpful bound on the error, but that is to be expected as the Taylor series does not converge at $x = 2$. (At $x = 2$, we are outside the interval of convergence.)

9. The maximum possible error for the nth degree Taylor polynomial about $x = 0$ approximating $\cos x$ is $|E_n| \le \frac{M \cdot |x-0|^{n+1}}{(n+1)!}$, where $|\cos^{(n+1)} x| \le M$ for $0 \le x \le 1$. Now the derivatives of $\cos x$ are simply $\cos x$, $\sin x$, $-\cos x$, and $-\sin x$. The largest magnitude these ever take is 1, so $|\cos^{(n+1)}(x)| \le 1$, and thus $|E_n| \le \frac{|x|^{n+1}}{(n+1)!} \le \frac{1}{(n+1)!}$. The same argument works for $\sin x$.

10. By the results of Problem 9, if we approximate $\cos 1$ using the nth degree polynomial, the error is at most $\frac{1}{(n+1)!}$. For the answer to be correct to four decimal places, the error must be less than 0.00005. Thus, the first n such that $\frac{1}{(n+1)!} < 0.00005$ will work. In particular, when $n = 7$, $\frac{1}{8!} = \frac{1}{40370} < 0.00005$, so the 7th degree Taylor polynomial will give the desired result. For six decimal places, we need $\frac{1}{(n+1)!} < 0.0000005$. Since $n = 9$ works, the 9th degree Taylor polynomial is sufficient.

11.
$$\sin x = x - \frac{x^3}{3!} + \frac{x^5}{5!} - \cdots$$

Write the error in approximating $\sin x$ by the Taylor polynomial of degree $n = 2k+1$ as E_n so that

$$\sin x = x - \frac{x^3}{3!} + \frac{x^5}{5!} - \cdots (-1)^k \frac{x^{2k+1}}{(2k+1)!} + E_n.$$

(Notice that $(-1)^k = 1$ if k is even and $(-1)^k = -1$ if k is odd.) We want to show that if x is fixed, $E_n \to 0$ as $k \to \infty$. Since $f(x) = \sin x$, all the derivatives of $f(x)$ are $\pm \sin x$ or $\pm \cos x$, so we have for all n and all x

$$|f^{(n+1)}(x)| \le 1.$$

Using the bound on the error given in the text on page 447, we see that

$$|E_n| \le \frac{1}{(2k+2)!}|x|^{2k+2}.$$

By the argument in the text on page 448, we know that for all x,

$$\frac{|x|^{2k+2}}{(2k+2)!} = \frac{|x|^{n+1}}{(n+1)!} \to 0 \quad \text{as} \quad n = 2k+1 \to \infty.$$

Thus the Taylor series for $\sin x$ does converge to $\sin x$ for every x.

12. The Taylor series for $\cos x$ is given by

$$\cos x = 1 - \frac{x^2}{2!} + \frac{x^4}{4!} - \frac{x^6}{6!} + \cdots$$

Write the error in approximating $\cos x$ by the Taylor polynomial of degree $n = 2k$ as E_n so that

$$\cos x = 1 - \frac{x^2}{2!} + \frac{x^4}{4!} - \frac{x^6}{6!} + \cdots (-1)^k \frac{x^{2k}}{(2k)!} + E_n.$$

(Notice that $(-1)^k = 1$ if k is even and $(-1)^k = -1$ if k is odd.) We want to show that if x is fixed, $E_n \to 0$ as $k \to \infty$. Since $f(x) = \cos x$, all the derivatives of $f(x)$ are $\pm \sin x$ or $\pm \cos x$, so we have for all n and all x

$$|f^{(n+1)}(x)| \le 1.$$

Using the bound on the error given in the text on page 447, we see that

$$|E_n| \leq \frac{1}{(2k+1)!}|x|^{2k+1}.$$

By the argument in the text on page 448, we know that for all x,

$$\frac{|x|^{2k+1}}{(2k+1)!} = \frac{|x|^{n+1}}{(n+1)!} \to 0 \quad \text{as} \quad n = 2k \to \infty.$$

Thus the Taylor series for $\cos x$ does converge to $\cos x$ for every x.

13. (a)

TABLE 9.1 $E_1 = \sin x - x$

x	$\sin x$	E
-0.5	-0.4794	0.0206
-0.4	-0.3894	0.0106
-0.3	-0.2955	0.0045
-0.2	-0.1987	0.0013
-0.1	-0.0998	0.0002

TABLE 9.2 $E_1 = \sin x - x$

x	$\sin x$	E
0	0	0
0.1	0.0998	-0.0002
0.2	0.1987	-0.0013
0.3	0.2955	-0.0045
0.4	0.3894	-0.0106
0.5	0.4794	-0.0206

(b) See answer to part (a) above.

(c)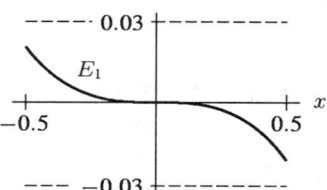

The fact that the graph of E_1 lies between the horizontal lines at ± 0.03 shows that $|E_1| < 0.03$ for $-0.5 \leq x \leq 0.5$.

14. (a)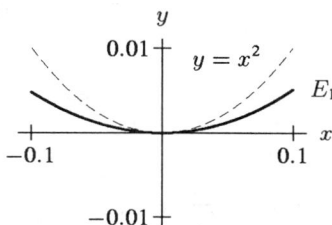

The graph of E_1 looks like a parabola. Since the graph of E_1 is sandwiched between the graph of $y = x^2$ and the x axis, we have

$$|E_1| \leq x^2 \quad \text{for} \quad |x| \leq 0.1.$$

(b)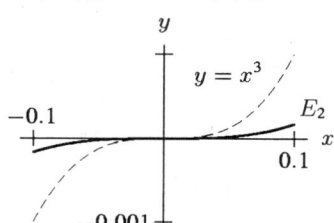

The graph of E_2 looks like a cubic, sandwiched between the graph of $y = x^3$ and the x axis, so

$$|E_2| \leq x^3 \quad \text{for} \quad |x| \leq 0.1.$$

(c) Using the Taylor expansion
$$e^x = 1 + x + \frac{x^2}{2!} + \frac{x^3}{3!} + \cdots$$
we see that
$$E_1 = e^x - (1+x) = \frac{x^2}{2!} + \frac{x^3}{3!} + \frac{x^4}{4!} + \cdots.$$
Thus for small x, the $x^2/2!$ term dominates, so
$$E_1 \approx \frac{x^2}{2!},$$
and so E_1 is approximately a quadratic.

Similarly
$$E_2 = e^x - (1 + x + \frac{x^2}{2}) = \frac{x^3}{3!} + \frac{x^4}{4!} + \cdots.$$
Thus for small x, the $x^3/3!$ term dominates, so
$$E_2 \approx \frac{x^3}{3!}$$
and so E_2 is approximately a cubic.

15.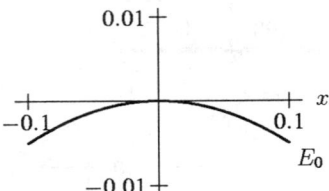

The graph of E_0 looks like a parabola, and the graph shows
$$|E_0| < 0.01 \quad \text{for} \quad |x| \leq 0.1.$$
(In fact $|E_0| < 0.005$ on this interval.) Since
$$\cos x = 1 - \frac{x^2}{2!} + \frac{x^4}{4!} - \frac{x^6}{6!} + \cdots,$$
$$E_0 = \cos x - 1 = -\frac{x^2}{2!} + \frac{x^4}{4!} - \frac{x^6}{6!} + \cdots.$$
So, for small x,
$$E_0 \approx -\frac{x^2}{2},$$
and therefore the graph of E_0 is parabolic.

16. We define $e^{i\theta}$ to be
$$e^{i\theta} = 1 + i\theta + \frac{(i\theta)^2}{2!} + \frac{(i\theta)^3}{3!} + \frac{(i\theta)^4}{4!} + \frac{(i\theta)^5}{5!} + \frac{(i\theta)^6}{6!} + \cdots$$
Suppose we consider the expression $\cos\theta + i\sin\theta$, with $\cos\theta$ and $\sin\theta$ replaced by their Taylor series:
$$\cos\theta + i\sin\theta = \left(1 - \frac{\theta^2}{2!} + \frac{\theta^4}{4!} - \frac{\theta^6}{6!} + \cdots\right) + i\left(\theta - \frac{\theta^3}{3!} + \frac{\theta^5}{5!} - \cdots\right)$$
Reordering terms, we have
$$\cos\theta + i\sin\theta = 1 + i\theta - \frac{\theta^2}{2!} - \frac{i\theta^3}{3!} + \frac{\theta^4}{4!} + \frac{i\theta^5}{5!} - \frac{\theta^6}{6!} - \cdots$$
Using the fact that $i^2 = -1$, $i^3 = -i$, $i^4 = 1$, $i^5 = i$, $\cdots$, we can rewrite the series as
$$\cos\theta + i\sin\theta = 1 + i\theta + \frac{(i\theta)^2}{2!} + \frac{(i\theta)^3}{3!} + \frac{(i\theta)^4}{4!} + \frac{(i\theta)^5}{5!} + \frac{(i\theta)^6}{6!} + \cdots$$
Amazingly enough, this series is the Taylor series for e^x with $i\theta$ substituted for x. Therefore, we have shown that
$$\cos\theta + i\sin\theta = e^{i\theta}.$$

CHAPTER TEN

Solutions for Section 10.1

1. (a) (III) An island can only sustain the population up to a certain size. The population will grow until it reaches this limiting value.
    (b) (V) The ingot will get hot and then cool off, so the temperature will increase and then decrease.
    (c) (I) The speed of the car is constant, and then decreases linearly when the breaks are applied uniformly.
    (d) (II) Carbon-14 decays exponentially.
    (e) (IV) Tree pollen is seasonal, and therefore cyclical.

2. If $Q = Ce^{kt}$, then
$$\frac{dQ}{dt} = Cke^{kt} = k(Ce^{kt}) = kQ.$$
We are given that $\frac{dQ}{dt} = -0.03Q$, so we know that $kQ = -0.03Q$. Thus we either have $Q = 0$ (in which case $C = 0$ and k is anything) or $k = -0.03$. Notice that if $k = -0.03$, then C can be any number.

3. If $P = P_0 e^t$, then
$$\frac{dP}{dt} = \frac{d}{dt}(P_0 e^t) = P_0 e^t = P.$$

4. If $y = \cos \omega t$, then
$$\frac{dy}{dt} = -\omega \sin \omega t, \qquad \frac{d^2 y}{dt^2} = -\omega^2 \cos \omega t.$$
Thus, if $\frac{d^2 y}{dt^2} + 9y = 0$, then
$$-\omega^2 \cos \omega t + 9 \cos \omega t = 0$$
$$(9 - \omega^2) \cos \omega t = 0.$$
Thus $9 - \omega^2 = 0$, or $\omega^2 = 9$, so $\omega = \pm 3$.

5. If $y = \sin 2t$, then $\frac{dy}{dt} = 2\cos 2t$, and $\frac{d^2 y}{dt^2} = -4 \sin 2t$.
Thus $\frac{d^2 y}{dt^2} + 4y = -4 \sin 2t + 4 \sin 2t = 0$.

6. (a) $P = \frac{1}{1+e^{-t}} = (1+e^{-t})^{-1}$
$\frac{dP}{dt} = -(1+e^{-t})^{-2}(-e^{-t}) = \frac{e^{-t}}{(1+e^{-t})^2}.$
Then $P(1-P) = \frac{1}{1+e^{-t}}\left(1 - \frac{1}{1+e^{-t}}\right) = \left(\frac{1}{1+e^{-t}}\right)\left(\frac{e^{-t}}{1+e^{-t}}\right) = \frac{e^{-t}}{(1+e^{-t})^2} = \frac{dP}{dt}.$
    (b) As t tends to ∞, e^{-t} goes to 0. Thus $\lim_{t \to \infty} \frac{1}{1+e^{-t}} = 1$.

7. If $y = Ax^\lambda$ then $y' = A\lambda x^{\lambda-1}$ and $y'' = A\lambda(\lambda-1)x^{\lambda-2}$. Substituting into the differential equation we get
$$x^2 A\lambda(\lambda-1)x^{\lambda-2} + 2xA\lambda x^{\lambda-1} - 6Ax^\lambda = 0,$$
or equivalently
$$Ax^\lambda(\lambda^2 + \lambda - 6) = 0.$$
Hence
$$(\lambda+3)(\lambda-2) = 0,$$
so that $\lambda = -3$ or 2.

8.
(I)	$y = 2\sin x,$	$dy/dx = 2\cos x,$	$d^2y/dx^2 = -2\sin x$
(II)	$y = \sin 2x,$	$dy/dx = 2\cos 2x,$	$d^2y/dx^2 = -4\sin 2x$
(III)	$y = e^{2x},$	$dy/dx = 2e^{2x},$	$d^2y/dx^2 = 4e^{2x}$
(IV)	$y = e^{-2x},$	$dy/dx = -2e^{-2x},$	$d^2y/dx^2 = 4e^{-2x}$

and so:

(a) (IV)
(b) (III)
(c) (III), (IV)
(d) (II)

9. The easiest way to approach this problem is to take the derivative dy/dx of all functions (I)–(V), and try to match the formulas obtained with the equations (A)–(E).

(I)
$$y = x^3$$
$$\frac{dy}{dx} = 3x^2 = 3\frac{x^3}{x} = 3\frac{y}{x},$$

thus (I) is a solution to (B).

(II)
$$y = 3x$$
$$\frac{dy}{dx} = 3 = \frac{y}{x},$$

thus (II) is a solution to (A).

(III)
$$y = e^{3x}$$
$$\frac{dy}{dx} = 3e^{3x} = 3y,$$

thus (III) is a solution to (E).

(IV)
$$y = 3e^x$$
$$\frac{dy}{dx} = 3e^x = y,$$

thus (IV) is a solution to (D).

(V)
$$y = x$$
$$\frac{dy}{dx} = 1 = \frac{y}{x},$$

thus (V) is a solution to (A).

We notice that none of the five functions is a solution to (C).

10. It is easiest to begin by writing down the first and second derivatives for each possible solution:

(I) $y = \cos x$, so $y' = -\sin x$, and $y'' = -\cos x$.
(II) $y = \cos(-x)$, so $y' = \sin(-x)$, and $y'' = -\cos(-x)$.
(III) $y = x^2$, so $y' = 2x$, and $y'' = 2$.
(IV) $y = e^x + e^{-x}$, so $y' = e^x - e^{-x}$, and $y'' = e^x + e^{-x}$.
(V) $y = \sqrt{2x}$, so $y' = \frac{1}{2}(2x)^{-1/2} \cdot 2 = 1/\sqrt{2x}$, and $y'' = -\frac{1}{2}(2x)^{-3/2} \cdot 2 = -(2x)^{-3/2}$.

By substituting these into the given differential equations, we get following solutions:

(a) (IV)
(b) None
(c) (V)
(d) (I), (II)
(e) (III)

11. (a) If $y = \frac{e^x + e^{-x}}{2}$, then $\frac{dy}{dx} = \frac{e^x - e^{-x}}{2}$, and $\frac{d^2y}{dx^2} = \frac{e^x + e^{-x}}{2}$.
If $k = 1$, then

$$k\sqrt{1 + \left(\frac{dy}{dx}\right)^2} = \sqrt{1 + \left(\frac{e^x - e^{-x}}{2}\right)^2} = \sqrt{1 + \frac{e^{2x}}{4} - \frac{1}{2} + \frac{e^{-2x}}{4}}$$
$$= \sqrt{\frac{e^{2x}}{4} + \frac{1}{2} + \frac{e^{-2x}}{4}} = \sqrt{\left(\frac{e^x + e^{-x}}{2}\right)^2}$$
$$= \left|\frac{e^x + e^{-x}}{2}\right| = \frac{e^x + e^{-x}}{2} \quad \text{(since } e^x + e^{-x} > 0\text{)}$$
$$= \frac{d^2y}{dx^2}.$$

(b) $y = \frac{e^{Ax} + e^{-Ax}}{2A}$, so

$$\frac{dy}{dx} = \frac{e^{Ax} - e^{-Ax}}{2} \quad \text{and} \quad \frac{d^2y}{dx^2} = A\left(\frac{e^{Ax} + e^{-Ax}}{2}\right).$$

Therefore we have

$$1 + \left(\frac{dy}{dx}\right)^2 = 1 + \left(\frac{e^{Ax} - e^{-Ax}}{2}\right)^2 = 1 + \frac{1}{4}\left(e^{2Ax} + e^{-2Ax} - 2\right)$$
$$= \frac{1}{4}\left(e^{2Ax} + e^{-2Ax} + 2\right) = \frac{1}{4}\left(e^{Ax} + e^{-Ax}\right)^2.$$

This means

$$k\sqrt{1 + \left(\frac{dy}{dx}\right)^2} = k\sqrt{\frac{1}{4}\left(e^{Ax} + e^{-Ax}\right)^2} = \frac{k}{2} \cdot \left|e^{Ax} + e^{-Ax}\right|$$
$$= k\frac{\left(e^{Ax} + e^{-Ax}\right)}{2} \quad \text{(since } e^{Ax} + e^{-Ax} > 0\text{)}.$$

Since we want $\frac{d^2y}{dx^2} = k\sqrt{1 + \left(\frac{dy}{dx}\right)^2}$, we must have $A = k$.

Solutions for Section 10.2

1. (a)

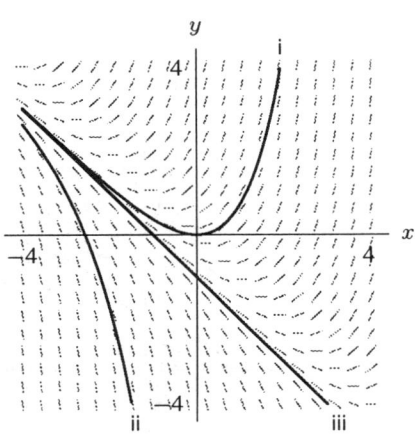

(b) The solution through $(-1, 0)$ appears to be linear, so its equation is $y = -x - 1$.
(c) If $y = -x - 1$, then $y' = -1$ and $x + y = x + (-x - 1) = -1$, so this checks as a solution.

2. (a)

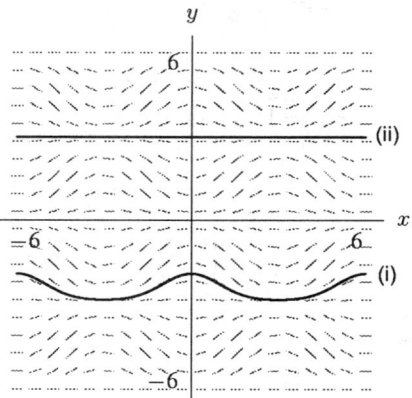

Figure 10.1

(b) We can see that the slope lines are horizontal when y is an integer multiple of π. We conclude from Figure 10.1 that the solution is $y = n\pi$ in this case.

To check this, we note that if $y = n\pi$, then $(\sin x)(\sin y) = (\sin x)(\sin n\pi) = 0 = y'$. Thus $y = n\pi$ is a solution to $y' = (\sin x)(\sin y)$, and it passes through $(0, n\pi)$.

3. (a) Since $y' = -y$, the slope is negative above the x-axis (when y is positive) and positive below the x-axis (when y is negative). The only slope field for which this is true is II.
    (b) Since $y' = y$, the slope is positive for positive y and negative for negative y. This is true of both I and III. As y get larger, the slope should get larger, so the correct slope field is I.
    (c) Since $y' = x$, the slope is positive for positive x and negative for negative x. This corresponds to slope field V.
    (d) Since $y' = \dfrac{1}{y}$, the slope is positive for positive y and negative for negative y. As y approaches 0, the slope becomes larger in magnitude, which correspond to solution curves close to vertical. The correct slope field is III.
    (e) Since $y' = y^2$, the slope is always positive, so this must correspond to slope field IV.

4. (a) II   (b) VI   (c) IV   (d) I   (e) III   (f) V

5. Notice that $y' = \dfrac{x+y}{x-y}$ is zero when $x = -y$ and is undefined when $x = y$. A solution curve will be horizontal (slope= 0) when passing through a point with $x = -y$, and will be vertical (slope undefined) when passing through a point with $x = y$. The only slope field for which this is true is slope field (b).

6. The slope fields in (I) and (II) appear periodic. (I) has zero slope at $x = 0$, so (I) matches $y' = \sin x$, whereas (II) matches $y' = \cos x$. The slope in (V) tends to zero as $x \to \pm\infty$, so this must match $y' = e^{-x^2}$. Of the remaining slope fields, only (III) shows negative slopes, matching $y' = xe^{-x}$. The slope in (IV) is zero at $x = 0$, so it matches $y' = x^2 e^{-x}$. This leaves field (VI) to match $y' = e^{-x}$.

7. (a), (b)

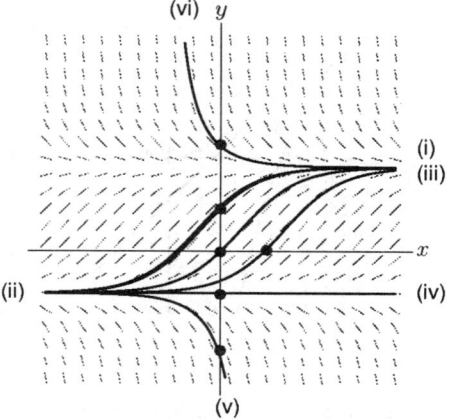

Figure 10.2

(c) Figure 10.2 shows that a solution will be increasing if its y-values fall in the range $-1 < y < 2$. This makes sense since if we examine the equation $y' = 0.5(1 + y)(2 - y)$, we will find that $y' > 0$ if $-1 < y < 2$. Notice that if the y-value ever gets to 2, then $y' = 0$ and the function becomes constant, following the line $y = 2$. (The same is true if ever $y = -1$.)

From the graph, the solution is decreasing if $y > 2$ or $y < -1$. Again, this also follows from the equation, since in either case $y' < 0$.

The curve has a horizontal tangent if $y' = 0$, which only happens if $y = 2$ or $y = -1$. This also can be seen on the graph in Figure 10.2.

8. (a) For $y = x^2 + 2x + 2$, we have $y - x^2 = 2x + 2$ and $y' = 2x + 2$; thus, $y' = y - x^2$ and the differential equation is satisfied.

For $y = x^2 + 2x + 2 - 2e^x$, we have $y - x^2 = 2x + 2 - 2e^x$ and $y' = 2x + 2 - 2e^x$ and so again $y' = y - x^2$.

The equation $y = x^2 + 2x + 2$ gives an upward opening parabola, so it corresponds to the top solution shown on the slope field.

(b)

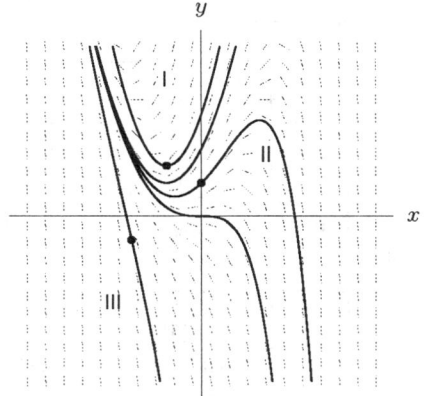

Figure 10.3

(c) The solution curves in Region I have one critical point each. They are concave up everywhere, decreasing at first, then increasing very rapidly.
(d) The curves in Region II have two critical points each and an inflection point. They are decreasing and concave up at first, then they become increasing, eventually becoming concave down and finally decreasing very rapidly.
(e) The curves in Region III are decreasing everywhere. They have no critical points.

Solutions for Section 10.3

1. (a)

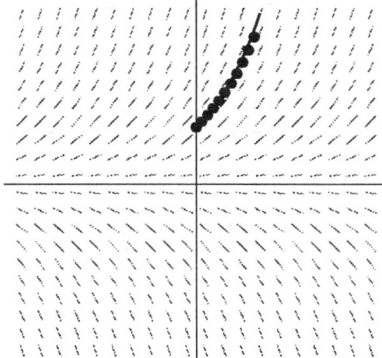

Figure 10.4

(b) $y(0) = 1$,
$y(0.1) \approx y(0) + 0.1y(0) = 1 + 0.1(1) = 1.1$
$y(0.2) \approx y(0.1) + 0.1y(0.1) = 1.1 + 0.1(1.1) = 1.21$
$y(0.3) \approx y(0.2) + 0.1y(0.2) = 1.21 + 0.1(1.21) = 1.331$
$y(0.4) \approx 1.4641$
$y(0.5) \approx 1.61051$
$y(0.6) \approx 1.77156$
$y(0.7) \approx 1.94872$
$y(0.8) \approx 2.14359$
$y(0.9) \approx 2.35795$
$y(1.0) \approx 2.59374$

(c) See Figure 10.4. A smooth curve drawn through the solution points seems to match the slopefield.
(d) For $y = e^x$, we have $y' = e^x = y$ and $y(0) = e^0 = 1$.

	Computed Solution	
x_n	Approx. $y(x_n)$	$y(x_n)$
0	1	1
0.1	1.1	1.10517
0.2	1.21	1.22140
0.3	1.331	1.34986
0.4	1.4641	1.49182
0.5	1.61051	1.64872
0.6	1.77156	1.82212
0.7	1.94872	2.01375
0.8	2.14359	2.22554
0.9	2.35795	2.45960
1.0	2.59374	2.71828

2. (a) (i)

TABLE 10.1 *Euler's method for* $y' = (\sin x)(\sin y)$, *starting at* $(0, 2)$

x	y	$\Delta y = $(slope)$\Delta x$
0	2	$0 = (\sin 0)(\sin 2)(0.1)$
0.1	2	$0.009 = (\sin 0.1)(\sin 2)(0.1)$
0.2	2.009	$0.018 = (\sin 0.2)(\sin 2.009)(0.1)$
0.3	2.027	

(ii)

TABLE 10.2 *Euler's method for* $y' = (\sin x)(\sin y)$, *starting at* $(0, \pi)$

x	y	$\Delta y = $(slope)$\Delta x$
0	π	$0 = (\sin 0)(\sin \pi)(0.1)$
0.1	π	$0 = (\sin 0.1)(\sin \pi)(0.1)$
0.2	π	$0 = (\sin 0.2)(\sin \pi)(0.1)$
0.3	π	

(b) The slope field shows that the slope of the solution curve through $(0, \pi)$ is always 0. Thus the solution curve is the horizontal line with equation $y = \pi$.

3. (a)

TABLE 10.3 Euler's method for $y' = x + y$ with $y(0) = 1$

x	y	$\Delta y =$(slope)Δx
0	1	$0.1 = (1)(0.1)$
0.1	1.1	$0.12 = (1.2)(0.1)$
0.2	1.22	$0.142 = (1.42)(0.1)$
0.3	1.362	$0.1662 = (1.662)(0.1)$
0.4	1.5282	

So $y(0.4) \approx 1.5282$.

(b)

TABLE 10.4 Euler's method for $y' = x + y$ with $y(-1) = 0$

x	y	$\Delta y =$(slope)Δx
-1	0	$-0.1 = (-1)(0.1)$
-0.9	-0.1	$-0.1 = (-1)(0.1)$
-0.8	-0.2	$-0.1 = (-1)(0.1)$
-0.7	-0.3	
$\vdots$	$\vdots$	Notice that y
0	-1	decreases by 0.1
$\vdots$	$\vdots$	for every step
0.4	-1.4	

So $y(0.4) = -1.4$. (This answer is exact.)

4. (a)

TABLE 10.5

	Computed Solution	
x_n	Approx. $y(x_n)$	$y(x_n)$
0	0	0
0.1	0	0.000025
0.2	0.0001	0.0004
0.3	0.0009	0.002025
0.4	0.0036	0.0064
0.5	0.01	0.015625
0.6	0.0225	0.0324
0.7	0.0441	0.060025
0.8	0.0784	0.1024
0.9	0.1296	0.164025
1.0	0.2025	0.25

(b) We guess
$$y(x) = \frac{x^4}{4} + C,$$
so that $y(0) = 0$ gives $C = 0$, and the required solution is therefore
$$y(x) = \frac{x^4}{4}.$$

This is shown in the 3rd column of Table 10.5 above.

(c) The computed solution underestimates the real solution since the solution is concave up and is approximated in every interval by the tangent which is beneath the curve. See Figure 10.5.

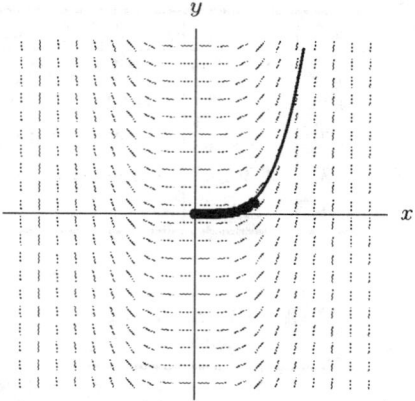

Figure 10.5

5. (a)

TABLE 10.6

t	y	slope = $\frac{1}{t}$	$\Delta y = $ (slope)$\Delta t = \frac{1}{t}(0.1)$
1	0	1	0.1
1.1	0.1	0.909	0.091
1.2	0.191	0.833	0.083
1.3	0.274	0.769	0.077
1.4	0.351	0.714	0.071
1.5	0.422	0.667	0.067
1.6	0.489	0.625	0.063
1.7	0.552	0.588	0.059
1.8	0.610	0.556	0.056
1.9	0.666	0.526	0.053
2	0.719		

(b) If $\frac{dy}{dt} = \frac{1}{t}$, then $y = \ln|t| + C$.
Starting at $(1, 0)$ means $y = 0$ when $t = 1$, so $C = 0$ and $y = \ln|t|$.
After ten steps, $t = 2$, so $y = \ln 2 \approx 0.693$.

(c) Approximate $y = 0.719$, Exact $y = 0.693$.
Thus the approximate answer is too big. This is because the solution curve is concave down, and so the tangent lines are above the curve. Figure 10.6 shows the slope field of $y' = 1/t$ with the solution curve $y = \ln t$ plotted on top of it.

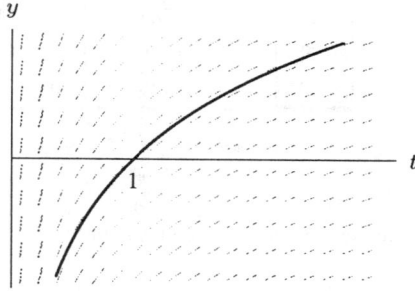

Figure 10.6

6. (a)

TABLE 10.7

x	y	$\Delta y = $ (slope)Δx
0	0	0
0.2	0	0.0016
0.4	0.0016	0.0128
0.6	0.0144	0.0432
0.8	0.0576	0.1024
1	0.1600	

At $x = 1$, $y \approx 0.16$.

(b)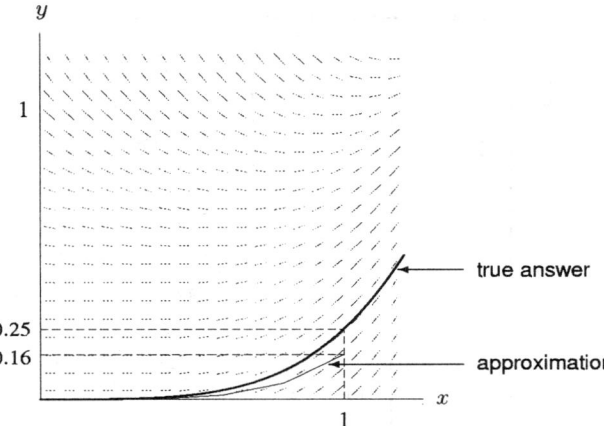

(c) Our answer to (a) appears to be an underestimate. This is as we would expect, since the curve is concave up.

7. (a) $\Delta x = 0.5$

TABLE 10.8 Euler's method for $y' = 2x$, with $y(0) = 1$

x	y	$\Delta y = $ (slope)Δx
0	1	$0 = (2 \cdot 0)(0.5)$
0.5	1	$0.5 = (2 \cdot 0.5)(0.5)$
1	1.5	

(b) $\Delta x = 0.25$

TABLE 10.9 Euler's method for $y' = 2x$, with $y(0) = 1$

x	y	$\Delta y = $ (slope)Δx
0	1	$0 = (2 \cdot 0)(0.25)$
0.25	1	$0.125 = (2 \cdot 0.25)(0.25)$
0.50	1.125	$0.25 = (2 \cdot 0.5)(0.25)$
0.75	1.375	$0.375 = (2 \cdot 0.75)(0.25)$
1	1.75	

(c) General Solution is $y = x^2 + C$, and $y(0) = 1$ gives $C = 1$. Thus, the solution is $y = x^2 + 1$. So the true value of y when $x = 1$ is $y = 1^2 + 1 = 2$.

(d) When $\Delta x = 0.5$, error $= 0.5$.
When $\Delta x = 0.25$, error $= 0.25$.
Thus, decreasing Δx by a factor of 2 has decreased the error by a factor of 2, as expected.

8. For $\Delta x = 0.2$, we get the following results.

$$y(1.2) \approx y(1) + 0.2\sin(1 \cdot y(1)) = 1.168294$$
$$y(1.4) \approx y(1.2) + 0.2\sin(1.2 \cdot y(1.2)) = 1.365450$$
$$y(1.6) \approx y(1.4) + 0.2\sin(1.4 \cdot y(1.4)) = 1.553945$$
$$y(1.8) \approx y(1.6) + 0.2\sin(1.6 \cdot y(1.6)) = 1.675822$$
$$y(2.0) \approx y(1.8) + 0.2\sin(1.8 \cdot y(1.8)) = 1.700779$$

Repeating this with $\Delta x = 0.1$ and 0.05 gives the results in Table 10.10 below

TABLE 10.10

	Computed Solution		
x-value	$\Delta x = 0.2$	$\Delta x = 0.1$	$\Delta x = 0.05$
1.0	1	1	1
1.1		1.084147	1.086501
1.2	1.168294	1.177079	1.181232
1.3		1.275829	1.280619
1.4	1.365450	1.375444	1.379135
1.5		1.469214	1.469885
1.6	1.553945	1.549838	1.546065
1.7		1.611296	1.602716
1.8	1.675822	1.650458	1.637809
1.9		1.667451	1.652112
2.0	1.700779	1.664795	1.648231

The computed approximations for $y(2)$ using step sizes $\Delta x = 0.2, 0.1, 0.05$ are $1.700779, 1.664795$, and 1.648231, respectively. Plotting these points we see that they lie approximately on a straight line.

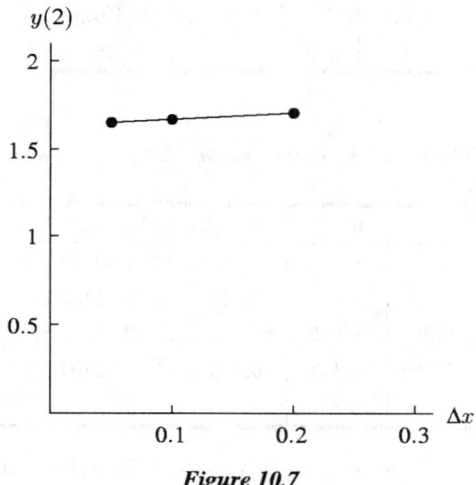

Figure 10.7

In the limit, as Δx tends to zero, the results produced by Euler's method should converge to the exact value of $y(2)$. This limiting value is the vertical intercept of the line drawn in Figure 10.7. This gives $y(2) \approx 1.632$.

9. (a) Using one step, $\frac{\Delta B}{\Delta t} = 0.05$, so $\Delta B = \left(\frac{\Delta B}{\Delta t}\right)\Delta t = 50$. Therefore we get an approximation of $B \approx 1050$ after one year.
    (b) With two steps, $\Delta t = 0.5$ and we have

    TABLE 10.11

t	B	$\Delta B = (0.05B)\Delta t$
0	1000	25
0.5	1025	25.63
1.0	1050.63	

    (c) Keeping track to the nearest hundredth with $\Delta t = 0.25$, we have

    TABLE 10.12

t	B	$\Delta B = (0.05B)\Delta t$
0	1000	12.5
0.25	1012.5	12.66
0.5	1025.16	12.81
0.75	1037.97	12.97
1	1050.94	

    (d) In part (a), we get our approximation by making a single increment, ΔB, where ΔB is just $0.05B$. If we think in terms of interest, ΔB is just like getting one end of the year interest payment. Since ΔB is 0.05 times the balance B, it is like getting 5% interest at the end of the year.
    (e) Part (b) is equivalent to computing the final amount in an account that begins with $1000 and earns 5% interest compounded twice annually. Each step is like computing the interest after 6 months. When $t = 0.5$, for example, the interest is $\Delta B = (0.05B) \cdot \frac{1}{2}$, and we add this to $1000 to get the new balance.
    Similarly, part (c) is equivalent to the final amount in an account that has an initial balance of $1000 and earns 5% interest compounded quarterly.

10. Assume that $x > 0$ and that we use n steps in Euler's method. Label the x-coordinates we use in the process $x_0, x_1, \ldots, x_n$, where $x_0 = 0$ and $x_n = x$. Then using Euler's method to find $y(x)$, we get

    TABLE 10.13

	x	y	$\Delta y = (\text{slope})\Delta x$
P_0	$0 = x_0$	0	$f(x_0)\Delta x$
P_1	x_1	$f(x_0)\Delta x$	$f(x_1)\Delta x$
P_2	x_2	$f(x_0)\Delta x + f(x_1)\Delta x$	$f(x_2)\Delta x$
$\vdots$	$\vdots$	$\vdots$	$\vdots$
P_n	$x = x_n$	$\sum_{i=0}^{n-1} f(x_i)\Delta x$	

    Thus the result from Euler's method is $\sum_{i=0}^{n-1} f(x_i)\Delta x$. We recognize this as the left-hand Riemann sum that approximates $\int_0^x f(t)\,dt$.

Solutions for Section 10.4

1. $\frac{dP}{dt} = 0.02P$ implies that $\frac{dP}{P} = 0.02\,dt$.

   $\int \frac{dP}{P} = \int 0.02\,dt$ implies that $\ln|P| = 0.02t + C$.

   $|P| = e^{0.02t+C}$ implies that $P = Ae^{0.02t}$, where $A = \pm e^C$.
   We are given $P(0) = 20$. Therefore, $P(0) = Ae^{(0.02) \cdot 0} = A = 20$. So the solution is $P = 20e^{0.02t}$.

2. $\frac{dQ}{dt} = \frac{Q}{5}$ implies that $\frac{dQ}{Q} = \frac{dt}{5}$.

   $\int \frac{dQ}{Q} = \int \frac{dt}{5}$ implies that $\ln|Q| = \frac{1}{5}t + C$.

   So $|Q| = e^{\frac{1}{5}t+C} = e^{\frac{1}{5}t}e^C$ implies that $Q = Ae^{\frac{1}{5}t}$, where $A = \pm e^C$. From the initial conditions we know that $Q(0) = 50$, so $Q(0) = Ae^{(\frac{1}{5}) \cdot 0} = A = 50$. Thus $Q = 50e^{\frac{1}{5}t}$.

3. $\frac{dm}{dt} = 3m$. As in problems 1 and 2, we get
   $$m = Ae^{3t}.$$
   Since $m = 5$ when $t = 1$, we have $5 = Ae^3$, so $A = \frac{5}{e^3}$. Thus $m = \frac{5}{e^3}e^{3t} = 5e^{3t-3}$.

4. $\frac{dI}{dx} = 0.2I$ implies that $\frac{dI}{I} = 0.2\,dx$ implies that $\int \frac{dI}{I} = \int 0.2\,dx$ implies that $\ln|I| = 0.2x + C$.
   So $I = Ae^{0.2x}$, where $A = \pm e^C$. According to the given boundary condition, $I(-1) = 6$. Therefore, $I(-1) = Ae^{0.2(-1)} = Ae^{-0.2} = 6$ implies that $A = 6e^{0.2}$. Thus $I = 6e^{0.2}e^{0.2x} = 6e^{0.2(x+1)}$.

5. $\frac{dy}{dx} + \frac{y}{3} = 0$ implies $\frac{dy}{dx} = -\frac{y}{3}$ implies $\int \frac{dy}{y} = -\int \frac{1}{3}\,dx$.
   Integrating and moving terms, we have $y = Ae^{-\frac{1}{3}x}$. Since $y(0) = A = 10$, we have $y = 10e^{-\frac{1}{3}x}$.

6. $\frac{1}{z}\frac{dz}{dt} = 5$ implies $\frac{dz}{z} = 5\,dt$.
   Integrating and moving terms, we have $z = Ae^{5t}$. Using the fact that $z(1) = 5$, we have $z(1) = Ae^5 = 5$, so $A = \frac{5}{e^5}$. Therefore, $z = \frac{5}{e^5}e^{5t} = 5e^{5t-5}$.

7. $\frac{dP}{dt} = P + 4$ implies that $\frac{dP}{P+4} = dt$.

   $\int \frac{dP}{P+4} = \int dt$ implies that $\ln|P+4| = t + C$.

   $P + 4 = Ae^t$ implies that $P = Ae^t - 4$. $P = 100$ when $t = 0$, so $P(0) = Ae^0 - 4 = 100$, and $A = 104$. Therefore $P = 104e^t - 4$.

8. $\frac{dy}{dx} = 2y - 4 = 2(y-2)$.

   Factoring out a 2 makes the integration easier: $\frac{dy}{y-2} = 2\,dx$ implies that $\int \frac{dy}{y-2} = \int 2\,dx$ implies that $\ln|y-2| = 2x + C$.
   $|y-2| = e^{2x+C}$ implies that $y - 2 = Ae^{2x}$ where $A = \pm e^C$. The curve passes through $(2,5)$, which means $3 = Ae^4$, so $A = \frac{3}{e^4}$. Thus, $y = 2 + \frac{3}{e^4}e^{2x} = 2 + 3e^{2x-4}$.

9. Factoring out the 0.1 gives $\frac{dm}{dt} = 0.1m + 200 = 0.1(m+2000)$.
   $\frac{dm}{m+2000} = 0.1\,dt$ implies that $\int \frac{dm}{m+2000} = \int 0.1\,dt$, so $\ln|m+2000| = 0.1t + C$. So $m = Ae^{0.1t} - 2000$. Using the initial condition, $m(0) = Ae^{(0.1) \cdot 0} - 2000 = 1000$, gives $A = 3000$. Thus $m = 3000e^{0.1t} - 2000$.

10. $\frac{dB}{dt} + 2B = 50$ implies $\frac{dB}{dt} = -2B + 50 = -2(B-25)$ implies $\int \frac{dB}{B-25} = -\int 2\,dt$.

    After integrating and doing some algebra, we have $B - 25 = Ae^{-2t}$. Using the initial condition, we have $75 = Ae^{-2}$, so $A = 75e^2$. Thus $B = 25 + 75e^2 e^{-2t} = 25 + 75e^{2-2t}$.

11. We know that the general solution to a differential equation of the form
    $$\frac{dy}{dt} = k(y - A)$$
    is
    $$y = Ce^{kt} + A.$$

Thus, in our case, we get
$$y = Ce^{t/2} + 200.$$
We know that at $t = 0$ we have $y = 50$, so solving for C we get
$$y = Ce^{t/2} + 200$$
$$50 = Ce^{0/2} + 200$$
$$-150 = Ce^0$$
$$C = -150.$$
Thus we get
$$y = 200 - 150e^{t/2}.$$

12. We know that the general solution to a differential equation of the form
$$\frac{dQ}{dt} = k(Q - A)$$
is
$$H = Ce^{kt} + A.$$
To get our equation in this form, we factor out a 0.3 to get
$$\frac{dQ}{dt} = 0.3\left(Q - \frac{120}{0.3}\right) = 0.3(Q - 400).$$
Thus, in our case, we get
$$Q = Ce^{0.3t} + 400.$$
We know that at $t = 0$ we have $Q = 50$ so solving for C we get
$$Q = Ce^{0.3t} + 400$$
$$50 = Ce^{0.3(0)} + 400$$
$$-350 = Ce^0$$
$$C = -350.$$
Thus we get
$$Q = 400 - 350e^{0.3t}.$$

13. Rearrange and write
$$\int \frac{1}{1-R} dR = \int dr$$
or
$$-\ln|1 - R| = r + C$$
which can be written as
$$1 - R = \pm e^{-C-r} = Ae^{-r}$$
or
$$R(r) = 1 - Ae^{-r}.$$
The initial condition $R(1) = 0.1$ gives $0.1 = 1 - Ae^{-1}$ and so
$$A = 0.9e.$$
Therefore
$$R(r) = 1 - 0.9e^{1-r}.$$

14. Write
$$\int \frac{1}{y} dy = \int \frac{1}{3+t} dt$$
and so
$$\ln|y| = \ln|3+t| + C$$
or
$$\ln|y| = \ln D|3+t|$$
where $\ln D = C$. Therefore
$$y = D(3+t).$$
The initial condition $y(0) = 1$ gives $D = \frac{1}{3}$ and so
$$y(t) = \frac{1}{3}(3+t).$$

15. $\frac{dz}{dt} = te^z$ implies $e^{-z} dz = t\, dt$ implies $\int e^{-z} dz = \int t\, dt$ implies $-e^{-z} = \frac{t^2}{2} + C$.
Since the solution passes through the origin, $z = 0$ when $t = 0$, we must have $-e^{-0} = \frac{0}{2} + C$, so $C = -1$. Thus $-e^{-z} = \frac{t^2}{2} - 1$, or $z = -\ln(1 - \frac{t^2}{2})$.

16. $dy/dx = 5y/x$ implies $\int dy/y = \int 5dx/x$. So $\ln|y| = 5\ln|x| + C = 5\ln x + C$ implies $|y| = e^{5\ln x}e^C$, and thus $y = Ax^5$ where $A = \pm e^C$. Since $y = 3$ when $x = 1$, so $A = 3$. Thus $y = 3x^5$.

17. $\frac{dy}{dt} = y^2(1+t)$ implies that $\int \frac{dy}{y^2} = \int (1+t)\, dt$ implies that $-\frac{1}{y} = t + \frac{t^2}{2} + C$ implies that $y = -\frac{1}{t + \frac{t^2}{2} + C}$.
Since $y = 2$ when $t = 1$, then $2 = -\frac{1}{1+\frac{1}{2}+C}$. So $2C + 3 = -1$, and $C = -2$. Thus $y = -\frac{1}{\frac{t^2}{2}+t-2} = -\frac{2}{t^2+2t-4}$.

18. $\frac{dz}{dt} = z + zt^2 = z(1+t^2)$ implies that $\int \frac{dz}{z} = \int (1+t^2)dt$ implies that $\ln|z| = t + \frac{t^3}{3} + C$ implies that $z = Ae^{t+\frac{t^3}{3}}$. $z = 5$ when $t = 0$, so $A = 5$ and $z = 5e^{t+\frac{t^3}{3}}$.

19. $\frac{dw}{d\theta} = \theta w^2 \sin\theta^2$ implies that $\int \frac{dw}{w^2} = \int \theta \sin\theta^2\, d\theta$ implies that $-\frac{1}{w} = -\frac{1}{2}\cos\theta^2 + C$. According to the initial conditions, $w(0) = 1$, so $-1 = -\frac{1}{2} + C$ and $C = -\frac{1}{2}$. Thus $-\frac{1}{w} = -\frac{1}{2}\cos\theta^2 - \frac{1}{2}$ implies that $\frac{1}{w} = \frac{\cos\theta^2+1}{2}$ implies that $w = \frac{2}{\cos\theta^2+1}$.

20. $x(x+1)\frac{du}{dx} = u^2$ implies $\int \frac{du}{u^2} = \int \frac{dx}{x(x+1)} = \int (\frac{1}{x} - \frac{1}{1+x})dx$ implies $-\frac{1}{u} = \ln|x| - \ln|x+1| + C$. $u(1) = 1$, so $-\frac{1}{1} = \ln|1| - \ln|1+1| + C$. So $C = \ln 2 - 1$. Solving for u yields $-\frac{1}{u} = \ln|x| - \ln|x+1| + \ln 2 - 1 = \ln\frac{2|x|}{|x+1|} - 1$, so $u = \frac{-1}{\ln|\frac{2x}{x+1}|-1}$.

21. Separating variables and integrating with respect to r gives
$$\int \frac{1}{z} dz = \int (1+r^2) dr$$
so that
$$\ln|z| = r + \frac{1}{3}r^3 + C.$$
The initial condition $z(0) = 1$ gives $C = 0$ so that
$$z(r) = e^{r+(1/3)r^3}.$$

22. Separating variables and integrating with respect to ψ gives
$$\int \frac{1}{w^2} dw = \int \psi \cos\psi^2 d\psi.$$
Now set $\psi^2 = t$, then this becomes
$$\int \frac{1}{w^2} dw = \frac{1}{2} \int \cos t\, dt$$

and so
$$-\frac{1}{w} = \frac{1}{2}\sin t + C$$
or
$$w = \frac{-2}{\sin(t) + D}$$
$$w = \frac{-2}{\sin\psi^2 + D}.$$

Using the initial conditions give $D = -2$, so the solution is
$$w = \frac{-2}{\sin\psi^2 - 2}.$$

23. $\frac{dR}{dt} = kR$ implies that $\frac{dR}{R} = k\,dt$ which implies that $\int \frac{dR}{R} = \int k\,dt$. Integrating gives $\ln|R| = kt + C$, so $|R| = e^{kt+C} = e^{kt}e^C$. $R = Ae^{kt}$, where $A = \pm e^C$.

24. $\frac{dQ}{dt} - \frac{Q}{k} = 0$ so $\frac{dQ}{dt} = \frac{Q}{k}$. This is now the same problem as Problem 25, except the constant factor on the right is $\frac{1}{k}$ instead of k. Thus the solution is $Q = Ae^{\frac{1}{k}t}$ for any constant A.

25. $\frac{dP}{dt} = P - a$, implying that $\frac{dP}{P-a} = dt$ so $\int \frac{dP}{P-a} = \int dt$. Integrating yields $\ln|P - a| = t + C$, so $|P - a| = e^{t+C} = e^t e^C$. $P = a + Ae^t$, where $A = \pm e^C$ or $A = 0$.

26. $\frac{dQ}{dt} = b - Q$ implies that $\frac{dQ}{b-Q} = dt$ which, in turn, implies $\int \frac{dQ}{b-Q} = \int dt$. Integrating yields $-\ln|b - Q| = t + C$, so $|b - Q| = e^{-(t+C)} = e^{-t}e^{-C}$. $Q = b - Ae^{-t}$, where $A = \pm e^{-C}$ or $A = 0$.

27. $\frac{dP}{dt} = k(P - a)$, so $\frac{dP}{P-a} = k\,dt$, so $\int \frac{dP}{P-a} = \int k\,dt$. Integrating yields $\ln|P - a| = kt + C$ so $P = a + Ae^{kt}$ where $A = \pm e^C$ or $A = 0$.

28. $\frac{dR}{dt} = aR + b$. If $a = 0$, then this is just $\frac{dR}{dt} = b$, where b is a constant. Thus in this case $R = bt + C$ is a solution for any constant C.
    If $a \ne 0$, then $\frac{dR}{dt} = a(R + \frac{b}{a})$.
    Now this is just the same as Problem 27, except here we have a in place of k and $-\frac{b}{a}$ in place of a, so the solutions are $R = -\frac{b}{a} + Ae^{at}$ where A can be any constant.

29. $\frac{dy}{dt} = y(2-y)$ which implies that $\frac{dy}{y(y-2)} = -dt$, implying that $\int \frac{dy}{(y-2)(y)} = -\int dt$, so $-\frac{1}{2}\int (\frac{1}{y} - \frac{1}{y-2})dy = -\int dt$. Integrating yields $\frac{1}{2}(\ln|y-2| - \ln|y|) = -t + C$, so $\ln\frac{|y-2|}{|y|} = -2t + 2C$.
    Exponentiating both sides yields $|1 - \frac{2}{y}| = e^{-2t+2C} \Rightarrow \frac{2}{y} = 1 - Ae^{-2t}$, where $A = \pm e^{2C}$. Hence $y = \frac{2}{1-Ae^{-2t}}$. But $y(0) = \frac{2}{1-A} = 1$, so $A = -1$, and $y = \frac{2}{1+e^{-2t}}$.

30. $t\frac{dx}{dt} = (1 + 2\ln t)\tan x$ implies that $\frac{dx}{\tan x} = (\frac{1+2\ln t}{t})\,dt$ which implies that $\int \frac{\cos x}{\sin x}\,dx = \int(\frac{1}{t} + \frac{2\ln t}{t})\,dt$.
    $\ln|\sin x| = \ln t + (\ln t)^2 + C$.
    $|\sin x| = e^{\ln t + (\ln t)^2 + C} = t(e^{\ln t})^{\ln t}e^C = t(t^{\ln t})e^C$. So $\sin x = At^{(\ln t + 1)}$, where $A = \pm e^C$. Therefore $x = \arcsin(At^{\ln t + 1})$.

31. $\frac{dx}{dt} = \frac{x\ln x}{t}$, so $\int \frac{dx}{x\ln x} = \int \frac{dt}{t}$ and thus $\ln|\ln x| = \ln|t| + C$, so $|\ln x| = e^C e^{\ln|t|} = e^C |t|$. Therefore $\ln x = At$, where $A = \pm e^C$, so $x = e^{At}$.

32. $\frac{dy}{dt} = -y\ln(\frac{y}{2})$ implies that $\frac{dy}{y\ln(\frac{y}{2})} = -dt$ implies that $\int \frac{dy}{y\ln(\frac{y}{2})} = \int(-dt)$.
    Substituting $w = \ln(\frac{y}{2})$, $dw = \frac{1}{y}dy$ gives:
    $\int \frac{dw}{w} = \int(-dt)$ implies that $\ln|w| = \ln|\ln(\frac{y}{2})| = -t + C$. Since $y(0) = 1$, we have $C = \ln|\ln\frac{1}{2}| = \ln|-\ln 2| = \ln(\ln 2)$. Thus $\ln|\ln(\frac{y}{2})| = -t + \ln(\ln 2)$, or
    $$|\ln(\frac{y}{2})| = e^{-t+\ln(\ln 2)} = (\ln 2)e^{-t}$$

    Again, since $y(0) = 1$, we see that $-\ln(y/2) = (\ln 2)e^{-t}$ and thus $y = 2(2^{-e^{-t}})$. (Note that $\ln(y/2) = (\ln 2)e^{-t}$ does not satisfy $y(0) = 1$.)

33. (a), (b)

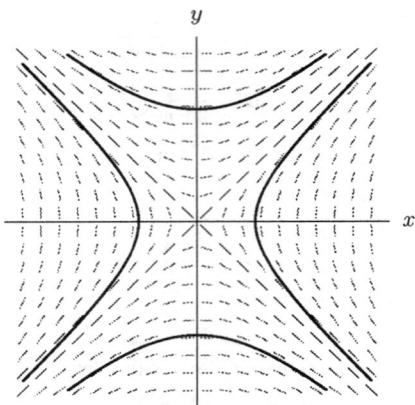

(c) $\frac{dy}{dx} = \frac{x}{y}$, so $\int y\,dy = \int x\,dx$ and thus $\frac{y^2}{2} = \frac{x^2}{2} + C$, or $y^2 - x^2 = 2C$. This is the equation of the hyperbolas in (b).

34. (a), (b)

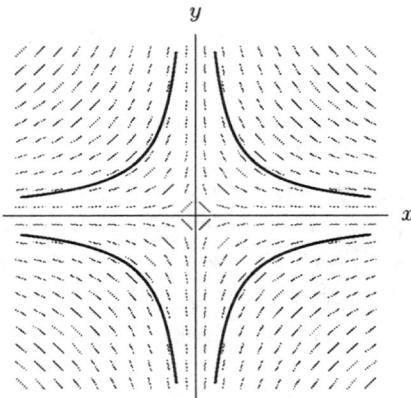

(c) $\frac{dy}{dx} = -\frac{y}{x}$, which implies that $\int \frac{dy}{y} = -\int \frac{dx}{x}$, so $\ln|y| = -\ln|x| + C$ implies that $|y| = e^{-\ln|x|+C} = (|x|)^{-1}e^C$. $y = \frac{A}{x}$, where $A = \pm e^C$.

35. By looking at the slope fields, we see that any solution curve of $y' = \frac{x}{y}$ intersects any solution curve to $y' = -\frac{y}{x}$. Now if the two curves intersect at (x, y), then the two slopes at (x, y) are negative reciprocals of each other, because $-\frac{1}{x/y} = -\frac{y}{x}$. Hence, the two curves intersect at right angles.

36. If $y(x) = xv(x)$, then $y' = v + xv'$ and the differential equation becomes
$$v + xv' = \frac{x + xv}{x - xv} = \frac{1 + v}{1 - v}.$$

Solving for v' gives
$$x\frac{dv}{dx} = \frac{1+v}{1-v} - v = \frac{1+v^2}{1-v}$$

which is a separable equation. Rearranging and integrating with respect to x gives

$$\int \frac{1-v}{1+v^2}dv = \int \frac{1}{x}dx$$

$$\int \frac{1}{1+v^2}dv - \frac{1}{2}\int \frac{2v}{1+v^2}dv = \int \frac{1}{x}dx$$

$$\arctan(v) - \frac{1}{2}\ln(1+v^2) = \ln|x| + C.$$

Now put $v = \frac{y}{x}$ to find

$$\arctan\left(\frac{y}{x}\right) - \frac{1}{2}\ln(x^2+y^2) = C.$$

37. Starting off with the homogeneous equation

$$f\left(\frac{y}{x}\right) = \frac{dy}{dx},$$

we replace $y(x)$ with $xv(x)$ and get

$$f(v) = f\left(\frac{xv}{x}\right) = f\left(\frac{y}{x}\right) = \frac{dy}{dx} = \frac{d(xv(x))}{dx} = v + x\frac{dv}{dx}.$$

Thus we have

$$f(v) - v = x\frac{dv}{dx}$$

which is clearly separable.

38. Making the substitution $s = x+a, t = y+b$. the differential equation becomes

$$\frac{dy}{dx} = \frac{x+y+a+b+5}{x-y+a-b-1}.$$

Now select a and b so that $a+b+5 = 0$ and $a-b-1 = 0$, that is, $a = -2, b = -3$. The differential equation becomes

$$\frac{dy}{dx} = \frac{x+y}{x-y}.$$

Now set $y(x) = xv(x)$, then $y' = v + xv'$ and the differential equation becomes

$$v + xv' = \frac{x+xv}{x-xv} = \frac{1+v}{1-v}.$$

Solving for v' gives

$$x\frac{dv}{dx} = \frac{1+v}{1-v} - v = \frac{1+v^2}{1-v}$$

which is a separable equation. Rearranging and integrating with respect to x gives

$$\int \frac{1-v}{1+v^2}dv = \int \frac{1}{x}dx$$

$$\int \frac{1}{1+v^2}dv - \frac{1}{2}\int \frac{2v}{1+v^2}dv = \int \frac{1}{x}dx$$

$$\arctan(v) - \frac{1}{2}\ln(1+v^2) = \ln|x| + C.$$

Now put $v = \frac{y}{x}$ to find

$$\arctan\left(\frac{y}{x}\right) - \frac{1}{2}\ln(x^2+y^2) = C.$$

Transforming back to the original variables $x = s+2$ and $y = t+3$ gives

$$\arctan\left(\frac{t+3}{s+2}\right) - \frac{1}{2}\ln((s+2)^2+(t+3)^2) = C.$$

Solutions for Section 10.5

1. (a) If the world's population grows exponentially, satisfying $\frac{dP}{dt} = kP$, and if the arable land used is proportional to the population, then we'd expect A to satisfy $\frac{dA}{dt} = kA$. One is, of course, also assuming that the total amount of arable land is large compared to the amount that is now being used.

   (b) We have $A(t) = A_0 e^{kt} = (1 \times 10^9)e^{kt}$, where t is the number of years after 1950. Since $2 \times 10^9 = (1 \times 10^9)e^{k(30)}$, we have $e^{30k} = 2$, so $k = \frac{\ln 2}{30} \approx 0.023$. Thus, $A \approx (1 \times 10^9)e^{0.023t}$. We want to find t such that $3.2 \times 10^9 = A(t) = (1 \times 10^9)e^{0.023t}$. Taking logarithms yields

   $$t = \frac{\ln(3.2)}{0.023} \approx 50.6 \text{ years.}$$

   Thus the arable land will have run out by the year 2001.

2. (a) Separating variables, we have $\frac{dH}{H-200} = -k\,dt$, so $\int \frac{dH}{H-200} = \int -k\,dt$, whence $\ln|H-200| = -kt + C$, and $H - 200 = Ae^{-kt}$, where $A = \pm e^C$. The initial condition is that the yam is $20°C$ at the time $t = 0$. Thus $20 - 200 = A$, so $A = -180$. Thus $H = 200 - 180e^{-kt}$.

   (b) Using part (a), we have $120 = 200 - 180e^{-k(30)}$. Solving for k, we have $e^{-30k} = \frac{-80}{-180}$, giving

   $$k = \frac{\ln \frac{4}{9}}{-30} \approx 0.027.$$

   Note that this k is correct if t is given in *minutes*. (If t is given in hours, $k = \frac{\ln \frac{4}{9}}{-\frac{1}{2}} \approx 1.62$.)

3. (a) (I)
   (b) (IV)
   (c) (II) and (IV)
   (d) (II) and (III)

4. (a) = (I), (b) = (IV), (c) = (III). Graph (II) represents an egg originally at $0°$ C which is moved to the kitchen table ($20°$ C) two minutes after the egg in part (a) is moved.

5. (a) We know that the equilibrium solutions are the functions satisfying the differential equation whose derivative everywhere is 0. Thus we have

   $$\frac{dy}{dt} = 0$$
   $$0.2(y-3)(y+2) = 0$$
   $$(y-3)(y+2) = 0.$$

   The solutions are $y = 3$ and $y = -2$.

   (b)

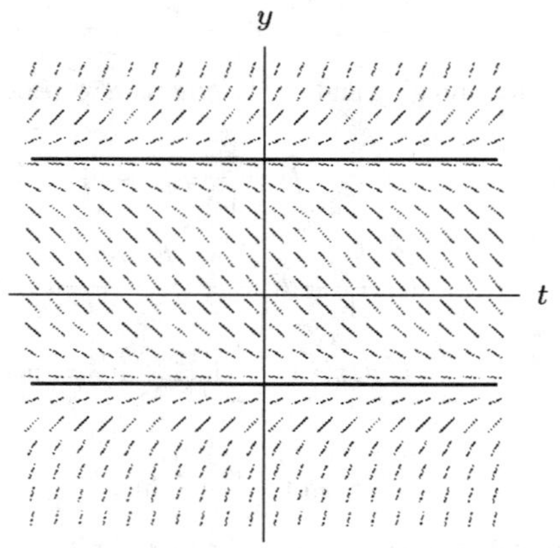

   Figure 10.8

   Looking at Figure 10.8, we see that the line $y = 3$ is an unstable solution, while the line $y = -2$ is a stable solution.

6. The equilibrium solutions of a differential equation are those functions satisfying the differential equation whose derivative is everywhere 0. Graphically, this means that a function is an equilibrium solution if it is a horizontal line that lies on the slope field. Looking at the figure in the problem, it appears that the equilibrium solutions for this problem are at $y = 1$ and $y = 3$. An equilibrium solution is stable if a small change in the initial value conditions gives a solution which tends toward equilibrium as $t \to \infty$. we see that $y = 3$ is a stable solution, while $y = 1$ is an unstable solution. See below.

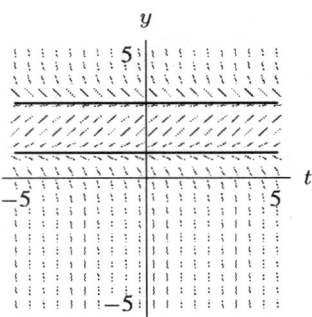

7. (a) We know that the equilibrium solution is the solution satisfying the differential equation whose derivative is everywhere 0. Thus we have
$$\frac{dy}{dt} = 0$$
$$0.5y - 250 = 0$$
$$y = 500.$$

(b) We use separation of variables. Since
$$\frac{dy}{dt} = 0.5y - 250,$$
we have
$$\int \frac{1}{0.5y - 250} dy = \int dt$$
$$2 \ln |0.5y - 250| = t + C$$
$$0.5y - 250 = e^{(t+C)/2}$$
$$y = Ae^{t/2} + 500,$$
where $A = 2e^{C/2}$.

(c) Using initial value $y(0) = 500$, we have $y = 500$, the equilibrium solution. Using initial value $y(0) = 400$, we have $A = -100$ and so $y = 500 - 100e^{t/2}$. Using initial value $y(0) = 600$, we have $A = 100$ and so $y = 500 + 100e^{t/2}$. These three solutions are shown below.

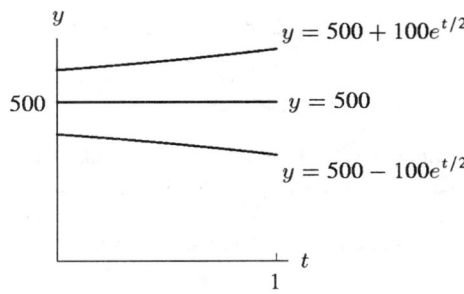

(d) We see above that the equilibrium solution $y = 500$ is unstable.

504 CHAPTER TEN /SOLUTIONS

8. (a) The rate of growth of the money in the account is proportional to the amount of money in the account. Thus
$$\frac{dM}{dt} = rM.$$

(b) Solving, we have $\frac{dM}{M} = r\,dt$.
$$\int \frac{dM}{M} = \int r\,dt$$
$$\ln|M| = rt + C$$
$$M = e^{rt+C} = Ae^{rt}, \qquad A = e^C.$$

When $t = 0$ (in 1970), $M = 1000$, so $A = 1000$ and $M = 1000e^{rt}$.

(c)

M

20000 — $M = 1000e^{0.10t}$

5000 — $M = 1000e^{0.05t}$

1000

$t = 0$ (1970), $t = 30$ (2000)

9. (a) $\frac{dB}{dt} = \frac{r}{100}B$. The constant of proportionality is $\frac{r}{100}$.
(b) Solving, we have
$$\frac{dB}{B} = \frac{r\,dt}{100}$$
$$\int \frac{dB}{B} = \int \frac{r}{100}\,dt$$
$$\ln|B| = \frac{r}{100}t + C$$
$$B = e^{(r/100)t+C} = Ae^{(r/100)t}, \qquad A = e^C.$$

A is the initial amount in the account, since A is the amount at time $t = 0$.

(c)

20,000 — $B = 1000e^{0.15t}$, $B = 1000e^{0.10t}$

10,000

1000 — $B = 1000e^{0.04t}$

15, 30

10. Since it takes 6 years to reduce the pollution to 10%, another 6 years would reduce the pollution to 10% of 10%, which is equivalent to 1% of the original. Therefore it takes 12 years for 99% of the pollution to be removed. (Note that the value of Q_0 does not affect this.) Thus the second time is double the first because the fraction remaining, 0.01, in the second instance is the square of the fraction remaining, 0.1, in the first instance.

11. Michigan:
$$\frac{dQ}{dt} = -\frac{r}{V}Q = -\frac{158}{4.9 \times 10^3}Q \approx -0.032Q$$
so
$$Q = Q_0 e^{-0.032t}.$$
We want to find t such that
$$0.1Q_0 = Q_0 e^{-0.032t}$$
so
$$t = \frac{-\ln(0.1)}{0.032} \approx 72 \text{ years}.$$

Ontario:
$$\frac{dQ}{dt} = -\frac{r}{V}Q = \frac{-209}{1.6 \times 10^3}Q = -0.131Q$$
so
$$Q = Q_0 e^{-0.131t}.$$
We want to find t such that
$$0.1Q_0 = Q_0 e^{-0.131t}$$
so
$$t = \frac{-\ln(0.1)}{0.131} \approx 18 \text{ years}.$$

Lake Michigan will take longer because it is larger (4900 km^3 compared to 1600 km^3) and water is flowing through it at a slower rate (158 km^3/year compared to 209 km^3/year).

12. Lake Superior will take the longest, because the lake is largest (V is largest) and water is moving through it most slowly (r is smallest). Lake Erie looks as though it will take the least time because V is smallest and r is close to the largest. For Erie, $k = r/V = 175/460 \approx 0.38$. The lake with the largest value of r is Ontario, where $k = r/V = 209/1600 \approx 0.13$. Since e^{-kt} decreases faster for larger k, Lake Erie will take the shortest time for any fixed fraction of the pollution to be removed.

For Lake Superior,
$$\frac{dQ}{dt} = -\frac{r}{V}Q = -\frac{65.2}{12{,}200}Q \approx -0.0053Q$$
so
$$Q = Q_0 e^{-0.0053t}.$$
When 80% of the pollution has been removed, 20% remains so $Q = 0.2Q_0$. Substituting gives us
$$0.2Q_0 = Q_0 e^{-0.0053t}$$
so
$$t = -\frac{\ln(0.2)}{0.0053} \approx 301 \text{ years}.$$
(Note: The 301 is obtained by using the exact value of $\frac{r}{V} = \frac{65.2}{12{,}200}$, rather than 0.0053. Using 0.0053 gives 304 years.)

For Lake Erie, as in the text
$$\frac{dQ}{dt} = -\frac{r}{V}Q = -\frac{175}{460}Q \approx -0.38Q$$
so
$$Q = Q_0 e^{-0.38t}$$
When 80% of the pollution has been removed
$$0.2Q_0 = Q_0 e^{-0.38t}$$
$$t = -\frac{\ln(0.2)}{0.38} \approx 4 \text{ years}.$$

So the ratio is
$$\frac{\text{Time for Lake Superior}}{\text{Time for Lake Erie}} \approx \frac{301}{4} \approx 75.$$
In other words it will take about 75 times as long to clean Lake Superior as Lake Erie.

13. (a)

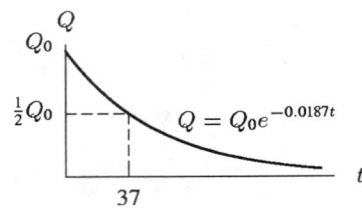

(b) $\frac{dQ}{dt} = -kQ$

(c) Since $25\% = 1/4$, it takes two half-lives $= 74$ hours for the drug level to be reduced to 25%. Alternatively, $Q = Q_0 e^{-kt}$ and $\frac{1}{2} = e^{-k(37)}$, we have

$$k = -\frac{\ln(1/2)}{37} \approx 0.0187.$$

Therefore $Q = Q_0 e^{-0.0187t}$. We know that when the drug level is 25% of the original level that $Q = 0.25Q_0$. Setting these equal, we get

$$0.25 = e^{-0.0187t}.$$

giving

$$t = -\frac{\ln(0.25)}{0.0187} \approx 74 \text{ hours} \approx 3 \text{ days}.$$

14. (a) We know that the rate at which morphine leaves the body is proportional to the amount of morphine in the body at that particular instant. If we let Q be the amount of morphine in the body, we get that

$$\text{Rate of morphine leaving the body} = kQ,$$

where k is the rate of proportionality. The solution is $Q = Q_0 e^{kt}$ (neglecting the continuously incoming morphine). Since the half-life is 2 hours, we have

$$\frac{1}{2}Q_0 = Q_0 e^{k \cdot 2}$$

and so

$$k = \frac{\ln(\frac{1}{2})}{2} = -0.347.$$

(b) $\frac{dQ}{dt} = -0.347Q + 2.5$

(c) Equilibrium will occur when $\frac{dQ}{dt} = 0$, i.e., when $0.347Q = 2.5$ or $Q = 7.2$ mg.

15. Let $C(t)$ be the current flowing in the circuit at time t, then

$$\frac{dC}{dt} = -\alpha C$$

where $\alpha > 0$ is the constant of proportionality between the rate at which the current decays and the current itself.

The general solution of this differential equation is $C(t) = Ae^{-\alpha t}$ but since $C(0) = 30$, we have that $A = 30$, and so we get the particular solution $C(t) = 30e^{-\alpha t}$.

When $t = 0.01$, the current has decayed to 11 amps so that $11 = 30e^{-\alpha 0.01}$ which gives $\alpha = -100\ln(11/30) = 100.33$ so that,

$$C(t) = 30e^{-100.33t}.$$

16. (a) Since the rate of change is proportional to the amount present, $dy/dt = ky$ for some constant k.

(b) Solving the differential equation, we have $y = Ae^{kt}$, where A is the initial amount. Since 100 grams become 54.9 grams in one hour, $54.9 = 100e^k$, so $k = \ln(54.9/100) \approx -0.5997$.
Thus, after 10 hours, there remains $100e^{(-0.5997)10} \approx 0.2486$ grams.

17. (a) If P = pressure and h = height, $\frac{dP}{dh} = -3.7 \times 10^{-5} P$, so $P = P_0 e^{-3.7 \times 10^{-5} h}$. Now $P_0 = 29.92$, since pressure at sea level (when $h = 0$) is 29.92, so $P = 29.92 e^{-3.7 \times 10^{-5} h}$. At the top of Mt. Whitney, the pressure is

$$P = 29.92 e^{-3.7 \times 10^{-5}(14500)} \approx 17.50 \text{ inches of mercury.}$$

At the top of Mt. Everest, the pressure is

$$P = 29.92 e^{-3.7 \times 10^{-5}(29000)} \approx 10.23 \text{ inches of mercury.}$$

(b) The pressure is 15 inches of mercury when

$$15 = 29.92 e^{-3.7 \times 10^{-5} h}$$

Solving for h gives $h = \frac{-1}{3.7 \times 10^{-5}} \ln(\frac{15}{29.92}) \approx 18{,}661.5$ feet.

18. (a) If I is intensity and l is the distance traveled through the water, then for some $k > 0$,

$$\frac{dI}{dl} = -kI.$$

(The proportionality constant is negative because intensity decreases with distance). Thus $I = Ae^{-kl}$. Since $I = A$ when $l = 0$, A represents the initial intensity of the light.

(b) If 50% of the light is absorbed in 10 feet, then $0.50A = Ae^{-10k}$, so $e^{-10k} = \frac{1}{2}$, giving

$$k = \frac{-\ln \frac{1}{2}}{10} = \frac{\ln 2}{10}.$$

In 20 feet, the percentage of light left is

$$e^{-\frac{\ln 2}{10} \cdot 20} = e^{-2\ln 2} = (e^{\ln 2})^{-2} = 2^{-2} = \frac{1}{4},$$

so $\frac{3}{4}$ or 75% of the light has been absorbed. Similarly, after 25 feet,

$$e^{-\frac{\ln 2}{10} \cdot 25} = e^{-2.5 \ln 2} = (e^{\ln 2})^{-\frac{5}{2}} = 2^{-\frac{5}{2}} \approx 0.177.$$

Approximately 17.7% of the light is left, so 82.3% of the light has been absorbed.

19. (a) $\frac{dT}{dt} = -k(T - A)$, where $A = 68°F$ is the temperature of the room, and t is time since 9 am.

(b)

$$\int \frac{dT}{T - A} = -\int k\, dt$$
$$\ln|T - A| = -kt + C$$
$$T = A + Be^{-kt}.$$

Using $A = 68$, and $T(0) = 90.3$, we get $B = 22.3$. Thus

$$T = 68 + 22.3 e^{-kt}.$$

At $t = 1$, we have

$$89.0 = 68 + 22.3 e^{-k}$$
$$21 = 22.3 e^{-k}$$
$$k = -\ln \frac{21}{22.3} \approx 0.06.$$

Thus $T = 68 + 22.3 e^{-0.06t}$.

We want to know when T was equal to 98.6°F, the temperature of a live body, so

$$98.6 = 68 + 22.3 e^{-0.06t}$$
$$\ln \frac{30.6}{22.3} = -0.06t$$
$$t = \left(-\frac{1}{0.06}\right) \ln \frac{30.6}{22.3}$$
$$t \approx -5.27.$$

The victim was killed approximately $5\frac{1}{4}$ hours prior to 9 am, at 3:45 am.

508 CHAPTER TEN /SOLUTIONS

20. (a) The differential equation is
$$\frac{dT}{dt} = -k(T - A),$$
where $A = 10°F$ is the outside temperature.

(b) Integrating both sides yields
$$\int \frac{dT}{T - A} = -\int k\, dt.$$
Then $\ln|T - A| = -kt + C$, so $T = A + Be^{-kt}$. Thus
$$T = 10 + 58e^{-kt}.$$
Since 10:00 pm corresponds to $t = 9$,
$$57 = 10 + 58e^{-9k}$$
$$\frac{47}{58} = e^{-9k}$$
$$\ln \frac{47}{58} = -9k$$
$$k = -\frac{1}{9}\ln\frac{47}{58} \approx 0.0234.$$
At 7:00 the next morning ($t = 18$) we have
$$T \approx 10 + 58e^{18(-0.0234)}$$
$$= 10 + 58(0.66)$$
$$\approx 48°F,$$
so the pipes won't freeze.

(c) We assumed that the temperature outside the house stayed constant at $10°F$. This is probably incorrect because the temperature was most likely warmer during the day (between 1 pm and 10 pm) and colder after (between 10 pm and 7 am). Thus, when the temperature in the house dropped from $68°F$ to $57°F$ between 1 pm and 10 pm, the outside temperature was probably higher than $10°F$, which changes our calculation of the value of the constant k. The house temperature will most certainly be lower than $48°F$ at 7 am, but not by much—not enough to freeze.

21. The rate of disintegration is proportional to the quantity of carbon-14 present. Let Q be the quantity of carbon-14 present at time t, with $t = 0$ in 1977. Then
$$Q = Q_0 e^{-kt},$$
where Q_0 is the quantity of carbon-14 present in 1977 when $t = 0$. Then we know that
$$\frac{Q_0}{2} = Q_0 e^{-k(5730)}$$
so that
$$k = -\frac{\ln(1/2)}{5730} = 0.000121.$$
Thus
$$Q = Q_0 e^{-0.000121t}.$$
The quantity present at any time is proportional to the rate of disintegration at that time so
$$Q_0 = c8.2 \quad \text{and} \quad Q = c13.5$$
where c is a constant of proportionality. Thus substituting for Q and Q_0 in
$$Q = Q_0 e^{-0.000121t}$$
gives
$$c13.5 = c8.2 e^{-0.000121t}$$
so
$$t = -\frac{\ln(13.5/8.2)}{0.000121} \approx -4120.$$
Thus Stonehenge was built about 4120 years before 1977, in about 2150 B.C.

22. (a) If $C' = -kC$, and then $C = C_0 e^{-kt}$. Since the half-life is 5730 years, $\frac{1}{2}C_0 = C_0 e^{-5730k}$. Solving for k, we have $-5730k = \ln(1/2)$ so $k = \frac{-\ln(1/2)}{5730} \approx 0.000121$.
    (b) From the given information, we have $0.91 = e^{-kt}$, where t is the age of the shroud. Solving for t, we have $t = \frac{-\ln 0.91}{k} \approx 779.4$ years.

23. (a) Since speed is the derivative of distance, Galileo's mistaken conjecture was $\frac{dD}{dt} = kD$.
    (b) We know that if Galileo's conjecture were true, then $D(t) = D_0 e^{kt}$, where D_0 would be the initial distance fallen. But if we drop an object, it starts out not having traveled any distance, so $D_0 = 0$. This would lead to $D(t) = 0$ for all t.

Solutions for Section 10.6

1. Let $D(t)$ be the quantity of dead leaves, in grams per square centimeter. Then $\frac{dD}{dt} = 3 - 0.75D$, where t is in years. We factor out -0.75 and then separate variables.

$$\frac{dD}{dt} = -0.75(D - 4)$$

$$\int \frac{dD}{D-4} = \int -0.75 \, dt$$

$$\ln|D-4| = -0.75t + C$$

$$|D-4| = e^{-0.75t+C} = e^{-0.75t} e^C$$

$$D = 4 + Ae^{-0.75t}, \text{ where } A = \pm e^C.$$

If initially the ground is clear, the solution looks like the following graph:

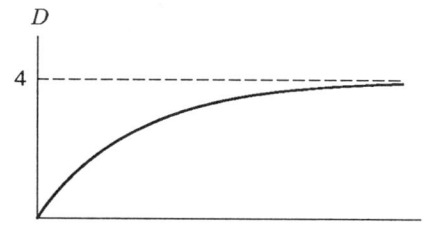

The equilibrium level is 4 grams per square centimeter, regardless of the initial condition.

2. (a) Since the rate of change of the weight is equal to

$$\frac{1}{3500}(\text{Intake} - \text{Amount to maintain weight})$$

we have

$$\frac{dW}{dt} = \frac{1}{3500}(I - 20W).$$

(b) Starting off with the equation

$$\frac{dW}{dt} = -\frac{2}{350}(W - \frac{I}{20}),$$

we separate variables and integrate:

$$\int \frac{dW}{W - \frac{I}{20}} = -\int \frac{2}{350} \, dt.$$

Thus we have

$$\ln\left|W - \frac{I}{20}\right| = -\frac{2}{350}t + C$$

so that

$$W - \frac{I}{20} = Ae^{-\frac{2}{350}t}$$

or in other words

$$W = \frac{I}{20} + Ae^{-\frac{2}{350}t}.$$

Let us call the person's initial weight W_0 at $t = 0$. Then $W_0 = \frac{I}{20} + Ce^0$, so $C = W_0 - \frac{I}{20}$. Thus

$$W = \frac{I}{20} + \left(W_0 - \frac{I}{20}\right)e^{-\frac{2}{350}t}.$$

(c) Using part (b), we have $W = 150 + 10e^{-\frac{2}{350}t}$. This means that $W \to 150$ as $t \to \infty$. See the following figure.

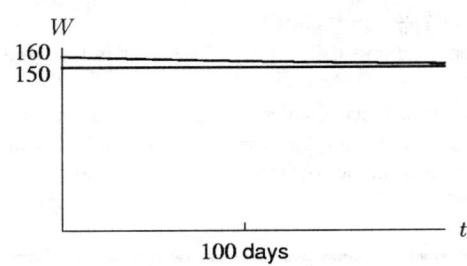

3. We are given that the rate of change of pressure with respect to volume, dP/dV is proportional to P/V, so that
$$\frac{dP}{dV} = k\frac{P}{V}.$$
Using separation of variables and integrating gives
$$\int \frac{dP}{P} = k \int \frac{dV}{V}.$$
Evaluating these integral gives
$$\ln P = k \ln V + c$$
or equivalently,
$$P = AV^k.$$

4. We are given that
$$BC = 2OC.$$
If the point A has coordinates (x, y) then $OC = x$ and $AC = y$. The slope of the tangent line, y', is given by
$$y' = \frac{AC}{BC} = \frac{y}{BC},$$
so
$$BC = \frac{y}{y'}.$$
Substitution into $BC = 2OC$ gives
$$\frac{y}{y'} = 2x,$$
so
$$\frac{y'}{y} = \frac{1}{2x}.$$
Separating variables to integrate this differential equation gives
$$\int \frac{dy}{y} = \int \frac{dx}{2x}$$
$$\ln|y| = \frac{1}{2} \ln|x| + C = \ln\sqrt{|x|} + \ln A$$
$$|y| = A\sqrt{|x|}$$
$$y = \pm(A\sqrt{x}).$$
Thus, in the first quadrant, the curve has equation $y = A\sqrt{x}$.

5. Let the depth of the water at time t be y. Then $\frac{dy}{dt} = -k\sqrt{y}$, where k is a positive constant. Separating variables,
$$\int \frac{dy}{\sqrt{y}} = -\int k\, dt,$$
so
$$2\sqrt{y} = -kt + C.$$
When $t = 0$, $y = 36$; $2\sqrt{36} = -k \cdot 0 + C$, so $C = 12$.
When $t = 1$, $y = 35$; $2\sqrt{35} = -k + 12$, so $k \approx 0.17$.
Thus, $2\sqrt{y} \approx -0.17t + 12$. We are looking for t such that $y = 0$; this happens when $t \approx \frac{12}{0.17} \approx 71$ hours, or about 3 days.

6. (a) $\frac{dM}{dt} = IM$, where I is the interest rate at time t. In this case, however, I is not constant, but depends on t. Using the beginning of 1975 as $t = 0$, and measuring time in years, we have $I = 0.50 + 0.25t$.
Thus our differential equation is $\frac{dM}{dt} = (0.50 + 0.25t)M$.

(b)
$$\frac{dM}{dt} = (0.50 + 0.25t)M$$
$$\int \frac{dM}{M} = \int (0.50 + 0.25t)\, dt$$
$$\ln|M| = 0.50t + 0.125t^2 + C$$
$$M = Ae^{0.50t + 0.125t^2}$$

Since the initial deposit is 100,000 cruzieros, we also know that $M = 100{,}000$ when $t = 0$, and so we have $A = 100{,}000$. Thus $M = 100{,}000 e^{0.50t + 0.125t^2}$.

(c) On January 1, 1980, we have $t = 5$. Thus $M = 100{,}000 e^{0.50(5) + 0.125(5)^2} \approx 27{,}727{,}000$ cruzieros.

7. (a) If the interest rate had been constant, the equation relating money and time would have been $M = 100{,}000 e^{rt}$, where r is the constant interest rate. Using Problem 6(c), we see that for the amounts of money to be equal after 5 years, we need
$$100{,}000 e^{0.50(5) + 0.125(5)^2} = 100{,}000 e^{5r}$$
$$5r = 0.50(5) + 0.125(5^2)$$
$$r = 1.125 = 112.5\%.$$

(b)
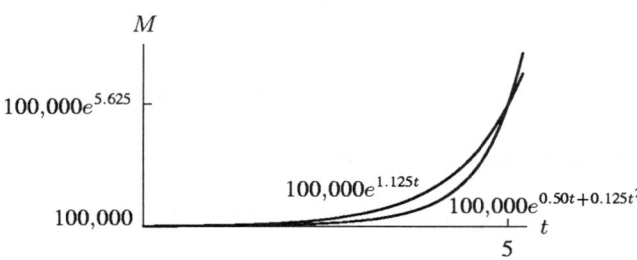

8. Since we know
$$ma = mg - kv,$$
differentiating gives
$$m\frac{da}{dt} = -k\frac{dv}{dt} = -ka.$$
Thus, the differential equation we want is
$$\frac{da}{dt} = -\frac{k}{m}a.$$
Solving for a gives
$$a = a_0 e^{-kt/m}.$$
At $t = 0$, we have $a = g$, the acceleration due to gravity. Thus, $a_0 = g$, so
$$a = g e^{-kt/m}.$$

9. (a) Quantity of A present at time t equals $(a - x)$.
Quantity of B present at time t equals $(b - x)$.
So
$$\text{Rate of formation of } C = k(\text{Quantity of } A)(\text{Quantity of } B)$$
gives
$$\frac{dx}{dt} = k(a - x)(b - x)$$

(b) Separating gives
$$\int \frac{dx}{(a-x)(b-x)} = \int k \, dt.$$

Rewriting the denominator as $(a-x)(b-x) = (x-a)(x-b)$ enables us to use Formula 26 in the Table of Integrals provided $a \neq b$. For some constant K, this gives
$$\frac{1}{a-b}\left(\ln|x-a| - \ln|x-b|\right) = kt + K.$$

Thus
$$\ln\left|\frac{x-a}{x-b}\right| = (a-b)kt + K(a-b)$$
$$\left|\frac{x-a}{x-b}\right| = e^{K(a-b)} e^{(a-b)kt}$$
$$\frac{x-a}{x-b} = Me^{(a-b)kt} \quad \text{where } M = \pm e^{K(a-b)}.$$

Since $x = 0$ when $t = 0$, we have $M = \frac{a}{b}$. Thus
$$\frac{x-a}{x-b} = \frac{a}{b} e^{(a-b)kt}.$$

Solving for x, we have
$$bx - ba = ae^{(a-b)kt}(x-b)$$
$$x(b - ae^{(a-b)kt}) = ab - abe^{(a-b)kt}$$
$$x = \frac{ab(1 - e^{(a-b)kt})}{b - ae^{(a-b)kt}} = \frac{ab(e^{bkt} - e^{akt})}{be^{bkt} - ae^{akt}}.$$

10. Quantity of A left at time $t =$ Quantity of B left at time t equals $(a-x)$.
Thus
$$\text{Rate of formation of } C = k(\text{Quantity of } A)(\text{Quantity of } B)$$
gives
$$\frac{dx}{dt} = k(a-x)(a-x) = k(a-x)^2.$$

Separating gives
$$\int \frac{dx}{(x-a)^2} = \int k \, dt$$

Integrating gives, for some constant K,
$$-(x-a)^{-1} = kt + K.$$

When $t = 0$, $x = 0$ so $K = a^{-1}$. Solving for x:
$$-(x-a)^{-1} = kt + a^{-1}$$
$$x - a = -\frac{1}{kt + a^{-1}}$$
$$x = a - \frac{a}{akt + 1} = \frac{a^2 kt}{akt + 1}$$

11. (a) The quantity and the concentration both increase with time. As the concentration increases, the rate at which the drug is excreted also increases, and so the rate at which the drug builds up in the blood decreases; thus the graph of concentration against time is concave down. The concentration rises until the rate of excretion exactly balances the rate at which the drug is entering; at this concentration there is a horizontal asymptote. (See Figure 10.9.)

10.6 SOLUTIONS   513

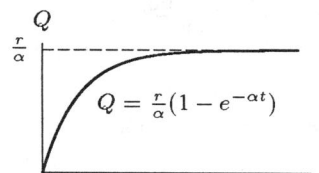

Figure 10.9

(b) Let's start by writing a differential equation for the quantity, $Q(t)$.

$$\begin{pmatrix} \text{Rate quantity} \\ \text{of drug changes} \end{pmatrix} = (\text{Rate in}) - (\text{Rate out})$$

$$\frac{dQ}{dt} = 43.2 - 0.082Q$$

where Q is measured in mg. We want an equation for concentration $c(t) = \frac{Q(t)}{v}$, where $c(t)$ is measured in mg/ml with $v = 35{,}000$ ml.

$$\frac{1}{v}\frac{dQ}{dt} = \frac{43.2}{v} - 0.082\frac{Q}{v},$$

giving

$$\frac{dc}{dt} = \frac{43.2}{35{,}000} - 0.082c.$$

(c) Factor out -0.082 and separate variables to solve.

$$\frac{dc}{dt} = -0.082(c - 0.015)$$

$$\int \frac{dc}{c - 0.015} = -0.082 \int dt$$

$$\ln|c - 0.015| = -0.082t + B$$

$$c - 0.015 = Ae^{-0.082t} \quad \text{where} \quad A = \pm e^B$$

Since $c = 0$ when $t = 0$, we have $A = -0.015$, so

$$c = 0.015 - 0.015e^{-0.082t} = 0.015(1 - e^{-0.082t}).$$

Thus $c \to 0.015$ mg/ml as $t \to \infty$.

12. (a)

$$\frac{dQ}{dt} = r - \alpha Q = -\alpha(Q - \frac{r}{\alpha})$$

$$\int \frac{dQ}{Q - r/\alpha} = -\alpha \int dt$$

$$\ln\left|Q - \frac{r}{\alpha}\right| = -\alpha t + C$$

$$Q - \frac{r}{\alpha} = Ae^{-\alpha t}$$

When $t = 0$, $Q = 0$, so $A = -\frac{r}{\alpha}$ and

$$Q = \frac{r}{\alpha}(1 - e^{-\alpha t})$$

So,

$$Q_\infty = \lim_{t \to \infty} Q = \frac{r}{\alpha}.$$

514 CHAPTER TEN /SOLUTIONS

(b) Doubling r doubles Q_∞. Since $Q_\infty = r/\alpha$, the time to reach $\frac{1}{2}Q_\infty$ is obtained by solving

$$\frac{r}{2\alpha} = \frac{r}{\alpha}(1 - e^{-\alpha t})$$

$$\frac{1}{2} = 1 - e^{-\alpha t}$$

$$e^{-\alpha t} = \frac{1}{2}$$

$$t = -\frac{\ln(1/2)}{\alpha} = \frac{\ln 2}{\alpha}.$$

So altering r doesn't alter the time it takes to reach $\frac{1}{2}Q_\infty$. See Figure 10.10.

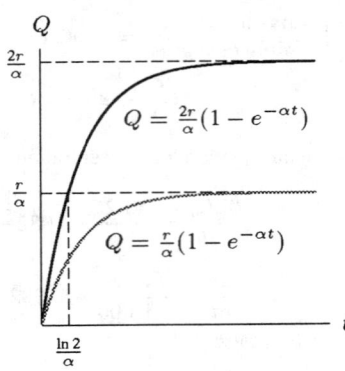

Figure 10.10

(c) Q_∞ is halved by doubling α, and so is the time, $t = \frac{\ln 2}{\alpha}$, to reach $\frac{1}{2}Q_\infty$.

13. (a) $\frac{dy}{dt} = -k(y - a)$, where $k > 0$ and a are constants.

(b) $\int \frac{dy}{y-a} = \int -k\,dt$, so $\ln|y-a| = \ln(y-a) = -kt + C$. Thus, $y - a = Ae^{-kt}$ where $A = e^C$. Initially nothing has been forgotten, so $y(0) = 1$. Therefore, $1 - a = Ae^0 = A$, so $y - a = (1-a)e^{-kt}$ or $y = (1-a)e^{-kt} + a$.

(c) As $t \to \infty$, $e^{-kt} \to 0$, so $y \to a$.
Thus, a represents the fraction of material which is remembered in the long run. The constant k tells us about the rate at which material is forgotten.

14. (a) $dp/dt = -k(p - p_0)$, where k is constant. Notice that $k > 0$, since if $p > p_0$ then dp/dt should be negative, and if $p < p_0$ then dp/dt should be positive.

(b) Separating variables, we have $\frac{dp}{p-p_0} = -k\,dt$.
Solving, we find $p = p_0 + (p - I)e^{-kt}$, where I is the initial price.

(c)

(d) As $t \to \infty$, $p \to p_0$. We see this in the solution in part (b), since as $t \to \infty$, $e^{-kt} \to 0$. (Remember $k > 0$.) In other words, as $t \to \infty$, p approaches the equilibrium price p_0.

15. (a) Concentration of carbon monoxide $= \frac{\text{Quantity in room}}{\text{Volume}}$.
If $Q(t)$ represents the quantity of carbon monoxide in the room at time t, $c(t) = Q(t)/60$.

$$\text{Rate quantity of carbon monoxide in room changes} = \text{rate in} - \text{rate out}$$

Now
$$\text{Rate in} = 5\%(0.002 \text{m}^3/\text{min}) = 0.05(0.002) = 0.0001 \text{m}^3/\text{min}.$$
Since smoky air is leaving at $0.002\text{m}^3/\text{min}$, containing a concentration $c(t) = Q(t)/60$ of carbon monoxide
$$\text{Rate out} = 0.002\frac{Q(t)}{60}$$
Thus
$$\frac{dQ}{dt} = 0.0001 - \frac{0.002}{60}Q$$
Since $c = Q/60$, we can substitute $Q = 60c$, giving
$$\frac{d(60c)}{dt} = 0.0001 - \frac{0.002}{60}(60c)$$
$$\frac{dc}{dt} = \frac{0.0001}{60} - \frac{0.002}{60}c$$

(b) Factoring the right side of the differential equation and separating gives
$$\frac{dc}{dt} = -\frac{0.0001}{3}(c - 0.05) \approx 3 \times 10^{-5}(c - 0.05)$$
$$\int \frac{dc}{c - 0.05} = -\int 3 \times 10^{-5} dt$$
$$\ln|c - 0.05| = -3 \times 10^{-5}t + K$$
$$c - 0.05 = Ae^{-3 \times 10^{-5}t} \quad \text{where } A = \pm e^K.$$

Since $c = 0$ when $t = 0$, we have $A = -0.05$, so
$$c = 0.05 - 0.05e^{-3 \times 10^{-5}t}$$

(c) As $t \to \infty$, $e^{-3 \times 10^{-5}t} \to 0$ so $c \to 0.05$.
Thus in the long run, the concentration of carbon monoxide tends to 5%, the concentration of the incoming air.

16. $c = 0.05 - 0.05e^{-3 \times 10^{-5}t}$
We want to solve for t when $c = 0.001$
$$0.001 = 0.05 - 0.05e^{-3 \times 10^{-5}t}$$
$$-0.049 = -0.05e^{-3 \times 10^{-5}t}$$
$$e^{-3 \times 10^{-5}t} = 0.98$$
$$t = \frac{-\ln(0.98)}{3 \times 10^{-5}} = 673 \text{ min} \approx 11 \text{ hours } 13 \text{ min}.$$

17. (a) Now
$$\frac{dS}{dt} = (\text{Rate at which salt enters the pool}) - (\text{Rate at which salt leaves the pool}),$$
and, for example,
$$\begin{pmatrix} \text{Rate at which salt} \\ \text{enters the pool} \end{pmatrix} = \begin{pmatrix} \text{Concentration of} \\ \text{salt solution} \end{pmatrix} \times \begin{pmatrix} \text{Flow rate of} \\ \text{salt solution} \end{pmatrix}$$
(grams/minute) = (grams/liter) $\times$ (liters/minute)

so
$$\text{Rate at which salt enters the pool} =$$
$$(10 \text{ grams/liter}) \times (60 \text{ liters/minute}) = (600 \text{ grams/minute})$$

The rate at which salt leaves the pool depends on the concentration of salt in the pool. At time t, the concentration is $\frac{S(t)}{2 \times 10^6 \text{ liters}}$, where $S(t)$ is measured in grams.
Thus

Rate at which salt leaves the pool =
$$\frac{S(t) \text{ grams}}{2 \times 10^6 \text{ liters}} \times \frac{60 \text{ liters}}{\text{minute}} = \frac{3S(t) \text{ grams}}{10^5 \text{ minutes}}.$$

Thus
$$\frac{dS}{dt} = 600 - \frac{3S}{100,000}.$$

(b) $\frac{dS}{dt} = -\frac{3}{100,000}(S - 20,000,000)$
$\int \frac{dS}{S - 20,000,000} = \int -\frac{3}{100,000}\,dt$
$\ln|S - 20,000,000| = -\frac{3}{100,000}t + C$
$S = 20,000,000 - Ae^{-\frac{3}{100,000}t}$
Since $S = 0$ at $t = 0$, $A = 20,000,000$. Thus $S(t) = 20,000,000 - 20,000,000 e^{-\frac{3}{100,000}t}$.

(c) As $t \to \infty$, $e^{-\frac{3}{100,000}t} \to 0$, so $S(t) \to 20,000,000$ grams. The concentration approaches 10 grams/liter. Note that this makes sense; we'd expect the concentration of salt in the pool to become closer and closer to the concentration of salt being poured into the pool as $t \to \infty$.

18. (a) Newton's Law of Motion says that
Force = (mass) × (acceleration).
Since acceleration, dv/dt, is measured upward and the force due to gravity acts downward,
$$-\frac{mgR^2}{(R+h)^2} = m\frac{dv}{dt}$$
so
$$\frac{dv}{dt} = -\frac{gR^2}{(R+h)^2}.$$

(b) Since $v = \frac{dh}{dt}$, the chain rule gives
$$\frac{dv}{dt} = \frac{dv}{dh} \cdot \frac{dh}{dt} = \frac{dv}{dh} \cdot v.$$
Substituting into the differential equation in part (a) gives
$$v\frac{dv}{dh} = -\frac{gR^2}{(R+h)^2}.$$

(c) Separating variables gives
$$\int v\,dv = -\int \frac{gR^2}{(R+h)^2}\,dh$$
$$\frac{v^2}{2} = \frac{gR^2}{(R+h)} + C$$
Since $v = v_0$ when $h = 0$,
$$\frac{v_0^2}{2} = \frac{gR^2}{(R+0)} + C \quad \text{gives} \quad C = \frac{v_0^2}{2} - gR,$$
so the solution is
$$\frac{v^2}{2} = \frac{gR^2}{(R+h)} + \frac{v_0^2}{2} - gR$$
$$v^2 = v_0^2 + \frac{2gR^2}{(R+h)} - 2gR$$

(d) The escape velocity v_0 ensures that $v^2 \geq 0$ for all $h \geq 0$. Since the positive quantity $\frac{2gR^2}{(R+h)} \to 0$ as $h \to \infty$, to ensure that $v^2 \geq 0$ for all h, we must have
$$v_0^2 \geq 2gR.$$
When $v_0^2 = 2gR$ so $v_0 = \sqrt{2gR}$, we say that v_0 is the escape velocity.

19. Since $R' > 0$ currently, R is increasing and R' is given by

$$R' = \sqrt{\frac{2GM_0}{R} - K},$$

where $C = -K$ and $K > 0$. Now R increases until $\frac{2GM_0}{R} = K$, giving $R_{\max} = \frac{2GM_0}{K}$. At this value of R, we have $R' = 0$. Since R' has decreased to 0, and since $R'' < 0$ always, R' will go on decreasing. (The original second order differential equation shows that $R'' < 0$.) Thus R' becomes negative, and is now given by

$$R' = -\sqrt{\frac{2GM_0}{R} - K}.$$

Since R' is now negative, R decreases, thereby making R' more and more negative – so the universe collapses. (See Figure 10.11.)

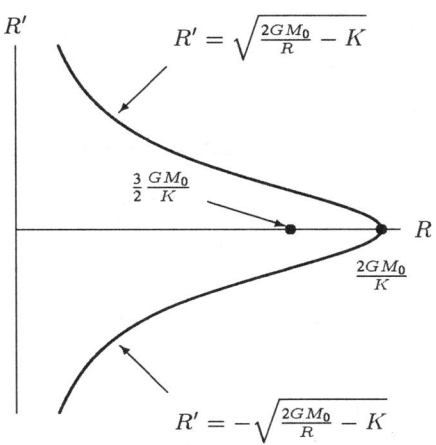

Figure 10.11

20. Separating variables gives

$$\frac{dR}{dt} = \frac{\sqrt{2GM_0}}{R^{1/2}}$$

$$\int R^{1/2}\, dR = \int \sqrt{2GM_0}\, dt$$

$$\frac{2}{3} R^{3/2} = \sqrt{2GM_0}\, t + A$$

Since $R = 0$ when $t = 0$, we have $A = 0$, so

$$R = \left(\frac{3\sqrt{2GM_0}}{2}\right)^{2/3} t^{2/3} = k t^{2/3} \quad \text{where } k = \left(\frac{3\sqrt{2GM_0}}{2}\right)^{2/3}, \text{ a constant.}$$

Therefore in the flat universe, $R = k t^{2/3}$ and $R' = \frac{2k}{3} t^{-1/3}$, meaning that the universe expands for all t, but the rate at which it expands goes to zero as $t \to \infty$.

21. (a) For a stable universe, we need $R' = 0$, so $R'' = 0$. However the differential equation for R'' shows that $R'' < 0$ for every R, so we never have $R'' = 0$. Thus R' and R must both be changing with time.

    (b) If the universe were expanding at a constant rate of $R'(t_0)$, then $R(t_0)/R'(t_0)$ would be the time it took for the radius to grow from 0 to $R(t_0)$ – a reasonable estimate for the age of the universe. Since in fact R' has been decreasing, in other words, the universe has actually been expanding faster than $R'(t_0)$, the Hubble constant is an overestimate (i.e. the universe is actually younger than the Hubble constant suggests.)

518 CHAPTER TEN /SOLUTIONS

Solutions for Section 10.7

1. A continuous growth rate of 0.2% means that

$$\frac{1}{P}\frac{dP}{dt} = 0.2\% = 0.002.$$

Separating variables and integrating gives

$$\int \frac{dP}{P} = \int 0.002\, dt$$

$$P = P_0 e^{0.002t} = (6.6 \times 10^6) e^{0.002t}.$$

2. (a)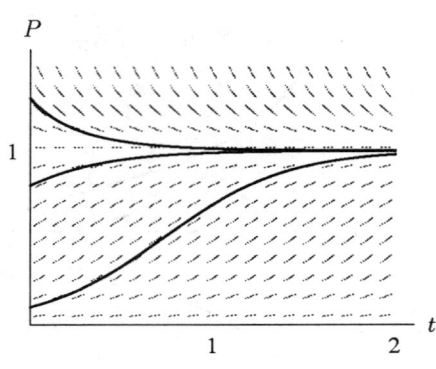

   (b) The value $P = 1$ is a stable equilibrium. (See part (d) below for a more detailed discussion.)
   (c) Looking at the solution curves, we see that P is increasing for $0 < P < 1$ and decreasing for $P > 1$. The values of $P = 0$, $P = 1$ are equilibria. In the long run, P tends to 1, unless you start with $P = 0$. The solution curves with initial populations of less than $P = \frac{1}{2}$ have inflection points at $P = \frac{1}{2}$. (This will be demonstrated algebraically in part (d) below.) At the inflection point, the population is growing fastest.
   (d)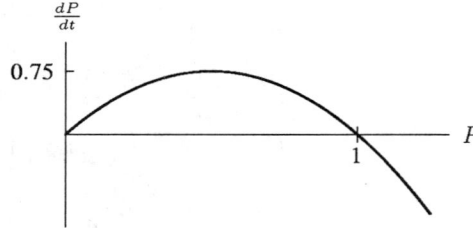

   Since $\frac{dP}{dt} = 3P - 3P^2 = 3P(1 - P)$, the graph of $\frac{dP}{dt}$ against P is a parabola, opening downwards with P intercepts at 0 and 1. The quantity $\frac{dP}{dt}$ is positive for $0 < P < 1$, negative for $P > 1$ (and $P < 0$). The quantity $\frac{dP}{dt}$ is 0 at $P = 0$ and $P = 1$, and maximum at $P = \frac{1}{2}$. The fact that $\frac{dP}{dt} = 0$ at $P = 0$ and $P = 1$ tells us that these are equilibria. Further, since $\frac{dP}{dt} > 0$ for $0 < P < 1$, we see that solution curves starting here will increase toward $P = 1$.

   If the population starts at a value $P < \frac{1}{2}$, it increases at an increasing rate up to $P = \frac{1}{2}$. After this, P continues to increase, but at a decreasing rate. The fact that $\frac{dP}{dt}$ has a maximum at $P = \frac{1}{2}$ tells us that there is a point of inflection when $P = \frac{1}{2}$. Similarly, since $\frac{dP}{dt} < 0$ for $P > 1$, solution curves starting with $P > 1$ will decrease to $P = 1$. Thus, $P = 1$ is a stable equilibrium.

3. (a) We know that a logistic curve can be modeled by the function
$$P = \frac{L}{1 + Ce^{-kt}}$$
where $C = (L - P_0)/(P_0)$ and P is the number of people infected by the virus at a particular time t. We know that L is the limiting value, or the maximal number of people infected with the virus, so in our case
$$L = 5000.$$
We are also told that initially there are only ten people infected with the virus so that we get
$$P_0 = 10.$$
Thus we have
$$C = \frac{L - P_0}{P_0}$$
$$= \frac{5000 - 10}{10}$$
$$= 499.$$
We are also told that in the early stages of the virus, infection grows exponentially with $k = 1.78$. Thus we get that the logistic function for people infected is
$$P = \frac{5000}{1 + 499e^{-1.78t}}.$$

(b)
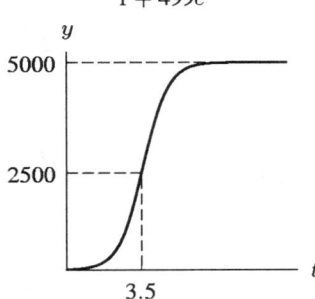

(c) Looking at the graph we see that the the point at which the rate changes from increasing to decreasing, the inflection point, occurs at roughly $t = 3.5$ giving a value of $P = 2500$. Thus after roughly 2500 people have been infected, the rate of infection starts dropping. See above.

4. (a) The logistic model is a reasonable one because at first very few houses have a VCR. As movie rentals become popular and as VCRs get cheaper, more people will buy VCRs. However, we know that the rate of VCR buying will start slowing down at some point as it is impossible for more than 100% of houses to have VCRs.

(b) To find the point of inflection, we must find the year at which the rate of VCR buying changes from increasing to decreasing. The following table shows the rate of change in the years from 1978 to 1990.

Year	1978	1979	1980	1981	1982	1983	1984
% Change per year	0.2	0.6	0.7	1.3	2.4	5.1	10.2
Year	1985	1986	1987	1988	1989	1990	1991
% Change per year	15.2	12.7	9.3	6.6	7.3	0	

Looking at the table, we see that the rate of percent change per year changes from increasing to decreasing in the year 1986. At this time 36% of households own VCRs giving $P = (1986, 36)$. Since at the inflection point we expect the vertical coordinate to be $L/2$, we get
$$L/2 = 36$$
$$L = 72\%.$$
Thus we expect the limiting value to be 72%. This fits in well with the data that we have for 1990 and 1991.

(c) Since the general form of a logistic equation is
$$P = \frac{L}{1 + Ce^{-kt}}$$
where L is the limiting value, we have that in our case $L = 75$ and the limiting value is 75%.

5. Rewriting the equation as $\frac{1}{P}\frac{dP}{dt} = \frac{(100-P)}{1000}$, we see that this is a logistic equation. Before looking at its solution, we explain why there must always be at least 100 individuals. Since the population begins at 200, $\frac{dP}{dt}$ is initially negative, so the population decreases. It continues to do so while $P > 100$. If the population ever reached 100, however, then $\frac{dP}{dt}$ would be 0. This means the population would stop changing – so if the population ever decreased to 100, that's where it would stay. The fact that $\frac{dP}{dt}$ will always be negative also shows that the population will always be under 200, as shown below.

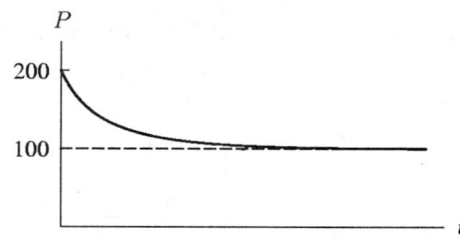

The solution, as given by the formula derived in the chapter, is

$$P = \frac{20000}{200 - 100e^{-t/10}}$$

6. (a) Let I be the number of informed people at time t, and I_0 the number who know initially. Then this model predicts that $\frac{dI}{dt} = k(M - I)$ for some positive constant k. Solving this, we find the solution is

$$I = M - (M - I_0)e^{-kt}.$$

We sketch the solution with $I_0 = 0$. Notice that $\frac{dI}{dt}$ is largest when I is smallest, so the information spreads fastest in the beginning, at $t = 0$. In addition, the graph below shows that $I \to M$ as $t \to \infty$, meaning that everyone gets the information eventually.

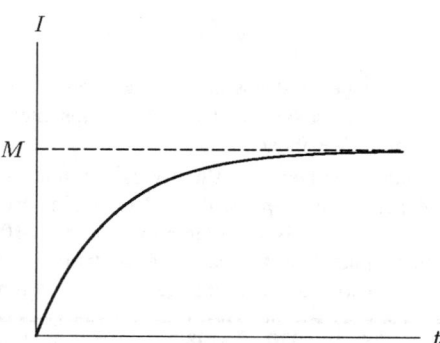

(b) In this case, the model suggests that $\frac{dI}{dt} = kI(M - I)$ for some positive constant k. This is a logistic model with carrying capacity M. We sketch the solutions for three different values of I_0 below.

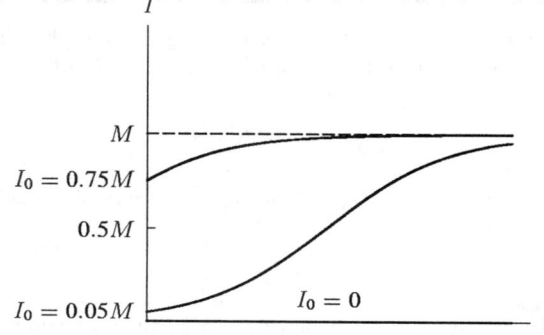

(i) If $I_0 = 0$ then $I = 0$ for all t. In other words, if nobody knows something, it doesn't spread by word of mouth!

(ii) If $I_0 = 0.05M$, then $\frac{dI}{dt}$ is increasing up to $I = \frac{M}{2}$. Thus, the information is spreading fastest at $I = \frac{M}{2}$.

(iii) If $I_0 = 0.75M$, then $\frac{dI}{dt}$ is always decreasing for $I > \frac{M}{2}$, so $\frac{dI}{dt}$ is largest when $t = 0$.

7. (a) Let the population at time t be $P(t)$ and the relative growth rate be $G = \alpha - \beta P$. When $P = 600, G = 35 - 15 = 20\%$, and when $P = 800, G = 30 - 20 = 10\%$ so

$$\alpha - 600\beta = 0.20$$

$$\alpha - 800\beta = 0.10.$$

Therefore, $\alpha = \frac{1}{2}$ and $\beta = \frac{1}{2000}$, hence

$$\frac{1}{P}\frac{dP}{dt} = \frac{1}{2} - \frac{1}{2000}P.$$

(b) The differential equation is a logistic equation

$$\frac{dP}{dt} = \frac{1}{2000}P(1000 - P)$$

and so the equilibrium population is $P = 1000$. We expect the population of 900 to increase to the equilibrium value of 1000.

(c) If the additional elk are added, the population of 1350 elk is above the equilibrium value, and the population will decrease to about 1000.

(d)

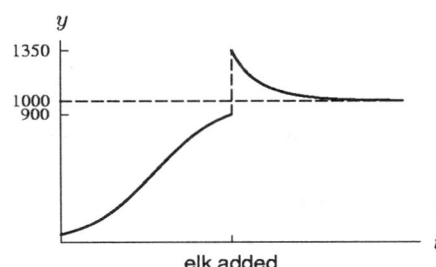

Importing more elk would be ecologically unsound, as the new population is in excess of the equilibrium population that Reading Island can support.

8. (a) $\frac{dp}{dt} = kp(B - p)$, where $k > 0$.

(b) To find when $\frac{dp}{dt}$ is largest, we notice that $\frac{dp}{dt} = kp(B - p)$, as a function of p, is a parabola opening downwards with the maximum at $p = \frac{B}{2}$, i.e. when $\frac{1}{2}$ the tin has turned to powder. This is the time when the tin is crumbling fastest.

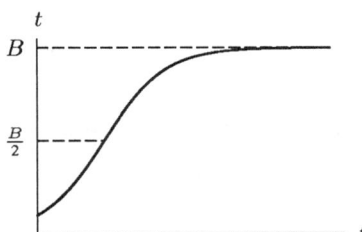

(c) If $p = 0$ initially, then $\frac{dp}{dt} = 0$, so we would expect p to remain 0 forever. However, since many organ pipes get tin pest, we must reconcile the model with reality. There are two possible ideas which solve this problem. First, we could assume that p is never 0. In other words, we assume that all tin pipes, no matter how new, must contain some small amount of tin pest. Assuming this means that all organ pipes must deteriorate due to tin pest eventually. Another explanation is that the powder forms at a slow rate even if there was none present to begin with. Since not all organ pipes suffer, it is possible that the conversion is catalyzed by some other impurities not present in all pipes.

9. (a) To estimate $\frac{dP}{dt}$ for 1810, for example, we'll take

$$\frac{(\text{pop. at } 1820)-(\text{pop. at } 1800)}{20 \text{ years}} = \frac{(9.6 - 5.3) \text{ million}}{20 \text{ years}} = 0.215 \frac{\text{million}}{\text{yr}}.$$

Thus $\frac{1}{P}\frac{dP}{dt} = \frac{0.215}{7.2} \approx 0.03$. We do this for several other points.

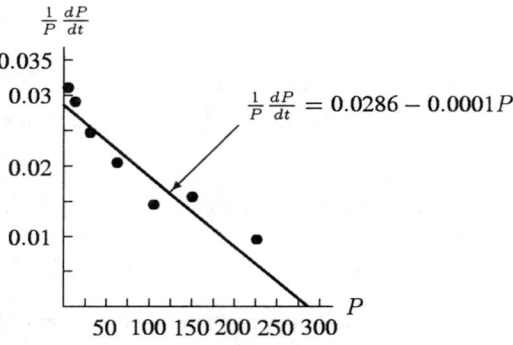

Year	P	$\frac{dP}{dt} \approx \frac{P(t+10) - P(t-10)}{20}$	$\frac{1}{P}\frac{dP}{dt}$
1800	5.3	$(7.2 - 3.9)/20 = 0.165$	0.0311
1830	12.9	$(17.1 - 9.6)/20 = 0.375$	0.0291
1860	31.4	$(38.6 - 23.1)/20 = 0.775$	0.0247
1890	62.9	$(76.0 - 50.2)/20 = 1.29$	0.0205
1920	105.7	$(122.8 - 92.0)/20 = 1.54$	0.0146
1950	150.7	$(179.0 - 131.7)/20 = 2.365$	0.0157
1980	226.5	$(248.7 - 205.0)/20 = 2.185$	0.0096

Plotting the data and fitting a line to it, we obtain

$$\frac{1}{P}\frac{dP}{dt} = 0.0286 - 0.0001P.$$

Thus $k \approx 0.0286$ and $a \approx 0.0001$.

(b) According to this model, $\frac{1}{P}\frac{dP}{dt} = 0.0001(286 - P)$. Thus, P will increase up to about 286 million, and then level off.

10. (a) Here we have, where $t =$ years since 1800:

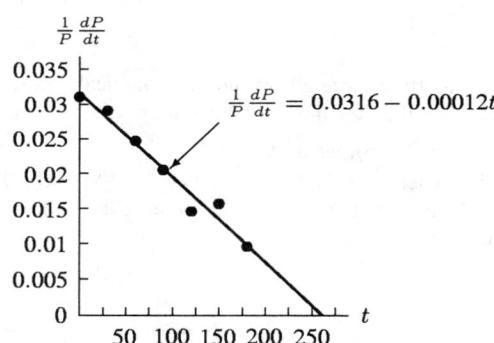

Year	t	$\frac{1}{P}\frac{dP}{dt}$
1800	0	0.0311
1830	30	0.0291
1860	60	0.0247
1890	90	0.0205
1920	120	0.0146
1950	150	0.0157
1980	180	0.0096

Graphing the data and fitting a line, we get $\frac{1}{P}\frac{dP}{dt} = 0.0316 - 0.00012t$ as our guess. So we have $a = 0.0316$ and $b = 0.00012$.

(b) dP/dt will be positive and P will increase until $0.0316 = 0.00012t$, i.e. until $t \approx 260$ or about the year 2060.

(c)
$$\frac{dP}{P} = (0.0316 - 0.00012t)\, dt.$$
$$\int \frac{dP}{P} = \int (0.0316 - 0.00012t)\, dt.$$
$$P = Ae^{0.0316t - 0.00006t^2}.$$

Using the fact that $P = 5.3$ when $t = 0$, we get $P = 5.3e^{0.0316t - 0.00006t^2}$.

11. (a) By the chain rule
$$\frac{dP}{dt} = \frac{d}{dt}\left(\frac{1}{u}\right) = \frac{d}{du}\left(\frac{1}{u}\right) \cdot \frac{du}{dt} = -\frac{1}{u^2}\frac{du}{dt}$$

(b) Substituting for $P = 1/u$ in the equation
$$\frac{dP}{dt} = kP\left(1 - \frac{P}{L}\right)$$

gives
$$-\frac{1}{u^2}\frac{du}{dt} = k\frac{1}{u}\left(1 - \frac{1}{Lu}\right).$$

Simplifying leads to
$$\frac{du}{dt} = -k\left(u - \frac{1}{L}\right)$$

and separating variables gives
$$\int \frac{du}{u - 1/L} = -\int k\,dt$$
$$\ln\left|u - \frac{1}{L}\right| = -kt + C$$
$$u - \frac{1}{L} = Ae^{-kt} \quad \text{where } A = \pm e^C$$
$$u = \frac{1}{L} + Ae^{-kt}$$

(c) Since $u = 1/P$, we have
$$\frac{1}{P} = \frac{1}{L} + Ae^{-kt} = \frac{1 + LAe^{-kt}}{L}$$

so
$$P = \frac{L}{1 + LAe^{-kt}} \quad \text{where } A \text{ is an arbitrary constant.}$$

12. (a)

TABLE 10.14

Year	P	$\frac{dP}{dt} \approx \frac{P(t+10) - P(t-10)}{20}$	$\frac{1}{P}\frac{dP}{dt}$
1790	3.9		
1800	5.3	$(7.2 - 3.9)/20 = 0.165$	0.0311
1810	7.2	$(9.6 - 5.3)/20 = 0.215$	0.0299
1820	9.6	$(12.9 - 7.2)/20 = 0.285$	0.0297
1830	12.9	$(17.1 - 9.6)/20 = 0.375$	0.0291
1840	17.1	$(23.2 - 12.9)/20 = 0.515$	0.0301
1850	23.2	$(31.4 - 17.1)/20 = 0.715$	0.0308
1860	31.4		

The method used in the text to calculate $k \approx 3.47\%$ is simply to average the values obtained for $\frac{1}{P}\frac{dP}{dt}$. Using this method on the values in Table 10.14 gives $k = 3.01\%$.

(b) To get $k \approx 2.98\%$, we assume k is the continuous growth constant satisfying $P = P_0 e^{kt}$. The data shows that, on average the population increased by 30.1% every ten years, meaning that if P_0 is the initial population and P is the population 10 years later:
$$\frac{P}{P_0} = 1.301 = e^{k(10)}.$$

Thus $k = \ln(1.301)/10 = 0.0263$.

13. Using a one-sided estimate for $f'(2)$, we get:

$$f'(2) \approx \frac{f(2+h) - f(2)}{h} = \frac{(2+h)^3 - (2)^3}{h}$$
$$= \frac{2^3 + 12h + 6h^2 + h^3 - 2^3}{h}$$
$$= 12 + 6h + h^2$$

If $h = 0.1$, we have $f'(2) \approx 12.61$.
If $h = 0.01$, we have $f'(2) \approx 12.0601$.
If $h = 0.001$, we have $f'(2) \approx 12.006001$.
As h decreases by a factor of ten, our approximation improves by one digit of accuracy.

Using a two-sided estimate for $f'(2)$, we get:

$$f'(2) \approx \frac{f(2+h) - f(2-h)}{2h} = \frac{(2+h)^3 - (2-h)^3}{2h}$$
$$= \frac{(2^3 + 12h + 6h^2 + h^3) - (2^3 - 12h + 6h^2 - h^3)}{2h}$$
$$= \frac{24h + 2h^3}{2h} = 12 + h^2$$

If $h = 0.1$, we have $f'(2) \approx 12.01$.
If $h = 0.01$, we have $f'(2) \approx 12.0001$.
If $h = 0.001$, we have $f'(2) \approx 12.000001$.
As h decreases by a factor of ten, the one-sided approximation improves by two digits of accuracy. The two-sided estimate is accurate to twice as many digits as the one-sided estimate.

14. If $f(x) = x^2$ then

$$\frac{f(x+h) - f(x-h)}{2h} = \frac{(x+h)^2 - (x-h)^2}{2h}$$
$$= \frac{(x^2 + 2xh + h^2) - (x^2 - 2xh + h^2)}{2h}$$
$$= \frac{4xh}{2h} = 2x = f'(x) \quad \text{for all } x.$$

15. The US population in 1860 was 31.4 million. If between 1860 and 1870 the population had increased at the same rate as previous decades, 34.7%, the population in 1870 would have been $(31.4 \text{ million})(1.347) = 42.3$ million. In actuality the US population in 1870 was only 38.6 million. This is a shortfall of 3.7 million people.

History records that about 618,000 soldiers died (total, both sides) during the Civil War (according to Collier's Encyclopedia, 1968). This accounts for only $\frac{1}{6}$ (roughly) of the shortfall. The rest of the shortfall can be attributed to civilian deaths and a decrease in the birth rate caused by absent males and an unwillingness to have babies under harsh economic conditions and political uncertainty.

16.

TABLE 10.15

Year	P	$\frac{dP}{dt} \approx \frac{P(t+10)-P(t-10)}{20}$
1790	3.9	
1800	5.3	$(7.2 - 3.9)/20 = 0.165$
1810	7.2	$(9.6 - 5.3)/20 = 0.215$
1820	9.6	$(12.9 - 7.2)/20 = 0.285$
1830	12.9	$(17.1 - 9.6)/20 = 0.375$
1840	17.1	$(23.2 - 12.9)/20 = 0.515$
1850	23.2	$(31.4 - 17.1)/20 = 0.715$
1860	31.4	$(38.6 - 23.2)/20 = 0.770$
1870	38.6	$(50.2 - 31.4)/20 = 0.940$
1880	50.2	$(62.9 - 38.6)/20 = 1.215$
1890	62.9	$(76.0 - 50.2)/20 = 1.290$
1900	76.0	$(92.0 - 62.9)/20 = 1.455$
1910	92.0	$(105.7 - 76.0)/20 = 1.485$
1920	105.7	$(122.8 - 92.0)/20 = 1.540$
1930	122.8	$(131.7 - 105.7)/20 = 1.300$
1940	131.7	$(150.7 - 122.8)/20 = 1.395$
1950	150.7	

According to these calculations, the largest value of dP/dt occurs in 1920 when the rate of change is $\frac{dP}{dt} = 1.540$ million people/year. The population in 1920 was 105.7 million. If we assume that the limiting value, L, is twice the population when it is changing most quickly, then $L = 2 \times 105.7 = 211.4$ million. This is greater than the estimate of 187 million computed in the text and closer to the actual 1990 population of 248.7 million.

17.

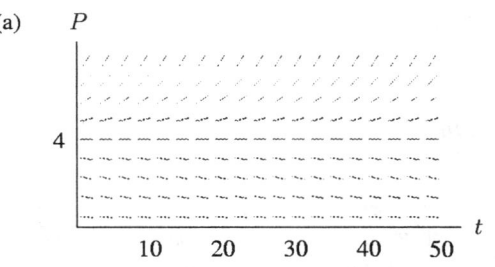

(c) There are two equilibrium values, $P = 0$, and $P = 4$. The first, representing extinction, is stable. The equilibrium value $P = 4$ is unstable because the populations increase if greater than 4, and decrease if less than 4. Notice that the equilibrium values can be obtained by setting $dP/dt = 0$:

$$\frac{dP}{dt} = 0.02P^2 - 0.08P = 0.02P(P - 4) = 0$$

so

$$P = 0 \text{ or } P = 4.$$

18. (a)

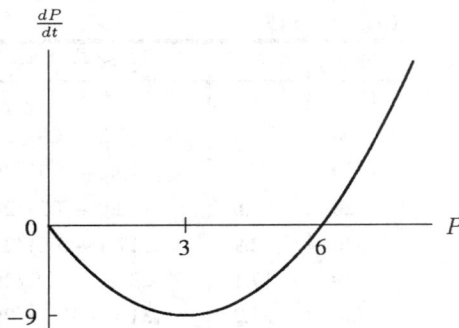

Figure 10.12

(b) Figure 10.12 shows that for $0 < P < 6$, the sign of dP/dt is negative. This means that P is decreasing over the interval $0 < P < 6$. As P decreases from $P(0) = 5$, the value of dP/dt gets more and more negative until $P = 3$. Thus the graph of P against t is concave down while P is decreasing from 5 to 3. As P decreases below 3, the slope of dP/dt increases toward 0, so the graph of P against t is concave up and asymptotic to the t-axis. At $P = 3$, there is an inflection point. See Figure 10.13.

(c) Figure 10.12 shows that for $P > 6$, the slope of dP/dt is positive and increases with P. Thus the graph of P against t is increasing and concave up. See Figure 10.13.

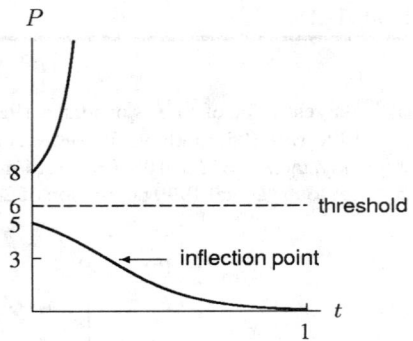

Figure 10.13

(d) For initial populations greater than the threshold value $P = 6$, the population increases without bound. Populations with initial value less than $P = 6$ decrease asymptotically towards 0, i.e. become extinct. Thus the initial population $P = 6$ is the dividing line, or threshold, between populations which grow without bound and those which die out.

19. (a)

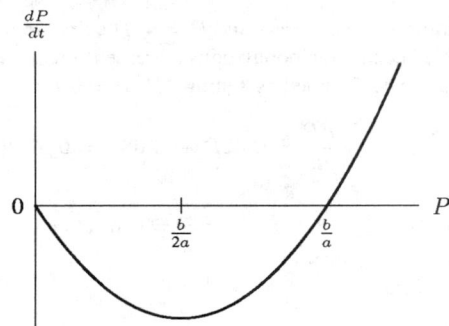

Figure 10.14

(b)

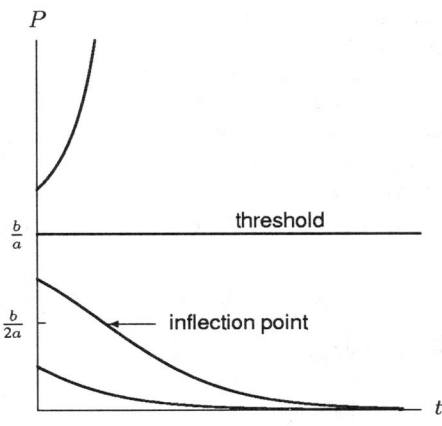

Figure 10.15

Figure 10.14 shows that dP/dt is negative for $P < \frac{b}{a}$, making P a decreasing function when $P(0) < \frac{b}{a}$. When $P > \frac{b}{a}$, the sign of dP/dt is positive, so P is an increasing function. Thus solution curves starting above $\frac{b}{a}$ are increasing, and those starting below $\frac{b}{a}$ are decreasing. See Figure 10.15.

For $P > \frac{b}{a}$, the slope, $\frac{dP}{dt}$, increases with P, so the graph of P against t is concave up. For $0 < P < \frac{b}{a}$, the value of P decreases with time. As P decreases, the slope $\frac{dP}{dt}$ decreases for $\frac{b}{2a} < P < \frac{b}{a}$, and increases towards 0 for $0 < P < \frac{b}{2a}$. Thus solution curves starting just below the threshold value of $\frac{b}{a}$ are concave down for $\frac{b}{2a} < P < \frac{b}{a}$ and concave up and asymptotic to the t-axis for $0 < P < \frac{b}{2a}$. See Figure 10.15.

(c) $P = \frac{b}{a}$ is called the threshold population because for populations greater than $\frac{b}{a}$, the population will increase without bound. For populations less than $\frac{b}{a}$, the population will go to zero, i.e. to extinction.

Solutions for Section 10.8

1. If $y = 2\cos t + 3\sin t$, then $y' = -2\sin t + 3\cos t$ and $y'' = -2\cos t - 3\sin t$. Thus, $y'' + y = 0$.

2. If $y = A\cos t + B\sin t$, then $y' = -A\sin t + B\cos t$ and $y'' = -A\cos t - B\sin t$. Thus, $y'' + y = 0$.

3. $y = A\cos\alpha t$
   $y' = -\alpha A \sin\alpha t$
   $y'' = -\alpha^2 A \cos\alpha t$
   If $y'' + 5y = 0$, then $-\alpha^2 A\cos\alpha t + 5A\cos\alpha t = 0$, so $A(5 - \alpha^2)\cos\alpha t = 0$. This is true for all t if $A = 0$, or if $\alpha = \pm\sqrt{5}$.
   We also have the initial condition: $y'(1) = -\alpha A \sin\alpha = 3$. Notice that this equation will not work if $A = 0$. If $\alpha = \sqrt{5}$, then
   $A = -\frac{3}{\sqrt{5}\sin\sqrt{5}} \approx -1.705$.
   Similarly, if $\alpha = -\sqrt{5}$, we find that $A \approx -1.705$. Thus, the possible values are $A = -\frac{3}{\sqrt{5}\sin\sqrt{5}} \approx -1.705$ and $\alpha = \pm\sqrt{5}$.

4. At $t = 0$, we find that $y = 2$, which is clearly the highest point since $-1 \leq \cos 3t \leq 1$. Thus, at $t = 0$ the mass is at its highest point. Since $y' = -6\sin 3t$, we see $y' = 0$ when $t = 0$. Thus, at $t = 0$ the object is at rest, although it will move down after $t = 0$.

5. At $t = 0$, we find that $y = 0$. Since $-1 \leq \sin 3t \leq 1$, y ranges from -0.5 to 0.5, so at $t = 0$ it is starting in the middle. Since $y' = -1.5\cos 3t$, we see $y' = -1.5$ when $t = 0$, so the mass is moving downward.

6. At $t = 0$, we find that $y = -1$, which is clearly the lowest point on the path. Since $y' = 3\sin 3t$, we see that $y' = 0$ when $t = 0$. Thus, at $t = 0$ the object is at rest, although it will move up after $t = 0$.

528 CHAPTER TEN /SOLUTIONS

7. All the differential equations have solutions of the form $s(t) = C_1 \sin \omega t + C_2 \cos \omega t$. Since for all of them, $s'(0) = 0$, we have $s'(0) = 0 = C_1 \omega \cos 0 - C_2 \omega \sin 0 = 0$, giving $C_1 \omega = 0$. Thus, either $C_1 = 0$ or $\omega = 0$. If $\omega = 0$, then $s(t)$ is a constant function, and since the equations represent oscillating springs, we don't want $s(t)$ to be a constant function. Thus, $C_1 = 0$, so all four equations have solutions of the form $s(t) = C \cos \omega t$.
   i) $s'' + 4s = 0$, so $\omega = \sqrt{4} = 2$. $s(0) = C \cos 0 = C = 5$. Thus, $s(t) = 5 \cos 2t$.
   ii) $s'' + \frac{1}{4}s = 0$, so $\omega = \sqrt{\frac{1}{4}} = \frac{1}{2}$. $s(0) = C \cos 0 = C = 10$. Thus, $s(t) = 10 \cos \frac{1}{2} t$.
   iii) $s'' + 6s = 0$, so $\omega = \sqrt{6}$. $s(0) = C = 4$, Thus, $s(t) = 4 \cos \sqrt{6} t$.
   iv) $s'' + \frac{1}{6}s = 0$, so $\omega = \sqrt{\frac{1}{6}}$. $s(0) = C = 20$. Thus, $s(t) = 20 \cos \sqrt{\frac{1}{6}} t$.

   (a) Spring (iii) has the shortest period, $\frac{2\pi}{\sqrt{6}}$. (Other periods are π, 4π, $2\pi\sqrt{6}$)

   (b) Spring (iv) has the largest amplitude, 20.

   (c) Spring (iv) has the longest period, $2\pi\sqrt{6}$.

   (d) Spring (i) has the largest maximum velocity. We can see this by looking at $v(t) = s'(t) = -C\omega \sin \omega t$. The velocity is just a sine function, so we look for the derivative with the biggest amplitude, which will have the greatest value. The velocity function for Spring i) has amplitude 10, the largest of the four springs. (The other velocity amplitudes are $10 \cdot \frac{1}{2} = 5$, $4\sqrt{6} \approx 9.8$, $\frac{20}{\sqrt{6}} \approx 8.2$)

8. (a) Since $\omega^2 = 9$, $\omega = 3$, and so the general solution is of the form
   $$y(t) = A \sin(3t) + B \cos(3t).$$

   (b) (i) $y(0) = 0$, gives $A \sin(0) + B \cos(0) = 0$ so that $B = 0$.
   $$y'(t) = 3A \cos(3t)$$
   $y'(0) = 1$ gives $3A = 1$ and so
   $$y(t) = \frac{1}{3} \sin(3t).$$
   (ii) $y(0) = 1$, gives $A \sin(0) + B \cos(0) = 1$ so that $B = 1$.
   $$y'(t) = 3A \cos(3t) - 3 \sin(3t)$$
   $y'(0) = 0$ gives $3A = 0$ and so
   $$y(t) = \cos(3t).$$
   (iii) $y(0) = 1$, gives $A \sin(0) + B \cos(0) = 1$ so that $B = 1$. $y(1) = 0$ gives $A \sin(3) + \cos(3) = 0$ and so $A = \frac{-\cos(3)}{\sin(3)}$, so
   $$y(t) = \frac{-\cos(3)}{\sin(3)} \sin(3t) + \cos(3t).$$
   Note that using the trigonometric identities, we can write this as:
   $$y(t) = \frac{-\cos(3)}{\sin(3)} \sin(3t) + \cos(3t)$$
   $$= \frac{1}{\sin(3)} (\sin(3) \cos(3t) - \cos(3) \sin(3t))$$
   $$= \frac{1}{\sin(3)} \sin(3 - 3t).$$
   (iv) $y(0) = 0$, gives $A \sin(0) + B \cos(0) = 0$ so that $B = 0$. $y(1) = 1$ gives $A \sin(3) = 1$ and so $A = \frac{1}{\sin(3)}$ so
   $$y(t) = \frac{1}{\sin(3)} \sin(3t).$$

(c)

(i)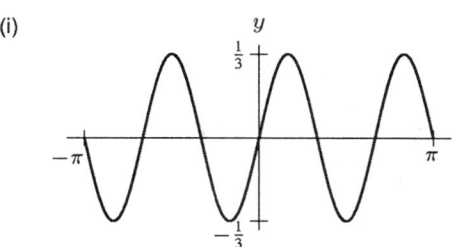

(ii)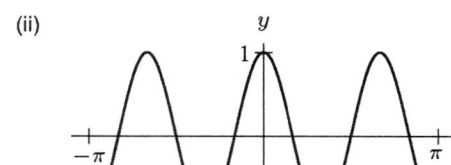

(iii)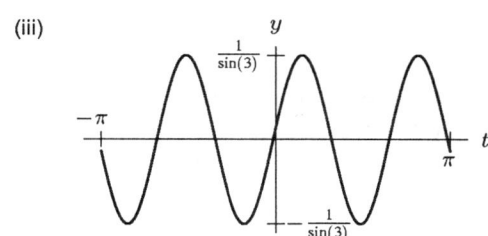

(iv)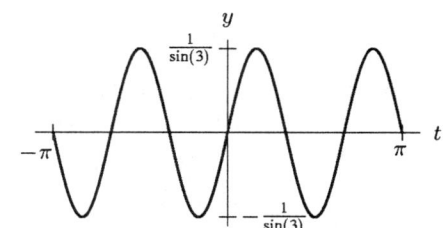

9. First, we note that the solutions of:
   (a) $x'' + x = 0$ are $x = A\cos t + B\sin t$;
   (b) $x'' + 4x = 0$ are $x = A\cos 2t + B\sin 2t$;
   (c) $x'' + 16x = 0$ are $x = A\cos 4t + B\sin 4t$.
   This follows from what we know about the general solution to $x'' + \omega^2 x = 0$.
   The period of the solutions to (a) is 2π, the period of the solutions to (b) is π, and the period of the solutions of (c) is $\frac{\pi}{2}$.
   Since the t-scales are the same on all of the graphs, we see that graphs (I) and (IV) have the same period, which is twice the period of graph (III). Graph (II) has twice the period of graphs (I) and (IV). Since each graph represents a solution, we have the following:

   - equation (a) goes with graph (II)
     equation (b) goes with graphs (I) and (IV)
     equation (c) goes with graph (III)
   - The graph of (I) passes through $(0,0)$, so $0 = A\cos 0 + B\sin 0 = A$. Thus, the equation is $x = B\sin 2t$. Since the amplitude is 2, we see that $x = 2\sin 2t$ is the equation of the graph. Similarly, the equation for (IV) is $x = -3\sin 2t$.
     The graph of (II) also passes through $(0,0)$, so, similarly, the equation must be $x = B\sin t$. In this case, we see that $B = -1$, so $x = -\sin t$.
     Finally, the graph of (III) passes through $(0,1)$, and 1 is the maximum value. Thus, $1 = A\cos 0 + B\sin 0$, so $A = 1$. Since it reaches a local maximum at $(0,1)$, $x'(0) = 0 = -4A\sin 0 + 4B\cos 0$, so $B = 0$. Thus, the solution is $x = \cos 4t$.

10. (a) We are given $\frac{d^2x}{dt^2} = -\frac{g}{l}x$, so $x = C_1\cos\sqrt{\frac{g}{l}}t + C_2\sin\sqrt{\frac{g}{l}}t$. We use the initial conditions to find C_1 and C_2.

    $$x(0) = C_1\cos 0 + C_2\sin 0 = C_1 = 0$$
    $$x'(0) = -C_1\sqrt{\frac{g}{l}}\sin 0 + C_2\sqrt{\frac{g}{l}}\cos 0 = C_2\sqrt{\frac{g}{l}} = v_0$$

    Thus, $C_1 = 0$ and $C_2 = v_0\sqrt{\frac{l}{g}}$, so $x = v_0\sqrt{\frac{l}{g}}\sin\sqrt{\frac{g}{l}}t$.

    (b) Again, $x = C_1\cos\sqrt{\frac{g}{l}}t + C_2\sin\sqrt{\frac{g}{l}}t$, but this time, $x(0) = x_0$, and $x'(0) = 0$.
    Thus, as before, $x(0) = C_1 = x_0$, and $x'(0) = C_2\sqrt{\frac{g}{l}} = 0$. In this case, $C_1 = x_0$ and $C_2 = 0$. Thus, $x = x_0\cos\sqrt{\frac{g}{l}}t$.

11. (a) If x_0 is increased, the amplitude of the function x is increased, but the period remains the same. In other words, the pendulum will start higher, but the time to swing back and forth will stay the same.

    (b) If l is increased, the period of the function x is increased. (Remember, the period of $x_0\cos\sqrt{\frac{g}{l}}t$ is $\frac{2\pi}{\sqrt{g/l}} = 2\pi\sqrt{l/g}$.) In other words, it will take longer for the pendulum to swing back and forth.

12. (a)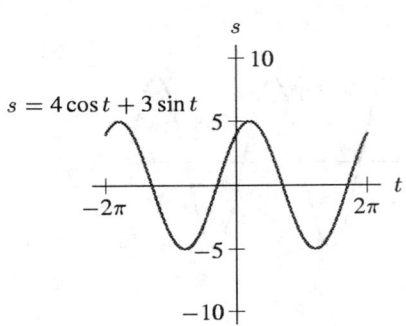

(b) Trace along the curve to the highest point; which has coordinates of about $(0.66, 5)$, so $A \approx 5$. If $s = 5\sin(t+\phi)$, then the maximum occurs where $t \approx 0.66$ and $t + \phi = \pi/2$, that is $0.66 + \phi \approx 1.57$, giving $\phi \approx 0.91$.

(c) Analytically
$$A = \sqrt{4^2 + 3^2} = 5$$
and
$$\tan\phi = \frac{4}{3} \quad \text{so} \quad \phi = \arctan\left(\frac{4}{3}\right) = 0.93.$$

13. The amplitude is $\sqrt{3^2 + 7^2} = \sqrt{58}$.

14. If we write $y = 3\sin(2t) + 4\cos(2t)$ in the form $y(t) = A\sin(2t + \phi)$, then $A = \sqrt{3^2 + 4^2} = 5$.

15. Take $\omega = 2$. The amplitude is $A = \sqrt{5^2 + 12^2} = \sqrt{169} = 13$. The phase shift is $\psi = \tan^{-1}\frac{12}{5}$.

16. The amplitude is $A = \sqrt{7^2 + 24^2} = \sqrt{625} = 25$.
The phase shift, ϕ, is given by $\tan\phi = \frac{24}{7}$, so $\phi = \arctan\frac{24}{7} \approx 1.287$ or $\phi \approx -1.855$.
Since $C_1 = 24 > 0$, we want $\phi = 1.287$, so the solution is $25\sin(\omega t + 1.287)$.

17. (a) Let $x = \omega t$ and $y = \phi$. Then
$$A\sin(\omega t + \phi) = A(\sin\omega t \cos\phi + \cos\omega t \sin\phi)$$
$$= (A\sin\phi)\cos\omega t + (A\cos\phi)\sin\omega t.$$

(b) If we want $A\sin(\omega t + \phi) = C_1 \cos\omega t + C_2 \sin\omega t$ to be true for all t, then by looking at the answer to part (a), we must have $C_1 = A\sin\phi$ and $C_2 = A\cos\phi$. Thus,
$$\frac{C_1}{C_2} = \frac{A\sin\phi}{A\cos\phi} = \tan\phi,$$
and
$$\sqrt{C_1^2 + C_2^2} = \sqrt{A^2\sin^2\phi + A^2\cos^2\phi} = A\sqrt{\sin^2\phi + \cos^2\phi} = A,$$
so our formulas are justified.

18. (a) Since the function executes 3 complete oscillations within 2π radians, we see $n = 3$. This graph is the shape of a cosine which has not been shifted to the right or left, so we have $\phi_0 = 0$. Since $\phi_0 = 0$, when $\phi = 0$ we have
$$U = \frac{V_0}{2}(1 + \cos 0) = V_0.$$

Thus, reading the graph we have $V_0 \approx 12$.

(b) The expression for U is
$$U = 6\left(1 + \cos(3\phi)\right) = 6 + 6\cos(3\phi).$$

We know that a differential equation of the form
$$\frac{d^2s}{dt^2} + \omega^2 s = 0$$

has solutions of the form
$$s = C_1\cos\omega t + C_2\sin\omega t.$$

Thus, if we consider the function
$$U - 6 = 6\cos(3\phi),$$

we have a function which satisfies a second order differential equation of the correct form. Thus, we take $K = 6$.
Differentiating $(U - 6)$ we have

$$\frac{d}{d\phi}(U - 6) = \frac{dU}{d\phi} = -18\sin(3\phi)$$

$$\frac{d^2}{d\phi^2}(U - 6) = \frac{d^2U}{d\phi^2} = -54\cos(3\phi) = -9\left(6\cos(3\phi)\right) = -9(U - 6).$$

Thus, we have $\omega^2 = 9$ since we have shown
$$\frac{d^2U}{d\phi^2} + 9(U - 6) = 0.$$

To find the values of A and B, substitute $\phi = 0$ giving
$$U(0) = 6 + 6\cos(0) = 12$$
$$U'(0) = -18\sin(0) = 0.$$

Thus, $A = 12$ and $B = 0$. So the differential equation is
$$\frac{d^2U}{d\phi^2} + 9(U - 6) = 0, \quad U(0) = 12, U'(0) = 0.$$

19. (a) Since the function executes 2 complete oscillations within 2π radians, we see that $n = 2$. The function shown is at its minimum when $\cos(2\phi)$ is at its maximum (namely, $\phi = 0$). In addition, the function shown is half a cycle, or π radians, out of phase with $\cos(2\phi)$. Thus, $\phi_0 = \pi$.
The maximum value of U occurs when the value of the cosine is 1, so
$$U = \frac{V_0}{2}(1 + 1) = V_0.$$

Thus, reading from the graph, $V_0 \approx 25$.

(b) The expression for U is
$$U = 25\left(1 + \cos(2\phi - \pi)\right) = 25 + 25\cos(2\phi - \pi).$$

We know that a differential equation of the form
$$\frac{d^2s}{dt^2} + \omega^2 s = 0$$

has solution of the form
$$s = C_1\cos\omega t + C_2\sin\omega t,$$

and that the solutions can be rewritten in the form $s = A\sin(\omega t + \alpha)$ (shown in the text) or in the form $S = A\cos(\omega t + \alpha)$. Thus, if we consider the function
$$U - 25 = 25\cos(2\phi - \pi),$$

we have a function which satisfies a differential equation of the correct form. Thus, we take $K = 25$.

Differentiating $U - 25$ we have

$$\frac{d}{d\phi}(U-25) = \frac{dU}{d\phi} = -50\sin(2\phi - \pi)$$

$$\frac{d^2}{d\phi^2}(U-25) = \frac{d^2U}{d\phi^2} = -100\cos(2\phi - \pi) = -4 \cdot 25\cos(2\phi - \pi)) = -4(U-25).$$

Thus we have $\omega^2 = 4$ since we have shown

$$\frac{d^2U}{d\phi^2} + 4(U - 25) = 0.$$

To find the values of A and B, substitute $\phi = 0$ giving

$$U(0) = 25 + 25\cos(-\pi) = 0$$
$$U'(0) = -50\sin(-\pi) = 0.$$

Thus, the differential equations is

$$\frac{d^2U}{d\phi^2} + 4(U-25) = 0, \quad U(0) = U'(0) = 0.$$

20. (a) $36\dfrac{d^2Q}{dt^2} + \dfrac{Q}{9} = 0$ so $\dfrac{d^2Q}{dt^2} = -\dfrac{Q}{324}.$
Thus,

$$Q = C_1 \cos\frac{1}{18}t + C_2 \sin\frac{1}{18}t.$$
$$Q(0) = 0 = C_1 \cos 0 + C_2 \sin 0 = C_1,$$
$$\text{so} \quad C_1 = 0.$$

So, $Q = C_2 \sin \dfrac{1}{18}t$, and

$$Q' = I = \frac{1}{18}C_2 \cos\frac{1}{18}t.$$
$$Q'(0) = I(0) = 2 = \frac{1}{18}C_2 \cos\left(\frac{1}{18}\cdot 0\right) = \frac{1}{18}C_2,$$
$$\text{so} \quad C_2 = 36.$$

Therefore, $Q = 36 \sin \dfrac{1}{18}t.$

(b) As in part (a), $Q = C_1 \cos \dfrac{1}{18}t + C_2 \sin \dfrac{1}{18}t.$
According to the initial conditions:

$$Q(0) = 6 = C_1 \cos 0 + C_2 \sin 0 = C_1,$$
$$\text{so} \quad C_1 = 6.$$

So $Q = 6\cos \dfrac{1}{18}t + C_2 \sin \dfrac{1}{18}t.$
Thus,

$$Q' = I = -\frac{1}{3}\sin\frac{1}{18}t + \frac{1}{18}C_2 \cos\frac{1}{18}t.$$
$$Q'(0) = I(0) = 0 = -\frac{1}{3}\sin\left(\frac{1}{18}\cdot 0\right) + \frac{1}{18}C_2 \cos\left(\frac{1}{18}\cdot 0\right) = \frac{1}{18}C_2,$$
$$\text{so} \quad C_2 = 0.$$

Therefore, $Q = 6\cos \dfrac{1}{18}t.$

21. The equation we have for the charge tells us that:
$$\frac{d^2Q}{dt^2} = -\frac{Q}{LC},$$
where L and C are positive.
If we let $\omega = \sqrt{\frac{1}{LC}}$, we know the solution is of the form:
$$Q = C_1 \cos \omega t + C_2 \sin \omega t.$$
Since $Q(0) = 0$, we find that $C_1 = 0$, so $Q = C_2 \sin \omega t$.
Since $Q'(0) = 4$, and $Q' = \omega C_2 \cos \omega t$, we have $C_2 = \frac{4}{\omega}$, so $Q = \frac{4}{\omega} \sin \omega t$.
But we want the maximum charge, meaning the amplitude of Q, to be $2\sqrt{2}$ coulombs. Thus, we have $\frac{4}{\omega} = 2\sqrt{2}$, which gives us $\omega = \sqrt{2}$.
So we now have: $\sqrt{2} = \frac{1}{\sqrt{LC}} = \frac{1}{\sqrt{10C}}$. Thus, $C = \frac{1}{20}$ farads.

22. We know that the general formula for Q will be of the form:
$$Q = C_1 \cos \omega t + C_2 \sin \omega t.$$
and
$$I = Q' = -C_1 \sin \omega t + C_2 \cos \omega t$$
Thus, as $t \to \infty$, neither one approaches a limit. Instead, they vary sinusoidally, with the same frequency but out of phase. We can think of the charge on the capacitor as being analogous to the displacement of a mass on a spring, oscillating from positive to negative. The current is then like the velocity of the mass, also oscillating from positive to negative. When the charge is maximal or minimal, the current is zero (just like when the spring is at the top or bottom of its motion), and when the current is maximal, the charge is zero (just like when the spring is at the middle of its motion).

Solutions for Section 10.9

1. The characteristic equation is $r^2 + 4r + 3 = 0$, so $r = -1$ or -3.
Therefore $y(t) = C_1 e^{-t} + C_2 e^{-3t}$.

2. The characteristic equation is $r^2 + 4r + 4 = 0$, so $r = -2$.
Therefore $y(t) = (C_1 t + C_2) e^{-2t}$.

3. The characteristic equation is $r^2 + 4r + 5 = 0$, so $r = -2 \pm i$.
Therefore $y(t) = C_1 e^{-2t} \cos t + C_2 e^{-2t} \sin t$.

4. The characteristic equation is $r^2 - 7 = 0$, so $r = \pm \sqrt{7}$.
Therefore $s(t) = C_1 e^{\sqrt{7}t} + C_2 e^{-\sqrt{7}t}$.

5. The characteristic equation is $r^2 + 7 = 0$, so $r = \pm \sqrt{7} i$.
Therefore $s(t) = C_1 \cos \sqrt{7} t + C_2 \sin \sqrt{7} t$.

6. If we try a solution $y(t) = Ae^{rt}$ then
$$r^2 - 3r + 2 = 0$$
which has the solutions $r = 2$ and $r = 1$ so that the general solution is of the form
$$y(t) = Ae^{2t} + Be^t$$

7. The characteristic equation is $4r^2 + 8r + 3 = 0$, so $r = -1/2$ or $-3/2$.
Therefore $z(t) = C_1 e^{-t/2} + C_2 e^{-3t/2}$.

8. The characteristic equation is $r^2 + 4r + 8 = 0$, so $r = -2 \pm 2i$.
Therefore $x(t) = C_1 e^{-2t} \cos 2t + C_2 e^{-2t} \sin 2t$.

9. The characteristic equation is $r^2 + r + 1 = 0$, so $r = -\frac{1}{2} \pm \frac{\sqrt{3}}{2} i$.
Therefore $p(t) = C_1 e^{-t/2} \cos \frac{\sqrt{3}}{2} t + C_2 e^{-t/2} \sin \frac{\sqrt{3}}{2} t$.

10. If we try a solution $z(t) = Ae^{rt}$ then
$$r^2 + 2 = 0$$
so that the general solution is of the form:
$$y(t) = A \sin \sqrt{2} t + B \cos \sqrt{2} t$$

11. If we try a solution $z(t) = Ae^{rt}$ then
$$r^2 + 2r = 0$$
which has solutions $r = 0$ and $r = -2$ so that the general solution is of the form
$$y(t) = A + Be^{-2t}$$

12. If we try a solution $P(t) = Ae^{rt}$ then
$$r^2 + 2r + 1 = 0$$
which has the repeated solution $r = -1$ so that the general solution is of the form
$$y(t) = (At + B)e^{-t}$$

13. The characteristic equation is $r^2 + 6r + 5 = 0$, so $r = -1$ or -5.
Therefore $y(t) = C_1 e^{-t} + C_2 e^{-5t}$.
$y'(t) = -C_1 e^{-t} - 5C_2 e^{-5t}$
$y'(0) = 0 = -C_1 - 5C_2$
$y(0) = 1 = C_1 + C_2$
Therefore $C_2 = -1/4$, $C_1 = 5/4$ and $y(t) = \frac{5}{4} e^{-t} - \frac{1}{4} e^{-5t}$.

14. The characteristic equation is $r^2 + 6r + 5 = 0$, so $r = -1$ or -5.
Therefore $y(t) = C_1 e^{-t} + C_2 e^{-5t}$.
$y'(t) = -C_1 e^{-t} - 5C_2 e^{-5t}$
$y'(0) = 5 = -C_1 - 5C_2$
$y(0) = 5 = C_1 + C_2$
Therefore $C_2 = -5/2$, $C_1 = 15/2$ and $y(t) = \frac{15}{2} e^{-t} - \frac{5}{2} e^{-5t}$.

15. The characteristic equation is $r^2 + 6r + 10 = 0$, so $r = -3 \pm i$.
Therefore $y(t) = C_1 e^{-3t} \cos t + C_2 e^{-3t} \sin t$.
$y'(t) = C_1 [e^{-3t}(-\sin t) + (-3e^{-3t}) \cos t] + C_2 [e^{-3t} \cos t + (-3e^{-3t}) \sin t]$
$y'(0) = 2 = -3C_1 + C_2$
$y(0) = 0 = C_1$
Therefore $C_1 = 0, C_2 = 2$ and $y(t) = 2e^{-3t} \sin t$.

16. The characteristic equation is $r^2 + 6r + 10 = 0$, so $r = -3 \pm i$.
Therefore $y(t) = C_1 e^{-3t} \cos t + C_2 e^{-3t} \sin t$.
$y'(t) = C_1 [e^{-3t}(-\sin t) + (-3e^{-3t}) \cos t] + C_2 [e^{-3t} \cos t + (-3e^{-3t}) \sin t]$
$y'(0) = 0 = -3C_1 + C_2$
$y(0) = 0 = C_1$
Therefore $C_1 = C_2 = 0$ and $y(t) = 0$.

17. The characteristic equation is $r^2 + 2r + 2 = 0$, so $r = -1 \pm i$.
Therefore $p(t) = C_1 e^{-t} \cos t + C_2 e^{-t} \sin t$.
$p(0) = 0 = C_1$ so $p(t) = C_2 e^{-t} \sin t$
$p(\pi/2) = 20 = C_2 e^{-\pi/2} \sin \frac{\pi}{2}$ so $C_2 = 20 e^{\pi/2}$
Therefore $p(t) = 20 e^{\frac{\pi}{2}} e^{-t} \sin t = 20 e^{\frac{\pi}{2} - t} \sin t$.

18. The characteristic equation is $r^2 + 4r + 5 = 0$, so $r = -2 \pm i$.
Therefore $p(t) = C_1 e^{-2t} \cos t + C_2 e^{-2t} \sin t$.
$p(0) = 1 = C_1$ so $p(t) = e^{-2t} \cos t + C_2 e^{-2t} \sin t$
$p(\pi/2) = 5 = C_2 e^{-\pi}$ so $C_2 = 5 e^{\pi}$.
Therefore $p(t) = e^{-2t} \cos t + 5 e^{\pi} e^{-2t} \sin t = e^{-2t} \cos t + 5 e^{\pi - 2t} \sin t$.

19. (a) $x'' + 4x = 0$ represents an undamped oscillator, and so goes with (IV).
    (b) $x'' - 4x = 0$ has characteristic equation $r^2 - 4 = 0$ and so $r = \pm 2$. The solution is $C_1 e^{-2t} + C_2 e^{2t}$. This represents non-oscillating motion, so it goes with (II).
    (c) $x'' - 0.2x' + 1.01x = 0$ has characteristic equation $r^2 - 0.2 + 1.01 = 0$ so $b^2 - 4ac = 0.04 - 4.04 = -4$, and $r = 0.1 \pm i$. So the solution is
    $$C_1 e^{(0.1+i)t} + C_2 e^{(0.1-i)t} = e^{0.1t}(A\sin t + B\cos t).$$
    The negative coefficient in the x' term represents an amplifying force. This is reflected in the solution by $e^{0.1t}$, which increases as t increases, so this goes with (I).
    (d) $x'' + 0.2x' + 1.01x$ has characteristic equation $r^2 + 0.2r + 1.01 = 0$ so $b^2 - 4ac = -4$. This represents a damped oscillator. We have $r = -0.1 \pm i$ and so the solution is $x = e^{-0.1t}(A\sin t + B\cos t)$, which goes with (III).

20. We solve the characteristic equation in each case to obtain solutions to the differential equation.
    (a) $r^2 + 5r + 6 = 0$, so $r = -2$ or -3. Then, $y = C_1 e^{-2t} + C_2 e^{-3t}$.
    (b) $r^2 + r - 6 = 0$, so $r = 2$ or -3. Then, $y = C_1 e^{2t} + C_2 e^{-3t}$.
    (c) $r^2 + 4r + 9 = 0$, so $r = -2 \pm \sqrt{5}i$. Then, $y = C_1 e^{-2t}\cos(\sqrt{5}t) + C_2 e^{-2t}\sin(\sqrt{5}t)$.
    (d) $r^2 = -9$, so $r = \pm 3i$. Then, $y = C_1 \cos(3t) + C_2 \sin(3t)$.

    Since (d) is undamped oscillations, it must be graph (I). Similarly, (c) is damped oscillations and so must be graph (II). Equation (a) is exponential decay, and so must be (IV). This leaves (III) to match with (b), which could be exponential growth or decay.

21. $0 = \frac{d^2}{dt^2}(e^{2t}) - 5\frac{d}{dt}(e^{2t}) + ke^{2t} = 4e^{2t} - 10e^{2t} + ke^{2t} = e^{2t}(k-6)$. Since $e^{2t} \neq 0$, we must have $k - 6 = 0$. Therefore $k = 6$.

    The characteristic equation is $r^2 - 5r + 6 = 0$, so $r = 2$ or 3. Therefore $y(t) = C_1 e^{2t} + C_2 e^{3t}$.

22. In the underdamped case, $b^2 - 4c < 0$ so $4c - b^2 > 0$. Since the roots of the characteristic equation are
    $$\alpha \pm i\beta = \frac{-b \pm \sqrt{b^2 - 4c}}{2} = \frac{-b \pm i\sqrt{4c - b^2}}{2}$$
    we have $\alpha = -b/2$ and $\beta = (\sqrt{4c - b^2})/2$ or $\beta = -(\sqrt{4c - b^2})/2$. Since the general solution is
    $$y = C_1 e^{\alpha t}\cos \beta t + C_2 e^{\alpha t}\sin \beta t$$
    and since α is negative, $y \to 0$ as $t \to \infty$.

23. Recall that $F_{\text{drag}} = -c\frac{ds}{dt}$, so to find the largest coefficient of damping we look at the coefficient of s'. Thus spring (iii) has the largest coefficient of damping.

24. The restoring force is given by $F_{\text{spring}} = -ks$, so we look for the smallest coefficient of s. Spring (iv) exerts the smallest restoring force.

25. The frictional force is $F_{\text{drag}} = -c\frac{ds}{dt}$. Thus spring (iv) has the smallest frictional force.

26. All of these differential equations have solutions of the form $C_1 e^{\alpha t}\cos \beta t + C_2 e^{\alpha t}\sin \beta t$. The spring with the longest period has the smallest β. Since $i\beta$ is the complex part of the roots of the characteristic equation, $\beta = \frac{1}{2}(\sqrt{4c - b^2})$. Thus spring (iii) has the longest period.

27. The stiffest spring exerts the greatest restoring force for a small displacement. Recall that by Hooke's Law $F_{\text{spring}} = -ks$, so we look for the differential equation with the greatest coefficient of s. This is spring (ii).

28. Recall that $s'' + bs' + c = 0$ is overdamped if the discriminant $b^2 - 4c > 0$, critically damped if $b^2 - 4c = 0$, and underdamped if $b^2 - 4c < 0$. Since $b^2 - 4c = 16 - 4c$, the circuit is overdamped if $c < 4$, critically damped if $c = 4$, and underdamped if $c > 4$.

29. Recall that $s'' + bs' + cs = 0$ is overdamped if the discriminant $b^2 - 4c > 0$, critically damped if $b^2 - 4c = 0$, and underdamped if $b^2 - 4c < 0$. Since $b^2 - 4c = 8 - 4c$, the solution is overdamped if $c < 2$, critically damped if $c = 2$, and underdamped if $c > 2$.

30. Recall that $s'' + bs' + cs = 0$ is overdamped if the discriminant $b^2 - 4c > 0$, critically damped if $b^2 - 4c = 0$, and underdamped if $b^2 - 4c < 0$. Since $b^2 - 4c = 36 - 4c$, the solution is overdamped if $c < 9$, critically damped if $c = 9$, and underdamped if $c > 9$.

31. The characteristic equation is $r^2 + r - 2 = 0$, so $r = 1$ or -2. Therefore $z(t) = C_1 e^t + C_2 e^{-2t}$. Since $e^t \to \infty$ as $t \to \infty$, we must have $C_1 = 0$. Therefore $z(t) = C_2 e^{-2t}$. Furthermore, $z(0) = 3 = C_2$, so $z(t) = 3e^{-2t}$.

536 CHAPTER TEN /SOLUTIONS

32. (a) If $r_1 = \frac{-b-\sqrt{b^2-4c}}{2}$ then $r_1 < 0$ since both b and $\sqrt{b^2-4c}$ are positive.
    If $r_2 = \frac{-b+\sqrt{b^2-4c}}{2}$, then $r_2 < 0$ because
    $$b = \sqrt{b^2} > \sqrt{b^2-4c}.$$

    (b) The general solution to the differential equation is of the form
    $$y = C_1 e^{r_1 t} + C_2 e^{r_2 t}$$
    and since r_1 and r_2 are both negative, y must go to 0 as $t \to \infty$.

33. The differential equation is $Q'' + 2Q' + \frac{1}{4}Q = 0$, so the characteristic equation is $r^2 + 2r + \frac{1}{4} = 0$. This has roots $\frac{-2 \pm \sqrt{3}}{2} = -1 \pm \frac{\sqrt{3}}{2}$. Thus, the general solution is

$$Q(t) = C_1 e^{(-1+\frac{\sqrt{3}}{2})t} + C_2 e^{(-1-\frac{\sqrt{3}}{2})t},$$
$$Q'(t) = C_1 \left(-1 + \frac{\sqrt{3}}{2}\right) e^{(-1+\frac{\sqrt{3}}{2})t} + C_2 \left(-1 - \frac{\sqrt{3}}{2}\right) e^{(-1-\frac{\sqrt{3}}{2})t}.$$

We have

(a)
$$Q(0) = C_1 + C_2 = 0$$
$$\text{and} \quad Q'(0) = \left(-1 + \frac{\sqrt{3}}{2}\right) C_1 + \left(-1 - \frac{\sqrt{3}}{2}\right) C_2 = 2.$$

Using the formula for $Q(t)$, we have $C_1 = -C_2$. Using the formula for $Q'(t)$, we have:
$$2 = \left(-1 + \frac{\sqrt{3}}{2}\right)(-C_2) + \left(-1 - \frac{\sqrt{3}}{2}\right) C_2 = -\sqrt{3} C_2$$
so, $\quad C_2 = -\frac{2}{\sqrt{3}}.$

Thus, $C_1 = \frac{2}{\sqrt{3}}$, and $Q(t) = \frac{2}{\sqrt{3}} \left(e^{(-1+\frac{\sqrt{3}}{2})t} - e^{(-1-\frac{\sqrt{3}}{2})t} \right).$

(b) We have
$$Q(0) = C_1 + C_2 = 2$$
$$\text{and} \quad Q'(0) = \left(-1 + \frac{\sqrt{3}}{2}\right) C_1 + \left(-1 - \frac{\sqrt{3}}{2}\right) C_2 = 0.$$

Using the first equation, we have $C_1 = 2 - C_2$. Thus,
$$\left(-1 + \frac{\sqrt{3}}{2}\right)(2 - C_2) + \left(-1 - \frac{\sqrt{3}}{2}\right) C_2 = 0$$
$$-\sqrt{3} C_2 = 2 - \sqrt{3}$$
$$C_2 = -\frac{2 - \sqrt{3}}{\sqrt{3}}$$
$$\text{and} \quad C_1 = 2 - C_2 = \frac{2 + \sqrt{3}}{\sqrt{3}}.$$

Thus, $Q(t) = \frac{1}{\sqrt{3}} \left((2 + \sqrt{3}) e^{(-1+\frac{\sqrt{3}}{2})t} - (2 - \sqrt{3}) e^{(-1-\frac{\sqrt{3}}{2})t} \right).$

34. In this case, the differential equation describing the charge is $Q'' + Q' + \frac{1}{4}Q = 0$, so the characteristic equation is $r^2 + r + \frac{1}{4} = 0$. This equation has one root, $r = -\frac{1}{2}$, so the equation for charge is

$$Q(t) = (C_1 + C_2 t)e^{-\frac{1}{2}t},$$
$$Q'(t) = -\frac{1}{2}(C_1 + C_2 t)e^{-\frac{1}{2}t} + C_2 e^{-\frac{1}{2}t}$$
$$= \left(C_2 - \frac{C_1}{2} - \frac{C_2 t}{2}\right)e^{-\frac{1}{2}t}.$$

(a) We have

$$Q(0) = C_1 = 0,$$
$$Q'(0) = C_2 - \frac{C_1}{2} = 2.$$

Thus, $C_1 = 0$, $C_2 = 2$, and

$$Q(t) = 2te^{-\frac{1}{2}t}.$$

(b) We have

$$Q(0) = C_1 = 2,$$
$$Q'(0) = C_2 - \frac{C_1}{2} = 0.$$

Thus, $C_1 = 2$, $C_2 = 1$, and

$$Q(t) = (2+t)e^{-\frac{1}{2}t}.$$

(c) The resistance was decreased by exactly the amount to switch the circuit from the overdamped case to the critically damped case. Comparing the solutions of parts (a) and (b) in Problems 33, we find that in the critically damped case the net charge goes to 0 much faster as $t \to \infty$.

35. In this case, the differential equation describing charge is $8Q'' + 2Q' + \frac{1}{4}Q = 0$, so the characteristic equation is $8r^2 + 2r + \frac{1}{4} = 0$. This quadratic equation has solutions

$$r = \frac{-2 \pm \sqrt{4 - 4 \cdot 8 \cdot \frac{1}{4}}}{16} = -\frac{1}{8} \pm \frac{1}{8}i.$$

Thus, the equation for charge is

$$Q(t) = e^{-\frac{1}{8}t}\left(A \sin \frac{t}{8} + B \cos \frac{t}{8}\right).$$
$$Q'(t) = -\frac{1}{8}e^{-\frac{1}{8}t}\left(A \sin \frac{t}{8} + B \cos \frac{t}{8}\right) + e^{-\frac{1}{8}t}\left(\frac{1}{8}A \cos \frac{t}{8} - \frac{1}{8}B \sin \frac{t}{8}\right)$$
$$= \frac{1}{8}e^{-\frac{1}{8}t}\left((A-B) \cos \frac{t}{8} + (-A-B) \sin \frac{t}{8}\right).$$

(a) We have

$$Q(0) = B = 0,$$
$$Q'(0) = \frac{1}{8}(A - B) = 2.$$

Thus, $B = 0$, $A = 16$, and

$$Q(t) = 16e^{-\frac{1}{8}t} \sin \frac{t}{8}.$$

(b) We have

$$Q(0) = B = 2,$$
$$Q'(0) = \frac{1}{8}(A - B) = 0.$$

Thus, $B = 2$, $A = 2$, and

$$Q(t) = 2e^{-\frac{1}{8}t}\left(\sin \frac{t}{8} + \cos \frac{t}{8}\right).$$

(c) By increasing the inductance, we have gone from the overdamped case to the underdamped case. We find that while the charge still tends to 0 as $t \to \infty$, the charge in the underdamped case oscillates between positive and negative values. In the over-damped case of Problem 33, the charge starts nonnegative and remains positive.

538 CHAPTER TEN /SOLUTIONS

36. The differential equation for the charge on the capacitor, given a resistance R, a capacitance C, and and inductance L, is

$$LQ'' + RQ' + \frac{Q}{C} = 0.$$

The corresponding characteristic equation is $Lr^2 + Rr + \frac{1}{C} = 0$. This equation has roots

$$r = -\frac{R}{2L} \pm \frac{\sqrt{R^2 - \frac{4L}{C}}}{2L}.$$

(a) If $R^2 - \frac{4L}{C} < 0$, the solution is

$$Q(t) = e^{-\frac{R}{2L}t}(A\sin\omega t + B\cos\omega t) \text{ for some } A \text{ and } B,$$

where $\omega = \frac{\sqrt{R^2 - \frac{4L}{C}}}{2L}$. As $t \to \infty$, $Q(t)$ clearly goes to 0.

(b) If $R^2 - \frac{4L}{C} = 0$, the solution is

$$Q(t) = e^{-\frac{R}{L}t}(A + Bt) \text{ for some } A \text{ and } B.$$

Again, as $t \to \infty$, the charge goes to 0.

(c) If $R^2 - \frac{4L}{C} > 0$, the solution is

$$Q(t) = Ae^{r_1 t} + Be^{r_2 t} \text{ for some } A \text{ and } B,$$

where

$$r_1 = -\frac{R}{2L} + \frac{\sqrt{R^2 - \frac{4L}{C}}}{2L}, \quad \text{and} \quad r_2 = -\frac{R}{2L} - \frac{\sqrt{R^2 - \frac{4L}{C}}}{2L}.$$

Notice that r_2 is clearly negative. r_1 is also negative since

$$\frac{\sqrt{R^2 - \frac{4L}{C}}}{2L} < \frac{\sqrt{R^2}}{2L} \quad (L \text{ and } C \text{ are positive})$$
$$= \frac{R}{2L}.$$

Since r_1 and r_2 are negative, again $Q(t) \to 0$, as $t \to \infty$.

Thus, for any circuit with a resistor, a capacitor and an inductor, $Q(t) \to 0$ as $t \to \infty$. Compare this with Problem 22 in Section 10.8, where we showed that in a circuit with just a capacitor and inductor, the charge varied along a sine curve.

37. In the overdamped case, we have a solution of the form

$$s = C_1 e^{r_1 t} + C_2 e^{r_2 t}$$

where r_1 and r_2 are real. We find a t such that $s = 0$, hence $C_1 e^{r_1 t} = -C_2 e^{r_2 t}$.

If $C_2 = 0$, then $C_1 = 0$, hence $s = 0$ for all t. But this doesn't match with Figure 37, so $C_2 \neq 0$. We divide by $C_2 e^{r_1 t}$, and get:

$$-\frac{C_1}{C_2} = e^{(r_2 - r_1)t}, \quad \text{where } -\frac{C_1}{C_2} > 0,$$

so the exponential is always positive. Therefore

$$(r_2 - r_1)t = \ln(-\frac{C_1}{C_2})$$

and

$$t = \frac{\ln(-\frac{C_1}{C_2})}{(r_2 - r_1)}.$$

So the mass passes through the equilibrium point only once, when $t = \frac{\ln(-\frac{C_1}{C_2})}{(r_2 - r_1)}$.

38. (a) $\frac{d^2y}{dt^2} = -\frac{dx}{dt} = y$ so $\frac{d^2y}{dt^2} - y = 0$.

(b) Characteristic equation $r^2 - 1 = 0$, so $r = \pm 1$.
The general solution for y is $y = C_1 e^t + C_2 e^{-t}$, so $x = C_2 e^{-t} - C_1 e^t$.

Solutions for Chapter 10 Review

1. Using the solution of the logistic equation given on page 500 in Section 10.7, and using $y(0) = 1$, we get $y = \frac{10}{1+9e^{-10t}}$.

2. $\frac{dP}{dt} = 0.03P + 400$ so $\int \frac{dP}{P + \frac{40000}{3}} = \int 0.03 dt$.
   $\ln|P + \frac{40000}{3}| = 0.03t + C$ giving $P = Ae^{0.03t} - \frac{40000}{3}$. Since $P(0) = 0$, $A = \frac{40000}{3}$, therefore $P = \frac{40000}{3}(e^{0.03t} - 1)$.

3. $\frac{dy}{dx} = \frac{y(3-x)}{x(\frac{1}{2}y-4)}$ gives $\int \frac{(\frac{1}{2}y-4)}{y} dy = \int \frac{(3-x)}{x} dx$ so $\int (\frac{1}{2} - \frac{4}{y}) dy = \int (\frac{3}{x} - 1) dx$. Thus $\frac{1}{2}y - 4\ln|y| = 3\ln|x| - x + C$.
   Since $y(1) = 5$, we have $\frac{5}{2} - 4\ln 5 = \ln|1| - 1 + C$ so $C = \frac{7}{2} - 4\ln 5$. Thus,
   $$\frac{1}{2}y - 4\ln|y| = 3\ln|x| - x + \frac{7}{2} - 4\ln 5.$$
   We cannot solve for y in terms of x, so we leave the equation in this form.

4. $\frac{dy}{dx} = e^{x-y}$ giving $\int e^y \, dy = \int e^x \, dx$ so $e^y = e^x + C$. Since $y(0) = 1$, we have $e^1 = e^0 + C$ so $C = e - 1$. Thus, $e^y = e^x + e - 1$, so $y = \ln(e^x + e - 1)$.
   [Note: $e^x + e - 1 > 0$ always.]

5. $\frac{df}{dx} = \sqrt{xf(x)}$ gives $\int \frac{df}{\sqrt{f(x)}} = \int \sqrt{x} \, dx$, so $2\sqrt{f(x)} = \frac{2}{3}x^{\frac{3}{2}} + C$. Since $f(1) = 1$, we have $2 = \frac{2}{3} + C$ so $C = \frac{4}{3}$.
   Thus, $2\sqrt{f(x)} = \frac{2}{3}x^{\frac{3}{2}} + \frac{4}{3}$, so $f(x) = (\frac{1}{3}x^{\frac{3}{2}} + \frac{2}{3})^2$.
   (Note: this is only defined for $x \geq 0$.)

6. $1 + y^2 - \frac{dy}{dx} = 0$ gives $\frac{dy}{dx} = y^2 + 1$, so $\int \frac{dy}{1+y^2} = \int dx$ and $\arctan y = x + C$. Since $y(0) = 0$ we have $C = 0$, giving $y = \tan x$.

7. $2\sin x - y^2 \frac{dy}{dx} = 0$ giving $2\sin x = y^2 \frac{dy}{dx}$. $\int 2\sin x \, dx = \int y^2 \, dy$ so $-2\cos x = \frac{y^3}{3} + C$. Since $y(0) = 3$ we have $-2 = 9 + C$, so $C = -11$. Thus, $-2\cos x = \frac{y^3}{3} - 11$ giving $y = \sqrt[3]{33 - 6\cos x}$.

8. $\frac{dk}{dt} = (1 + \ln t)k$ gives $\int \frac{dk}{k} = \int (1 + \ln t) dt$ so $\ln|k| = t\ln t + C$. $k(1) = 1$, so $0 = 0 + C$, or $C = 0$. Thus, $\ln|k| = t\ln t$ and $|k| = e^{t \ln t} = t^t$, giving $k = \pm t^t$.
   But recall $k(1) = 1$, so $k = t^t$ is the solution.

9. $\frac{dy}{dx} + xy^2 = 0$ means $\frac{dy}{dx} = -xy^2$, so $\int \frac{dy}{y^2} = \int -x \, dx$ giving $-\frac{1}{y} = -\frac{x^2}{2} + C$. Since $y(1) = 1$ we have $-1 = -\frac{1}{2} + C$ so $C = -\frac{1}{2}$. Thus, $-\frac{1}{y} = -\frac{x^2}{2} - \frac{1}{2}$ giving $y = \frac{2}{x^2+1}$.

10. $\frac{dy}{dx} = e^{x+y} = e^x e^y$ implies $\int e^{-y} dy = \int e^x dx$ implies $-e^{-y} = e^x + C$. Since $y = 0$ when $x = 1$, we have $-1 = e + C$, giving $C = -1 - e$. Therefore $-e^{-y} = e^x - 1 - e$ and $y = -\ln(1 + e - e^x)$.

11. This equation is separable and so we write it as
    $$\frac{1}{z(z-1)} \frac{dz}{dt} = 1.$$
    We integrate with respect to t, giving
    $$\int \frac{1}{z(z-1)} dz = \int dt$$
    $$\int \frac{1}{z-1} dz - \int \frac{1}{z} dz = \int dt$$
    $$\ln|z-1| - \ln|z| = t + C$$
    $$\ln\left|\frac{z-1}{z}\right| = t + C,$$
    so that
    $$\frac{z-1}{z} = e^{t+C} = ke^t.$$
    Solving for z gives
    $$z(t) = \frac{1}{1 - ke^t}.$$
    The initial condition $z(0) = 10$ gives
    $$\frac{1}{1-k} = 10$$
    or $k = 0.9$. The solution is therefore
    $$z(t) = \frac{1}{1 - 0.9e^t}.$$

540 CHAPTER TEN /SOLUTIONS

12. $\frac{dy}{dx} = \frac{y(100-x)}{x(20-y)}$ gives $\int (\frac{20-y}{y}) dy = \int (\frac{100-x}{x}) dx$. Thus, $20 \ln |y| - y = 100 \ln |x| - x + C$. The curve passes through $(1, 20)$, so $20 \ln 20 - 20 = -1 + C$ giving $C = 20 \ln 20 - 19$. Therefore, $20 \ln |y| - y = 100 \ln |x| - x + 20 \ln 20 - 19$. We cannot solve for y in terms of x, so we leave the equation in this form.

13. $\frac{dy}{dt} = 2^y \sin^3 t$ implies $\int 2^{-y} dy = \int \sin^3 t \, dt$. Using Integral Table Formula 17, we have

$$-\frac{1}{\ln 2} 2^{-y} = -\frac{1}{3} \sin^2 t \cos t - \frac{2}{3} \cos t + C.$$

According to the initial conditions: $y(0) = 0$ so $-\frac{1}{\ln 2} = -\frac{2}{3} + C$, and $C = \frac{2}{3} - \frac{1}{\ln 2}$. Thus,

$$-\frac{1}{\ln 2} 2^{-y} = -\frac{1}{3} \sin^2 t \cos t - \frac{2}{3} \cos t + \frac{2}{3} - \frac{1}{\ln 2}.$$

Solving for y gives:

$$2^{-y} = \frac{\ln 2}{3} \sin^2 t \cos t + \frac{2 \ln 2}{3} \cos t - \frac{2 \ln 2}{3} + 1.$$

Taking natural logs, (Notice the right side is always > 0.)

$$y \ln 2 = -\ln \left(\frac{\ln 2}{3} \sin^2 t \cos t + \frac{2 \ln 2}{3} \cos t - \frac{2 \ln 2}{3} + 1 \right),$$

so

$$y = \frac{-\ln \left(\frac{\ln 2}{3} \sin^2 t \cos t + \frac{2 \ln 2}{3} \cos t - \frac{2 \ln 2}{3} + 1 \right)}{\ln 2}.$$

14. $e^{-\cos \theta} \frac{dz}{d\theta} = \sqrt{1 - z^2} \sin \theta$ implies $\int \frac{dz}{\sqrt{1-z^2}} = \int e^{\cos \theta} \sin \theta \, d\theta$ implies $\arcsin z = -e^{\cos \theta} + C$. According to the initial conditions: $z(0) = \frac{1}{2}$, so $\arcsin \frac{1}{2} = -e^{\cos 0} + C$, therefore $\frac{\pi}{6} = -e + C$, and $C = \frac{\pi}{6} + e$. Thus $z = \sin(-e^{\cos \theta} + \frac{\pi}{6} + e)$.

15. $(1+t^2) y \frac{dy}{dt} = 1 - y$ implies that $\int \frac{y \, dy}{1-y} = \int \frac{dt}{1+t^2}$ implies that $\int (-1 + \frac{1}{1-y}) dy = \int \frac{dt}{1+t^2}$. Therefore $-y - \ln|1-y| = \arctan t + C$. $y(1) = 0$, so $0 = \arctan 1 + C$, and $C = -\frac{\pi}{4}$, so $-y - \ln|1-y| = \arctan t - \frac{\pi}{4}$. We cannot solve for y in terms of t.

16. $\frac{dy}{dx} = \frac{0.2y(18+0.1x)}{x(100+0.5y)}$ giving $\int \frac{(100+0.5y)}{0.2y} dy = \int \frac{18+0.1x}{x} dx$, so

$$\int \left(\frac{500}{y} + \frac{5}{2} \right) dy = \int \left(\frac{18}{x} + \frac{1}{10} \right) dx.$$

Therefore, $500 \ln |y| + \frac{5}{2} y = 18 \ln |x| + \frac{1}{10} x + C$. Since the curve passes through $(10,10)$, $500 \ln 10 + 25 = 18 \ln 10 + 1 + C$, so $C = 482 \ln 10 + 24$. Thus, the solution is

$$500 \ln |y| + \frac{5}{2} y = 18 \ln |x| + \frac{1}{10} x + 482 \ln 10 + 24.$$

We cannot solve for y in terms of x, so we leave the answer in this form.

17. $\frac{dQ}{dt} = -t^2 Q^2 - Q^2 + 4t^2 + 4 = -Q^2(t^2+1) + 4(t^2+1) = (t^2+1)(4-Q^2)$. Separating variables yields $\frac{dQ}{4-Q^2} = (t^2+1) dt$, so

$$-\int \frac{dQ}{(Q-2)(Q+2)} = -\frac{1}{4} \int \left(\frac{1}{Q-2} - \frac{1}{Q+2} \right) dQ = \int (t^2 + 1) dt.$$

Integrating, we obtain $-\frac{1}{4}(\ln |Q-2| - \ln |Q+2|) = \frac{t^3}{3} + t + C$, so $\ln \left| \frac{Q-2}{Q+2} \right| = -\frac{4t^3}{3} - 4t - 4C$. Exponentiating yields $\left| \frac{Q-2}{Q+2} \right| = e^{-\frac{4t^3}{3} - 4t} e^{-4C}$. $\frac{Q-2}{Q+2} = Ae^{-\frac{4t^3}{3} - 4t}$ where $A = \pm e^{4C}$. Solving for Q, $Q = \frac{4}{1 - Ae^{-\frac{4t^3}{3} - 4t}} - 2$. Notice that A could be any constant, including 0. In fact, we also lost the solution $Q = -2$ when we divided both sides by $4 - Q^2$. (The solution $Q = 2$ corresponds to $A = 0$, but $Q = -2$, another valid solution, is lost by our division.)

18. Separation of variables yields $\int \frac{dy}{y \ln y} = \int \frac{dt}{t^2}$, so $\ln |\ln y| = -\frac{1}{t} + C$.
    Exponentiating both sides gives:

$$|\ln y| = e^{-1/t + C} = e^{-1/t} e^C.$$

So, $\ln y = Ae^{-1/t}$, where $A = \pm e^C$. Exponentiating once more gives $y = e^{Ae^{-1/t}}$.

19. $\frac{x}{y}\frac{dx}{dy} = e^{(\frac{x}{a})^2}\ln y \Rightarrow xe^{-\frac{x^2}{a^2}}dx = y\ln y\, dy \Rightarrow \int xe^{-\frac{x^2}{a^2}}dx = \int y\ln y\, dy$. So $-\frac{a^2}{2}e^{-\frac{x^2}{a^2}} = \frac{y^2}{2}\ln y - \frac{y^2}{4} + C$. We could solve for x, but it would be much messier than this convenient form for the solution.

20. $(y\sqrt{x^3+1})\frac{dy}{dx} + x^2y^2 + x^2 = 0$ is equivalent to $(y\sqrt{x^3+1})\frac{dy}{dx} = -x^2y^2 - x^2 = -x^2(y^2+1)$. Separating variables yields $\frac{y\,dy}{y^2+1} = -\frac{x^2}{\sqrt{x^3+1}}dx$. Integrating, we obtain $\int \frac{y\,dy}{y^2+1} = -\int \frac{x^2}{\sqrt{x^3+1}}dx$. This implies $\frac{1}{2}\ln|y^2+1| = -\frac{2}{3}\sqrt{x^3+1} + C$, whence $y^2 + 1 = Ae^{-\frac{4}{3}\sqrt{x^3+1}}$ where $A = \pm e^{2C}$. So $y = \pm\sqrt{Ae^{-\frac{4}{3}\sqrt{x^3+1}} - 1}$.
Note that A cannot be 0; in fact A must be greater than 1.

21. (a) $\Delta x = \frac{1}{5} = 0.2$.
At $x = 0$:
$y_0 = 1, y' = 4$; so $\Delta y = 4(0.2) = 0.8$. Thus, $y_1 = 1 + 0.8 = 1.8$.
At $x = 0.2$:
$y_1 = 1.8, y' = 3.2$; so $\Delta y = 3.2(0.2) = 0.64$. Thus, $y_2 = 1.8 + 0.64 = 2.44$.
At $x = 0.4$:
$y_2 = 2.44, y' = 2.56$; so $\Delta y = 2.56(0.2) = 0.512$. Thus, $y_3 = 2.44 + 0.512 = 2.952$.
At $x = 0.6$:
$y_3 = 2.952, y' = 2.048$; so $\Delta y = 2.048(0.2) = 0.4096$. Thus, $y_4 = 3.3616$.
At $x = 0.8$:
$y_4 = 3.3616, y' = 1.6384$; so $\Delta y = 1.6384(0.2) = 0.32768$. Thus, $y_5 = 3.68928$. So $y(1) \approx 3.689$.

(b)

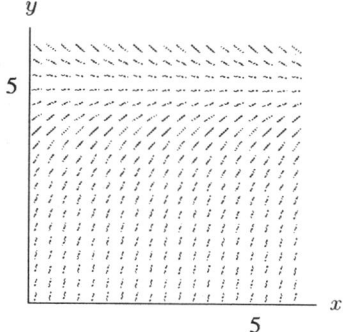

Since solution curves are concave down for $0 \le y \le 5$, and $y(0) = 1 < 5$, the estimate from Euler's method will be an overestimate.

(c) Solving by separation:
$$\int \frac{dy}{5-y} = \int dx, \text{ so } -\ln|5-y| = x + C.$$
Then $5 - y = Ae^{-x}$ where $A = \pm e^{-C}$. Since $y(0) = 1$, we have $5 - 1 = Ae^0$, so $A = 4$.
Therefore, $y = 5 - 4e^{-x}$, and $y(1) = 5 - 4e^{-1} \approx 3.528$.
(Note: as predicted, the estimate in (a) is too large.)

(d) Doubling the value of n will probably halve the error and, therefore, give a value half way between 3.528 and 3.689, which is approximately 3.61.

22. (a) 1 step: $\Delta y = \frac{1}{(\cos x)(\cos y)}\Delta x = \frac{1}{(\cos 0)(\cos 0)}\cdot\frac{1}{2} = \frac{1}{2}$.
Thus, using 1 step, we get $(\frac{1}{2}, \frac{1}{2})$ as our approximation.

(b) 2 steps: $\Delta x = \frac{1}{4}$.

x	y	$\Delta y = \frac{1}{(\cos x)(\cos y)}\Delta x$
0	0	0.25
0.25	0.25	0.266
0.5	0.516	

Thus, using 2 steps, we get $(0.5, 0.516)$ as our approximation.

(c) 4 steps: $\Delta x = \frac{1}{8}$

x	y	$\Delta y = \frac{1}{(\cos x)(\cos y)}\Delta x$
0	0	0.125
0.125	0.125	0.127
0.25	0.252	0.133
0.375	0.385	0.145
0.5	0.530	

Thus, using 4 steps, we get (0.5,0.530) as our approximation.

(d) We have
$$\frac{dy}{dx} = \frac{1}{(\cos x)(\cos y)},$$
so
$$\int \cos y \, dy = \int \frac{dx}{\cos x},$$
whence
$$\sin y = \frac{1}{2} \ln \left| \frac{(\sin x) + 1}{(\sin x) - 1} \right| + C.$$

Our curve passes through (0,0), so, $0 = 0 + C$, and $C = 0$. Therefore
$$y = \arcsin \left(\frac{1}{2} \ln \left| \frac{(\sin x) + 1}{(\sin x) - 1} \right| \right).$$

When $x = \frac{1}{2}$, $y \approx 0.549$. Our answers in parts (a)-(c) are all underestimates. In each case, the error is about $\frac{1}{n+1}$, where n is the number of steps. We expect the error to be approximately proportional to $\frac{1}{n}$, so this seems reasonable.

23. The characteristic equation of $y'' + 6y' + 8y = 0$ is
$$r^2 + 6r + 8 = 0.$$

We have that
$$b^2 - 4c = 6^2 - 4(8) = 4 > 0.$$

This indicates overdamped motion. Since the roots of the characteristic equation are $r_1 = -2$ and $r_2 = -4$, the general solution is
$$y(t) = C_1 e^{-2t} + C_2 e^{-4t}.$$

24. The characteristic equation of $9z'' + z = 0$ is
$$9r^2 + 1 = 0$$
If we write this in the form $r^2 + br + c = 0$, we have that $r^2 + 1/9 = 0$ and
$$b^2 - 4c = 0 - (4)(1/9) = -4/9 < 0$$
This indicates underdamped motion and since the roots of the characteristic equation are $r = \pm \frac{1}{3} i$, the general equation is
$$y(t) = C_1 \cos \left(\frac{1}{3} t \right) + C_2 \sin \left(\frac{1}{3} t \right)$$

25. The characteristic equation of $9z'' - z = 0$ is
$$9r^2 - 1 = 0.$$
If this is written in the form $r^2 + br + c = 0$, we have that $r^2 - 1/9 = 0$ and
$$b^2 - 4c = 0 - (4)(-1/9) = 4/9 > 0$$
This indicates overdamped motion and since the roots of the characteristic equation are $r = \pm 1/3$, the general solution is
$$y(t) = C_1 e^{\frac{1}{3} t} + C_2 e^{-\frac{1}{3} t}.$$

26. The characteristic equation of $x'' + 2x' + 10x = 0$ is

$$r^2 + 2r + 10 = 0$$

We have that
$$b^2 - 4c = 2^2 - 4(10) = -36 < 0$$

This indicates underdamped motion and since the roots of the characteristic equation are $r = -1 \pm 3i$, the general solution is

$$y(t) = C_1 e^{-t} \cos 3t + C_2 e^{-t} \sin 3t$$

27. Recall that $s'' + bs' + cs = 0$ is overdamped if the discriminant $b^2 - 4c > 0$, critically damped if $b^2 - 4c = 0$, and underdamped if $b^2 - 4c < 0$. Since $b^2 - 4c = b^2 - 20$, the solution is overdamped if $b > 2\sqrt{5}$ or $b < -2\sqrt{5}$, critically damped if $b = \pm 2\sqrt{5}$, and underdamped if $-2\sqrt{5} < b < 2\sqrt{5}$.

28. Recall that $s'' + bs' + cs = 0$ is overdamped if the discriminant $b^2 - 4c > 0$, critically damped if $b^2 - 4c = 0$, and underdamped if $b^2 - 4c < 0$. This has discriminant $b^2 - 4c = b^2 + 64$. Since $b^2 + 64$ is always positive, the solution is always overdamped.

29. (a) Since the amount leaving the blood is proportional to the quantity in the blood,

$$\frac{dQ}{dt} = -kQ \quad \text{for some positive constant } k.$$

Thus $Q = Q_0 e^{-kt}$, where Q_0 is the initial quantity in the bloodstream. Only 20% is left in the blood after 3 hours. Thus $0.20 = e^{-3k}$, so $k = \frac{\ln 0.20}{-3} \approx 0.5365$. Therefore $Q = Q_0 e^{-0.5365t}$.

(b) Since 20% is left after 3 hours, after 6 hours only 20% of that 20% will be left. Thus after 6 hours only 4% will be left, so if the patient is given 100 mg, only 4 mg will be left 6 hours later.

30. Let $V(t)$ be the volume of water in the tank at time t, then

$$\frac{dV}{dt} = k\sqrt{V}$$

This is a separable equation which has the solution

$$V(t) = \left(\frac{kt}{2} + C\right)^2$$

Since $V(0) = 200$ this gives $200 = C^2$ so

$$V(t) = \left(\frac{kt}{2} + \sqrt{200}\right)^2.$$

However, $V(1) = 180$ therefore

$$180 = \left(\frac{k}{2} + \sqrt{200}\right)^2,$$

so that $k = 2\left(\sqrt{180} - \sqrt{200}\right) = -1.45146$. Therefore,

$$V(t) = (-0.726t + \sqrt{200})^2.$$

The tank will be half-empty when $V(t) = 100$, so we solve

$$100 = (-0.726t + \sqrt{200})^2$$

to obtain $t = 5.7$ days. The tank will be half empty in 5.7 days.

The volume after 4 days is $V(4)$ which is approximately 126.32 liters.

31. (a) A very hot cup of coffee cools faster than one near room temperature. The differential equation given says that the rate at which the coffee cools is proportional to the difference between the temperature of the surrounding air and the temperature of the coffee. Since $\frac{dT}{dt} < 0$ (the coffee is cooling) and $T - 20 > 0$ (the coffee is warmer than room temperature), k must be positive.

(b) Separating variables gives

$$\int \frac{1}{T-20} dT = \int -k\, dt$$

and so

$$\ln|T-20| = -kt + C$$

and

$$T(t) = 20 + Ae^{-kt}.$$

If the coffee is initially boiling (100° C), then $A = 80$ and so

$$T(t) = 20 + 80e^{-kt}.$$

When $t = 2$, the coffee is at $90°C$ and so $90 = 20 + 80e^{-2k}$ so that $k = \frac{1}{2} \ln \frac{8}{7}$.
Let the time when the coffee reaches $60°C$ be T_d, so that

$$60 = 20 + 80e^{-kT_d}$$

$$e^{-kT_d} = \frac{1}{2}.$$

Therefore, $T_d = \frac{1}{k} \ln 2 = \frac{2\ln 2}{\ln \frac{8}{7}} \approx 10$ minutes.

32. According to Newton's Law of Cooling, the temperature, T, of the roast as a function of time, t, satisfies

$$T'(t) = k(350 - T)$$
$$T(0) = 40.$$

Solving this differential equation, we get that $T = 350 - 310e^{-kt}$ for some $k > 0$. To find k, we note that at $t = 1$ we have $T = 90$, so

$$90 = 350 - 310e^{-k(1)}$$
$$\frac{260}{310} = e^{-k}$$
$$k = -\ln\left(\frac{260}{310}\right)$$
$$\approx 0.17589.$$

Thus, $T = 350 - 310e^{-0.17589t}$. Solving for t when $T = 140$, we have

$$140 = 350 - 310e^{-0.17589t}$$
$$\frac{210}{310} = e^{-0.17589t}$$
$$t = \frac{\ln(210/310)}{-0.17589}$$
$$t \approx 2.21 \text{ hours}.$$

33. (a) For this situation,

$$\begin{pmatrix} \text{Rate money added} \\ \text{to account} \end{pmatrix} = \begin{pmatrix} \text{Rate money added} \\ \text{via interest} \end{pmatrix} + \begin{pmatrix} \text{Rate money} \\ \text{deposited} \end{pmatrix}$$

Translating this into an equation yields

$$\frac{dB}{dt} = 0.1B + 1200.$$

(b) Solving this equation via separation of variables gives

$$\frac{dB}{dt} = 0.1B + 1200$$
$$= (0.1)(B + 12000)$$

So
$$\int \frac{dB}{B + 12000} = \int 0.1\, dt$$
and
$$\ln|B + 12000| = 0.1t + C$$
solving for B,
$$|B + 12000| = e^{(0.1)t+C} = e^C e^{(0.1)t}$$
or
$$B = Ae^{0.1t} - 12000, \text{ (where } A = e^c\text{)}$$
We may find A using the initial condition $B_0 = f(0) = 0$
$$A - 12000 = 0 \quad \text{or} \quad A = 12000$$

(c) After 5 years, the balance is
$$B = f(5) = 12{,}000(e^{(0.1)(5)} - 1)$$
$$\approx 7784.66 \text{ dollars}.$$

34. (a) The balance in the account at the beginning of the month is given by the following sum

$$\begin{pmatrix} \text{balance in} \\ \text{account} \end{pmatrix} = \begin{pmatrix} \text{previous month's} \\ \text{balance} \end{pmatrix} + \begin{pmatrix} \text{interest on} \\ \text{previous month's balance} \end{pmatrix} + \begin{pmatrix} \text{monthly deposit} \\ \text{of \$100} \end{pmatrix}$$

Denote month i's balance by B_i. Assuming the interest is compounded continuously, we have

$$\begin{pmatrix} \text{previous month's} \\ \text{balance} \end{pmatrix} + \begin{pmatrix} \text{interest on previous} \\ \text{month's balance} \end{pmatrix} = B_{i-1}e^{0.1/12}.$$

Since the interest rate is $10\% = 0.1$ per year, interest is $\frac{0.1}{12}$ per month. So at month i, the balance is

$$B_i = B_{i-1}e^{\frac{0.1}{12}} + 100$$

Explicitly, we have for the five years (60 months) the equations:

$$B_0 = 0$$
$$B_1 = B_0 e^{\frac{0.1}{12}} + 100$$
$$B_2 = B_1 e^{\frac{0.1}{12}} + 100$$
$$B_3 = B_2 e^{\frac{0.1}{12}} + 100$$
$$\vdots \quad \vdots$$
$$B_{60} = B_{59} e^{\frac{0.1}{12}} + 100$$

In other words,

$$B_1 = 100$$
$$B_2 = 100 e^{\frac{0.1}{12}} + 100$$
$$B_3 = (100 e^{\frac{0.1}{12}} + 100) e^{\frac{0.1}{12}} + 100$$
$$= 100 e^{\frac{(0.1)2}{12}} + 100 e^{\frac{0.1}{12}} + 100$$
$$B_4 = 100 e^{\frac{(0.1)3}{12}} + 100 e^{\frac{(0.1)2}{12}} + 100 e^{\frac{(0.1)}{12}} + 100$$
$$\vdots \quad \vdots$$
$$B_{60} = 100 e^{\frac{(0.1)59}{12}} + 100 e^{\frac{(0.1)58}{12}} + \cdots + 100 e^{\frac{(0.1)1}{12}} + 100$$
$$B_{60} = \sum_{k=0}^{59} 100 e^{\frac{(0.1)k}{12}}$$

(b) The sum $B_{60} = \sum_{k=0}^{59} 100e^{\frac{(0.1)k}{12}}$ can be written as $B_{60} = \sum_{k=0}^{59} 1200e^{\frac{(0.1)k}{12}}\left(\frac{1}{12}\right)$ which is the left Riemann sum for $\int_0^5 1200e^{0.1t}\,dt$, with $\Delta t = \frac{1}{12}$ and $N = 60$. Evaluating the sum on a calculator gives $B_{60} = 7752.26$.

(c) The situation described by this problem is almost the same as that in Problem 33, except that here the money is being deposited once a month rather than continuously; however the nominal yearly rates are the same. Thus we would expect the balance after 5 years to be approximately the same in each case. This means that the answer to part (b) of this problem should be approximately the same as the answer to part (c) to Problem 33. Since the deposits in this problem start at the end of the first month, as opposed to right away, we would expect the balance after 5 years to be slightly smaller than in Problem 33, as is the case.

Alternatively, we can use the Fundamental Theorem of Calculus to show that the integral can be computed exactly

$$\int_0^5 1200e^{0.1t}\,dt = 12000(e^{(0.1)5} - 1) = 7784.66$$

Thus $\int_0^5 1200e^{0.1t}\,dt$ represents the exact solution to Problem 33. Since $1200e^{0.1t}$ is an increasing function, the left hand sum we calculated in part (b) of this problem underestimates the integral. Thus the answer to part (b) of this problem should be less than the answer to part (c) of Problem 33.

35. Let I be the number of infected people. Then, the number of healthy people in the population is $M - I$. The rate of infection is

$$\text{Infection rate} = \frac{0.01}{M}(M - I)I.$$

and the rate of recovery is

$$\text{Recovery rate} = 0.009I.$$

Therefore,

$$\frac{dI}{dt} = \frac{0.01}{M}(M - I)I - 0.009I$$

or

$$\frac{dI}{dt} = 0.001I\left(1 - 10\frac{I}{M}\right).$$

This is a logistic differential equation, and so the solution will look like the following graph:

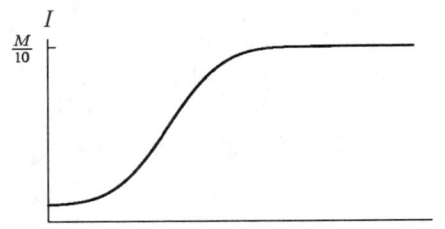

The limiting value for I is $\frac{1}{10}M$, so $1/10$ of the population is infected in the long run.

36. (a) The equilibrium population will be reached when dP/dt approaches zero. Solving $1 - 0.0004P = 0$ gives $P = 2500$ fish as the equilibrium population.

(b) The solution of the differential equation is

$$P(t) = \frac{2500}{(1 + Ae^{-0.25t})}$$

subject to $P(-10) = 1000$ if $t = 0$ represents the present time. So we have

$$1000 = \frac{2500}{(1 + Ae^{2.5})}$$

from which $A = 0.123127$ and

$$P(0) = \frac{2500}{(1 + 0.123127)} \approx 2230.$$

Therefore, the current population is approximately 2230 fish.

(c) The effect of losing 10% of the fish each year gives the revised differential equation

$$\frac{dP}{dt} = (0.25 - 0.0001P)P - 0.1P$$

or

$$\frac{dP}{dt} = (0.15 - 0.0001P)P.$$

The revised equilibrium population is therefore about 1500 fish.

37. (a) When Juliet loves Romeo (i.e. $j > 0$), Romeo's love for her decreases (i.e. $\frac{dr}{dt} < 0$). When Juliet hates Romeo ($j < 0$), Romeo's love for her grows ($\frac{dr}{dt} > 0$). So j and $\frac{dr}{dt}$ have opposite signs, corresponding to the fact that $-B < 0$. When Romeo loves Juliet ($r > 0$), Juliet's love for him grows ($\frac{dj}{dt} > 0$). When Romeo hates Juliet ($r < 0$), Juliet's love for him decreases ($\frac{dj}{dt} < 0$). Thus r and $\frac{dj}{dt}$ have the same sign, corresponding to the fact that $A > 0$.

(b) Since $\frac{dr}{dt} = -Bj$, we have

$$\frac{d^2r}{dt^2} = \frac{d}{dt}(-Bj) = -B\frac{dj}{dt} = -ABr.$$

Rewriting the above equation as $r'' + ABr = 0$, we see that the characteristic equation is $R^2 + AB = 0$. Therefore $R = \pm\sqrt{AB}i$ and the general solution is

$$r(t) = C_1 \cos\sqrt{AB}t + C_2 \sin\sqrt{AB}t.$$

(c) Using $\frac{dr}{dt} = -Bj$, and differentiating r to find j, we obtain

$$j(t) = -\frac{1}{B}\frac{dr}{dt} = -\frac{\sqrt{AB}}{B}(-C_1 \sin\sqrt{AB}t + C_2 \cos\sqrt{AB}t).$$

Now, $j(0) = 0$ gives $C_2 = 0$ and $r(0) = 1$ gives $C_1 = 1$. Therefore, the particular solutions are

$$r(t) = \cos\sqrt{AB}t \quad \text{and} \quad j(t) = \sqrt{\frac{A}{B}}\sin\sqrt{AB}t$$

(d) Consider one period of the graph of $j(t)$ and $r(t)$:

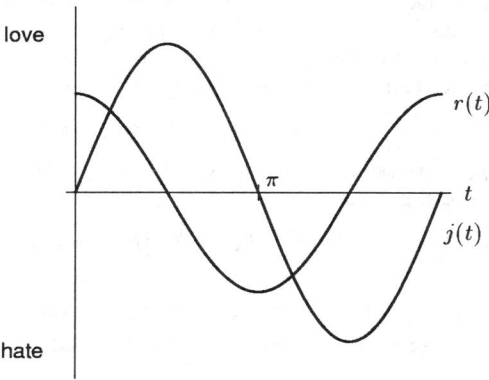

From the graph, we see that they both love each other only a quarter of the time.

38. (a) We have $\Psi = C_1 \cos(\omega x) + C_2 \sin(\omega x)$, and we want $\Psi(0) = \Psi(l) = 0$.

$$\Psi(0) = C_1 = 0 \quad \text{so } C_1 = 0.$$
$$\Psi(l) = C_2 \sin(\omega l) = 0 \quad \text{so } \omega l = n\pi \text{ for some positive integer } n.$$

Thus, $\omega = (n\pi)/l$, so

$$\Psi = C_2 \sin\left(\frac{n\pi x}{l}\right).$$

(b) Using this formula for Ψ, we have

$$\frac{d\Psi}{dx} = \frac{n\pi}{l} C_2 \cos\left(\frac{n\pi x}{l}\right)$$

$$\frac{d^2\Psi}{dx^2} = -\frac{n^2\pi^2}{l^2} C_2 \sin\left(\frac{n\pi x}{l}\right).$$

Thus, substituting for $d^2\Psi/dx^2$ and $\Psi = C_2 \sin(n\pi x/l)$, we have

$$\frac{-h^2}{8\pi^2 m} \frac{d^2\Psi}{dx^2} = \frac{h^2}{8\pi^2 m} \cdot \frac{n^2\pi^2}{l^2} C_2 \sin\left(\frac{n\pi x}{l}\right) = \frac{h^2 n^2}{8ml^2}\Psi,$$

so

$$E = \frac{h^2 n^2}{8ml^2}.$$

(c) Since n must be a positive integer, so $n = 1, 2, 3, 4, ...$, the possible values of E are

$$E_1 = \frac{h^2}{8ml^2}, \quad E_2 = \frac{4h^2}{8ml^2}, \quad E_3 = \frac{9h^2}{8ml^2}, \quad E_4 = \frac{16h^2}{8ml^2}, \quad ....$$

The lowest energy level is $E_1 = h^2/(8ml^2)$, and we see that other energy levels are multiples of E_1:

$$E_2 = 4E_1, \quad E_3 = 9E_1, \quad E_4 = 16E_1, \quad ....$$

Solutions to the Projects and Computer Algebra Investigations

1. (a)
$$p(x) = \text{the number of people with incomes} \geq x.$$
$$p(x + \Delta x) = \text{the number of people with incomes} \geq x + \Delta x.$$

So the number of people with incomes between x and $x + \Delta x$ is

$$p(x) - p(x + \Delta x) = -\Delta p.$$

Since all the people with incomes between x and $x + \Delta x$ have incomes of about x (if Δx is small), the total amount of money earned by people in this income bracket is approximately $x(-\Delta p) = -x\Delta p$.

(b) Pareto's law claims that the average income of all the people with incomes $\geq x$ is kx. Since there are $p(x)$ people with income $\geq x$, the total amount of money earned by people in this group is $kxp(x)$.

The total amount of money earned by people with incomes $\geq (x + \Delta x)$ is therefore $k(x + \Delta x)p(x + \Delta x)$. Then the total amount of money earned by people with incomes between x and $x + \Delta x$ is

$$kxp(x) - k(x + \Delta x)p(x + \Delta x).$$

Since $\Delta p = p(x + \Delta x) - p(x)$, we can substitute $p(x + \Delta x) = p(x) + \Delta p$. Thus the total amount of money earned by people with incomes between x and $x + \Delta x$ is

$$kxp(x) - k(x + \Delta x)(p(x) + \Delta p).$$

Multiplying out, we have

$$kxp(x) - kxp(x) - k(\Delta x)p(x) - kx\Delta p - k\Delta x \Delta p$$

Simplifying and dropping the second order term $\Delta x \Delta p$ gives the total amount of money earned by people with incomes between x and $x + \Delta x$ as

$$-kp\Delta x - kx\Delta p.$$

(c) Setting the answers to parts (a) and (b) equal gives

$$-x\Delta p = -kp\Delta x - kx\Delta p.$$

Dividing by Δx, and letting $\Delta x \to 0$ so that $\frac{\Delta p}{\Delta x} \to p'$, we have

$$x\frac{\Delta p}{\Delta x} = kp + kx\frac{\Delta p}{\Delta x}$$
$$xp' = kp + kxp'$$

so

$$(1-k)xp' = kp.$$

(d) We solve this equation by separating variables

$$\int \frac{dp}{p} = \int \frac{k}{(1-k)} \frac{dx}{x}$$

$$\ln p = \frac{k}{(1-k)} \ln x + C \quad \text{(no absolute values needed since } p, x > 0\text{)}$$

$$\ln p = \ln x^{k/(1-k)} + \ln A \quad \text{(writing } C = \ln A\text{)}$$

$$\ln p = \ln[Ax^{k/(1-k)}] \quad \text{(using } \ln(AB) = \ln A + \ln B\text{)}$$

$$p = Ax^{k/(1-k)}$$

(e) We take $A = 1$. For $k = 10, p = x^{-10/9} \approx x^{-1}$. For $k = 1.1, p = x^{-11}$. The functions are graphed in Figure 10.16. Notice that the larger the value of k, the less negative the value of $k/(1-k)$ (remember $k > 1$), and the slower $p(x) \to 0$ as $x \to \infty$.

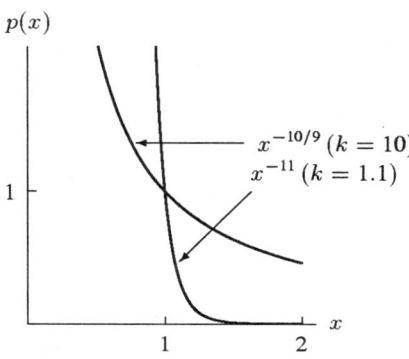

Figure 10.16

2. (a) Writing $F = b\left(\frac{a^2 - ar}{r^3}\right) = 0$ shows $F = 0$ when $r = a$, so $r = a$ gives the equilibrium position.

(b) Expanding $1/r^3$ about $r = a$ gives

$$\frac{1}{r^3} = \frac{1}{(a+r-a)^3} = \frac{1}{a^3}\left(1 + \frac{r-a}{a}\right)^{-3}$$

$$= \frac{1}{a^3}\left(1 - 3\left(\frac{r-a}{a}\right) + \frac{(-3)(-4)}{2!}\left(\frac{r-a}{a}\right)^2 - \cdots\right)$$

$$= \frac{1}{a^3}\left(1 - \frac{3(r-a)}{a} + \frac{6(r-a)^2}{a^2} - \cdots\right).$$

Similarly, expanding $1/r^2$ about $r = a$ gives

$$\frac{1}{r^2} = \frac{1}{(a+r-a)^2} = \frac{1}{a^2}\left(1 + \frac{r-a}{a}\right)^{-2}$$

$$= \frac{1}{a^2}\left(1 - 2\left(\frac{r-a}{a}\right) + \frac{(-2)(-3)}{2!}\left(\frac{r-a}{a}\right)^2 - \cdots\right)$$

$$= \frac{1}{a^2}\left(1 - 2\left(\frac{r-a}{a}\right) + 3\left(\frac{r-a}{a}\right)^2 - \cdots\right).$$

Thus, combining gives

$$F = b\left(\frac{1}{a}\left(1 - \frac{3(r-a)}{a} + \frac{6(r-a)^2}{a^2} - \cdots\right) - \frac{1}{a}\left(1 - \frac{2(r-a)}{a} + \frac{3(r-a)^2}{a^2} - \cdots\right)\right)$$

$$= \frac{b}{a}\left(-\frac{(r-a)}{a} + \frac{3(r-a)^2}{a^2} - \cdots\right)$$

$$= \frac{b}{a^2}\left(-(r-a) + \frac{3(r-a)^2}{a} - \cdots\right).$$

(c) Setting $x = r - a$ gives
$$F \approx \frac{b}{a^2}\left(-x + \frac{3x^2}{a}\right).$$

(d) For small x, we discard the quadratic term in part (c), giving
$$F \approx \frac{-b}{a^2}x.$$

The acceleration is d^2x/dt^2. Thus, using Newton's Second Law:

$$\text{Force} = \text{Mass} \cdot \text{Acceleration}$$

we get
$$\frac{-bx}{a^2} = m\frac{d^2x}{dt^2}.$$

So
$$\frac{d^2x}{dt^2} + \frac{b}{a^2m}x = 0.$$

This differential equation represents an oscillation of the form $x = C_1 \cos \omega t + C_2 \sin \omega t$, where $\omega^2 = b/(a^2m)$ so $\omega = \sqrt{b/(a^2m)}$. Thus, we have
$$\text{Period} = \frac{2\pi}{\omega} = 2\pi a\sqrt{\frac{m}{b}}.$$

3. (a) Equilibrium values are $N = 0$ (unstable) and $N = 200$ (stable). The graphs are shown in Figures 10.17 and 10.18.

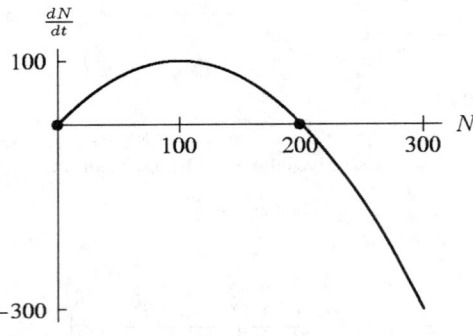

Figure 10.17: $dN/dt = 2N - 0.01N^2$

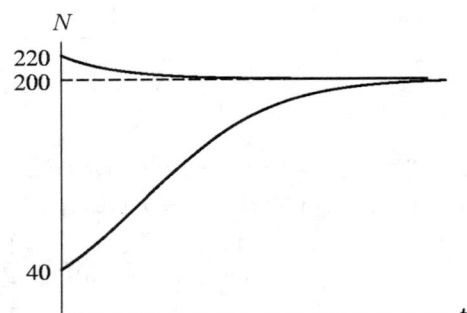

Figure 10.18: Solutions to $dN/dt = 2N - 0.01N^2$

(b) When there is no fishing the rate of population change is given by $\frac{dN}{dt} = 2N - 0.01N^2$. If fishermen remove fish at a rate of 75 fish/year, then this results in a decrease in the growth rate, $\frac{dP}{dt}$, by 75 fish/year. This is reflected in the differential equation by including the -75.

(c)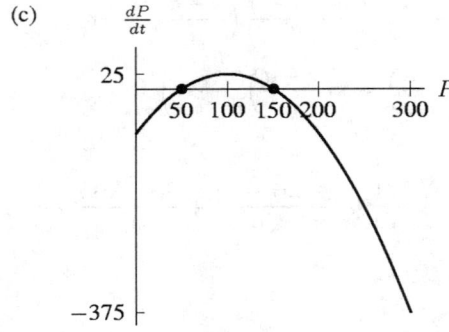

Figure 10.19: $dP/dt = 2P - 0.01P^2 - 75$

(d)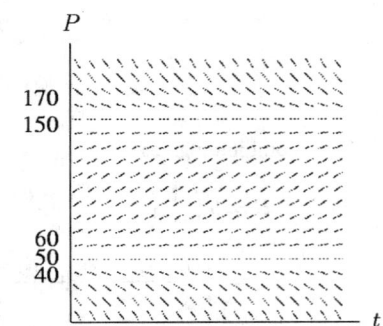

Figure 10.20

(e)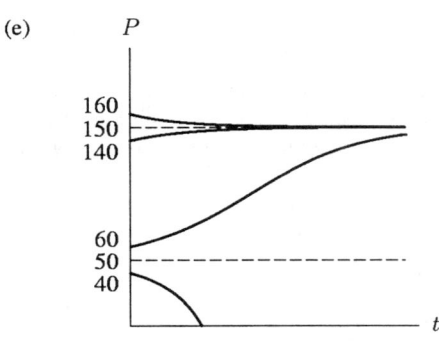

Figure 10.21: Solutions to $dP/dt = 2P - 0.01P^2 - 75$

(f)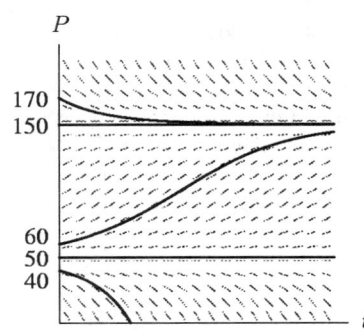

Figure 10.22

(g) The two equilibrium populations are $P = 50, 150$. The stable equilibrium is $P = 150$, while $P = 50$ is unstable. Notice that $P = 50$ and $P = 150$ are solutions of $dP/dt = 0$:

$$\frac{dP}{dt} = 2P - 0.01P^2 - 75 = -0.01(P^2 - 200P + 7500) = -0.01(P - 50)(P - 150).$$

(h) (i)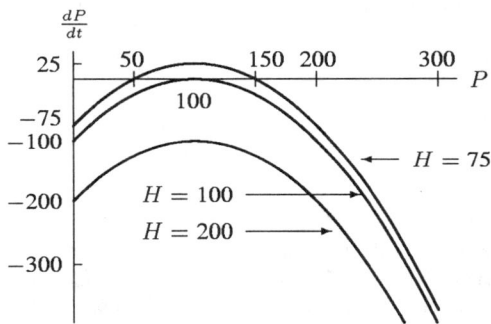

(ii) For $H = 75$, the equilibrium populations (where $dP/dt = 0$) are $P = 50$ and $P = 150$. If the population is between 50 and 150, dP/dt is positive. This means that when the initial population is between 50 and 150, the population will increase until it reaches 150, when $dP/dt = 0$ and the population no longer increases or decreases. If the initial population is greater than 150, then dP/dt is negative, and the population decreases until it reaches 150. Thus 150 is a stable equilibrium. However, 50 is unstable.

For $H = 100$, the equilibrium population (where $dP/dt = 0$) is $P = 100$. For all other populations, dP/dt is negative and so the population decreases. If the initial population is greater than 100, it will decrease to the equilibrium value, $P = 100$. However, for populations less than 100, the population decreases until the species dies out.

For $H = 200$, there are no equilibrium points where $dP/dt = 0$, and dP/dt is always negative. Thus, no matter what the initial population, the population always dies out eventually.

(iii) If the population is not to die out, looking at the three cases above, there must be an equilibrium value where $dP/dt = 0$, i.e. where the graph of dP/dt crosses the P axis. This happens if $H \leq 100$. Thus provided fishing is not more than 100 fish/year, there are initial values of the population for which the population will not be depleted.

(iv) Fishing should be kept below the level of 100 per year.

4. (a) We find the equilibrium solutions by setting $dP/dt = 0$, that is, $P(P-1)(2-P) = 0$, which gives three solutions, $P = 0, P = 1$, and $P = 2$.
   (b) To get your computer algebra system to check that P_1 and P_2 are solutions, substitute one of them into the equation and form an expression consisting of the difference between the right and left hand sides, then ask the CAS to simplify that expression. Do the same for the other function. In order to avoid too much typing, define P_1 and P_2 as functions in your system.
   (c) Substituting $t = 0$ gives

$$P_1(0) = 1 - \frac{1}{\sqrt{4}} = 1/2$$

$$P_2(0) = 1 + \frac{1}{\sqrt{4}} = 3/2.$$

We can find the limits using a computer algebra system. Alternatively, setting $u = e^t$, we can use the limit laws to calculate

$$\lim_{t \to \infty} \frac{e^t}{\sqrt{3 + e^{2t}}} = \lim_{u \to \infty} \frac{u}{\sqrt{3 + u^2}} = \lim_{u \to \infty} \sqrt{\frac{u^2}{3 + u^2}}$$

$$= \sqrt{\lim_{u \to \infty} \frac{u^2}{3 + u^2}} = \sqrt{\lim_{u \to \infty} \frac{1}{\frac{3}{u^2} + 1}}$$

$$= \sqrt{\frac{1}{\lim_{u \to \infty} \frac{3}{u^2} + 1}} = \sqrt{\frac{1}{0 + 1}} = 1.$$

Therefore, we have

$$\lim_{t \to \infty} P_1(t) = 1 - 1 = 0$$

$$\lim_{t \to \infty} P_2(t) = 1 + 1 = 2.$$

To predict these limits without having a formula for P, looking at the original differential equation. We see if $0 < P < 1$, then $P(P-1)(2-P) < 0$, so $P' < 0$. Thus, if $0 < P(0) < 1$, then $P'(0) < 0$, so P is initially decreasing, and tends toward the equilibrium solution $P = 0$. On the other hand, if $1 < P < 2$, then $P(P-1)(2-P) > 0$, so $P' > 0$. So, if $1 < P(0) < 2$, then $P'(0) > 0$, so P is initially increasing and tends towards the equilibrium solution $P = 2$.

5. (a) Using the integral equation with $n + 1$ replaced by n, we have

$$y_n(a) = b + \int_a^a (y_{n-1}(t)^2 + t^2) \, dt = b + 0 = b.$$

   (b) We have $a = 1$ and $b = 0$, so the integral equation tells us that

$$y_{n+1}(s) = \int_1^s (y_n(t)^2 + t^2) \, dt.$$

With $n = 0$, since $y_0(s) = 0$, the CAS gives

$$y_1(s) = \int_1^s 0 + t^2 \, dt = -\frac{1}{3} + \frac{s^3}{3}.$$

Then

$$y_2(s) = \int_1^s (y_1(t)^2 + t^2) \, dt = -\frac{17}{42} + \frac{s}{9} + \frac{s^3}{3} - \frac{s^4}{18} + \frac{s^7}{63},$$

and

$$y_3(s) = \int_1^s (y_2(t)^2 + t^2) \, dt$$

$$= -\frac{157847}{374220} + \frac{289\,s}{1764} - \frac{17\,s^2}{378} + \frac{82\,s^3}{243} - \frac{17\,s^4}{252} + \frac{s^5}{42} - \frac{s^6}{486} + \frac{s^7}{63} - \frac{11\,s^8}{1764} +$$

$$\frac{5\,s^9}{6804} + \frac{2\,s^{11}}{2079} - \frac{s^{12}}{6804} + \frac{s^{15}}{59535}.$$

(c) The solution y, and the approximations y_1, y_2, y_3 are graphed in Figure 10.23. The approximations appear to be accurate on the range $0.5 \leq s \leq 1.5$.

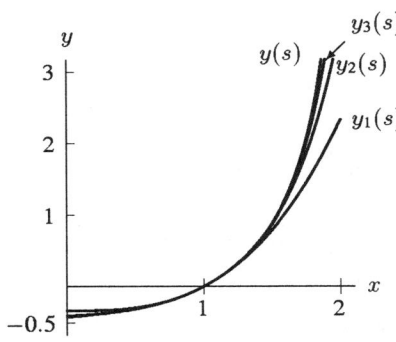

Figure 10.23

6. (a) See Figure 10.24.

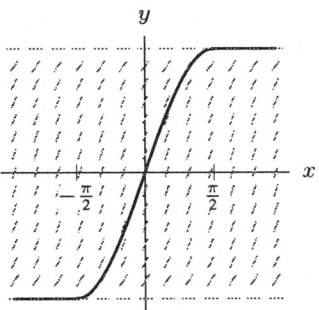

Figure 10.24

(b) Different CASs give different answers, for example they might say $y = \sin x$, or they might say
$$y = \sin x, \quad -\frac{\pi}{2} \leq x \leq \frac{\pi}{2}.$$

(c) Both the sample CAS answers in part (b) are wrong. The first one, $y = \sin x$, is wrong because $\sin x$ starts decreasing at $x = \pi/2$, where the slope field clearly shows that y should be increasing at all times. The second answer is better, but it does not give the solution outside the range $-\pi/2 \leq x \leq \pi/2$. The correct answer is the one sketched in Figure 10.24, which has formula
$$y = \begin{cases} -1 & x \leq -\frac{\pi}{2} \\ \sin x & -\frac{\pi}{2} \leq x \leq \frac{\pi}{2} \\ 1 & x \leq \frac{\pi}{2} \leq x. \end{cases}$$

Solutions to Problems on Systems of Differential Equations

1. Since
$$\frac{dS}{dt} = -aSI,$$
$$\frac{dI}{dt} = aSI - bI,$$
$$\frac{dR}{dt} = bI$$

we have
$$\frac{dS}{dt} + \frac{dI}{dt} + \frac{dR}{dt} = -aSI + aSI - bI + bI = 0.$$

Thus $\frac{d}{dt}(S + I + R) = 0$, so $S + I + R =$ constant.

2. Here x and y both increase at about the same rate.

3. Initially $x = 0$, so we start with only y. Then y decreases while x increases. Then x continues to increase while y starts to increase as well. Finally y continues to increase while x decreases.

4. x decreases quickly while y increases more slowly.

5. The closed trajectory represents populations which oscillate repeatedly.

6. This is an example of a predator-prey relationship. Normally, we would expect the worm population, in the absence of predators, to increase without bound. As the number of worms w increases, so would the rate of increase dw/dt; in other words, the relation $dw/dt = w$ might be a reasonable model for the worm population in the absence of predators.

    However, since there are predators (robins), dw/dt won't be that big. We must lessen dw/dt. It makes sense that the more interaction there is between robins and worms, the more slowly the worms are able to increase their numbers. Hence we lessen dw/dt by the amount wr to get $dw/dt = w - wr$. The term $-wr$ reflects the fact that more interactions between the species means slower reproduction for the worms.

    Similarly, we would expect the robin population to decrease in the absence of worms. We'd expect the population decrease at a rate related to the current population, making $dr/dt = -r$ a reasonable model for the robin population in absence of worms. The negative term reflects the fact that the greater the population of robins, the more quickly they are dying off. The wr term in $dr/dt = -r + wr$ reflects the fact that the more interactions between robins and worms, the greater the tendency for the robins to increase in population.

7. If there are no worms, then $w = 0$, and $\frac{dr}{dt} = -r$ giving $r = r_0 e^{-t}$, where r_0 is the initial robin population. If there are no robins, then $r = 0$, and $\frac{dw}{dt} = w$ giving $w = w_0 e^t$, where w_0 is the initial worm population.

8. There is symmetry across the line $r = w$. Indeed, since $\frac{dr}{dw} = \frac{r(w-1)}{w(1-r)}$, if we switch w and r we get $\frac{dw}{dr} = \frac{w(r-1)}{r(1-w)}$, so $\frac{dr}{dw} = \frac{r(1-w)}{w(r-1)}$. Since switching w and r changes nothing, the slope field must be symmetric across the line $r = w$. The slope field shows that the solution curves are either spirals or closed curves. Since there is symmetry about the line $r = w$, the solutions must in fact be closed curves.

9. If $w = 2$ and $r = 2$, then $\frac{dw}{dt} = -2$ and $\frac{dr}{dt} = 2$, so initially the number of worms decreases and the number of robins increases. In the long run, however, the populations will oscillate; they will even go back to $w = 2$ and $r = 2$.

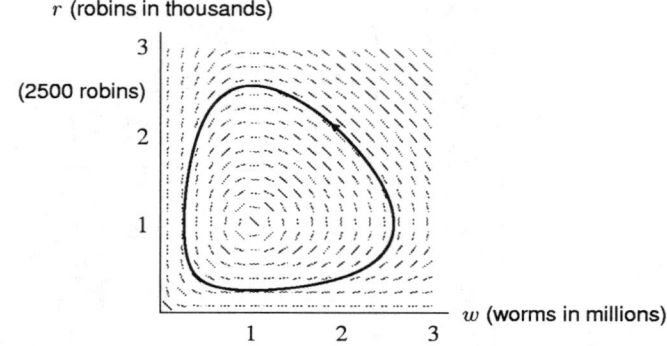

10. Sketching the trajectory through the point $(2,2)$ on the slope field given shows that the maximum robin population is about 2500, and the minimum robin population is about 500. When the robin population is at its maximum, the worm population is about 1,000,000.

11.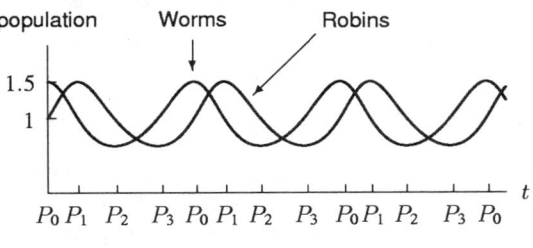

Figure 10.25

12. It will work somewhat; the maximum number the robins reach will increase. However, the minimum number the robins reach will decrease as well. (See graph of slope field.) In the long term, the robin-worm populations will again fall into a cycle. Notice, however, if the extra robins are added during the part of the cycle where there are the fewest robins, the new cycle will have smaller variation.

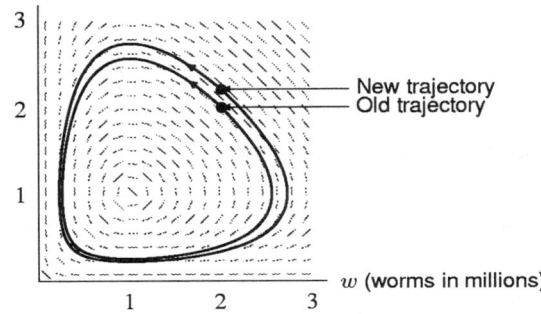

Note that if too many robins are added, the minimum number may get so small the model may fail, since a small number of robins are more susceptible to disaster.

13. The numbers of robins begins to increase while the number of worms remains approximately constant.

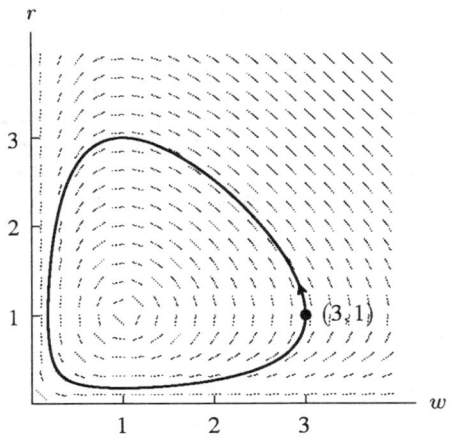

The numbers of robins and worms oscillate periodically between 0.2 and 3, with the robin population lagging behind the worm population.

14. (a) If B were not present, then we'd have $A' = 2A$, so company A's net worth would grow exponentially. Similarly, if A were not present, B would grow exponentially. The two companies restrain each other's growth, probably by competing for the market.
    (b) To find equilibrium points, find the solutions of the pair of equations
    $$A' = 2A - AB = 0$$
    $$B' = B - AB = 0$$
    The first equation has solutions $A = 0$ or $B = 2$. The second has solutions $B = 0$ or $A = 1$. Thus the equilibrium points are (0,0) and (1,2).
    (c) In the long run, one of the companies will go out of business. Two of the trajectories in the figure below go towards the A axis; they represent A surviving and B going out of business. The trajectories going towards the B axis represent A going out of business. Notice both the equilibrium points are unstable.

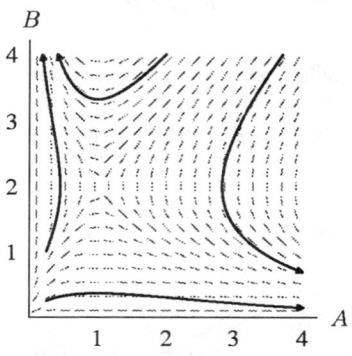

15. (a) Thinking of y as a function of x and x as a function of t, then by the chain rule: $\dfrac{dy}{dt} = \dfrac{dy}{dx}\dfrac{dx}{dt}$, so:
    $$\frac{dy}{dx} = \frac{dy/dt}{dx/dt} = \frac{-0.01x}{-0.05y} = \frac{x}{5y}$$

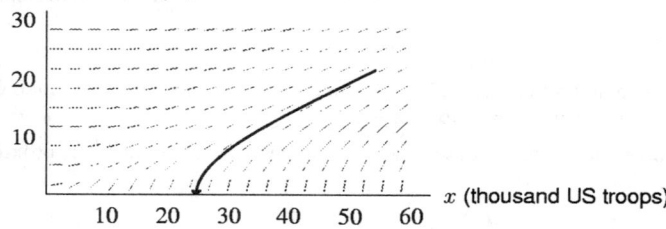

   (b) The figure above shows the slope field for this differential equation and the trajectory starting at $x_0 = 54$, $y_0 = 21.5$. The trajectory goes to the x-axis, where $y = 0$, meaning that the Japanese troops were all killed or wounded before the US troops were, and thus predicts the US victory (which did occur). Since the trajectory meets the x-axis at $x \approx 25$, the differential equation predicts that about 25,000 US troops would survive the battle.
   (c) The fact that the US got reinforcements, while the Japanese did not, does not alter the predicted outcome (a US victory). The US reinforcements have the effect of changing the trajectory, altering the number of troops surviving the battle. See the graph below.

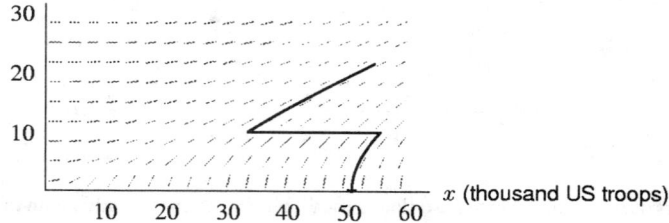

16. (a) Thinking of y as a function of x and x as a function of t, then by the chain rule: $\dfrac{dy}{dt} = \dfrac{dy}{dx}\dfrac{dx}{dt}$, so:

$$\frac{dy}{dx} = \frac{dy/dt}{dx/dt} = \frac{-bx}{-ay} = \frac{bx}{ay}$$

(b) Separating variables,

$$\int ay\,dy = \int bx\,dx$$

$$a\frac{y^2}{2} = b\frac{x^2}{2} + k$$

$$ay^2 - bx^2 = C \quad \text{where } C = 2k$$

17. (a) Lanchester's square law for the battle of Iwo Jima is

$$0.05y^2 - 0.01x^2 = C.$$

If we measure x and y in thousands, $x_0 = 54$ and $y_0 = 21.5$, so $0.05(21.5)^2 - 0.01(54)^2 = C$ giving $C = -6.0475$. Thus the equation of the trajectory is

$$0.05y^2 - 0.01x^2 = -6.0475$$

giving

$$x^2 - 5y^2 = 604.75.$$

(b) Assuming that the battle did not end until all the Japanese were dead or wounded, that is, $y = 0$, then the number of US soldiers remaining is given by $x^2 - 5(0)^2 = 604.75$. This gives $x = 24.59$, or about 25,000 troops. This is approximately what happened.

18. (a) Since the guerrillas are hard to find, the rate at which they are put out of action is proportional to the number of chance encounters between a guerrilla and a conventional soldier, which is in turn proportional to the number of guerrillas and to the number of conventional soldiers. Thus the rate at which guerrillas are put out of action is proportional to the product of the strengths of the two armies.

(b)
$$\frac{dx}{dt} = -xy$$
$$\frac{dy}{dt} = -x$$

(c) Thinking of y as a function of x and x a function of of t, then by the chain rule: $\dfrac{dy}{dt} = \dfrac{dy}{dx}\dfrac{dx}{dt}$ so:

$$\frac{dy}{dx} = \frac{dy/dt}{dx/dt} = \frac{-x}{-xy} = \frac{1}{y}$$

Separating variables:

$$\int y\,dy = \int dx$$

$$\frac{y^2}{2} = x + C$$

The value of C is determined by the initial strengths of the two armies.

(d) The sign of C determines which side wins the battle. Looking at the general solution $\dfrac{y^2}{2} = x + C$, we see that if $C > 0$ the y-intercept is at $\sqrt{2C}$, so y wins the battle by virtue of the fact that it still has troops when $x = 0$. If $C < 0$ then the curve intersects the axes at $x = -C$, so x wins the battle because it has troops when $y = 0$. If $C = 0$, then the solution goes to the point $(0, 0)$, which represents the case of mutual annihilation.

(e) We assume that an army wins if the opposing force goes to 0 first. Figure 10.26 shows that the conventional force wins if $C > 0$ and the guerrillas win if $C < 0$. Neither side wins if $C = 0$ (all soldiers on both sides are killed in this case).

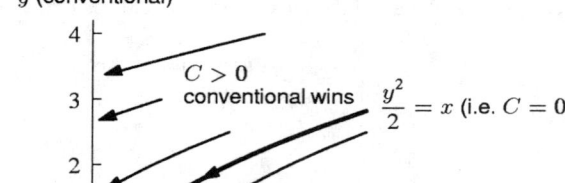

Figure 10.26

19. (a) Taking the constants of proportionality to be a and b, with $a > 0$ and $b > 0$, the equations are
$$\frac{dx}{dt} = -axy$$
$$\frac{dy}{dt} = -bxy$$

(b) $\frac{dy}{dx} = \frac{dy/dt}{dx/dt} = \frac{-bxy}{-axy} = \frac{b}{a}$. Solving the differential equation gives $y = \frac{b}{a}x + C$, where C depends on the initial sizes of the two armies.

(c) The sign of C determines which side wins the battle. Looking at the general solution $y = \frac{b}{a}x + C$, we see that if $C > 0$ the y-intercept is at C, so y wins the battle by virtue of the fact that it still has troops when $x = 0$. If $C < 0$ then the curve intersects the axes at $x = -\frac{a}{b}C$, so x wins the battle because it has troops when $y = 0$. If $C = 0$, then the solution goes to the point $(0, 0)$, which represents the case of mutual annihilation.

(d) We assume that an army wins if the opposing force goes to 0 first.

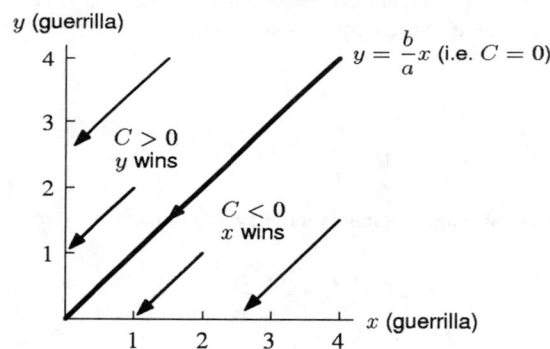

20. (a) We have
$$\frac{\frac{dy}{dt}}{\frac{dx}{dt}} = \frac{dy}{dx} = \frac{-3y - xy}{-2x - xy} = \frac{y(x+3)}{x(y+2)}.$$

Thus,
$$\left(\frac{y+2}{y}\right) dy = \left(\frac{x+3}{x}\right) dx$$

so
$$\int \left(1 + \frac{2}{y}\right) dy = \int \left(1 + \frac{3}{x}\right) dx.$$

So,
$$y + 2\ln|y| = x + 3\ln|x| + C.$$

Since x and y are non-negative,
$$y + 2\ln y = x + 3\ln x + C.$$

This is as far as we can go with this equation – we cannot solve for y in terms of x, for example. We can, however, put it in the form
$$e^{y + 2\ln y} = e^{x + 3\ln x + C}, \quad \text{or} \quad y^2 e^y = Ax^3 e^x.$$

(b) An equilibrium state satisfies
$$\frac{dx}{dt} = -2x - xy = 0 \quad \text{and} \quad \frac{dy}{dt} = -3y - xy = 0.$$

Solving the first equation, we have
$$-x(y+2) = 0, \quad \text{so} \quad x = 0 \quad \text{or} \quad y = -2.$$

The second equation has solutions
$$y = 0 \quad \text{or} \quad x = -3.$$

Since $x, y \geq 0$, the only equilibrium point is $(0, 0)$.

(c) We can use either of our forms for the solution. Looking at
$$y^2 e^y = Ax^3 e^x,$$

we see that if x and y are very small positive numbers, then
$$e^x \approx e^y \approx 1.$$

Thus,
$$y^2 \approx Ax^3, \quad \text{or} \quad \frac{y^2}{x^3} \approx A, \text{ a constant.}$$

Looking at
$$y + 2\ln y = x + 3\ln x + C,$$

we note that if x and y are small, then they are negligible compared to $\ln y$ and $\ln x$. Thus,
$$2\ln y \approx 3\ln x + C,$$

giving
$$\ln y^2 - \ln x^3 \approx C,$$

so
$$\ln \frac{y^2}{x^3} \approx C$$

and therefore
$$\frac{y^2}{x^3} \approx e^C, \text{ a constant.}$$

(d) If
$$x(0) = 4 \quad \text{and} \quad y(0) = 8,$$

then
$$8 + 2\ln 8 = 4 + 3\ln 4 + C.$$

Note that
$$2\ln 8 = 3\ln 4 = \ln 64,$$

giving
$$4 = C.$$

So the phase trajectory is
$$y + 2\ln y = x + 3\ln x + 4.$$

(Or equivalently, $y^2 e^y = e^4 x^3 e^x = x^3 e^{x+4}$.)

(e) If the concentrations are equal, then
$$y + 2\ln y = y + 3\ln y + 4,$$

giving
$$-\ln y = 4 \quad \text{or} \quad y = e^{-4}.$$

Thus, they are equal when $y = x = e^{-4} \approx 0.0183$.

(f) Using part (c), we have that if x is small,
$$\frac{y^2}{x^3} \approx e^4.$$

Since $x = e^{-10}$ is certainly small,
$$\frac{y^2}{e^{-30}} \approx e^4, \quad \text{and} \quad y \approx e^{-13}.$$

21. (a) Symbiosis, because both populations decrease while alone but are helped by the presence of the other.
    (b)

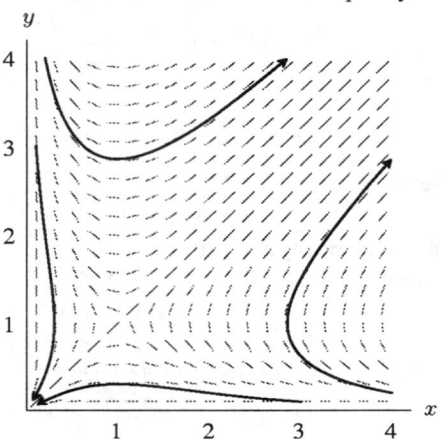

Both populations tend to infinity or both tend to zero.

22. (a) Competition, because both populations grow logistically when alone, but are harmed by the presence of the other.
    (b)

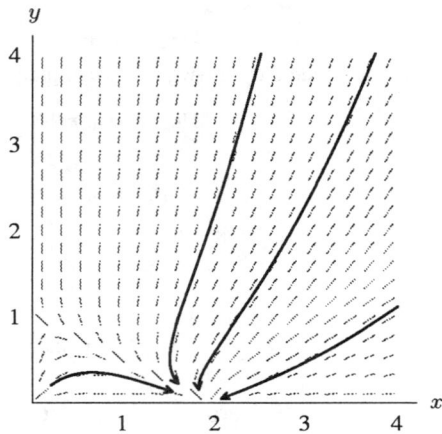

In the long run, $x \to 2$, $y \to 0$. In other words, y becomes extinct.

23. (a) Predator-prey, because x decreases while alone, but is helped by y, whereas y increases logistically when alone, and is harmed by x. Thus x is predator, y is prey.
    (b)

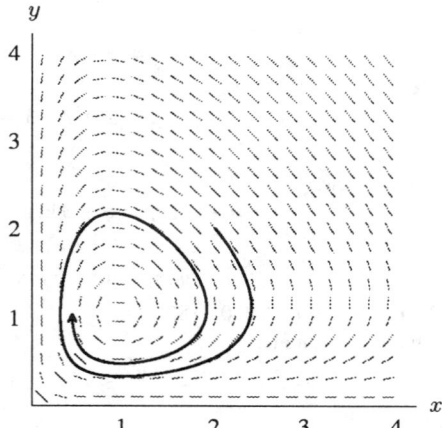

Provided neither initial population is zero, both populations tend to about 1. If x is initially zero, but y is not, then $y \to \infty$. If y is initially zero, but x is not, then $x \to 0$.

Solutions to Problems on Analyzing the Phase Plane

1. (a) $dS/dt = 0$ where $S = 0$ or $I = 0$ (both axes).
   $dI/dt = 0.0026I(S - 192)$, so $dI/dt = 0$ where $I = 0$ or $S = 192$.
   Thus every point on the S axis is an equilibrium point (corresponding to no one being sick).

   (b) In region I, where $S > 192$, $\dfrac{dS}{dt} < 0$ and $\dfrac{dI}{dt} > 0$.
   In region II, where $S < 192$, $\dfrac{dS}{dt} < 0$ and $\dfrac{dI}{dt} < 0$. See Figure 10.27.

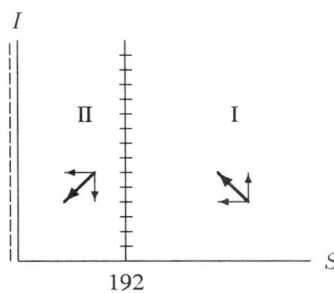

Figure 10.27

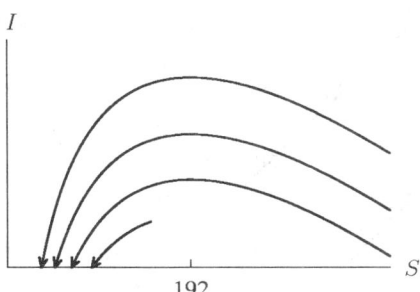

Figure 10.28

   (c) If the trajectory starts with $S_0 > 192$, then I increases to a maximum when $S = 192$. If $S_0 < 192$, then I always decreases. See Figure 10.27. Regardless of the initial conditions, the trajectory always goes to a point on the S-axis (where $I = 0$). The S-intercept represents the number of students who never get the disease. See Figure 10.28.

2. The nullclines are where $\frac{dw}{dt} = 0$ or $\frac{dr}{dt} = 0$.
   $\frac{dw}{dt} = 0$ when $w - wr = 0$, so $w(1 - r) = 0$ giving $w = 0$ or $r = 1$.
   $\frac{dr}{dt} = 0$ when $-r + rw = 0$, so $r(w - 1) = 0$ giving $r = 0$ or $w = 1$.

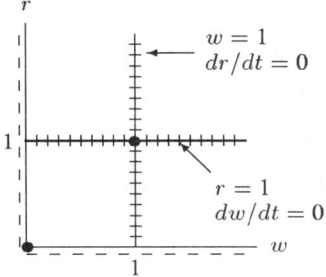

Figure 10.29: Nullclines and equilibrium points (dots)

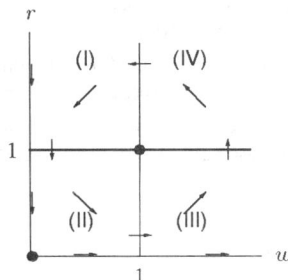

Figure 10.30

The equilibrium points are where the nullclines intersect: $(0, 0)$ and $(1, 1)$. The nullclines split the first quadrant into four sectors. See Figure 10.29. We can get a feel for how the populations interact by seeing the direction of the trajectories in each sector. See Figure 10.30. If the populations reach an equilibrium point, they will stay there. If the worm population dies out, the robin population will also die out, too. However, if the robin population dies out, the worm population will continue to grow.

Otherwise, it seems that the populations cycle around the equilibrium $(1, 1)$. The trajectory moves from sector to sector: trajectories in sector (I) move to sector (II); trajectories in sector (II) move to sector (III); trajectories in sector (III) move to sector (IV); trajectories in sector (IV) move back to sector (I). The robins keep the worm population down by feeding on them, but the robins need the worms (as food) to sustain the population. These conflicting needs keep the populations moving in a cycle around the equilibrium.

3. We first find the nullclines. Again, we assume $x, y \geq 0$.
Vertical nullclines occur where $dx/dt = 0$, which happens when $\frac{dx}{dt} = x(2 - x - y) = 0$, i.e. when $x = 0$ or $x + y = 2$.
Horizontal nullclines occur where $dy/dt = 0$, which happens when $\frac{dy}{dt} = y(1 - x - y) = 0$, i.e. when $y = 0$ or $x + y = 1$.
These nullclines are shown in Figure 10.31.

Equilibrium points (also shown in Figure 10.31) occur where both dy/dt and dx/dt are 0, i.e. at the intersections of vertical and horizontal nullclines. There are three such points for these equations: $(0, 0)$, $(0, 1)$, and $(2, 0)$.

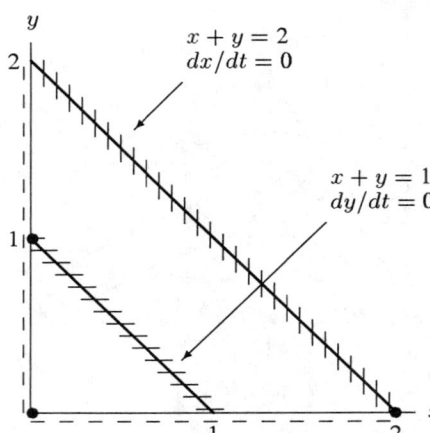

Figure 10.31: Nullclines and equilibrium points (dots)

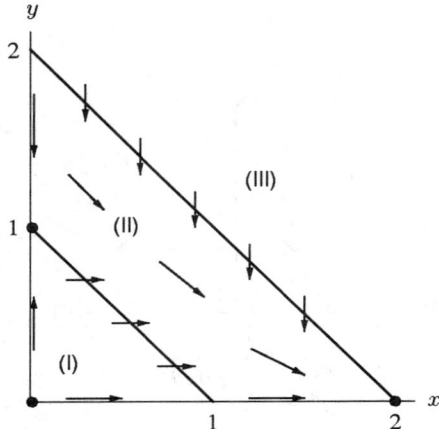

Figure 10.32: General directions of trajectories and equilibrium points (dots)

Looking at sectors in Figure 10.32, we see that no matter in what sector the initial point lies, the trajectory will head toward the equilibrium point $(2, 0)$.

4. We first find the nullclines. Vertical nullclines occur where $\frac{dx}{dt} = 0$, which happens when $x = 0$ or $y = \frac{1}{3}(2 - x)$. Horizontal nullclines occur where $\frac{dy}{dt} = y(1 - 2x) = 0$, which happens when $y = 0$ or $x = \frac{1}{2}$. These nullclines are shown in Figure 10.33.

Equilbrium points (also shown in Figure 10.33) occur at the intersections of vertical and horizontal nullclines. There are three such points for this system of equations; $(0, 0)$, $(\frac{1}{2}, \frac{1}{2})$ and $(2, 0)$.

The nullclines divide the positive quadrant into four regions as shown in Figure 10.33. Trajectory directions for these regions are shown in Figure 10.34.

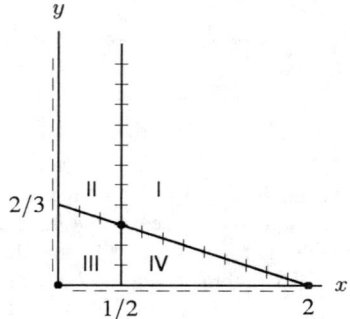

Figure 10.33: Nullclines and equilibrium points (dots)

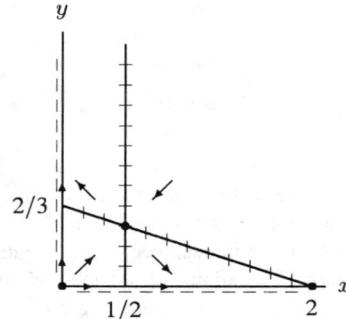

Figure 10.34: General directions of trajectories and equilibrium points (dots)

5. We first find nullclines. Vertical nullclines occur where $\frac{dx}{dt} = x(2 - x - 2y) = 0$, which happens when $x = 0$ or $y = \frac{1}{2}(2 - x)$. Horizontal nullclines occur where $\frac{dy}{dt} = y(1 - 2x - y) = 0$, which happens when $y = 0$ or $y = 1 - 2x$. These nullclines are shown in Figure 10.35.

   Equilibrium points (also shown in the figure below) occur at the intersections of vertical and horizontal nullclines. There are three such points for this system; $(0, 0)$, $(0, 1)$, and $(2, 0)$.

   The nullclines divide the positive quadrant into three regions as shown in the figure below. Trajectory directions for these regions are shown in Figure 10.36.

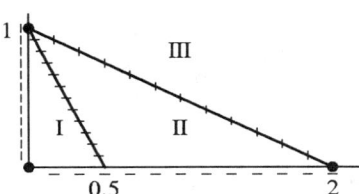

Figure 10.35: Nullclines and equilibrium points (dots)

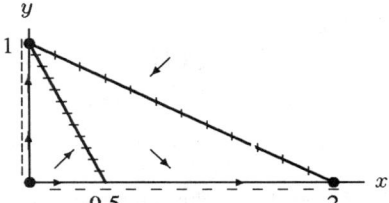

Figure 10.36: General directions of trajectories and equilibrium points (dots)

6. We first find the nullclines. Vertical nullclines occur where $\frac{dx}{dt} = x(1 - y - \frac{x}{3}) = 0$, which happens when $x = 0$ or $y = 1 - \frac{x}{3}$. Horizontal nullclines occur where $\frac{dy}{dt} = y(1 - \frac{y}{2} - x) = 0$, which happens when $y = 0$ or $y = 2(1 - x)$. These nullclines are shown in Figure 10.37.

   Equilibrium points (also shown in Figure 10.37) occur at the intersections of vertical and horizontal nullclines. There are four such points for this system: $(0, 0)$, $(0, 2)$, $(3, 0)$, and $(\frac{3}{5}, \frac{4}{5})$.

   The nullclines divide the positive quadrant into four regions as shown in Figure 10.37. Trajectory directions for these regions are shown in Figure 10.38.

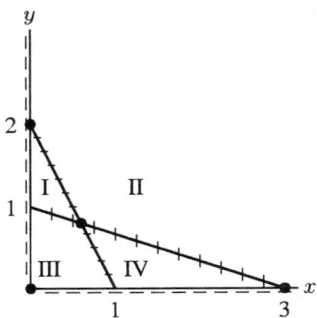

Figure 10.37: Nullclines and equilibrium points (dots)

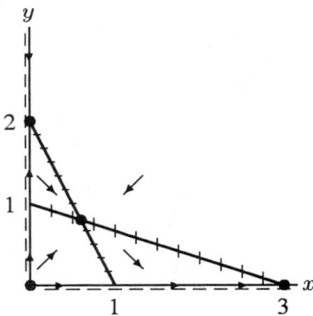

Figure 10.38: General directions of trajectories and equilibrium points (dots)

7. We first find the nullclines. Again, we assume $x, y \geq 0$.
   $\frac{dx}{dt} = x(1 - x - \frac{y}{3}) = 0$ when $x = 0$ or $x + y/3 = 1$.
   $\frac{dy}{dt} = y(1 - y - \frac{x}{2}) = 0$ when $y = 0$ or $y + x/2 = 1$.
   These nullclines are shown in Figure 10.39. There are four equilibrium points for these equations. Three of them are the points, $(0, 0)$, $(0, 1)$, and $(1, 0)$. The fourth is the intersection of the two lines $x + y/3 = 1$ and $y + x/2 = 1$. This point is $(\frac{4}{5}, \frac{3}{5})$.

564 CHAPTER TEN /SOLUTIONS

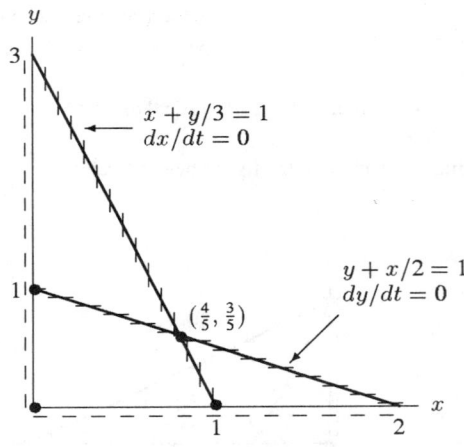

Figure 10.39: Nullclines and equilibrium points (dots)

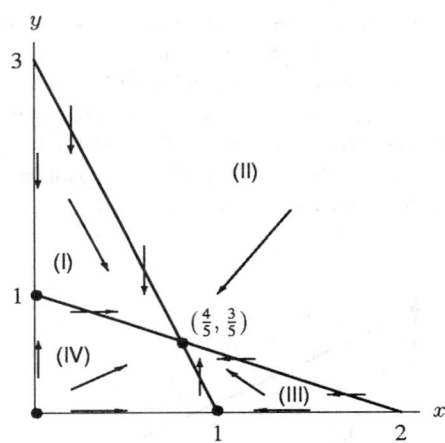

Figure 10.40: General directions of trajectories and equilibrium points (dots)

Looking at sectors in Figure 10.40, we see that no matter in what sector the initial point lies, the trajectory will head toward the equilibrium point $\left(\frac{4}{5}, \frac{3}{5}\right)$. Only if the initial point lies on the x- or y-axis, will the trajectory head towards the equilibrium points at $(1, 0)$, $(0, 1)$, or $(0, 0)$. In fact, the trajectory will go to $(0, 0)$ only if it starts there, in which case $x(t) = y(t) = 0$ for all t. From direction of the trajectories in Figure 10.40, it appears that if the initial point is in sectors (I) or (III), then it will remain in that sector as it heads towards the equilibrium.

8. We assume that x, $y \geq 0$ and then find the nullclines. $\frac{dx}{dt} = x(1 - \frac{x}{2} - y) = 0$ when $x = 0$ or $y + \frac{x}{2} = 1$. $\frac{dy}{dt} = y(1 - \frac{y}{3} - x) = 0$ when $y = 0$ or $x + \frac{y}{3} = 1$.
We find the equilibrium points. They are $(2, 0)$, $(0, 3)$, $(0, 0)$, and $\left(\frac{4}{5}, \frac{3}{5}\right)$. The nullclines and equilibrium points are shown in Figure 10.41.

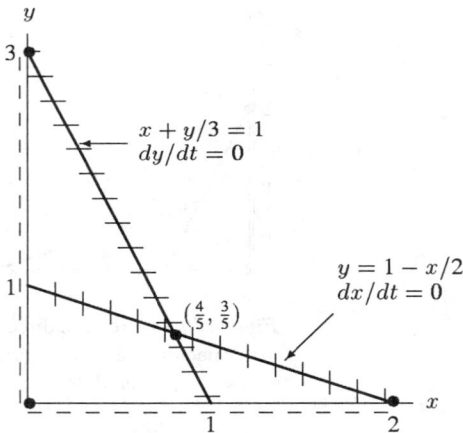

Figure 10.41: Nullclines and equilibrium points (dots)

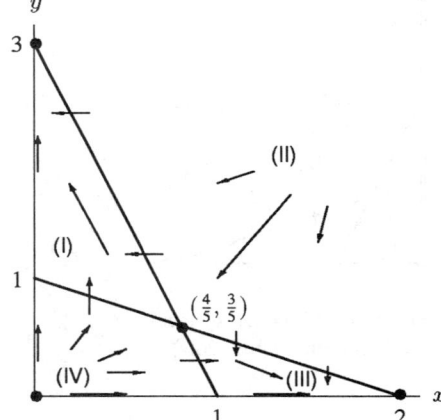

Figure 10.42: General directions of trajectories and equilibrium points (dots)

Figure 10.42 shows that if the initial point is in sector (I), the trajectory heads towards the equilibrium point $(0, 3)$. Similarly, if the trajectory begins in sector (III), then it heads towards the equilibrium $(2, 0)$ over time. If the trajectory begins in sector (II) or (IV), it can go to any of the three equilibrium points $(2, 0)$, $(0, 3)$, or $\left(\frac{4}{5}, \frac{3}{5}\right)$.

9. (a) The nullclines are $P = 0$ or $P_1 + 3P_2 = 13$ (where $dP_1/dt = 0$) and $P = 0$ or $P_2 + 0.4P_1 = 6$ (where $dP_2/dt = 0$).
   (b) The phase plane in Figure 10.43 shows that P_2 will eventually exclude P_1 regardless of where the experiment starts so long as there were some P_2 originally. Consequently, the data points would have followed a trajectory that starts at the origin, crosses the first nullcline and goes left and upwards between the two nullclines to the point $P_1 = 0$, $P_2 = 6$.

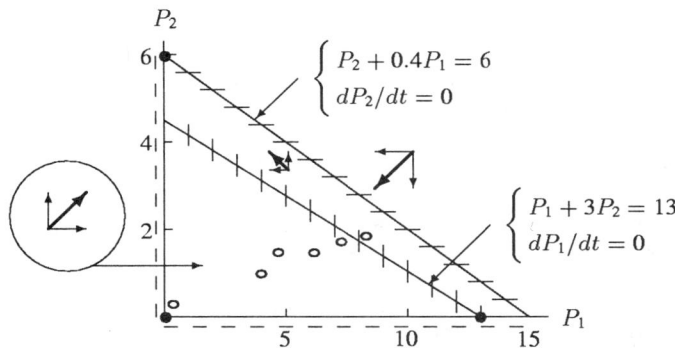

Figure 10.43: Nullclines and equilibrium points (dots) for Gauses's yeast data (hollow dots)

10. (a) In the equation for dx/dt, the term involving x, namely $-0.2x$, is negative meaning that as x increases, dx/dt decreases. This corresponds to the statement that the more a country spends on armaments, the less it wants to increase spending.

   On the other hand, since $+0.15y$ is positive, as y increases, dx/dt increases, corresponding to the fact that the more a country's opponent arms, the more the country will arm itself.

   The constant term, 20, is positive means that if both countries are unarmed initially, (so $x = y = 0$), then dx/dt is positive and so the country will start to arm. In other words, disarmament is not an equilibrium situation in this model.

   (b) The nullclines are shown in Figure 10.44. When $dx/dt = 0$, the trajectories are vertical (on the line $-0.2x + 0.15y + 20 = 0$); when $dy/dt = 0$ the trajectories are horizontal (on $0.1x - 0.2y + 40 = 0$). There is only one equilibrium point, $x = y = 400$.

   (c) In region I, try $x = 400, y = 0$, giving

$$\frac{dx}{dt} = -0.2(400) + 0.15(0) + 20 < 0$$
$$\frac{dy}{dt} = 0.1(400) - 0.2(0) + 4 - 0 > 0$$

   In region II, try $x = 500, y = 500$, giving

$$\frac{dx}{dt} = -0.2(500) + 0.15(500) + 20 < 0$$
$$\frac{dy}{dt} = 0.1(500) - 0.2(500) + 40 < 0$$

   In region III, try $x = 0, y = 400$, giving

$$\frac{dx}{dt} = -0.2(0) + 0.15(400) + 20 > 0$$
$$\frac{dy}{dt} = 0.1(0) - 0.2(400) + 40 < 0$$

   In region IV, try $x = 0, y = 0$, giving

$$\frac{dx}{dt} = -0.2(0) + 0.15(0) + 20 > 0$$
$$\frac{dy}{dt} = 0.1(0) - 0.2(0) + 40 > 0$$

   See Figure 10.44.

(d) The one equilibrium point is stable.

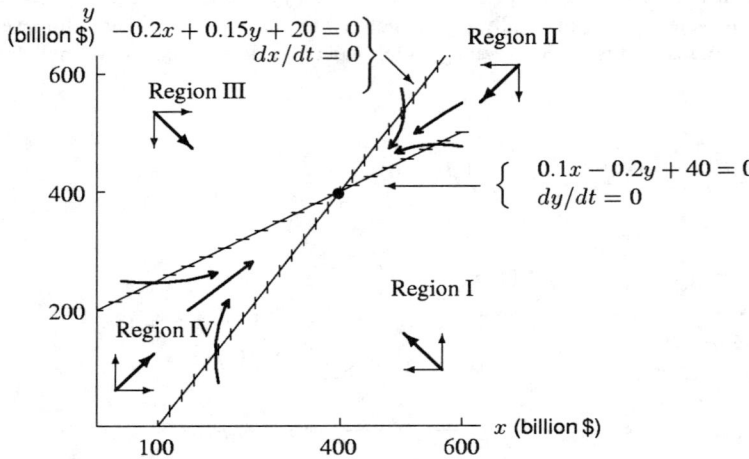

Figure 10.44: Nullclines and equilibrium point(dot) for arms race

(e) If both sides disarm, then both sides spend $0. Thus initially $x = y = 0$, and $dx/dt = 20$ and $dy/dt = 40$. Since both dx/dt and dy/dt are positive, both sides start arming. Figure 10.44 shows that they will both arm until each is spending about $400 billion.

(f) If the country spending y billion is unarmed, then $y = 0$ and the corresponding point on the phase plane is on the x-axis. Any trajectory starting on the x-axis tends towards the equilibrium point $x = y = 400$. Similarly, a trajectory starting on the y-axis represents the other country being unarmed; such a trajectory also tends to the same equilibrium point.

Thus, if either side disarms unilaterally, that is, if we start out with one of the countries spending nothing, then over time, they will still both end up spending roughly $400 billion.

(g) This model predicts that, in the long run, both countries will spend near to $400 billion, no matter where they start.

11. (a)

$$\frac{dx}{dt} = 0 \text{ when } x = \frac{10.5}{0.45} = 23.3$$

$$\frac{dy}{dt} = 0 \text{ when } 8.2x - 0.8y - 142 = 0$$

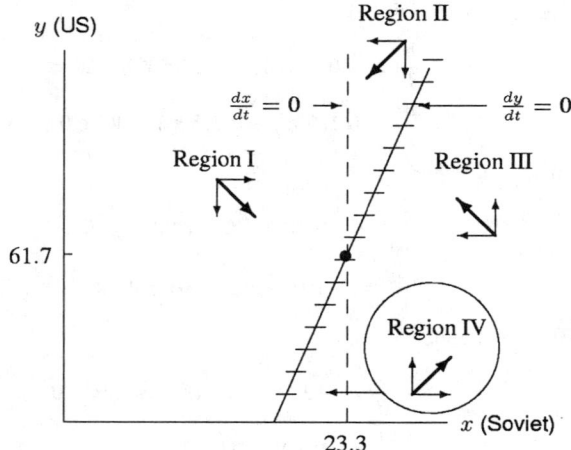

Figure 10.45: Nullclines and equilibrium point (dot) for US-Soviet arms race

There is an equilibrium point where the trajectories cross at $x = 23.3, y = 61.7$
In region I, $\frac{dx}{dt} > 0, \frac{dy}{dt} < 0$.
In region II, $\frac{dx}{dt} < 0, \frac{dy}{dt} < 0$.
In region III, $\frac{dx}{dt} < 0, \frac{dy}{dt} > 0$.
In region IV, $\frac{dx}{dt} > 0, \frac{dy}{dt} > 0$.

(b)

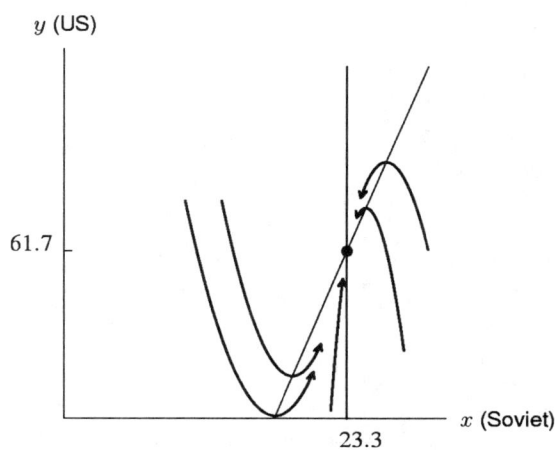

Figure 10.46: Trajectories for US-Soviet arms race.

(c) All the trajectories tend towards the equilibrium point $x = 23.3, y = 61.7$. Thus the model predicts that in the long run the arms race will level off with the Soviet Union spending 23.3 billion dollars a year on arms and the US 61.7 billion dollars.

(d) As the model predicts, yearly arms expenditure did tend towards 23 billion for the Soviet Union and 62 billion for the US.

APPENDIX

Solutions for Section A

1. The graph is

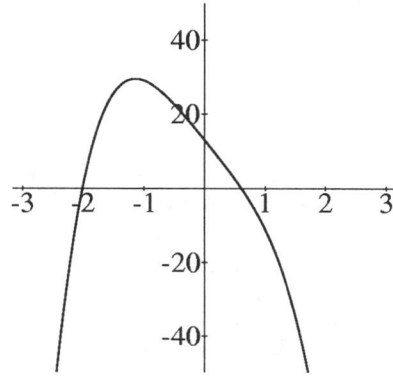

   (a) The range appears to be $y \leq 30$.
   (b) The function has two zeros.

2. (a) The root is between 0.3 and 0.4, at about 0.35.
   (b) The root is between 1.5 and 1.6, at about 1.55.
   (c) The root is between -1.8 and -1.9, at about -1.85.

3. The root occurs at about -1.05

4. The root is between -1.7 and -1.8, at about -1.75.

5. The largest root is at about 2.5.

6. There is one root at $x = -1$ and another at about $x = 1.35$.

7. There is one real root at about $x = 1.05$.

8. The root occurs at about 0.9, since the function changes sign between 0.8 and 1.

9. Using a graphing calculator, we see that when x is around 0.45, the graphs intersect.

10. The root occurs between 0.6 and 0.7, at about 0.65.

11. The root occurs between 1.2 and 1.4, at about 1.3.

12. Zoom in on graph: $t = \pm 0.824$. [Note: t must be in radians; one must zoom in two or three times.]

13. (a) Only one real zero, at about $x = -1.15$.
    (b) Three real zeros: at $x = 1$, and at about $x = 1.41$ and $x = -1.41$.

14. First, notice that $f(3) \approx 0.5 > 0$ and that $f(4) \approx -0.25 < 0$.
    1st iteration: $f(3.5) > 0$, so a zero is between 3.5 and 4.
    2nd iteration: $f(3.75) < 0$, so a zero is between 3.5 and 3.75.
    3rd iteration: $f(3.625) < 0$, so a zero is between 3.5 and 3.625.
    4th iteration: $f(3.588) < 0$, so a zero is between 3.5 and 3.588.
    5th iteration: $f(3.545) > 0$, so a zero is between 3.545 and 3.588.
    6th iteration: $f(3.578) < 0$, so a zero is between 3.567 and 3.578.
    7th iteration: $f(3.572) > 0$, so a zero is between 3.572 and 3.578.
    8th iteration: $f(3.575) > 0$, so a zero is between 3.575 and 3.578.

    Thus we know that, rounded to two places, the value of the zero must be 3.58. We know that this is the largest zero of $f(x)$ since $f(x)$ approaches -1 for larger values of x.

15. (a) Let $F(x) = \sin x - 2^{-x}$. Then $F(x) = 0$ will have a root where $f(x)$ and $g(x)$ cross. The first positive value of x for which the functions intersect is $x \approx 0.7$.
    (b) The functions intersect for $x \approx 0.4$.

16. The graph is

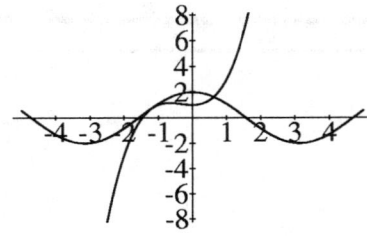

We find one zero at about 0.6. It looks like there might be another one at about -1.2, but zoom in close... closer... closer, and you'll see that though the graphs are very close together, they do not touch, and so there is no zero near -1.2. Thus the zero at about 0.6 is the only one. (How do you know there are no other zeros off the screen?)

17. (a) Since f is continuous, there must be one zero between $\theta = 1.4$ and $\theta = 1.6$, and another between $\theta = 1.6$ and $\theta = 1.8$. These are the only clear cases. We might also want to investigate the interval $0.6 \leq \theta \leq 0.8$ since $f(\theta)$ takes on values close to zero on at least part of this interval. Now, $\theta = 0.7$ is in this interval, and $f(0.7) = -0.01 < 0$, so f changes sign twice between $\theta = 0.6$ and $\theta = 0.8$ and hence has two zeros on this interval (assuming f is not *really* wiggly here, which it's not). There are a total of 4 zeros.

(b) As an example, we find the zero of f between $\theta = 0.6$ and $\theta = 0.7$. $f(0.65)$ is positive; $f(0.66)$ is negative. So this zero is contained in $[0.65, 0.66]$. The other zeros are contained in the intervals $[0.72, 0.73]$, $[1.43, 1.44]$, and $[1.7, 1.71]$.

(c) You've found all the zeros. A picture will confirm this; see Figure A.1.

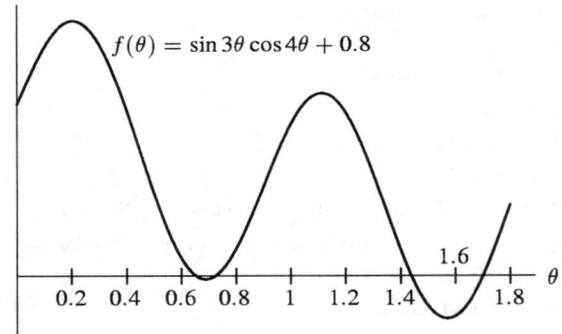

Figure A.1

18. (a) There appear to be two solutions: one on the interval from 1.13 to 1.14 and one on the interval from 1.08 to 1.09. From 1.13 to 1.14, $\frac{x^3}{\pi^3}$ increases from 0.0465 to 0.0478 while $(\sin 3x)(\cos 4x)$ decreases from 0.0470 to 0.0417, so they must cross in between. Similarly, going from 1.08 to 1.09, $\frac{x^3}{\pi^3}$ increases from 0.0406 to 0.0418 while $(\sin 3x)(\cos 4x)$ increases from 0.0376 to 0.0442. Thus the difference between the two changes sign over that interval, so their difference must be zero somewhere in between.

(b) Reasonable estimates are $x = 1.085$ and $x = 1.131$.

19. (a) The first ten results are:

n	0	1	2	3	4	5	6	7	8
1	3.14159	5.05050	5.50129	5.56393	5.57186	5.57285	5.57297	5.57299	5.57299

(b) The solution is $x \approx 5.573$. We started with an initial guess of 1, and kept repeating the given procedure until our values converged to a limit at around 5.573. For each number on the table, the procedure was in essence asking the question "Does this number equal 4 times the arctangent of itself?" and then correcting the number by repeating the question for 4 times the arctangent of the number.

(c) P_0 represents our initial guess of $x = 1$ (on the line $y = x$). P_1 is 4 times the arctangent of 1. If we now use take this value for P_1 and slide it horizontally back to the line $y = x$, we can now use this as a new guess, and call it P_2. P_3, of course, represents 4 times the arctangent of P_2, and so on. Another way to make sense of this diagram is to consider the function $F(x) = 4 \arctan x - x$. On the diagram, this difference is represented by the vertical lines connecting P_0 and P_1, P_2 and P_3 and so on. Notice how these lines (and hence the difference between $\arctan x$ and x) get smaller as we approach the intersection point, where $F(x) = 0$.

(d) For an initial guess of $x = 10$, the procedure gives a decreasing sequence which converges (more quickly) to the same value of about 5.573. Graphically, our initial guess of P_0 will lie to the right of the intersection on the line $y = x$. The iteration procedure gives us a sequence of $P_1, P_2, \ldots$ that zigzags to the left, toward the intersection point. For an initial guess of $x = -10$, the procedure gives an increasing sequence converging to the other intersection point of these two curves at $x \approx -5.573$. Graphically, we get a sequence which is a reflection through the origin of the sequence we got for an initial guess of $x = 10$. This is so because both $y = x$ and $y = \arctan x$ are odd functions.

20. Starting with $x = 0$, and repeatedly taking the cosine, we get the numbers below. Continuing until the first three decimal places remain fixed under iteration, we have this list and diagram:

x	$\cos x$
0	0.735069
1	0.7401473
0.5403023	0.7356047
0.8575532	0.7414251
0.6542898	0.7375069
0.7934804	0.7401473
0.7013688	0.7383692
0.7639597	0.7395672
0.7221024	0.7387603
0.7504178	0.7393039
0.7314043	0.7389378
0.7442374	etc.

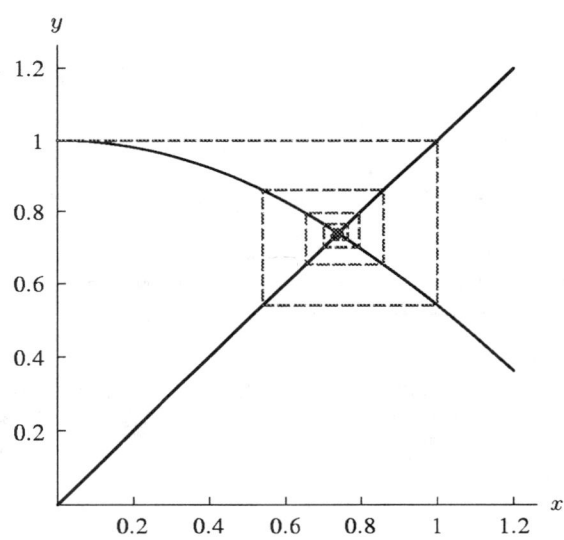

21.

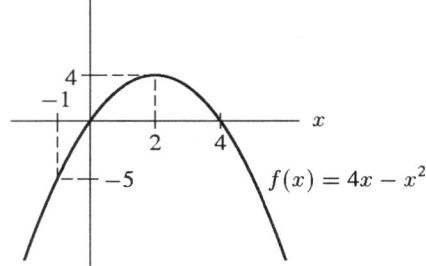

Bounded and $-5 \leq f(x) \leq 4$.

22.

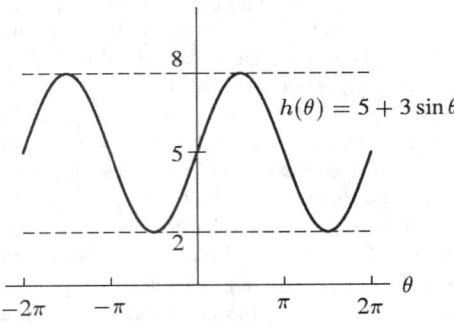

Bounded and $2 \leq h(\theta) \leq 8$.

23.

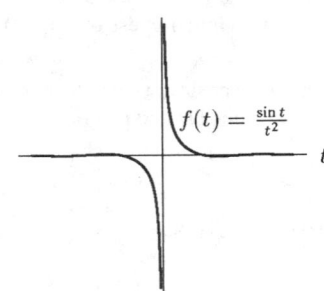

Not bounded because $f(t)$ goes to infinity as t goes to 0.

Solutions for Section B

1. For $20 \leq x \leq 100$, $0 \leq y \leq 1.2$, this function looks like a horizontal line at $y \approx 1.0725$ (In fact, the graph approaches this line from below.) Now, $e^{0.07} \approx 1.0725$, which strongly suggests that, as we already know, as $x \to \infty$, $\left(1 + \frac{0.07}{x}\right)^x \to e^{0.07}$.

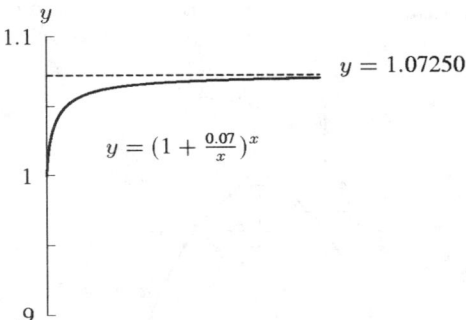

2. We use the equation $B = Pe^{rt}$, where P is the initial principal, t is the time in years the deposit is in the account, r is the annual interest rate and B is the balance. After $t = 5$ years, we will have

$$B = (10{,}000)e^{(0.08)5} = (10{,}000)e^{0.4} \approx \$14{,}918.25.$$

3. (a) (i)
$$\left(1 + \frac{0.05}{1000}\right)^{1000} = 1.05126978\ldots,$$
so the effective annual yield is $5.126978\ldots\%$.

(ii)
$$\left(1 + \frac{0.05}{10000}\right)^{10000} = 1.05127096\ldots,$$
so the effective annual yield is $5.127096\ldots\%$.

(iii)
$$\left(1 + \frac{0.05}{100000}\right)^{100000} = 1.05127108\ldots,$$
so the effective annual yield is $5.127108\ldots\%$.

(b) The effective annual rates in part (a) are closing in on 5.127%, so this is the effective annual yield for a 5% annual rate compounded continuously.

(c) $e^{0.05} = 1.05127109\ldots$ Since continuous compounding is equivalent to multiplying by $e^{0.05}$, the effective annual yield for continuous compounding is $0.05127109\ldots \approx 5.127\%$.

4. (a)
$$\left(1 + \frac{0.04}{10000}\right)^{10000} \approx 1.0408107$$
$$\left(1 + \frac{0.04}{100000}\right)^{100000} \approx 1.0408108$$
$$\left(1 + \frac{0.04}{1000000}\right)^{1000000} \approx 1.0408108$$

Effective annual yield: 4.08108%

(b) $e^{0.04} \approx 1.048108$ as expected.

5. $e^{0.06} \approx 1.0618365$, so the effective annual yield $\approx 6.18365\%$.

6. (a) $e^{0.06} = 1.0618365$ which means the bank balance has increased by approximately 6.18%.
(b) $e^{0.06t} = 2$, so $t = \frac{\ln 2}{0.06} = 11.55$ years.
(c) $e^{rt} = 2$ so $t = \frac{\ln 2}{r}$.

7. (a) Since we have 6% annual interest compounded continuously,
$$P = P_0 e^{0.06t}$$
$$2000 = 1000 e^{0.06t}$$
$$2 = e^{0.06t}$$
$$0.06t = \ln 2$$
$$t = \frac{\ln 2}{0.06} = 11.55 \text{ years}.$$

(b) The investment doubles approximately every 11.55 years, so
$$P = P_0(2)^{t/11.55}.$$

8. $e^{0.12} \approx 1.1274969$, so the effective annual yield $\approx 12.74969\%$.

9. (a) Using the formula $A = A_0(1 + \frac{r}{n})^{nt}$, we have $A = 10^6(1 + \frac{1}{12})^{12} \approx 10^6(2.61303529) \approx 2{,}613{,}035$ zaïre after one year.

(b) Compounding daily, $A = 10^6(1 + \frac{1}{365})^{365} \approx 10^6(2.714567) \approx 2{,}714{,}567$ zaïre. Compounding hourly, $A = 10^6(1 + \frac{1}{8760})^{8760} \approx 10^6(2.7181267) \approx 2{,}718{,}127$ zaïre. Compounding each minute, $A = 10^6(1 + \frac{1}{525600})^{525600} \approx 10^6(2.718280) \approx 2{,}718{,}280$ zaïre.

(c) The amount does not seem to be increasing without bound, but rather it seems to level off at a value just over $2{,}718{,}000$ zaïre. A close upper limit might be $2{,}718{,}300$ (amounts may vary). In fact, the limit is $(e \times 10^6)$ zaïre.

10. We know that for a given annual rate, the higher the frequency of compounding, the higher the effective annual yield. So the effective yield of (a) will be greater than that of (c) which is greater than that of (b). Also, the effective annual yield of (e) will be greater than that of (d). Now the effective annual yield of (e) will be less than the effective annual yield of 5.5% annual rate, compounded twice a year, and the latter will be less than the yield from (b). Thus $d < e < b < c < a$. Matching these up with our choices, we get
(d) I, (e) II, (b) III, (c) IV, (a) V.

11. (a) At the end of the year, the landlord's investment amounts to $\$1000e^{0.06} \approx \$1061.84$, and so has earned $61.84 in interest. Each year, the landlord pays the tenant $\$1000(0.05) = \$50$. Therefore the landlord made $11.84 in interest.

    (b) At the end of the year, the landlord's investment amounts to $\$1000e^{0.04} = \$1040.81$, and so has earned $40.81 in interest. By paying the tenant $50, the landlord loses $9.19.

Solutions for Section C

1. (1,0)
2. (0,0)
3. (−2, 0)
4. (−1, −1)
5. $(\frac{5\sqrt{3}}{2}, -\frac{5}{2})$
6. (0, 3)
7. (cos 1, sin 1)
8. $r = \sqrt{1^2 + 0^2} = 1$, $\theta = 0$.
9. $r = \sqrt{0^2 + 2^2} = 2$, $\theta = \pi/2$.
10. $r = \sqrt{1^2 + 1^2} = \sqrt{2}$.
    $\tan\theta = 1/1 = 1$. Since the point is in the first quadrant, $\theta = \pi/4$.
11. $r = \sqrt{(-1)^2 + 1^2} = \sqrt{2}$.
    $\tan\theta = (-1)/1 = -1$. Since the point is in the second quadrant, $\theta = 3\pi/4$.
12. $r = \sqrt{(-3)^2 + (-3)^2} = 4.2$.
    $\tan\theta = (-3/-3) = 1$. Since the point is in the third quadrant, $\theta = 5\pi/4$.
13. $r = \sqrt{(0.2)^2 + (-0.2)^2} = 0.28$.
    $\tan\theta = 0.2/(-0.2) = -1$. Since the point is in the fourth quadrant, $\theta = 7\pi/4$. (Alternatively $\theta = -\pi/4$.)
14. $r = \sqrt{3^2 + 4^2} = 5$, $\tan\theta = 4/3$. The point is in the first quadrant, so $\theta = 0.92$.
15. $r = \sqrt{(-3)^2 + 1^2} = 3.16$, $\tan\theta = 1/(-3)$. Since the point is in the second quadrant $\theta = 2.82$.
16. Substituting $r = \sqrt{x^2 + y^2}$ into $r = 1$ gives $\sqrt{x^2 + y^2} = 1$. This means $x^2 + y^2 = 1$, which is a circle of radius 1 centered at the origin. See figure below.

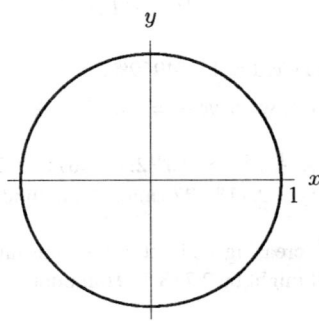

APPENDIX C SOLUTIONS **575**

17. Putting $\theta = \pi/3$ into $\tan\theta = y/x$ gives $\sqrt{3} = y/x$, or $y = \sqrt{3}x$. This is a line through the origin of slope $\sqrt{3}$.

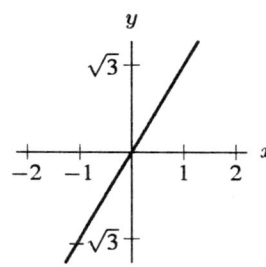

18. For $r = \theta/10$, the radius r increases as the angle θ winds around the origin, so this is a spiral.

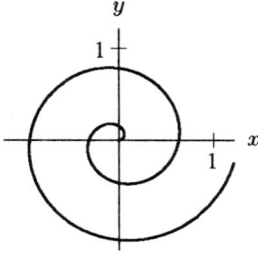

19. We can rewrite $r = 2/\cos\theta$ as $2 = r\cos\theta = x$. So this is a vertical line at $x = 2$.

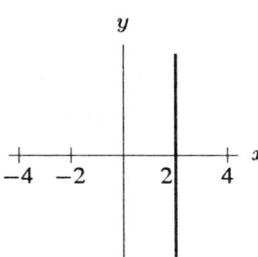

20. For $r = 2\cos\theta$, we multiply by r and get $r^2 = 2r\cos\theta$. Substituting $r^2 = x^2 + y^2$ and $x = r\cos\theta$ gives $x^2 + y^2 = 2x$. We now complete the square:

$$x^2 - 2x + 1 + y^2 = 1, \quad \text{so} \quad (x-1)^2 + y^2 = 1.$$

This is a circle of radius 1 centered at the point $(1, 0)$.

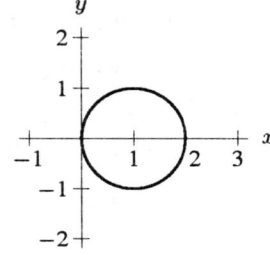

21. We multiply $r = \sin\theta/\cos^2\theta$ by $r\cos^2\theta$ and get $r^2\cos^2\theta = r\sin\theta$. Since $r\cos\theta = x$ and $r\sin\theta = y$, we have $y = x^2$, a parabola.

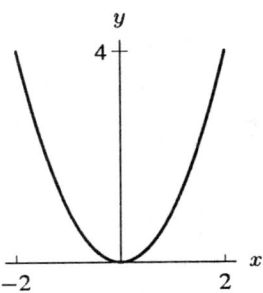

22. For each pair of Cartesian coordinates, there is more than one pair of polar coordinates for that point. For example, if $(x, y) = (1, 0)$ then $(r, \theta) = (1, 0)$, $(r, \theta) = (1, 2\pi)$, and $(r, \theta) = (1, 4\pi)$ all represent the same point.

Solutions for Section D

1. $2e^{i\frac{\pi}{2}}$
2. $5e^{i\pi}$
3. $\sqrt{2}e^{i\frac{\pi}{4}}$
4. $5e^{i4.069}$
5. $0e^{i\theta}$, for any θ.
6. $e^{3\pi i/2}$
7. $\sqrt{10}e^{i\theta}$, where $\theta = \arctan(-3) \approx -1.249 + \pi = 1.893$ is an angle in the second quadrant.
8. $13e^{i\theta}$, where $\theta = \arctan(-\frac{12}{5}) \approx -1.176$ is an angle in the fourth quadrant.
9. $-3 - 4i$
10. $-11 + 29i$
11. $-5 + 12i$
12. $1 + 3i$
13. $\frac{1}{4} - \frac{9i}{8}$
14. $3 - 6i$
15. $\cos\frac{2\pi}{3} + i\sin\frac{2\pi}{3} = -\frac{1}{2} + i\frac{\sqrt{3}}{2}$
16. $\cos\frac{\pi}{6} + i\sin\frac{\pi}{6} = \frac{\sqrt{3}}{2} + \frac{i}{2}$ is one solution.
17. $5^3(\cos\frac{3\pi}{2} + i\sin\frac{3\pi}{2}) = -125i$
18. $\sqrt[4]{10}\cos\frac{\pi}{8} + i\sqrt[4]{10}\sin\frac{\pi}{8}$ is one solution.
19. One value of $\sqrt{i}$ is $\sqrt{e^{i\frac{\pi}{2}}} = (e^{i\frac{\pi}{2}})^{\frac{1}{2}} = e^{i\frac{\pi}{4}} = \cos\frac{\pi}{4} + i\sin\frac{\pi}{4} = \frac{\sqrt{2}}{2} + i\frac{\sqrt{2}}{2}$
20. One value of $\sqrt{-i}$ is $\sqrt{e^{i\frac{3\pi}{2}}} = (e^{i\frac{3\pi}{2}})^{\frac{1}{2}} = e^{i\frac{3\pi}{4}} = \cos\frac{3\pi}{4} + i\sin\frac{3\pi}{4} = -\frac{\sqrt{2}}{2} + i\frac{\sqrt{2}}{2}$
21. One value of $\sqrt[3]{i}$ is $\sqrt[3]{e^{i\frac{\pi}{2}}} = (e^{i\frac{\pi}{2}})^{\frac{1}{3}} = e^{i\frac{\pi}{6}} = \cos\frac{\pi}{6} + i\sin\frac{\pi}{6} = \frac{\sqrt{3}}{2} + \frac{i}{2}$
22. One value of $\sqrt{7i}$ is $\sqrt{7e^{i\frac{\pi}{2}}} = (7e^{i\frac{\pi}{2}})^{\frac{1}{2}} = \sqrt{7}e^{i\frac{\pi}{4}} = \sqrt{7}\cos\frac{\pi}{4} + i\sqrt{7}\sin\frac{\pi}{4} = \frac{\sqrt{14}}{2} + i\frac{\sqrt{14}}{2}$
23. $(1 + i)^{100} = (\sqrt{2}e^{i\frac{\pi}{4}})^{100} = (2^{\frac{1}{2}})^{100}(e^{i\frac{\pi}{4}})^{100} = 2^{50} \cdot e^{i \cdot 25\pi} = 2^{50}\cos 25\pi + i2^{50}\sin 25\pi = -2^{50}$
24. One value of $(1 + i)^{2/3}$ is $(\sqrt{2}e^{i\frac{\pi}{4}})^{2/3} = (2^{\frac{1}{2}}e^{i\frac{\pi}{4}})^{\frac{2}{3}} = \sqrt[3]{2}e^{i\frac{\pi}{6}} = \sqrt[3]{2}\cos\frac{\pi}{6} + i\sqrt[3]{2}\sin\frac{\pi}{6} = \sqrt[3]{2} \cdot \frac{\sqrt{3}}{2} + i\sqrt[3]{2} \cdot \frac{1}{2}$
25. One value of $(-4 + 4i)^{2/3}$ is $[\sqrt{32}e^{(i3\pi/4)}]^{(2/3)} = (\sqrt{32})^{2/3}e^{(i\pi/2)} = 2^{5/3}\cos\frac{\pi}{2} + i2^{5/3}\sin\frac{\pi}{2} = 2i\sqrt[3]{4}$

26. One value of $(\sqrt{3}+i)^{1/2}$ is $(2e^{i\frac{\pi}{6}})^{1/2} = \sqrt{2}e^{i\frac{\pi}{12}} = \sqrt{2}\cos\frac{\pi}{12} + i\sqrt{2}\sin\frac{\pi}{12} \approx 1.366 + 0.366i$

27. One value of $(\sqrt{3}+i)^{-1/2}$ is $(2e^{i\frac{\pi}{6}})^{-1/2} = \frac{1}{\sqrt{2}}e^{i(-\frac{\pi}{12})} = \frac{1}{\sqrt{2}}\cos(-\frac{\pi}{12}) + i\frac{1}{\sqrt{2}}\sin(-\frac{\pi}{12}) \approx 0.683 - 0.183i$

28. Since $\sqrt{5} + 2i = 3e^{i\theta}$, where $\theta = \arctan\frac{2}{\sqrt{5}} \approx 0.730$, one value of $(\sqrt{5}+2i)^{\sqrt{2}}$ is $(3e^{i\theta})^{\sqrt{2}} = 3^{\sqrt{2}}e^{i\sqrt{2}\theta} = 3^{\sqrt{2}}\cos\sqrt{2}\theta + i3^{\sqrt{2}}\sin\sqrt{2}\theta \approx 3^{\sqrt{2}}(0.513) + i3^{\sqrt{2}}(0.859) \approx 2.426 + 4.062i$

29. Substituting $A_1 = 2 - A_2$ into the second equation gives

$$(1-i)(2-A_2) + (1+i)A_2 = 0$$

so

$$2iA_2 = -2(1-i)$$
$$A_2 = \frac{-(1-i)}{i} = \frac{-i(1-i)}{i^2} = i(1-i) = 1+i$$

Therefore $A_1 = 2 - (1+i) = 1 - i$.

30. Substituting $A_2 = 2 - A_1$ into the second equation gives

$$(i-1)A_1 + (1+i)(2-A_1) = 0$$
$$iA_1 - A_1 - A_1 - iA_1 + 2 + 2i = 0$$
$$-2A_1 = -2 - 2i$$
$$A_1 = 1 + i$$

Substituting, we have

$$A_2 = 2 - A_1 = 2 - (1+i) = 1 - i.$$

31. (a)
$$z_1 z_2 = (-3 - i\sqrt{3})(-1 + i\sqrt{3}) = 3 + (\sqrt{3})^2 + i(\sqrt{3} - 3\sqrt{3}) = 6 - i2\sqrt{3}.$$
$$\frac{z_1}{z_2} = \frac{-3 - i\sqrt{3}}{-1 + i\sqrt{3}} \cdot \frac{-1 - i\sqrt{3}}{-1 - i\sqrt{3}} = \frac{3 - (\sqrt{3})^2 + i(\sqrt{3} + 3\sqrt{3})}{(-1)^2 + (\sqrt{3})^2} = \frac{i \cdot 4\sqrt{3}}{4} = i\sqrt{3}.$$

(b) We find (r_1, θ_1) corresponding to $z_1 = -3 - i\sqrt{3}$:

$$r_1 = \sqrt{(-3)^2 + (\sqrt{3})^2} = \sqrt{12} = 2\sqrt{3};$$
$$\tan\theta_1 = \frac{-\sqrt{3}}{-3} = \frac{\sqrt{3}}{3}, \text{ so } \theta_1 = \frac{7\pi}{6}.$$

(See Figure D.1.) Thus,
$$-3 - i\sqrt{3} = r_1 e^{i\theta_1} = 2\sqrt{3}\, e^{i\frac{7\pi}{6}}.$$

We find (r_2, θ_2) corresponding to $z_2 = -1 + i\sqrt{3}$:

$$r_2 = \sqrt{(-1)^2 + (\sqrt{3})^2} = 2;$$
$$\tan\theta_2 = \frac{\sqrt{3}}{-1} = -\sqrt{3}, \text{ so } \theta_2 = \frac{2\pi}{3}.$$

(See Figure D.2.) Thus,
$$-1 + i\sqrt{3} = r_2 e^{i\theta_2} = 2 e^{i\frac{2\pi}{3}}.$$

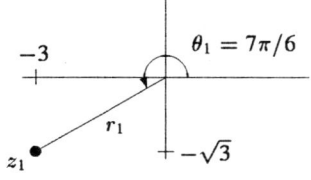

Figure D.1

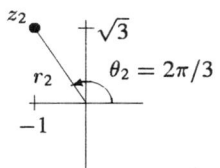

Figure D.2

We now calculate $z_1 z_2$ and $\dfrac{z_1}{z_2}$.

$$z_1 z_2 = \left(2\sqrt{3}e^{i\frac{7\pi}{6}}\right)\left(2e^{i\frac{2\pi}{3}}\right) = 4\sqrt{3}e^{i(\frac{7\pi}{6}+\frac{2\pi}{3})} = 4\sqrt{3}e^{i\frac{11\pi}{6}}$$

$$= 4\sqrt{3}\left[\cos\frac{11\pi}{6} + i\sin\frac{11\pi}{6}\right] = 4\sqrt{3}\left[\frac{\sqrt{3}}{2} - i\frac{1}{2}\right] = 6 - i2\sqrt{3}.$$

$$\frac{z_1}{z_2} = \frac{2\sqrt{3}e^{i\frac{7\pi}{6}}}{2e^{i\frac{2\pi}{3}}} = \sqrt{3}e^{i(\frac{7\pi}{6}-\frac{2\pi}{3})} = \sqrt{3}e^{i\frac{\pi}{2}}$$

$$= \sqrt{3}\left(\cos\frac{\pi}{2} + i\sin\frac{\pi}{2}\right) = i\sqrt{3}.$$

These agrees with the values found in (a).

32. If the roots are complex numbers, we must have $(2b)^2 - 4c < 0$ so $b^2 - c < 0$. Then the roots are

$$x = \frac{-2b \pm \sqrt{(2b)^2 - 4c}}{2} = -b \pm \sqrt{b^2 - c}$$

$$= -b \pm \sqrt{-1(c - b^2)}$$

$$= -b \pm i\sqrt{c - b^2}.$$

Thus, $p = -b$ and $q = \sqrt{c - b^2}$.

33. True, since $\sqrt{a}$ is real for all $a \geq 0$.
34. True, since $(x - iy)(x + iy) = x^2 + y^2$ is real.
35. False, since $(1 + i)^2 = 2i$ is not real.
36. False. Let $f(x) = x$. Then $f(i) = i$ but $f(\bar{i}) = \bar{i} = -i$.
37. True. We can write any nonzero complex number z as $re^{i\beta}$, where r and β are real numbers with $r > 0$. Since $r > 0$, we can write $r = e^c$ for some real number c. Therefore, $z = re^{i\beta} = e^c e^{i\beta} = e^{c+i\beta} = e^w$ where $w = c + i\beta$ is a complex number.
38. False, since $(1 + 2i)^2 = -3 + 4i$.
39.
$$1 = e^0 = e^{i(\theta-\theta)} = e^{i\theta}e^{i(-\theta)}$$
$$= (\cos\theta + i\sin\theta)(\cos(-\theta) + i\sin(-\theta))$$
$$= (\cos\theta + i\sin\theta)(\cos\theta - i\sin\theta)$$
$$= \cos^2\theta + \sin^2\theta$$

40. Using Euler's formula, we have:

$$e^{i(2\theta)} = \cos 2\theta + i\sin 2\theta$$

On the other hand,

$$e^{i(2\theta)} = (e^{i\theta})^2 = (\cos\theta + i\sin\theta)^2 = (\cos^2\theta - \sin^2\theta) + i(2\cos\theta\sin\theta)$$

Equating imaginary parts, we find

$$\sin 2\theta = 2\sin\theta\cos\theta.$$

41. Using Euler's formula, we have:

$$e^{i(2\theta)} = \cos 2\theta + i\sin 2\theta$$

On the other hand,

$$e^{i(2\theta)} = (e^{i\theta})^2 = (\cos\theta + i\sin\theta)^2 = (\cos^2\theta - \sin^2\theta) + i(2\cos\theta\sin\theta)$$

Equating real parts, we find

$$\cos 2\theta = \cos^2\theta - \sin^2\theta.$$

APPENDIX E SOLUTIONS   579

42. $\dfrac{d}{d\theta}(e^{i\theta}) = ie^{i\theta} = i(\cos\theta + i\sin\theta) = -\sin\theta + i\cos\theta$

Since in addition $\dfrac{d}{d\theta}(e^{i\theta}) = \dfrac{d}{d\theta}(\cos\theta + i\sin\theta) = \dfrac{d}{d\theta}(\cos\theta) + i\dfrac{d}{d\theta}(\sin\theta)$, by equating imaginary parts, we conclude that $\dfrac{d}{d\theta}\sin\theta = \cos\theta$.

43. $\dfrac{d^2}{d\theta^2}(e^{i\theta}) = \dfrac{d^2}{d\theta^2}(\cos\theta + i\sin\theta) = \dfrac{d^2}{d\theta^2}(\cos\theta) + i\dfrac{d^2}{d\theta^2}(\sin\theta)$

But $\dfrac{d^2}{d\theta^2}(e^{i\theta}) = i^2 e^{i\theta} = -e^{i\theta} = -\cos\theta - i\sin\theta$

Equating real parts, we find $\dfrac{d^2}{d\theta^2}(\cos\theta) = -\cos\theta$.

Solutions for Section E

1. (a) $f'(x) = 3x^2 + 6x + 3 = 3(x+1)^2$. Thus $f'(x) > 0$ everywhere except at $x = -1$, so it is increasing everywhere except perhaps at $x = -1$. The function is in fact increasing at $x = -1$ since $f(x) > f(-1)$ for $x > -1$, and $f(x) < f(-1)$ for $x < -1$.
   (b) The original equation can have at most one root, since it can only pass through the x-axis once if it never decreases. It must have one root, since $f(0) = -6$ and $f(1) = 1$.
   (c) The root is in the interval $[0, 1]$, since $f(0) < 0 < f(1)$.
   (d) Let $x_0 = 1$.

$$x_0 = 1$$
$$x_1 = 1 - \dfrac{f(1)}{f'(1)} = 1 - \dfrac{1}{12} = \dfrac{11}{12} \approx 0.917$$
$$x_2 = \dfrac{11}{12} - \dfrac{f\left(\tfrac{11}{12}\right)}{f'\left(\tfrac{11}{12}\right)} \approx 0.913$$
$$x_3 = 0.913 - \dfrac{f(0.913)}{f'(0.913)} \approx 0.913.$$

Since the digits repeat, they should be accurate. Thus $x \approx 0.913$.

2. Let $f(x) = x^3 - 50$. Then $f(\sqrt[3]{50}) = 0$, so we can use Newton's method to solve $f(x) = 0$ to obtain $x = \sqrt[3]{50}$. Since $f'(x) = 3x^2$, f' is always positive, and f is therefore increasing. Consequently, f has only one zero. Since $3^3 = 27 < 50 < 64 = 4^3$, let $x_0 = 3.5$. Then

$$x_0 = 3.5$$
$$x_1 = 3.5 - \dfrac{f(3.5)}{f'(3.5)} \approx 3.694$$

Continuing, we find

$$x_2 \approx 3.684$$
$$x_3 \approx 3.684.$$

Since the digits repeat, x_3 should be correct, as can be confirmed by calculator.

3. Let $f(x) = x^4 - 100$. Then $f(\sqrt[4]{100}) = 0$, so we can use Newton's method to solve $f(x) = 0$ to obtain $x = \sqrt[4]{100}$. $f'(x) = 4x^3$. Since $3^4 = 81 < 100 < 256 = 4^4$, try 3.1 as an initial guess.

$$x_0 = 3.1$$
$$x_1 = 3.1 - \dfrac{f(3.1)}{f'(3.1)} \approx 3.164$$
$$x_2 = 3.164 - \dfrac{f(3.164)}{f'(3.164)} \approx 3.162$$
$$x_3 = 3.162 - \dfrac{f(3.162)}{f'(3.162)} \approx 3.162$$

Thus $\sqrt[4]{100} \approx 3.162$.

4. Let $f(x) = x^3 - \frac{1}{10}$. Then $f(10^{-1/3}) = 0$, so we can use Newton's method to solve $f(x) = 0$ to obtain $x = 10^{-1/3}$. $f'(x) = 3x^2$. Since $\sqrt[3]{\frac{1}{27}} < \sqrt[3]{\frac{1}{10}} < \sqrt[3]{\frac{1}{8}}$, try $x_0 = \frac{1}{2}$. Then $x_1 = 0.5 - \frac{f(0.5)}{f'(0.5)} \approx 0.467$. Continuing, we find $x_2 \approx 0.464. x_3 \approx 0.464$. Since $x_2 \approx x_3$, $10^{-1/3} \approx 0.464$.

5. Let $f(x) = \sin x - 1 + x$; we want to find all zeros of f, because $f(x) = 0$ implies $\sin x = 1 - x$.

   Graphing $\sin x$ and $1 - x$ in Figure E.1, we see that $f(x)$ has one solution at $x \approx \frac{1}{2}$.

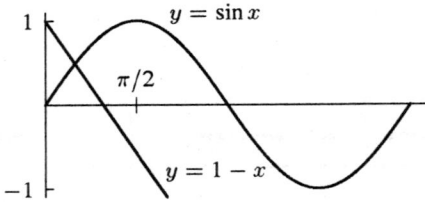

Figure E.1

Letting $x_0 = 0.5$, and using Newton's method, we have $f'(x) = \cos x + 1$, so that

$$x_1 = 0.5 - \frac{\sin(0.5) - 1 + 0.5}{\cos(0.5) + 1} \approx 0.511,$$

$$x_2 = 0.511 - \frac{\sin(0.511) - 1 + 0.511}{\cos(0.511) + 1} \approx 0.511.$$

Thus $\sin x = 1 - x$ has one solution at $x \approx 0.511$.

6. Let $f(x) = \cos x - x$. We want to find all zeros of f, because $f(x) = 0$ implies that $\cos x = x$. Since $f'(x) = -\sin x - 1$, f' is always negative (as $-\sin x$ never exceeds 1). This means f is always decreasing and consequently has at most 1 root. We now use Newton's method. Since $\cos 0 > 0$ and $\cos \frac{\pi}{2} < \frac{\pi}{2}$, $\cos x = x$ for $0 < x < \frac{\pi}{2}$. Thus, try $x_0 = \frac{\pi}{6}$.

$$x_1 = \frac{\pi}{6} - \frac{\cos \frac{\pi}{6} - \frac{\pi}{6}}{-\sin \frac{\pi}{6} - 1} \approx 0.7519,$$

$$x_2 \approx 0.7391,$$

$$x_3 \approx 0.7390.$$

$x_2 \approx x_3 \approx 0.739$. Thus $x \approx 0.739$ is the solution.

7. Let $f(x) = e^{-x} - \ln x$. Then $f'(x) = -e^{-x} - \frac{1}{x}$. We want to find all zeros of f, because $f(x) = 0$ implies that $e^{-x} = \ln x$. Since e^{-x} is always decreasing and $\ln x$ is always increasing, there must be only 1 solution. Since $e^{-1} > \ln 1 = 0$, and $e^{-e} < \ln e = 1$, then $e^{-x} = \ln x$ for some x, $1 < x < e$. Try $x_0 = 1$. We now use Newton's method.

$$x_1 = 1 - \frac{e^{-1} - 0}{-e^{-1} - 1} \approx 1.2689,$$

$$x_2 \approx 1.309,$$

$$x_3 \approx 1.310.$$

Thus $x \approx 1.310$ is the solution.

8. Let $f(x) = e^x \cos x - 1$. Then $f'(x) = -e^x \sin x + e^x \cos x$. Now we use Newton's method, guessing $x_0 = 1$ initially.

$$x_1 = 1 - \frac{f(1)}{f'(1)} \approx 1.5725$$

Continuing: $x_2 \approx 1.364$, $x_3 \approx 1.299$, $x_4 \approx 1.293$, $x_5 \approx 1.293$. Thus $x \approx 1.293$ is a solution. Looking at a graph of $f(x)$ suffices to convince us that there is only one solution.

9. Let $f(x) = \ln x - \frac{1}{x}$, so $f'(x) = \frac{1}{x} + \frac{1}{x^2}$.
   Now use Newton's method with an initial guess of $x_0 = 2$.
   $$x_1 = 2 - \frac{\ln 2 - \frac{1}{2}}{\frac{1}{2} + \frac{1}{4}} \approx 1.7425,$$
   $$x_2 \approx 1.763,$$
   $$x_3 \approx 1.763.$$

   Thus $x \approx 1.763$ is a solution. Since $f'(x) > 0$ for positive x, f is increasing: it must be the only solution.

10. (a) One zero in the interval $0.6 < x < 0.7$.
    (b) Three zeros in the intervals $-1.55 < x < -1.45$, $x = 0$, $1.45 < x < 1.55$.
    (c) Two zeros in the intervals $0.1 < x < 0.2$, $3.5 < x < 3.6$.

11. $f'(x) = 3x^2 + 1$. Since f' is always positive, f is everywhere increasing. Thus f has only one zero. Since $f(0) < 0 < f(1)$, $0 < x_0 < 1$. Pick $x_0 = 0.68$.
    $$x_0 = 0.68,$$
    $$x_1 = 0.6823278,$$
    $$x_2 \approx 0.6823278.$$

    Thus $x \approx 0.682328$ (rounded up) is a root. Since $x_1 \approx x_2$, the digits should be correct.

12. Let $f(x) = x^2 - a$, so $f'(x) = 2x$.
    Then by Newton's method, $x_{n+1} = x_n - \frac{x_n^2 - a}{2x_n}$
    For $a = 2$:
    $x_0 = 1, x_1 = 1.5, x_2 \approx 1.416, x_3 \approx 1.414215, x_4 \approx 1.414213$ so $\sqrt{2} \approx 1.4142$.
    For $a = 10$:
    $x_0 = 5, x_1 = 3.5, x_2 \approx 3.17857, x_3 \approx 3.162319, x_4 \approx 3.162277$ so $\sqrt{10} \approx 3.1623$.
    For $a = 1000$:
    $x_0 = 500, x_1 = 251, x_2 \approx 127.49203, x_3 \approx 67.6678, x_4 \approx 41.2229, x_5 \approx 32.7406, x_6 \approx 31.6418, x_7 \approx 31.62278, x_8 \approx 31.62277$ so $\sqrt{1000} \approx 31.6228$.
    For $a = \pi$:
    $x_0 = \frac{\pi}{2}, x_1 \approx 1.7853, x_2 \approx 1.7725$ $x_3 \approx 1.77245, x_4 \approx 1.77245$ so $\sqrt{\pi} \approx 1.77245$.

Solutions for Section F

1. Between times $t = 0$ and $t = 1$, x goes at a constant rate from 0 to 1 and y goes at a constant rate from 1 to 0. So the particle moves in a straight line from $(0, 1)$ to $(1, 0)$. Similarly, between times $t = 1$ and $t = 2$, it goes in a straight line to $(0, -1)$, then to $(-1, 0)$, then back to $(0, 1)$. So it traces out the diamond shown in Figure F.1.

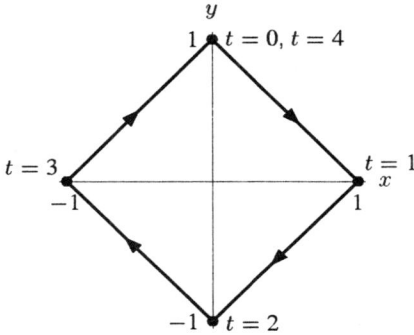

Figure F.1

2. This is like Example 2, except that the x-coordinate goes all the way to 2 and back. So the particle traces out the rectangle shown in Figure F.2.

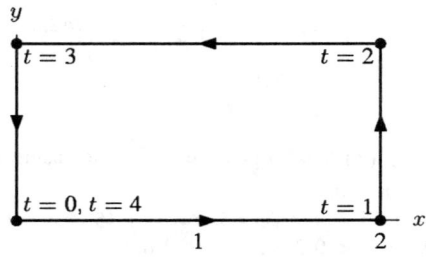

Figure F.2

3. Between times $t = 0$ and $t = 1$, x goes from -1 to 1, while y stays fixed at 1. So the particle goes in a straight line from $(-1, 1)$ to $(1, 1)$. Then both the x- and y-coordinates decrease at a constant rate from 1 to -1. So the particle goes in a straight line from $(1, 1)$ to $(-1, -1)$. Then it moves across to $(1, -1)$, then back diagonally to $(-1, 1)$. See Figure F.3.

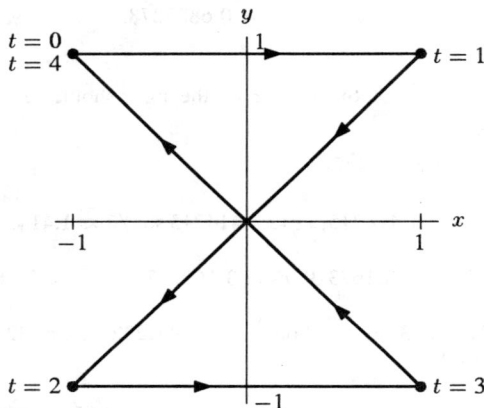

Figure F.3

4. As the x-coordinate goes at a constant rate from 2 to 0, the y-coordinate goes from 0 to 1, then down to -1, then back to 0. So the particle zigs and zags from $(2, 0)$ to $(1.5, 1)$ to $(1, 0)$ to $(.5, -1)$ to $(0, 0)$. Then it zigs and zags back again, forming the shape in Figure F.4.

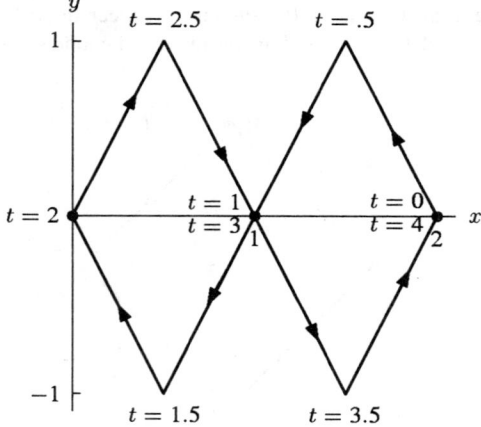

Figure F.4

5. The particle moves clockwise: For $0 \leq t \leq \frac{\pi}{2}$, we have $x = \cos t$ decreasing and $y = -\sin t$ decreasing. Similarly, for the time intervals $\frac{\pi}{2} \leq t \leq \pi$, $\pi \leq t \leq \frac{3\pi}{2}$, and $\frac{3\pi}{2} \leq t \leq 2\pi$, we see that the particle moves clockwise.

6. For $0 \leq t \leq \frac{\pi}{2}$, we have $x = \sin t$ increasing and $y = \cos t$ decreasing, so the motion is clockwise for $0 \leq t \leq \frac{\pi}{2}$. Similarly, we see that the motion is clockwise for the time intervals $\frac{\pi}{2} \leq t \leq \pi$, $\pi \leq t \leq \frac{3\pi}{2}$, and $\frac{3\pi}{2} \leq t \leq 2\pi$.

7. Let $f(t) = t^2$. The particle is moving clockwise when $f(t)$ is decreasing, that is, when $f'(t) = 2t < 0$, so when $t < 0$. The particle is moving counterclockwise when $f'(t) = 2t > 0$, so when $t > 0$.

8. Let $f(t) = t^3 - t$. The particle is moving clockwise when $f(t)$ is decreasing, that is, when $f'(t) = 3t^2 - 1 < 0$, and counterclockwise when $f'(t) = 3t^2 - 1 > 0$. That is, it moves clockwise when $-\sqrt{\frac{1}{3}} < t < \sqrt{\frac{1}{3}}$, between $(\cos((-\sqrt{\frac{1}{3}})^3 + \sqrt{\frac{1}{3}}), \sin((-\sqrt{\frac{1}{3}})^3 + \sqrt{\frac{1}{3}}))$ and $(\cos((\sqrt{\frac{1}{3}})^3 - \sqrt{\frac{1}{3}}), \sin((\sqrt{\frac{1}{3}})^3 - \sqrt{\frac{1}{3}}))$, and counterclockwise when $t < -\sqrt{\frac{1}{3}}$ or $t > \sqrt{\frac{1}{3}}$.

9. Let $f(t) = \ln t$. Then $f'(t) = \frac{1}{t}$. The particle is moving counterclockwise when $f'(t) > 0$, that is, when $t > 0$. Any other time, when $t \leq 0$, the position is not defined.

10. Let $f(t) = \cos t$. Then $f'(t) = -\sin t$. The particle is moving clockwise when $f'(t) < 0$, or $-\sin t < 0$, that is, when
$$2k\pi < t < (2k+1)\pi,$$
where k is an integer. The particle is otherwise moving counterclockwise, that is, when
$$(2k-1)\pi < t < 2k\pi,$$
where k is an integer. Actually, the particle does not fully trace out a circle. The range of $f(t)$ is $[-1, 1]$ so the particle oscillates between the points $(\cos(-1), \sin(-1))$ and $(\cos 1, \sin 1)$.

11. In all three cases, $y = x^2$, so that the motion takes place on the parabola $y = x^2$.

    In case (a), the x-coordinate always increases at a constant rate of one unit distance per unit time, so the equations describe a particle moving to the right on the parabola at constant horizontal speed.

    In case (b), the x-coordinate is never negative, so the particle is confined to the right half of the parabola. As t moves from $-\infty$ to $+\infty$, $x = t^2$ goes from ∞ to 0 to ∞. Thus the particle first comes down the right half of the parabola, reaching the origin $(0, 0)$ at time $t = 0$, where it reverses direction and goes back up the right half of the parabola.

    In case (c), as in case (a), the particle traces out the entire parabola $y = x^2$ from left to right. The difference is that the horizontal speed is not constant. This is because a unit change in t causes larger and larger changes in $x = t^3$ as t approaches $-\infty$ or ∞. The horizontal motion of the particle is faster when it is farther from the origin.

12. One possible answer is $x = 3\cos t, y = -3\sin t, 0 \leq t \leq 2\pi$.

13. One possible answer is $x = -2, y = t$.

14. One possible answer is $x = 2 + 5\cos t, y = 1 + 5\sin t, 0 \leq t \leq 2\pi$.

15. The parameterization $x = 2\cos t, y = 2\sin t, 0 \leq t \leq 2\pi$, is a circle of radius 2 traced out counterclockwise starting at the point $(2, 0)$. To start at $(-2, 0)$, put a negative in front of the first coordinate
$$x = -2\cos t \quad y = 2\sin t, \qquad 0 \leq t \leq 2\pi.$$
Now we must check whether this parameterization traces out the circle clockwise or counterclockwise. Since when t increases from 0, $\sin t$ is positive, the point (x, y) moves from $(-2, 0)$ into the second quadrant. Thus, the circle is traced out clockwise and so this is one possible parameterization.

16. The slope of the line is
$$m = \frac{3 - (-1)}{1 - 2} = -4.$$
The equation of the line with slope -4 through the point $(2, -1)$ is $y - (-1) = (-4)(x - 2)$, so one possible parameterization is $x = t$ and $y = -4t + 8 - 1 = -4t + 7$.

17. The ellipse $x^2/25 + y^2/49 = 1$ can be parameterized by $x = 5\cos t, y = 7\sin t, 0 \leq t \leq 2\pi$.

18. The parameterization $x = -3\cos t, y = 7\sin t, 0 \leq t \leq 2\pi$, starts at the right point but sweeps out the ellipse in the wrong direction (the y-coordinate becomes positive as t increases). Thus, a possible parameterization is $x = -3\cos(-t) = -3\cos t, y = 7\sin(-t) = -7\sin t, 0 \leq t \leq 2\pi$.

19. (a) If $t \geq 0$, we have $x \geq 2, y \geq 4$, so we get the part of the line to the right of and above the point $(2, 4)$.
    (b) When $t = 0, (x, y) = (2, 4)$. When $t = -1, (x, y) = (-1, -3)$. Restricting t to the interval $-1 \leq t \leq 0$ gives the part of the line between these two points.
    (c) If $x < 0$, giving $2 + 3t < 0$ or $t < -2/3$. Thus $t < -2/3$ gives the points on the line to the left of the y-axis.

584 APPENDIX /SOLUTIONS

20. (I) has a positive slope and so must be l_1 or l_2. Since its y-intercept is negative, these equations must describe l_2. (II) has a negative slope and positive x-intercept, so these equations must describe l_3.

21. (a) C_1 has center at the origin and radius 5, so $a = b = 0, k = 5$ or -5.
    (b) C_2 has center at $(0, 5)$ and radius 5, so $a = 0, b = 5, k = 5$ or -5.
    (c) C_3 has center at $(10, -10)$, so $a = 10, b = -10$. The radius of C_3 is $\sqrt{10^2 + (-10)^2} = \sqrt{200}$, so $k = \sqrt{200}$ or $k = -\sqrt{200}$.

22. It is a straight line through the point $(3, 5)$ with slope -1. A linear parameterization of the same line is $x = 3 + t$, $y = 5 - t$.

23. For $0 \leq t \leq 2\pi$

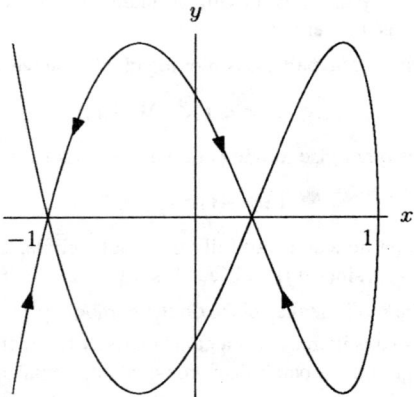

24. For $0 \leq t \leq 2\pi$

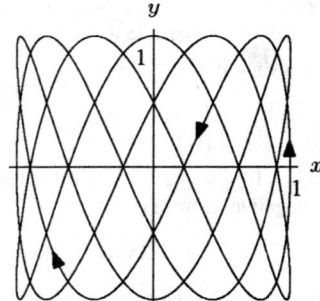

25. For $0 \leq t \leq 2\pi$

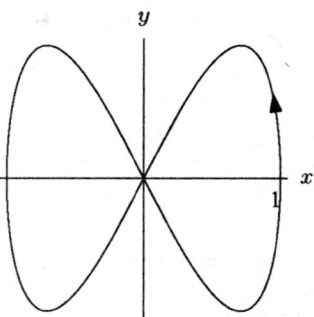

26. This curve never closes on itself. The plot for $0 \leq t \leq 8\pi$ is in Figure F.5.

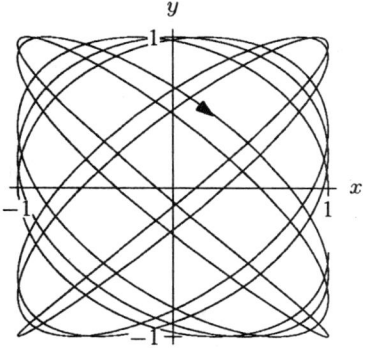

Figure F.5

27.

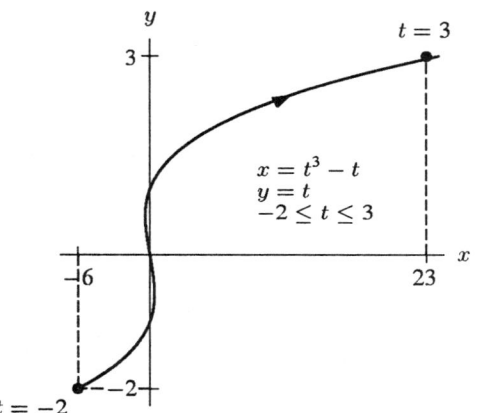

Figure F.6

The particle starts moving from left to right, then reverses its direction for a short time, then continues motion left to right. See Figure F.6.

28.

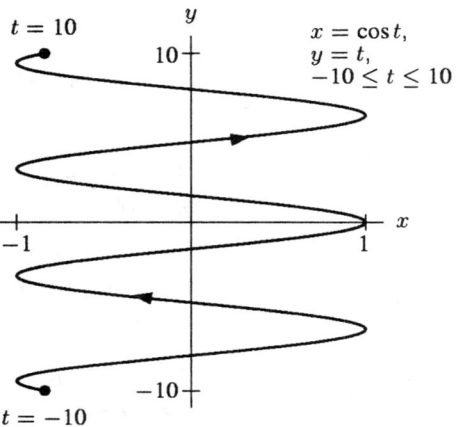

The particle moves back and forth between -1 and 1.

29.

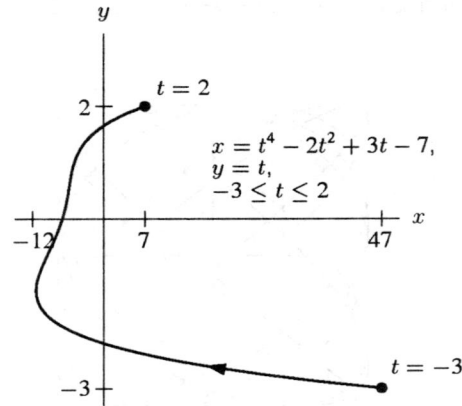

The particle starts moving to the left, reverses direction three times, then ends up moving to the right.

30.

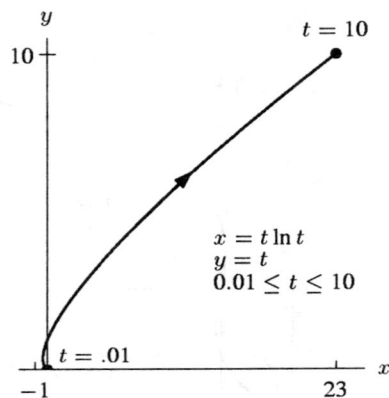

After a short move to the left, the particle moves steadily to the right.

Solutions for Section G

1. (a) Eliminating t between
$$x = 2 + t, \quad y = 4 + 3t$$
gives
$$y - 4 = 3(x - 2),$$
$$y = 3x - 2.$$

Eliminating t between
$$x = 1 - 2t, \quad y = 1 - 6t$$
gives
$$y - 1 = 3(x - 1),$$
$$y = 3x - 2.$$

Since both parametric equations give rise to the same equation in x and y, they both parameterize the same line.

(b) Slope = 3, y-intercept = -2.

2. (a) We get the part of the line with $x < 10$ and $y < 0$.
   (b) We get the part of the line between the points $(10, 0)$ and $(11, 2)$.

3. (a) The curve is a spiral as shown in Figure G.1.

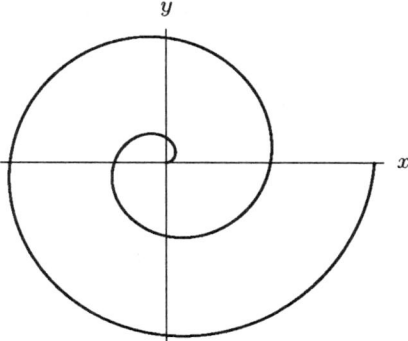

Figure G.1: The spiral $x = t \cos t, y = t \sin t$ for $0 \le t \le 4\pi$

(b) At $t = 2$, the position is $(2\cos 2, 2 \sin 2) = (-0.8323, 1.8186)$, and at $t = 2.01$ the position is $(2.01 \cos 2.01, 2.01 \sin 2.01) = (-0.8546, 1.8192)$. The distance between these points is

$$\sqrt{(-0.8546 - (-0.8323))^2 + (1.8192 - 1.8186)^2} \approx 0.022.$$

Thus the speed is approximately $0.022/0.01 \approx 2.2$.

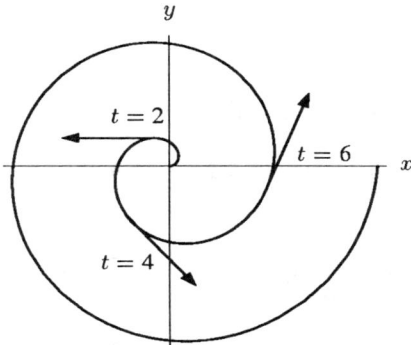

Figure G.2: The spiral $x = t \cos t, y = t \sin t$ and three velocity vectors

(c) Evaluating the exact formula

$$v = \sqrt{(\cos t - t \sin t)^2 + (\sin t + t \cos t)^2}$$

gives:

$$v(2) = \sqrt{(-2.235)^2 + (0.077)^2} = 2.2363.$$

4. We have $dx/dt = 2t$ and $dy/dt = 3t^2$. Therefore, the speed of the particle is

$$v = \sqrt{(\frac{dx}{dt})^2 + (\frac{dy}{dt})^2} = \sqrt{((2t)^2 + (3t^2)^2)} = |t| \cdot \sqrt{(4 + 9t^2)}.$$

The particle comes to a complete stop when its speed is 0, that is, if $t\sqrt{4 + 9t^2} = 0$, and so when $t = 0$.

5. We have $dx/dt = -2t\sin(t^2)$ and $dy/dt = 2t\cos(t^2)$. Therefore, the speed of the particle is given by

$$v = \sqrt{(-2t\sin(t^2))^2 + (2t\cos(t^2))^2}$$
$$= \sqrt{4t^2(\sin(t^2))^2 + 4t^2(\cos(t^2))^2}$$
$$= 2|t|\sqrt{\sin^2(t^2) + \cos^2(t^2)}$$
$$= 2|t|.$$

The particle comes to a complete stop when speed is 0, that is, if $2|t| = 0$, and so when $t = 0$.

6. We have
$$\frac{dx}{dt} = -2\sin 2t, \quad \frac{dy}{dt} = \cos t.$$

The speed is
$$v = \sqrt{4\sin^2(2t) + \cos^2 t}.$$

Thus, $v = 0$ when $\sin(2t) = \cos t = 0$, and so the particle stops when $t = \pm\pi/2, \pm 3\pi/2, \ldots$ or $t = (2n+1)\frac{\pi}{2}$, for any integer n.

7. We have
$$\frac{dx}{dt} = (2t - 2), \quad \frac{dy}{dt} = (3t^2 - 3).$$

The speed is given by:
$$v = \sqrt{(2t-2)^2 + (3t^2-3)^2}.$$

The particle stops when $2t - 2 = 0$ and $3t^2 - 3 = 0$. Since these are both satisfied only by $t = 1$, this is the only time that the particle stops.

8. At $t = 2$, the position is $(2^2, 2^3) = (4, 8)$, the velocity in the x-direction is $2 \cdot 2 = 4$, and the velocity in the y-direction is $3 \cdot 2^2 = 12$. So we want the line going through the point $(4, 8)$ at the time $t = 2$, with the given x- and y-velocities:

$$x = 4 + 4(t-2), \quad y = 8 + 12(t-2).$$

9. The length is
$$\int_1^2 \sqrt{(x'(t))^2 + (y'(t))^2 + (z'(t))^2}\, dt = \int_1^2 \sqrt{5^2 + 4^2 + (-1)^2}\, dt = \sqrt{42}.$$

This is the length of a straight line from the point $(8, 5, 2)$ to $(13, 9, 1)$.

10. We have
$$D = \int_0^1 \sqrt{(-e^t\sin(e^t))^2 + (e^t\cos(e^t))^2}\, dt$$
$$= \int_0^1 \sqrt{e^{2t}}\, dt = \int_0^1 e^t\, dt$$
$$= e - 1.$$

This is the length of the arc of a unit circle from the point $(\cos 1, \sin 1)$ to $(\cos e, \sin e)$—in other words between the angles $\theta = 1$ and $\theta = e$. The length of this arc is $(e - 1)$.

11. We have
$$D = \int_0^{2\pi} \sqrt{(-3\sin 3t)^2 + (5\cos 5t)^2}\, dt.$$

We cannot find this integral symbolically, but numerical methods show $D \approx 24.6$.

12. (a) In order for the particle to stop, its velocity $\vec{v} = (dx/dt)\vec{i} + (dy/dt)\vec{j}$ must be zero, so we solve for t such that $dx/dt = 0$ and $dy/dt = 0$, that is

$$\frac{dx}{dt} = 3t^2 - 3 = 3(t-1)(t+1) = 0,$$

$$\frac{dy}{dt} = 2t - 2 = 2(t-1) = 0.$$

The value $t = 1$ is the only solution. Therefore, the particle stops when $t = 1$ at the point $(t^3 - 3t, t^2 - 2t)|_{t=1} = (-2, -1)$.

(b) In order for the particle to be traveling straight up or down, the x-component of the velocity vector must be 0. Thus, we solve $dx/dt = 3t^2 - 3 = 0$ and obtain $t = \pm 1$. However, at $t = 1$ the particle has no vertical motion, as we saw in part (a). Thus, the particle is moving straight up or down only when $t = -1$. Since the velocity at time $t = -1$ is

$$\vec{v}(-1) = \frac{dx}{dt}\bigg|_{t=-1}\vec{i} + \frac{dy}{dt}\bigg|_{t=-1}\vec{j} = -4\vec{j},$$

the motion is straight down. The position at that time is $(t^3 - 3t, t^2 - 2t)|_{t=-1} = (2, 3)$.

(c) For horizontal motion we need $dy/dt = 0$. That happens when $dy/dt = 2t - 2 = 0$, and so $t = 1$. But from part (a) we also have $dx/dt = 0$ also at $t = 1$, so the particle is not moving at all when $t = 1$. Thus, there is no time when the motion is horizontal.

13. (a) The parametric equation describing Emily's motion is

$$x = 10\cos\left(\frac{2\pi}{20}t\right) = 10\cos\left(\frac{\pi}{10}t\right), \quad y = 10\sin\left(\frac{2\pi}{20}t\right) = 10\sin\left(\frac{\pi}{10}t\right) \quad z = \text{constant.}$$

Her velocity vector is

$$\vec{v} = \frac{dx}{dt}\vec{i} + \frac{dy}{dt}\vec{j} + \frac{dz}{dt}\vec{k} = -\pi\sin\left(\frac{\pi}{10}t\right)\vec{i} + \pi\cos\left(\frac{\pi}{10}t\right)\vec{j}.$$

Her speed is given by:

$$\|\vec{v}\| = \sqrt{\left(-\pi\sin\left(\frac{\pi}{10}t\right)\right)^2 + \left(\pi\cos\left(\frac{\pi}{10}t\right)\right)^2 + 0^2}$$

$$= \pi\sqrt{\sin^2\left(\frac{\pi}{10}t\right) + \cos^2\left(\frac{\pi}{10}t\right)}$$

$$= \pi\sqrt{1} = \pi \text{ m/sec},$$

which is independent of time (as we expected). This is certainly the long way to solve this problem though, since we could have simply divided the circumference of the circle (20π) by the time taken for a single rotation (20 seconds) to arrive at the same answer.

(b) When Emily drops the ball, it initially has Emily's velocity vector, but it immediately begins accelerating in the z-direction due to the force of gravity. The motion of the ball will then be tangential to the merry-go-round, curving down to the ground. In order to find the tangential component of the ball's motion, we must know Emily's velocity at the moment she dropped the ball. Then we can integrate the velocity and obtain the position of the ball. Assuming Emily drops the ball at time $t = 0$, her position and velocity vector are

$$\vec{r}(0) = 10\vec{i} + 3\vec{k} \text{ and } \vec{v}(0) = \pi\vec{j}.$$

Thus, the ball has velocity only in the y-direction when it is dropped. In the z-direction, we have

$$\text{Acceleration} = \frac{d^2z}{dt^2} = -9.8 \text{ m/sec}^2.$$

Since the initial velocity 0 and initial height 3, we have

$$z = 3 - 4.9t^2.$$

The ball touches the ground when $z = 0$, that is, when $t = 0.78$ sec. In that time, the ball also travels $\pi(0.78) = 2.45$ meters in the y-direction. So, the final position is $(10, 2.45, 0)$. The distance between this point and $P = (10, 0, 0)$ is 2.45 meters.

(c) The distance of the ball from Emily when it hits the ground is found by finding Emily's position at $t = 0.78$ sec and using the distance formula. Emily's position when the ball hits the ground is $(10\cos(0.078\pi), 10\sin(0.078\pi), 3) = (9.70, 2.43, 3)$. The distance between this point and the point where the ball struck the ground is:

$$d \approx \sqrt{(10 - 9.70)^2 + (2.45 - 2.43)^2 + (0 - 3)^2} = 3.01 \text{ meters.}$$

Note that the merry-go-round doesn't rotate very much in the 0.78 sec needed for the ball to reach the ground, so our answer makes sense.

14. (a) To find the equations of the moon's motion relative to the star, you must first calculate the equation of the planet's motion relative to the star, and then the moon's motion relative to the planet, and then add the two together.

The distance from the planet to the star is R, and the time to make one revolution is one unit, so the parametric equations for the planet relative to the star are $x = R\cos t$, $y = R\sin t$.

The distance from the moon to the planet is 1, and the time to make one revolution is twelve units, therefore, the parametric equations for the moon relative to the planet are $x = \cos 12t$, $y = \sin 12t$.

Adding these together, we get:

$$x = R\cos t + \cos 12t,$$
$$y = R\sin t + \sin 12t.$$

(b) For the moon to stop completely at time t, the velocity of the moon must be equal to zero. Therefore,

$$\frac{dx}{dt} = -R\sin t - 12\sin 12t = 0,$$
$$\frac{dy}{dt} = R\cos t + 12\cos 12t = 0.$$

There are many possible values to choose for R and t that make both of these equations equal to zero. We choose $t = \pi$, and $R = 12$.

(c) The graph with $R = 12$ is shown below.

15. Substituting $x = t$, and $y = t^2$ into the equation gives us $F = 1/(t^2 + t^4 + 1)$. To find the rate of change of temperature at time t, we differentiate F with respect to t:

$$\text{Rate of change of temperature} = \frac{dF}{dt} = -(t^2 + t^4 + 1)^{-2}(2t + 4t^3).$$

16. (a) If $\Delta t = t_{i+1} - t_i$ is small enough so that C_i is approximately a straight line, then we can make the linear approximations

$$x(t_{i+1}) \approx x(t_i) + x'(t_i)\Delta t,$$
$$y(t_{i+1}) \approx y(t_i) + y'(t_i)\Delta t,$$

and so

$$\text{Length of } C_i \approx \sqrt{(x(t_{i+1}) - x(t_i))^2 + (y(t_{i+1}) - y(t_i))^2}$$
$$\approx \sqrt{x'(t_i)^2(\Delta t)^2 + y'(t_i)^2(\Delta t)^2}$$
$$= \sqrt{x'(t_i)^2 + y'(t_i)^2}\Delta t.$$

(b) From point (a) we obtain the approximation

$$\text{Length of } C = \sum \text{length of } C_i$$
$$\approx \sum \sqrt{x'(t_i)^2 + y'(t_i)^2}\,\Delta t.$$

The approximation gets better and better as Δt approaches zero, and in the limit the sum becomes a definite integral:

$$\text{Length of } C = \lim_{\Delta t \to 0} \sum \sqrt{x'(t_i)^2 + y'(t_i)^2}\,\Delta t$$
$$= \int_a^b \sqrt{x'(t)^2 + y'(t)^2}\,dt.$$

Solutions for Section H

1. AN ORBITING SATELLITE: SOLUTION

Mathematics skills required: Writing formulas for piecewise defined functions, integration, using estimation techniques to evaluate integrals of functions given graphically, working with units of amperes (current), ampere-minutes and ampere-hours (total discharge).

Comments: This is a fairly straightforward problem, involving finding integrals of functions given in different ways.

Solution:

(a) For Operation 1, we have the following:
  (i) A plot of current use versus time is given in Figure H.1.

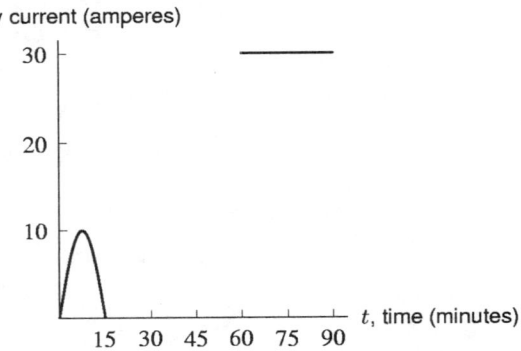

Figure H.1: Operation 1

The battery current function is given by the formulas

$$D(t) = \begin{cases} 10 \sin \dfrac{2\pi t}{30} & 0 \leq t \leq 15 \\ 0 & 15 \leq t \leq 60 \\ 30 & 60 \leq t \leq 90. \end{cases}$$

  (ii) The battery current function gives the rate at which the current is flowing. Thus, the total discharge is given by the integral of the battery current function:

$$\text{Total discharge} = \int_0^{90} D(t)\,dt = \int_0^{15} 10 \sin \frac{2\pi t}{30}\,dt + \int_{15}^{60} 0\,dt + \int_{60}^{90} 30\,dt$$
$$\approx 95.5 + 900$$
$$= 995.5 \text{ ampere-minutes} = 16.6 \text{ ampere-hours}.$$

  (iii) The battery can discharge up to 40% of 50 ampere-hours, which is 20 ampere-hours, without damage. Since Operation 1 can be performed with just 16.6 ampere-hours, it is safe.

(b) For Operation 2, we have the following:
  (i) The total battery discharge is given by the area under the battery current curve in Figure H.2. The area under the right-most portion of the curve, (when the satellite is shadowed by the earth), is easily calculated as 30 amps· 30 minutes = 900 ampere-minutes = 15 ampere-hours. For the other part we estimate by trapezoids, which are the average of left and right rectangles on each subinterval. Estimated values of the function are in Table H.1.

TABLE H.1 *Estimated values of the battery current*

Time	0	5	10	15	20	25	30
Current	5	16	18	12	5	12	0

Using $\Delta t = 5$, we see

$$\begin{aligned}\text{Total discharge} &= \frac{1}{2}(5+16)\cdot 5 + \frac{1}{2}(16+18)\cdot 5 + \frac{1}{2}(18+12)\cdot 5 \\ &\quad + \frac{1}{2}(12+5)\cdot 5 + \frac{1}{2}(5+12)\cdot 5 + \frac{1}{2}(12+0)\cdot 5 \\ &\approx 330 \text{ ampere-minutes} = 5.5 \text{ ampere-hours.}\end{aligned}$$

The total estimated discharge is 20.5 ampere-hours.

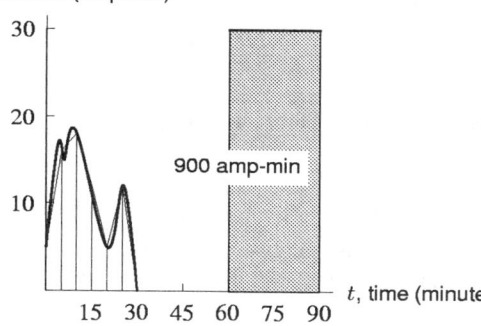

Figure H.2: Operation 2

(ii) Since the estimated discharge appears to be an underestimate, Operation 2 probably should not be performed.

2. THE EARTH'S PATH: SOLUTION

Mathematics skills required: Numerical integration.

Comments: This is a fairly short project which requires numerical integration and introduces some of the properties of an ellipse.

Solution:

(a) For a circle, both a and b equal the radius of the circle. So $c = \sqrt{a^2 - b^2} = 0$, and then $e = c/a = 0$.

(b) Since we have

$$e = \frac{c}{a} = \frac{\sqrt{a^2 - b^2}}{a} = \sqrt{1 - \left(\frac{b}{a}\right)^2},$$

the larger e is, the smaller b is relative to a, which means the ellipse is more "squashed."

(c) Looking at Figure H.3, we see

$$a = \frac{94.5 + 91.5}{2} = 93 \text{ million miles.}$$
$$c = 94.5 - 93 = 1.5 \text{ million miles.}$$
$$b = \sqrt{a^2 - c^2} \approx 92.99 \text{ million miles.}$$
$$e = \frac{c}{a} = \frac{1.5}{93} = 0.0161.$$

The fact that e is so small reflects the fact that the earth's orbit is nearly a circle.

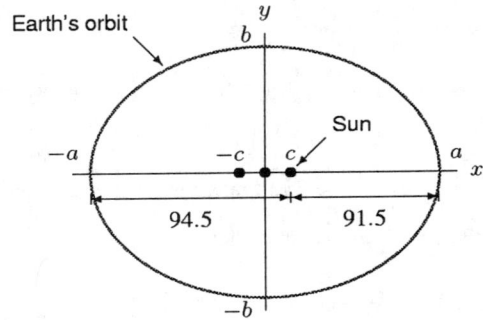

Figure H.3

(d) The distance traveled is the perimeter of the ellipse, with $a = 93$, $e = 0.0161$:

$$\text{Distance} = \int_0^{\pi/2} 4 \cdot 93(1 - (0.0161 \sin \theta)^2)^{1/2} d\theta \approx 584.30 \text{ million miles.}$$

This integral was approximated numerically.

(e) It takes the earth 1 year to go once round its orbit, so

$$\text{Average speed} = \frac{584.3 \text{ million miles}}{1 \text{ year}} = \frac{584.3 \times 10^6}{1 \times 365 \times 24 \times 60 \times 60} = 18.528 \text{ miles/sec.}$$

(f) If the radius of the circle is r, we need $2\pi r = 584.30$, so $r \approx 92.99$ million miles.

(g) For Halley's comet, we have

$$a = 1674 \text{ million miles}$$
$$e = \frac{c}{a} = 0.97, \text{ so } c = 1623.78$$
$$\frac{\text{Minimum distance}}{\text{from sun}} = 1674 - 1623.78 = 50.22 \text{ million miles}$$
$$\text{Distance around orbit} = \int_0^{\pi/4} 4(1674)(1 - (0.97 \sin \theta)^2)^{1/2} d\theta \approx 7159.00 \text{ million miles}$$
$$\text{Average speed} = \frac{7159 \times 10^6 \text{ miles}}{76 \text{ years}} = \frac{7159 \times 10^6}{76 \times 365 \times 24 \times 60 \times 60} = 2.987 \frac{\text{miles}}{\text{sec}}$$

Halley's comet travels much more slowly than the earth.

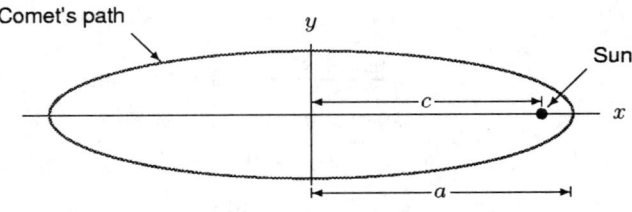

Figure H.4

(h) Halley must have seen the comet in $1758 - 76 = 1682$. Since Halley died before 1758 he did not live to see the comet's return and never knew whether his prediction came true.

3. GAS TANKS: SOLUTION

Mathematics skills required: Area interpretation of definite integral.

Comments: An understanding of how the sign of the integrand affects the value of the integral is needed in part (e).

Solution:

(a) The volume of the cylindrical tank is $\pi \cdot \left(\frac{d}{2}\right)^2 \cdot l$. The volume of the rectangular tank is $w \cdot d \cdot l$. Setting these two volumes equal to each other gives $w = \frac{\pi \cdot d}{4}$.

If the gas is at a depth of h in the rectangular tank, then the tank is exactly h/d full, since

$$\frac{\text{Volume left in rectangular tank}}{\text{Total volume of rectangular tank}} = \frac{w \cdot h \cdot l}{w \cdot d \cdot l} = \frac{h}{d}.$$

(b) For the same height h in the two tanks, (see Figure H.5), the error is given by

$$\begin{aligned}
\text{Error} &= \begin{array}{c}\text{Fraction of cylindrical}\\ \text{tank which is full}\end{array} - \frac{h}{d}\\
&= \frac{A_1 l}{\pi(d/2)^2 l} - \frac{h\pi(d/4)l}{d\pi(d/4)l}\\
&= \frac{A_1 l}{\pi(d/2)^2 l} - \frac{A_2 l}{\pi(d^2/4)l}\\
&= \frac{A_1 - A_2}{\pi(d^2/4)}
\end{aligned}$$

Since the length has canceled out, we can ignore it in calculating the error.

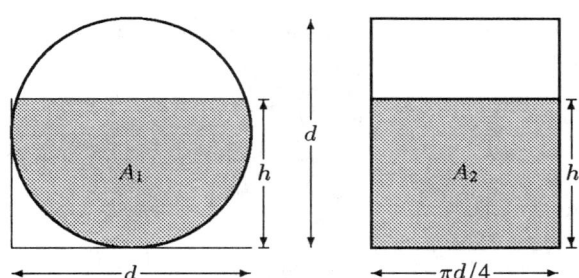

Figure H.5

(c)

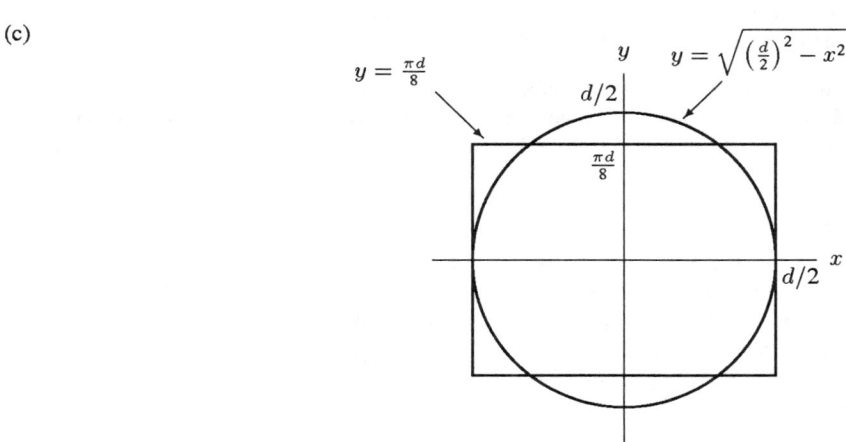

Figure H.6

(d)

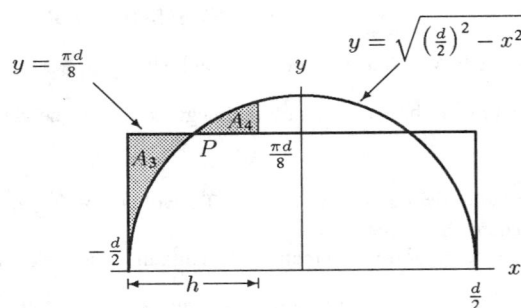

Figure H.7

For any $0 \leq h \leq d$, Figure H.7 shows that the error in measuring the fraction of gas remaining, assuming that the tank is rectangular rather than cylindrical, is given by

$$E(h) = 2 \cdot \frac{(A_3 - A_4) \cdot l}{\pi(d^2/4)l} = 2 \cdot \frac{\int_{-(d/2)}^{-(d/2)+h} \left(\frac{\pi d}{8} - \sqrt{\left(\frac{d}{2}\right)^2 - x^2}\right) dx}{\pi(d^2/4)}$$

(e) The width of the rectangular tank was chosen so that its volume equals the volume of the cylindrical tank. In addition, the volume of half the rectangular tank equals half the volume of the cylindrical tank. This means that the error $E(h) = 0$ when $h = 0$, $h = d/2$, and $h = d$. Let us consider the value of $E(h)$ as h increases from 0 to $d/2$. When h is slightly above 0, the fact that the line $y = \pi d/8$ is above the circle means that $E(h)$ increases as h increases. After h passes the value corresponding to the point P in Figure H.7, the error starts to decrease as the circle is above the line.

This means the maximum value of $E(h)$ occurs where the circle and the line cross. This happens when

$$\frac{\pi d}{8} = \sqrt{\left(\frac{d}{2}\right)^2 - x^2}$$

$$x = \pm\sqrt{\frac{d^2}{4} - \frac{\pi^2 d^2}{64}} = \pm\frac{d}{8}\sqrt{16 - \pi^2}$$

Thus,

$$h = \frac{d}{2} + \frac{d}{8}\sqrt{16 - \pi^2} \quad \text{or} \quad h = \frac{d}{2} - \frac{d}{8}\sqrt{16 - \pi^2}.$$

4. **THE GRAPHS OF** $y = b^x$ **AND** $y = x^b$**: SOLUTION**

Mathematics skills required: Ability to find points of intersection of graphs. Derivatives of exponential functions and using derivatives to see where curves touch. Optional part: finding local and global maxima/minima.

Comments: This project demonstrates an interesting property of e, namely that the graphs $y = e^x$ and $y = x^e$ meet only at (e, e^e). It is best done with a graphing calculator which has with a zoom feature. However, it can be hard to see that $y = e^x$ and $y = x^e$ intersect only once. The analytical answer to part (g) is the only way to be sure that $b = e$ is the value making graphs touch.

Solution:

(a) The points of intersection in the first quadrant are $(2, 4)$ and $(4, 16)$.
(b) The points of intersection in the first quadrant are $(3, 27)$ and approximately $(2.48, 15.22)$.
(c) The points of intersection in the first quadrant are $(2, 16)$ and $(4, 64)$.
(d) The point (b, b^b) is a crossing point for all b.
(e) Parts (a)-(c) suggest that there are, in general, two crossing points, but the other point can be either to the left or right of the point (b, b^b).

(f) Looking at parts (a)-(c) suggests that as b goes from 2 to 4 the second cross-over moves from the right of (b, b^b) to the left of (b, b^b). Somewhere between $b = 2$ and $b = 4$, the two cross-over points match up. Thus, we expect there to be a value of b giving just one crossing point. Since if $b = 3$, the second crossing point is just to the left of (b, b^b), we expect this value of b to be between $b = 2$ and $b = 3$, and closer to $b = 3$. Graphs of $y = b^x$ and $y = x^b$ with $b = 2.7$, for example, look as though the two crossing points have pretty much merged into one. This may lead you to guess that if $b = e$ there is only one crossing point. Repeated zooming in on the graphs of $y = e^x$ and $y = x^e$ shows that there seems to be only one crossing point.

(g) To confirm that when $b = e$ the two graphs touch, we show that the two functions have the same y-values and the same slopes. The y-values of both are $y = e^e$. The slopes of each are

$$\frac{dy}{dx} = e^x \quad \text{and} \quad \frac{dy}{dx} = ex^{e-1}.$$

At $x = e$, the slopes are equal, since

$$\left.\frac{dy}{dx}\right|_{x=e} = e^x\Big|_{x=e} = e^e \quad \text{and} \quad \left.\frac{dy}{dx}\right|_{x=e} = ex^{e-1}\Big|_{x=e} = e \cdot e^{e-1} = e^e.$$

(h) When $b^x = x^b$, taking natural logs gives

$$x \ln b = b \ln x$$

so

$$f(x) = x \ln b - b \ln x = 0.$$

This means that crossing points of $y = b^x$ and $y = x^b$ correspond exactly to zeros of $f(x)$. We will show that $f(x)$ has exactly two zeros for $x > 0$, except in the case $b = e$, when it has one. Differentiating, we have

$$f'(x) = \ln b - \frac{b}{x} = \frac{x \ln b - b}{x}.$$

Provided $b > 1$, we have $\ln b > 0$, so

$$f'(x) > 0 \quad \text{for} \quad x > b/\ln b \quad \text{and} \quad f'(x) < 0 \quad \text{for} \quad x < b/\ln b.$$

So $x = b/\ln b$ gives a local minimum. Since $\ln x \to -\infty$ as $x \to 0^+$ and x dominates $\ln x$ as $x \to \infty$ (that is, x is much larger than $\ln x$ as $x \to \infty$), we have

$$\lim_{x \to 0^+} f(x) = \infty \quad \text{and} \quad \lim_{x \to \infty} f(x) = \infty.$$

We know $f(x)$ has one zero (at $x = b$), and so, unless the zero occurs at the minimum, $f(x)$ must have two zeros. The only case in which $f(x)$ has one zero occurs when the minimum and the zero coincide, that is, when

$$b = \frac{b}{\ln b}.$$

This occurs when $\ln b = 1$, so $b = e$.

5. THE ROAD AND THE RIVER: SOLUTION

Mathematics skills required: Taking derivatives of polynomials, solving simultaneous equations, optimization, using calculators or computers to find roots, and computing areas using integrals.

Comments: A fairly straightforward problem. The most difficult aspect for students not familiar with using a calculator or computer may be realizing what computations they have to do by machine and what they can do by hand.

Solution: We are given that the river follows the course $f(x) = x^2$ and the road follows the course $g(x) = ax^3 + bx^2 + cx + d$.

(a) To find the values for the coefficients of $g(x) = ax^3 + bx^2 + cx + d$, we begin by substituting the coordinates of the three bridges into the equation for the road:

$$0 = g(0) = a(0)^3 + b(0)^2 + c(0) + d = d$$
$$1 = g(-1) = a(-1)^3 + b(-1)^2 + c(-1) + d = -a + b - c + d$$
$$1 = g(1) = a(1)^3 + b(1)^2 + c(1) + d = a + b + c + d$$

From the first equation we know $d = 0$ and so $-a + b - c = 1$ and $a + b + c = 1$. Adding these two equations together we get $2b = 2$ or $b = 1$. Putting $b = 1$ into the second equation gives $a + c = 0$ or $c = -a$. So we have

$$g(x) = ax^3 + x^2 - ax.$$

To determine a we use the fact that the bridge (and therefore the road) crosses the river at right angles at $(1, 1)$. Since $f'(x) = 2x$, the slope of the river at $(1, 1)$ is 2, so the slope of the road must be $-\frac{1}{2}$. Given that $g'(x) = 3ax^2 + 2x - a$, we must have $-\frac{1}{2} = g'(1) = 2a + 2$. So $a = -\frac{5}{4}$. So the equation for the road is

$$g(x) = -\frac{5}{4}x^3 + x^2 + \frac{5}{4}x.$$

See Figure H.8.

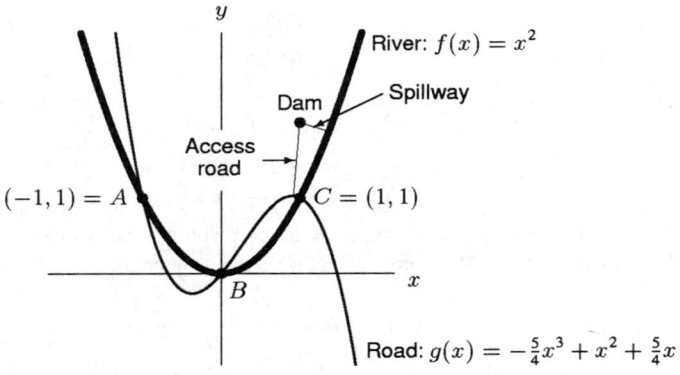

Figure H.8

(b) The distance between Bridge A at $(-1, 1)$ and Bridge C at $(1, 1)$ is 2 km. Between Bridge A and Bridge B the distance is $\sqrt{(-1)^2 + (1)^2} = \sqrt{2}$ km. The distance between Bridge B and Bridge C is $\sqrt{(1)^2 + (1)^2} = \sqrt{2}$ km.

(c) We already know Bridge C crosses at an angle of $90°$ or $\pi/2$ radians. To determine the other angles, we need to determine the slopes of the road and the river at the bridges. First the river: $f'(x) = 2x$. So $f'(0) = 0$ and $f'(-1) = -2$. For the road $g'(x) = -\frac{15}{4}x^2 + 2x + \frac{5}{4}$. So $g'(0) = \frac{5}{4}$ and $g(-1) = -\frac{18}{4}$.

At Bridge B the angle is the angle between the road and the x-axis. So $\theta = \arctan(5/4)$; giving $\theta \approx 0.9$ radians or $51°$. At Bridge A, the road is at an angle of $\arctan(18/4) \approx 1.3$ radians $= 77.5°$ with the x-axis. The angle of the river with the x-axis is $\arctan 2 \approx 1.1$ radians $\approx 63.4°$. So the angle between the road and river at Bridge A is 0.2 radians $\approx 14.1°$.

(d) Let the spillway from $(1, 2)$ intersect the river at $(x, f(x)) = (x, x^2)$. This means the length of the spillway is $\sqrt{(x^2 - 2)^2 + (x - 1)^2}$. To find x, we must minimize this length. Since the square root function is strictly increasing, we may as well minimize $D = (x^2 - 2)^2 + (x - 1)^2$. (It is easier to minimize D because D has no square root.) This function can be minimized either by tracing on a calculator or by using a computer algebra system.

Alternatively, we can take the derivative and set it equal to zero:

$$D' = 2(x^2 - 2)(2x) + 2(x - 1)$$
$$= 4x^3 - 8x + 2x - 2$$
$$= 4x^3 - 6x - 2.$$

Setting D' equal to 0 and using a calculator or computer gives roots at $x = -1, \frac{1}{2} \pm \frac{1}{2}\sqrt{3}$ so $x = -1, -0.37, 1.37$. There are several ways to determine which of these critical points is the global minimum (graphing $y = (x^2 - 2)^2 + (x - 1)^2$, or using first or second derivatives, for example.) The minimum is $x = 1.37$, so the point of intersection with the river is approximately $(1.37, 1.88)$.

(e) The access road goes from a point $(x, -\frac{5}{4}x^3 + x^2 + \frac{5}{4}x)$ on the original road to the point $(1, 2)$. We want to minimize its length $\sqrt{(-\frac{5}{4}x^3 + x^2 + \frac{5}{4}x - 2)^2 + (x - 1)^2}$. Again, we get rid of the square root and actually minimize

$$D = \left(-\frac{5}{4}x^3 + x^2 + \frac{5}{4}x - 2\right)^2 + (x - 1)^2.$$

The derivative is

$$D' = 2\left(-\frac{5}{4}x^3 + x^2 + \frac{5}{4}x - 2\right)\left(-\frac{15}{4}x^2 + 2x + \frac{5}{4}\right) + 2(x - 1).$$

Using a calculator or computer, we find that $x \approx -1.12, -0.50,$ and 0.92 are the only real zeros of D'. We evaluate D at these critical points:

$$D(-1.12) \approx 4.65$$
$$D(-0.50) \approx 7.17$$
$$D(0.92) \approx 0.96$$

Thus, the closest point on the road to the dam is $(0.92, 1.02)$.

(f) We want to compute the area between the two curves from $x = -1$ to $x = 1$. Note that from $x = -1$ to $x = 0$ the river is higher on the graph, but that from $x = 0$ to $x = 1$ the road is higher. Thus, evaluating the integral numerically or using the Fundamental Theorem of Calculus, we have

$$\text{Area} = \int_{-1}^{0} \left[x^2 - \left(-\frac{5}{4}x^3 + x^2 + \frac{5}{4}x\right)\right] dx + \int_{0}^{1} \left[\left(-\frac{5}{4}x^3 + x^2 + \frac{5}{4}x\right) - x^2\right] dx$$

$$= \frac{5}{8} \text{km}^2.$$

6. INTERSTATE TRUCKING: SOLUTION

Mathematics skills required: Constructing a mathematical model, optimization.

Comments: The optimization problem in part (d) is unusual in that the optimum value is known (55 mph), and the problem is to find the conditions which lead to this optimum. A variant of this project is to ask what group of people in the real world might be interested in each of the questions asked. A possible answer is owners of trucking companies for parts (b) and (c), traffic police for part (d), and Interstate Commerce Commission for parts (e) and (f).

Solution:

(a) The total cost per mile is the cost of the driver plus the cost of fuel. We let

w be the driver's hourly wage in dollars/hour,

v be the average speed in miles/hour,

m be the weight of the truck in thousands of pounds,

f the cost of fuel in dollars/gallon.

The cost per mile of the driver's wages is w/v. The cost of fuel per mile will be one over the "mileage per gallon" times the cost of fuel per gallon—i.e. f/mpg. The mileage per gallon is $6 - (m - 25)(0.02) - (v - 45)(0.1)$ for velocities over 45 and $6 - (m - 25)(0.02)$ for velocities under 45. So the total cost per mile, c, is

$$c = \begin{cases} \dfrac{w}{v} + \dfrac{f}{6 - (m - 25)(0.02)} & 0 < v \leq 45 \\ \dfrac{w}{v} + \dfrac{f}{6 - (m - 25)(0.02) - (v - 45)(0.1)} & 45 < v. \end{cases}$$

Note that there is an upper limit to the velocity in this last expression given when

$$6 - (m - 25)(0.02) - (v - 45)(0.1) = 0.$$

(b) We are now given the values

$$w = 15.00 \text{ dollars/hour}$$
$$m = 75 \text{ thousand pounds}$$
$$f = 1.25 \text{ dollars/gallon}.$$

We have
$$c = \begin{cases} \dfrac{15}{v} + \dfrac{1.25}{6 - (75 - 25)(0.02)} & 0 < v \le 45 \\ \dfrac{15}{v} + \dfrac{1}{6 - (75 - 25)(0.02) - (v - 45)(0.1)} & 45 < v, \end{cases}$$

which simplifies to
$$c = \begin{cases} \dfrac{15}{v} + \dfrac{1}{4} & 0 < v \le 45 \\ \dfrac{15}{v} + \dfrac{1.25}{5 - (v - 45)(0.1)} & 45 < v < 95. \end{cases}$$

The upper limit for v occurs when $5 - (v - 45)(0.1) = 0$, that is, $v = 95$.

To initiate our search for a minimum, note that the function $c = 15/v + 1/4$ is strictly decreasing. So we only need find the minimum of the function
$$c = \dfrac{15}{v} + \dfrac{1.25}{5 - (v - 45)(0.1)}$$

on the interval $45 \le v < 95$. Rearranging this slightly, we get
$$c = \dfrac{15}{v} + \dfrac{1.25}{9.5 - 0.1v}.$$

Then differentiating gives
$$\dfrac{dc}{dv} = -\dfrac{15}{v^2} + \dfrac{(1.25)(0.1)}{(9.5 - 0.1v)^2}.$$

Setting this to zero and solving, we get
$$0 = -\dfrac{15}{v^2} + \dfrac{(1.25)(0.1)}{(9.5 - 0.1v)^2}$$
$$15(9.5 - 0.1v)^2 = 0.125 v^2$$
$$3.87(9.5 - 0.1v) \approx \pm 0.354 v$$
$$36.8 - 0.387 v \approx \pm 0.354 v$$
$$36.8 \approx 0.741 v \text{ or } 36.8 \approx 0.033 v$$
$$v \approx 49.7 \text{ or } v \approx 1100.$$

This last value is not in the domain, so we only consider the critical point $v = 49.7$ and the endpoints of $v = 45$ and $v = 95$. We evaluate the cost function:
$$c(45) = 0.333 + 0.25 = 58.3¢/\text{mile}$$
$$c(49.7) = 0.302 + 0.276 = 57.8¢/\text{mile}$$
$$c(95) = \infty.$$

So $v = 49.7$ is a minimum; the cheapest speed is 49.7 mph.

(c) Evaluating the cost at $v = 55$ mph, $v = 60$ mph, and the minimum $v = 49.7$ mph gives
$$c(49.7) = 57.8¢$$
$$c(55) = 58.5¢$$
$$c(60) = 60.7¢.$$

Notice that the cost per mile does not rise very quickly. A produce hauler often gets extra revenue for getting there fast. Increasing speed from 50 to 60 mph decreases the transit time by over 15% but increases the costs by only 5%. Thus, many produce haulers will choose a speed above 49.7 mph.

(d) Now we are not given the price of fuel, but we want the minimum to be at $v = 55$ mph. We find the value of f making $v = 55$ the minimum. The function we want to minimize is
$$c = \dfrac{15}{v} + \dfrac{f}{9.5 - 0.1v}.$$

Differentiating gives
$$\frac{dc}{dv} = -\frac{15}{v^2} + \frac{0.1f}{(9.5 - 0.1v)^2}$$

Setting this equal to 0, we have
$$0 = -\frac{15}{v^2} + \frac{0.1f}{(9.5 - 0.1v)^2}$$
$$0 = -15(9.5 - 0.1v)^2 + 0.1fv^2.$$

Substituting $v = 55$ and solving for f gives
$$0 = -15(4)^2 + 0.1(55)^2 f$$
$$f \approx 80¢/\text{gallon}.$$

(e) Now we are not told the driver's wages, w, or the fuel cost, f. We want to find the relationship between w and f making the minimum cost occur at $v = 55$ mph. We have
$$c = \frac{w}{v} + \frac{f}{9.5 - 0.1v}$$
$$\frac{dc}{dv} = -\frac{w}{v^2} + \frac{0.1f}{(9.5 - 0.1v)^2}.$$

We need this to equal 0 when $v = 55$, so
$$0 = -\frac{w}{3025} + \frac{0.1f}{16}.$$

This means
$$\frac{w}{f} = \frac{(3025)(0.1)}{16} = 18.9,$$

that is, the fuel cost per gallon should be 1/18.9 that of the driver's hourly wage. If the Interstate Commerce Commision wants truck drivers to keep to a speed of 55 mph, they should consider taxing fuel or driver's wages so that they remain in the relation $w = 18.9f$.

(f) Now we assume $w = 18.9f$ and that m is variable. We want to minimize cost, getting a relationship between m and the optimal v. The function we want to minimize is
$$c = \frac{18.9f}{v} + \frac{f}{6 - (m - 25)(0.02) - (v - 45)(0.1)}$$
$$= \frac{18.9f}{v} + \frac{f}{11 - 0.02m - 0.1v}.$$

Differentiating gives
$$\frac{dc}{dv} = \frac{-18.9f}{v^2} + \frac{0.1f}{(11 - 0.02m - 0.1v)^2}.$$

We are interested in when $dc/dv = 0$:
$$-\frac{18.9f}{v^2} + \frac{0.1f}{(11 - 0.02m - 0.1v)^2} = 0.$$

Solving gives
$$v = 63.7 - 0.116m \quad \text{or} \quad v = 403.5 - 0.734m.$$

Only the first gives plausible speeds (and gives $v = 55$ when $m = 75$), so we conclude the optimal speed varies linearly with weight according to the equation $v = 63.7 - 0.116m$. This means that every 10,000 increase in weight reduces the optimal speed by just over 1 mph.

7. APPROXIMATING π WITH INTEGRALS: SOLUTION

Mathematics skills required: Integrating by parts. Observing inverse proportionality in a table of values.

Comments: The history of various approximations to π (especially those that are derived from calculus) would make an excellent student research project.

Solution: Write

$$\int \sin^n x\, dx = \int (\sin^{n-2} x)(1 - \cos^2 x)\, dx$$
$$= \int \sin^{n-2} x\, dx - \int \sin^{n-2} x \cos x \cos x\, dx.$$

Using integration by parts on the second integral with $u = \cos x$ and $v' = \sin^{n-2} x \cos x\, dx$, we have for $n > 1$:

$$\int_0^{\pi/2} \sin^n x\, dx = -\frac{1}{n} \sin^{n-1} x \cos x \Big|_0^{\pi/2} + \frac{n-1}{n} \int_0^{\pi/2} \sin^{n-2} x\, dx.$$

Since $\cos \frac{\pi}{2} = 0$, and $\sin^{n-1} 0 = 0$, this gives

$$\int_0^{\pi/2} \sin^n x\, dx = \frac{n-1}{n} \int_0^{\pi/2} \sin^{n-2} x\, dx.$$

(a) Suppose n is a positive even integer. Then

$$\int_0^{\pi/2} \sin^n x\, dx = \frac{n-1}{n} \int_0^{\pi/2} \sin^{n-2} x\, dx.$$

If $n > 2$, then

$$\int_0^{\pi/2} \sin^n x\, dx = \frac{n-1}{n} \cdot \frac{n-3}{n-2} \int_0^{\pi/2} \sin^{n-4} x\, dx.$$

We can continue this process until we have

$$\int_0^{\pi/2} \sin^n x\, dx = \frac{n-1}{n} \cdot \frac{n-3}{n-2} \cdots \frac{3}{4} \cdot \frac{1}{2} \cdot \int_0^{\pi/2} \sin^0 x\, dx.$$

Notice that the last numerical factor is $\frac{1}{2}$. This is because the last step involves the integration of $\int \sin^2 x\, dx$ and so n was equal to 2. Since

$$\int_0^{\pi/2} \sin^0 x\, dx = \int_0^{\pi/2} dx = \pi/2,$$

we have, for n even and positive,

$$\int_0^{\pi/2} \sin^n x\, dx = \frac{n-1}{n} \cdot \frac{n-3}{n-2} \cdots \frac{3}{4} \cdot \frac{1}{2} \cdot \pi/2.$$

(b) If n is odd, we go through the same process, but in the end we have

$$\int_0^{\pi/2} \sin^n dx = \frac{n-1}{n} \cdot \frac{n-3}{n-2} \cdots \frac{4}{5} \cdot \frac{2}{3} \cdot \int_0^{\pi/2} \sin x\, dx.$$

But

$$\int_0^{\pi/2} \sin x\, dx = -\cos x \Big|_0^{\pi/2} = 1.$$

Thus

$$\int_0^{\pi/2} \sin^n x\, dx = \frac{n-1}{n} \cdot \frac{n-3}{n-2} \cdots \frac{4}{5} \cdot \frac{2}{3}$$

for n odd and greater than 1.

(c) If $0 < x < \pi/2$ then $0 < \sin x < 1$, so for any integer k,

$$\sin^{2k+1} x < \sin^{2k} x < \sin^{2k-1} x.$$

It follows that the area under the graph of $\sin^{2k-1} x$ must be greater than the area under the graph of $\sin^{2k} x$ because its values are larger. Similarly, the area under $\sin^{2k} x$ must be larger than the area under $\sin^{2k+1} x$. So

$$\int_0^{\pi/2} \sin^{2k+1} x \, dx < \int_0^{\pi/2} \sin^{2k} x \, dx < \int_0^{\pi/2} \sin^{2k-1} x \, dx.$$

Thus, we have:

$$\frac{(2k)(2k-2)\cdots 4\cdot 2}{(2k+1)(2k-1)\cdots 5\cdot 3} < \frac{(2k-1)(2k-3)\cdots 3\cdot 1}{(2k)(2k-2)\cdots 4\cdot 2}\cdot\frac{\pi}{2} < \frac{(2k-2)(2k-4)\cdots 4\cdot 2}{(2k-1)(2k-3)\cdots 5\cdot 3}.$$

Multiplying all sides of the inequality by

$$2\cdot\frac{(2k)(2k-2)\cdots 4\cdot 2}{(2k-1)(2k-3)\cdots 3\cdot 1}$$

gives

$$2\cdot\frac{2\cdot 2\cdot 4\cdot 4\cdots(2k-2)(2k-2)(2k)(2k)}{1\cdot 3\cdot 3\cdot 5\cdot 5\cdots(2k-1)(2k-1)(2k+1)} < \pi < 2\cdot\frac{2\cdot 2\cdot 4\cdot 4\cdots(2k-2)(2k-2)(2k)}{1\cdot 3\cdot 3\cdot 5\cdot 5\cdots(2k-1)(2k-1)},$$

which is our desired inequality.

(d) A table for various values of k follows:

k	Lower bound	Upper bound	Gap length
2	2.8444	3.5556	0.7112
3	2.9258	3.4134	0.4876
4	2.9722	3.3436	0.3714
5	3.0022	3.3024	0.3002
6	3.0232	3.2752	0.2520
10	3.0678	3.2210	0.1532
20	3.1036	3.1812	0.0776
40	3.1222	3.1612	0.0390
80	3.1318	3.1514	0.0196
100	3.1338	3.1494	0.0156

We see that as k doubles (for example, as k goes from 5 to 10 to 20 and so on), the gap length approximately halves each time. These values suggest that the gap length is approximately proportional to $1/k$. We can understand why this is so. The gap is

$$\text{Gap} = 2\cdot\frac{2\cdot 2\cdot 4\cdot 4\cdots(2k-2)(2k-2)(2k)}{1\cdot 3\cdot 3\cdots(2k-1)(2k-1)}$$

$$-2\cdot\frac{2\cdot 2\cdot 4\cdot 4\cdots(2k)(2k)}{1\cdot 3\cdot 3\cdots(2k-1)(2k-1)(2k+1)}$$

$$= \left(2\cdot\frac{2\cdot 2\cdot 4\cdot 4\cdots(2k-2)(2k-2)(2k)}{1\cdot 3\cdot 3\cdots(2k-1)(2k-1)}\right)\left(1 - \frac{2k}{2k+1}\right)$$

$$= \underbrace{\left(2\cdot\frac{2\cdot 2\cdot 4\cdot 4\cdots(2k-2)(2k-2)(2k)}{1\cdot 3\cdot 3\cdots(2k-1)(2k-1)}\right)}_{\text{upper limit for }\pi}\left(\frac{1}{2k+1}\right)$$

$$\approx \pi\cdot\frac{1}{2k} \quad \text{because} \quad \frac{1}{2k+1}\approx\frac{1}{2k}$$

$$= \frac{\pi}{2}\cdot\frac{1}{k}.$$

Assuming we take the average of the upper and lower bounds as the approximation, we know that the error bound is also proportional to $1/k$, since the error bound is half the gap size.

8. SURFACE AREA OF AN UNPAINTABLE CAN OF PAINT: SOLUTION

Mathematics skills required: Arclength, an understanding of Riemann sums, finding volumes of solids of revolution, computing improper integrals. For part (e) only, solving differential equations.

Comments: This project extends, to finding surface areas, the methods used to find volumes of solids of revolution and arclengths. In particular, we demonstrate an interesting paradox: we find a surface that holds a finite volume but has an infinite surface area. Part (e) requires the solution of a differential equation, but may be omitted without affecting the remainder of the problem.

Solution:

(a) Revolving the semi-circle $y = \sqrt{r^2 - x^2}$ around the x-axis yields the sphere of radius r. See Figure H.9. Differentiating yields:
$$\frac{dy}{dx} = \frac{-1}{\sqrt{r^2 - x^2}} \cdot x = -\frac{x}{y}.$$

Thus, substituting $-x/y$ for $f'(x)$, we get

$$\text{Surface area} = 2\pi \int_{-r}^{r} y\sqrt{1 + \frac{x^2}{y^2}}\, dx = 2\pi \int_{-r}^{r} \sqrt{x^2 + y^2}\, dx$$
$$= 2\pi r \int_{-r}^{r} dx = 4\pi r^2.$$

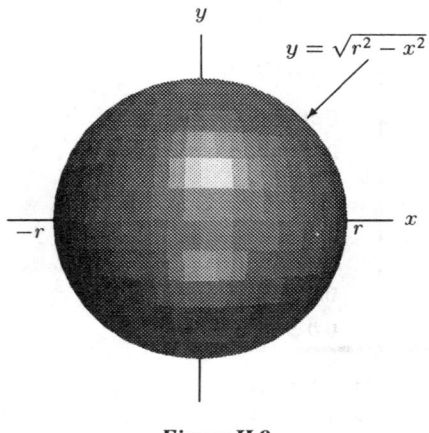

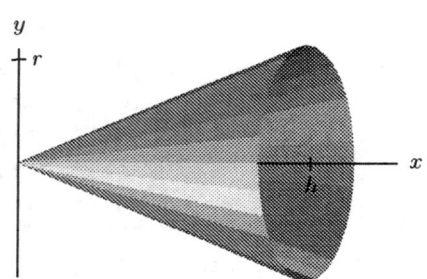

Figure H.9 **Figure H.10**

(b) Revolving the line $y = rx/h$ around the x-axis yields the cone. The base of the cone is a circle with area πr^2. See Figure H.10. The area of the rest of the cone is

$$\text{Surface area} = 2\pi \int_0^h y\sqrt{1 + \frac{r^2}{h^2}}\, dx = 2\pi\sqrt{1 + \frac{r^2}{h^2}}\left(\frac{r}{h}\int_0^h x\, dx\right)$$
$$= 2\pi \frac{r}{h}\frac{h^2}{2}\sqrt{1 + \frac{r^2}{h^2}} = \pi r \sqrt{r^2 + h^2}$$

Adding the area of the base, we get

$$\text{Total surface area of cone} = \pi r^2 + \pi r \sqrt{r^2 + h^2}.$$

(c) We find the volume of $y = 1/x$ revolved about the x-axis as x runs from 1 to ∞. See Figure H.11.

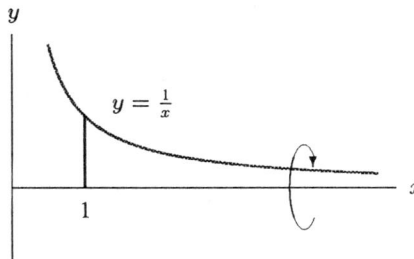

Figure H.11

$$\text{Volume} = \int_1^\infty \pi y^2 \, dx = \pi \int_1^\infty \frac{1}{x^2} \, dx = \pi \lim_{b \to \infty} \int_1^b \frac{dx}{x^2} = \pi \lim_{b \to \infty} \left. \frac{-1}{x} \right|_1^b = \pi$$

Thus, the volume of this solid is finite and equal to π.

(d) Now we show the surface area of this solid is unbounded. We have

$$\text{Surface area} = 2\pi \int_1^\infty y \sqrt{1 + \left(\frac{dy}{dx}\right)^2} \, dx = 2\pi \int_1^\infty \frac{1}{x} \sqrt{1 + \frac{1}{x^4}} \, dx$$

We cannot easily compute the antiderivative of $\frac{1}{x}\sqrt{1 + \frac{1}{x^4}}$, so we bound the integral from below by noticing that

$$\sqrt{1 + \frac{1}{x^4}} \geq 1.$$

Thus we see that

$$\text{Surface area} \geq 2\pi \int_1^\infty \frac{1}{x} \, dx = 2\pi \lim_{b \to \infty} \int_1^b \frac{dx}{x} = 2\pi \lim_{b \to \infty} \ln x \Big|_1^b.$$

Since $\ln x$ goes to infinity as x goes to infinity, the surface area is unbounded.

Alternatively, we can try calculating

$$2\pi \int_1^b \frac{1}{x} \sqrt{1 + \frac{1}{x^4}} \, dx$$

for larger and larger values of b. We would see that the integral seems to diverge.

(e) For a solid generated by the revolution of a curve $y = f(x)$ for $a \leq x \leq b$,

$$\text{Volume} = \int_a^b \pi y^2 \, dx$$

and

$$\text{Surface area} = \int_a^b 2\pi y \sqrt{1 + (f'(x))^2} \, dx.$$

The volume and the surface area will be equal if

$$f(x) = 2\sqrt{1 + (f'(x))^2}.$$

We find a function $y = f(x)$ which satisfies this relation:

$$y = 2\sqrt{1 + \left(\frac{dy}{dx}\right)^2}$$

$$\frac{y^2}{4} = 1 + \left(\frac{dy}{dx}\right)^2$$

$$\frac{dy}{dx} = \sqrt{\frac{y^2}{4} - 1}$$

$$\frac{dy}{\sqrt{y^2 - 4}} = \frac{1}{2} dx$$

$$\int \frac{dy}{\sqrt{y^2 - 4}} = \int \frac{1}{2} dx$$

$$\ln|y + \sqrt{y^2 - 4}| = \frac{x}{2} + C$$

$$y + \sqrt{y^2 - 4} = Ae^{x/2}$$

Notice in the third line we have used the fact that $dy/dx \geq 0$. Any function, $y = f(x)$, which satisfies this relationship has the required property.

9. QUININE: SOLUTION

Mathematics skills required: Ability to solve elementary differential equations, to work with piecewise defined functions and to find average values of a function.

Comments: This project shows students that there is important mathematics behind how drugs are prescribed. The difficulty of this problem is that the function describing the process is given by different analytic expressions over different intervals.

Solution:

(a) Let k by the relative rate of decay, per minute, of quinine. Since quinine's half-life is 11.5 hours, we have

$$\frac{1}{2} = e^{-k(11.5)(60)},$$

so

$$k = \frac{\ln 2}{(11.5)(60)} \approx 0.001.$$

Hence, $k = 0.1\%$/min.

(b) Just prior to 8 am of the first day the patient has no quinine in her body. Assuming the drug mixes rapidly in the patient's body, she has about $50/70 \approx 0.714$ mg/kg of the drug soon after 8 am. Suppose we represent the concentration of quinine in the patient (in mg/kg) by x and represent time since 8 am (in minutes) by t. Then

$$x = Ae^{-0.001t},$$

where A is the initial concentration and $k = -0.001$ is the rate at which quinine is metabolized per minute. There are $24 \cdot 60 = 1440$ minutes in a day. On the first day, the patient begins with 0.714 mg/kg in her system, so just before 8 am of the second day the patient's system holds

$$0.714 e^{-0.001 \cdot 1440} \approx 0.169 \text{ mg/kg}.$$

After the patient's second dose of quinine, her system contains $0.714 + 0.169 = 0.883$ mg/kg of quinine.

(c) By continuing in a similar manner, we see that just prior to 8 am on the third day, she has $0.883 e^{-0.001 \cdot 1440} \approx 0.209$ mg/kg; just after 8 am, she has $0.209 + 0.714 = 0.923$ mg/kg. Just prior to 8 am on the fourth day, she has $0.923 e^{-0.001 \cdot 1440} \approx 0.218$ mg/kg; just after 8 am, she has $0.228 + 0.714 = 0.932$ mg/kg. We can keep going with these calculations: just prior to 8 am on the fifth day, the concentration is 0.221 mg/kg; on the sixth day, it is 0.222 mg/kg; on the seventh day, it is 0.222 mg/kg, and so on forever.

We find a formula for the concentration just after the nth dose as follows. The last dose contributes 0.714 mg/kg. The previous dose contributes $0.714 e^{-0.001(1440)}$ mg/kg. The dose before that contributes $0.714 e^{-0.001(2)(1440)}$ mg/kg, and so on, back to $0.714 e^{-0.001(n-1)(1440)}$ mg/kg from the initial dose. So

$$\text{Concentration just after } n \text{ doses} = 0.714 + 0.714 e^{-1.44} + 0.714 \left(e^{-1.44}\right)^2 + \cdots + 0.714 \left(e^{-1.44}\right)^{n-1}.$$

We notice that this is a geometric series, with sum given by

$$\text{Concentration just after } n \text{ doses} = 0.714\left(\frac{1-e^{-1.44n}}{1-e^{-1.44}}\right) = 0.936(1-e^{-1.44n}).$$

Although the concentration of quinine does not reach an equilibrium it does fall into a steady-state pattern which repeats over and over again. This makes sense; at some point the patient must metabolize the daily dosage exactly. If we let $n \to \infty$ in our formula, we have $e^{-1.44n} \to 0$, which means that the concentration just after the nth dose gets very close to 0.936. So the concentration just before the nth dose is $0.936 - 0.714 = 0.222$, as we found in our calculations for the first few days.

(d)

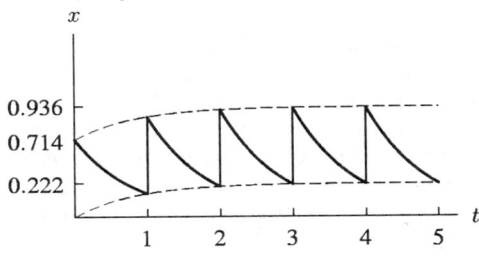

Figure H.12

If we keep setting the clock back to 0 minutes each day at 8 am, then we have that at $t = 0$ each day, the concentration (starting on the fifth day or so) is 0.936 mg/kg. As the day progresses, we have

$$x = 0.936 e^{-0.001 \cdot t}.$$

(e) The average concentration of quinine in the patient is given by the integral of the concentration over a day, divided by the time in a day:

$$\text{Average concentration} = \frac{1}{1440}\int_0^{1440} x\, dt = \frac{1}{1440}\int_0^{1440} 0.936 e^{-0.001t}\, dt$$

$$= \frac{0.936}{1440}\left(\frac{-e^{-0.001t}}{0.001}\right)\bigg|_0^{1440} = \frac{0.936}{1.44}(1 - e^{-1.44})$$

$$\approx 0.496 \text{ mg/kg}.$$

(f) Since the average concentration is 0.496 mg/kg and the minimum effective average concentration is 0.4 mg/kg, this treatment is effective. It is also safe—the highest concentration (0.936 mg/kg, achieved shortly after 8 am) is less than the toxic concentration of 3.0 mg/kg.

(g) Each dose of 25 mg corresponds to $25/70 = 0.357$ mg/kg. Let x_s be the steady-state concentration just before each 0.357 mg/kg dose. Then $x_s + 0.357$ will be the concentration just after the dose. Since we are in a steady-state, this concentration decays to exactly x_s just before the next dose. So

$$x_s = (x_s + 0.357)e^{-0.001(12)(60)}.$$

This means

$$x_s = \frac{0.357 e^{-0.001(12)(60)}}{1 - e^{-0.001(12)(60)}} \approx 0.339 \text{ mg/kg},$$

so $x_s + 0.357 = 0.696$ mg/kg is the concentration just after each dose. At t minutes after a dose, for $0 \le t \le (12)(60)$, there is a steady-state concentration of

$$x = 0.696 e^{-0.001t} \text{ mg/kg}.$$

This means

$$\text{Average concentration} = \frac{1}{720}\int_0^{720} x\, dt \approx \frac{1}{720}\int_0^{720} 0.696 e^{-0.001t}\, dt$$

$$= \frac{0.696}{720}\left[\frac{-e^{-0.001t}}{0.001}\right]\bigg|_0^{720} = \frac{0.696}{0.72}[1 - 0.487]$$

$$\approx 0.496 \text{ mg/kg}.$$

This treatment is also effective and safe. The average concentration of 0.496 mg/kg is greater than 0.4 mg/kg, and the highest concentration of 0.696 mg/kg is less than 3 mg/kg.

(h) For an exponentially decaying function, the average value between two points (x_0, y_0) and (x_1, y_1) is $\frac{(y_0 - y_1)}{(x_1 - x_0)r}$, where r is the relative rate of decay and A_0 is the initial concentration. The reason is as follows.

$$\text{Average} = \frac{1}{x_1 - x_0} \int_{x_0}^{x_1} A_0 e^{-rt} dt$$

$$= \frac{A_0}{x_1 - x_0} \left[\frac{e^{-rt}}{r} \right]_{x_0}^{x_1}$$

$$= \frac{y_0 - y_1}{(x_1 - x_0) \cdot r}$$

(i) Since a steady state has been reached, y_0 is the concentration right after a dose and y_1 is the concentration just prior to a dose. Thus, $y_0 - y_1$ represents the increase in concentration from each dose. Furthermore, $x_1 - x_0$ is the time between doses. When we go to the new protocol, we halve both the numerator and the denominator of the equation for the average concentration, and so the average remains unchanged. Similarly, if we were to double the dose to 100 mg and give it every 48 hours we would simply be doubling both the numerator and the denominator; again the average concentration would not change.

(j) We want the final concentration to be 10^{-10} kg/kg = 10^{-4} mg/kg. We therefore need to solve for t in $10^{-4} = 0.883 \cdot e^{-0.001 \cdot t}$. Doing so yields $t \approx 9086$ min ≈ 6.3 days.

10. U.S. ELECTRICITY CONSUMPTION: SOLUTION

Mathematical skills required: Fitting lines to plotted data, solving elementary differential equations.

Comments: This project introduces many of the issues which arise in making predictions from numerical data: Converting data to a form expected to lead to linear relationship; fitting a line to a scatterplot. Lines can be fitted either by eye using a scatterplot on paper or using a calculator or computer. Students should be encouraged to think about how each mathematical model relates to the physical problem.

Solution: Note: Your estimates for a, b, c are highly dependent on the type of approximations and line fitting you use, so your estimates may differ significantly from those presented here.

(a) In order to generate the necessary plots we need $\frac{dE}{dt}$, $\frac{1}{E}\frac{dE}{dt}$, E and t. For 1912, we approximate

$$\frac{dE}{dt} \approx \frac{\Delta E}{\Delta t} = \frac{25 - 12}{5} \approx 2.6;$$

for 1917,

$$\frac{dE}{dt} \approx \frac{\Delta E}{\Delta t} = \frac{39 - 25}{3} \approx 4.7$$

and so on. All the other values of the following chart can then be directly computed.

TABLE H.2

Year(t)	Electricity Consumption (E)	$\frac{\Delta E}{\Delta t} \approx \frac{dE}{dt}$	$\frac{1}{E}\frac{\Delta E}{\Delta t} \approx \frac{1}{E}\frac{dE}{dt}$
1912	12	2.6	0.217
1917	25	4.7	0.187
1920	39	5.9	0.151
1929	92	2.4	0.026
1936	109	12.6	0.115
1945	222	32.5	0.146
1955	547	41.6	0.076
1960	755	60.0	0.079
1965	1055	95.2	0.090
1970	1531	75.5	0.049
1980	2286	24.1	0.011
1987	2455		

We rewrite equation (i) as
$$\frac{dE}{dt} = cE.$$

In Figure H.13 we plot $\frac{dE}{dT}$ versus E. The best line through these data points that passes through the origin (which can be found, for instance, by the least squares method) has a slope of about 0.036, so $c = 0.036$. Equation (ii) is of the form
$$\frac{dE}{dt} = a - bE,$$
so we use the same plot, but allow lines which do not go through the origin. The slope of the best fitting line in this case is about 0.024, so $b = -0.024$ and the $\frac{dE}{dt}$ intercept is 18.1, so $a = 18.1$.

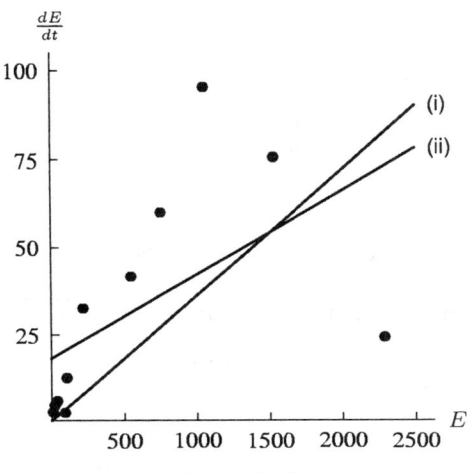

Figure H.13

Since equation (iii) is
$$\frac{1}{E}\frac{dE}{dt} = a - bE,$$
to check equation (iii) we plot $\frac{1}{E}\frac{dE}{dt}$ versus E, as in Figure H.14. The line shown has slope $m = -6.1 \cdot 10^{-5}$ and $\frac{1}{E}\frac{dE}{dt}$ intercept at 0.14. So $b = 6.1 \cdot 10^{-5}$ and $a = 0.14$.

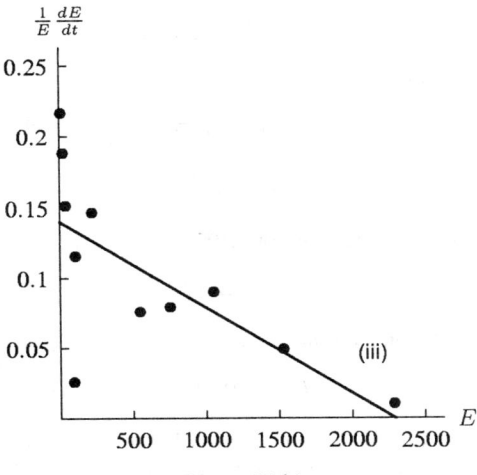

Figure H.14

Since equation (iv) is
$$\frac{1}{E}\frac{dE}{dt} = a - bt,$$
for equation (iv) we graph $\frac{1}{E}\frac{dE}{dt}$ versus t, where t is measured since 1900; we get Figure H.15. The best line has a slope of -0.002 and a $\frac{1}{E}\frac{dE}{dt}$ intercept of 0.2. So $a = 0.2$ and $b = 0.002$.

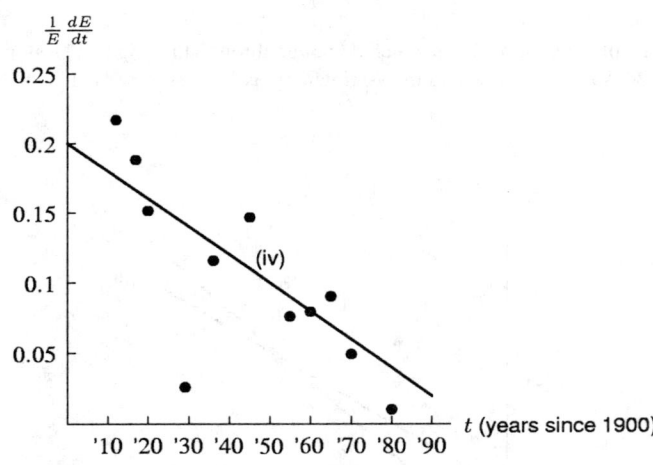

Figure H.15

(b) (i) We have
$$\frac{dE}{dt} = 0.036E,$$
so we get
$$E = E_0 e^{0.036t}.$$
This is exponential growth at a continuous rate of 3.6%. To estimate E in the year 2000, we measure time from 1987, and so $E_0 = 2455$ and
$$E \approx 2455 e^{0.036(13)} \approx 3920.$$
This model predicts that growth will continue at 3.6% forever. This is not reasonable. For instance, it predicts that in the year 2920 the US energy consumption will equal the entire energy output of the sun.

(ii) We have
$$\frac{dE}{dt} = 18.1 + 0.024E.$$
Separating variables, this is
$$\int \frac{dE}{E + 754} = \int 0.024 dt.$$
Solving, we get
$$E = Ae^{0.024t} + 754.$$
Again assuming that we measure time from 1987, this becomes
$$E = 1701 e^{0.024t} + 754.$$
So this model predicts that in the year 2000,
$$E = 1701 e^{0.024(13)} + 754 = 3078.$$
Again, this growth pattern does not seem reasonable because, although it is a slower growth (2.4% versus 3.6%) than the last example, it is still forever exponential. This model predicts that it will take longer for US consumption to reach the total output of the sun, but it is still predicted to happen (sometime around 3400).

(iii) The third equation is
$$\frac{1}{E}\frac{dE}{dt} = 0.14 - (6.1 \times 10^{-5})E.$$

This is solved by partial fractions:
$$\frac{1}{E}dE = 6.1 \cdot 10^{-5}(2295 - E)dt$$
$$\int \frac{1}{E(2295-E)}dE = \int 6.1 \cdot 10^{-5} dt$$
$$\int \left(\frac{1}{E} + \frac{1}{2295-E}\right)dE = \int 0.14 dt$$
$$\ln|E| - \ln|2295 - E| = 0.14t + C$$
$$\frac{|E|}{|E-2295|} = Ke^{0.14t}$$

Solving for E, this is
$$E = \frac{-2295Ke^{0.14t}}{1 - Ke^{0.14t}}.$$

Measuring time from 1987, we get $K \approx 2455/(2455 - 2295) \approx 15.3$, so
$$E = \frac{-35,100e^{0.14t}}{1 - 15.3e^{0.14t}}.$$

Thus the predicted consumption in the year 2000 is
$$E = \frac{-35,100e^{0.14(13)}}{1 - 15.3e^{0.14(13)}} \approx 2318.$$

This model predicts logistic growth leveling off at 2295 billion kilowatt hours per year. In some ways this model is more satisfactory than the previous ones because it acknowledges that energy consumption will not grow indefinitely. However, this model is problematic in that the 1987 value for E of 2455 is bigger than the leveling off value of 2295. (Your numerical values may differ, depending on your estimating method.)

(iv) The equation here is
$$\frac{1}{E}\frac{dE}{dt} = 0.2 - 0.002t.$$

Integrating this gives
$$\ln|E| = 0.2t - 0.001t^2 + C$$
or
$$E = Ke^{0.2t - 0.001t^2}.$$

Since t is measured from 1900 we know that $E = 2455$ when $t = 87$. This gives $K = 0.132$, so the predicted consumption in the year 2000 is
$$E = 0.132e^{0.2(100) - 0.001(100)^2} \approx 2910.$$

This model predicts that energy consumption reaches a maximum in the year 2000 (this is when the maximum of $0.2t - 0.001t^2$ occurs).

11. TERMINAL VELOCITY: SOLUTION

Mathematics skills required: Using slope field diagrams, finding equilibrium solutions of differential equations, approximating functions using local linearity, setting up Riemann sums, evaluating improper integrals, integrating by substitution and by partial fractions.

Comments: This problem is a challenging investigation of various models for free fall with air resistance. Parts (d) and (e) will be difficult for many students as they involve setting up an unfamiliar Riemann sum. Part (g) is also likely to be difficult, since it requires the student to view the upper limit of an integral as a variable. Hints involving determining how long it takes to reach particular velocities when the acceleration is given might be helpful.

Solution:

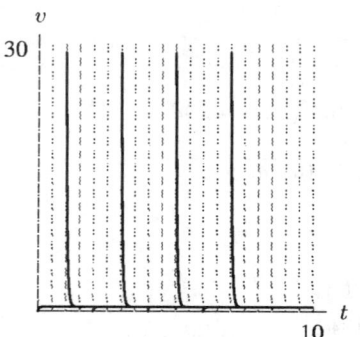

Figure H.16: Slopefields and several solutions for $\frac{dv}{dt} = 9.8 - 20v$.

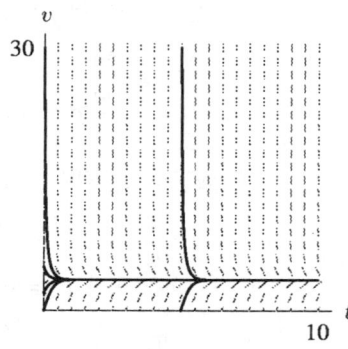

Figure H.17: Slopefields and several solutions for $\frac{dv}{dt} = 9.8 - 0.8v^2$.

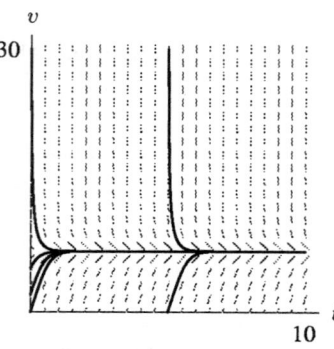

Figure H.18: Slopefields and several solutions for $\frac{dv}{dt} = 9.8 - 0.03v^3$.

(a) The concavity of the solutions will be measured by the second derivative d^2v/dt^2. Since

$$\frac{dv}{dt} = g - \frac{1}{m}f(v),$$

we have

$$\frac{d^2v}{dt^2} = -\frac{1}{m}f'(v)\frac{dv}{dt}.$$

We chose $f(v)$ to be increasing so $f'(v) > 0$.

- When $f(v) < mg$, we have $\dfrac{dv}{dt} > 0$. So $\dfrac{d^2v}{dt^2} < 0$, meaning the graph is concave down.
- When $f(v) > mg$, on the other hand, $\dfrac{d^2v}{dt^2} > 0$ and the solution is concave up.

(b) If $f(v) = kv$, then

$$\frac{dv}{dt} = g - \frac{k}{m}v.$$

The constant solution is obtained by solving $dv/dt = 0$, which gives

$$v = \frac{mg}{k}.$$

If $f(v) = kv^2$, the constant solution is again obtained by solving $dv/dt = 0$, which gives

$$g - \frac{k}{m}v^2 = 0, \quad \text{or} \quad v = \sqrt{\frac{mg}{k}}.$$

For general f, the constant solution v is obtained when $f(v) = mg$. Since f is increasing, $f(0) = 0$ and $\lim_{v \to \infty} f = \infty$, there will be exactly one value of v satisfying the above condition, and we can write $v = f^{-1}(mg)$. This velocity is called the terminal velocity because this velocity is a stable equilibrium for these differential equations. Thus any object falling through the atmosphere eventually gets close to this velocity.

(c) We are assuming the terminal velocity is reached and that air resistance is proportional to a power of the velocity. We have two equations

$$0 = 50g - k(24)^p \quad \text{and} \quad 0 = 100g - k(33.5)^p$$

and we want to solve for k and p. Multiplying the first equation by 2, we get

$$2k(24)^p = k(33.5)^p$$

$$2 = \left(\frac{33.5}{24}\right)^p = (1.4)^p$$

$$\ln 2 = p \ln 1.4$$

$$p = \frac{\ln 2}{\ln 1.4} \approx \frac{0.693}{0.336} \approx 2.06 \approx 2.$$

Then solving $50g - k(24)^2 = 0$ for k gives $k \approx 0.85$. So the correct form for f is

$$f(v) = 0.85v^2.$$

(d) We have
$$\frac{dv}{dt} = 9.8 - \frac{0.85}{50}v^2.$$
Using local linearity we have
$$\frac{\Delta v}{\Delta t} \approx 9.8 - \frac{0.85}{50}v^2$$
or
$$\Delta t \approx \frac{1}{9.8 - 0.017v^2}\Delta v.$$
We have $\Delta v = 13 - 12 = 1$. The smaller value of v is 12 and the larger value is 13. Substituting we have
$$\frac{1}{9.8 - 0.017(12)^2} \le \Delta t \le \frac{1}{9.8 - 0.017(13)^2},$$
or
$$0.13 \text{ sec} \le \Delta t \le 0.15 \text{ sec}.$$
During this time the distance traveled by the ball is at least $(0.13 \text{ sec})(12 \text{ m/sec})$ and no more than $(0.15 \text{ sec})(13 \text{ m/sec})$, so
$$1.56 \text{ m} < \text{Distance traveled} < 1.95 \text{ m}.$$

(e) Continuing to use $m = 50$ gm and $f(v) = 0.85v^2$, we have
$$\Delta t \approx \frac{1}{9.8 - 0.017v^2}\Delta v.$$
If we split the whole velocity interval $v_1 \le v \le v_2$ into n segments of length Δv, then the total time to change the velocity from v_1 to v_2 is given approximately by the Riemann sum
$$t \approx \sum \frac{\Delta v}{9.8 - 0.017v^2},$$
and exactly by
$$t = \lim_{\Delta v \to 0} \sum \frac{\Delta v}{9.8 - 0.017v^2} = \int_{v_1}^{v_2} \frac{dv}{9.8 - 0.017v^2}.$$
As the ball accelerates by Δv, it travels $\Delta s \approx v\Delta t \approx \frac{v}{9.8-0.017v^2}\Delta v$. So the total distance, s, traveled in accelerating from speed v_1 to v_2 is given approximately by
$$s \approx \sum \frac{v\Delta v}{9.8 - 0.017v^2},$$
and given exactly by
$$s = \lim_{n\to\infty} \sum_{i=0}^{n-1} \frac{v_i \Delta v}{9.8 - 0.017v_i^2} = \int_{v_1}^{v_2} \frac{v\,dv}{9.8 - 0.017v^2}.$$

(f) In part (b) we found the terminal velocity to be $v = \sqrt{mg/k}$. So we want to evaluate
$$\int_0^{\sqrt{mg/k}} \frac{dv}{g - \frac{k}{m}v^2} = \frac{1}{g}\int_0^{\sqrt{mg/k}} \frac{dv}{1 - \frac{k}{mg}v^2}.$$
Since the denominator of the integrand is zero at $v = \sqrt{mg/k}$, this is an improper integral. We wish to evaluate
$$\lim_{b \to \sqrt{\frac{mg}{k}}^-} \int_0^b \frac{dv}{1 - \frac{k}{mg}v^2}.$$
Rewriting the integral by partial fractions, or using the table of integrals, we have
$$\int_0^b \frac{dv}{1 - \frac{k}{mg}v^2} = \frac{1}{2}\int_0^b \frac{dv}{1 + \sqrt{\frac{k}{mg}}v} + \frac{1}{2}\int_0^b \frac{dv}{1 - \sqrt{\frac{k}{mg}}v}$$
$$= \left[\frac{1}{2}\sqrt{\frac{mg}{k}}\ln\left(1 + \sqrt{\frac{k}{mg}}v\right) - \frac{1}{2}\sqrt{\frac{mg}{k}}\ln\left(1 - \sqrt{\frac{k}{mg}}v\right)\right]_0^b$$

(since both $1 + \sqrt{\frac{k}{mg}}v > 0$ and $1 - \sqrt{\frac{k}{mg}}v > 0$ on the interval $0 \leq v \leq b < \sqrt{\frac{mg}{k}}$)

$$= \frac{1}{2}\sqrt{\frac{mg}{k}} \ln\left(\frac{1 + \sqrt{\frac{k}{mg}}b}{1 - \sqrt{\frac{k}{mg}}b}\right).$$

So

$$\lim_{b \to \sqrt{\frac{mg}{k}}^-} \int_0^b \frac{dv}{1 - \frac{k}{mg}v^2} = \lim_{b \to \sqrt{\frac{mg}{k}}^-} \frac{1}{2}\sqrt{\frac{mg}{k}} \ln\left(\frac{1 + \sqrt{\frac{k}{mg}}b}{1 - \sqrt{\frac{k}{mg}}b}\right) = \infty.$$

Thus this integral diverges. Physically, this means it takes forever to reach terminal velocity and so a falling object can never exceed it.

(g) There are several ways to solve this problem. They all involve first finding the velocity of the ball when it hit the ground, and then determining how long it took to attain that velocity. The solution given here is analytic; though by doing numerical integrations on a calculator, good approximate solutions can be found. The solution is attained in two steps. First, use the integral which tells us the distance traveled in accelerating from $v_1 = 0$ to v_2 to determine how fast the ball is going when it has fallen 160 m. And then use the integral for time to determine how long it took to accelerate that far.

Step 1: Find v. The distance traveled is

$$s = \int_0^{v_1} \frac{v\,dv}{9.8 - 0.017v^2}.$$

This integral is evaluated by substitution $w = 9.8 - 0.017v^2$, $dw = -0.034v\,dv$:

$$s = -\frac{1}{0.034} \int_{9.8}^{9.8 - 0.017v_2^2} \frac{dw}{w} = -\frac{1}{0.034} \ln w \Big]_{9.8}^{9.8 - 0.017v_2^2}$$

$$= -\frac{1}{0.034} \ln(9.8 - 0.017v_2^2) + \frac{1}{0.034} \ln 9.8$$

$$\approx -30 \ln(9.8 - 0.017v_2^2) + 67.1$$

Substituting $s = 160$ and solving for v_2 gives $v_2 \approx 23.95$ m/sec. (This tells us that the ball was pretty close to terminal velocity).

Step 2: Find t. We use the integral for time to determine how long it takes to accelerate to this speed. We evaluate

$$t = \int_0^{23.95} \frac{dv}{9.8 - 0.017v^2} \approx 8.2 \text{ sec}.$$

(h) To compute the terminal velocity of the baseball, we use

$$mg - 0.85v^2 = 145(9.8) - 0.85v^2 = 0.$$

This gives

$$v \approx 40.1 \text{ m/sec}.$$

To determine the velocity of the baseball after falling 160 m, we begin by evaluating the integral

$$s = \int_0^{v_1} \frac{v\,dv}{9.8 - 0.006v^2} = -\frac{1}{0.006} \int_{9.8}^{9.8 - 0.006v_1^2} \frac{du}{u}$$

$$= -\frac{1}{0.006} \ln u \Big]_{9.8}^{9.8 - 0.006v_1^2} = -\frac{1}{0.006}(\ln(9.8 - 0.006v_1^2) - \ln 9.8)$$

$$\approx -\frac{1}{0.006} \ln(9.8 - 0.006v_1^2) + 380.$$

Substituting $s = 160$, we get $v_1 = 31.8$ m/sec.